REVIEWS IN MINERALOGY — VOLUME 23

MINERAL-WATER INTERFACE GEOCHEMISTRY

M. F. HOCHELLA, JR. AND ART F. WHITE, editors

Authors:

G. MICHAEL BANCROFT
Department of Chemistry
University of Ontario
London, Ontario, Canada N6A 5B7

GORDON E. BROWN, JR.
MICHAEL F. HOCHELLA, JR.
Department of Geology
Stanford University
Stanford, California 94305

BRUCE BUNKER
Electronic Ceramics
Sandia National Laboratories
Albuquerque, New Mexico 87185

WILLIAM H. CASEY
Geochemical Research, Division 6233
Sandia National Laboratories
Albuquerque, New Mexico 87185

JAMES A. DAVIS
DOUGLAS B. KENT
U. S. Geological Survey, MS 465
345 Middlefield Road
Menlo Park, CA 94025

JANET G. HERING
WERNER STUMM
EAWAG, Swiss Federal Institute of
Technology Zürich
CH-8600 Dübendorf, Switzerland

MARGARET M. HYLAND
Dept. of Chemical & Materials Engineering
University of Auckland, Private Bag
Auckland, New Zealand

ANTONIO C. LASAGA
Department of Geology & Geophysics
Yale University
New Haven, CT 06511

GEORGE NANCOLLAS
JING-WU ZHANG
Department of Chemistry
State University of NY at Buffalo
Buffalo, New York 14214

GEORGE A. PARKS
Department of Applied Earth Sciences
Stanford University
Stanford, California 94305

PAUL W. SCHINDLER
Department of Inorganic Chemistry
University of Bern
CH-3012 Bern, Switzerland

GARRISON SPOSITO
Department of Soil Science
University of California
Berkeley, California 94720

T. DAVID WAITE
Australian Nuclear Sciences
 & Technology Organization
Private Mail Bag 1, Menai
New South Wales 2234, Australia

ART F. WHITE
U. S. Geological Survey, MS 420
345 Middlefield Road
Menlo Park, California 94025

Series Editor: PAUL H. RIBBE
Department of Geological Sciences
Virginia Polytechnic Institute and State University
Blacksburg, Virginia 24061

PUBLISHED BY THE MINERALOGICAL SOCIETY OF AMERICA
1130 SEVENTEENTH STREET, N.W., SUITE 330, WASHINGTON, D.C. 20036

COPYRIGHT: 1990

MINERALOGICAL SOCIETY of AMERICA

Printed by BookCrafters, Inc., Chelsea, Michigan 48118

REVIEWS in MINERALOGY
(Formerly: SHORT COURSE NOTES)
ISSN 0275-0279

VOLUME 23, MINERAL-WATER
INTERFACE GEOCHEMISTRY

ISBN 0-939950-28-6

ADDITIONAL COPIES of this volume as well as those listed below
may be obtained from the MINERALOGICAL SOCIETY OF AMERICA,
1130 Seventeenth St., N.W., Suite 330, Washington, D.C. 20036 U.S.A.

Volume 1: Sulfide Mineralogy, 1974; P. H. Ribbe, Ed. 284 pp.
Six chapters on the structures of sulfides and sulfosalts; the crystal chemistry and chemical bonding of sulfides, synthesis, phase equilibria, and petrology. ISBN# 0-939950-01-4.

Volume 2: Feldspar Mineralogy, 2nd Edition, 1983; P. H. Ribbe, Ed. 362 pp. Thirteen chapters on feldspar chemistry, structure and nomenclature; Al,Si order/disorder in relation to domain textures, diffraction patterns, lattice parameters and optical properties; determinative methods; subsolidus phase relations, microstructures, kinetics and mechanisms of exsolution, and diffusion; color and interference colors; chemical properties; deformation. ISBN# 0-939950-14-6.

Volume 4: Mineralogy and Geology of Natural Zeolites, 1977; F. A. Mumpton, Ed. 232 pp. Ten chapters on the crystal chemistry and structure of natural zeolites, their occurrence in sedimentary and low-grade metamorphic rocks and closed hydrologic systems, their commercial properties and utilization. ISBN# 0-939950-04-9.

Volume 5: Orthosilicates, 2nd Edition, 1982; P. H. Ribbe, Ed. 450 pp. Liebau's "Classification of Silicates" plus 12 chapters on silicate garnets, olivines, spinels and humites; zircon and the actinide orthosilicates; titanite (sphene), chloritoid, staurolite, the aluminum silicates, topaz, and scores of miscellaneous orthosilicates. Indexed. ISBN# 0-939950-13-8.

Volume 6: Marine Minerals, 1979; R. G. Burns, Ed. 380 pp. Ten chapters on manganese and iron oxides, the silica polymorphs, zeolites, clay minerals, marine phosphorites, barites and placer minerals; evaporite mineralogy and chemistry. ISBN# 0-939950-06-5.

Volume 7: Pyroxenes, 1980; C. T. Prewitt, Ed. 525 pp. Nine chapters on pyroxene crystal chemistry, spectroscopy, phase equilibria, subsolidus phenomena and thermodynamics; composition and mineralogy of terrestrial, lunar, and meteoritic pyroxenes. ISBN# 0-939950-07-3.

Volume 8: Kinetics of Geochemical Processes, 1981; A. C. Lasaga and R. J. Kirkpatrick, Eds. 398 pp. Eight chapters on transition state theory and the rate laws of chemical reactions; kinetics of weathering, diagenesis, igneous crystallization and geochemical cycles; diffusion in electrolytes; irreversible thermodynamics. ISBN# 0-939950-08-1.

Volume 9A: Amphiboles and Other Hydrous Pyriboles—Mineralogy, 1981; D. R. Veblen, Ed. 372 pp. Seven chapters on biopyribole mineralogy and polysomatism; the crystal chemistry, structures and spectroscopy of amphiboles; subsolidus relations; amphibole and serpentine asbestos—mineralogy, occurrences, and health hazards. ISBN# 0-939950-10-3.

Volume 9B: Amphiboles: Petrology and Experimental Phase Relations, 1982; D. R. Veblen and P. H. Ribbe, Eds. 390 pp. Three chapters on phase relations of metamorphic amphiboles (occurrences and theory); igneous amphiboles; experimental studies. ISBN# 0-939950-11-1.

Volume 10: Characterization of Metamorphism through Mineral Equilibria, 1982; J. M. Ferry, Ed. 397 pp. Nine chapters on an algebraic approach to composition and reaction spaces and their manipulation; the Gibbs' formulation of phase equilibria; geologic thermobarometry; buffering, infiltration, isotope fractionation, compositional zoning and inclusions; characterization of metamorphic fluids. ISBN# 0-939950-12-X.

Volume 11: Carbonates: Mineralogy and Chemistry, 1983; R. J. Reeder, Ed. 394 pp. Nine chapters on crystal chemistry, polymorphism, microstructures and phase relations of the rhombohedral and orthorhombic carbonates; the kinetics of CaCO₃ dissolution and precipitation; trace elements and isotopes in sedimentary carbonates; the occurrence, solubility and solid solution behavior of Mg-calcites; geologic thermobarometry using metamorphic carbonates. ISBN# 0-939950-15-4.

Volume 12: Fluid Inclusions, 1984; by E. Roedder. 644 pp. Nineteen chapters providing an introduction to studies of all types of fluid inclusions, gas, liquid or melt, trapped in materials from the earth and space, and their application to the understanding of geological processes. ISBN# 0-939950-16-2.

Volume 13: Micas, 1984; S. W. Bailey, Ed. 584 pp. Thirteen chapters on structures, crystal chemistry, spectroscopic and optical properties, occurrences, paragenesis, geochemistry and petrology of micas. ISBN# 0-939950-17-0.

Volume 14: Microscopic to Macroscopic: Atomic Environments to Mineral Thermodynamics, 1985; S. W. Kieffer and A. Navrotsky, Eds. 428 pp. Eleven chapters attempt to answer the question, "What minerals exist under given constraints of pressure, temperature, and composition, and why?" Includes worked examples at the end of some chapters. ISBN# 0-939950-18-9.

Volume 15: Mathematical Crystallography, 1985; by M. B. Boisen, Jr. and G. V. Gibbs. 406 pp. A matrix and group theoretic treatment of the point groups, Bravais lattices, and space groups presented with numerous examples and problem sets, including solutions to common crystallographic problems involving the geometry and symmetry of crystal structures. ISBN# 0-939950-19-7.

Volume 16: Stable Isotopes in High Temperature Geological Processes, 1986; J. W. Valley, H. P. Taylor, Jr., and J. R. O'Neil, Eds. 570 pp. Starting with the theoretical, kinetic and experimental aspects of isotopic fractionation, 14 chapters deal with stable isotopes in the early solar system, in the mantle, and in the igneous and metamorphic rocks and ore deposits, as well as in magmatic volatiles, natural water, seawater, and in meteoric-hydrothermal systems. ISBN #0-939950-20-0.

Volume 17: Thermodynamic Modelling of Geological Materials: Minerals, Fluids, Melts, 1987; H. P. Eugster and I. S. E. Carmichael, Eds. 500 pp. Thermodynamic analysis of phase equilibria in simple and multi-component mineral systems, and thermodynamic models of crystalline solutions, igneous gases and fluid, ore fluid, metamorphic fluids, and silicate melts, are the subjects of this 14-chapter volume. ISBN # 0-939950-21-9.

Volume 18: Spectroscopic Methods in Mineralogy and Geology, 1988; F. C. Hawthorne, Ed. 698 pp. Detailed explanations and encyclopedic discussion of applications of spectroscopies of major importance to earth sciences. Included are IR, optical, Raman, Mossbauer, MAS NMR, EXAFS, XANES, EPR, Auger, XPS, luminescence, XRF, PIXE, RBS and EELS. ISBN # 0-939950-22-7.

Volume 19: Hydrous Phyllosilicates (exclusive of micas), 1988; S. W. Bailey, Ed. 698 pp. Seventeen chapters covering the crystal structures, crystal chemistry, serpentine, kaolin, talc, pyrophyllite, chlorite, vermiculite, smectite, mixed-layer, sepiolite, palygorskite, and modulated type hydrous phyllosilicate minerals.

Vol. 20: Modern Powder Diffraction, 1989; D.L. Bish & J.E. Post, Eds. 369 pp.
Vol. 21: Geochemistry and Mineralogy of Rare Earth Elements, 1989; B.R. Lipin & G.A. McKay, Eds. 348 pp.
Vol. 22: The Al2SiO5 Polymorphs, 1990; D.M. Kerrick (monograph) 406 pp.
Vol. 23: Mineral-Water Interface Geochemistry, 1990; M.F. Hochella, Jr. & A.F. White, Eds.

To the pioneers of mineral-water interface geochemistry

P. Berthier, M. Daubree, J. Ebelmen, and J. Forchhammer
who studied mineral dissolution in the early to mid-1800's

and

C. Newbury, W. Skey, and C. Wilkinson
who studied oxidation-reduction reactions at mineral-solution interfaces
in the 1860's and 1870's.

MINERAL-WATER INTERFACE GEOCHEMISTRY

FOREWORD

The Mineralogical Society of America (MSA) began publishing books in this series in 1974 under the title *Short Course Notes*. From the beginning the volumes were very much more than "notes," and MSA renamed this serial publication *Reviews in Mineralogy* in 1980.

As discussed in the Preface, this particular volume resulted from the collective efforts of the eighteen authors who presented a short course on Mineral-Water Interface Geochemistry at Tanglewood Resort near Dallas, Texas, October 25-28, 1990, just prior to the annual meeting of MSA (in conjunction with the Geological Society of America).

As editor of *Reviews* I thank Mike Hochella for a splendid job of coercing manuscripts from authors (most of them on time!) and, with Art White, doing an excellent job of technical editing. Under a great deal of "deadline pressure" my editorial assistant, Marianne Stern, did most of the paste-up for the camera-ready copy, and Margie Sentelle rendered highly skilled assistance in manuscript preparation.

Paul H. Ribbe
Blacksburg, VA
September 20, 1990

PREFACE AND ACKNOWLEDGMENTS

This book and accompanying MSA short course was first considered in 1987 in response to what seemed to be a growing interest in the chemical reactions that take place at mineral-water interfaces. Now, in 1990, this area of work is firmly established as one of the major directions in mineralogical and geochemical research (see Chapter 1). We believe that there are two major reasons for this. The first is that there is a growing awareness within various earth science disciplines that interface chemistry is *very* important in many natural processes, i.e., these processes cannot be adequately described, much less understood, unless the role of interface chemistry is carefully considered. Perhaps the best illustration of this increase in awareness is the diverse backgrounds of the scientists who will be attending the short course. Participants have research interests in aqueous and environmental geochemistry, mineralogy, petrology, and crystallography. In the final list of participants, one-quarter are from outside the United States, and include scientists from Australia, Canada, England, France, Israel, The Netherlands, Sweden, and Switzerland.

The second reason that this field is one of the major new research directions in the earth sciences is because many methods, both experimental and theoretical, have relatively recently become available to study mineral surfaces and mineral-water interfaces. Many important spectroscopic techniques now used routinely to characterize surfaces and interfaces were not available twenty years ago, and some were not available just five years ago. To emphasize the importance of these methods, two Nobel prizes were awarded in the 1980's to the developers of x-ray photoelectron spectroscopy (XPS) and scanning tunneling microscopy (STM).

We have directed ourselves and the other authors of this book to follow the general guidelines of writing for "Reviews in Mineralogy". However, for the subject of mineral-water interface geochemistry, this is not easy because the field is far from mature. Several chapters are not reviews in the traditional sense in that they cover research that is relatively recent for which a considerable amount of work remains. In any case, we believe that this book describes most of the important concepts and contributions that have driven mineral-water interface geochemistry to its present state. We begin in Chapter 1 with examples of the global importance of mineral-water interface reactions and a brief review of the contents of the entire book. Thereafter, we have divided the book into four sections, including atomistic approaches (Chapters 2-3), adsorption (Chapters 4-8), precipitation and dissolution (Chapters 9-11), and oxidation-reduction reactions (Chapters 11-14).

We gratefully acknowledge the Mineralogical Society of America for agreeing to sponsor a short course on interface geochemistry and to publish this book. We are fortunate to be able to help produce what we believe to be a first-rate scientific book that is also quite affordable and widely distributed, and to offer a short course with speakers from around the world. MSA has given us, and many before us, a convenient vehicle on which to accomplish these things. In addition, we greatly appreciate the outstanding production talent of series editor Paul Ribbe, without whom this "Reviews in Mineralogy" series would probably not exist. Paul and his staff always seem to come through with a professional product which is on-time despite those of us who do our best to make it late. Susan Myers and her staff at the MSA business office in Washington, D.C. have also been indispensable and have worked tirelessly to arrange logistics, handle participant correspondence, and keep the meeting within our budget.

The short course was financially supported by the following three corporate sponsors (listed in alphabetical order):

Digital Instruments, Inc., manufacturers of scanning tunneling and atomic force microscopes,

Physical Electronics Division of Perkin Elmer, manufacturers of vacuum-based surface analytic instrumentation, and

VG Scientific, manufacturers of vacuum-based surface analytic instrumentation.

Their generous contributions were used to provide scholarships to short course participants who otherwise could not have attended the course. Certainly, the innovative and sophisticated instrumentation that these companies produce plays a vital role in pushing the frontiers of surface science forward.

Finally, books like this are meant ultimately to combat unwise advise such as: "A few months in the laboratory can save you a few hours in the library." From the perspective of this interesting(?) thought, we hope that this book will be of use.

Michael F. Hochella, Jr.
Stanford, California

Art F. White
Menlo Park, California
September 1, 1990

v

TABLE OF CONTENTS

Page

ii   Copyright; Additional Copies
iii   Dedication
iv   Foreword; Preface and Acknowledgments

Chapter 1   M. F. Hochella, Jr. & A. F. White
MINERAL-WATER INTERFACE GEOCHEMISTRY: AN OVERVIEW

1   INTRODUCTION
3   THE IMPORTANCE OF MINERAL-WATER INTERFACE GEOCHEMISTRY
3   Metals in aquatic systems
4   Ore deposit formation
4   *Sorption-concentration of ore metals*
6   *Hydrothermal leaching as a metal source for ore*
6   Low-temperature chemical weathering
9   A REVIEW OF THIS BOOK
9   Atomistic approaches
10   Adsorption
12   Precipitation and dissolution
13   Oxidation-reduction reactions
15   ACKNOWLEDGMENTS
15   REFERENCES

Chapter 2   A. C. Lasaga
ATOMIC TREATMENT OF MINERAL-WATER SURFACE REACTIONS

17   INTRODUCTION
23   Definition and measurement of surfaces – crystal habits
26   SURFACE TOPOGRAPHY AND KINETICS
31   REACTION MECHANISMS AND TRANSITION STATE THEORY
36   Application of TST to quartz dissolution
39   *AB INITIO* METHODS IN MINERAL SURFACE REACTIONS
39   *Ab initio* theory
43   *Ab initio* studies of adsorption
56   *Ab initio* studies of mechanisms of water-rock kinetics
57   Transition state of the hydrolysis reaction
59   MOLECULAR DYNAMICS METHODS IN SURFACE STUDIES
63   Applications to surfaces
70   MONTE CARLO METHODS IN SURFACE STUDIES
80   REFERENCES

Chapter 3   M. F. Hochella, Jr.
ATOMIC STRUCTURE, MICROTOPOGRAPHY, COMPOSITION, AND REACTIVITY OF MINERAL SURFACES

87   INTRODUCTION
87   EXPERIMENTAL TECHNIQUES
88   Spectroscopies for determining surface composition
88   *X-ray photoelectron spectroscopy (XPS)*
89   *Auger electron spectroscopy (AES)*
90   *Secondary ion mass spectroscopy (SIMS)*
90   *Surface analysis by laser ionization (SALI)*

90	*Resonant ionization mass spectroscopy (RIMS)*
90	*Rutherford backscattering (RBS)*
92	*Resonant nuclear reaction (RNR) analysis*
92	*Scanning tunneling spectroscopy (STS)*
93	Microscopies for determining surface microtopography
93	*Optical methods*
94	*Scanning electron microscopy (SEM)*
94	*Transmission electron microscopy (TEM)*
94	*Scanning tunneling microscopy (STM)*
95	*Atomic force microscopy (AFM)*
95	Tools for determining surface structure
95	*Low energy electron diffraction (LEED)*
96	*Scanning tunneling microscopy (STM)*
96	*Atomic force microscopy (AFM)*
97	MINERAL SURFACE COMPOSITION
97	Adventitious modification of mineral surfaces
98	Geochemically modified mineral surface compositions
98	*Sorption modification*
99	*Detachment modification*
99	*Redox modification*
101	Compositional inhomogeneities
103	MINERAL SURFACE MICROTOPOGRAPHY
103	Microtopography models
103	Growth surfaces
105	Cleavage surfaces
105	Dissolution surfaces
108	MINERAL SURFACE ATOMIC STRUCTURE
108	General description of surface atomic structure
108	Surface structures in vacuum
109	Adsorption-induced surface structural modification
109	Structure of various mineral surfaces
111	*Galena {001}*
111	*Hematite {001}*
113	*Rutile {110}*
113	*Olivine {010}*
113	*Albite {010}*
116	*Calcite {101}*
116	REACTIVITY OF SURFACES
116	Some general concepts concerning surface reaction
116	*The two-dimensional phase approximation*
117	*Heterogeneous (surface) catalysis*
117	Examples of the effect of surface microtopography on reactivity
117	*Carbon monoxide dissociation on Pt*
117	*Ethylene decomposition on various metals*
119	*Adsorption of gases on metals*
119	*Dissolution of minerals*
120	*^{235}U sorption on sheet silicates*
120	Examples of the effect of surface composition on reactivity
120	*Bonding modifiers on the surfaces of catalysts*
121	*Dissolution rates across the plagioclase series*
122	Examples of the effect of surface atomic structure on reactivity
122	*Catalytic reactions on single crystal melts*
122	*Mineral and glass dissolution reactions*
123	*Sorption reactions on minerals*
123	Surface reactivity observed atom by atom
126	CONCLUSIONS
128	ACKNOWLEDGMENTS
128	REFERENCES

Chapter 4   G. A. Parks
SURFACE ENERGY AND ADSORPTION AT MINERAL/WATER
INTERFACES: AN INTRODUCTION

133   INTRODUCTION
133   PARTICLE SIZE, SHAPE, AND SURFACE AREA
135   THERMODYNAMICS
135   Surface free energy
135   Gibbs' definition of surface excess properties: the dividing surface
137   Contribution of surfaces and interfaces to thermodynamic criteria of
           equilibrium
137   The Laplace (or Young-Laplace) equation: Curved surfaces imply a
           pressure gradient
138   The Kelvin effect: Equilibrium constants depend on A_S
139   The Kelvin equation and vapor pressures
140   The Freundlich-Ostwald equation
141   The Gibbs equation: Adsorption reduces surface free energy
142   REACTIONS WITH WATER
142   Fracture surfaces
142   Hydroxylation
142   High field gradients enhance dissociation of electrolytes
143   Hydration
146   Immersion, surface ionization, and electrified interfaces
147   Surface charge
147   Origins of charge
149   Electrified interfaces
150   SORPTION: REACTIONS WITH AQUEOUS SOLUTES
150   pH and ionic strength dependence
152   Non-specific or physical adsorption
152   Concentration dependence
153   Specific or chemical adsorption
154   Concentration dependence
157   Hydrolysis, polymerization, and precipitation
158   Competition and synergism between adsorbate and complexing
                ligands
158   Identifying sorption reactions: Clues to the composition and
                structure of surface complexes
159   Surface charge
162   Proton stoichiometry
162   The last analysis
163   SURFACE AND INTERFACIAL FREE ENERGIES OF QUARTZ
163   Fracture surface energy
165   APPLICATIONS
165   Fracture and crack propagation
167   Ostwald ripening and the Ostwald step rule
168   Earthquake prediction
169   ACKNOWLEDGMENTS
169   REFERENCES

Chapter 5   J. A. Davis and D. B. Kent
SURFACE COMPLEXATION MODELING IN AQUEOUS
GEOCHEMISTRY

177   INTRODUCTION
178   SURFACE FUNCTIONAL GROUPS
178   Surface functional groups and mineral types
179   Oxides and aluminosilicates without permanent charge

179	*Types of surface hydroxyl groups*
179	*Density of surface hydroxyl groups*
183	*Site density determined by adsorption*
184	Organic matter
184	Minerals with permanent structural charge
184	*Phyllosilicate minerals*
185	*Kaolinite*
185	*Smectites, vermiculites, and illitic micas*
187	*Manganese oxides*
187	Carbonate minerals
188	Sulfide minerals
189	SURFACE AREA AND POROSITY
189	Physical methods
190	Gas adsorption
190	*Adsorption isotherms on minerals*
192	*BET analysis*
192	*Problems caused by microporosity*
193	*Low surface area materials*
193	*Evaluation of microporosity*
195	*Surface area of clay minerals and soils*
197	ADSORPTION OF IONS AT HYDROUS OXIDE SURFACES IN WATER
197	Adsorption of cations
199	Adsorption of anions
199	Surface site heterogeneity and competitive adsorption of ions
202	Kinetics of sorption reactions
203	*Reversibility of sorption processes*
204	Effect of solution speciation on ion adsorption
204	Adsorption of hydrophobic molecules
204	THE ELECTRIFIED MINERAL-WATER INTERFACE
204	Definition of mineral surface charge
205	Classical electrical double layer models
206	The electrical double layer at oxide surfaces
206	*The Nernst equation and proton surface charge*
209	*The zero surface charge condition*
211	*Zeta potential*
212	Early developments of surface coordination theory
213	MODELS FOR ADSORPTION-DESORPTION EQUILIBRIUM
213	Empirical adsorption models
214	*Distribution coefficients*
214	*Langmuir isotherm*
215	*Freundlich and other isotherms*
215	*General partitioning equation*
217	Surface complexation models
218	*Properties of solvent water at the interface*
218	*Surface acidity of hydrous oxides*
219	*Surface coordination reactions*
219	*The Constant Capacitance Model (CCM)*
220	*The Diffuse Double Layer Model (DDLM)*
222	*Triple Layer Model (TLM)*
225	*Four layer models*
225	*The non-electrostatic surface complexation model*
225	Proton stoichiometry in surface complexation reactions
226	Parameter estimation
228	Comparison of the performance of surface complexation models
230	APPLICATIONS IN AQUEOUS GEOCHEMISTRY
231	Surface area and functional groups of soils and sediments
233	Observations of sorption phenomena in complex mineral-water systems
233	*Effect of aqueous composition*

ix

234	*Identification of dominant sorptive mineral components in composite materials*
235	*Interactive effects of mineral phases*
236	*Special problems in sorption experiments with natural composite materials*
236	The electrical double layer of soils and sediments
237	Use of empirical adsorption models for soils and sediments
237	*Distribution coefficients*
238	*Modeling based on the partitioning equation*
239	Use of surface complexation models with soils and sediments
239	*Modeling with the non-electrostatic SCM*
240	*Modeling with electrical double layer corrections*
240	*SCM modeling to dominant adsorptive components of composite materials*
242	Guidelines for surface complexation modeling with natural composite materials
243	CONCLUDING REMARKS
245	ACKNOWLEDGMENTS
245	LIST OF TERMS AND SYMBOLS
246	APPENDIX A. DETAILS OF SURFACE AREA MEASUREMENT
246	Gas adsorption
246	*Sample drying*
247	*t- and α_S plots*
247	Adsorption from solution
248	REFERENCES

Chapter 6   G. Sposito

MOLECULAR MODELS OF ION ADSORPTION
ON MINERAL SURFACES

261	INTRODUCTION
262	DIFFUSE DOUBLE LAYER MODELS
262	Modified Gouy-Chapman theory
264	Accuracy of MGC theory
266	Counterion condensation
267	SURFACE COORDINATION MODELS
267	Types of surface coordination
269	The Bragg-Williams approximation
271	Surface complexation equilibria
272	APPLICATIONS
272	Proton adsorption
274	Metal cation adsorption
276	Coion exclusion
278	CONCLUDING REMARKS
278	ACKNOWLEDGMENTS
278	REFERENCES

Chapter 7   P. W. Schindler

CO-ADSORPTION OF METAL IONS AND ORGANIC LIGANDS:
FORMATION OF TERNARY SURFACE COMPLEXES

281	INTRODUCTION
282	Adsorption of organic compounds
282	*Hydrophobic expulsion*
283	*Electrostatic attraction*

x

285   *Surface complexation*
286   Adsorption of metal ions
286   *Electrostatic attraction*
287   *How to distinguish between inner sphere and outer sphere*
           complexes
288   Co-adsorption of metal ions and organic ligands
289   THERMODYNAMIC STABILITY
289   Conditional and intrinsic constants
291   Evaluating intrinsic stability constants
291   *Studies at low surface coverage*
291   *Extrapolation techniques*
292   *Double layer techniques*
292   Stability constants of ternary surface complexes
292   *Definitions*
293   *Predictions from statistics*
296   *Experimental results*
298   *Charge effects*
298   *Ternary surface complexes with π-acceptor ligands*
299   SPECTROSCOPY
299   Methods
300   Results
300   THE ROLE OF TERNARY SURFACE COMPLEXES IN NATURE AND TECHNOLOGY
300   Effect of organic ligands upon the fate of trace metals in aquatic
           environments
302   The structure of the clay-organic interface
304   Ternary surface complexes in froth flotation
304   Ternary surface complexes in heterogeneous redox reactions
305   ACKNOWLEDGMENTS
305   REFERENCES

Chapter 8   Gordon E. Brown, Jr.

SPECTROSCOPIC STUDIES OF CHEMISORPTION REACTION
MECHANISMS AT OXIDE-WATER INTERFACES

309   INTRODUCTION AND OVERVIEW
310   Chemisorption versus physiosorption
311   The need for molecular-level information about chemisorption reaction
           mechanisms
313   The structure of "clean" and "real" surfaces
319   OVERVIEW OF STRUCTURAL METHODS PROVIDING MOLECULAR-LEVEL
       INFORMATION ABOUT CHEMISORBED SPECIES
320   X-ray absorption spectroscopy (XAS)
324   *Production of x-ray absorption spectra*
328   *XANES and pre-edge spectra*
328   *EXAFS spectra*
329   *Analysis of EXAFS spectra*
333   Other in situ spectroscopic methods and selected applications
333   *Magnetic resonance spectroscopies*
335   *FTIR and Raman spectroscopy*
335   *Mössbauer spectroscopy*
336   XAS STUDIES OF CHEMISORPTION REACTION MECHANISMS AT SOLID/LIQUID
       INTERFACES
337   Co(II) on γ-Al$_2$O$_3$ and TiO$_2$ (rutile)
341   Co(II) on kaolinite and α-SiO$_2$ (quartz)
342   Co(II) on calcite
343   Pb(II) on γ-Al$_2$O$_3$ and α-FeOOH (goethite)

343   Np(V) on α-FeOOH and U(VI) on ferric oxide-hydroxide gels
347   Se oxyanions on α-FeOOH
347   Other XAS studies of sorption complexes at solid/water or solid/air
       interfaces
348   SEXAFS STUDIES OF MOLECULAR CHEMISORPTION AT SOLID/VACUUM
       INTERFACES
352   CONCLUSIONS AND OUTLOOK
353   ACKNOWLEDGMENTS
353   REFERENCES

Chapter 9   J.-W. Zhang & G. H. Nancollas
MECHANISMS OF GROWTH AND DISSOLUTION
OF SPARINGLY SOLUBLE SALTS

365   INTRODUCTION
365   THE DRIVING FORCES FOR GROWTH AND DISSOLUTION
365   Definition
368   Calculation
369   NUCLEATION
372   THE CRYSTAL-SOLUTION INTERFACE
372   Kink densities
372   Steps from surface nucleation
373   Steps from screw dislocations
375   CRYSTAL GROWTH RATE LAWS
375   Volume diffusion
375   Adsorption and surface diffusion
376   Integration
377   Surface nucleation
378   Combined mechanisms
379   CRYSTAL DISSOLUTION RATE LAWS
379   INFLUENCE OF ADDITIVES ON THE RATES
379   Inhibitory effect of additives
380   Inactivation of kink sites
380   Retardation of step movement
381   Reduction of the concentration of growth units on a terrace
381   Dual effects of additives
382   EXPERIMENTAL METHODS
382   Experimental approaches
382   Free drift method
382   Potentiostatic method
382   Constant composition method
383   Titrant composition for CC experiments
383   Systems containing only lattice ions
384   Systems involving supporting electrolytes
384   Systems involving acid or base addition
385   Instrumentation for CC method
387   Rate determination
387   EXPERIMENTAL DETERMINATION OF REACTION MECHANISMS
389   GROWTH AND DISSOLUTION OF SOME ALKALINE EARTH SALTS
389   Calcium phosphate
391   Alkaline earth fluoride
392   Calcium carbonate
392   ACKNOWLEDGMENT
393   REFERENCES

Chapter 10   W. H. Casey & B. Bunker
           LEACHING OF MINERAL AND GLASS SURFACES
                      DURING DISSOLUTION

397   INTRODUCTION
398   THE STRUCTURE OF MIXED-OXIDE MINERALS AND GLASSES
398   The simplified structure of oxide minerals and glasses
399   Reactive and unreactive sites in a structure
404   The effect of temperature
404   GENERAL REACTION MECHANISMS
405   Hydration
407   Ion-exchange reactions
408   Hydrolysis reactions
412   The pH-dependence of leaching rates
414   EXAMPLES
414   Phosphate oxynitride glass
416   Plagioclase
417   Beryl
419   PROPERTIES OF THE LEACHED LAYER
419   Changes in cation coordination with leaching
421   Repolymerization of hydroxyl groups subsequent to leaching
421   Crazing and spallation of the leached layer
423   Chemisorption in the leaching layer
423   CONCLUSIONS
424   ACKNOWLEDGMENTS
424   REFERENCES

Chapter 11   J. G. Hering & W. Stumm
    OXIDATION AND REDUCTIVE DISSOLUTION OF MINERALS

427   INTRODUCTION
428   Objectives
428   BACKGROUND
428   Redox processes in natural systems
428   Major redox couples and the distribution of redox-active species
431   Specific reductants and oxidants
434   Role of the biota
434   Reductive and oxidative dissolution of minerals
434   Some theoretical background
434   Dependence of the rate of surface-controlled mineral dissolution on
                  surface structure
435   Surface-controlled dissolution kinetics and the surface complexation
                  model
437   Reactivity of surface species and redox reactions at mineral surfaces
439   Application of surface complexation model to surface-controlled
                  dissolution: model assumptions
440   Application of the surface complexation model: a generalized rate
                  law for dissolution
440   Application of the surface complexation model: a descriptive
                  example
443   Catalysis of redox reactions by mineral surfaces
445   CASE STUDIES

                              xiii

445   Reductive dissolution of iron oxides
445   *Reaction with a reductant*
447   *Reaction with a reduced metal and a ligand*
449   *Reaction with a (non-metal) reductant and a ligand*
452   Reductive dissolution of manganese oxides
452   Oxidative dissolution of pyrite
453   Oxidative dissolution of uranium(IV) oxides
453   Oxidative dissolution of iron (II) silicates
454   DISCUSSION
454   Applicability and limitations of the surface complexation model
454   *Laboratory studies*
454   *Applicability to field observations*
455   Dissolution and its reverse: precipitation
456   Some geochemical implications
458   CONCLUDING REMARKS
459   ACKNOWLEDGMENTS
459   REFERENCES

Chapter 12   A. F. White

HETEROGENEOUS ELECTROCHEMICAL REACTIONS ASSOCIATED WITH OXIDATION OF FERROUS OXIDE AND SILICATE SURFACES

467   INTRODUCTION
469   SOLID STATE ELECTROCHEMISTRY
471   OXIDATION OF FERROUS-CONTAINING OXIDES
472   Solid state electron transfer
472   Solid state oxidation
473   Oxide electrode processes
480   Heterogeneous redox reactions
480   *Reductive dissolution involving transition metals*
482   *Oxidative electron transfer*
483   OXIDATION OF FERROUS ORTHO- AND CHAIN SILICATES
484   Solid state electron transfer
485   Solid state oxidation
487   Dissolution processes
487   Coupled electron-cation transport
489   *Dissolved oxygen reduction*
491   *Actinide reduction on olivine and basalt*
491   *Nitrate reduction*
493   OXIDATION OF MICAS
494   Oxidation of structural Fe
495   Dissolution
496   Heterogeneous redox reactions
498   *Reduction of transition metals*
498   *Reduction of organics*
500   OXIDATION OF CLAY MINERALS
500   Structural oxidation
501   Dissolution
501   Heterogeneous reduction reactions
501   *Reduction of transition metals*
503   CONCLUSIONS
505   ACKNOWLEDGMENTS
505   REFERENCES

Chapter 13   G. M. Bancroft & M. M. Hyland
SPECTROSCOPIC STUDIES OF ADSORPTION/REDUCTION REACTIONS OF AQUEOUS METAL COMPLEXES ON SULPHIDE SURFACES

511   INTRODUCTION
513   BRIEF REVIEW OF LABORATORY STUDIES
514   SURFACE STUDIES OF METAL COMPLEX/SULPHIDE SYSTEMS
514   An introduction to X-ray photoelectron spectroscopy (XPS)
520   An introduction to Auger electron spectroscopy (AES) and a comparison of techniques
522   The chemistry of the surface from XPS and Raman spectroscopies
535   Spatial distribution of metal species from SEM and Auger studies
542   Other techniques for obtaining surface information
546   Summary of mechanisms
552   GEOCHEMICAL IMPLICATIONS OF THE LABORATORY STUDIES
552   The nature of Au in natural sulphides from SEM, SIMS, and Mössbauer studies
555   ACKNOWLEDGMENTS
555   REFERENCES

Chapter 14   T. D. Waite
PHOTO-REDOX PROCESSES AT THE MINERAL-WATER INTERFACE

559   INTRODUCTION
560   PHOTO-REDOX PROCESSES INVOLVING ABSORPTION BY MINERALS
560   Electronic structure and optical properties of minerals
560   Molecular orbital theory
563   Band theory
566   Semiconducting minerals
568   Effects of illumination
568   Charge carrier mobility
569   Interfacial electron transfer
569   Surface states
571   Space charge layer, band bending and electron transfer
576   Factors influencing semiconductor reactivity
576   Morphology
577   Size
579   Substitutional doping
579   Surface modification
580   Attachment to supports
580   Intercalation of foreign species into semiconductors
581   Applications of geochemical significance
581   Photodissolution of semiconducting minerals
584   Hydrogen peroxide production
586   Degradation of organic and inorganic pollutants
587   PHOTO-REDOX PROCESSES INVOLVING INTERFACIAL CHARGE TRANSFER TO THE MINERAL SUBSTRATE
587   Charge injection into semiconductors
588   Charge transfer via adsorbed chromophores
589   Charge transfer via photoactive surface complexes

590   PHOTO-REDOX PROCESSES INVOLVING CHROMOPHORES ADSORBED ON OR
           INCORPORATED IN MINERAL SUBSTRATES
591   Photoprocesses of chromophores adsorbed to silica and clay surfaces
591   *Photoprocesses on particulate silica and silica gel*
593   *Photoprocesses on clay minerals*
593   Photoprocesses in zeolites
593   *General features of zeolites*
594   *Photochemistry of inorganic ions exchanged into zeolites*
595   *Photochemistry of organic molecules in zeolites*
595   CONCLUSIONS
597   REFERENCES

CHAPTER 1   M. F. HOCHELLA, JR. & A. F. WHITE

MINERAL-WATER INTERFACE GEOCHEMISTRY: AN OVERVIEW

INTRODUCTION

Mineral-water interface geochemistry, the subject of this book, is arguably the most important subdiscipline within the general field of surface science because it affects many of the fundamental aspects of the way we live and the world around us. It plays critical roles in, for example, the quality of the world's fresh water, the development of soils and the distribution of plant nutrients within them, the integrity of underground waste repositories, the genesis of certain types of ore and hydrocarbon deposits, and in a more global sense, the geochemical cycling of the elements. The scientific development of this subdiscipline has to a large degree depended on other areas relevant to the extremely diverse field of surface science such as crystal growth, electrochemistry, biochemistry, catalytic chemistry, and corrosion science. Applications of surface science to these areas have been important in materials fabrication and durability, adhesion and lubrication research, battery development, cell biology, and industrial processes which utilize heterogeneous catalytic reactions, just to name a few.

The crust of planet Earth can be thought of as a catalytic bed of enormous diversity consisting of trillions of square kilometers of surface area. Most of this surface is in contact with water or its vapor. Despite this gigantic scope and the obvious complexity which accompanys it, mineral-water interface geochemistry has a foundation which consists of only two fundamental mechanisms, chemical species attachment to and detachment from mineral surfaces. Throughout this book, we will refer to these processes as sorption and desorption, respectively. In addition, modifiers to and synonyms of sorption and desorption will be used to designate different specific mechanisms of attachment and detachment reactions. Table 1 lists some of the important consequences of these two basic phenomena in natural systems.

Mineral-water interface geochemistry is not a new field. For example, observations concerning silicate dissolution were made more than 150 years ago (see the dedication page of this volume). But it has been in just the last twenty years, and mostly in the last ten, that means have been developed to directly study mineral surfaces, fluids near mineral surfaces, and the solid-fluid interface. This has lead to a tremendous resurgence in geochemical interface science and a reconsideration of the importance of mineral-water interface reactions within the global geochemical picture. Unfortunately, with vast numbers of possible mineral assemblages, and with an infinite variety of fluid compositions, the detailed understanding and prediction of mineral-water interaction, along with the consequences of such processes, is exceedingly difficult. However, with advances in experimentation-observation and modeling, there remains a great deal of hope that a more comprehensive predictive capability in this field is within reach. This book discusses major parts of both of these approaches, and reviews specific developments in these fields up to the present time.

In the following section, we first present several diverse examples of important natural and perturbed systems in which interface chemistry plays a key roll. After this, we present a technical review of this volume, the contents of which form a basis on

TABLE 1. PROCESSES OCCURRING AT MINERAL-WATER INTERFACES

Fundamental Processes	sorption	desorption
Possible Results	attachment of cations and anions, electron transfer, thin-film growth, mineral growth	detachment of cations and anions, electron transfer, congruent and incongruent dissolution
Possible Consequences	sediment, rock or secondary mineral formation, reduction in permeability, solute immobility, heterogeneous catalysis, oxidation-reduction	weathering, increase in permeability, solute mobility, oxidation-reduction
Consequences Compounded (Examples)	geochemical recycling of elements soil formation water quality hydrothermal alteration diagenesis and metamorphism ore deposit formation	

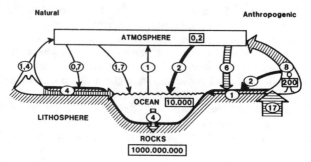

Figure 1. Global cadmium cycling from natural and anthropogenic sources. All numbers are relative. Numbers in boxes and ovals are reservoirs and fluxes, respectively. From Salomons and Forstner (1984); modified from Brunner and Baccini (1981).

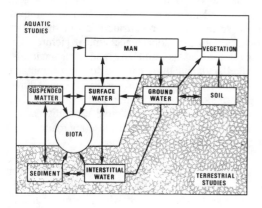

Figure 2. Schematic diagram of metal reservoirs and possible lines of interaction in terrestrial and aquatic systems. From Salomons and Forstner (1984).

which current problems in mineral-water interface geochemistry are being addressed.

THE IMPORTANCE OF MINERAL-WATER INTERFACE GEOCHEMISTRY

Metals in aquatic systems

Metals are naturally introduced into aquatic systems from many sources including water-rock interaction, biodegradation, atmospheric deposition, and volcanic activity. However, anthropogenic addition of metals into surface waters is becoming more and more significant. On a local, regional, and even global scale, this can outweigh natural introduction, sometimes by dramatic proportions. Anthropogenic introduction can result from the direct dumping of industrial waste derived from manufacturing and ore extraction processes, from the chemical weathering of landfills and mine tailings, and from the atmospheric deposition of materials resulting from burning fossil fuels and smelting ore. Looking at all industrialized metals, it has been estimated that 7 to 70×10^3 metric tons are released to the aquatic environment each year from milling and smelting operations alone (see Moore and Luoma, 1990, and references therein). Metals from industrial waste and combustion by-products would add considerably more to this amount.

As an example of metal introduction and distribution on a global basis, Figure 1 shows a schematic diagram which gives estimates of natural and anthropogenic cadmium fluxes. There is considerable interest in this because cadmium and solutions of its compounds are known to be highly toxic in humans (e.g., Yasumura et al., 1980, and references therein), and this element has a strong tendency to bioaccumulate in the food web. As shown, industrialized societies release over 5 times as much cadmium to the atmosphere as natural sources. Most of the cadmium released by human activity is deposited on the continents where it enters various aquatic systems. Within aquatic systems, its transport and eventual distribution are critically influenced by interactions with minerals, and particularly with soil particles, clay minerals, and hydroxides of iron and manganese (e.g., see Astruc, 1986; Forstner, 1986; Bewers et al., 1987; and references therein). A more complete picture of the lines of interaction of metals on the Earth's surface is shown in Figure 2.

A site specific example of the importance of mineral-water interface reactions in the release and eventual distribution of metal contaminants can be found in the Clark Fork River basin of western Montana, USA. Due to over 125 years of copper and related metal mining and smelting in this area, it is today one of the largest hazardous waste sites in the world with over 1,600 km^2 of contaminated land (see review by Moore and Luoma, 1990, and references therein). Although aquatic contamination due to airborne dust from smelting is not currently a factor in the area, massive amounts of tailings act as the primary source of metals which are released to rivers and groundwater (Fig. 3). These tailings contain hundreds to thousands of times natural levels of As, Cd, Cu, Pb, and Zn. The fine grain size of these tailings, and their exposure in dump sites, results in a metal desorption rate which far exceeds natural weathering processes. The problem is compounded as the metal sulfides in the tailings react with oxygen-rich water. A series of oxidation-reduction reactions at the sulfide-water interface results in the introduction of hydrogen and metal ions, as well as sulfate, to solution (e.g., see Nordstrom, 1982). The increased acidity due to these processes further increases desorption rates, again compounding the problem which has come to be

4

Figure 3. Tailings from copper mining near Anaconda, Montana, USA. The tailings contain up to a few percent Cu and Zn, thousands of ppm As, and hundreds of ppm Cd. A portion of the ruins of the original workings of the Anaconda mine, started in 1881, can be seen on the side of the hill in the left background. Photograph by M.F. Hochella, Jr.

known as acid mine drainage. In the Clark Fork drainage basin, such processes are ultimately responsible for elevated concentrations of toxic metals more than 560 km downstream from the primary source area.

Ore deposit formation

There are many ways in which chemical reactions at the mineral-water interface can influence or directly result in ore deposit formation. For example, reactions occurring at mineral surfaces can either preferentially release certain metals, providing a metal source for an ore deposit, and preferentially exclude other metals, potentially resulting in residual enrichment. In addition, specific sorption and/or oxidation-reduction reactions can concentrate a metal solute as it passes through a host body. Some examples of these processes are given below.

Sorption-concentration of ore metals. Models which describe the emplacement of ore-forming minerals in a host have traditionally invoked changes in the temperature or fluid pressure of the parent solution, or change in the fugacity or activity of a component or components in solution which cause the ore-forming phase to reach saturation and precipitate. However, it has also been known for some time that a solution component that is below bulk saturation can "precipitate" in the presence of a surface. Newbury, Skey, and Wilkinson first showed this by plating out precious metals on various metal sulfide surfaces (see Skey, 1871, and Liversidge, 1893, for reviews of this work). Recently, the extensive experimentation and spectroscopic measurements of Jean, Hyland, and Bancroft (see Bancroft and Hyland, this volume) has shown that this phenomenon is driven by adsorption-reduction reactions at the sulfide-aqueous solution interface. In addition, James and Parks (1975) and Morse et al. (1987), among others, have recognized the importance of these metal solute-sulfide surface reactions in reducing aqueous environments.

Figure 4. Scanning tunneling micrograph of Au accumulation on a galena surface due to a reduction-adsorption interface reaction. The width of the image is 1000 Å. The rounded mounds on the left and top sides of the image, all of which are less than 30 Å in height, are gold. Most of the lower right quarter of the image is the original galena surface. Micrograph taken by C.M. Eggleston.

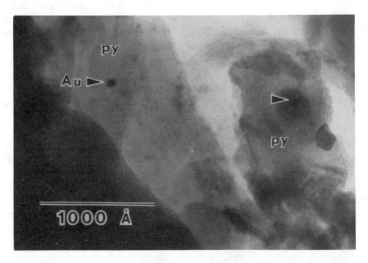

Figure 5. Transmission electron micrograph of two gold particles (arrows) within pyrite grains from the Carlin gold mine, Nevada, USA. The gold particles, each about 50 Å in diameter, were identified by x-ray dispersive analysis. From Bakken et al. (1989).

Figure 4 shows what happens when a piece of galena is dipped into an aqueous solution of 50 ppm $KAuCl_4$ at room temperature for just a few minutes (C.M. Eggleston, unpublished data). Au^{3+} is reduced to Au^0 at the sulfide-water interface, and adsorption occurs resulting in the formation of bulk metallic gold on the sulfide surface as shown. Recently, Bakken et al. (1989), Starling et al. (1989), and Bakken (1990) have found natural examples of associated precious metals and sulfides where the concentrating process described above *may* have had a controlling influence in the formation of the ore deposit. For example, Figure 5 is a transmission electron micrograph of gold in pyrite from the highly-disseminated, sediment-hosted gold-bearing ore of the Carlin mine, Nevada, USA. As Bakken (1990) points out, there are several mechanisms which realistically could have acted to precipitate gold in this deposit, including a change in the activity of sulfur or other dissolved species in solution, a change in oxygen fugacity, or a decrease in temperature. On the other hand, reduction-adsorption of gold on pyrite surfaces during pyrite precipitation is another possibility which does not require that the ore-forming fluid reach bulk saturation with respect to gold. In addition, the textural similarity between what is seen in laboratory experiments (see Bancroft and Hyland, this volume) and TEM images from the Carlin mine is intriguing.

Hydrothermal leaching as a metal source for ore. Perhaps the best examples of this process are found in volcanic-associated (volcanogenic) massive sulfide deposits. These deposits have been extensively reviewed by Franklin et al. (1981). The metal sources for these deposits are generally considered to be either the rocks stratigraphically below the deposit itself through which hydrothermal solutions have flown, or the volatile components extracted from the underlying magma system. Although both possibilities have supporting evidence, the majority of the evidence points to hydrothermal leaching of metal from the underlying rocks. The strongest evidence supporting this hypothesis (Franklin et al., 1981, and references therein) comes from (1) the analysis of natural hydrothermal fluids in active systems like the Salton Sea geothermal brines where a close link with igneous activity can be ruled out, (2) the known ability of hydrothermal solutions to effectively leach metals from rocks in laboratory experiments, (3) the relationship of the chemistry of the deposit with the underlying lithology, and (4) various stable and radiogenic isotopic studies.

Two particularly interesting and actively forming volcanogenic sulfide deposits whose metals may have been leached from surrounding rocks are in the Atlantis II Deep area of the Red Sea and on the East Pacific Rise. Pottorf and Barnes (1983) describe the Red Sea deposit as one in which two compositionally distinct hydrothermal fluids have leached mostly Fe, Zn, and Cu from surrounding and underlying mafic volcanic rocks and evaporite-shale units. They suggest that this may be a massive sulfide orebody in the process of formation. The sulfide deposits on the East Pacific Rise near 21°N were first extensively described by Hekinian et al. (1980). Figure 6 shows the now-classic picture of a "black smoker" hydrothermal vent flanked by other vents (some inactive), all resting on a basal mound made up mostly of Fe, Zn and Cu sulfides and their oxidation products. Although the source of these metals has not been definitively determined, Bischoff and Dickson (1975), Bischoff and Seyfried (1978), and Mottl et al. (1979), Seyfried and Janecky (1985), Bischoff and Rosenbauer (1987), and Seyfried (1987), among others, have shown experimentally how effectively heated seawater leaches and mobilizes metals during hydrothermal alteration processes at mid-ocean ridges. One of the latest models proposes two vertically nested seawater circulation cells between the sea-floor and the underlying magma chamber (Bischoff and Rosenbauer, 1989). The water entering the upper cell is fresh seawater, and the lower cell consists of a recirculating highly saline brine which leaches oceanic crust even more effectively than heated seawater. This model explains the high metal content, the variations in salinity, and the temperature of the exiting hydrothermal fluids at spreading centers, all without invoking a significant magmatic metal contribution.

Low-temperature chemical weathering

The importance of mineral-water interface reactions within the context of chemical weathering and the geochemical cycling of the elements cannot be overstated. All of the global-scale phenomena discussed in this section begin with atomic-scale disso-

Figure 6. Hot hydrothermal fluids jetting from a vent on the crest of the East Pacific Rise spreading center at 21°N. The fluid is dark due to suspended sulfides and is >350°C upon exiting the vent. The instrumentation in the foreground is from the deep-diving submersible Alvin. Photograph by D. Foster; from Hekinian et al., 1980; fee paid to *Science*; copyright 1980 by the American Association for the Advancement of Science.

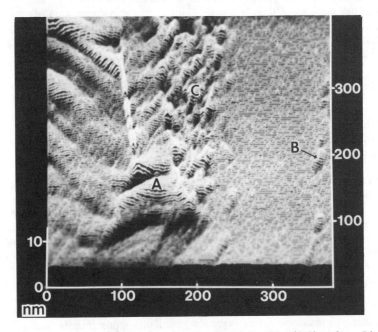

Figure 7. A high-resolution atomic force microscopic view of the albite {010} surface. Dissolution reactions on this and other mineral surfaces ultimately control most of the natural solute load which is transported to the oceans. In this image, the relief around A is 30 Å, and around B and C, only 5 Å. Loss of Na by interaction with water would initially be relatively rapid from this surface in the rough area to the left compared to the mostly atomically flat area to the right. From Hochella et al. (1990).

lution reactions at the mineral-water interface (Fig. 7). These reactions ultimately control most of the dissolved load of cations and anions that feed the oceans via rivers and groundwater, and various aspects of dissolution reactions are described by Nancollas, Casey and Bunker, and Hering and Stumm in this volume.

Chemical weathering is one of the major components of the Earth's *exogenic subcycle* in which continental source material finds its way to the oceans. This subcycle ends with precipitation and sedimentation in the ocean. The other half of the complete cycle, called the *endogenic subcycle*, puts this material back onto the continents by accretion or subduction and volcanism, from which point the entire cycle can start again. Intriguing insights into the exogenic part of the cycle can be found in Drever et al. (1988), Lerman (1988), and references therein.

A great deal of work has gone into attempts to understand global fluxes from the continents to the oceans and the specific role that chemical weathering has on this process since the pioneering insights of Garrels and Mackenzie (1971). This work starts with large-scale weathering studies which usually involve actual river systems in various geologic settings. Many of these studies have been reviewed by Drever (1988), and several generalizations can be made from them. Waters draining from igneous and metamorphic rocks are generally very dilute and contain Na and Ca as the dominant cations and bicarbonate as the major anion (Mg is also important in waters draining mafic rocks). This is because silicate minerals are relatively insoluble, and alkaline and alkaline earth elements are initially leached preferentially compared with Al and Si at the surfaces of these minerals. However, the concentration of solutes added by the atmosphere may be more than that added by dissolution, especially in areas of high discharge. Importantly, relatively minor phases in silicic rocks (e.g., pyroxenes) may be the dominant source of dissolved species in the outflowing waters.

Carbonate minerals are much more soluble in water than silicates, and it is observed, not surprisingly, that waters draining from carbonate rocks carry a higher solute load than waters draining igneous and siliceous metamorphic rocks. The principal solute species are Ca, Mg, and bicarbonate. Salts in evaporite deposits are the most soluble of the common minerals, and they obviously give rise to waters with quickly changing compositions and high salinity.

Detrital rocks offer mixed possibilities. The compositions of waters that percolate through arkoses and other sandstones can be compositionally similar to that percolating through fractured felsic igneous rocks. However, waters draining from shales can be quite variable, with Na and Ca being the common cations and sulfate, chloride, and bicarbonate as common anions. Relative proportions of these, and the possibilities of other species, is dependent on the bulk composition and mineralogy of the shale.

A major result of the observations discussed above is that the relative proportions of solute load entering the oceans resulting from the weathering of various rock types can be approximated (e.g., see Meybeck, 1987, and references therein). Carbonate minerals from sedimentary rocks contribute approximately half of the total solute load to the oceans. Crystalline rocks contribute a disproportionally small amount to the total solute load, accounting for only approximately 10% while representing one-third of the exposed rock on the continents. Evaporites contribute, as expected, a disproportionally large amount to the total solute load, perhaps over 15%, while only accounting for approximately 1% of exposed crustal rocks.

A REVIEW OF THIS BOOK

The organization of the following chapters vary considerably due both to specific topics reviewed and the approach adopted by individual authors. Some topics are generally recognized as critical geochemical issues and have been subjected to intense levels of research. The challenge in reviewing these subjects lies in critically evaluating a large literature base, sometimes containing contradictory and unreconciled results, and in synthesizing the important processes into a format useful to current researchers. Other topics in the volume deal with specific subjects for which more limited research has been conducted, but which also hold promise for significant future advances. In any case, it is difficult to subdivide this book due to fundamental interrelationships among most mineral-water interface processes. Given this, we have chosen to organize the volume into four broadly defined sections: (1) general microscopic or atomistic theories and observations, (2) adsorption processes and models, (3) dissolution and precipitation mechanisms, and (4) surficial heterogeneous oxidation-reduction reactions. The following brief summary will attempt to highlight and interrelate some of the central concepts presented within these four general topics in this volume.

Atomistic approaches

Ultimately, all geochemical processes at mineral-water interfaces involve interactions on the atomic scale. Unfortunately due to inherent complexities, geochemical interfaces have in the past eluded characterization by many of the "atomistic" methodologies development in the fields of surface chemistry, physics, and material science. However, recent developments in the application of statistical and quantum mechanical studies and in analytical techniques have permitted significant insights into atomistic processes occurring at mineral surfaces. The following two chapters by Lasaga (Chapter 2) and Hochella (Chapter 3) review theoretical and observational approaches, respectively, to understanding atomistic interactions at the mineral-water interface.

Four major theoretical approaches to describing fundamental atomic processes at the mineral-water interface are reviewed by Lasaga. First, transition state theory permits calculation of kinetic rate constants based on identification of activated complexes whose formation and destruction control reaction rates. Such an approach is rigorously correct only in cases where the elementary reactions can be defined and partition coefficients and vibrational properties deduced. Second, ab-initio calculations evaluate the interatomic forces within molecular clusters based on first principal assumptions in quantum mechanics and can be applied to the determination of the energetics associated with chemisorption processes and dissolution and precipitation kinetics. The final two approaches, molecular dynamic and Monte Carlo calculations, are complementary methods permitting dynamic modeling of the physical and chemical characteristics of surfaces for given time steps. In the former case, the calculation of the position and velocity of individual particles or atoms are calculated as functions of time based on equilibrium and non-equilibrium interatomic potentials within set initial and boundary conditions. Monte Carlo calculations are a stochastic approach which simulates the dynamic rates of ion attachment, detachment, and diffusion at the surface based on the probability of interaction determined by the potential energy involved. A major contribution of these theories is the ability to link atomistic or microscopic simulations with the macroscopic thermodynamics of the mineral-water interface.

The recent major advances in analytical methodology have permitted detailed structural and chemical observations of mineral surfaces and the mineral-water interface down to the atomic scale. Such observations form a basis on which the preceding theories can be developed and validated. Important results, reviewed by Hochella in Chapter 3, demonstrate that the chemical composition of the solid near the mineral-water interface is not generally representative of the bulk solid and is most often heterogeneous. The atomic structure of the top few monolayers of a mineral is also found not to be strictly representative of the bulk structure. Further, mineral surface microtopography is highly complex for even seemingly simple surfaces such as well developed cleavage planes. All of these factors affect the reactivity and catalytic activity of the mineral surface with respect to species in solution. Some experiments have been conducted which specifically show the change in mineral surface reactivity at mineral-water interfaces as the composition, atomic structure, and microtopography of the solid surface changes. Using these results and the results from other materials systems, it is becoming more clear how surface roughness, structure, and composition affect reactivity. In addition, it has also become clear that mineral surfaces are not structurally, chemically, or morphologically inert, but they are dynamic systems responding to changes in ambient chemical and physical conditions, this significantly influencing reaction mechanisms and rates.

Adsorption

Adsorption, reviewed in the second part of this volume (Chapters 4-8), is a subject which has been extensively studied in the fields of soil science, mineral processing, water treatment, as well as in general geochemistry. Fundamental thermodynamics of mineral surfaces in the presence of water is discussed (Parks, Chapter 4) followed by reviews of surface complexation models (Davis and Kent, Chapter 5) and molecular adsorption models (Sposito, Chapter 6). The last chapters in this section deal with more specific adsorption topics. Schindler (Chapter 7) discusses co-adsorption of metal ions and organic ligands, and Brown (Chapter 8) discusses recent advances in spectroscopic methods for characterizing adsorbed complexes at the mineral-water interface *in situ.*

The physical properties of the mineral-water interface place unique constraints on the thermodynamics of the sorption process. As addressed in the reviews by Parks as well as Davis and Kent, a number of methods, including geometric estimates and sorption isotherms, are employed to define surface areas, porosity, and roughness factors related to surface microtopography. The reactive surface areas determined by sorption methods may vary depending on the size and chemical properties of the ions or molecules used. As discussed by Parks, surface area and curvature also can be related to excess surface free energies by the Laplace and Kelvin effects, respectively, and to solid phase solubility by the Freundlich-Ostwald equation.

The excess surface free energy is also strongly affected by chemical adsorption and interaction with solvent and solute components. As summarized by Schindler, specific modes of adsorptive interaction can be classified as (1) polarity, (2) surface charge, (3) Lewis acidity, and (4) Lewis basicity. Parks first reviews these mechanisms in a fundamental system involving the mineral substrate and water. Polarity controls the interactions with the polar solvent water and hydrophilic mineral surfaces. The differences in the properties of adsorbed water and bulk water decrease as the adsorption density or surface coverage increases.

Charge interaction arises from strong electrostatic forces. Positive and negative charge is generated by broken bonds at the oxide surface in addition to chemical substitutions and defects in the mineral structure. As pointed out by Parks, the dissociation constant for water is much greater on mineral surfaces than in solution, producing hydroxylation of the surface. The net surface charge in a simple water-substrate system is determined by the distribution of these H^+ and OH^- groups. Solvation of metal cations by water forms a Lewis acid site. These reactive sites are unstable due to the positively-charged water and can either deprotonate or exchange. Deprotonated surface groups behave as Lewis bases.

The charged hydroxylated surfaces can interact with a myriad of solute species which compete for the finite number of sorption sites. The need to characterize the interaction of such species often under complex natural conditions has led to the development of a number of models describing adsorption-desorption equilibrium in multicomponent systems. As documented in the review by Davis and Kent, these models can be classified as either empirical or surface complexation approaches. The empirical approach, using simple distribution coefficients, Langmuir and Freundlich isotherms, and general partitioning coefficients, describe adsorption based on a specific set of chemical conditions such as pH and ionic composition. This approach has had considerable appeal due to its conceptual and numerical simplicity, but it is seriously constrained by the inability to describe chemical or physical perturbations in a geochemical system.

Surface complexation models are essentially extensions of the ion association models in aqueous chemistry with the tenets that: (1) the surface is composed of specific functional groups that react with dissolved solutes to form surface complexes, (2) the equilibrium of surface complexation and ionization can be described via mass law equations with correlation factors applied for variable electrostatic conditions and boundaries, (3) surface charge is treated as necessary consequences of chemical reaction of the functional groups, and (4) the empirically derived equilibrium constants are related to thermodynamic properties via the activity coefficient for the surface species. As described by Davis and Kent, both conceptual and numerical differences in surface complexation models stem principally from assumptions concerning the charge and ion distributions in the fluid phase adjacent to the mineral surface and whether these distributions are described by constant capacitance, diffuse layer, triple layer, or four layer models. Calculations using different surface complexation models often produce comparable results which successfully model adsorption in experimental and natural systems.

Molecular adsorption models based on statistical mechanics, as reviewed by Sposito, represent important approaches to understanding atomic and charge distributions at the mineral-water interface. Such approaches also provide chemical constraints which can be verified by quantum chemistry and molecular-level experiments, notably surface spectroscopy. These methodologies are mechanistically more sensitive than data from tradition adsorption experiments. The Modified Gouy-Chapman model relates the distribution of diffuse ions in a uniform liquid continuum to the surface charge distribution density using the Posisson-Boltzman expression. Applications of the model include description of the ionic distributions and counterion condensation occurring at the surface of charged colloids. From the viewpoint of statistical mechanics, the molecular theory of surface complexes is a special case of the general molecular models discussed by Lasaga. In the case of surface complexes, the calculation of the partition functions can be simplified by the Bragg-Williams approximation which assumes a lack of specific surface site interaction in which an average potential energy field of each surface species is determined collectively. As pointed out by Sposito, surface complexation models of the type reviewed by Davis and Kent can be directly derived from such statistical models.

Surface compositional analyses reviewed by Hochella indicate that most minerals that have been exposed to air or aqueous solutions have hydrocarbons and possibly other organic compounds on their surfaces. Therefore, co-adsorption processes involving metals and organic ligands as reviewed by Schindler may be of paramount importance in many geochemical systems. Three important points which are discussed in detail are (1) hydrophobic expulsion assisted by van der Waals interaction is the dominant mode of absorption of amphipathic compounds on nonpolar surfaces, (2) electrostatic attraction is the principal mode of adsorption of ionizable organics on polarizable surfaces, and (3) surface complexation most commonly involves the surface hydroxyl sites and deprotonation of the organic ligand. Concurrent adsorption of metals onto a substrate may either proceed independently of organic adsorption or may proceed by co-adsorption. The species formed by complexing a metal-ligand with surface hydroxyls are called mixed or ternary complexes. The review by Schindler discusses conventions adapted to describe the thermodynamic stability of surface ternary complexes, the experimental methods used to evaluate the stability constants of these complexes, and factors effecting their stability.

Brown discusses spectroscopic studies used to characterize chemisorption reactions at oxide-water interfaces as well as mineral surface atomic structure and how sorbates may perturb this structure. The key to the spectroscopic approaches discussed is that they have the ability to probe the sorbed complex *in situ*, thus giving direct and often quantitative information. Perhaps the most important method used for this type of study is X-ray absorption spectroscopy (XAS) using a high intensity source (e.g., synchrotron-based XAS). XAS is not a surface or interface sensitive technique, but it is element specific, and if the element of interest is found only at the mineral-water interface in the system being studied, the technique can provide information concerning the chemical environment of this element. Specifically, XAS provides average interatomic distances between the probed atom and its first and second nearest neighbors. It also gives the average number and identity of atoms in each of these shells, although this information is slightly more qualitative in nature. Nevertheless, this information has been used successfully to discriminate among various types of sorption complexes (inner-sphere vs. outer-sphere, mono-dentate vs. bi-dentate, etc.) in a number of systems, including Co^{2+} (aq) on γ-Al_2O_3 and rutile, Pb^{2+} (aq) on γ-Al_2O_3 and goethite, and aqueous selenate and selenite ions on goethite.

Precipitation and dissolution

Precipitation and dissolution at the mineral-water interfaces are reviewed in Chapters 9-11. These opposing processes involve multiple adsorption and desorption reactions at surface sites and surface and near-surface diffusion. As discussed by Zhang and Nancollas in Chapter 9, classical theories of precipitation and crystal growth stem from consideration of surface fine structure, principally kinks resulting from either surface nucleation or the development of screw dislocations. Both the critical nucleus diameter producing precipitation and the distance between consecutive spiral steps are related to the degree of solution saturation. At low supersaturation, surface nucleation is negligibly slow and the precipitation rate will be determined by available spiral dislocation sites. However at sufficiently high supersaturation, surface nucleation may dominate the overall growth rate. Foreign aqueous species normally act as inhibitors to mineral growth by poisoning kink sites through adsorption. Foreign ions of the same size as the lattice ions in more limited cases can increase nucleation rates and mineral growth by lowering the energy of adsorption. Zhang and Nancollas discuss experimental free drift, potentiostatic, and constant composition methodologies in evaluating the above surface reaction mechanisms and rates.

Casey and Bunker in Chapter 10 review the effects of bulk mineral and glass structure on dissolution and leaching. Considering these phases as inorganic polymers is helpful in distinguishing between different dissolution reaction sites. The most important variations between structures include the number and covalency of bridges or cross-links between polymer chains and the number of metal ions or network modifiers in the structure which can produce ionic rather than covalent bonds. Mineral structures with low densities of cross-links tend to dissolve congruently while minerals with a high proportion of cross-links, such as tectosilicates, dissolve incongruently which produces leached surface layers. Incongruency and selective leaching result from three processes: hydration, hydrolysis, and ion-exchange which are reviewed in detail in this chapter.

As discussed by Casey and Bunker, as well as Hering and Stumm in Chapter 11, an important recent development in surface geochemistry has been the ability to explain observed dissolution rates on surface complexation models which incorporate the principles of coordination chemistry and electric layer theory. As reviewed by Hering and Stumm, application of the surface complexation model to surface dissolution is based on the concept that the dissolution rate is proportional to the concentration of a precursor complex controlling decomposition of an activated complex (also discussed by Lasaga). This precursor is a surface metal center which has been destabilized either by formation of a surface complex, protonation of neighboring hydroxyl groups, or oxidation to a more liable oxidation state. Corollary assumptions are that formation of the precursor is rapid, the precursor is regenerated, and its concentration is constant but small relative to the total concentration of surface sites. The distribution of reactive sites with different activation energies, and thus reaction rates, can be represented by the conventional kink and step model discussed by Zhang and Nancollas. Hering and Stumm demonstrate the applicability of a surface complexation model for reductive dissolution and discuss limitations and environmental ramifications.

Oxidation-reduction reactions

Electrochemical processes which occur at the mineral-water interface have only been investigated recently in significant detail by the geochemical community. Such heterogeneous redox processes involve electron transfer between surface sites on the mineral and multivalent aqueous species. Electrochemical reactions involving reductive dissolution are discussed by Hering and Stumm in Chapter 11. Reductive dissolution results from the enhanced solubility of the mineral structure due to reduction of oxidized metal species in the structure. In the examples of manganese and iron oxyhydroxides, effects of pH and other solution parameters on rates of reductive dissolution can be explained by surface complexation. In addition, the effects of organic ligands on these rates can be explained by an increase in the adsorption of electron donors.

The opposing reaction to that discussed above, involving oxidation of the mineral surface, is discussed by White in Chapter 12 and Bancroft and Hyland in Chapter 13. White discusses the role of structural oxidation of ferrous oxides and silicates and resultant coupled reduction of aqueous species. The oxidation energetics of these substrates are favorably enhanced by the formation of smaller ferric atoms in the surface structure. Charge balance constraints require concurrent coupled electron-cation transfer from the surface to the solution. The rate controlling step in solid state oxidation can either be the rate of dehydroxylation and proton transfer in the case of hydrated silicates or the rate of tetrahedral or octahedral cation loss and vacancy site formation in other silicates and oxides. Some oxide minerals can undergo either reductive dissolution or structural oxidation depending on the electrochemical potential and therefore act as effective redox buffers to multivalent aqueous metal species. Oxidation on the surface of iron silicates enhances surface structural stability and suppresses hydration reactions by competing for available structural cations.

Comparable oxidation processes for sulfides are addressed by Bancroft and Hyland, in which case sulfide surfaces are oxidized to polysulfides and sulfate. When such processes occur in the presence of adsorbed oxidized metal ions, reduction and precipitation of the metal species occurs. Bancroft and Hyland review the application of a number of surface sensitive analytical techniques, including X-ray photoelectron spectroscopy and Raman spectroscopy, in characterizing the sulfur oxidation products and reduced metal species on the mineral surfaces. Results indicate the following reaction steps: (1) diffusion of ions to the sulfide surface, (2) preferential adsorption at defect and kink sites, (3) electron transfer accompanying partial or complete loss of ligands, resulting in so-called ad-atoms, (4) surface diffusion of ad-atoms with clustering to form critical nuclei, and (5) development of crystallographic and morphological characteristics of the bulk metal.

The semiconducting aspects of oxide minerals in relation to electron transfer during oxidation and reduction processes are discussed by both White and Waite and the application of electrode processes and methodologies addressed. Waite (Chapter 14) reviews both solid state and solution state processes related to photo-redox systems in which adsorption in the ultraviolet-visible wavelength region leads to electronic excitation of the absorber. The wavelength dependency of adsorption, the nature of the excited species, and subsequent redox reactions of these excited species are dependent on the nature of the chromophore. Three major chromophore types involving minerals are considered in the review by Waite: (1) adsorption of light by the mineral bulk resulting in production of positive and negative charges (holes and electrons) which subsequently induce electron transfer at the mineral-water interface; (2) light adsorbing species sorbed at mineral surfaces for which the mineral is intimately involved in photo-induced (or enhanced) electron transfer; and (3) light adsorbing species sorbed to the mineral surface for which the underlying mineral modifies the photochemistry of the chromophore but for which the mineral does not directly participate in the electron transfer process. Mechanistic aspects of photoprocesses associated with each of these chromophore types are considered in the chapter and examples discussed.

ACKNOWLEDGMENTS

We would like to thank Barbara Bakken, Francois Farges, and Pat Johnsson (Stanford University), Alex Blum (USGS), and Johnnie Moore (University of Montana) for assisting with various aspects of this chapter.

REFERENCES

Astruc, M. (1986) Evaluation of methods for the speciation of cadmium. In: Cadmium in the Environment (H. Mislin and O. Ravera, eds.). Birkhauser Verlag, Basel, pp. 12-17.

Bakken, B.M. (1990) Gold mineralization, wall-rock alteration, and the geochemical evolution of the hydrothermal system in the main orebody, Carlin mine, Nevada. Ph.D. dissertation, Stanford University, Stanford, CA, 236 p.

Bakken, B.M., Hochella, M.F., Jr., Marshall, A.F., and Turner, A.M. (1989) High resolution microscopy of gold in unoxidized ore from the Carlin mine, Nevada. Econ. Geol., 84, 171-179.

Bewers, J.M., Barry, P.J., and MacGregor, D.J. (1987) Distribution and cycling of cadmium in the environment. In: Cadmium in the Aquatic Environment (J. Nriagu and J. Sprague, eds.), Wiley-Interscience, New York, pp. 1-18.

Bischoff, J.L., and Dickson, F.W. (1975) Seawater-basalt interaction at 200°C and 500 bars: Implications for the origins of sea-floor heavy metal deposits and regulation of seawater chemistry. Earth Planet. Sci. Lett., 25, 385-397.

Bischoff, J.L., and Seyfried, W.E. (1978) Hydrothermal chemistry of seawater from 25 to 350°C. Amer. J. Sci., 278, 838-860.

Bischoff, J.L. and Rosenbauer, R.J. (1987) Phase separation in seafloor geothermal systems: An experimental study of the effects on metal transport. Am. J. Sci., 287, 953-978.

Bischoff, J.L. and Rosenbauer, R.J. (1989) Salinity variations in submarine hydrothermal systems by layered double-diffusive convection. J. Geol., 97, 613-623.

Brunner, P.H., and Baccini, P. (1981) Die schwermetalle, sorgenkinder der entsorgung? Neue Zurcher Zeitung, 25 Marz Nr 70.

Drever, J.I. (1988) The Geochemistry of Natural Waters (2nd Ed.). Prentice Hall, Englewood Cliffs, NJ, 437 p.

Drever, J.I., Li, Y.-H., and Maynard, J.B. (1988) Geochemical cycles: The continental crust and the oceans. In: Chemical Cycles in the Evolution of the Earth (C.B. Gregor, R.M. Garrels, F.T. Mackenzie, and J.B. Maynard, eds.). Wiley, New York, 276 p.

Forstner, U. (1986) Cadmium in sediments. In: Cadmium in the Environment (H. Mislin and O. Ravera, eds.). Birkhauser Verlag, Basel, pp. 40-46.

Franklin, J.M., Lydon, J.W., and Sangster, D.F. (1981) Volcanic-associated massive sulfide deposits. Econ. Geol., 75th Anniv. Vol., 485-627.

Garrels, R.M., and Mackenzie, F.T. (1971) Evolution of Sedimentary Rock. Norton, New York, 397 p.

Hekinian, R., Fevrier, M., Bischoff, J.L., Picot, P., and Shanks, W.C. (1980) Sulfide deposits from the East Pacific Rise near 21°N. Science, 207, 1433-1444.

Hochella, M.F., Jr., Eggleston, C.M., Elings, V.B., and Thompson, M.S. (1990) Atomic structure and morphology of the albite (010) surface: An atomic-force microscope and electron diffraction study. Am. Mineral., 75, 723-730.

James, R.O., and Parks, G.A. (1975) Adsorption of zinc(II) at the cinnabar (HgS)/H_2O interface. Amer. Inst. Chem. Eng. Sympos. Series 150, 71, 157-164.

Lerman A. (1988) Weathering rates and major transport processes: An introduction. In: Physical and Chemical Weathering in Geochemical Cycles (A. Lerman and M. Meybeck, eds.). Kluwer Academic Publishers, Dordrecht, The Netherlands, 375 p.

16

Liversidge, A. (1893) On the origin of gold nuggets. J. Royal Soc. New South Wales, 27, 303-343.

Meybeck, M. (1987) Global chemical weathering of surficial rocks estimated from river dissolved loads. Amer. J. Sci., 287, 401-428.

Moore, J.N., and Luoma, S.N. (1990) Hazardous wastes from large-scale metal extraction: A case study. Environ. Sci. Technol., 24 (in press).

Morse, J.W., Millero, F.J., Cornwell, J.C., and Rickard, D. (1987) The chemistry of the hydrogen sulfide and iron sulfide systems in natural waters. Earth Sci. Rev., 24, 1-42.

Mottl, M.J., Holland, H.D., and Corr, R.F. (1979) Chemical exchange during hydrothermal alteration of basalts by seawater. II. Experimental results of Fe, Mn, and sulfur species. Geochim. Cosmochim. Acta, 43, 869-884.

Nordstrom, D.K. (1982) Acid sulfate weathering. Special Publ. #10, Soil Sci. Soc. Amer., Madison, WI, 37 p.

Pottorf, R.J., and Barnes, H.L. (1983) Mineralogy, geochemistry, and ore genesis of hydrothermal sediments from the Atlantis II Deep, Red Sea. Econ. Geol. Monograph 5, The Kuroko and Related Volcanogenic Massive Sulfide Deposits (H. Ohmoto and B.J. Skinner, eds.), 198-223.

Salomons, W., and Forstner, U. (1984) Metals in the Hydrocycle. Springer-Verlag, Berlin, 349 p.

Seyfried, W.E. (1987) Experimental and theoretical constraints on hydrothermal alteration processes at mid-ocean ridges. Ann. Rev. Earth Planet. Sci., 15, 317-335.

Seyfried, W.E., and Janecky, D.R. (1985) Heavy metal and sulfur transport during subcritical and supercritical hydrothermal alteration of basalt: Influence of fluid pressure and basalt composition and crystallinity. Geochim. Cosmochim. Acta, 49, 2545-2560.

Skey, W. (1871) On the reduction of certain metals from their solutions by metallic sulphides, and the relation of this to the occurrence of such metals in a native state. Trans. Proc. New Zealand Inst., vol. III, 225-235.

Starling, A., Gilligan, J.M., Carter, A.H.C., Foster, R.P., and Sanders, R.A. (1989) High temperature hydrothermal precipitation of precious metals on the surface of pyrite. Nature, 340, 298-300.

Yasumura, S., Vartsky, D., Ellis, K.J., and Cohn, S.H. (1980) Cadmium in human beings. In: Cadmium in the Environment (J. Nriagu, ed.), Wiley-Interscience, New York, pp. 12-34.

ATOMIC TREATMENT OF MINERAL-WATER SURFACE REACTIONS

INTRODUCTION

Most geochemical processes involve the transfer of chemical elements between a fluid phase and a solid mineral phase. The dynamics controlling all such processes, whether growth, dissolution, oxidation/reduction, adsorption, absorption, ion-exchange, heterogeneous catalysis etc., are governed by the detailed structure and chemical bonding of the mineral surface in contact with the fluid.

In developing the treatment of surface processes, it is possible to use either a phenomenological and macroscopic description or a microscopic and atomic theory. The thermodynamic description of surfaces is based primarily on the classic work of Gibbs. This chapter will emphasize the atomic theory and discuss the relation between the atomic details of the mineral surface and their macroscopic manifestations in the thermodynamic and kinetic properties of the natural systems. After developing the link between basic kinetic theory and surface processes, several theories based on atomic physics of fundamental importance to surface studies will be discussed: transition state theory, quantum mechanical approaches, molecular dynamics and Monte Carlo methods.

At the outset, it is important to highlight the role of mineral surfaces in the bigger picture of geochemistry. To carry this out, the surface processes must be related to geochemical processes such as weathering, the dynamics of geothermal systems, environmental pollution, the formation of ore deposits, the chemical evolution of sedimentary basins, the migration and containment of toxic and radioactive wastes and diagenesis, to name only a few. Among the various physical and chemical processes taking place in the systems just mentioned, dissolution/precipitation reactions of minerals are among the most important. The kinetics of these reactions, as well as adsorption of tracer elements or stable isotope exchange, depend on two simultaneous processes: 1) the transport of chemical species to and from the mineral surfaces and 2) the adsorption/dissolution/precipitation reaction at the active surface sites. Quantitative models of many of these natural phenomena, therefore, require deciding which of the two processes dominates the kinetics.

Figure 1 illustrates the basic processes during crystal growth or dissolution. We can summarize the steps as

1) Transport of atoms through the parent phase(s)
2) Attachment/detachment of atoms to the surface
3) Movement of adsorbed atoms on or into the surface
4) Attachment/detachment of atoms to edges or kinks

Processes 2-4, which themselves may involve several elementary steps as we will elaborate in this chapter, are usually labeled surface processes to distinguish them from process 1, the long-range transport of atoms and/or the possible movement through a diffusion boundary layer. The latter processes are labeled transport processes. Figure 1 includes a detailed description of the topography of the surface, such as steps and kinks, which will be further discussed later in this chapter. The first task, however, is to differentiate surface process from transport processes and decide which, if any, controls the rate of crystal growth or dissolution.

18

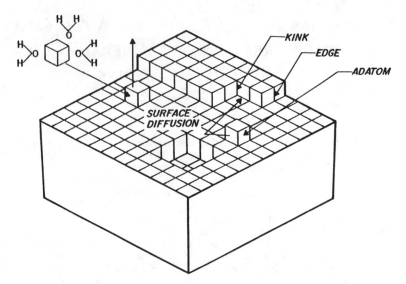

Figure 1. Basic surface processes during crystal growth or dissolution. Model used in Monte Carlo simulations. Note the use of blocks of size a for the molecular species. Detailed dynamics of surfaces must account for the changes in atomic topography such as steps, kinks, edges etc. (from Blum and Lasaga, 1987).

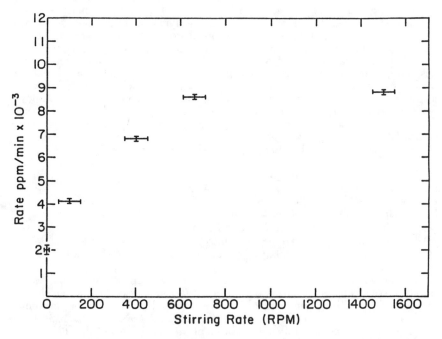

Figure 2. Variation of the nepheline dissolution rate with the stirring rate of the solution at pH 3 and 80°C (from Tole et al., 1985). The plateau region corresponds to the surface reaction rate.

Table 1. Dissolution mechanism for various substances arranged in
order of solubilities in pure water (from Berner, 1978).

Substance	Solubility
Surface reaction control	
$KAlSi_3O_8$	3×10^{-7}
$NaAlSi_3O_8$	6×10^{-7}
$BaSO_4$	1×10^{-5}
$SrCO_3$	3×10^{-5}
$CaCO_3$	6×10^{-5}
Ag_2CrO_4	1×10^{-4}
$SrSO_4$	9×10^{-4}
Opaline SiO_2	2×10^{-3}
Mixed control	
$PbSO_4$	1×10^{-4}
Transport control	
$AgCl$	1×10^{-5}
$Ba(IO_3)_2$	8×10^{-4}
$CaSO_4 \cdot 2\,H_2O$	5×10^{-3}
$Na_2SO_4 \cdot 10\,H_2O$	2×10^{-1}
$MgSO_4 \cdot 7\,H_2O$	3×10^{0}
$Na_2CO_3 \cdot 10\,H_2O$	3×10^{0}
KCl	4×10^{0}
$NaCl$	5×10^{0}
$MgCl_2 \cdot 6\,H_2O$	5×10^{0}

The interaction of mass transfer and surface reaction kinetics in quantitatively determining the overall rate has been discussed by Berner (1978), and Lasaga (1984), (1986). The major conclusion at low temperatures (e.g., 25 °C) is that surface processes constitute the rate-limiting or "controlling step" in the overall reaction. However, in the case of highly reactive minerals, there are situations where diffusion control becomes important. For example, dissolution of calcite at low pH (pH $\leq$ 3) (Berner and Morse, 1974) or dissolution of nepheline at low pH and higher temperatures (e.g., pH = 3 and 60-80 °C) (Tole et al. 1986) are transport controlled. We use transport controlled to indicate transport through the fluid medium; diffusion through an altered solid surface is considered, therefore, a "surface" process.

Much has been written on or sometimes simply assumed about whether a particular crystal growth rate in a geologic process is transport or surface controlled. Noting that the transport and surface reactions occur normally in series would indicate that the slowest process would control the rate of the overall growth reaction (Lasaga, 1981a). If the surface reactions could incorporate the atomic units into the crystal at a very fast rate, the crystal would grow only as fast as reactant atoms could get to the surface. This extreme case is termed **transport control growth** in the literature. On the other hand, in the extreme case that the surface rates are exceedingly slow compared to the transport rates, the overall growth rate would be determined by the rate-limiting surface reaction. This latter case is termed, appropiately enough, **surface controlled growth**. It is important to note that both cases are only appropiate as an extreme limiting case. Often neither case is valid and the growth rate is a more complex function of both the transport and the surface kinetics.

There are some clues that help to decipher what the dominant mode (if any) of crystal growth is. Experimentally, if the rate of growth (or dissolution) is surface controlled, there should be nearly horizontal concentration gradients of the reactants within the medium surrounding the growing phase. In growth studies involving silicate melts, such evidence can be examined by measuring composition profiles of the quenched glass surrounding the growing crystal using an electron microprobe (Muncill and Lasaga, 1988; Zhang et al., 1989). In aqueous solutions, it is usually difficult to measure concentration gradients; however, in this case, a method to determine if the surface controls the reaction is to vary the stirring rate of the sample cell during the growth or dissolution experiment. Changing the stirring rate varies the thickness of the stationary boundary layer through which molecules must diffuse from the homogeneous bulk solution to the mineral surface. If the rate is independent of stirring rate, it is likely that the reaction rate is surface-controlled. Figure 2 illustrates this procedure for dissolution rate data on the mineral nepheline.

Other methods to verify surface control in aqueous media are high activation energies and crystallographic control of the growth or dissolution (e.g., Berner, 1978). The activation energy, E_a, for diffusion in aqueous media is around 5 kcal/mole (e.g., the energy of a hydrogen bond). Therefore, activation energies that are much higher than 5 kcal/mole are indicative of surface control. In fact, surface controlled dissolution of silicates in aqueous solutions have activation energies that cluster around 15-20 kcal/mole (Lasaga, 1984). A molecular rationale for the magnitude of this activation energy will be given in a later section. A theoretical rule of thumb proposed by Berner (1978) states that for aqueous solutions, minerals with low solubility dissolve by surface control whereas high solubility minerals dissolve by transport control. His results are given in Table 1.

In the field, Fisher (1978) has given a qualitative means of distinguishing the two extreme modes of growth. This method is based on analysis of the re-distribution of reactants surrounding a growing crystal during a geochemical reaction. In the case of transport control, the components needed for growth that are near the crystal are quickly exhausted, leading to the formation of a depletion halo in some of the reactants. In the case of surface control, however, the reaction is slow enough that the depletion of growth components is fairly uniform throughout the spatial area, because transport can act fast enough to ensure such homogeneity. Therefore, in this case there will be no depletion halo. Such arguments can be applied to the analysis of sedimentary or metamorphic processes.

Generally speaking, one would expect that transport processes should become more limiting than surface processes at higher temperatures. The rationale for this conclusion is based on the difference in activation energies between transport processes and surface processes. The variation with temperature of both the viscosity of water (relevant to fluid flow in Darcy's law) and the diffusion of components in solution is small, resulting in activation energies around 5 kcal/mole. Using the value of E_a for surface reactions discussed above (15 kcal/mole), one can readily conclude that

$$\frac{k_{200C}^{surf}}{k_{25C}^{surf}} = 11760 , \tag{1}$$

whereas using an activation energy of 5 kcal/mole,

$$\frac{k_{200C}^{diff}}{k_{25C}^{diff}} = 23 . \tag{2}$$

Obviously, the rates of surface chemical reactions increase much more rapidly with temperature than those of diffusion in solution. However, recent experimental data on metamorphic reactions (Schramke et al., 1987; Lasaga, 1986) have shown that even at temperatures around 600 °C, there are systems which can be described as surface controlled, e.g., the muscovite dehydration reaction,

$$KAl_3Si_3O_{10}(OH)_2 + SiO_2 \Longrightarrow KAlSi_3O_8 + Al_2SiO_5 + H_2O, \tag{3}$$

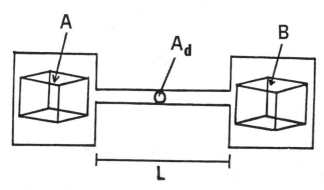

Figure 3. Schematic diagram illustrating the model used to quantify the breakdown of mineral B, the transport of components a distance L through the grain boundary network (area A_d) and the growth of mineral A.

or the reactions:

$$1\,tremolite \;+\; 11\,dolomite \;\rightleftharpoons\; 8\,forsterite \;+\; 13\,calcite \;+\; 9\,CO_2 \;+\; H_2O,$$

$$1\,tremolite \;+\; 3\,calcite \;+\; 2\,quartz \;=\; 5\,diopside \;+\; 3\,CO_2 \;+\; H_2O.$$

Heinrich et al. (1989), Dachs and Metz (1988). Nonetheless, other reactions, e.g., the calcite + quartz reaction,

$$CaCO_3 + SiO_2 \;\Longrightarrow\; CaSiO_3 + CO_2, \tag{4}$$

are indeed diffusion controlled (Tanner et al., 1985).

The same rule of thumb as proposed by Berner (1978) seems to hold for metamorphic reactions. For example, the data of Schramke et al. (1987) indicate that the dehydration of muscovite is controlled by surface reaction at the surface of andalusite (Al_2SiO_5), a low solubility mineral. Likewise, the reaction tremolite + calcite + quartz is controlled by surface reaction at the surface area of tremolite (Dachs and Metz, 1988). On the other hand, the reaction of calcite + quartz, involving relatively soluble minerals, is transport controlled (Tanner et al., 1985).

At this point, it is very important to emphasize that, regardless of the "controlling" processes, after a microscopically small time period, the rate of transport of components away from the surface of any mineral (which could be by diffusion, fluid flow or both) must necessarily **equal** the reaction rate at the surface. This equality of the various rates is required to ensure that the overall reaction takes place according to its stoichiometry. The discussion about whether diffusion or surface processes "control" the rate really refers to the question: "are the concentrations in the fluid adjacent to mineral surfaces the equilibrium concentrations?" In the case of slow surface rates, we expect the surface concentration to reflect closely the bulk chemical composition and not the equilibrium concentration. On the other hand, for fast surface reaction rates, the surface concentration will quickly approach the equilibrium concentration to slow down the surface reaction rate and maintain equality of the transport and surface rates. The proper understanding of the interplay between surface reaction and diffusion in geological reactions is so fundamental that I have developed a general (see Fig. 3), though simplified, model to show the kinetic relationships in detail (Lasaga, 1986). In that model, it is shown that in the general case that one mineral, B, is breaking down and transporting elements to the site of another mineral, A, which is forming (e.g., Reactions (3) and (4) above), either the area of A, A_A, or the area of B, A_B, or the effective area for diffusion, A_d, will control the rate, depending on which rate is slowest. The key parameters, which control the form of the rate law are the dimensionless quantities:

$$\gamma_A = \frac{k_A L}{D A_d} A_A ,$$ (5)

and

$$\gamma_B = \frac{k_B L}{D A_d} A_B ,$$ (6)

where k_A and k_B are the rate constants for the surface reactions in minerals A and B (with rate law $kA(C - C_{eq})$ so that k has units of cm/sec), L is the effective diffusion distance between A and B, D the diffusion coefficient for mass transport. The surface reaction rates will depend on the composition of the "fluid" at the surface of mineral A and B. By requiring that the surface reaction rates of a common component at the surfaces of mineral A and B be the same as the rate of transport of the component, the two surface compositions can be solved for and hence the overall rate can be determined. The overall rate is given by (Lasaga, 1986):

$$Rate = \frac{D A_d}{L} \frac{\gamma_A \gamma_B (C_{eq}^B - C_{eq}^A)}{(\gamma_B + 1)(\gamma_A + 1) - 1} .$$ (7)

This rate law includes the transport control and surface control rate laws as limiting cases. If the surface rate constants are very large (i.e., γ_A and $\gamma_B \gg 1$), then Equation (7) reduces to simply

$$Rate = \frac{D}{L} A_d (C_{eq}^B - C_{eq}^A) ,$$ (8)

i.e., a transport controlled rate. However, if k_A becomes very small ($\gamma_A \ll 1$ and $\gamma_B \gg 1$), then the rate will approach

$$Rate = k_A A_A (C_{eq}^B - C_{eq}^A)$$ (9)

or a similar equation if k_B is small. If both γ's are much less than 1, then the rate is given by:

$$Rate = \frac{k_A A_A k_B A_B}{k_A A_A + k_B A_B} (C_{eq}^B - C_{eq}^A) .$$ (10)

If $\gamma_A \ll \gamma_B$ then Equation (10) reduces to Equation (9). If $\gamma_B \ll \gamma_A$ then,

$$Rate = k_B A_B (C_{eq}^B - C_{eq}^A)$$ (11)

If the various rates are similar, then **no** simple dependence on surface area will be found.

As a rule, Equation (7) does predict that even in cases where the overall reaction involves several minerals (either reactants or products), the overall rate will tend to be determined by the surface area of the mineral with the lowest γ. Most of the work on metamorphic reactions quoted above, have indeed found a strong dependence of the rate on the surface area of **one** mineral in the reaction (which could be a reactant or a product), in agreement with the limiting case of this model. It is important to stress that which mineral controls the rate, while usually following a rule such as given by Berner (1978), will vary depending on the overall abundance (e.g., the area) of the mineral. Therefore, if a reactant is being consumed in a reaction, its area may decrease until the reactant's γ value is the lowest one and at this point the diminishing area will control the overall rate. As a result, the rate law may vary as a coupled system evolves. Once we understand more about the behavior of mineral surface reaction rates, we will be in a position to link this important concept to the observed textural relations of reaction progress in nature. Estimating reactive surface areas is certainly an important aspect of quantifying overall rates (including approach to equilibrium or lack thereof) in either sedimentary or metamorphic environments.

Definition and measurement of surfaces - crystal habits

In general, rates of surface reactions will be reported as moles/unit area/time (e.g., moles/cm^2/sec). That is, the rate measured in the laboratory (usually total moles released or grown or adsorbed etc over some period of time) is divided by some surface area to obtain the kinetic rate constant. Often the total surface area, defined generally as the area measured in a BET analysis, is used. Much has been written about the difference between total surface area and **reactive** surface area (e.g., Aagaard and Helgeson, 1982; Helgeson et al., 1984). In this chapter some of the fundamental atomic details of the "reactive" surface area will be discussed. At this point, however, it is useful to emphasize that many of the reactions discussed will involve some precursor adsorbed complexes (in the Monte Carlo section, these adsorbed complexes will be differentiated by whether they occur on steps, kinks, etc.). In fact, it is almost certain that there will be a number of different precursor reactive adsorbed atomic structures contributing to each geochemical reaction. If we label the concentration of these species as, C_i^{ads}, in units of moles/cm^2 total surface area, then the complete rate (remember these are parallel reactions) will be given by the sum of the contribution from each of these reactive sites. Using an Arrhenius equation for each individual reaction, then we may write,

$$Rate \ = \ \sum_i C_i^{ads} \, k_i^r \ = \ \sum_i C_i^{ads} \, A_i \, exp(-E_i/RT) \tag{12}$$

where the individual k_i^r have units of sec^{-1} and represent the rate of crossing of the activated barrier for the particular adsorbed precursor. A_i is the pre-exponential factor for the ith individual rate constant and E_i is the activation energy. Note that the different activation energies in Eqn.(12) can often lead to a change in the dominant reaction mechanism when the kinetics are studied over wide temperature changes. Generally speaking, C_i^{ads} can be estimated from the mineral structure (Lasaga, 1981b). For a well covered mineral surface, the values will be around 10^{15} sites/cm^2. (the reader may want to check this out for a simple structure such as NaCl). If we can estimate the k_i^r from theory, then it may be possible to relate the extent of coverage of the surface by an adsorbed precursor to the experimental rate data. It is this kind of information that will be forthcoming in the next decade and that will advance our fundamental understanding of surface reactions. This approach will also be taken up in the transition state section.

The previous section has emphasized the central role which surfaces play in many of earth's chemical processes. It is important, therefore, to focus much more carefully on the definition and characterization of mineral surfaces, especially at the atomic level. A plane is mathematically defined as all the points perpendicular to a given line and passing through any one point. If $\vec{p}$ is a point on the plane and the perpendicular direction is given by the vector, $\vec{n}$, all the points on the plane, $\vec{x}$ satisfy the equation:

$$\vec{n} \bullet (\vec{x} - \vec{p}) \ = \ 0 \, , \tag{13}$$

i.e., the vector $\vec{x} - \vec{p}$ is perpendicular to $\vec{n}$. Equivalently Equation (13) can be written

$$\vec{n} \bullet \vec{x} \ = \ k \, , \tag{14}$$

where k is some fixed scalar. Crystallographers have long used the Miller indices (hkl) to define planes in mineral structures. Such indices can also be used to define the variety of crystal surfaces possible in a mineral. In an orthorhombic system, the (hkl) plane is defined as the normal to the [uvw] direction, where [uvw] is defined by

$$[uvw] = u\vec{a} + v\vec{b} + w\vec{c}, \tag{15}$$

and $\vec{a}$, $\vec{b}$, $\vec{c}$ are the unit cell axes. In other crystallographic systems, the (hkl) plane can be written in the usual geometric formula above, by solving the equations:

$$n_1 a_1 + n_2 a_2 + n_3 a_3 = h$$

$$n_1 b_1 + n_2 b_2 + n_3 b_3 = k \tag{16}$$

$$n_1 c_1 + n_2 c_2 + n_3 c_3 = l$$

for the three components of the vector $\vec{n}$, which can be made into a unit vector, i.e., $\vec{n}/\mid \vec{n} \mid$. In Equation (16), (a_1, a_2, a_3) are the x,y,z components of the general unit cell vector, $\vec{a}$, in some chosen Cartesian coordinate system. Similarly for b_i and c_i. In crystallography, (hkl) does not stand for any one plane but for a whole system of parallel planes i.e., those defined by the vector $\vec{n}$. Usually, the only other constraint is that it pass through lattice points in the crystal; therefore, picking some $\vec{p}$ in the lattice will specify an allowed plane (e.g., Eqn.(13)). Some simple surfaces for close-packed structures are given in Figure (4). Surfaces with complex (hkl), e.g., (521), can usually be built out of terraces or steps in simpler surfaces, with a spacing between steps such as to yield the desired orientation.

Different crystal faces (labeled in the usual crystallographic manner by the Miller indices (hkl)), have different bonding properties and hence different surface free energies (see later chapter), σ, as well as different adsorption properties and different growth (or dissolution) rates. The different growth rates of different crystallographic faces are the basis for the establishment of the most obvious aspect of growth, the **crystal habit**. The habit of a crystal is dominated by the slowly growing faces. As an example, Figure 5 shows three types of faces on a simple substance. Faces, S, (stepped) and K (kinked) provide ample defects for growth and therefore grow very quickly. As a result, these faces are rarely, if ever, observed. Instead, the slowly growing F faces dominate the crystal habit. To see this phenomenon clearly, we can look at Figure 6, which shows a growing simple cubic phase with two faces present, (111) and (100), in the morphology of a truncated octahedron. If the (100) face grows faster than the (111) face, then the movement of the (100) face with the (111) faces moving very slightly, will make the crystal evolve as shown in Figure 6a. Ultimately, the fast growing (100) face will disappear leaving an octahedral crystal with all (111) faces. Note that to mantain the habit shown in Figure 6b, the (111) faces would need to have a growth rate similar to the (100) faces.

The different rates of growth of these faces can be explained from the atomic structures of the various crystal surfaces. If an atom adsorbs onto a (111) face of a cubic crystal (Figure 4b), the adsorbed atom forms 3 bonds with the atoms lying beneath it. However, an atom adsorbed onto a (100) face (Figure 4b) will form 4 bonds and, thus, be far more stable. Finally, an atom adsorbed onto a (110) face will form 5 bonds to the atoms underneath. This surface produces the most stable adsorbed atoms. Because the stability of the adsorbed atoms is critical to the subsequent growth of the face (e.g., by enabling attachment of the atoms to surface steps), it is expected that the growth rates of the various faces should reflect the differences in the stability of the atoms adsorbed onto them. These expectations are substantiated by Monte Carlo simulations of growth as a function of supersaturation, $\Delta\mu$, as can be seen in Figure 7, which plots the Monte Carlo computed growth rates of the (111), (100) and (110) faces (Gilmer, 1980). Note

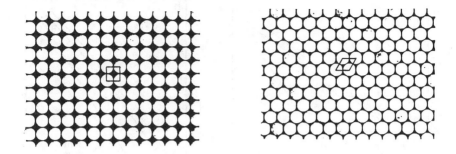

Figure 4. Ball models of a) an ideal (100) surface and b) an ideal (111) surface of a face-centered cubic monatomic crystal (from Blakely, 1973).

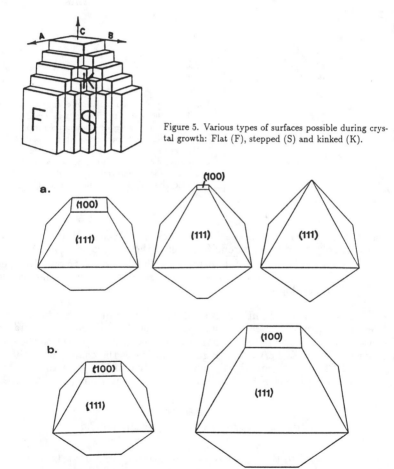

Figure 5. Various types of surfaces possible during crystal growth: Flat (F), stepped (S) and kinked (K).

Figure 6. Two different evolutions during the growth of a simple cubic crystal initially bounded by (100) and (111) faces. (a) Case where the growth rate of the (100) face is much faster than that of the (111) face. (b) Case with similar growth rates for the (100) and (111) faces.

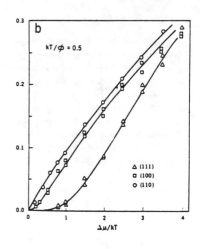

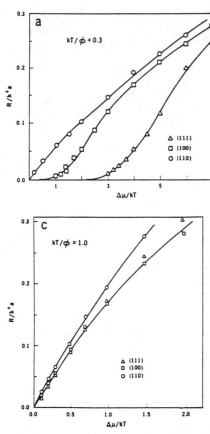

Figure 7. Monte Carlo simulations of the growth rates of the (111), (100) and (110) faces of a face-centered cubic crystal. $\Delta\mu$ is the supersaturation of the solution ($\Delta\mu = 0$ at equilibrium). The parameter ϕ is the bonding energy of the lattice as discussed later in the chapter. The calculated speed of the interface is normalized by the factor k^+a, where a is the unit cell length and k^+ is an atom arrival rate constant (see discussion in text) (from Gilmer, 1980).

that the slowest growing surface is the densely packed (111) face. However, note that the difference in growth rates is a <u>strong</u> function of the supersaturation. At low $\Delta\mu$, the (111) face grows so slowly that only octahedra (fig 6a) will develop. However, at high $\Delta\mu$, the (111) and (100) growth rates become similar enough so that truncated octahedra (Fig 6b) develop.

Preferential adsorption of impurities on different crystallographic faces can significantly alter the growth/dissolution surface rates of the different faces, and hence alter the morphology of crystals. For example, Pb^{2+} has a strong effect on the growth rate and morphology of KCl crystals. Figure 8 illustrates the change from the usual cubic habit of KCl grown from slightly supersaturated solutions to the octahedral habit obtained at high levels of Pb^{2+} and higher supersaturations (Li et al., 1990).

SURFACE TOPOGRAPHY AND KINETICS

To proceed further into the theory of crystal surfaces, we must take a closer look at the energetics and structure of crytal surfaces. Figure 9 illustrates some of the surface features important for either adsorption, growth or dissolution. A loosely adsorbed atom (or molecule), such as species 3 in Figure 9, is commonly termed an **adatom**. The mobility plays a prominent role in much of the growth literature. The

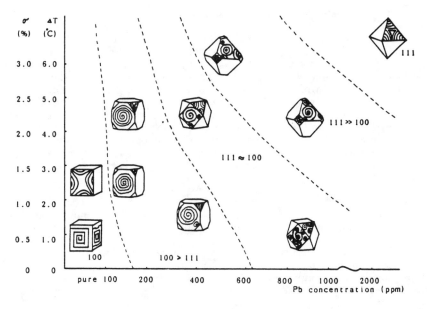

Figure 8. Variation in the crystal habits of KCl in relation to the supersaturation and impurity Pb^{2+} concentrations (from Li et al., 1990).

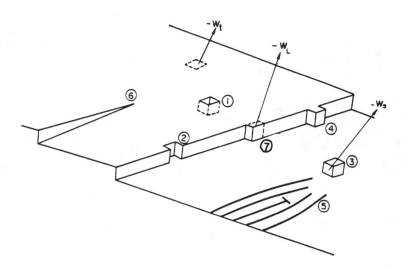

Figure 9. Diagram illustrating some of the defects which may occur at a free surface. The examples shown are a vacancy on a terrace (1), a vacancy at a step (2), an adatom on a terrace (3), an adatom at a ledge or step (4), an edge dislocation-surface intersection (5), a screw dislocation-surface intersection (6) and a kink site (7). Note that in the removal of an atom from a kink site the kink site is regenerated.

counterpart to adatoms are **vacancy defects** (e.g., species 1 in Fig. 9). Impurity ions can also play an essential role in surface dynamics and can lead to either catalytic or inhibitory effects on the kinetics (e.g., the effect of phosphate ions on calcite growth, Berner and Morse, 1974). Another important defect is the **surface step**, illustrated in Figure 9. Obviously, atoms attached to a step (such as species 4) are much more stable than simple adatoms because there are more bonds formed with other surface atoms. The propagation of steps is basic to orderly crystal growth or dissolution. The central focus of step growth occurs at sites, such as number 2 in Figure 9, which have <u>half</u> the number of bonds present as in the bulk phase. Such sites are special because they self-generate (i.e., an atom arriving at such a site creates a new similar site); they are given the name **kinks**, because of their obvious appearance. Kink sites are important sites for modeling growth and, in particular, we can show that <u>at</u> equilibrium the impingement rate of atoms to kink sites, k_+, is equal to the kink site dissolution rate, k_-,

$$k_+^{kink} = k_-^{kink} . \tag{17}$$

Note that Equation (17) is **not** true of other sites (e.g., adatoms) at equilibrium. The proof of (17) uses microscopic reversibility and will be given later in this chapter.

Two other defects that can greatly affect growth or dissolution are the outcropping of edge or screw dislocations at the surface, as displayed in Figure 9. Because these outcrops provide high energy sites as well as (in the case of screw dislocations) self-generating steps, they play a key role especially in the classic Burton-Cabrera-Frank (BCF) theory.

A feature of crystal faces that is of general interest is whether the surface is rough (high degree of disorder) or smooth. A smooth surface is associated with the slow growth of euhedral crystals, while rough surfaces are more akin to fast growth from highly supersaturated solutions or to transport-controlled growth (e.g., dendritic or spherulitic textures). Examples of smooth and rough surfaces are given in Figure 10 from Monte Carlo simulations. Both simulations have "added" the same number of atoms to the surface. However, the topography has changed drastically from one case to the other. Obviously, there seems to be a "phase transition" between an ordered and a disordered surface, quite similar to the magnetic transitions in ferromagnetic systems. In fact, if we consider spin up to be an occupied site and spin down to be an empty site, an isomorphism can be set up between magnetic problems and surface problems.

Jackson (1967) has developed a simple model that illustrates nicely the transition from ordered growth (clean surfaces) to rough or continuous growth (bumpy surfaces) as the entropy of dissolution or melting decreases. The model begins by taking into account the atomic bonding between species on the surface. For simplicity, let us assume that all the species are the same (e.g., a monatomic crystal), that the bonding energy between nearest neighbors in the solid is given by ϕ and that in the bulk crystal each atom has ν neighbors. If we assume that the solid bond energies are much greater than those with the fluid or between fluid species (this assumption will be removed later in the chapter), then the energy required to dissolve or melt the mineral, is simply the bonding energy of the lattice. Ignoring PV effects, this energy can be set equal to the macroscopically observed enthalpy of dissolution or melting per atom in the solid,

$$\Delta H = \frac{1}{2} \nu \phi . \tag{18}$$

In computing the lattice energy, Equation (18) has inserted a 1/2 factor, which corrects for the sharing of a bond between two atoms. Alternatively, the 1/2 can

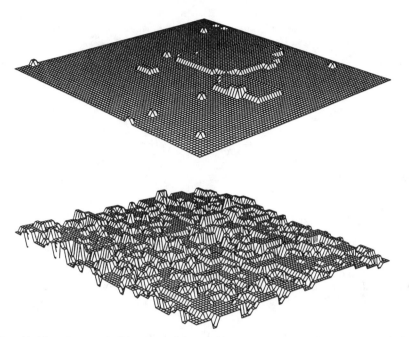

Figure 10. Monte Carlo simulations of growth using a bonding energy of $\Phi/kT = 4.0$. a) ordered growth observed for low supersaturation $\Delta\mu/kT = 1.5$ b) disordered growth (rough surface) observed for high supersaturation, $\Delta\mu/kT = 5.0$. Results obtained using the MC program of Alex Blum.

arise if we consider the atom to be removed from the unique site where the number of bonds is exactly half that in the bulk crystal. This special site is, of course, the "kink" site discussed earlier. Now we need to look at the change in free energy of the surface upon the addition of N atoms. There will be two contributions, one from energy changes and the other from entropy changes (omitting volume changes here). Specifically,

$$\Delta G = \Delta E_0 + \Delta E_1 - T\,\Delta S_0 - T\,\Delta S_1 . \tag{19}$$

We will analyze each term in detail. ΔE_0 is the energy gained (a negative quantity) by putting N single adatoms on the surface. If there are η_0 nearest neighbors to each adatom on the plane *below*, then ΔE_0 is equal to $-N\,\eta_0\,\phi$. In addition, some of the N atoms on the surface will be adjacent to each other. ΔE_1 is the energy gained by the bonds between surface atoms. Assuming that each atom has η_1 possible adjacent nearest neighbors, then the probability of observing an adjacent atom is N/N_{tot}, where N_{tot} is the total number of surface sites. Therefore, each atom on average will have a total of $\eta_1\,(N/N_{tot})$ adjacent atoms and so $\Delta E_1 = -1/2\,N\,\eta_1\,(N/N_{tot})\,\phi$. Again, the 1/2 arises because each atom-atom bond is counted twice. ΔS_0 is the entropy loss in going from the liquid to the solid. If T_{eq} is the temperature at which the crystal and the fluid are in equilibrium (i.e., saturation temperature or melting point), then ΔS_0 $= -N\,\frac{\Delta H}{T_{eq}}$, where the minus sign reflects that ΔH, defined by Equation (18) , refers to dissolution, while ΔS_0 refers to the precipitation of N atoms. Finally, ΔS_1 is the configurational entropy of the surface configuration of N adatoms. Assuming *random* mixing of the N surface atoms and the $N_{tot} - N$ surface vacancies on the surface sites (this is in the spirit of so-called "mean field" theories), then (e.g., Hill, 1962)

$$\Delta S_1 = -k\left[(N_{tot} - N)\,ln(\frac{N_{tot} - N}{N_{tot}}) + Nln(\frac{N}{N_{tot}})\right] \tag{20}$$

where k is Boltzmann's factor. Equation (20) is the usual equation for the entropy of mixing of ideal solutions. Combining, all these terms yields

$$\Delta G = -N\eta_0\phi \ - \ \frac{1}{2}N^2\frac{\eta_1}{N_{tot}}\phi + TN\frac{\Delta H}{T_{eq}} + kT\left[(N_{tot} - N)\,ln\left(\frac{N_{tot} - N}{N_{tot}}\right) + N\,ln\left(\frac{N}{N_{tot}}\right)\right]$$

(21)

If this equation is divided by N_{tot} k T_{eq}, and if we define the following ratios

$$x \equiv \frac{N}{N_{tot}} \ ; \quad \xi \equiv \frac{\eta_1}{\nu} \ ,$$

then Equation(21) becomes:

$$\Delta G \ = \ - \frac{\phi\eta_0}{kT_{eq}}\,x \ - \ \frac{\eta_1\phi}{2kT_{eq}}\,x^2 + \frac{\nu\phi}{2kT_{eq}}\frac{T}{T_{eq}}\,x + \frac{T}{T_{eq}}\left[(1-x)\,ln(1-x) + x\,ln\,x\right] \quad (22)$$

where the Relation (19) involving ΔH has been used. To obtain the total number of neighbors for any one atom in the crystal one adds up all those neighbors adjacent to the atom and the number of neighbors in the plane above and in the plane below the particular surface of the atom. Hence, using the definition of ν, η_1 and η_0,

$$\nu = \eta_1 + 2\,\eta_0 \ . \tag{23}$$

Combining the two terms linear in x in Equation (22) yields

$$\frac{\phi}{kT_{eq}}(-\eta_0 + \nu/2)x \ ,$$

where we have approximated T/T_{eq} as close to 1. Using the relation between ν and η_0, we have that the x term is

$$\frac{\phi}{kT_{eq}}(\eta_1/2)\,x \ .$$

Both the x and x^2 terms have the same coefficient, α; therefore, combining the x and x^2 terms we can write

$$\frac{\Delta G}{N_{tot}kT_{eq}} \ = \ \alpha\,x(1 - x) + x\,ln\,x + (1 - x)\,ln(1-x) \ , \tag{24}$$

where α is defined by

$$\alpha \ = \ \frac{\eta_1\,\phi}{2\,kT_{eq}} \ . \tag{25}$$

Now we want to investigate the possible minima or maxima of this ΔG function and thereby obtain the expected stable states of the surface. To do this, we need to set the derivative of the right side of Equation (24) with respect to x to zero. The solution needs to be done numerically for each value of α. Table 2 gives the results. Obviously for low values of α, ΔG has only a minimum at a value of x = 0.5. Such a surface has half its sites empty and half its sites occupied, i.e., it is quite

Table 2. Degree of Surface Roughness.
Minima in ΔG (Eqn. (24)) as a function of α

α	x_{min}	α	x_{min}
1	0.5	2.7	0.11, 0.89
1.5	0.5	2.9	0.08, 0.92
2.0	0.5	3.0	0.07, 0.93
2.1	0.31, 0.69	3.5	0.04, 0.96
2.2	0.25, 0.75	4.0	0.02, 0.98
2.3	0.20, 0.80		
2.5	0.14, 0.86		

disordered (e.g., Figure 10). Once α increases past 2.0, two minima appear in the ΔG curve and the minimum at x = 0.5 becomes a maximum (i.e., the ΔG versus x curve has two minima with a maximum in between). This value of α, then, represents the equivalent of a phase transition temperature. Again, the surface roughness problem is isomorphic with ferromagnetic transition problems. Thus the value $\alpha = 2.0$ marks the equivalent of a Curie temperature or a critical temperature. The two minima in ΔG quickly move apart so that for α values greater than 2.5, x_{min} is nearly 0.0 or nearly 1.0. In this case, the surface is either all empty or all filled (see Figure 10), i.e., the surface is quite ordered. Equation (25) shows that small values of α correspond to small bonding parameters. Therefore, systems with little change in bonding between the surface and the fluid, will exhibit surfaces during growth or dissolution that are quite disordered and rough. These systems should also have low interfacial free energies, σ. On the other hand, systems with strong surface bonding will have high σ and relatively clean surfaces.

REACTION MECHANISMS AND TRANSITION STATE THEORY

Understanding the atomic nature of surface reactions is necessary to unravel the kinetic mechanisms of geochemical reactions. Other chapters of this book deal with the experimental probes of surface structure and chemistry and with the analysis of adsorption processes at mineral surfaces. The next two sections will address the *ab initio* and molecular dynamics approaches and their application to surface processes. In this section, I would like to discuss the use of transition state theory in these studies.

Numerous recent papers have invoked transition state theory in developing the mechanisms of low temperature water-rock reactions (e.g., see Lasaga, 1981b; Aagaard and Helgeson, 1982). One of the central themes in understanding heterogeneous kinetics is the elucidation of the important activated complexes. The variation in the atomic structure and bonding of the activated complex as a function of the surface properties, the composition of the solution near the surface, temperature and pressure will determine many of the kinetic properties such as catalysis, changes in activation energy, kinetic isotope effects and the overall rate law.

Contrary to some usage in the literature, transition state complexes are **not** structures representing true free energy minima and hence are not normal (in the thermodynamic sense) chemical species. For example, the attack by H^+ on the surface of feldspars will lead to a surface which has exchanged alkalies for SiOH or $SiOH_2^+$ groups. While these latter species are surface complexes, **none** of these

Reaction in water

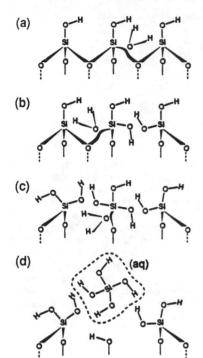

(a)

(b)

(c)

(d)

Figure 11. Sequence of steps in the proposed mechanism for the overall reaction in the dissolution of quartz (from Dove and Crerar, 1990).

chemical moieties is an activated complex. Further attack by H^+ or by H_2O will hydrolyze the Si-O-Al and Si-O-Si bridging bonds as shown in the reaction:

$$H_2O + \ \equiv Si - O - Si \equiv \ \Longrightarrow \ 2 \equiv Si - O - H \ . \tag{26}$$

Again, neither side of Reaction (26) represents the activated complex. Both sides represent "stable" surface species.

Activated complexes are specific to a particular **elementary reaction**. One cannot discuss the activated complex without specifying the precise elementary reaction taking place; one cannot speak of the activated complex of an overall reaction. For example, we expect the dissolution and precipitation of silica to proceed with several elementary reactions, as illustrated in Figure 11 (see Lasaga and Gibbs, 1990b; Dove and Crerar, 1990). Note also that for the complex many-variable potential surfaces that are of interest to us, it is very possible to find several "activated complexes" in different regions of the surface. The task, then, is to find the relevant elementary pathway through one (or more) of these activated complexes that controls the overall geochemical reaction being studied.

Once the activated complex has been found, transition state theory can be used to quantify the kinetics of each elementary reaction. For example, the partition function calculation (e.g., see Lasaga, 1981b) can be used to calculate adsorption equilibrium constants. According to statistical mechanics, if C_l is the concentration of molecules in the liquid or gas phase (molecules/cm^3) and C_{ads}, C_s are the concentrations of adsorbed molecules and of bare surface sites respectively (both in number/cm^2) then the equilibrium constant is given by:

$$K_{ads} = \frac{C_{ads}}{C_l\,C_s} = \frac{q_{ads}}{(q_l/V)\,q_s}\,e^{-\Delta E_{ads}/RT} \qquad (27)$$

(e.g., Hill, 1962). The term ΔE_{ads} refers to the adsorption energy, which is usually a negative number. In Equation (27), q_{ads} stands for the partition function of the adsorbed complex, q_l is the partition function of the molecule in the liquid phase and q_s is the partition function of the solid. The partition functions can be obtained if one knows the geometry and vibrational frequencies of the isolated and adsorbed species, as well as the vibrational frequencies of the surface. Under normal circumstances, the partition functions can be broken down into contributions from the translational, rotational and vibrational energies, i.e., as a product:

$$q = q_{trans}\,q_{rotn}\,q_{vib} \quad . \qquad (28)$$

The vibrational partition function, q_{vib}, in turn, can be broken down into a contribution from each vibrational mode i.e., q_{vib} can be calculated from the product of the vibrational partition functions of each individual vibrational frequency:

$$q_{vib} = \prod_i q_{i,vib} \quad . \qquad (29)$$

Table 3 gives the formulas for the translational and rotational partition functions as well as for the vibrational partition functions of an individual vibrational frequency. Note that these partition functions are dimensionless. Usually partition functions (which add up the terms exp(-ϵ/kT) for all the possible energies, ϵ), will not contain contributions from energies that are much greater than kT. Therefore, while changes in translational or rotational properties (low energies) will affect the value of q and hence the equilibrium constant, K (Eqn.(27)), vibrations will contribute significantly to the vibrational partition function only if the vibrational frequencies are low e.g., on the order of several hundred cm^{-1} or less (RT = 600 cal/mole at 25 C and the vibrational energy, hν, is less than RT only if the vibrational frequency, ν, is less than 209 cm^{-1}, using the conversion 1 cm^{-1} = 2.859 cal/mole). If the frequencies are much bigger than several hundred wavenumbers then the partition function will simply take the value 1 (see Table 3). It is customary, for example, to set the vibrational partition function for the surface, q$_s$, to unity (i.e., ignore contributions from the vibrations). The formulas for the translation and rotational partition functions can also be found in Table 3; for further discussion see Hill (1962) or Lasaga (1981b). The ratio of partition functions is related to the ratio of the number of molecules. Because concentration is usually in molecules per unit volume, the partition functions must be divided by the volume V of the system to obtain equilibrium constants. This is done in Equation (27) for the concentration, C_l, i.e., we use q/V. Note that the formula for the translational partition function in table 3 already has divided q by the system volume V (and hence is not a dimensionless formula). For the concentration of adsorbed species, we require units of molecules/unit area. Thus the partition functions q_{ads} and q_s should both be divided by the solid surface A. However, because these two partition functions occur in a ratio, the areas cancel out so that the total partition functions can be used in Equation (27).

The fraction of molecules adsorbed on the surface, θ, can be obtained from

$$\frac{\theta}{1 - \theta} = \frac{C_{ads}}{C_s} \quad .$$

Table 3. Partition functions

Motion	Degrees of Freedom	Partition function	Order of Magnitude
Translational[a]	3	$(2\pi\,m\,k\,T)^{3/2}/h^3$	10^{24}-10^{25}
Translational[a]	2	$(2\pi\,m\,k\,T)/h^2$	10^{16}
Rotational (linear molecule)	2	$(8\,\pi^2\,I\,k\,T)/h^2$	10-100
Rotational (non-linear)	3	$\dfrac{8\pi^2(8\pi^3 ABC)^{1/2}(kT)^{3/2}}{h^3}$	100-1000
Vibrational	1	$1/(1 - \exp(-h\nu/kT)\,)$	1-10

m - mass of molecule
k - Boltzmann's constant
T - Temperature (K)
h - Planck's constant
I - moment of inertia of linear molecule
A,B,C - principal moments of inertia of non-linear molecule
ν - normal mode frequency of vibration
a - translational partition functions are per unit volume (3 degrees of freedom) or per unit area (2 degrees of freedom)

Table 4. Observed and Calculated rates for zero-order reactions (from Laidler, 1965).

Decomposition of	Surface	E (kcal)	Temp (°K)	Calc.	Obs.
NH_3	W	38.0	904	8.0×10^{18}	4×10^{17}
NH_3	W	41.5	1316	3.4×10^{21}	2×10^{19}
NH_3	Mo	53.2	1228	8.5×10^{18}	$5\text{-}20 \times 10^{17}$
HI	Au	25.0	978	5.2×10^{22}	1.6×10^{17}
$HCOOCH(CH_3)_2$	Glass	35.0	714	7.5×10^{17}	5.8×10^{17}

Therefore, using Equation (27) we have

$$\frac{\theta}{1 - \theta} = C_l \frac{q_{ads}}{(q_l/V) \, q_s} \, e^{-\Delta E_{ads}/RT} \quad . \tag{30}$$

Equation (30) will be discussed again in the *ab initio* section below.

Proceeding to the kinetics, the transition state theory (TST) can be applied to the rate of adsorption, the rate of desorption or the rate of unimolecular or bimolecular reaction on the surface. For example, for unimolecular reactions species A will adsorb onto the surface, S, to produce an adsorbed complex, AS. Given sufficient energy, this complex will reach an activated complex, $AS^{\ddagger}$, and proceed to form products:

$$A + S \rightarrow AS \rightarrow AS^{\ddagger} \rightarrow products \quad . \tag{31}$$

In this case, the TST expression for the rate of formation of products is

$$Rate = C_{ads} \frac{kT}{h} \frac{q^{\ddagger}}{q_{ads}} \, exp(-E^{\ddagger}/RT) \quad , \tag{32}$$

where $E^{\ddagger}$ is the energy change between the activated complex and the adsorbed complex, $q^{\ddagger}$, refers to the partition function of the activated complex and q_{ads} refers to the partition function of the adsorbed complex. Note that although the activated complex is not a true minimum, it is a minimum in all degrees of freedom except the reaction coordinate. Hence, the partition function, $q^{\ddagger}$, is calculated in TST with the same formulas as in Table 3, using the usual vibrational frequencies of the activated complex and the rotational and translational properties of the complex geometry. However, one of the vibrational frequencies will be imaginary (corresponding to the reaction coordinate) and it is implicit that the partition function excludes this one degree of freedom.

In the case that the surface coverage is fairly extensive, C_{ads} will approach a constant value (on the order of 10^{15} molecules/cm^2) and the rate will be zeroth order with respect to the concentration of A in the fluid. As a further simplification one may assume that the two partition functions in (32) are unity, because they contain mostly vibrational contributions (this will be wrong if there are significant low frequencies of vibration). Thus we could write

$$Rate = 10^{15} \frac{kT}{h} \, exp(-E_{expt}/RT) \quad , \tag{33}$$

where E_{expt} is the experimental activation energy and combines the effect of both the reaction activation energy and the adsorption energy change. For example, table 4 gives results from simple applications of Equation (33) to some unimolecular rate kinetics. Even at this extremely simple level, the TST results are encouraging, i.e., the results yield the correct order of magnitude for the various rates.

It is important to stress that TST can be used not only to say something about the activation energies but also about the full rate constant, e.g., the pre-exponential factor. Furthermore, in silicate studies (e.g., Dove and Crerar, 1990), it has sometimes been found that an increase in the rate occurs mostly in the pre-exponential. Such results can be studied from the specific formulas for the pre-exponential using the formulas for the partition functions. In the next section, we look at an important geochemical reaction from the TST point of view.

Application of TST to quartz dissolution

The formulas given above can be applied to an analysis of the kinetics of quartz dissolution. Based on the suggestion in Lasaga and Gibbs (1990a,b) we will postulate that the rate of dissolution depends on the rate of hydrolysis of the SiOSi unit following a water "acceptor" adsorption (see the *ab initio* section). Figure 11 gives the proposed mechanism. If we break it into each step, the reaction mechanism would be:

$$H_2O + \; \equiv Si- \; \overset{K_1}{\rightleftharpoons} \; H_2O_{ads} \bullet Si \overset{k_1}{\rightarrow} \; \equiv Si - OH$$

$$H_2O + \; \equiv Si - OH \overset{K_2}{\rightleftharpoons} \; H_2O_{ads} \bullet Si - OH \overset{k_2}{\rightarrow} = Si(OH)_2$$

$$H_2O + \; = Si(OH)_2 \overset{K_3}{\rightleftharpoons} \; H_2O_{ads} \bullet Si(OH)_2 \overset{k_3}{\rightarrow} \; -Si(OH)_3 \qquad (34)$$

$$H_2O + \; -Si(OH)_3 \overset{K_4}{\rightleftharpoons} \; H_2O_{ads} \bullet Si(OH)_3 \overset{k_4}{\rightarrow} \; H_4SiO_4^{ads}$$

$$H_4SiO_4^{ads} \overset{k_5}{\rightarrow} \; H_4SiO_4(aq)$$

The K's refer to the equilibrium constants for H_2O adsorption onto each particular site. The slow hydrolysis steps have individual rate constants given by k_i. Applying the principle of steady state for the various intermediates (see Lasaga, 1981a), which sets the rate of change of the concentration of each adsorbed species to zero, the reader can verify that

$$k_1[H_2O_{ads} \bullet Si] = k_2[H_2O_{ads} \bullet Si(OH)] = k_3[H_2O_{ads} \bullet Si(OH)_2] \qquad (35)$$

$$= k_4[H_2O_{ads} \bullet Si(OH)_3] = k_5[H_4SiO_4^{ads}] \; .$$

The total concentration of reactive surface sites, S_{tot}, will be given by:

$$S_{tot} = [\equiv Si] + [\equiv Si(OH)] + [= Si(OH)_2] + [-Si(OH)_3] \qquad (36)$$

$$+ [H_4SiO_4^{ads}] + [H_2O_{ads} \bullet Si] + [H_2O_{ads} \bullet Si(OH)]$$

$$+ [H_2O_{ads} \bullet Si(OH)_2] + [H_2O_{ads} \bullet Si(OH)_3] \; .$$

We can use Equation (35) and the equilibrium assumption for the H_2O adsorbed species to rewrite Equation (36) as,

$$S_{tot} = [\equiv Si] \, D \qquad (37)$$

where the parameter D is given by:

$$D = 1 + \frac{k_1 K_1}{k_2 K_2} + \frac{k_1 K_1}{k_3 K_3} + \frac{k_1 K_1}{k_4 K_4} + K_1 a_{H_2O} + \frac{k_1}{k_2} K_1 a_{H_2O} + \qquad (38)$$

$$\frac{k_1}{k_3} K_1 a_{H_2O} + \frac{k_1}{k_4} K_1 a_{H_2O} + \frac{k_1}{k_5} K_1 a_{H_2O} \; .$$

The overall rate law (i.e., the release rate of H_4SiO_4 to solution) is given by

$$Rate = k_5[H_4SiO_4^{ads}] = k_1[H_2O_{ads} \bullet Si]$$

$$= k_1 \, K_1 a_{H_2O} \, [\equiv Si] \; ,$$

where Equation (35) and the equilibrium assumption were used. Substituting from Equation (37) yields the final result:

$$Rate = k_1 K_1 a_{H_2O} \frac{\alpha A_{tot}}{D} \ , \tag{39}$$

where A_{tot} is the total surface area of "active" sites (i.e., where the water can adsorb) and is related to S_{tot} by a simple geometric factor, α. If the adsorption equilibrium constants and the rates are similar and if k_5 is fast, D is of the order of magnitude of unity. Equation (39), then, simplifies to the product of the hydrolysis rate constant (assumed similar in all steps) and the equilibrium constant for adsorption. The activation energy, E_{act}, in this case, will be given by:

$$E_{act} = E^{\ddagger}_{hydrolysis} + \Delta H_{ads} \ , \tag{40}$$

where $E^{\ddagger}$ is the activation energy for the hydrolysis reaction activated complex and ΔH_{ads} is the standard state enthalpy of the adsorption equilibrium constant for water onto silica surface. Let us now use the simplified rate law in a transition state calculation. It will be more convenient to replace $K_1 a_{H_2O}$ by the surface concentration of adsorbed water molecules and to write this concentration as a product of the surface sites, C_s, and a mole fraction X. In this case, applying the transition state formula for the hydrolysis rate constant, k_1, (e.g., Eqn.(32)), the rate will be:

$$Rate = C_s X_{H_2O} \frac{kT}{h} \frac{q^{\ddagger}}{q^{ads}} exp(-E^{\ddagger}_{hydrolysis}/RT) \ , \tag{41}$$

where C_s refers to the **total** number of Si sites on the surface (in moles/cm^2), X_{H_2O} is the mole fraction of sites occupied by adsorbed water molecules, q are the partition functions (see Table 3) and $E^{\ddagger}_{hydrolysis}$ is the activation energy (including the zero-point energies of vibration). We can estimate C_s from the geometry of the crystal lattice. In general, C_s for most minerals will be around 10^{15} sites per cm^2 or 1.6×10^{-9} moles per cm^2. For example, the total number of silanol (SiOH) groups in a fully hydroxylated silica surface is around 1.1×10^{-9} moles per cm^2 (Iler, 1979). *Ab initio* and experimental activation energies are around 20 kcal/mole. We need to correct the experimental E for the energy of adsorption of water (a few kcal/mole; Iler p. 627, 1979). For our purposes, we will take E as 24 kcal/mole.

The partition functions for the silica dissolution model require knowledge of the vibrational frequencies of the adsorption complex and the activated complex. (There are no translational or rotational modes to consider in this model). In fact, given the vibrational frequencies, ν_i, the partition function ratio is given by:

$$\frac{q^{\ddagger}}{q^{ads}} = \frac{\prod_i (1 - e^{-h\nu_i^{ads}/kT})}{\prod_i (1 - e^{-h\nu_i^{\ddagger}/kT})} \ . \tag{42}$$

(e.g., see Table 3). These vibrational frequencies can be obtained from *ab initio* calculations (see next section) on disiloxane (Casey et al., 1990). Table 5 gives the frequencies. At 200 °C, the results of inserting the frequencies into Equation (42) are

$$\frac{q^{\ddagger}}{q^{ads}} = 3.3 \times 10^{-3} \ .$$

Table 5. 6-31G* *ab initio* frequencies for the water-disiloxane
adsorption complex and the hydrolysis transition state

Adsorption complex	Transition state	Adsorption complex	Transition state
4066.1	4134.7	819.3	870.9
4180.0	2324.5	775.5	790.9
2377.6	2399.5	766.9	782.2
2422.3	2431.9	437.4	728.5
2411.2	2416.0	175.0	483.4
2414.1	2163.4	204.3	574.4
2397.7	2194.1	117.2	560.0
2389.5	2427.3	133.1	405.6
1831.8	1365.5 i	58.5	142.6
1125.2	1239.6	95.4	203.6
1150.2	1332.3	27.0	48.1
1062.8	1116.7	47.1	72.6
1042.5	1182.0		
1057.0	1022.2		
1049.3	1061.8		
1051.8	1039.8		
837.7	959.4		
627.0	1047.9		

Note that the partition function ratio, in this case, is **not** unity. This deviation arises because both the adsorption complex and the transition state have a number of low vibrational frequencies. In fact, the adsorption complex has the water molecule at a distance of 3.5 Å and hence has even more low frequencies than the transition state, thereby giving rise to a ratio that is less than one. Note that the biggest contributions to the partition function come from vibrational frequencies less than $300\ cm^{-1}$. Therefore, the contribution from the high frequency of vibration of SiH bonds (or equivalently the SiO bonds in quartz) would be negligible. The size of the partition function is governed by the low frequency modes directly related to either the adsorption or the activated complex. Combining these numbers we have that at 200 °C:

$$Rate\ =\ 4.4 \, x \, 10^{-10} \, X_{H_2O}\ \ moles/cm^2/sec \ . \tag{43}$$

We can compare this result with the experimental data of Dove and Crerar (1990), for dissolution of quartz in deionized water at 200 C:

$$Rate\ =\ 2.14 \, x \, 10^{-12}\ \ moles/cm^2/sec \ . \tag{44}$$

The conclusion from this calculation is that the fraction of sites occupied by water molecules is only around 0.02. This is not surprising, taking into account that the adsorption is not as favorable as the adsorption onto a silanol group (see *ab initio* section below).

Once the complex has been found, the familiar formulae of TST can be used to learn something about $\Delta S^{\ddagger}$, $\Delta H^{\ddagger}$. $\Delta H^{\ddagger}$ can be obtained from

$$\Delta H^{\ddagger} \ = \ \Delta E^{\ddagger} + \frac{1}{2} \sum_i h\nu_i^{\ddagger} - \frac{1}{2} \sum_i h\nu_i^{react} \ , \tag{45}$$

where the zero-point energy of vibration has been added to the energy term. $\Delta S^{\ddagger}$ is obtained by first rewriting k as

$$k = \frac{kT}{h} \frac{\gamma^{react}}{\gamma^{\ddagger}} e^{\Delta s^{\ddagger}/R} e^{-\Delta H^{\ddagger}/RT} \quad , \tag{46}$$

which yields,

$$\Delta S^{\ddagger} = R \ln \left(\frac{q^{\ddagger}/V}{q^{react}/V} \right) \quad . \tag{47}$$

The Arrhenius parameters can be obtained from $\Delta H^{\ddagger}$ and $\Delta S^{\ddagger}$:

$$E_a = RT + \Delta H^{\ddagger} \quad , \tag{48}$$

and

$$A = e \frac{kT}{h} \frac{\gamma^{react}}{\gamma^{\ddagger}} e^{\Delta s^{\ddagger}/R} \tag{49}$$

(Lasaga, 1981b).

Ab initio METHODS IN MINERAL SURFACE REACTIONS

A first-principles investigation of the bonding and atomic dynamics of minerals and their interaction with both hydrated ions and water or with melts can treat such important problems as, for instance, the link between local defects or impurities and the bulk kinetic properties or the nature of the bonds and forces involved in surface reactions or the association of a local adsorption complex with a particular catalytic or inhibitory effect. The task at hand is to evaluate the interatomic forces from the fundamental laws of physics and a handful of universal constants (such as Planck's constant, the electron charge, the speed of light, the mass of nuclei and electrons) - hence the word *ab initio*. No empirical or semi-empirical constants are introduced into the calculations. On the other hand, this is not to say that there are no approximations used in the *ab initio* calculations. These approximations will become clearer below. In many cases *Ab initio* methods have become as good as or better than experiments for the accurate description of structures and vibrational frequencies of new gas-phase molecules. Fortunately, current computational levels enable us to extend the applications of *ab initio* methods to mineral surface reactions. As will be emphasized again later, an important component of successfully carrying out *ab initio* calculations on silicates and their surface reactions is the **local** nature of the chemical forces shaping the dynamics of silicates. Earlier work (Gibbs, 1982; Lasaga and Gibbs, 1987, 1988, 1989) has shown that *ab initio* calculations on molecular clusters are capable of predicting the crystal structure and the equations of state of many silicates quite successfully.

Ab Initio theory

The evaluation of potential surfaces is based on the use of the **Born-Oppenheimer approximation** (e.g., see Lasaga and Gibbs, 1990a). In the Born-Oppenheimer approximation the positions of the nuclei are fixed and the wave equation is solved for the wavefunction of the electrons. This separation of electron and nuclear motion is allowed because the nuclear masses are much greater than the mass of the electrons and, as a result, nuclei move much more slowly. In the Born-Oppenheimer approximation the eigenvalue, E, will then be a function of the atomic positions i.e., $E(\vec{R})$. The Schrodinger equation now becomes:

$$\hat{H}^{elec} \Psi^{elec}(\mathbf{r}, \mathbf{R}) = E(\mathbf{R}) \Psi^{elec}(\mathbf{r}, \mathbf{R}) \quad , \tag{50}$$

where the electronic Hamiltonian is given by the sum of the kinetic and potential energies of the electrons and the nuclear-nuclear electrostatic repulsion. $\mathbf{r}$ stands for the spatial coordinates of all the electrons in the system.

The eigenvalue, $E(\mathbf{R})$, in Equation (50) yields the Born-Oppenheimer potential surface if the nuclear positions, $\mathbf{R}$, are all varied. Sometimes this surface is also called the adiabatic (for slow motion) potential surface. If we know $E(\mathbf{R})$ accurately, we can predict the detailed atomic forces and the chemical behavior of the entire system.

A usual solution to the Schrodinger equation assumes that the electrons can be approximated as independent particles that interact mainly with the nuclear charges and with an average potential from the other electrons. In this case, Ψ^{elec} can be written as a product of functions of only one electron coordinates, $\psi_i(x, y, z)$, an approximation that is called the **Hartree-Fock approximation**. This separation of variables allows us to write:

$$\Psi = \sum_P (-1)^P \mathbf{P}[\psi_1(1)\alpha(1) \ \psi_1(2)\beta(2) \ \cdots \ \psi_n(2n)\beta(2n)] \quad . \tag{51}$$

The electronic wavefunction, Ψ, must be anti-symmetric with respect to interchange of any two electrons. Therefore, we sum over all the possible permutations, $\mathbf{P}$, of the 2n individual functions, except that a +1 or -1 is inserted in front of each product, depending on whether the permutation is even or odd. The product in Equation (51) assumes that each spatial function can accomodate two electrons, one with spin up (α) and one with spin down (β). Thus Equation (51) is correct for closed shell systems, which form the majority of systems relevant to our discussion.

The one-electron functions, ψ_i, in Equation (51) are called **molecular orbitals**. These molecular orbitals form the basis for the conceptual treatment of bonding in molecules (see Lasaga and Gibbs, 1990a). In practice, the molecular orbitals are expanded as a sum over some set of prescribed atomic orbitals, ϕ_μ (i.e., the usual 1s, 2s, 2p ... functions),

$$\psi_i = \sum_{\mu=1}^{N} c_{\mu i} \phi_\mu \quad . \tag{52}$$

The set of coefficients, $c_{\mu i}$, are obtained from optimizing the solution to the Schrodinger equation. One problem with this scheme is that the unknown coefficients, $c_{\mu i}$, appear also in the potential energy term of the one-electron differential equation. The usual method is to iterate the solution for the coefficients until the new coefficients equal the old coefficients. Hence, after a set of coefficients is solved for, this set is input into the potential term to update the differential Equation and obtain a new matrix. Then this new equation is used to obtain a new set of coefficients. This process is repeated until the set of coefficients does not change and we reach what is termed a "self-consistent field" (SCF) solution . These Equations are termed the **Hartree-Fock SCF** equations.

The set of atomic orbitals, { ϕ_μ }, used to obtain the molecular orbitals (Eqn.(52)) is termed the **basis set**. The size of the atomic orbital set, N, varies with the accuracy demanded of the calculation. A **minimal basis set** is one which merely uses the minimum atomic orbitals needed to accomodate all the electrons in the system up to the valence electrons. For example, a minimal basis set on silicon would have one 1s function, one 2s function, three 2p functions, one 3s function and three 3p functions. Bigger basis sets are labeled **extended basis sets** (see Lasaga and Gibbs, 1990b). Mathematically, if the number of "different" atomic orbitals increases to infinity (i.e., this set forms what is termed a "complete" set), then the orbitals, ψ_i, obtained by the coefficients $c_{\mu i}$ will be the exact solutions to the one-electron differential equation. In turn, this limit would yield the best possible solution to the full Born-Oppenheimer Schrodinger equation <u>within</u> the "separation of variables" scheme represented by Equation (51). Such a solution is termed the **Hartree-Fock limit**.

To go beyond the Hartree-Fock limit and obtain the full solution to the Schrodinger equation (in the non-relativistic and Born-Oppenheimer limit), one would have to combine various solutions of the type used in Equation (51). In general, the n molecular orbitals with the <u>lowest</u> molecular orbital energies are used in the Hartree-Fock solution, Equation (51), for the ground state of a 2n electron system. The rest of the molecular orbitals obtained will be **excited molecular orbitals**. Other possible wavefunctions of the type given in Equation (51) can be formed by using excited molecular orbitals in the product. The set of all possible products can now be used as a basis set to solve the full Schrodinger equation:

$$\Psi = \sum_i C_i \sum_P (-1)^P \mathbf{P}[\psi_{i_1}(1)\alpha(1)\cdots\psi_{i_n}(2n)\beta(2n)] \quad , \tag{53}$$

where $i = \{i_1, i_2, ..., i_n\}$ stands for the set of n molecular orbitals used in the particular ith product and the n orbitals are picked from the (in principle) infinite set of molecular orbitals allowable in the system. Such a calculation corrects the energy for what is termed **electron correlation** (i.e., it corrects for the assumption of average motion used in the one-electron differential equation for the ψ_i). The method of Equation (53) is termed **configuration interaction**, because the sum in (53) is one over many electronic configurations involving excited electronic states For most chemical systems, the solutions obtained by extensive application of Equation (53) (e.g., high N) become equivalent to the exact solution of the potential energy surface discussed earlier.

A faster convergence is achieved if both the molecular orbital coefficient and the coefficients C_i are optimized at the same time. Such a scheme is labeled the multiconfiguration self-consistent field or MCSCF for short.

Atomic orbitals, ϕ_{μ}, are usually expanded in terms of a number M of gaussian functions:

$$\phi_\mu = \sum_{s=1}^{M} d_{\mu s}\, g_s \quad . \tag{54}$$

The gaussian functions, themselves, would have the appropiate behavior as the particular atomic orbital being approximated. For example for p_x, p_y, p_z orbitals, the general gaussians would all be:

$$g_{p_x} = (\frac{128\alpha^5}{\pi^3})^{1/4}\, x\, exp(-\alpha\, r^2) \quad ,$$

$$g_{p_y} = (\frac{128\alpha^5}{\pi^3})^{1/4}\, y\, exp(-\alpha\, r^2) \quad ,$$

$$g_{p_z} = (\frac{128\alpha^5}{\pi^3})^{1/4} z \, exp(-\alpha \, r^2) \quad .$$

Note that in each case the only difference between the gaussians of a given type is in the exponential, α. The coefficient in front simply normalizes the function so that the integral over all space is unity.

The coefficients, $d_{\mu s}$, in Equation (54) are chosen so as to minimize the difference between the atomic orbitals given by (53) and by (54). Once chosen, these coefficients are fixed in all subsequent ab initio calculations i.e., ϕ_μ is completely specified by (54). The size of the exponent α in the gaussian determines how close to the nucleus the electron charge is or conversely how "diffuse" the electron charge is. For higher level basis sets, two sets of valence atomic orbitals are used (these are called **split valence** basis sets). One set lies close to the nucleus and mimics closely the true valence atomic orbitals. The other set is more diffuse (smaller α) and enables the molecular orbital to respond to electron cloud deformation away from the nucleus due to chemical bond formation (see Lasaga and Gibbs, 1990b).

For example, the basis set 3-21G for the oxygen **valence** orbitals would have the following split functions:

$$2p_{x,O} = 0.245N_\alpha \, x \, e^{-7.403r^2} + 0.854N_\alpha \, x \, e^{-1.576r^2} \quad (inner)$$

and

$$2p'_{x,O} = N_\alpha \, x \, e^{-0.374r^2} \quad (outer) \quad .$$

The N_α in front of each gaussian refers to the normalization factor (which depends on the size of α) given earlier. Note the much higher values of α for the $2p_x$ orbital of oxygen than the $2p'_x$, which keeps the $2p_x$ orbital closer to the nucleus. Thus, the $2p'_{x,O}$ orbital is much more diffuse than the $2p_{x,O}$. This added flexibility is needed to describe major changes in the bonding of minerals.

The size of the M used in expanding each atomic orbital in terms of gaussians (Eqn.(54)) is usually included in the basis set description. Thus a STO-3G set is a minimal basis set with each atomic orbital (i.e., each STO) expanded by three gaussian functions in Equation (54) (i.e., M = 3). For extended sets, one normally uses more gaussians to describe the inner (core) atomic orbitals; therefore, more numbers are given in the label. For example, a 3-21G basis set is an extended basis set with three gaussians used to expand the core atomic orbitals, two gaussians used to expand one set of valence atomic orbitals and one gaussian used to expand a more "diffuse" set of atomic orbitals. (This is a **split-valence basis set**). If, in addition, orbitals of higher angular momemtum than required by the electrons in a given atom are used (termed **polarization functions**), an asterisk is added. For example, the 3-21G* basis would add 3d orbitals on all second row atoms.

As an illustration, a Gaussian 3-21G* basis set for silicon will consist of the following atomic orbitals:

1s, 2s, $2p_x$, $2p_y$, $2p_z$ − all described with 3 Gaussians in (10)

3s, $3p_x$, $3p_y$, $3p_z$ inner orbitals − 2 Gaussians

3s, $3p_x$, $3p_y$, $3p_z$ outer orbitals − 1 Gaussian

$3d_{x^2}$, $3d_{y^2}$, $3d_{z^2}$, $3d_{xy}$, $3d_{xz}$, $3d_{yz}$ − 1 Gaussian (polarization)

Altogether, 19 atomic orbitals will be inputted into the *ab initio* calculations per Si atom, requiring 33 gaussian functions to describe them. For a calculation on H_3SiOH, there would be 36 atomic orbitals (e.g., 19 on Si, 9 on O and 2 on each H) input to obtain the molecular orbitals. The 36 atomic orbitals would be expanded by 60 gaussians.

Once an accurate wavefunction has been obtained in *ab initio* calculations, the forces on all the atoms in a cluster can be computed exactly and analytically using well-developed quantum mechanical techniques. This ability enables us to carry out a full *ab initio* minimization of the cluster geometry and extract the optimal equilibrium geometry. In general, the optimization algorithm searches for a **stationary point**, i.e., a molecular structure such that for all atomic coordinates the force is zero. Mathematically, this means that

$$\frac{\partial E}{\partial x_i} = 0 \quad i = 1, ..., 3N \tag{55}$$

for a cluster of N atoms, where E is the potential energy as a function of the atomic nuclear positions i.e., the Born-Oppenheimer energy, $E(\mathbf{R})$, in Equation (50). Having reached a stationary point, it is important to ascertain whether the structure is a **true minimum**. This test is achieved by analyzing the eigenvalues of the second derivative or **Hessian matrix**:

$$\mathbf{H}_{ij} \equiv \frac{\partial^2 E}{\partial x_i \partial x_j} \quad . \tag{56}$$

For a true minimum, the eigenvalues of $\mathbf{H}$ must be all positive except for the six zeroes corresponding to three translations and three rotations of the cluster (which don't change the energy in our case). This test is important because it can enable us to check postulated atomic structures to see if they are stable and also enables us to distinguish minima from saddle points and other more complex stationary points to be discussed in the next section.

The kind of accuracy possible in solving the Schrodinger equation today can be clearly seen by the impressive agreement of many *ab initio* results with experimental data. As an example, Figure 12 gives a number of results comparing *ab initio* interatomic bond lengths with experimental values. There is no question that with the present capabilities we can accurately calculate the potential surfaces, which describe the chemical behavior of a wide range of systems. Our next task is to apply this capability to the dynamics of geochemical reactions.

Ab Initio studies of adsorption

Ab initio studies of adsorption have been numerous during the past few years. Generally, the studies have focused on the chemical bonding of species on a variety of surfaces, usually by chemisorption. This work can be subdivided into studies of chemisorption onto semiconductor surfaces, e.g., Si or Ge, chemisorption onto metals, chemisorption onto simple ionic crystals, or chemisorption onto silica, aluminosilicates and zeolites.

44

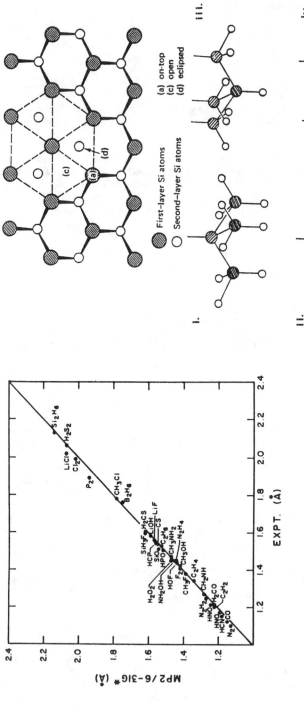

Figure 12. Comparison of experimental results with high-level *ab initio* calculations (from Hehre et al., 1986).

Figure 13. a) Schematic of the Si (111) surface showing first and second layer Si atoms. Two kinds of threefold sites, open and eclipsed, are indicated by the dashed lines. Representative on-top, open, and eclipsed sites are labeled with letters a,c, and d. b) The molecular clusters used to model Si chemisorption onto Si (111). i) The two-layer Si_4H_9 cluster that models the on-top site of Si(111) ii) The four layer $Si_{10}H_{15}$ cluster used to model the on-top site iii) the two-layer Si_4H_7 cluster used to model the eclipsed site iv) the four-layer $Si_{10}H_{13}$ cluster used to model the open site of Si (111). The dangling bonds of the first layer Si atoms are indicated. (from Seel ang Bagus, 1983).

In general, the assumption of all these *ab initio* studies is that local (i.e., nearest neighbor or second nearest neighbor) chemical forces determine most of the dynamics and energetics of chemisorption processes. The undisputed success of these studies in a large number of cases has proven the validity of that claim.

Chemisorption processes onto Si surfaces are of obvious interest to the electronics industry and it is no surprise that the majority of works have focused on the Si surface. Seel and Bagus (1983) have carried out a study of the interaction of fluorine and chlorine with the Si (111) surface. The cluster model of the Si (111) surface is illustrated in Figure 13. Notice the commonly used embedding procedure where hydrogen atoms are used to truncate the lattice. Three adsorption sites were studied (1) on-top adsorption (2) deeper adsorption onto the second layer Si atoms in the open and (3) eclipsed geometries (see Fig.13). The *ab initio* calculations used a special basis, which is probably in the range of a 6-31G level basis. An important result compares the on-top adsorption (the most stable chemisorption found) of F with two clusters of different size, Si_4H_9 and $Si_{10}H_{15}$. The results were nearly identical for both clusters, thus establishing the local nature of the adsorption (i.e., the basic premise of all these methods). The calculated valence bandwidth and the position of the surface dangling-bond states are in satisfactory agreement with experiment. A significant difference between F and Cl is that the F atom has another minimum adsorption site 1.4 $\overset{\circ}{A}$ below the surface with an energy barrier of 1 eV. On the other hand, Cl with its larger size has a significant 13 eV barrier to penetrate the surface. Similar results were found with higher level calculations by Illas et al. (1985).

Calculations of adsorption on metals are also becoming increasingly common. These *ab initio* results are now being incorporated into MD simulations (see next section). For example, Panas et al.(1988) modelled adsorption of C or N onto a Ni surface using a Ni_5X cluster (X = C, N, etc.). Their calculations used a high level multireference CCI method. Nakatsuji et al. (1988) looked at the reaction of H_2 with a Pt surface.

Adsorption onto ionic surfaces has been modelled using point charges for the crystal lattice. For example, Bourg (1984) has modelled adsorption of Li_2 onto LiF by carrying out STO-6G *ab initio* calculations on + and - point charges laid out in the (100) surface. The Li_2 adsorbs stably only on top of fluorine sites, as would be expected, and the binding energy is calculated to be small, i.e., -0.019 eV (relative to the Li_2 energy). Considering that thermal energy, RT, is on the order of 0.026 eV, most of the adatoms will evaporate. On the other hand, calculations carried out in the presence of a step yield a binding energy that is around -0.054 eV. Here we see a first principles approach to the stability question of attachment to steps, a topic that plays such a central role in all atomic theories of crystal growth (see Monte Carlo section).

Ab initio studies of adsorption onto silica and aluminosilicate surfaces have been done in a number of earlier papers (Sauer 1987; 1989; Geerlings et al., 1984; Hobza et al., 1981; Mortier et al., 1984). The 1-4 orders of magnitude difference in catalytic activity between zeolites and aluminosilica has fueled a number of studies on the role of hydroxyl surface groups in the surface chemistry of these materials. Central to these studies has been the relative acidity of different OH sites on these surfaces. In fact the question of relative acidity of surface groups forms an important part of other chapters in this volume. Two particularly important OH groups are the terminal OH and the bridging OH groups. These are depicted in Figure 14. Usually the *ab initio* calculations on terminal OH's have used H_3SiOH or H_4SiO_4 (see Fig.14) as clusters, again simulating the rest of the lattice by the H atoms. The bridging OH model is usually based on the disiloxane molecule as shown in Figure 14. For example, Geerlings et al. (1984) used H_3SiOH and $H_3Si - OH - SiH_3$ to

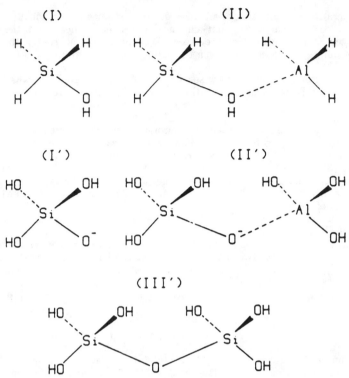

Figure 14. Typical clusters used in *ab initio* calculations of the relative acidity of surface groups on silica and aluminosilicate surfaces. I. Simple model for a terminal OH group on silica . II. Simple model for a bridging OH group on aluminosilicate surfaces. I′. Higher level cluster for a dissociated terminal OH group. II′. Cluster used to model a dissociated bridging OH group on aluminosilicates. III′. Cluster model of a bridging oxygen site on silica.

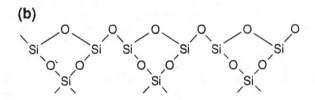

Figure 15. a) Hydroxylated silica surface showing the usual adsorbed water onto silanol groups. b) Hydrophobic dehydroxilated silica surface comprised essentially of siloxane bridging bonds.

Table 6. O-H stretching vibrational frequencies in cm^{-1}, 3-21G calculations[a]

molecule	without scaling	scaling[b]	observed
H$_3$COH	3868	3690	3681
H$_2$O	3876	3698	3707
fig 14 II'	3931	3750	3665
fig 14 I'	3995	3811	3745

a - from Mortier et al. (1984)
b - frequencies scaled by 0.9102 based on earlier *ab initio* work.

study the adsorption of CO, H_2O and NH$_3$. They employed a 3-21G split-valence basis set and observed the weakening of the OH bond from the terminal to the bridging type. Comparing the binding energies of water molecule on both types of OH surface groups, it is clear that the *ab initio* results predict that adsorption onto a terminal group is much more stable than adsorption onto a bridging group. These results have been confirmed in the other works (e.g., Hobza et al., 1981; Mortier et al., 1984; Sauer, 1987). Once the silanol groups are eliminated during dehydroxylation (e.g., heating to several hundred degrees for several hours; Iler, 1979), and the surface is comprised of essentially Si-O-Si siloxane bonds (see Fig.15b), the silica surface has a **hydrophobic** character, which illustrates the major role of silanol groups in water adsorption.

The adsorption of water molecule onto a terminal OH group can occur by either donation of a proton or reception of a proton (see Fig.16). Although earlier semi-empirical calculations had the proton-donor being the most stable, all subsequent higher level *ab initio* calculations have shown that the most stable adsorption has the oxygen of the water molecule forming a H-bond with the hydrogen of the OH surface group Figure (16) Lasaga and Gibbs, 1990a,b; Hobza et al., 1981).

Sauer (1987) uses a higher level 6-31G* basis to compare the relative acidities of the OH surface groups, i.e., the relative energies of the reaction:

$$X - OH \rightarrow XO^- + H^+ \ . \tag{57}$$

In addition to the groups discussed above, he includes calculations on $H_3Si - OH - AlH_3$, which are relevant to aluminosilicates (including feldspars) and zeolites. His results, shown in Figure 17, show that the Si-OH-Al group is by far the most acidic, in agreement with experiment. Mortier et al. (1984) have also carried out similar calculations with the same conclusions. The SiOHAl moiety is of great importance in the hydrolysis of feldspar (see below). In addition, Table 6 from their work shows the good agreement between the *ab initio* ν_{OH} frequencies and experimental work.

As discussed by Hobza et al. (1981) in studying the adsorption energetics of water on silica surfaces, the appropiate reaction to use is:

$$H_2O \cdots H_2O + M \leftrightarrow M \cdots H_2O + H_2O \ , \tag{58}$$

where M stands for the appropiate silica surface group. In other words, the energy gained in adsorbing the water must be weighed against the energy lost in breaking a hydrogen bond. In fact, a study of the hydrophilic or hydrophobic nature of a surface group can be done by comparing the ΔH_0 for adsorption to the ΔH_0 for the reaction

$$H_2O + H_2O \Longrightarrow H_2O \cdots H_2O \ ,$$

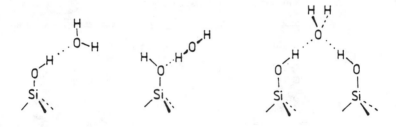

Figure 16. Different possible geometries for the adsorption of water onto a silanol group.

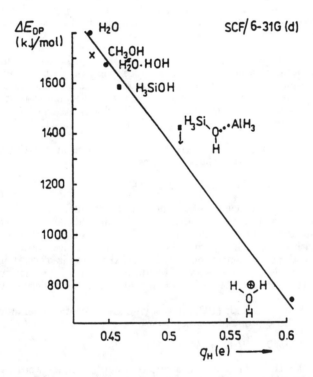

Figure 17. Calculated deprotonation energies and their dependence on the net atomic charge on hydrogen. The asterisk in *H₂O•HOH indicates the acidic proton considered. The arrow shows for H₃SiOH•AlH₃ the decrease of deprotonation energy expected on relaxation of the geometry of the anions (from Sauer, 1987).

the product being the water dimer (e.g., Hobza et al., 1981).

Normally the surface of silica is covered by numerous silanol (SiOH) groups reaching about 6-7 OH groups per 100 $\mathring{A}^2$ (Iler, 1979). By far the most favorable mode of adsorption of water on silica is by "donor" adsorption onto a silanol group i.e., by the hydrogen bonding of the H in the silanol group to the oxygen of the incoming water molecule so that water "sits oxygen down on the SiOH groups" (Iler 1979, p. 627). This is illustrated in Figure 15a.

Figure 18 gives the geometry of the adsorbed water molecule in the case of donor adsorption onto the silanol group. Note the strong hydrogen bond formed. The oxygen-oxygen distance in Figure 18 is 2.66 $\mathring{A}$, which is actually shorter than the oxygen-oxygen distance (2.80 $\mathring{A}$) calculated (using 3-21G*) for the water dimer. Analysis of the Hessian matrix verifies that the adsorbed structure is a local minimum. In fact, Table 7 confirms that water adsorption onto the hydrogens of silanol groups on silica surfaces is energetically favorable; i.e., the energy change for the reaction:

$$\equiv SiOH + H_2O \cdots H_2O \rightarrow \equiv SiOH \cdots OH_2 + H_2O \qquad (59)$$

from Table 7 is $\Delta E = $ -3.5 kcal/mole.

The key step in the kinetics of silica dissolution, however, requires a special type of water adsorption, i.e., the acceptor adsorption onto disiloxane shown in Figure 19 (Lasaga and Gibbs, 1990a). This type of adsorption is required by the need to form a new SiO bond between the oxygen of the water molecule and the surface silicon atom in step 2 of Figure 20. Note that the kind of adsorption postulated in Figure 20 (and exhibited further below) leads to a surface silicon that is 5-fold coordinated. The existence of 5-fold coordinated silicon in glasses and melts has been postulated in several MD papers (Angell et al., 1982; Kubicki and Lasaga, 1988), in spectroscopic studies (Xue et al., 1990) and in other works (Liebau, 1984). Table 7 gives our results for the adsorption of water on H_3SiOH and disiloxane. In agreement with experiment, acceptor adsorption onto a disiloxane group is not as stable as adsorption onto the silanol group. However, an analysis of the Hessian matrix indeed shows that the geometry in Figure 19 corresponds to a full minimum in the Born-Oppenheimer surface. Note that for the 6-31G* calculation the distance from the silicon to the oxygen of the adsorbed water molecule is 3.538 $\mathring{A}$, which is even longer than the OO distance in the water dimer. It is interesting to point out that the OO distance in the channels along the c-axis of quartz is only around 3.56 $\mathring{A}$. In other words, a water molecule would have to "squeeze" to adsorb within the spiraling network of silica tetrahedra. On the other hand, the distance between the hydrogen of the adsorbed water and the bridging oxygen is 2.197 $\mathring{A}$, only slightly larger than the OH distance (2.028 $\mathring{A}$) in the 6-31G* water dimer. In addition, the original SiO bond length increases slightly upon adsorption of the water molecule from 1.626 $\mathring{A}$ to 1.648 $\mathring{A}$, a precursor state to the ultimate rupture of the bond in the subsequent dynamics.

The energetics of reaction (59) for the water acceptor adsorption depicted in Figure 19 are not favorable (i.e., $\Delta E > 0$, see Table 7). Because this adsorption is precisely the type of adsorption onto siloxane bonds, the unfavorable ΔE confirms the hydrophobic nature of the dehydroxylated disiloxane-rich surface of silica.

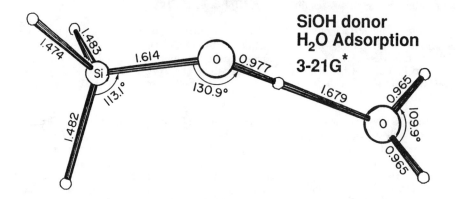

Figure 18. *Ab initio* calculation of the optimized geometry of water adsorption onto a terminal silanol group. Big circles are oxygens, intermediate circles are silicons and small circles are hydrogens (from Lasaga and Gibbs, 1990b).

Table 7. Water Adsorption energies[a]

3-21G* Adsorption energies onto H_3SiOH

	ΔE^a	ΔE^b	ΔH^a	ΔH^b
Donor Adsorption	-14.46	-3.51	-11.94	-3.96
Acceptor Adsorption	-10.91	+0.04	-7.76	+0.22
Water Dimer	-10.96	-	-7.98	-

6-31G* Adsorption energies onto disiloxane

	ΔE^a	ΔE^b	ΔH^a	ΔH^b
Acceptor Adsorption	-2.786	+2.812	-1.073	+2.375
Water Dimer	-5.598	-	-3.448	-

a - energy change of reaction $A + H_2O \rightarrow A \cdot H_2O$
b - energy change relative to the water-dimer (reaction (20))

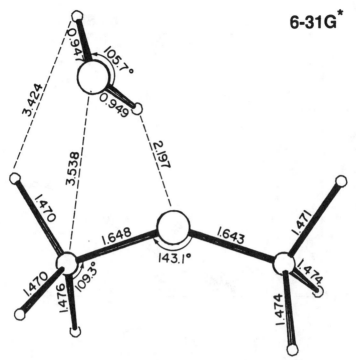

6-31G*

Figure 19. *Ab initio* 6-31G* fully optimized structure for the adsorption of water onto a disiloxane group. Big circles are oxygens, intermediate circles are silicons and small circles are hydrogens (from Lasaga and Gibbs, 1990b).

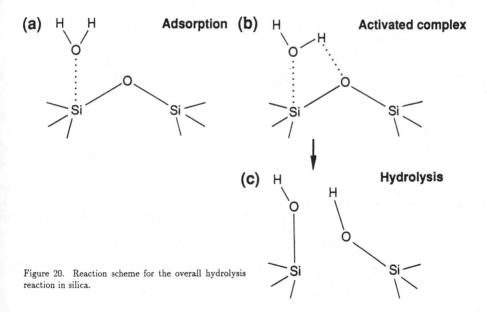

(a) Adsorption   **(b)** Activated complex

(c) Hydrolysis

Figure 20. Reaction scheme for the overall hydrolysis reaction in silica.

52

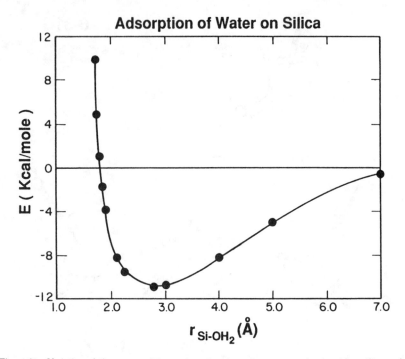

Adsorption of Water on Silica

Figure 21. Variation of the energy of interaction of an incoming water molecule with a silica surface as a function of distance. Calculations used a 3-21G* basis set and H_3SiOH.

The energetics of the H_2O acceptor adsorption are of interest in monitoring the molecular dynamics of the adsorption process. We can obtain an interesting profile by fixing the bond distance between the silicon atom and the oxygen of the incoming water molecule and then optimizing **all** the remaining degrees of freedom. Figure 21 shows the results for several SiO bond distances, in the case of acceptor adsorption using H_3SiOH. Note that the energy rise to the right of the minimum is quite gradual so that the chemical interaction between a water molecule and the silica surface would extend out to nearly 7 $\overset{\circ}{A}$ away from the surface. This result has important implications for the treatment of the adsorbed layer in silicates. In particular, it suggests that the so-called Stern layer may be several water molecules thick, even in "pure" water.

The energetics given in Table 7 must be corrected for zero-point energy of vibration to convert the energies to enthalpies i.e.,

$$\Delta H_{0K} = \Delta E + \sum_i \frac{1}{2} h \nu_i^P - \sum_i \frac{1}{2} h \nu_i^R , \qquad (60)$$

where P and R stand for product and reactant respectively, i.e., the adsorbed complex and the infinitely separated water + surface in one case or the water dimer and the adsorbed water complex in the other. To calculate this correction, a full normal mode analysis of each stable configuration is carried out. The frequencies are obtained in a fully *ab initio* manner by using the analytically derived Hessian matrix from the electronic wavefunctions. The corrected ΔH are given in Table 7. Note that part of the correction to the ΔE^a for the case with the infinitely separated reactants arises from the conversion of a translational degree of freedom into a vibrational degree

Table 8. BSSE Counterpoise calculation
Water donor adsorption onto H_3SiOH 3-21G*

	E^a_{SCF}	E^a_{BSSE}	ΔE^b	ΔE^b_{BSSE}
H_3SiOH	-364.293375	-364.294344	-	-
H_2O	-75.585960	-75.592741	-	-
$H_3SiOH \bullet H_2O$ (donor)	-439.902384	-439.902384	-14.46	-9.60

a - energies in Hartree
b - adsorption energy change in kcal/mole

of freedom. The more relevant corrections are the corrections to the ΔE^b shown in Table 7 (which don't have this loss of degree of freedom). The donor adsorption is stabilized (lower ΔH, -3.96) by the zero-point correction. The acceptor adsorption onto disiloxane is also stabilized by 0.5 kcal/mole, while the acceptor adsorption onto H_3SiOH is destabilized by the correction (ΔH increases to 0.22 kcal/mole). Note, however, that the 6-31G* results yield a hydrogen bond energy, which is in much better agreement with experiment (i.e., 5 kcal/mole).

Hobza et al. (1981) and Sauer and Zahradnik (1984) have emphasized the need to correct the usual *ab initio* energies for several effects 1) the correlation energy 2) the basis set superposition error, BSSE and 3) the changes in entropy occurring during adsorption. The entropy corrections amount to calculating the partition functions (as discussed in the TST section) and therefore the equilibrium constants(Hill, 1962 p.128). Corrections 1) and 2) for the enthalpy change of any of the adsorption reactions lead to:

$$\Delta H_0 = \Delta E_{SCF} + \Delta E_{ZPE} + \Delta E_{corr} + \Delta E_{BSSE} \quad , \tag{61}$$

where the 0 reminds us that this is the ΔH at 0K. The symbols SCF and ZPE stand for the self-consistent field *ab initio* energy (Hartree-Fock) and for the zero-point vibrational energy respectively; E_{corr} is the correlation energy. In addition, Hobza et al. add a correction for dispersion energy (i.e., a C/r^6 term), but at this point semi-empirical terms are brought in. Corrections to higher temperatures can be made within the same framework as the entropy calculations.

The BSSE arises because the basis set used to compute the properties of the adsorbed complex is bigger than the basis set used to compute the properties of the individual surface and adsorbing molecules respectively. If the basis set were sufficiently flexible, this would not cause any problems. However, for small basis sets, this error could be important. The simplest way to correct for BSSE (Boys and Bernardi, 1971) is to introduce a calculation involving "ghost" atoms. For example, in the case of water adsorbing onto H_3SiOH, rather than carry out an energy calculation for the H_3SiOH precursor (at say 3-21G* basis), one would now carry out the calculation using the $H_3SiOH - H_2O$ adsorption complex but setting the nuclear charge on the H_2O part to zero. In this way, the calculation is using the basis set from water but is not including any nuclear charges nor any extra electrons, i.e., it truly is calculating the energy of H_3SiOH but with a bigger basis set, equal in size to the basis set used for the full adsorption complex calculation. This method is called the **counterpoise** method. As shown in Table 8, the net effect is to drop the energies of both H_2O and H_3SiOH relative to the complex. Because the H_3SiOH basis set is a significant addition to the water molecule basis set (not viceversa), the drop in the water energy is greater than the change in the H_3SiOH energy. Because of the lowering of the energies of the reactants, the BSSE reduces the stability of the adsorption complex from 14.5 kcal/mole to 9.6 kcal/mole. Note that the correction would also shift the other adsorption energies similarly. Hence the *relative* energies of the adsorption complexes would be much less affected. Nonetheless the absolute value changes in a non-trivial way. For higher level basis sets (more flexible), e.g., the 6-31G* or the 6-311G** basis, the BSSE would become a much smaller correction.

The information obtained from *ab initio* calculations include the optimized geometries and the curvature of the potential surface at the minimum, i.e., the vibrational frequencies. With this type of information, it is possible to add the entropy corrections to the enthalpy changes and obtain ΔG's snd thus to calculate equilibrium constants for the adsorption process. The theory was already discussed in the earlier section on transition state theory (e.g., Eqn.(27)). As discussed by Sauer and Zahradnik (1984) and Hobza et al. (1981), for gas adsorption a rigid cluster model may be invoked whereby a water molecule with three rotations and three translational degrees of freedom approaches the surface and forms a complex which now has

6 low frequencies of vibration. The internal degrees of freedom of the surface clusters are ignored in their calculations. There is no question that there should be no rotation or translational component to the surface cluster, however future work should incorporate the internal vibrational modes (see TST section) . However, only the low frequency vibrational modes will contribute to the partition function (see TST section).

One can analyze experimental rate data of heterogeneous reactions in terms of adsorption complexes on the surface. For example, the dissolution rate of quartz, can be broken down into:

$$k_{diss} = C_{H_2O}^{ads} \, k_{hydrolysis} \, , \qquad (62)$$

where $C_{H_2O}^{ads}$ is the number of water molecules adsorbed in the reactive position (see Fig.19) per cm^2 of quartz surface area. We can estimate $k_{hydrolysis}$ from the *ab initio* data and transition state theory, as was done earlier.

Understanding the acid and base catalysis of silicate surface reactions is a major effort in the kinetics of geochemical processes. A key step in such a study is knowledge of the types of adsorption complexes formed by H$^+$ and OH$^-$ on mineral surfaces. For example, it is known that the dissolution of quartz or silica is catalyzed by OH$^-$ in the basic pH region. The effect of OH$^-$ adsorption on silica can be studied by a cluster such as in Figure 22. Preliminary *ab initio* calculations indicate the type of geometry shown in the figure. Note the significant elongation of the SiO bridging bond in disiloxane. Clearly, this distortion will reduce the energetics of hydrolysis (compare the structure in Fig.22 to the water adsorption structure in Fig.19).

Likewise, the acid catalysis of feldspar dissolution can be studied by a cluster such as in Figure 23. Once more, the adsorption of H$^+$ onto the bridging oxygen of a disiloxane group significantly alters the bonding properties of the nearby bonds. In particular, the adsorption complex computed *ab initio* leads to quite elongated Si-O and Al-O bonds, a condition that leads to an obvious lowering of the activation energy.

There has been recent interest in developing proton adsorption models for metal oxides and hydroxides that incorporate multisite adsorption (e.g., Hiemstra et al., 1989a,b; Stumm and Wollast, 1990; Schindler and Stumm, 1987). Usually these models include electrostatic interactions between the surface and adsorbed molecules as well as a coulombic treatment of the electrical double layer (Westall, 1987). The models have been successful in systematizing the effect of changes in coordination number of metal or oxygen ions on the adsorption. On the other hand, the new approaches discussed here can expand these important earlier treatments to include a much better handling of the chemical bonding involved at these surfaces. Ultimately, it will be very useful to combine both approaches.

Calculations involving much bigger clusters have all been carried out using semiempirical methods, e.g., the CNDO or MINDO or MNDO methods (e.g., Zhidomirov and Kazansky, 1986; Takahashi and Tanaka, 1986). In the latter case, the adsorption

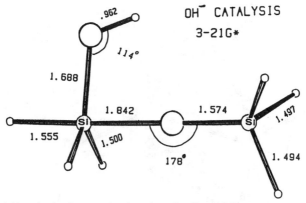

Figure 22. *Ab initio* molecular cluster used to investigate the effect of OH- adsorption on the catalysis of SiOSi hydrolysis in silica. Note the elongated SiO bond and the formation of a five-fold coordinated silicon.

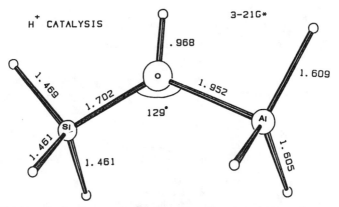

Figure 23. *Ab initio* molecular cluster used to investigate the acid catalysis of feldspar dissolution. Note the elongated SiO and AlO bonds as a result of the H$^+$ adsorption onto the bridging oxygen.

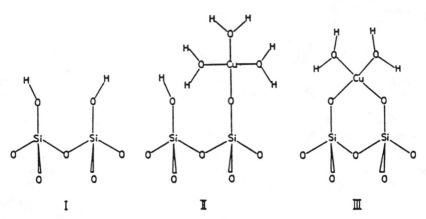

Figure 24. Models used to study the adsorption of Cu^{2+} onto the silica surface (from Zhidomirov and Kasansky, 1986).

56

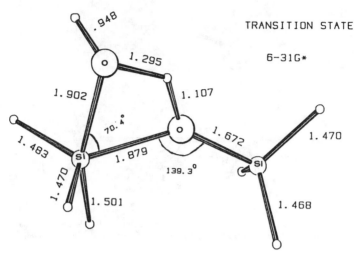

TRANSITION STATE

6-31G*

Figure 25. Full transition state geometry using a 6-31G* basis for the hydrolysis reaction of water on a disiloxane group (from Casey et al., 1990).

Table 9. Hydrolysis transition state (kcal/mole)

	ΔE^a	ΔE^b	ΔE^c	ΔE^d	ΔE^{expt}
$\Delta E_a^{\ddagger}$	22.74	16.70	45.95	32.31	
Vibrational Zero-Point	-1.30	-1.30	-1.42	-1.42	
ΔH_0	21.44	15.40	44.53	30.89	19-21^e

---

a - 3-21G* results for H_3SiOH hydrolysis
b - MP2/3-21G* results for H_3SiOH hydrolysis
c - 6-31G* results for disiloxane hydrolysis
d - MP2/6-31G* results for disiloxane hydrolysis
e - quartz dissolution data (Rimstidt and Barnes, 1980; Dove and Crerar, 1990)

of hydrated Cu ($Cu(H_2O)_4)^{2+}$) on silica is monitored and it is found that adsorption of Cu onto a single silica site is better than adsorption onto two sites (see Fig.24). In addition, processes such as the oxidation of the surface of silicon have been studied (Hagon et al., 1987).

Ab initio studies of mechanisms of water-rock kinetics

With the vast increases in computational power, it is now feasible to calculate *ab initio* potential surfaces that incorporate bond-breaking and bond-forming processes. In particular, the reaction pathways and energetics of **surface** reactions involving oxides and silicates can be studied from the atomic point of view with these techniques.

One of the most important potential energy surfaces to investigate from the point of view of geochemical kinetics, petrology, ceramics and rock mechanics is the hydrolization of the Si-O-Si structural unit at the surface of silicates, i.e., the reaction

$$H_2O + \ \equiv Si - O - Si \equiv \ \Longrightarrow \ 2 \equiv Si - O - H \ . \tag{63}$$

This reaction is a key step in the dissolution processes occurring at the surfaces of silicates. Equation (63) is also one of the essential aspects of the widely studied hydrolytic weakening in the area of rock mechanics. One of the current areas of active research in that field involves determining the molecular details of water or hydrogen defects in minerals. The incorporation of water by reactions, such as Equation (63), can affect the rates of diffusion or creep by many orders of magnitude. In addition, Equation (63) is the basic step by which water is initially incorporated into silicate melts, thereby significantly affecting both the phase diagrams and the transport properties such as viscosity and diffusion.

Figure 11 illustrates the elementary reactions that govern the dissolution of silica and similar silicates. The sequence depicts the sequential hydrolysis by water molecules of the three to four Si-O-Si bridges anchoring a surface silicon atom . At the end of the four hydrolysis steps, the silicon is surrounded by 4 OH groups and leaves the surface as orthosilicic acid. **None** of the steps illustrated in Figure 11 depict the activated complex. Figure 20 focuses on one of the hydrolysis steps. The attack of water molecules is decomposed into two molecular steps in Figure 20. The first step involves the **adsorption** of water near a Si-O-Si group. The second step involves the formation of a new Si-O bond by the oxygen of the adsorbed water and the cleavage of the Si-O-Si group. **It is the activated complex associated with this step that we believe accounts for the energetics of silica dissolution.**

Transition State of the hydrolysis reaction

Having obtained the adsorption data, the next step is to ascertain the kinetics of the hydrolysis reaction, which requires computation of the activated complex. Returning to Figure 20, it is clear that the hydrolysis reaction proceeds in such a manner that the original SiO bond is broken and new Si-OH and OH bonds are formed.

The true activated complex using disiloxane, is given in Figure 25. That this is a true activated complex follows from the exact fulfilment of the two key requirements (1) the structure in Figure 25 is a stationary point for all degrees of freedom (Eqn.(55)) and (2) the eigenvalues of the second derivative matrix (Hessian) are all positive except for one negative and the usual 6 zeroes. Interestingly enough, the symmetry of the two SiO bonds is closely maintained in the full transition state complex. However, the two OH distances (1.107 Å and 1.295 Å) are not symmetrical. This assymetry arises from the symmetry breaking induced by the other groups in the activated complex and also manifests itself in the malleable nature of the lengthened OH bonds.

Again to compare the ΔE's to experimental quantities we must include the zero-point vibrational energies.

$$\Delta H^{\ddagger} = \Delta E^{\ddagger} + \sum_{i=1}^{3N-7} \frac{1}{2} h\nu_i^{\ddagger} - \sum_{i=1}^{3N-6} \frac{1}{2} h\nu_i^{ads} \quad . \tag{64}$$

This correction is non-trivial and reduces the barrier by 1.22 kcal/mole for the hydrolysis of SiOH and by 1.42 kcal/mole for the hydrolysis of disiloxane (see Table 9). Note that part of the loss in the barrier is due to the fact that the activated complex is "loose" compared to the adsorption complex. Structures with many elongated bonds tend to have lower frequencies of vibration and hence smaller $\frac{1}{2}h\nu$ corrections. Therefore, the increase in zero-point energy of vibration is bigger for the stable adsorption complex than for the activated complex. This correction has important consequences for the isotope effect (Lasaga and Gibbs, 1990a).

Table 10. Reaction coordinate normal mode for the   disiloxane-water transition state ($\nu = 1365$ i).

Internal Coordinate[a]	TST value[b]	Normal Mode value[c]
OH^d bond	1.107 Å	1.221 Å
OH^d bond	1.295 Å	1.156 Å
H^wOH^d angle	119.7⁰	124.6⁰
OH^dO angle	130.1⁰	133.1⁰
$SiOH^d$ angle	82.3⁰	77.6⁰

a - only internal coordinates that change by 0.02 Å or 2⁰ are listed
b - see geometry in Figure 25
c - value of internal coordinate after vibration is allowed
d - H atom being exchanged in Figure 25
w - H atom (unreactive) in water molecule

The new values for $\Delta E^{\ddagger}$, $\Delta(\frac{1}{2}h\nu)$, and $\Delta H^{\ddagger}$ are given in Table 9. The activation energy predicted from our exact activated complex for the hydrolysis of H_3SiOH is in excellent agreement with experimental values. The results for the disiloxane hydrolysis using the higher level calculation (adding electron correlation at the MP2/6-31G* level) are in reasonable agreement with experiment. One should add that the correct comparison, given the kinetic mechanism in Figures 11 and 20, should add the true enthalpy of adsorption for the water molecule (Lasaga, 1981a; p. 37) and the RT correction in Equation (64)

$$E_a^{expt} = RT + \Delta H^{\ddagger} + \Delta H_{H_2O}^{ads} \ .$$ (65)

However, we expect ΔH^{ads} to be either close to zero or a small negative number (e.g., several kcal/mole - see Table 7). The net effect of the corrections in (65) would most likely be to increase slightly the agreement with experiment of the calculated activation energy for disiloxane hydrolysis case (i.e., to values of 26-28 kcal/mole versus 19-21 kcal/mole for experiment).

It is instructive to analyze the normal mode of the activated complex which has the negative eigenvalue (i.e., an imaginary frequency because $(2\pi\nu)^2 = \lambda$; for disiloxane, $\nu^{\ddagger} = 1365$ i cm^{-1}). As mentioned earlier, this normal mode is, in fact, the **reaction coordinate** for the hydrolysis reaction. One can analyze this normal mode by distorting the molecular geometry of the transition state by the cartesian coordinate vector corresponding to the particular normal mode and observing which interanl coordinates change. If only bonds that change more than 0.02 Å and angles that change more than 2⁰ are picked, then the results are as given in Table 10. This analysis indicates that the motion involves predominantly the transfer of the H-atom, i.e., the "vibration" has one OH distance increasing and the other OH distance decreasing from the corresponding value at the transition state geometry. Interestingly, the two long SiO bonds (which ultimately must also get involved in the reaction) are **not** very active in the reaction coordinate normal mode. This result states that the H transfer dominates the hydrolysis reaction in this mechanism.

The agreement of some of the calculated energies with experiment strongly indicates that we are indeed obtaining for the first time a look at the key chemical moieties driving the dissolution of quartz and related minerals. Nonetheless, while the value of $\Delta H^{\ddagger}$ is in the vicinity of the observed energetics of silicate reactions, it is important to stress that we are just beginning our investigation of the possible surface reactions which govern the dynamics of mineral-water interactions. Future experimental work (isotopic, structural, spectroscopic, kinetic) will undoubtedly uncover numerous pathways where different chemical species, e.g., OH^- complexes, H^+ complexes, simultaneous bonds to several SiOH or AlOH groups and so on, will play an important role. Furthermore, different reaction mechanisms may be operative at different temperature ranges as well as different pH or ionic strength conditions. The message emphasized here is that the *ab initio* calculations are now capable of providing critical results to help decipher the atomic details of these processes.

MOLECULAR DYNAMICS METHODS IN SURFACE STUDIES

Given the complex collective motions possible in the dynamics of mineral surfaces and the surrounding fluids, it is only natural that scientists have turned to molecular dynamics to provide some deep insight into the nature of surface dynamics. The basis of molecular dynamics (MD) rests on the *ergodic* hypothesis of statistical mechanics which states that the time-averaged property of one system will be equivalent to the instantaneous ensemble-average over many systems (see Ciccotti et al., 1987 for a good review of the methods). Molecular dynamics results are based on trajectories of particles over time to simulate a macroscopic system on an atomic level. Molecular dynamics systems usually consist of a few hundred to a few thousand particles (e.g., hard spheres, atoms, ions, or molecules) within a finite cell and the number of particles during the simulation remains constant. To eliminate surface effects, periodic boundary conditions are imposed in all three spatial directions. If a particle leaves the central cell during solution of the equations of motion, then an identical particle is introduced on the opposite side of the central cell according to the imposed boundary conditions. In effect, this boundary condition treats all the particles as if they were at the center of an infinite volume. Once the trajectories are obtained, statistical theories enable us to calculate both equilibrium and non-equilibrium properties. First, let us discuss briefly the methods used to obtain the actual trajectories.

Most molecular dynamics systems are treated using classical Newtonian mechanics. Calculation of the position and velocity of each particle in the system rests on the numerical solution of Newton's equations of motion:

$$\frac{d\vec{r}_i}{dt} = \vec{v}_i \quad i = 1, ..., N \ , \tag{66}$$

$$m_i \frac{d\vec{v}_i}{dt} = \vec{F}_i \quad i = 1, ..., N \ , \tag{67}$$

where m_i, $\vec{r}_i$, $\vec{v}_i$, and $\vec{F}_i$ are the mass, position, velocity and force vectors of particle i. These equations are solved with finite time steps. Time steps in the numerical solution must be short enough to accurately reproduce the "true" trajectory a particle would follow given the interatomic potential and the initial conditions, but as

60

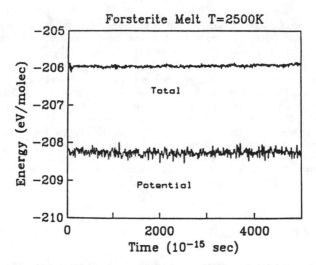

Figure 26. Constant total energy and variable potential energy as a function of time for a MD simulation of Mg_2SiO_4 melt at 2500K and 2.75 g/cm^3 (from Kubicki and Lasaga, 1990).

large as possible to efficiently simulate the greatest amount of real time per unit of computer time. To model silicates, time steps of 10^{-15} seconds are common. For reference, high-frequency vibrations in Si-O bonds (≈ 1000 cm^{-1}) have a vibrational period of approximately 3×10^{-14} seconds; therefore, tens of time steps will be calculated during one vibrational period.

Molecular dynamics simulations may be carried out under a variety of conditions and constraints. Perhaps the most common and simple ensemble is the **microcanonical** ensemble. In this ensemble, the number of particles, N, the volume, V, and the total energy, E, remain constant during the simulation. V is easily fixed by holding the edges of the periodic cell constant within the simulation. N is fixed by the three-dimensional periodic boundary conditions which allow atoms to move freely without affecting the number of particles within the central cell. The total energy of the system is calculated at every time step. Both the potential and kinetic energy will fluctuate with time during the simulation, but the total energy must remain constant (see Fig.26) as long as the chosen time step is small enough, there are no external perturbations, and the potential energy function between particles is conservative (i.e., the potential energy depends only on the positions of the particles).

A more convenient ensemble than the microcanonical ensemble is the **isothermal-isobaric** ensemble. In this ensemble, the number of particles, pressure, and temperature (N, P, T) remain constant. Several papers (e.g., Andersen, 1980; Nosé, 1984; Nosé and Yonezawa, 1986; Hoover, 1985) have focused on MD techniques for calculating NPT-ensemble MD trajectories.

The choice of interatomic potential for a given system is a key ingredient in the molecular dynamics. For a reference on the wide variety of potentials used to model gases, water, ionic solutions, and organic molecules see Ciccotti et al. (1987).

Ionic models were first used for silicates because MD simulations of silicate melts evolved from studies on molten salts (Woodcock et al., 1976). The simplicity of this potential was also an attractive feature in the early years of computer development when computation time was expensive and less available than at present. Nonetheless, simple ionic models have been surprisingly successful in in reproducing experimental data on complex structures in highly covalent systems such as SiO_2 glass (Woodcock et al., 1976; Mitra, 1982; Mitra et al., 1981; Kubicki and Lasaga, 1988; Feuston and Garofalini, 1988) as well as in $MgSiO_3$-perovskite (Matsui, 1988) and forsterite, Mg_2SiO_4, (Matsui et al., 1981; Kubicki and Lasaga, 1990; Kubicki et al., 1989), both important phases in the earth's mantle (see also Angell et al. 1983, 1988; Dempsey and Kawamura, 1984).

A typical form for the interionic potential, V_{ionic}, between two atoms, i and j, at a distance, r_{ij}, is

$$V_{ionic} = \frac{Z_i Z_j e^2}{r_{ij}} + A_{ij} exp(-r_{ij}/\rho_{ij}) + C_{ij} r_{ij}^{-6} \quad , \tag{68}$$

where Z_i is the ionic charge of ion i (formal or partial), A_{ij} and ρ_{ij} are the Born-Mayer repulsion parameters between ions i and j, and C_{ij} is a Van der Waal's term.

In molecular dynamics simulations of surfaces, the most common potential used is the Lennard-Jones "6-12" potential, $V(r)$:

$$V(r) = -4\epsilon[\frac{\sigma}{r^6} - \frac{\sigma}{r^{12}}] \quad , \tag{69}$$

where σ is such that the minimum in the potential occurs at a distance of $2^{1/6} \sigma$ and the value of the potential at the minimum is given by $-\epsilon$. If this potential is integrated over all half-space (e.g., a surface), the result is the potential of interaction of a molecule in a fluid with all the atoms in a solid surface. If the density of atoms in the surface is ρ, then the integrated potential for a molecule a distance z away from the surface will be given by (another exercise for the reader):

$$V_{surface}(z) = \frac{4\epsilon\sigma^{12}\rho\pi}{45z^9} - \frac{2\epsilon\sigma^6\rho\pi}{3z^3} \quad . \tag{70}$$

Hence, it is common to see reference to the 3-9 potential in the literature.

An alternative approach to develop interatomic potentials is *ab initio* calculation (e.g., see Lasaga and Gibbs, 1987). In this method, a large amount of computer time is necessary to generate the theoretical data on which the interatomic potential is based. However, increasing availability of computational power and the Gaussian 86 program (Frisch, 1983) allow a researcher to perform *ab initio* calculations on molecules conveniently. In general, *ab initio* calculations may be carried out over a larger range of coordination states, structures, and bond distances than are found in crystals and glasses. Hence, the *ab initio*-derived potential may be able to reproduce accurately interatomic forces far from the equilibrium positions where the dynamics of bond-breaking and formation actually take place.

The link between the microscopic motions and the macroscopic properties of systems is made with the theorems of statistical mechanics (Hill, 1962). Woodcock (1975) has discussed many of the formulas for calculating thermodynamic properties from MD simulation data. These relationships between the simulated microscopic systems and the macroscopic properties are of primary importance for testing the accuracy of a given potential model. Further, they may be used to predict thermodynamic quantities of materials at pressure and temperature conditions that are not attainable from experiment.

Some of the important thermodynamic quantities derived from MD simulations at constant N and V are the internal energy, E, temperature, T, and pressure, P. The internal energy of the system is calculated from the sum of kinetic and potential energies (Woodcock, 1975)

$$E = \langle \frac{1}{2} \sum_{i=1}^{N} m_i v_i^2 \rangle + \langle \Phi(\mathbf{r_i}) \rangle \quad , \tag{71}$$

with the "$\langle \ \rangle$" denoting an ensemble- or time-average of the system. In molecular dynamics, by virtue of the ergodic hypothesis, the thermodynamic average of a property, $\langle B \rangle$, is given by the time average of the instantaneous value of B at time t, B(t):

$$\langle B \rangle = lim_{T \to \infty} \frac{1}{T} \int_0^T B(\tau) d\tau \quad , \tag{72}$$

or using the discrete time steps in an MD simulation:

$$\langle B \rangle = \frac{1}{M} \sum_{j=1}^{M} B_j \quad , \tag{73}$$

where B_j is the value of B at the jth time step and the average is taken over M time steps. In the microcanonical ensemble, E (in Eqn.(71)) should remain constant (within numerical error) over the duration of the run.

Temperature is obtained from the average kinetic energy over time (usually averaged over 100 or more time steps i.e., $M \geq 100$ in Eqn.(73)). For a system with N particles,

$$T = \frac{1}{3Nk_B} \langle \sum_{i=1}^{N} m_i \vec{v}_i^2 \rangle \quad , \tag{74}$$

where k_B is Boltzmann's constant. Temperature fluctuations in the microcanonical ensemble are proportional to the inverse square root of the number of particles (i.e., $N^{-\frac{1}{2}}$; Hill, 1962).

Pressure calculations are based on the thermodynamic relation, $\partial E/\partial V = -P$ with E given by Equation (71). The resulting expression, derived from the virial theorem (Woodcock, 1975), is given by Equation (73),

$$P = \frac{Nk_BT}{V} - (\frac{1}{3V}) \langle \sum_{i=1}^{N} (\vec{r}_i \cdot \vec{F}_i) \rangle . \tag{75}$$

The $(\vec{r}_i \cdot \vec{F}_i)$ term is the dot product of the position and force vectors of each particle. The second term in silicate simulations is typically at least an order of magnitude bigger than the first term.

In surfaces, just as in liquids and glasses, there is structural disorder. Therefore, the structures of surfaces must be described in a statistical sense. Short-range order

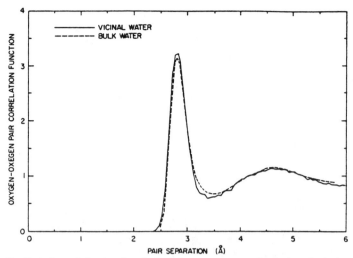

Figure 27. Typical correlation function obtained for oxygen-oxygen distributions for both water near a surface and bulk water. Note the approach to 1 as long range order is lost in either case (from Mulla et al., 1984).

in MD simulations is commonly analyzed with the statistical correlation function defined as

$$g_{ij}(r) \equiv \frac{1}{4\pi\rho_j r^2} \frac{d\langle N_{ij}(r)\rangle}{dr}, \qquad (76)$$

where $N_{ij}(r)$ is the number of atoms of type j inside a sphere of radius , r, around a selected atom of type i, and ρ_j is the bulk density of the atoms of type j. As usual, time-averages are implied by the "$\langle \; \rangle$." Note that as $r \to \infty$, short-range order and structure are lost and $N_{ij}(r)$ approaches the bulk value (i.e., $\frac{4}{3}\pi r^3\rho_j$). Therefore, $g_{ij} \to 1$ as $r \to \infty$ (see Fig.27). From MD simulations, it is possible to obtain directly individual pair correlation functions (e.g., g_{Si-O}). To calculate g_{ij}, each atom is selected over 100's of time steps and the number of atoms of type j in thin spherical shells with radii of (r) and (r + Δr) around these atoms (i.e., dN(r)/dr) are counted out to 10 Å. The correlation functions are averaged over the number of time steps and plotted to show the short-range surface or liquid structure.

Applications to surfaces

Most MD studies on surfaces have dealt with either the mineral surface itself or with the interaction of an aqueous fluid near an interface. In all these cases, the study of a surface requires a reconsideration of the periodic boundary conditions. In some cases, the surface was created by removing the periodic boundary condition in the z-direction. This creates a thin film extending semi-infinitely in the x and y directions (which still retain the periodic boundary conditions, e.g., Heyes, 1983). This approach has been taken in studying the mineral surfaces themselves. Alternatively, when dealing with a fluid in contact with a surface, all periodic boundary conditions are retained, which leads to an infinite set of systems with fluid trapped between two walls (see Fig.28).

Some recent work has focused on delineating the surface structure, without ex-

64

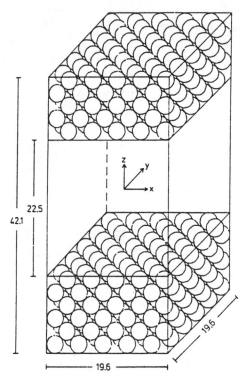

Figure 28. Sketch of the basic tetragonal simulation cell with top and bottom containing an ordered solid surface (in this case platinum atoms). The water molecules are located in the center of the box. Distances are in Å (from Spohr and Heinzinger, 1986).

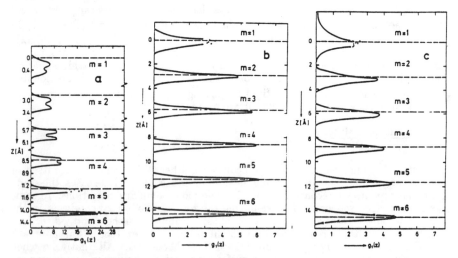

Figure 29. Single-particle distribution function for MD simulations of the solid Kr (100) surface (no fluid) at several temperatures. a) T = 7 K b) T = 70 K c) T = 102 K. Solid lines and solid circles are the correlation function, g(z) at the surface (z is the distance perpendicular to the (100) surface). Crosses are g(z) in the bulk. Dashed lines are the mean positions of the Kr layers in the bulk of the crystal (from Schommers and Von Blanckenhagen, 1985).

plicit regard for a fluid, e.g., in vacuum or low density gas. The most general result has been that the surface structure of solids deviates, sometimes strongly, from the structure in the bulk due to surface relaxation; however, this deviation decreases strongly with distance of the atomic layers from the surface. Therefore, usually the changes take place in only the top 2 or 3 atomic planes of the material (e.g., Toxvaerd and Praestgaard, 1977).

The next emphasis has been on understanding the topography of the atomic layers at or near the surface. As an example, Schommers and Von Blanckenhagen (1985) studied the properties of the (100) surface of solid Kr as a function of temperature. They employed 500 atoms in their MD simulation and found evidence for surface roughening or surface "premelting" below the melting temperature of the crystal. The structure was analyzed using pair correlation functions calculated at each atomic (100) layer from the surface (m = 1) down to the sixth layer (m = 6). At low temperatures (T = 7 K), the pair correlation function, g(z) as a function of distance from the surface, exhibited a double peak for low values of m (see Fig.29). This double peak is characteristic of a classical oscillator (i.e., it spends most of its time at the ends of the trajectory). This double peak (vibrational) motion of surface atoms was lost as more disordered motion occurred at higher temperatures. At the higher temperatures (T = 102 K) the pair correlation function in the x-y plane (g(r)) showed quite a bit of disorder for m near 1 (Fig.29). This disorder was taken as an indication of a roughening transition, a topic that has been raised in the petrologic melt literature and was discussed earlier in this chapter.

Pontikis and Rosato (1985) looked at the roughening transition in the (110) surface of solid Ar. The MD simulations employed a common potential, the 6-12 Lennard Jones (LJ) potential (Eqn.(58)). This potential has been the most widely used in molecular dynamics calculations and even many common complex potentials, such as the ST2 potential of water (see below) use the LJ in their formulation. Pontikis and Rosato (1985) found that the number of vacancies in the surface increased exponentially with temperature (as expected from an Arrhenius behavior) until a temperature T = 0.7 T_m, where T_m is the melting temperature. Thereafter, the number of vacancies was high and independent of temperature. They associated this change with a roughening transition at 0.7 T_m. A recent paper by Lutsko et al. (1988), however, has questioned whether the transitions seen by these earlier papers were not due to the periodic boundary conditions used to carry out the surfac studies.

Other studies of surface structure have focused on the phenomena of surface diffusion, a topic of importance in the Monte Carlo simulations (next section) and in the general theory of mineral growth and dissolution. Most of these papers have used simple faces in fcc materials (Doll and McDowell, 1982; Ghaleb, 1984; Tully et al., 1979). The work of Tully et al.(1979) actually used the idea of "ghost" particles to simulate the vibrational motion of non-diffusing surface and bulk atoms. Each diffusing atom would interact explicitly with all other diffusing atoms but its entire interaction with the rest of the lattice was embedded in the interaction with the "ghost" particle associated with it. This "ghost" had a vibrational force, a frictional force and white random force component acting on it. An interesting conclusion from the work is that while adatoms indeed are the fastest "diffusers" on the surface, the diffusion of clusters of 2-6 atoms could not be ignored. This result should be studied from the point of view of Monte Carlo simulation (next section).

The most studied geological material from the surface MD point of view has been silica. Garofalini and coworkers (Garofalini, 1982, 1983; Levine and Garofalini; 1987) have used a modified Born Mayer potential,

$$V_{ij}(r) = A_{ij}e^{-\frac{r}{\rho_{ij}}} + \frac{z_iz_je^2}{r} \, erfc(\frac{r}{B_{ij}}) \, , \tag{77}$$

to look at the surface structure of silica glass. The complementary error function, erfc(u), in Equation (77) arises from the Ewald summation of all the Coulomb charges in the system (see Lasaga, 1981c). The MD calculations employed 800 atoms and after initially equilibrating the glass (using 3D periodic boundary conditions), the boundary condition in the z-direction was removed allowing the surface atoms to relax, except for the bottom 50-100 atoms, which remained fixed. In analyzing the correlation function for the surface of the glass, a new peak at 1.54 Å has appeared, which is related to the formation of non-bridging SiO^{NBO} bonds. Furthermore, a new peak at 2.2 Å (compared to the g(r) for the bulk silica glass) signaled the formation of edge shared SiOSi linkages (see Fig.30). The formation of 5-fold coordinated silicon (see last section) was also observed. Finally, and of interest to the water-rock mechanisms, the formation of "channels" 15 Å deep from the surface were observed. Figure 31 gives such a channel. Perhaps these channels are playing a role in determining how far the initial penetration of water molecules can get during silica (and maybe aluminosilicate) reactions.

The next step in the use of MD in surfaces depends on the simulation of a fluid in contact with a surface. The trick here is to incorporate the solid-fluid interface within the framework of the periodic boundary conditions normally employed in molecular dynamics. Usually one begins by equilibrating the solid and "fluid" separately in MD simulations. Then the two are joined as in Figure 28 with the solid bounding the fluid on both sides (e.g., along the z-direction). Now one applies the periodic boundary conditions in the x and y directions (thereby obtaining an infinite film). At this point, either, the atoms in the solid near the ends are frozen (so that a single thin film is obtained) or periodic boundary conditions in the z-axis are reinstated, creating an infinite set of thin films in the z-direction. For the purposes of Ewald coulombic sums, such periodicity is desirable. Obviously, one of the key questions is the effect of these boundary conditions or of the proximity of the two walls on the MD results.

Of particular interest to chemists as well as geochemists is the interaction of water and aqueous solutions with mineral surfaces. The early studies of surface phenomena, limited the "structure" of the surface to that of a wall with a simple potential, usually a simple repulsive potential or a 6-12 Lennard Jones potential (Eqn.(69)). For example, Marchesi (1983) has simulated water in contact with a wall. 150 molecules of water were used in the MD simulations within two walls 20 Å apart. The interaction of the water molecules with the wall were completely determined by the distance, z, from the wall and given by a repulsive potential:

$$V_{wall} = \alpha \, e^{\beta z} \, . \tag{78}$$

The interactions between the water molecules themselves, was determined by the commonly used ST2 potential of Rahman and Stillinger (1971) and Stillinger and Rahman (1974), (1978). In this potential, the water molecule is treated as a tetrahedral set of positive and negative point charges (see Figure 32). The charge on the protons is +0.2357 e and that on the "lone pairs" is -0.2357 e. The distance to the oxygen is 1 Å for the protons and 0.8 Å for the electrons. The interaction between pairs of water molecules then is given by a Lennard Jones potential (Eqn.(69)), depending on the oxygen-oxygen distance and by the Coulomb interaction between all 16 pairs (4 on each water) of charges. This Coulomb interaction, V_{Coul}, is then smoothed so that for distances less than 2.01 Å it is zero and for distances greater than 3.1287 Å it is the full value. Table 11 from Mulla (1986) shows how well the different potentials for water predict observed properties. Other studies include Evans et al. (1988).

Marchesi began with the ice structure, then melted it and equilibrated it for 8000 time steps. This was followed by insertion of the walls and 54000 time steps of

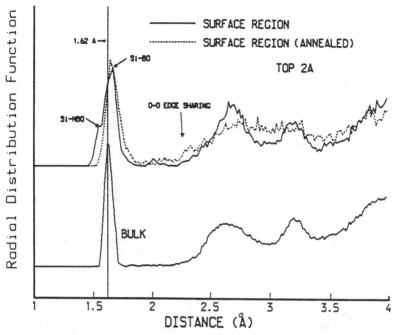

Figure 30. Radial distribution functions from MD simulations of the silica glass surface showing the effect of anealing on surface structure. Bulk refers to bulk silica glass (from Garofalini, 1987).

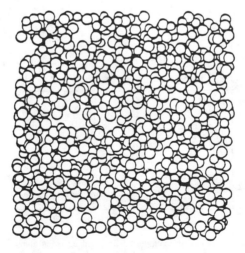

Figure 31. Snapshot of the surface of an 800 atom MD simulation of the silica glass surface. Note the "channels" (from Garofalini, 1987).

68

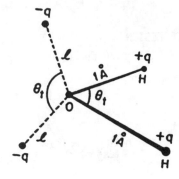

Figure 32. ST2 model of a water molecule. The electrons and protons are replaced by electrostatic charges for the MD simulations.

Table 11. Comparison of water properties for the ST2, MCY and CF simulations models and bulk water at approximately 298 K (from Mulla, 1986).

Property	ST2	MCY	CF	bulk water
-U (kJ/mol)	34	28.5	33	34
C_v (J/K/mol)	71	79	--	75
u (Debye units)	2.35	2.26	1.86	1.86
PV/NkT (V/N=1)	0.09	8.5 (29)	0.1	0.05
D (10^{-9} m²/sec)	3.1 (30)	2.3 (29)	1.10 (31)	2.85 (32)

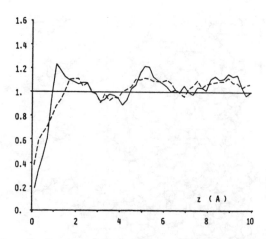

Figure 33. Density profiles for oxygen atoms (solid line) and hydrogen atoms (dotted line) from MD simulations as a function of distance from the wall, z. The value for bulk water is 1.0 in each case. Note strong repulsion near walls and approach to bulk values at distances of 10 Å (from Marchesi, 1983).

simulations. Analysis of the structure shows that the repulsion lasts for 1 Å away from the surface (see Fig.33). The density of water molecules increases near the walls to values higher than bulk. Finally by distances greater than 8 Å the structure of the water was indistinguishable from that of the bulk (the small fluctuations at larger distances seen by Marchesi (1983) have been attributed to the short duration of the MD run (Sonnenschein and Heinzinger, 1983)). This last observation has been generally true of most of the MD results (e.g., Sonnenschein and Heinzinger, 1983; Barabino et al., 1984), which find beyond a few angstrons little difference structurally between water near the walls and bulk water (see Fig.27). The main difference between the two types of water molecules arises from dynamical properties, e.g., the water orientational relaxation time. In either case, most of the studies find essentially bulk water behavior beyond 10-20 Å. Barabino et al. (1984) conclude that "water properties near a wall are determined more by changes in geometric structure of water induced by the presence of a wall rather than by direct influence of the wall potential on water molecules". In this regard the behavior near walls, is similar to the behavior seen in water near non-polar solutes. The fact that the effect of walls on water structure is generally lost past distances of 10-20 Å has important implications for the question of the applicability of phase equilibrium studies (or kinetic studies for that matter) to natural systems (e.g., metamorphic reactions), where the fluid is in a tightly constrained grain boundary network. If the grain boundaries are bigger than 20 Å in radius, then the MD work on surfaces, suggests that the fluid will behave in large part as bulk water at the same P and T. Of course, these MD studies have been done at lower P and T. Future work, should check the applicability of the lower P and T results to temperatures of 100-1000 °C and pressures of 1-20 kbar.

More recent work (e.g., Hautman et al., 1989; Spohr and Heinzinger, 1986, 1987, 1988a,b; Foster et al., 1989) have included more realistic *ab initio* potentials between the atoms comprising the metal walls, in particular Pt metal, and between the metal atoms and water molecules. The interactions between water molecules is still basically an ST2 model or something very similar. *Ab initio* studies of the interaction of water with Pt atoms indicate that the water molecule adsorbs head on with the oxygen directly above the Pt atom in the surface and the dipole moment of the water pointing in a direction perpendicular to the surface. These new MD results also obtain a transition region between the surface and the bulk water, around 5-7 Å. However, with these potentials, there is now strong adsorption of the water on the surface, as evidenced by oxygen-oxygen correlation functions, g_{OO}, with peaks close to the surface much bigger than the OO peaks in bulk water. Spohr and Heinzinger (1988a) show (Fig.34) that as the amount of coverage (defined as ratio of water molecules on the surface to Pt atoms) increases, g_{OO} first has one peak (i.e., monolayer of water) and then picks up a second broader peak (next adsorption layer).

A particular discrepancy between MD simulations on the one hand and experimental and *ab initio* results on the other is the predicted orientation of the dipole moment of the water molecules. The MD simulations are all in agreement with a dipole moment of the water near the surface that is oriented **parallel** to the surface plane. This ordering of the water dipoles is contrary to experimental work on the adsorption of water on Pt, Ni and Ag (see Heinzinger and Spohr, 1989), which, in agreement with *ab initio*, predict a dipole moment perpendicular to the surface.

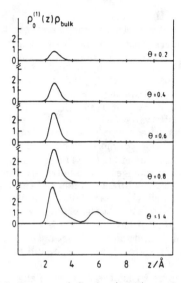

Figure 34. Oxygen atom density profiles from 5 MD simulations showing variation with increasing coverage, Θ, of water on the surface of Pt. z is the distance from the average position of the surface layer of platinum atoms. Note the development of a secondary layer of adsorbed water at high coverage (from Spohr and Heinzinger, 1988a).

Heinzinger and Spohr (1989) used a more realistic water-metal potential to see if a better potential (based on *ab initio*) would ameliorate the situation; however, the results (Figure 35) were still the same (e.g., the g_{HH} and g_{OO} peaks coincide at 2.5 Å, yielding a parallel orientation of the water molecules on the surface). They conclude that the water-water interactions predominantly determine the structure even of the water molecules in the adsorption layer.

Applications of water MD on mineral surfaces have been few. In fact, MD work on both aqueous solutions of geochemical interest and on fluid-mineral dynamics should be a high priority for future geochemists. Recent work includes that of Gruen et al. (1981) on water within mica walls, Grivtsov et al. (1988) on adsorption of water on β-tridymite and of Mulla et al. (1984) on an analogue pyrophyllite surface. Grivtsov et al (1988) used 9 water molecules and did not use Ewald-like techniques to sum up the charges, hence needed about 15000 atom centers to obtain 5% accuracy. Nonetheless, they found that the adsorbed waters are dominantly either between hydroxyl groups of the surface (deeper penetration) or on silanol groups (see *ab initio* section). Mulla (1986) has a nice review of the earlier work, including the work on water itself (see Table 12).

MONTE CARLO METHODS IN SURFACE STUDIES

The Monte Carlo method, a complementary method to the molecular dynamics technique, models surface structure and dynamics by using statistical sampling with random numbers (Gilmer and Bennema, 1972; Gilmer 1976, 1977, 1980; Broughton and Gilmer, 1983; Lasaga and Blum, 1986; Blum and Lasaga, 1987; Wehrli, 1989a,b). From the point of view of surface reactions, three processes dominate the dynamics of adsorption, growth and dissolution:

(1) Attachment of an ion or molecule from solution.
(2) Surface diffusion of an ion or molecule from one
    surface site to an adjacent surface site.

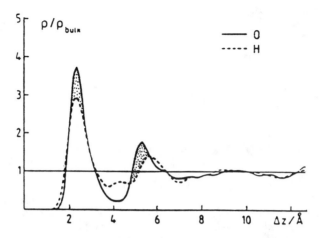

Figure 35. Normalized oxygen and hydrogen atom densities as a function of distance from the Pt (100) surface. The shaded areas indicate an excess of positive or negative charges (from Spohr and Heinzinger, 1988a).

Table 12. Results from selected molecular dynamics studies of water near hydrophobic surfaces (from Mulla, 1986).

# molecules	150	216	150	256
cell volume (nm³)	4.5	8.787	5.198	9.156
cell density (g/cc)	1.0	0.74	0.87	0.84
temperature (K)	301	287	304	286
trajectory time (ps)	25	20	14	0.75
water potential	ST2	ST2	MCY	ST2
surface potential	hard wall	L-J 12-6	L-J 4-2	L-J 12-6
range of density oscillations (g/cc)	0.9 to 1.0	0.9 to 1.0	0.5 to 3.2	0.8 to 1.1
density trend towards surfaces	decreases	decreases	increases	decreases
hydrogen bonding trend towards surfaces	---	---	---	decreases
preferred dipolar orientations	yes	yes	yes	yes
self-diffusion coeff. near surfaces (m²/s)	3.3×10^{-9}	4.8×10^{-9}	3.1×10^{-9}	2.1×10^{-9}
near midplane (m²/s)	4.2×10^{-9}	3.3×10^{-9}	3.7×10^{-9}	2.7×10^{-9}
dipole relaxation time near surfaces (ps)	3.1	---	2.3	---
near midplane (ps)	2.1	---	2.0	---

(3) Detachment of an ion or molecule back to solution.

These three individual atomic processes have rate constants that vary with the local details of the surface topography. Furthermore, any one of these elementary processes will be assumed to be an activated process so that the rate constants can be adequately described by an Arrhenius formulation. The details of these rate constants will be discussed below. For now, assume that there is an arrival rate constant, k^+, (the number of molecules per total area per unit time), a detachment rate constant, which depends on the number of bonds, n, of a given atom on the surface to adjacent atoms, k_n^-, and let us ignore surface diffusion. The Monte Carlo treatment turns these rates into probabilities and the whole surface dynamics into a stochastic process. For example, suppose that we are modeling a surface with 100 sites (with a periodic boundary condition in two dimensions, so that the atoms at the edge "see" the atoms on the other side) and that k^+ is 5 atoms/total surface/unit time and $k_n^- = 0.8, 0.4$, 0.2, 0.1 (atoms/site/unit time) for n = 1,2,3,4 respectively. Note that the k^- are given in units of atoms per site of a given kind per unit time (so that the total rate is given by k_n X_n, where X_n is the fraction of surface sites that have an atom with n bonds). The Monte Carlo scheme would then pick an interval of time of $\Delta t = 0.1$ unit times so that the number of atoms arriving is 0.5 in one Δt. Now this rate becomes a probability; therefore, we pick a site at random, then pick a random number, r (0 $\leq$ r $\leq$ 1), if r is less than 0.5 a new species is added to that site; if r is greater than 0.5, no new species is added. After adding an atom (or doing nothing if r $\geq$ 0.5), the method proceeds to pick another site at random for possible dissolution. Once a site is picked, the method checks how many bonds the atom on the surface has. If n = 2 (for example), then using the appropiate Δt, the probability of detachment is 0.04. Therefore, a new random number, r, is picked and if r is less than 0.04, the atom is removed. This scheme is continued for as long as necessary to establish the kinetics of the system. Blum and Lasaga (1987) also discuss methods to speed up the low probability events in the simulation.

At this point, it is imperative to give a clear explanation of the eneergetics involved in the atomic processes and their relation to the individual atomic rate constants used in Monte Carlo simulations. In particular, the importance of **kink** sites in surface theories and the role of microscopic reversibility in linking the various rate constants needs to be discussed. Let us take the simple case of a solid containing N atoms, with each atom having s nearest neighbors in the bulk solid. Focusing only on the short range interactions let us define the following parameters:

- ϕ_{ss} = solid-solid interaction energy
- ϕ_{sf} = solid-fluid interaction energy
- ϕ_{ff} = fluid-fluid interaction energy

For simplicity, let us now assume that the fluid molecules also have the same number of nearest neighbors, s (this is not too critical, because the solid bonds dominate the energetics, and it can be corrected in more sophisticated models). Let us assume that there are M molecules of fluid. The energy of the initial system (i.e., crystal + fluid) is then

$$E_{before} = -\frac{N}{2} s \phi_{ss} - \frac{M}{2} s \phi_{ff} . \tag{79}$$

Equation (79) ignores surface effects (i.e., large V/A ratio). The factors of 1/2 in the equation take into account that each bond is counted twice in the interactions of each solid species or fluid species with each other. This is an important point. Let us now, dissolve the crystal completely. There are now N solid molecules surrounded by fluid molecules. and we have lost some of the fluid-fluid interactions. The new energy of the system is

$$E_{after} = -\frac{M - N}{2} s \, \phi_{ff} - s \, N \, \phi_{sf} \ , \tag{80}$$

where the loss of N fluid-fluid interactions leads to the M - N term. The net change in energy, therefore, is

$$\Delta E = \frac{s}{2} N \left(\phi_{ss} + \phi_{ff} - 2 \phi_{sf} \right) \ , \tag{81}$$

$$\Delta E = \frac{s}{2} N \, \Phi \ ,$$

where Φ has been defined by the last equation. Note the similarity to solid solution models (e.g., regular solutions). The energy per solid particle is then given by

$$\Delta E = \frac{s}{2} \, \Phi \ . \tag{82}$$

Let us now contrast this calculation with the process whereby a particle on the **surface** with n solid bonds (i.e., n nearest solid neighbors and $(s - n)$ fluid neighbors) dissolves. The dissolution process can be thought of as an exchange of a fluid molecule with the molecule on the surface. In this case, the energy before the dissolution (focusing only on the particles to be exchanged) is given by

$$E_{before} = -s \, \phi_{ff} - n \, \phi_{ss} - (s - n) \, \phi_{sf} \ .$$

After the exchange, the energy is given by:

$$E_{after} = -s \, \phi_{sf} - n \, \phi_{sf} - (s - n) \, \phi_{ff} \ .$$

Therefore, the net change in energy on dissolution is given by

$$\Delta E = n \left(\phi_{ff} + \phi_{ss} - 2\phi_{sf} \right) , \tag{83}$$

$$\Delta E = n \, \Phi \ , \tag{84}$$

Comparing Equations (82) and (84), it is clear that the energy change upon bulk dissolution of the solid equals the change upon detachment of a surface species, **only** when the number of bonds, n, equals s/2. This surface species is precisely what is termed a **kink** site. For example, for a cubic material (s = 6), the kink site would have 3 neighbor bonds (see Fig. 1 above). This is the fundamental link of kink sites to the thermodynamics of surfaces. Basically, the 1/2 comes into play because in the bulk we count bonds twice. This is not true of dissolution of surface species and so a kink site will lead to the same energy change as a bulk dissolution.

We can take this treatment further and also understand the role of kink sites in the kinetics by introducing the principle of **microscopic reversibility** (see Lasaga, 1981a). Let us return to the atomic rate constants, k^+ and k_n^-. Imagine a surface in contact with a solution with which it is not in bulk thermodynamic equilibrium, i.e., the $\Delta\mu$ of the reaction is not zero ($\Delta\mu \geq 0$ means the solution is supersaturated and $\Delta\mu \leq 0$ means the solution is undersaturated). In a short time, equilibrium with the surface topography (not with the bulk) will be established so that the rate of arrival of atoms on a site with n bonds will equal the rate of detachment. Because the total rate of detachment from the surface (remember k^+ is defined with respect to total area) depends upon the fraction of the surface that has sites with n bonds, X_n, we have

$$k^+ = k_n^- X_n \ . \tag{85}$$

74

Therefore,

$$\frac{k^+}{k_n^-} = X_n = exp(-\frac{\Delta G_n^0}{RT}) .$$ (86)

The change in free energy upon dissolution of a site with n bonds must correct the bulk free energy change, $\Delta\mu$, for the difference in energy between the bulk dissolution and that of the surface species. Comparing Equations (82) and (84), it is clear that the correction is $(n - s/2)$ Φ. Therefore,

$$\Delta G_n^0 = -\Delta\mu - (n - s/2)\,\Phi , $$ (87)

and so

$$k^+ = k_n^- \, e^{\frac{\Delta\mu}{RT}} \, e^{(n-s/2)\Phi/RT} .$$ (88)

Note that if the solid and the liquid are at bulk thermodynamic equilibrium, $(\Delta\mu = 0)$ then

$$k^+ = k_{s/2}^- = k_{kink}^- .$$ (89)

This result is the rationale behind the often made statement that at equilibrium the rate of attachment and detachment of atoms at **kink** sites is equal (this is not true for the other surface sites as one can see from Eqn.(88)). Equation (88) also justifies the Monte Carlo treatment where if the arrival rate is set to:

$$k^+ = k_{kink}^- \, e^{\Delta\mu/RT} = k_{eq}^+ \, e^{\Delta\mu/RT} ,$$ (90)

then the detachment rate is given by

$$k_n^- = \nu \, e^{-n\Phi/RT} ,$$ (91)

where ν is defined by

$$\nu = k_{kink}^- \, e^{(s/2)\Phi/RT} .$$

Equations (90) and (91) form the core of the various Monte Carlo simulations (Blum and Lasaga, 1987).

The same treatment as carried out in Equations (82)-(91) can be extended to solid solutions or the incorporation of impurities. First, Equation (82) would need to be generalized to take into account the solid solution composition and/or the impurity content of the bulk solid. This would redefine the base expression for the energy term in $\Delta\mu$. Then the changes in free energy ΔG^0 (e.g., needed in Eqn.(86)) can be properly corrected for variations in the energy change due to local variations in the bonding of the particular surface species being investigated. For example, in more complex cases there will be several kink sites possible. Note that even in these cases, Equation (89) holds except that now the "kink" site meant in (89) is an "average" kink site i.e., a kink site with the bonding being half the average number of bonds to each solid solution atom or impurity atom in the bulk solid. The bottom line is to obtain a new formula for ΔH of the bulk solid and use this formula to correct the energies of surface processes from that given in $\Delta\mu$. These corrected energies are used in the microscopic reversibility equations.

Equations (90) and (91) can also be generalized to include changes in surface energy due to structural defects such as dislocations (Blum and Lasaga, 1987). In the case of an additional strain energy term, u(r), at a site a distance r away from a dislocation line, the dissolution rate constants discussed above can be modified to become:

$$k_n^- = \nu \, exp(-n(\phi - \frac{s}{2}u(r))/kT) .$$ (92)

It is this ability to readily model quite complex (and hopefully more realistic) cases, that makes the Monte Carlo method such a powerful tool to investigate a wide variety of surface phenomena.

It is important to link the values of the MC parameters to the macroscopic observables of minerals. As discussed in Lasaga and Blum (1986), the size of the Monte Carlo "block", a (Fig.1 above), can be related to the molecular volume,

$$\bar{V} = a^3 . \tag{93}$$

Likewise, the bonding parameter Φ can be related to the surface free energy by

$$\sigma = \frac{\Phi}{a^2} . \tag{94}$$

Of course, Φ can also be related to the $\Delta H_{dissoln}$ (Eqn.(82)). For example, for quartz $\bar{V}$ is 22.7 cm^3/mole, $\Delta H_{dissoln}$ is 5.34 kcal/mole (Robie et al., 1978) and the surface free energy is 350 ergs/cm^2 (Parks, 1984). If $\Phi/kT = 10$ and a = 3.36 Å then $\bar{V} =$ 22.7 cm^3/mole, $\sigma = 367$ ergs/cm^2 and $\Delta H_{dissoln} = 4.47$ kcal/mole. Blum and Lasaga (1987) used slightly different values of Φ/kT and a (4 and 2-3) based mostly on data for σ and the Burgers vector for dislocations in quartz. In either case, values can be tailored to ensure that the macroscopic variables of interest are properly predicted by the MC parameters.

One of the interesting applications of the Monte Carlo method is in delineating the transition from an atomically smooth surface to a rough surface in crystal growth (or dissolution). The simple analytic model for such a transition was discussed in the earlier section. It was shown there that a key parameter is the ΔH of the transition, which is related to the bonding parameter in the surface. Such a transition can be seen in the Monte Carlo results shown in Figure 10 above. For a $\phi/kT = 4$ and $\Delta\mu/kT = 1.5$, the growth occurs in an orderly planar configuration. However, as the supersaturation increases to $\Delta\mu/kT = 5.0$, the growth is now leading to the formation of a rough surface, i.e., we have a roughening transition. Note that in the figures, several layers have already been added to the initial surface, i.e., the planar growth in Figure 10a is **not** an artifact of initial conditions. The layers are all laid down one after the other in an orderly fashion. This type of transition from orderly (euhedral crystals) to rough growth as a function of the supersaturation has been seen in melt growth (Kirkpatrick, 1981; Miller, 1977).

Another application of the Monte Carlo approach follows the influence that a step (screw dislocation) on the surface has on the growth or dissolution process (for an example of natural occurrences in silicates see the work on kaolinite by Yonebayashi and Hattori, 1979). For example, if a screw dislocation is introduced into the Monte Carlo simulation (we do this by changing which atoms are called "bonded" in the vicinity of the dislocation), the dissolution is seen to spiral downward due to the presence of the step in Figure 36. Note that the strain field of the dislocation has not been included in the simulation and therefore no etch pit is being formed. One of the nice properties of the Monte Carlo method is to enable us to unravel different effects, e.g., the effect of the step on the dissolution versus the effect of the stress.

An interesting application of the Monte Carlo method has been to the prediction of the role of dislocations in mineral dissolution rates (Blum and Lasaga, 1987). Using the parameters for quartz, Blum and Lasaga (1987) calculated the relative rates of dissolution of defect free quartz and of an etch pit forming at a screw dislocation. If there are enough dislocations, the rate should be controlled by the etch pit dissolution and the rate should be proportional to the dislocation density.

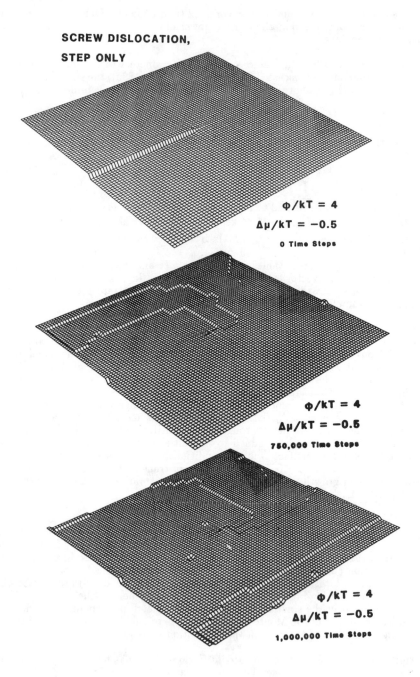

Figure 36. Monte Carlo simulations of dissolution in the presence of a screw dislocation. Only the step of the dislocation is included i.e. the strain energy is set to zero. The MC parameters used were Φ/kT = 4 and $\Delta\mu/kT$ = -0.5 (undersaturation). a) 0 time steps b) 750,000 time steps c) 1,000,000 time steps (Program by Alex Blum).

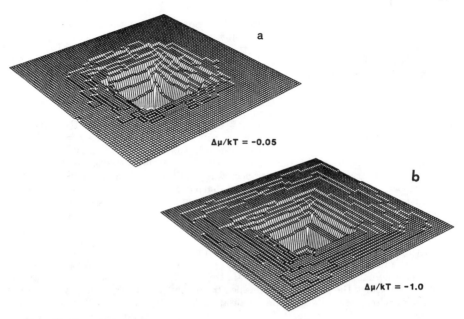

a

Δμ/kT = -0.05

b

Δμ/kT = -1.0

Figure 37. Monte Carlo simulations of etch pit formation. a) $\Phi/kT = 4$ and $\Delta\mu/kT = $ -0.05. Note formation of hollow core b) $\Phi/kT = 4$ and $\Delta\mu/kT = $ -1.0. In this case (for the same amount of units dissolved as in a), the hollow core is expanding to form an etch pit (from Lasaga and Blum, 1986).

The Monte Carlo results indicated that this "cross-over" would not occur until the dislocation density reached values of 10^9-10^{10} dislocations/cm^2. In other words, the dissolution rate of most natural quartz would not be sensitive to the dislocation density. These high values have been confirmed in subsequent experiments (Blum and Lasaga, 1990), which showed that the dissolution rate of quartz did not vary significantly, when the dislocation density was varied from 10^6 to 10^{10}.

On the other hand, Blum and Lasaga (1987) showed that the earlier theory of Lasaga and Blum (1986) and experimental work of Brantley et al. (1986) quantifying the formation of dissolution etch pits as a result of the strain field of a dislocation, were in very good agreement with detailed MC simulations. Figure 37 depicts the two cases a) low undersaturation and the formation of a hollow core b) high undersaturation and the formation of an etch pit. Much more can be done with MC in this regard, for example studying the effect of bonding anisotropy on the shape of the etch pit or the interaction between dislocations.

One of the useful applications of the Monte Carlo method is in understanding the various rate laws possible for surface reactions. This is a critical field, where the geochemical data are sparse. In general, any geochemical reaction (e.g., the dissolution of albite, the precipitation of gibbsite or kaolinite, etc.) will have a rate that depends on the ΔG of the reaction (as well as other parameters). For a surface reaction one may write:

$$Rate = k_0\, e^{E/RT} \prod_i a_i^{n_i}\, f(\Delta G)\ , \qquad (95)$$

where the E is an "activation" energy (operationally defined because it may consist of contributions from true activated complexes and from ΔH of equilibrium reactions (see earlier in this chapter or Lasaga, 1981a). The product term takes into account the possible catalysis or inhibition of the reaction by different chemical species. Activities are used (although one could rephrase this in terms of concentrations) and a power law dependence is assumed (for simplicity). The majority of geochemical research to date has focused on obtaining an E and on ascertaining the values of n_i for various important species such as H^+, OH^-, Na^+ etc. Little has been done on the last term in Equation (95), $f(\Delta G)$, which reflects the important dependence of the overall rate on the proximity to equilibrium. This term can vary the rate by orders of magnitude as equilibrium is approached. Therefore, unless we know the shape of the f function, kinetic values that may be way too high will be used in modeling natural systems. One may erroneously conclude that equilibrium may be reached in some cases based on such calculations. Furthermore, the shape of the f functions (for example, if a severe drop in rate occurs as equilibrium is approached, at what supersaturation or undersaturation does the sharp drop appear?) will have non-trivial implications if there is significant coupling between several reactions in the system. A most important question depends on whether the function f is a nearly linear or highly non-linear function of ΔG.

Little is known about the shape of this function. One certain constraint is that the value of the function is 0 when ΔG is zero, i.e., the kinetics must be fully consistent with thermodynamics. The requirement that $f(0) = 0$, embodies for overall reactions what is often called the principle of "detailed balancing" or "microscopic reversibility" in elementary reactions. Recent experimental work by Nagy et al. (1990) has begun to unravel the shape of this important function. Monte Carlo simulations in conjunction with experiments can guide us in what to expect. Blum and Lasaga (1987) have carried out some simulations to explore the shape of possible f functions. For example, Figure 38 shows the $f(\Delta G)$ curve for a surface that is defect free and for the case where no surface diffusion is allowed. Two cases are compared with different values of the bonding parameter, Φ/kT. Clearly, the strength of the bonding has a dramatic effect on the shape of the curves. Furthermore, the presence of an induction region, where no reaction takes place, is also readily observable. Figure 39 shows the effect of adding the presence of a screw dislocation step and also compares the effect of surface diffusion on $f(\Delta G)$. An interesting result follows if the precipitation rate is fitted to an equation of the type:

$$Rate = A \left(\frac{\Delta \mu}{kT}\right)^n , \tag{96}$$

where A and n are adjustable parameters. The MC data yields values of n between 2 and 3. The BCF crystal growth theory predicts a value of 2. Experiments on the dehydration of muscovite (see earlier) yields values betweeen 2.0 and 2.6 (Schramnke et al., 1985). Recent data by Metz and co-workers also yield values in this range (Lasaga, unpublished). This agreement suggest that the Monte Carlo simulations may be important in our understanding of the surface kinetics involved in many of the geochemical mineral reactions.

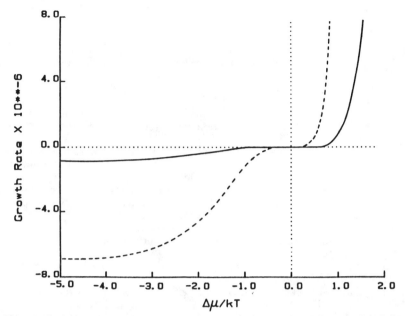

Figure 38. Surface reaction rate laws for dislocation-free surfaces. No surface diffusion is allowed. Rate in units of a ν (see text). Solid line uses $\Phi/kT = 3.5$; Dashed line uses $\Phi/kT = 3.0$ (from Blum and Lasaga, 1987).

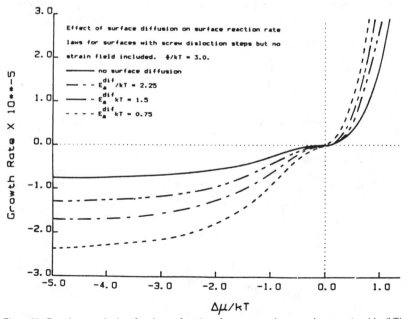

Figure 39. Reaction rate (units of a ν) as a function of supersaturation or undersaturation ($\Delta\mu/kT$) for various conditions. $\Phi/kT = 3.0$ in all cases. Solid line: dislocation-free surface and no surface diffusion. -.. : screw dislocation step (no strain) and no surface diffusion. -.- : Dislocation-free surface and surface diffusion included (with activation energy $E/kT = 1.5$). — : Screw-dislocation step (no strain) and surface diffusion included (with activation energy $E/kT = 1.5$) (From Blum and Lasaga, 1987)

REFERENCES

Aagard, P. and H.C. Helgeson, 1982, Thermodynamic and kinetic constraints on reaction rates among minerals and aqueous solutions, I. Theoretical Considerations, Am. J. Sci., **282**, 237-285.

Andersen, H.C., 1980, Molecular dynamics simulations at constant pressure and/or temperature, J. of Chem. Phys., **72**, 2384-2393.

Angell, C.A., P.A. Cheeseman, and S. Tammaddon, 1982, Pressure enhancement of ion mobilities in liquid silicates from computer simulation studies to 800 kbar, Science, **218**, 885-887.

Angell, C.A., P. Cheeseman, and S. Tammaddon, 1983, Water-like transport property anomalies in liquid silicates investigated at high T and P by computer simulation techniques, Bull. Mineral.,**106**, 87-97.

Angell, C.A., C.A. Scamehorn, C.C. Phifer, R.R. Kadiyala, and P.A. Cheeseman, 1988, Ion dynamics studies of liquid and glassy silicates, and gas-in-liquid solutions, Phys. and Chemistry of Minerals, **15**, 221-227.

Barabino, G., C., Gavotti, M., Marchesi, 1984, Molecular dynamics simulation of water near walls using an improved wall-water interaction potential, Chem. Phys. Lett., **104**, 478-483.

Berner, R.A., 1978, Rate control of mineral dissolution under Earth surface conditions, Am. J. Sci., **278**, 1235-1252.

Berner, R.A. and J.W., Morse, 1974, Dissolution kinetics of calcium carbonate .n sea water: IV. Theory of calcite dissolution, Am. J. Sci., **274**, 108-134.

Blakely, J.M., 1973, Introduction to the Properties of Crystal Surfaces, Pergamon Press, Oxford, 261 pp.

Blum, A.E. and A.C. Lasaga, 1987, Monte Carlo simulations of surface reaction rate laws, in Aquatic Surface Chemistry: Chem. processes at the particle-water interface, W. Stumm (ed.), John Wiley and Sons, New York, 255-292.

Blum, A.E. and A.C. Lasaga, 1990, The effect of dislocation density on the dissolution rate of quartz, Geochim. Cosmochim. Acta, **54**, 283-297.

Bourg, M., 1984, Preliminary results of a theoretical *ab initio* model study of adsorption on ionic crystals, Int'l. J. Quantum Chem., **26**, 775-781.

Boys, S.F. and F., Bernardi, 1970, The calculation of small molecular interactions by the differences of separate total energies. Some procedures with reduced errors, Mol. Phys., **19**, 553-566.

Brantley, S.L., S.R., Crane, D.A. Crerar, R., Hellmann, and R., Stallard, 1986, Dissolution at dislocation etch pits in quartz, Geochim. Cosmochim. Acta, **50**, 2349-2378.

Broughton, J.Q. and G.H. Gilmer, 1983, Molecular dynamics investigation of the crystal-fluid interface. I. Bulk properties, J. Chem. Phys., **79**, 5095-5104.

Casey, W.H., A.C., Lasaga, and G.V., Gibbs, 1990, Mechanisms of silica dissolution as inferred from the kinetic isotope effect, Geochim. Cosmochim. Acta, in press.

Ciccotti, G., D. Frenkel, I.R. McDonald, eds., 1987, Simulation of Liquids and Solids - Molecular Dynamics and Monte Carlo Methods in Statistical Mechanics, North-Holland, Amsterdam, 481 pp.

Dachs, E. and P. Metz, 1988, The mechanism of the reaction: 1 tremolite + 3 calcite + 2 quartz ↔ 5 diopside + 3 CO_2 + 1 H_2O: Results of powder experiments, Contrib. Mineral. Petrol., **100**, 542-551.

Dempsey, M.J., and K., Kawamura, 1984, Molecular dynamics simulation of the structure of aluminosilicate melts. In: Progress in Experimental Petrology, Natural Environ. Res. Council Pub. Series D, **25**, 49-55.

Doll, J.D. and H.K., McDowell, 1982, Theoretical studies of surface diffusion: Self-diffusion in the fcc (111) system, J. Chem. Phys., **77**, 479-490.

Dove, P.M. and D.A., Crerar, 1990, Kinetics of quartz dissolution in electrolyte solutions using a hydrothermal mixed flow reactor, Geochim. Cosmochim. Acta, **54**, 955-969.

81

Dovesi, R., C., Pisani, C., Roetti, and B., Silvi, 1987, The electronic structure of α-quartz: a periodic Hartree-Fock calculation, J. Chem. Phys., **86**, 6967-6971.

Evans, M.W., G.C. Lie, and E. Clementi, 1988, Molecular dynamics simulation of water from 10 to 1273 K, J. Chem. Phys., **88**, 5157-5165.

Feuston, B.P. and S.H. Garofalini, 1988, Empirical three-body potential for vitreous silica, J. Chem. Phys., **89**, 5818-5824.

Fisher, G.W., 1978, Rate laws in metamorphism, Geochim. Cosmochim. Acta, **42**, 1035-1050.

Foster, K., K., Raghavan, and M. Berkowitz, 1989, A molecular dynamics study of the effect of temperature on the structure and dynamics of water between Pt walls, Chem. Phys. Lett., **162**, 32-38.

Frisch, M.J., J.S. Binkley, H.B., Schlegel, K., Raghavachari, C.F., Melius, L., Martin, J.J.P., Stewart, F.W., Bobrowicz, C.M., Rohlfing, L.R., Kahn, D.J., Defrees, R., Seeger, R.A., Whiteside, D.J., Fox, E.M., Fleuder, J.A., Pople, 1983 **Gaussian 86**, Carnegie-Mellon Quantum Chemistry Publishing Unit, Pittsburgh, PA.

Garofalini, S.H., 1982, Molecular dynamics simulation of the frequency spectrum of amorphous silica, J. Chem. Phys., **76**, 3189-3192.

Garofalini, S.H., 1983, A molecular dynamics simulation of the vitreous silica surface, J. Chem. Phys., **78**, 2069-2072.

Geerlings, P., N., Tariel, A., Botrel, R., Lissillour, and W.J., Mortier, 1984, Interaction of surface hydroxyls with adsorbed molecules. A quantum chemical study, J. Phys. Chem., **88**, 5752-5759.

Ghaleb, D., 1984, Diffusion of adatom dimers on (111) surface of face centred crystals: a Molecular Dynamics study, Surf. Sci., **137**, L103-L108.

Gibbs, G.V., 1982, Molecules as models for bonding in silicates, Amer. Mineral., **67**, 421-450.

Gilmer, G.H., 1976, Growth on imperfect crystal faces. I. Monte Carlo growth rates, J. Cryst. Growth, **35**, 15-28.

Gilmer, G.H., 1977, Computer simulation of crystal growth, J. Cryst. Growth, **42**, 3-10.

Gilmer, G.H., 1980, Computer models of crystal growth, Science, **208**, 355-363.

Gilmer, G.H. and P., Bennema, 1972, Simulation of crystal growth with surface diffusion, J. Appl. Phys., **43**, 1347-1360.

Grivtsov, A.G., L.T., Zhuravlev, G.A., Gerasimova, L.G., Khazin, 1988, Molecular dynamics of water: adsorption of water on β-tridymite, J. Colloid Interface Sci., **126**, 397-406.

Gruen, D.W.R., S., Marcelja, and B.A., Pailthorpe, 1981, Theory of polarization profiles and the "hydration force", Chem. Phys. Lett., **82**, 315-322.

Hagon, J.P., A.M., Stoneham, and M., Jaros, 1987, Transport processes in silicon oxidation I. Dry oxidation, Phil. Mag. B, **55**, 211-224.

Hautman, J., J.W., Halley, and Y.J., Rhee, 1989, Molecular dynamics simulation of water between two ideal classical metal walls, J. Chem. Phys., **91**, 467.

Hehre, W.J., L., Radom, P.R., Schleyer, and J.A., Pople, 1986, *Ab initio* Molecular Orbital Theory, John Wiley and Sons, New York

Heinrich, W., P., Metz, M., Gottschalk, 1989, Experimental investigation of the kinetics of the reaction: 1 tremolite + 11 dolomite ↔ 8 forsterite + 13 calcite + 9 CO_2 + 1 H_2O, Contrib. Mineral. Petrol., **102**, 163-173.

Heinzinger, K. and E., Spohr, 1989, Computer simulations of water-metal interfaces, Electrochim. Acta, **34**, 1849-1856.

82

Helgeson, H.C., W.M., Murphy, and P., Aagaard, 1984, Thermodynamic and kinetic constraints on reaction rates among minerals and aqueous solutions. II. Rate constants, effective surface area, and the hydrolysis of feldspar, Geochim. Cosmochim. Acta, **48**, 2405-2432.

Heyes, D.M., 1983, Molecular dynamics simulations of ionic crystal films, J. Chem. Phys., **79**, 4010-4027.

Hiemstra, T., W.H., Van Riemsdijk, and G.H., Bolt, 1989, Multisite proton adsorption modeling at the solid/solution interface of (hydr)oxides: a new approach, J. Colloid Interface Sci., **133**, 91-104.

Hill, T.L., 1962, An Introduction to Statistical Thermodynamics, Addison-Wesley Pub. Co., New York, 508 pp.

Hobza, P., J., Sauer, C., Morgeneyer, J., Hurych, and R., Zahradnik, 1981, Bonding ability of surface sites on silica and their effect on hydrogen bonds. A quantum-chemical and statistical thermodynamic treatment, J. Phys. Chem., 85, 4061-4067.

Hoover, W.G., 1985, Canonical dynamics: Equilibrium phase-space distributions, Phys. Rev. A, **31**, 1695-1697.

Iler, R. K., 1979, The Chemistry of Silica, John Wiley and Sons, New York, 835 pp.

Illas, F., J., Rubio, and J.M., Ricart, 1985, *Ab initio* cluster-model study of the on-top chemisorption of F and Cl on Si(111) and Ge(111) surfaces, Phys Rev. B, **31**, 8068-8082.

Jackson, K.A., 1967, Current concepts in crystal growth from the melt, Prog. Solid State Chem., **4**, 53-80.

Jonsson, B., 1981, Monte Carlo simulations of liquid water between two rigid walls, Chem. Phys. Lett., **82**, 520-529.

Karim, O.A. and A.D.J. Haymet, 1988, The ice/water interface: A molecular dynamics simulation study, J. Chem. Phys., **89**, 6889-6896.

Kirkpatrick, R.J., 1981, Kinetics of crystallization of igneous rocks. In: Kinetics of Geochemical Processes, Reviews in Mineralogy 8, A.C. Lasaga and R.J. Kirkpatrick (eds.), 321-395

Kubicki, J.D. and A.C. Lasaga, 1988, Molecular dynamics simulations of SiO_2 melt and glass: Ionic and covalent models, Am. Mineral., **73**, 945-955.

Kubicki, J.D. and A.C. Lasaga, 1990, Molecular dynamics simulation of pressure and temperature effects on $MgSiO_3$ and Mg_2SiO_4 melts and glasses, Phys. Chem. Minerals, in press.

Kubicki, J.D., A.C. Lasaga, and R.J. Hemley, 1989, Ab-initio ,molecular dynamics simulations of forsterite and $MgSiO_3$-perovskite, EOS, Trans. Am. Geophys. Union, **70**, 349.

Laidler, K.J., 1965, Chem. Kinetics, McGraw-Hill Co., New York, 425 pp.

Landman, U., W.D. Luedtke, R.N. Barnett, C.L. Cleveland, M.W. Ribarsky, E. Arnold, S. Ramesh, H. Baumgart, A. Martinez, and B. Khan, 1986, Faceting at the silicon (100) crystal-melt interface: Theory and experiment, Phys. Rev. Lett., **56**, 155-158.

Lasaga, A.C., 1981a, Rate Laws of geochemical reactions. In: Kinetics of Geochemical Processes, Rev. Mineral., 8, A.C. Lasaga and R.J. Kirkpatrick (eds.), 1-81

Lasaga, A.C., 1981b, Transition state theory. In: Kinetics of Geochemical Processes, Rev. Mineral., 8, A.C. Lasaga and R.J. Kirkpatrick (eds.), 135-169.

Lasaga, A.C., 1981c, The atomistic basis of kinetics: defects in minerals. In: Kinetics of Geochemical Processes, Rev. Mineral., 8, A.C. Lasaga and R.J. Kirkpatrick (eds.), 261-319.

Lasaga, A.C., 1984, Chemical kinetics of water-rock interactions, J. Geophys. Res., B6, 4009-4025.

Lasaga, A.C., 1986, Metamorphic reaction rate laws and the development of isograds, Mineral.. Mag., **50**, 359-373.

Lasaga, A.C. and A.E., Blum, 1986, Surface chemistry, etch pits and mineral-water reactions, Geochim. Cosmochim. Acta, **50**, 2363-2379.

Lasaga, A.C., and G.V. Gibbs, 1987, Applications of quantum mechanical potential surfaces to mineral Phys. calculations. Phys. Chem. Minerals, **14**, 107-117.

Lasaga, A.C. and G.V. Gibbs, 1988, Quantum mechanical potential surfaces and calculations on minerals and molecular clusters, Phys. Chem. Minerals, **16**, 29-41.

Lasaga, A.C., and G. V. Gibbs, 1989, *Ab-initio* calculations on hydroxyacid silicate molecules and implications for the structure defects and spectra of SiO_2 glass, Phys. Chem. Minerals, submitted.

Lasaga, A.C. and G.V., Gibbs, 1990a, *Ab initio* quantum mechanical calculations of surface reactions - A new era?. In: Aquatic Chemical Kinetics, W. Stumm (ed.), John Wiley and Sons, New York.

Lasaga, A.C. and G.V., Gibbs, 1990b, *Ab Initio* quantum mechanical calculations of water-rock interactions: Adsorption and hydrolysis reactions, Am. J. Science, **290**, 263-295.

Levine, S.M. and S.H., Garofalini, 1987, A structural analysis of the vitreous silica surface via a molecular dynamics computer simulation, J. Chem. Phys., **86**, 2997-3002.

Li, L., K., Tsukamoto, I., Sunagawa, 1990, Impurity adsorption and habit changes in aqueous solution grown KCl crystals, J. Cryst. Growth, **99**, 150-155

Liebau, F., 1984, Pentacoordinate silicon intermediate states during silicate condensation and decondensation. Crystallographic suport, Theoretica Chim. Acta, **89**, 1-7.

Lutsko, J.F., D., Wolf, S., Yip, S.R., Phillpot, and T., Nguyen, 1988, Molecular dynamics method for the simulation of bulk-solid interfaces at high temperatures, Phys. Rev. B, **38**, 11572-11581.

Marchesi, M., 1983, Molecular Dynamics simulation of liquid water between two walls, Chem. Phys. Lett., **97**, 224-230.

Matsui, M., 1988, Molecular dynamics study of $MgSiO_3$ perovskite, Phys. and Chemistry of Minerals, **16**, 234-238.

Matsui, Y., K. Kawamura, and Y. Syono, 1981, Molecular dynamics calculations applied to silicate systems: Molten and vitreous $MgSiO_3$ and Mg_2SiO_4. In: S. Akimoto and M.H. Manghnani, (eds.), High Pressure Research in Geophysics, Advances in Earth and Planetary Science, **12**, 511-524. Reidel, Boston, MA.

Miller, C.E., 1977, Faceting transition in melt grown crystals, J. Cryst. Growth, **42**, 357-363.

Mitra, S.K., M. Amini, D. Fincham, and R.W. Hockney, 1981, Molecular dynamics simulation of silicon dioxide glass, Phil. Mag., B, **43**, 365-372.

Mitra, S.K., 1982, Molecular dynamics simulation on silicon dioxide glass, Phil. Mag., B, **45**, 529-548.

Mortier, W.J., J., Sauer, J.A., Lercher, and H., Noller, 1984, Bridging and terminal hydroxyls. A structural chemical and quantum chemical discussion, J. Phys. Chem., **88**, 905-912.

Mulla, D.J., P.F., Low, J.H., Cushman, and D.J., Diestler, 1984, A molecular dynamics study of water near silicate surfaces, J. Colloid Interface Sci., **100**, 576-580.

Mulla, D.J., 1986, Simulating liquid water near mineral surfaces: current methods and limitations. In: Geochemical Processes at Mineral Surfaces, Am. Chem. Society, Washington, D.C., pp. 20-36.

Muncill, G.E. and A.C., Lasaga, 1988, Crystal-growth kinetics of plagioclase in igneous systems: Isothermal H_2O-saturated experiments and extension of a growth model to complex silicate melts, Am. Mineral., **73**, 982-992.

Nagy, K.L., C.I., Steefel, A.E., Blum, A.C., Lasaga, 1990, Dissolution and precipitation kinetics of kaolinite: Initial results at $80^{\circ}C$ with application to porosity evolution in a sandstone, AAPG Memoir, in press.

Nakatsuji, H., Y., Matsuzaki, and T., Yonezawa, 1988, *Ab initio* theoretical study on the reactions of a hydrogen molecule with small platinum clusters: A model for chemisorption on a Pt surface, J. Chem. Phys., **88**, 5759-5772.

Nosé, S., 1984, A unified formulation of the constant temperature molecular dynamics methods, J. Chem. Phys., **81**, 511-519.

Nosé, S. and F. Yonezawa, 1986, Isothermal-isobaric computer simulations of melting and crystallization of a Lennard-Jones system, J. Chem. Phys., **84**, 1803-1814.

Panas, I., J., Schule, U., Brandemark, P., Siegbahn, and U., Wahlgren, 1988, Comparison of the binding of carbon, nitrogen, and oxygen atoms to single nickel atoms and to nickel surfaces, J. Phys. Chem., **92**, 3079-3086.

Parks, G.A., 1984, Surface and interfacial free energies of quartz, J. Geophys. Res., **89**, 3997-4008.

Pontikis, V., and Rosato, V., 1985, Roughening transition on the (110) face of Argon: a Molecular Dynamics study, Surf. Sci., **162**, 150-155.

Rahman, A. and F.H. Stillinger, 1971, Molecular dynamics study of liquid water, J. Chem. Phys., **55**, 3336-3359.

Rimstidt, J.D. and H.L., Barnes, H.L., 1980, The kinetics of silica-water reactions, Geochim. Cosmochim. Acta, 44, 1683-1699.

Robie, R.A., B.S., Hemingway, and J.R., Fisher, 1978, Thermodynamic properties of minerals and related substances at 298.15 K and 1 bar (10^5 Pascals) pressure and at higher temperatures, U.S. Geol. Surv. Bull., 1452, 456 pp.

Sanders, M.J., M., Leslie, and C.R.A., Catlow, 1984, Interatomic potentials for SiO_2, J. Chem. Soc., Chem Commun., 1271-1274.

Sauer, J., 1987, Molecular structure of orthosilicic acid, silanol, and H_3SiOH • AlH_3 complex: Models of surface hydroxyls in silica and zeolites, J. Phys. Chem., **91**, 2315-2319.

Sauer, J., 1989, Molecular Models in *ab Initio* studies of solids and surfaces: From ionic crystals and semiconductors to catalysts, Chem. Rev., **89**, 199-255.

Sauer, J. and R., Zahradnik, 1984, Quantum chemical studies of zeolites and silica, Int'l. J. Quant. Chem., **26**, 793-822.

Schindler, P.W. and W., Stumm, 1987, The surface chemistry of oxides, hydroxides and oxide minerals. In: Aquatic Surface Chemistry, W. Stumm (ed.), Wiley Interscience, New York, p. 311-338.

Schommers, W., P., von Blanckenhagen, 1985, Study of the surface structure by means of molecular dynamics, Surf. Sci., **162**, 144-149.

Schramke, J. A., D.M., Kerrick, and A.C., Lasaga, 1987, The reaction Mmuscovite + quartz → andalusite + K-feldspar + water. Part I. Growth kinetics and mechanism, Am. J. Sci., **287**, 517-559.

Seel, M., and P.S., Bagus, 1983, *Ab initio* cluster study of the interaction of fluorine and chlorine with the Si(111) surface, Phys. Rev. B, **28**, 2023-2039.

Sonnenschein, R., and K., Heinzinger, 1983, a Molecular Dynamics study of water between Lennard-Jones walls, Chem. Phys. Lett., **102**, 550-558.

Spohr, E., and K., Heinzinger, 1986, Molecular Dynamics simulation of a water/metal interface, Chem. Phys. Lett., **123**, 218-231.

Spohr, E., and K., Heinzinger, 1988a, A molecular dynamics study on the water/metal interfacial potential, Ber. Busenges. Phys. Chem., **92**, 1358-1363.

Spohr, E., and K., Heinzinger, 1988b, Computer simulations of water and aqueous electrolyte solutions at interfaces, Electrochim. Acta, **33**, 1211-1222.

Sprik, M. and M.L. Klein, 1988, A polarizable model for water using distributed charge sites, J. Chem. Phys., **89**, 7556-7560.

Stillinger, F.H. and A., Rahman, 1974, Improved simulation of liquid water by molecular dynamics, J. Chem. Phys., **60**, 1545-1557.

Stillinger, F.H. and A., Rahman, 1978, Revised central force potentials for water, J. Chem. Phys., **68**, 666-670.

Stumm, W. and R., Wollast, 1990, Coordination chemistry of weathering: Kinetics of the surface-controlled dissolution of oxide minerals, Rev. Geophys., **28**, 53-69.

Takahashi, K. and K., Tanaka, 1986, Adsorption of copper ion and alkyldithiocarbonate at silica by a molecular orbital method, J. Colloid Interface Sci., **113**, 21-31.

Tanner, S.B., D.M., Kerrick, A.C., Lasaga, 1985, Experimental kinetic study of the reaction: Calcite + Quartz = Wollastonite + Carbon Dioxide, from 1 to 3 kilobars and 500° to 850°C, Am. J. Sci., **285**, 577-620.

Tole, M.P., A.C., Lasaga, C., Pantano, W.B., White, 1986, The kinetics of dissolution of nepheline ($NaAlSiO_4$), Geochim. Cosmochim. Acta, **50**, 379-392.

Toxvaerd, S. and E., Praestgaard, 1977, Molecular dynamics calculation of the liquid structure up to a solid surface, J. Chem. Phys., **67**, 5291-5304.

Tully, J.C., G.H., Gilmer, and M., Shugard, 1979, Molecular dynamics of surface diffusion. I. The motion of adatoms and clusters, J. Chem. Phys., **71**, 1630-1644.

Ugliengo, P., V., Saunders, and E., Garrone, 1990, Silanol as a model for the free hydroxyl of amorphous silica: *Ab initio* calculations of the interaction with water, J. Phys. Chem., **94**, 2260-2267.

Wehrli, B., 1989a, Monte Carlo simulations of surface morphologies during mineral dissolution, J. Colloid Interface Sci., **132**, 230-242.

Wehrli, B., 1989b, Surface structure and mineral dissolution kinetics: A Monte Carlo study. In: Water-Rock Interaction, D.L. Miles (ed.), A.A. Balkema, Rotterdam, Netherlands, pp. 751-753.

Westall, J.C., 1987, Adsorption mechanisms in aquatic surface chemistry. In: Aquatic Surface Chemistry, W. Stumm,(ed.), John Wiley and Sons, New York, 520 pp.

Woodcock, L.V., 1975, Molecular dynamics calculations on molten ionic salts. In: J. Braunstein, G. Mamantov, and G. P. Smith, (eds.), Advances in Molten Salt Chemistry, **3**, 1-74.

Woodcock, L.V., C.A., Angell, P., Cheeseman, 1976, Molecular dynamics studies of the vitreous state: Ionic systems and silica, J. Chem. Phys., **65**, 1565-1567.

Xue, X., J.F., Stebbins, M., Kanzaki, R.G. Tronnes, 1989, Silicon coordination and speciation changes in a silicate liquid at high pressures, Science, **245**, 962-964.

Yonebayashi, K. and T., Hattori, 1979, Growth spirals on etched kaolinite crystals, Clay Science, **5**, 177-188.

Zhang, Y., D., Walker, C.E., Lesher, 1989, Diffusive crystal dissolution, Contrib. Mineral. Petrol., **102**, 492-513.

Zhidomirov, G.M. and V.B., Kazansky, 1986, Quantum-chemical cluster models of acid-base sites of oxide catalysts, Adv. Catalysis, **34**, 131-204.

ATOMIC STRUCTURE, MICROTOPOGRAPHY, COMPOSITION, AND REACTIVITY OF MINERAL SURFACES

Reactivity is one of the most important chemical properties of any material, and because reactions of a solid with another material must take place starting on the surface of that solid, understanding *surface reactivity* is fundamentally important in chemistry. Atoms at the surface (the top atomic layer) of a solid reside in a markedly different environment than atoms just below the surface (the near-surface), whose environment in turn is different from atoms representative of the bulk. There is at least one bond direction, if not several, from a surface atom in which the coordination chemistry is not the same as is characteristic of like atoms in the bulk. "Foreign" atoms contacting a solid surface, whether themselves from gaseous, liquid, or solid phases, may under- or overbond the original surface atom compared to its typical bulk environment(s). The near-surface atoms are influenced indirectly by this disruption of the surface atoms. This influence may go several atomic layers into the solid, but it diminishes with depth. All of this results in one basic and underlying chemical principle: *The chemical and physical properties of a bulk material are different from the chemical and physical properties of its surface.*

The surface reactivity of a mineral, as with any solid, is ultimately dependent on three surface characteristics. They are (1) chemical composition, (2) atomic structure, and (3) fine-scale morphology, or microtopography. These three items, along with how each affects mineral surface reactivity, are the principal subjects of this chapter. The objective of this chapter is to show that detailed knowledge of surface composition, structure, and microtopography allow for a much more complete understanding of mineral surface reactivity. This in turn can be applied to a more thorough and systematic understanding of mineral surface geochemistry.

This chapter starts with a brief review of the principal diffraction, spectroscopic, and microscopic techniques used to study the physical and chemical characteristics of mineral surfaces. After this, mineral surface and near-surface composition and atomic structure, as well as surface microtopography, are discussed. It should be emphasized that this discussion mostly focuses on surface characteristics as they are directly observed, rather than on theoretical or modeling points of view. Next, specific examples of how the reactivity of surfaces is dependent on composition, atomic structure, and microtopography are discussed. Not all of the examples given will be from the mineralogic or geochemical literature. Unfortunately, the state of mineral surface chemistry lags significantly behind several aspects of surface science in other disciplines, but the non-mineralogic literature represents a valuable resource of information whose basic principles are applicable to all materials, including minerals under geologically relevant conditions.

EXPERIMENTAL TECHNIQUES

In this section, the methods that are most commonly used to study the composition, microtopography, and atomic structure of mineral surfaces are briefly reviewed (see also Table 1). This coverage is not comprehensive, but it does review the

Table 1. Tools for determining mineral surface composition, microtopography, and atomic structure which are discussed in this chapter.

Composition

	input	output
X-ray photoelectron spectroscopy (XPS)	photons	electrons
Auger electron spectroscopy (AES)[a]	electrons	electrons
secondary ion mass spectroscopy (SIMS)	ions	ions
surface analysis by laser ionization (SALI)	heat, photons, electrons, ions	ions
resonance ionization mass spectroscopy (RIMS)	heat, photons, electrons, ions	ions
Rutherford backscattering spectroscopy (RBS)	usually He	same as input
resonant nuclear reaction (RNR)	ions	gamma rays
scanning tunneling spectroscopy (STS)	tunneling electrons	electrons

Microtopography

	microscope type
phase contrast	visible light
differential interference contrast	visible light
multiple-beam interferometry	visible light
scanning electron microscopy (SEM)	electron
transmission electron microscopy (TEM)	electron
scanning tunneling microscopy (STM)	electron tunneling
atomic force microscopy (AFM)[b]	tip-sample repulsion

Atomic structure

	technique type
low energy electron diffraction (LEED)	electron diffraction
scanning tunneling microscopy (STM)	electron tunneling
atomic force microscopy (AFM)[b]	tip-sample repulsion

[a] Electron stimulated AES. X-ray stimulated AES is also possible.
[b] Repulsive-mode AFM. Attractive-mode AFM is also possible.

methods that are in most common use today or methods that show particular promise. In this regard, some of the methods listed are well established and have been available for decades, while others are so new that their full potential is not yet known. The description of each method is very brief and they are meant simply as an introduction for those not familiar with them. References are given if the reader would like additional information.

Spectroscopies for determining surface composition

X-ray photoelectron spectroscopy (XPS). XPS is the most widely used surface analytic technique. This is so because it allows relatively straight-forward analysis of the near-surface of materials. Analysis depths range between a few to over 100 Å depending on the surface analyzed and the instrument conditions. XPS can also give useful chemical state information, including oxidation and structural state parameters. Both conductors and insulators can be analyzed, and quantitative near-surface compositions can usually be calculated easily. The latest XPS technology, among other things, has provided for high spatial resolution surface analysis, with some instruments now capable of collecting data on areas of 1 μm diameter.

This technique utilizes soft x-rays which impinge on a surface, ejecting photoelectrons from valence and core levels of the surface and near-surface atoms. An electron energy analyzer is used to accurately measure the kinetic energy, E_K, of the ejected photoelectrons. Photoelectrons which have not lost energy on the way out of the sample have characteristic energies according to the classic relationship

$$E_K = h\nu - E_B - \phi_{sp} \ ,\tag{1}$$

where $h\nu$ is the photon energy which ejects the electron, E_B is the binding energy by which the electron was held in the subatomic orbital of the parent atom, and ϕ_{sp} is the work function of the spectrometer (a machine constant). XPS spectra are plotted as E_B (calculated from Eqn. 1) vs. photoelectron signal intensity, and surface and near-surface atoms are easily identified using the characteristic set of binding energies of each element. The probability (cross-section) for photoelectron ejection from H and He atoms at the photon energies used for XPS are extremely low, and these are the only elements that cannot be practically detected with the technique.

Detailed reviews of XPS written specifically for geologic materials are given in Hochella (1988) and Perry et al. (1990).

Auger electron spectroscopy (AES). Electron stimulated AES has been a popular surface analytic technique, especially when used to analyze conductors and semiconductors, due to its high spatial resolution capabilities. The surface sensitivity of the technique is equivalent to XPS described above. Analyses over areas down to a few hundred Angstroms in diameter have been successful with this technique on non-insulators; submicron surface analysis has been accomplished on insulators. Charging problems on insulators can usually be overcome, especially on relatively flat surfaces. However, electron beam damage of the surface being analyzed is always a potential problem, and the chemical state information from AES is generally more difficult to interpret and less useful than that from XPS.

Auger electrons are generated whenever a core vacancy is created in an atom. The process that generates Auger electrons and gives them their characteristic energies can be summarized in the equation

$$E(VXY) = E(V) - E(X) - E(Y') \ ,\tag{2}$$

where $E(VXY)$ is the energy of the Auger electron, and V, X, and Y are the three subatomic levels involved in the Auger process. $E(V)$ and $E(X)$ are the energies of the subatomic levels from which the hole is created and the filling electron originates, respectively. $E(Y)$ is the energy of the level from which the Auger electron comes, and $E(Y')$ is the energy of this level with the atom in an Auger transition imposed (doubly ionized) final state. (Note that due to the number and configuration of electrons on H and He, these elements cannot be detected with AES.) If the atom is near a surface, the Auger electron may escape without energy loss and its energy can be analyzed as with XPS. X-rays or electrons are most often used to create the core hole. Electrons are used whenever surface analysis with high spatial resolution is required. Because electron stimulated Auger spectra characteristically have a very high background and weak peak intensities, the spectra, like those shown in this chapter, are typically differentiated.

Reviews of Auger spectroscopy and microscopy specifically for geologic materials can be found in Hochella (1988), Browning and Hochella (1990), and Mogk (1990).

Secondary ion mass spectroscopy (SIMS). SIMS is an outstanding surface analytic technique that has found wide acceptance in a number of fields. The strengths of the technique lie in the fact that all elements can be analyzed to trace levels, high quality depth profiles are readily obtained, and isotopic ratios can be measured on certain instruments. Insulating samples are often a challenge to analyze, but there are a number of techniques available to overcome these problems. High spatial resolution analyses, down to a few microns in diameter and sometimes less, can also be performed. Quantitative analysis is possible but generally challenging, especially without standards, because of potentially complex primary ion - sample interactions.

SIMS works by bombarding the surface to be analyzed with an ion beam (commonly used ions are Cs^+, O_2^+, and O^-) which sputters atomic and molecular ions from the sample into the vacuum of the instrument. Positive or negative ions are then extracted into a mass spectrometer, which on the most advanced instruments consists of magnetic and electrostatic sector analyzers.

An excellent recent review of the SIMS technique, including the development of quantitative analysis and examples of and references to its many uses in the earth sciences, is given in Metson (1990).

Surface analysis by laser ionization (SALI). SALI is a relatively new surface analytic technique which works by chemically analyzing desorbed or sputtered neutral atoms or molecules from the surface of interest. Desorption can be accomplished by heating, or using an incident electron, ion, or laser beam. SALI operates by the ionization of the emitted neutral atoms or molecules by intense laser light which passes just above and parallel to the sample surface. The atomic or molecular ions are then mass analyzed by high resolution, high transmission time-of-flight mass spectrometry. The result is an extremely sensitive analytic technique, capable of obtaining analyses down to ppm levels on a single monolayer. The other major advantage of this technique is that the ionization step is completely separate from the desorption step. Therefore, a laser wavelength is usually chosen which universally ionizes desorbed species, making quantitative analysis considerably easier.

General descriptions of the SALI technique are given in Becker and Gillen (1984) and Schuhle et al. (1988). In the earth sciences to date, SALI has been used for the analysis of thin films of organic molecules on mineral surfaces (Tingle et al., 1990a,b).

Resonant ionization mass spectroscopy (RIMS). This is a very similar technique to SALI, but uses tunable laser light to selectively ionize specific emitted neutral atoms or molecules of interest instead of the more universal ionization approach used by SALI. The RIMS technique is developing rapidly (e.g., Willis et al., 1989) and has already, among other things, opened up the study of the Re-Os isotopic system in rocks (e.g., Walker and Fassett, 1986; Walker et al., 1989).

Rutherford backscattering (RBS). Although not as popular as XPS, AES, or SIMS, RBS is one of the few techniques that can produce relatively non-destructive depth profiles (relative to ion sputtering). A disadvantage of the technique is that the depth resolution of an RBS depth-profile is typically not better than several tens to a few hundred Angstroms. Therefore, a compositional change within a few tens of Angstroms of the surface, for example, may not be detected.

RBS works by using a particle accelerator to generate a high energy beam of helium (usually several MeV) which is channeled toward the surface of interest. The helium will backscatter from the target material as a convoluted function of the helium energy and mass, the mass of the specific atom in the sample doing the backscattering, and the scattering angle. Scattering cross-sections of He for the elements are known, so when the constituent elements and density of the target are known, a depth profile of these constituents in the near-surface of the sample can be established from the energy distribution of the backscattered He.

The RBS technique is described in detail in Chu et al. (1978), and examples of its use for the analysis of mineral surfaces can be found in Casey et al. (1988, 1989).

Resonant nuclear reaction (RNR) analysis. Resonant nuclear reactions can also be used, like RBS, to relatively non-destructively depth profile for elements in the near-surface of a solid, and also like RBS, the depth resolution is no better than several tens of Angstroms. Unlike RBS, however, only one element can be depth profiled at a time depending on the nuclear reaction used. For example, hydrogen can be depth profiled by bombarding the sample with high energy ^{15}N ions. If the ^{15}N energy is exactly 6.385 MeV (the resonant energy in this case), there is a large probability that the reaction ^{15}N + ^{1}H → ^{12}C + ^{4}He + 4.43 MeV gamma-ray will occur. At ^{15}N energies higher or lower than this, the reaction probability drops off dramatically. Also, as ^{15}N penetrates the sample to deeper and deeper levels, it looses energy at a predictable rate. Therefore, if a ^{15}N energy of greater than the resonant energy for this reaction is used, the N will get to a certain depth before it obtains the resonant energy, at which point easily detectable 4.43 MeV gamma-rays will be produced depending on the amount of H at that depth in the sample. In general, depth profiles are produced by varying the energy of the incoming ions and measuring the intensities of the characteristic gamma-rays which are emitted.

Examples of the use of RNR analysis for geologically important materials can be found, for example, in Lanford et al. (1979) and Petit et al. (1987).

Scanning tunneling spectroscopy (STS). STS is still in its early stages of development and is listed here because of its enormous future potential. Ultimately, it is hoped that STS, in conjunction with scanning tunneling microscopy (STM) which is discussed below, will be capable of identifying individual atoms on the surfaces of conducting and semiconducting materials such as certain sulfides, oxides, and native elements.

Briefly, STS works by bringing a sharp tip within electronic tunneling distance (several Angstroms) of a surface directly over the atom of interest (see the section describing STM for an explanation of how this can be done). With the tip stationary in this position and tunneling current established, the bias voltage between the tip and sample is ramped, and the changes in the tunneling current are recorded. This information can be used to map out the density and energy of electron states in the atom in question according to the following formalism:

$$I_T = \int_{E_F}^{E_F+eV} \rho_s(E,\vec{r}) \, \rho_T(E - eV,\vec{r}) \, T(E,eV,\vec{r}) \, dE \quad , \tag{3}$$

where I_T is the tunneling current, E_F is the energy of the Fermi level, eV is the bias voltage between the sample and tip, ρ_s is the density of electronic states at energy E (with respect to E_F) at the sample surface under the tip position $\vec{r}$, ρ_T is the density of electronic states at energy $E - eV$ (with respect to E_F), and T is a transmission function where

$$T = e^{-ks} \quad . \tag{4}$$

In this function, s is the tip to sample distance, and

$$k = \left[\frac{4\pi m}{h^2} \left[\frac{\phi_s + \phi_T}{2} + \frac{eV}{2} - E \right] \right]^{1/2} , \qquad (5)$$

where m is the electron rest mass, h is Planck's constant, and ϕ_s and ϕ_T are the electronic work functions of the sample and tip, respectively. If enough is already known about the density of states from other spectroscopies like ultraviolet photoelectron spectroscopy (UPS), tunneling spectroscopy, in certain cases, may be sufficient for atomic identification.

A simple form of STS, used on sulfide surfaces, has been recently described in Eggleston and Hochella (1990) (see also Fig. 20, this chapter). An example of the use of STS on a non-geologic material is discussed in the section entitled "Reactivity of Surfaces" in this chapter (see also Fig. 32, this chapter). In the next several years, major important advances should be made in this field.

Microscopies for determining surface microtopography

Optical methods. There have been a number of optical methods available for some time that are relatively easy to use and which can provide excellent detail of the microtopography of surfaces down to the molecular level. Each of the three techniques briefly described below, including phase contrast, differential interference contrast, and multiple beam interferometry microscopies, give a lateral view of a surface at normal optical magnification, but can dramatically enhance features which have only minute relief.

Phase contrast microscopy works on the principle of shifting the phase of diffracted light beams by using a high absorption phase plate (e.g., see Sunagawa, 1961). Features with a relief of just a few Angstroms can be seen with this technique. The method is so sensitive because a large amplitude difference can be obtained from two rays having only a minute initial phase difference as caused by, for example, a step one unit cell high. With this method, the high and low sides of steps can be unambiguously identified. However, for a minute step to be seen, its face must be sharp and perpendicular to the terraces that it divides. The method works in either the transmission or reflection mode.

Differential interference contrast microscopy utilizes a Wollaston prism or a Savart plate to split the ordinary and extraordinary rays from an image, and interference is created between these rays using an analyzer (e.g., see Sunagawa, 1987). Interference colors are generated which highlight microtopographic surface features as small as a few Angstroms in height. Steps do not have to be sharp and perpendicular to the surface in order to be seen as with phase contrast microscopy, but the high and low side of steps cannot be unambiguously identified.

Multiple-beam interferometry is a optical microscopic technique that was developed into its present state of use by Tolansky (1948). Like phase contrast and differential interference contrast microscopies, this technique gives a lateral view of a surface at normal optical magnification, but a series of interference fringes also present in the image are sensitive to topography on the surface of only a few Angstroms. The surface to be studied is coated with several hundred Angstroms of a highly reflecting layer (usually silver). A smooth and similarly coated piece of glass is placed on the study surface, silver sides in, and the pair are illuminated at normal incidence with parallel monochromatic light. Interference fringes will form from the recombination of multiply reflected beams, and these can be used to measure height variations on the study surface down to the molecular scale. Details of the physical mechanisms of interferometry can be found in the book by Tolansky (1973), and many examples of its use in the study of the microtopography of surfaces can be found in Tolansky (1968).

Scanning electron microscopy (SEM). SEM is one of the most universal and useful techniques available for the study of the physical characteristics of solid surfaces. The primary reasons for this are its ease of usage, its very high imaging resolution, and its great depth of field. Imaging resolution down to a few tens of Angstroms is possible with advanced research SEM's equipped with field emission electron sources.

In the SEM technique, a highly focused electron beam is rastered over the sample surface, and secondary, backscattered, and Auger electrons are generated very near the point of impact of the beam. Secondary electrons are defined as those whose origin is the sample, and their energies can be anything up to the energy of the incident beam. However, SEM images are formed by low energy secondary electrons (energies less than 50 eV) which generally originate from the valence and conduction bands of the near-surface atoms. Secondary electron generation is primarily a function of composition and the angle at which the incident beam strikes the sample (more secondary electrons are produced at glancing angles). This latter phenomenon, along with shadowing effects on rough surfaces and the depth of field, are what give SEM photomicrographs such a striking appearance and why they are similar to viewing an object by eye.

Comprehensive reviews of SEM and related techniques can be found in the books by Goldstein et al. (1981) and Newbury et al. (1985). More recently, a review of SEM for geologic materials have been written by Blake (1990).

Transmission electron microscopy (TEM). Standard TEM microscopy can be used to study surface microtopography by what has been called the TEM decoration technique. The technique was developed by Bethge and Krohn (1965) (see also Krohn and Bethge, 1977) as a way to use the TEM to routinely study surface morphology. The technique works by vapor depositing a thin film of gold onto the surface of interest which has been cleaned and heated. Next, the surface is carbon-coated. Finally, the original specimen is etched away with an acid, leaving the thin carbon-gold film which is effectively a molecular-scale mold of the original surface and amenable to TEM imaging. An example of this technique, used to image molecular-scale growth spirals on kaolin group minerals, can be found in Sunagawa and Koshino (1975).

Scanning tunneling microscopy (STM). STM has helped drive a revolution in surface science since its introduction less than 10 years ago (Binnig et al., 1982a,b, 1983). STM allows for the imaging of conducting and semiconducting surfaces and fluid-solid interfaces down to the atomic level. The imaging of microtopography with STM is becoming more and more routine.

The tunneling microscope utilizes a sharp metallic tip that is rastered over the surface of interest with a piezoelectric translator. The ceramic translator is controllable to within 0.1 Å in x, y, and z. When the tip is brought to within several Angstroms of the surface and a small bias is applied (usually in the mV range), electrons can tunnel across the gap as described in the STS section above and Equations 3-5. The measured tunneling current is extremely sensitive to the tip-sample separation, s, as described by

$$I_T \equiv \exp\left[\left[\frac{-4s\pi}{h}\right](2m\phi)^{\frac{1}{2}}\right] , \qquad (6)$$

where ϕ is the tunneling barrier height which is proportional to the sample and tip electronic work functions. Therefore, if the tip is rastered across the surface at a constant tunneling current, it will follow the physical contours of the surface down to the atomic level. This produces a three-dimensional "image" of the surface which can look similar to SEM images, except that the resolution capabilities with STM are far greater. Also, the dimensions of any feature on a surface, down to the atomic-scale in x, y, and z, can be obtained.

The application of STM to conducting and semiconducting mineral surfaces is just beginning (Zheng et al., 1988; Hochella et al., 1989; Eggleston and Hochella, 1990).

Atomic force microscopy (AFM). This is a recently developed technique (Binnig et al., 1986) which is a result of the invention of the STM. AFM has demonstrated atomic resolution capabilities on both conducting and insulating surfaces, potentially making it a very important tool for the surface study of a wide range of geologic materials. As with STM, it is quite versatile in that it can be operated in vacuum, air, or at a solid-fluid interface.

AFM works by rastering a sample, by means of a single tube piezoelectric translator, under a sharp tip mounted on or part of a flexible microcantilever. The most useful cantilever to date is a Si_3N_4 sheet, the flexible (thin) part being about 100-200 µm long and approximately 0.5 µm thick. In the repulsive mode of operation, the tip is actually "touching" the surface with an exceptionally small force ($< 10^{-7}$ N) provided by the flexing microcantilever. In the attractive mode of operation, the tip does not physically touch the surface, but is positioned close enough to it to be deflected slightly downward by van der Waals forces. In either mode, as the sample is rastered under the tip, the up and down deflection of the tip can be measured by, for example, minute deflections of a laser beam reflecting off the back side of the microcantilever into a photodiode array. As with STM, the piezoelectric translator has controlled movements to the Angstrom level, and vertical tip deflections can also be measured on the Angstrom level. Although microtopography of surfaces can be relatively easily observed down to the atomic scale with AFM operating in the repulsive mode, image details are dependent on the tip shape. In comparison, attractive mode AFM currently has lower spatial resolution, but it is in an earlier stage of development.

Hochella et al. (1990) and Hartman et al. (1990) give examples of the applications of AFM to mineral surface science. In the next few years, AFM should gain wide acceptance among scientists who study mineral surfaces and mineral-water interactions.

Tools for determining surface structure

Low energy electron diffraction (LEED). Historically, LEED has been the primary tool for determining the atomic structure of surfaces. In this technique, a low energy beam of electrons (usually between 30 and 200 eV) impinges perpendicular to a surface. Conducting or insulating surfaces can be analyzed. Elastically backscattered primary electrons will undergo Bragg diffraction if the surface is atomically ordered (i.e. crystalline). These electrons, whose penetration into the surface is a few monolayers or less, can be viewed on a fluorescent screen over the sample. For qualitative LEED, this pattern is photographed and measured, from which the unit cell size and shape can be determined. Repeat distances in direct space can be determined with the equation

$$d_{hk} = \frac{D \lambda}{p_{hk}} , \tag{7}$$

where d_{hk} is the perpendicular spacing of hk lines on the surface in Angstroms, D is the sample surface to screen distance in millimeters in the LEED instrument, p_{hk} is the distance between spots on the screen, and λ is the wavelength of the incident electron beam in Angstroms given by the relativistically corrected deBroglie equation

$$\lambda = \frac{h}{\left[2m_0E\,e\left[1 + \dfrac{E\,e}{2m_0c^{\frac{1}{2}}}\right]\right]^{\frac{1}{2}}} , \tag{8}$$

where m_0 is the electron rest mass, E is the acceleration voltage of the primary electron beam, e is the electron charge, and c is the velocity of light. With the substitution of constants, and at the limit of low accelerating voltages used for LEED, Equation 6 simplifies to

$$\lambda = \left[\frac{150.4}{E}\right]^{\frac{1}{2}} . \tag{9}$$

Finally, d_{hk} is converted into a unit cell edge depending on the cell shape. Besides measuring the size and shape of the cell in this way, systematic absences can also be used to determine or narrow the possibilities for the two-dimensional space group (plane group) symmetry for the surface atomic arrangement. Alternatively, the diffracted electron beam intensities can be measured, each as a function of incident beam energy. With this, it is often possible to solve the atomic structure of the surface, although this is generally a complex and not always unambiguous task.

Examples of the application of qualitative LEED to mineral surface structure analysis can be found in Hochella et al. (1989, 1990). A full description of LEED atomic structure analysis can be found in Clarke (1985) and Van Hove et al. (1986).

Scanning tunneling microscopy (STM). STM can now be thought of as an important auxiliary tool to LEED for surface crystal structure analysis. The fundamentals of STM, especially pertaining to the imaging of surface microtopography, have already been discussed. Here, atomic level STM imaging as it pertains to surface structure analysis is briefly discussed.

When the sample is biased positively with respect to the tip, electrons tunnel from the tip into unoccupied conduction band levels of the sample surface directly beneath it. When the sample is biased negatively, electrons tunnel to the tip from occupied valence bands of the atom directly beneath the tip on the sample surface. With the tip easily controlled down to subatomic dimensions, the above give a basis for the imaging of individual atomic sites. Combined with the possibility of atomic identification with STS as discussed above, the potential of STM as a surface atomic structure analysis tool is clear. At this time, however, atomic resolution STM can be very challenging, especially on atomically rough surfaces were atoms in valleys can be difficult to "see". Also, the electronic structure of surfaces with more complicated compositions may be difficult to unravel. STM will continue to develop into a more complete surface structure analyzer in the near future.

Atomic force microscopy (AFM). The use of AFM to determine surface atomic structure is in a very early stage of development (see AFM section above), and there is some controversy as to how an atomic resolution image is created in the first place with AFM (e.g., see Abraham and Batra, 1989). To date, the AFM has demonstrated apparent atomic resolution on graphite (e.g., Binnig et al., 1987; Albrecht and Quate, 1988), boron nitride (e.g., Binnig et al, 1987; Albrecht and Quate, 1987), as well as a few other insulating materials (e.g., Hansma et al., 1988). It remains to be seen how atomic resolution imaging with AFM will proceed, and how useful it will become.

MINERAL SURFACE COMPOSITION

The surface composition of a mineral, as with all materials, is rarely representative of the bulk. There are only two cases where a mineral surface might have a composition which is the same as the bulk. First, one must consider the few minerals which are essentially inert. Presumably, one might find a surface of a natural gold or platinum grain to be essentially pure, but even this will not always be the case. Second, a mineral surface created by fracture in an ultra-high vacuum, as in a surface analytic instrument, will have a composition representative of the bulk. However, in this latter case, the clean condition is usually short-lived. Contaminants on the original specimen surface near the freshly fractured surface can quickly migrate onto this fresh surface. In addition, one can calculate, using kinetic theory, that even in the ultra-high vacuum state of 10^{-9} torr, the frequency of collisions of the residual gas molecules with the clean surface will result in a measurable amount of contamination in less than 20 minutes. This can easily be confirmed by experiment. Therefore, given the fact that nearly all minerals in natural environments will have modified surface compositions compared to their bulk, it is obviously important to characterize the nature of these modifications when trying to understand mineral-water interactions. Below, the way in which mineral surface compositions can be geochemically modified is discussed, and the important and seldom studied aspects of surface chemical inhomogeneities is also addressed.

Adventitious modification of mineral surfaces

For all mineral surfaces except the most inert, exposure to air or aqueous solutions will immediately result in the uptake of H, O, and usually C. These are generally known as "adventitious" elements in the field of surface science due to their chemical reactivity and ubiquitous nature. (Parks, as well as Davis and Kent, both in this volume, present of detailed look at the uptake of H and O on mineral surfaces.) As exposure continues, more and more adventitious layers can continue to accumulate, although the process quickly slows, or "passivates". Excluding gross contamination, adventitious layers can easily obtain a thickness of many monolayers. Although additional layers may not be firmly held on the mineral surface, they generally will have some effect on the reactive behavior of a mineral surface with respect to other species in the contacting atmosphere or solution (see section entitled "Reactivity of Surfaces" below).

Figure 1 shows an example of the adventitious C on the surface of calcite after exposure to air for several days. The XPS spectra taken over the C 1s binding energy region shows one C photopeak originating from the CO_3 groups of the bulk carbonate, and another line due to the adventitious C on the surface. Calculations (Stipp and Hochella, 1990) show that the adventitious C from this surface is the equivalent of approximately 10 monolayers thick. It is interesting to note, however, that the chemical shift of this photopeak indicates that the adventitious species is not CO_2 as one might expect, but species with C-H, C-C, and C-O bonds (remember that CO_2 has C=O bonding; the C 1s photopeak for CO_2 is centered at 292 eV). Although CO_2 is generally more abundant in air and many aqueous solutions than hydrocarbons and molecules with carbonyl groups, it is apparently less reactive with this surface, or it dissociates in contact with it. In fact, we have never seen molecular CO_2 adsorbed on a mineral surface using vacuum-based surface analysis techniques. However, we consistently find spectral evidence for the sorption of OH^-, H_2O and hydrocarbons. These are by far the most common adventitious species.

More information on the nature of adventitious species on mineral surfaces can be gained by analysis with SALI using laser VUV light at 118 nm (10.5 eV) for the photoionization step (Tingle et al., 1990a). This energy is optimized for efficient ionization of most organic compounds. Figure 2 shows a mass analysis of the species coming off a heater strip with no sample attached as it is heated inside the instrument.

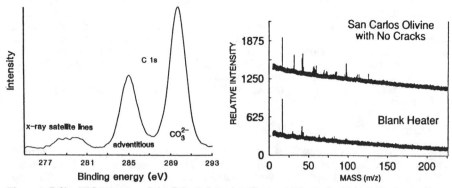

Figure 1 (left). XPS spectrum of the C 1s region taken from a calcite surface which has been exposed to air for several days. The peaks at approximately 285.0 eV and 289.7 eV are due to adventitious carbon from air exposure and the carbon in the carbonate groups of the calcite, respectively. The adventitious line can be eliminated by low energy ion sputtering. See text for additional details. From Hochella (1988).

Figure 2 (right). Mass spectra generated by SALI (using 118 nm laser light which gives efficient organic ionization) during the heating of a small strip heater to several hundred °C with (top) and without (bottom) an attached olivine crystal in ultra-high vacuum. The olivine fragment used has been broken out of the center of a large single crystal and exposed to air for less than 5 min. The difference between the spectra represent minute quantities of organic compounds that have become attached to the olivine surface during air exposure. From Tingle et al. (1990a).

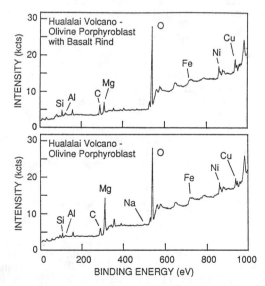

Figure 3. XPS survey spectra of the {010} crack surfaces of two olivine xenolithic porphyroblasts from the 1801 eruption of the Hualalai Volcano, Hawaii. Besides olivine constituents (Mg, Fe, Si, and O), Na, Al, Cu, Ni, and C (the latter in excess of adventitious amounts) are observed on the crack surface. These additional elements were vapor-transported to these surfaces in the late stages of eruption and reside in thin discontinuous films (less than a few nanometers thick) and minute particles (smaller than a few tens of nanometers in diameter). From Tingle et al. (1990b).

The upper mass spectrum is the same experiment except that a small olivine crystal has been attached to the heater strip. This olivine fragment was broken out of a gem-quality, essentially flawless crystal in air using clean tools and introduced into the vacuum of the SALI instruments in a matter of minutes. The difference between the two spectra represent the organic compounds that attached themselves to the olivine surface from a few minutes of air exposure. SALI is an extremely sensitive technique, and the amount of organics present is estimated to be less than a monolayer, but this example serves to illustrate how quickly surfaces can pick up trace organics just from brief air exposure.

Geochemically modified mineral surface compositions

The term "geochemically modified" is used to denote a change due to a natural process other than adventitious contamination as discussed above. Geochemical modification of mineral surfaces involves surface reactions which may only affect the top monolayer of atoms, or it may modify the mineral to considerable depths. For convenience, geochemical modification of mineral surface compositions has been divided into three categories. Elements can be added to the surface (sorption), elements can be lost (desorption), and as a special subset of these two, oxidation-reduction processes (redox modification) is also included. Illustrative examples of these three processes are given below.

Sorption modification. The best known example of mineral surface compositional modification in geochemistry is the sorption of species from aqueous solution onto mineral surfaces. This phenomenon has been widely studied for some time (see Parks, Davis and Kent, Schindler, and Brown, this volume). Recently, the structural analysis of these sorbed complexes has become quite sophisticated, allowing for in-situ structural assessment (see Brown, this volume). Below, an example of sorption modification that has not been as extensively studied in geochemistry is given, that of vapor transport and deposition on mineral surfaces. (Tingle and Fenn, 1984, and references therein give other interesting natural examples of this phenomenon.)

Figure 3 shows XPS spectra from two olivine porphyroblasts from xenoliths collected from the 1801 eruption of the Hualalai Volcano, Hawaii (Tingle et al., 1990b). The olivine is approximately Fo_{90}, and the pieces analyzed were in pristine condition. The actual surfaces analyzed with XPS are {010} crack surfaces which appear fresh and show a very faint purple iridescence in reflected light. These unhealed cracks are presumably very late stage features formed by extreme temperature gradients and/or rapid cooling during the eruptive event (Mathez, 1987; Tingle et al., 1990a). Therefore, it is thought that these olivine surfaces generated by late-stage fracture are immediately exposed to volcanic gases. There is also strong evidence that suggests that these cracks are somewhat self-sealing (Tingle et al., 1990a), and there is no evidence that these crack surfaces have been weathered subsequent to eruption and emplacement.

The XPS spectra of the olivine {010} crack surfaces from Hualalai show all of the olivine constituents (O, Mg, Si, and Fe). However, other elements are present, including C, Na, Al, Cu, and Ni. The C line intensity is well in excess of what one would expect for the few minutes of air exposure that these surfaces have seen before introduction into the instrument. SALI analysis of similar samples show a rich array of organic molecules as described in Tingle et al. (1990a,b). The possibility that these organics are laboratory contaminants has been eliminated, and there is very strong evidence that they are not from environmental biogenic contamination. It has been shown that it is possible that these organics are produced by the catalytic interaction of simple C and H-containing gases interacting with the fresh olivine surface. The other inorganic species (Na, Al, Cu, and Ni) have been vapor-transported to these surfaces. Finally, ion sputtering experiments on these surfaces suggests that the organic-containing carbonaceous films are less than 3-4 nm thick, and that the Ni and Cu are present as particles less than 20 nm in diameter.

Although the example of surface compositional modification given above does not deal with a mineral-water interaction, it does serve to illustrate that mineral surface chemistry is extremely variable. Solutions passing over a mineral surface may not interact with the typically simpler bulk composition for that mineral, and the elements it does interact with will depend on the exact history of each surface.

Detachment modification. Chemical weathering of minerals (see Zhang and Nancollas, Casey and Bunker, and Hering and Stumm, this volume), even at neutral pH's, generally results in a modification of surface composition. This modification is usually in the form of a leached layer, a layer from which components of the original mineral have been removed. (Note that the net result of chemical weathering is the removal of material from the mineral, but species from solution can also be added to the leached layer.) The thickness of this leached layer can be a few Angstroms thick (essentially the top 1 or 2 monolayers); on the other hand, it can range to many hundreds to thousands of Angstroms, or even deeper, depending on the mineral and the weathering conditions (e.g., see Petit et al., 1987; Schott and Petit, 1987; Nesbitt and Muir, 1988; Mogk and Locke, 1988; Casey et al., 1988, 1989; Hellmann et al., 1990). Also, there is evidence (direct as well as indirect) that the leached layer is typically not uniform in thickness or in composition (see "Compositional Inhomogeneous" section below).

An example of the kind of surface compositional modification which can result from chemical weathering is shown in Figure 4. This plot shows what happens to the surface and near-surface composition of the {010} surface of albite when it is subjected to deionized water acidified with HCl to a pH of 2.47 and run in a single-pass flow-through hydrothermal vessel at 225°C for 7.5 h (Hellmann et al., 1990). The plot is the result of XPS analyses collected between ion-sputtering intervals, resulting in a compositional depth profile. Because all atoms do not sputter away at the same rate (Hochella et al., 1988b), the solid data points are obtained by sputtering into an unreacted homogeneous albite crystal. Any deviation from the curves defined by the solid data points indicates that a leached layer is present, and the depth of the layer can be obtained by noting where the two curves coincide. The open data points in Figure 4 are for the reacted crystal. They show that, under the conditions stated above, Al is depleted with respect to Si to a depth of about 100 Å, and O and Na are depleted with respect to Si to a depth of about 50 Å. Cl does not penetrate this leached layer at a pH of 2.47, but it does penetrate to about the depth of Al depletion at pH 2.26. (The reason for this has to do with the zero point of charge on this surface as explained in Hellmann et al., 1990.)

Another example of a leached layer on albite is given in Figure 5 (also from Hellmann et al., 1990). The conditions of this experiment where 225°C for 4 h using only deionized water equilibrated with air (pH 5.7). In this case, the Na depleted layer is only between 10 and 30 Å thick, and would be easily missed by most other surface analytic depth profiling techniques. The O (not shown) and Al concentrations with respect to Si are actually increased in the top several tens of Angstroms. This example has been included because chemical weathering in nature often occurs near neutral pH. It has been suggested in the past that leached layers may not form under these conditions. However, it is clear from Figure 5 and other studies, such as that of Muir et al. (1989), that leached layers can quickly form under non-drastic weathering conditions, although they may be very thin. However, even the thinnest leached layer results in a modification of the composition of the true surface, and this is important in the reactivity of minerals (see section entitled "Reactivity of Surfaces" below).

Redox modification. As mentioned above, this type of surface compositional modification is typically a subset of both sorption and desorption modification, as oxygen and perhaps other elements are added to or removed from mineral surfaces

100

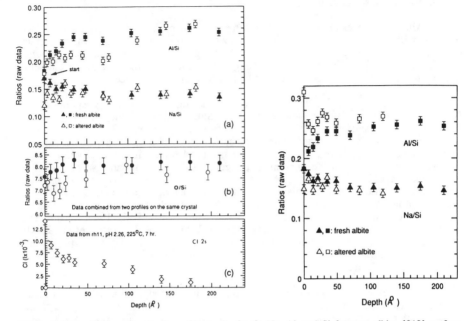

Figure 4 (left). XPS sputter depth profile results for O, Na, Al, and Si from an albite {010} surface reacted at pH 2.47, 225°C for 7.5 h (open symbols) compared to results from a fresh albite (filled symbols, see Hochella et al., 1988b, for an explanation of this type of profiling). The Cl profile is from a similar run at pH 2.26. See text for an interpretation of this data. From Hellmann et al. (1990).

Figure 5 (right). XPS sputter depth profile results for Na, Al, and Si from an albite {010} surface reacted at pH 5.7, 225°C for 4 h (open symbols) compared to results from a fresh albite (filled symbols, see Hochella et al., 1988b, for an explanation of this type of profiling). See text for an interpretation of this data. From Hellmann et al. (1990).

undergoing redox reactions (see Hering and Stumm, White, and Bancroft and Hyland, this volume). We include an example of redox modification of a mineral surface here simply to show the dramatic effects that it can have on surface composition.

The example that has been chosen here is from the gold-bearing carbonate strata of the Roberts Mountain Formation near Carlin, Nevada. A detailed description of this ore body can be found in Bakken and Einaudi (1986). Using scanning Auger microscopy and spectroscopy, the surfaces of a number of sulfide grains in samples that contain high concentrations of gold were analyzed (Hochella et al., 1986a,b; Bakken et al., 1989). These samples were also taken from the "unoxidized" portion of the ore body, showing little evidence of supergene alteration or weathering. In these specimens, fresh appearing sulfides are well in excess of any amorphous iron oxides. Figure 6 shows the Auger spectra of a typical euhedral pyrite (probably diagenetic) in these rocks both before and after ion sputtering. The surface composition of the apparently fresh pyrite grain is far from that of FeS_2. The S Auger line, considering the very high sensitivity that Auger spectroscopy has for S, is very weak, and the oxygen line is very strong. The Si, K, and Ca lines, and a portion of the O line, are from micron-sized grains of various silicates dotting the surface of this grain. Adventitious C is also clearly apparent. The major component of the surface of this pyrite grain is an Fe-oxide. Semiquantitative analysis of the peak intensities indicate that the composition is, not surprisingly, Fe_2O_3 (Hochella et al., 1986a,b). There is only a minor amount of sulfate on the surface. After sputtering off approximately 400 Å of the surface, the S Auger line dramatically increases, and the oxygen left is mostly associated with the silicate grains remaining on the surface.

The above is an illustrative example because the mineralogy and petrography of this rock indicates that it is unoxidized, yet aqueous solutions percolating through this rock will not, at least initially, interact with any sulfides. Instead, the solutions will interact with transition metal oxides that coat the sulfide grains. In others words, solutions will not "see" any sulfides in this rock, at least initially, even though they are an important part of the bulk mineralogy.

There are a great many papers that deal with the surface modification of redox sensitive minerals using surface sensitive spectroscopies. Interested readers may want to obtain, for example, Hagstrom and Fahlman (1978), Brion (1980), Buckley and Woods (1984), White and Yee (1985), Jean and Bancroft (1985), and Hyland and Bancroft (1989).

Compositional inhomogeneities

Compositional inhomogeneities on mineral surfaces resulting from sorption and desorption surface modifications are not often considered. Considering desorption first, a few silicate dissolution studies have suggested, using indirect evidence, that the thickness of leached layers on a surface may be highly variable (e.g., Berner et al., 1985; Chou and Wollast, 1985; Hellmann et al., 1990). This could imply, although not necessarily so, that the leached layer itself may be compositionally inhomogeneous. There is now direct evidence that this is indeed the case. Hochella et al. (1988) used scanning Auger microscopy and spectroscopy to obtain the surface composition of hydrothermally altered labradorite with a lateral resolution of from 2 to 10 μm. The results clearly show that the surface composition of the dissolving labradorite under the conditions used (deionized water at 300°C and 300 bars for 57 days) is highly variable. For example, Ca was nearly completely leached from the surface over certain portions of certain grains, while it was still present in significant quantities on different portions of the same grains.

We have recently been able to perform even higher lateral resolution surface analysis on feldspar surfaces. Figure 7 shows an SEM photomicrograph of a labradorite grain that has undergone the same hydrothermal treatment described in the proceeding paragraph. The accompanying Auger map in this figure is of the same area as the photomicrograph and shows Ca Auger peak intensities. The intensities, displayed in gray-scale tones (darker gray, higher intensity), have been corrected for background and topographic effects. The pixel size is only 0.25 × 0.25 μm, but the actual analysis comes from an area in the center of the pixel about 0.1 μm (1000 Å) in diameter and less than 10 Å deep. The extremely shallow depth of analysis is due to the fact that the Ca LMM Auger line measured has a relatively low energy (290 eV), and therefore a low escape depth. On this lateral scale, the variation in Ca surface concentration is rather dramatic. While the absolute Ca concentration is difficult to obtain for these spectra, the variation is about the same as described in the Hochella et al. (1988a) study (preceding paragraph). This analysis clearly illustrates how complex mineral dissolution can be on a micro-scale. At least in this case, the variability in the surface chemistry is probably controlled by the starting microtopography of the surface, the local density and distribution of crystal defect and twin plane outcrops, and the local flow patterns of reacting solution over the surface.

Surface compositional inhomogeneities resulting from sorption modification is more intuitively obvious, and more evidence, both direct and circumstantial, exists for this in the literature. It is clear that different sites on the surface should have different reaction potential with, for example, species in aqueous solution. A discussion of this topic will be given in the section entitled "Reactivity of Surfaces" below.

102

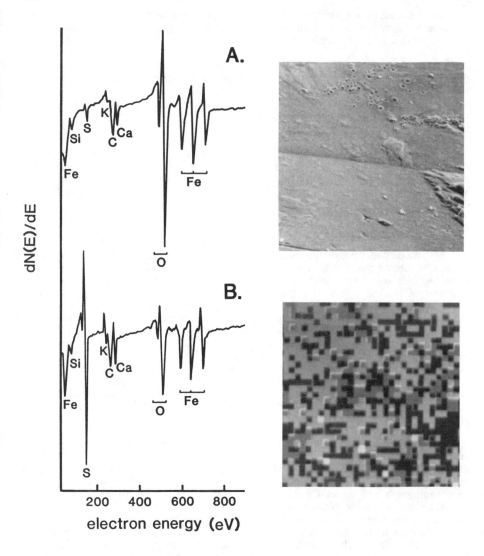

Figure 6 (left). Electron-stimulated AES spectra of a pyrite grain from the gold-bearing carbonate strata of the Robert Mountain Formation both before (A) and after (B) ion sputtering. The grain surface consists of an Fe-oxide and small amounts of sulfate. The oxidized coating is removed after several hundred Angstroms of ion sputtering. The peaks due to Si, K, and Ca (and a portion of the O line) are due to small silicate grains sticking to the pyrite surface. From Bakken et al. (1989).

Figure 7 (right). SEM photomicrograph and Ca Auger map over the same area of a labradorite grain after exposure to deionized water at 300°C and 300 bars pressure for 57 days. The field of view is 8 μm. The pixel size in the Auger map is 0.25 × 0.25 μm, and the shade of gray represents Ca concentration within the top 10 Å of the surface (darker gray, higher concentrations). See text for additional details.

MINERAL SURFACE MICROTOPOGRAPHY

As mentioned in the introduction to this chapter, if one is to understand surface reactivity, it is important to attempt to characterize the microtopography of these surfaces. Here, microtopography is defined as surface morphological features on the general dimensional scale of the chemical reactions that will take place at that site. Therefore, it is most useful to consider features in the range of a few Angstroms to perhaps a few tens of Angstroms. Obviously, the appearance of a surface as seen with the unaided eye can be highly deceiving. For example, the {001} parting surface of specular hematite appears to have a mirror-like flatness (which is enhanced by its metallic luster), yet there is a great deal of microtopography on this surface (see below). In fact, surfaces can appear perfectly flat using normally configured petrographic microscopes or standard SEM's, and still the surface in question may have considerable microtopography. It is becoming more and more apparent that surfaces are very rarely flat over even relatively small areas. If an imaging method shows a surface to be perfectly flat, it is most likely that the method is not sensitive enough to see the microtopography present.

Surface microtopography studies, especially on mineral surfaces, are not very common, although Sunagawa and colleagues have a number of excellent papers on the subject (e.g., see Sunagawa, 1987, and references therein). Below, we briefly review general surface microtopography models, and also give a few examples of observed microtopography on mineral surfaces.

Microtopography models

Models for the microtopography of surfaces have been developed from many studies over the last few decades (e.g., Somorjai, 1981). The general model that has been developed from growth, dissolution, and other surface studies is shown in Figure 8. At the microtopographic scale and as depicted in this figure, the model implies that surfaces consist of flat areas, called terraces, which are separated by steps. In their simplest form, steps are one atomic layer high, although they can be considerably higher than this. Kink sites appear where there is a corner on a step, that is where a step changes directions. An atomic or molecular-size hole in a terrace is called a vacancy. The opposite of this, that is a atom or molecule sitting on top of a terrace, is called an adatom or admolecule, respectively. Also, atoms which protrude from an atomically rough surface are often called adatoms. Microtopographic imaging with phase contrast microscopy, interferometry, TEM, STM, and AFM have generally supported the above model.

Considering only atoms on the surface, atoms that make up the terraces have the greatest number of surface neighbors. At the edge of a step, the number of nearest-neighbors is reduced, and atoms at the outer corners of kink sights have even fewer neighbors. Adatoms generally have the fewest nearest-neighbors of all surface sites, and adatom sites are potentially, but not always, the most reactive sites on a surface.

Growth surfaces

There exists a number of excellent reviews on the growth of crystals in nature (e.g., Kirkpatrick, 1975, 1981; Tiller, 1977; Sunagawa, 1984; and references therein). The two principle growth mechanisms are spiral growth around a dislocation, and surface nucleation two-dimensional growth (Fig. 9). This figure clearly shows that both growth mechanisms create steps and potentially all of the features described in the model above. Consequently, it is of little surprise that microtopographic imaging techniques commonly show steps and related features on growth surfaces. For example, Figure 10 shows spiral steps on a natural hematite {001} growth surface using phase contrast microscopy (Sunagawa and Bennema, 1979). Various steps in the image are

104

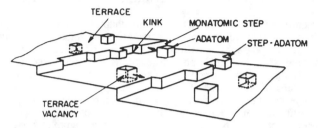

Figure 8. Standard block model of a heterogeneous surface. In this depiction, each block represents an atom.

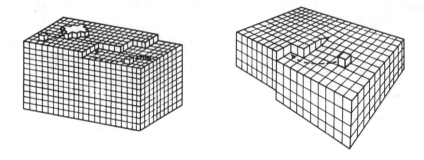

Figure 9. Standard schematic representation of crystal growth models, including surface nucleation-controlled growth (left), and screw dislocation-controlled growth (right). Each block represents an atom or a small molecule. Both models require surface sorption and migration as shown. From Kirkpatrick (1981).

Figure 10. Phase contrast micrograph showing unit-cell scale spiral steps on a natural hematite {001} surface. From Sunagawa and Bennema (1979).

calculated to be 7, 14, or 21 Å high, corresponding to 1/2, 1, and 1 1/2 unit cells in height along the c crystallographic axis (see also Sunagawa, 1961).

Growth steps on mineral surfaces can take on a seemingly endless number of different forms. Relatively small growth steps can group together to give the appearance of macroscopic steps visible to the unaided eye. Steps may also be straight or curved, regular or highly irregular, and they may bunch to form gently rising or steep hillocks. The overall characteristics will depend on the particular crystal face in question and the condition under which it has grown.

Cleavage surfaces

As with growth surfaces, mineral cleavage surfaces are rarely perfectly flat. For example, portions of a {001} cleavage surface of galena may appear perfectly flat to the unaided eye and perhaps even to moderate resolution SEM. However, these supposedly perfect surfaces can be exceedingly rough on the microtopographic scale as shown in Figure 11. In this STM image which shows moderate vertical exaggeration, the wide arrows show a step which varies between 30 and 50 Å in height over the length imaged. Along this step, several kink sites can be seen (e.g., point 1), as well as protrusions from the step (e.g., point 2) which appear to be step ad-features. The hillocks that are seen over much of the image are between 5 and 30 Å in height and do not seem to show any particular pattern. A closer inspection of these hillocks may reveal that they are actually formed by closely bunched steps, although this has not been investigated as yet.

Atomic resolution STM images on a relatively flat area of the galena surface shows other features in the surface microtopography model described above. Figure 12 was collected under tunneling conditions where the S atoms appear as high mounds, and the Pb atoms are very low and can just be distinguished in a portion of the image between the S atoms. This image shows several S vacancies and an atomic step in the lower left portion of the image.

Finally, Figure 13 is an AFM image of the {010} albite cleavage surface showing a vertical exaggeration of approximately 3:1. The main feature of this image is the curving 250 to 300 Å step, although there are several other steps in this image with heights of less than 50 Å (see arrow, Fig. 13). The non-vertical drop-off of the steps in this image is the result of an imaging artifact of the AFM and is a function of the tip shape (see Hochella et al., 1990, for more details). The large bumps on the upper terrace are particles of albite fracture-dust which are not connected structurally to the surface. In other areas on this surface (not shown), the AFM has measured steps down to 5 Å in height, as well as wide terraces over 1000 Å in width which seem to be atomically flat.

Dissolution surfaces

When a mineral dissolves, the fundamental reactions, as with growth processes, are taking place on the atomic and molecular scale as emphasized by Hellmann et al. (1990) and Casey and Bunker (this volume). In general, the dissolution microtopography will depend on the original microtopography of the surface, the atomic structure and composition of the crystal, the nature of the crystal defects, the nature of the dissolving medium, and the rate and duration of the dissolution event. Dissolution will preferentially occur at sharp kink sites and edges, and also at defect outcrops. Generally, etch pits with pointed bottoms will occur at line defects, and etch pits with flat bottoms will occur at point defects (e.g., Sunagawa, 1987). Etch pits are known to follow lines of concentrated strain along defects, and also to follow trains of impurities. Etch pits can also initiate at impurity sites. Sungwal (1987) can be consulted for an extensive discussion concerning all of these subjects.

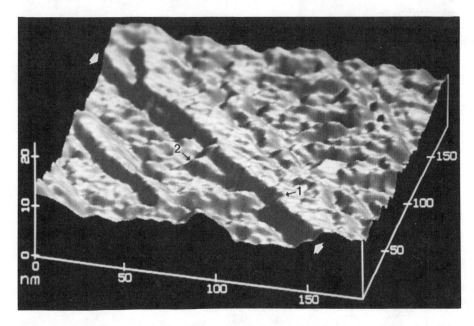

Figure 11. STM image of a galena {001} cleavage surface showing moderate vertical exaggeration. The wide arrows mark a cleavage step 30 to 50 Å in height. Also denoted are a kink site (labeled 1) and a half-step protrusion (labeled 2). From Hochella et al. (1989).

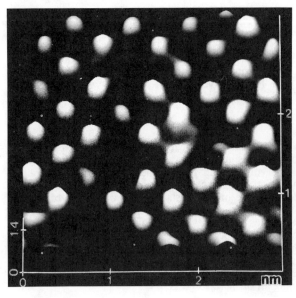

Figure 12. Atomically-resolved STM image of the galena {001} surface taken at +200 mV sample bias and a tunneling current of 1.1 nA. The large mounds represent S sites, with the galena unit cell represented by four S atoms in a square (cell edge 5.9 Å) with one in the center. Note the S vacancies and the atomic step in the lower left corner. From Eggleston and Hochella (1990).

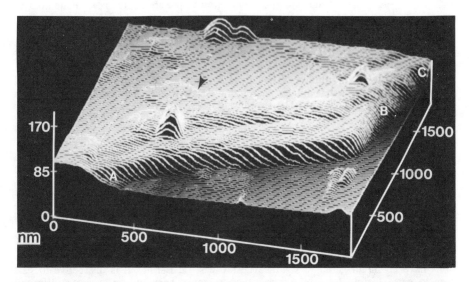

Figure 13. AFM image of the albite {010} surface showing a vertical exaggeration of approximately 3:1. The large bumps on the upper plateau are ultrafine particles of albite left after fracture, and the arrow points to a 40 Å step on this plateau. The major curving step crossing the lower third of the field of view ranges between 25 to 30 nm in height. Between points A and B, the step appears to slope down over 60 nm laterally. The step shape is probably more square, and this slope is a function of the shape of the end of the tip. Between points B and C, the apparent lateral extent of the step increases to 150-200 nm, probably reflecting a true change in the step shape. From Hochella et al. (1990).

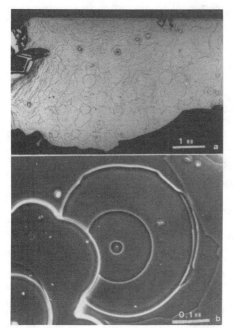

Figure 14. Hematite {001} two-dimensional step dissolution. (a) Low magnification optical micrograph showing circular depressions. (b) Phase contrast micrograph showing retreating circular steps from defect or impurity sites at the center. From Sunagawa (1987).

It is particularly interesting, as with growth processes, to look at the role of steps in dissolution processes. Whether dissolving on a relatively flat surface or along the walls of an etch pit, steps will generally be present and dissolution will proceed in a two-dimensional sense laterally from them. As an example, dissolution steps on the {001} surface of hematite are shown in Figure 14 (Sunagawa, 1987). The top part of the figure is a normal optical micrograph showing circular patterns of depressions due to two-dimensional dissolution. The lower portion of the figure is a phase contrast photomicrograph showing minute dissolution steps receding from a central defect which in this case is believed to be an impurity (Sunagawa, 1987).

MINERAL SURFACE ATOMIC STRUCTURE

General description of surface atomic structure

Just as with composition, it is typical for the atoms on the surface, and sometimes those as deep as several layers down, to assume positions different from their equilibrium positions in the bulk. These structural differences, if they exist, can be subtle or dramatic. A good place to start in understanding surface structure is to look at unreacted, structurally simple surfaces exposed by fracture in vacuum (most of these studies have been performed on non-minerals). Although studies using these conditions have no direct geologic significance (at least on Earth!), it is instructive to start here to get a feeling for how atoms can be arranged near and on surfaces. We will then consider how surface structures can change when atoms or molecules are slowly adsorbed from vacuum. Finally, the surface atomic structure of six minerals will be examined, including one sulfide, two oxides, two silicates, and a carbonate, all after exposure to air, and for the carbonate, also after exposure to aqueous solutions.

Surface structures in vacuum

Physicists, chemists, and materials scientists have determined the surface structure of a large number of metals and a somewhat smaller number of oxides in vacuum. (Note that, unlike bulk structure analysis, there may be several surface structures to solve for each material depending on the number of crystal faces of interest. Different faces of the same material can have dramatically different properties. See section entitled "Reactivity of Surfaces" below.) All of these surface structure analyses, performed mostly with LEED, have resulted in an interesting underlying theme: Most clean surfaces (i.e., in vacuum) relax inward toward the bulk, with the spacing between the first and second atomic layers being reduced by up to 15% (see Somorjai, 1981, 1990, and references therein). With nothing to bond to on the vacuum side of the interface, some of the electron density in the dangling bonds is redistributed to the remaining underlying bonds. With this the overlap population is increased and the bonds contract. In many cases, the lateral positions of the surface atoms also change as they seek equilibrium positions to better satisfy bonding preferences, including coordination number, bond lengths and bonding angles. As a result of even minor shifts, the surface unit cell dimensions may undergo incremental increases depending on new surface repeat distances. Another generality that can be drawn from previous work is as follows: The more structurally open or atomically rough the surface, the more likely it is to have a greater relaxation inward and/or lateral reconstruction.

The Si(100) surface is a good example of a relatively simple surface reconstruction resulting from exposure of a face by fracture in vacuum. The left side of Figure 15 shows the unreconstructed Si(100) surface. Silicon has the diamond structure (each Si is bonded to four other Si's), and the top layer of Si atoms on the surface (heavy cross-hatch in Fig. 15) have two dangling bonds each. The Si atoms just below have their normal complement of 4 bonds. In an attempt to reduce the dangling bond population, these uppermost Si atoms lean towards one another (right side of Fig. 15) to within bonding distance. With this, the hypothetically unreconstructed unit cell on the surface, designated 1 × 1, becomes a 2 × 1 cell as drawn on the figure (i.e., one surface cell edge doubles in length, and the other remains the same). More recently, it has been shown that the formation of these surface Si-dimers actually results in a somewhat buckled or uneven surface as shown in Figure 16, and that relaxation extends 4 or 5 layers down into the crystal (Tromp et al., 1983; Holland et al., 1984).

It should also be noted that, just as bulk atomic structures undergo structural transformations as a function of temperature, surface structures do the same, quite independently from the bulk. For example, the Si(111) surface adopts a 2 × 1 unit cell at room temperature in vacuum. At 400°C, the surface reverts back to the 1 × 1 bulk-like state, but between 600 and 700°C, enough thermal energy is available to drive a major surface reconstruction which reduces the number of dangling bonds in the 1 × 1 structure by 61% (Takayanagi et al., 1985). The unit cell size of this complex reconstruction is 7 × 7. Temperature driven reconstructions can occur on any surface, including those of any mineral, resulting in a change in reactivity (see "Reactivity of Surfaces" below).

Adsorption-induced surface structural modification

One of the immensely useful attributes of vacuum-based surface science, even for understanding geochemically important surface reactions, is that it allows one to simplify and control surface interactions so that one has some chance of understanding fundamental mechanisms. In this section, we will see how residual gases in a vacuum can modify surface structure when they sorb to the surface. In geologic systems, the chemistry will be more convoluted, and it will be difficult to observe individual reactions, but the overall fundamental processes as revealed by in-vacuum work will likely be applicable.

A particularly illustrative example of adsorption-induced surface structural modification is the adsorption of C gas from an ultra-high vacuum onto the Ni(100) surface (Onuferko et al., 1979). The fresh Ni(100) surface is a simple square array of Ni atoms, with one Ni atom per surface unit cell. As C atoms interact with this surface, they move to the surface site between 4 Ni atoms. To increase the C bonding energy further, thus providing for a more thermodynamically stable situation, the 4 Ni atoms bonded to one C twist and spread slightly, allowing the C to sink down and come within bonding distance of a Ni in the next layer down. This results in a 2 × 2 unit cell as shown in Figure 17.

A few generalities can be obtained by studying the results of a number of adsorption-induced structural modification studies (Somorjai and Van Hove, 1989). First, the inward relaxation so often seen in vacuum surface analysis work is reduced, eliminated, or sometimes even reversed when species are allowed to sorb to the surface. Second, as seen in the example presented above, substrate surface atoms that are nearest to the sorbate may shift in position to better accommodate it. However, this generally only occurs when the the sorbate-surface interaction is strong, i.e., the sorbate is chemically reactive with the surface and strong bonding occurs.

Structure of various mineral surfaces

The surface structures of galena and hematite (Hochella et al., 1989), olivine (Tingle et al., 1990a), albite (Hochella et al., 1990), and calcite (Stipp and Hochella, 1990) described below were obtained from cleavage surfaces exposed by fracture in air and introduced to ultra-high vacuum within 5 minutes for LEED analysis. Some of the calcite surfaces analyzed were also exposed to aqueous solutions before introduction into the LEED vacuum chamber. The surface structure reported below for rutile was obtained by LEED after in-vacuum fracture (Henrich and Kurtz, 1981).

Adventitious carbon and oxygen resulting from air and/or solution exposure on

110

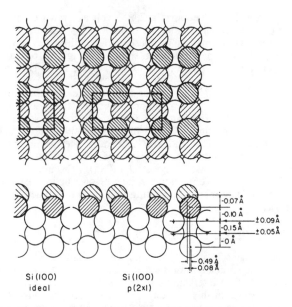

Si (100)
ideal

Si (100)
p(2x1)

Figure 15. Top view (upper drawing) and side view (lower drawing) of the Si (100) surface. Atom shading designates different layers. The left sides of both drawings represent the ideal structure (the 1 × 1 surface cell is shown). The right side shows top layer Si atom-pairs leaning towards each other, resulting in the 2 × 1 unit cell shown. The measurements to the lower right show the displacement of atoms from the ideal structure down to the fifth layer. This is known as near-surface relaxation due to surface reconstruction. Modified from Somorjai (1981).

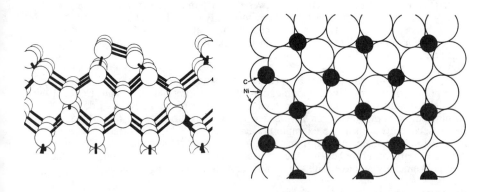

Figure 16 (left). Ball and stick side view model of the Si (100) surface, which more accurately depicts the true surface structure shown on the right side of Figure 15. Extensive LEED and ion scattering experiments indicate that the Si atoms which lean toward each other on the (100) surface actually result in a buckled dimer configuration as shown. From Somorjai (1990).

Figure 17 (right). The Ni (100) surface after sorption of C atoms (Ni - open circles, C - solid circles). Before sorption, the Ni atoms form a simple square array. After sorption, each four-atom square rotates slightly as shown, resulting in a new 2 × 2 cell. From Somorjai (1990); modified from Onuferko et al. (1979).

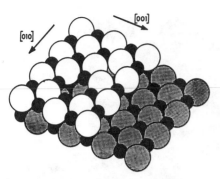

Figure 18. Model of the galena {001} surface with one atomic step. The top layer shows S atoms (large open spheres), Pb atoms (small solid spheres), and a S and Pb vacancy. The second layer shows S (stippled pattern) and Pb (solid). From Tossell and Vaughan (1987).

these surfaces (except for rutile) has been characterized with XPS. Adventitious hydrogen is presumably always present, but it is not detectable with XPS. The adventitious atoms on these samples are generally three monolayers or less in thickness. The contaminant overlayers do not contribute to the LEED pattern except for diffuse scattering due to their disordered nature. However, because of the very short attenuation lengths of low energy electrons in condensed matter (e.g., see Hochella, 1988, Hochella and Carim, 1988, and references therein), these thin overlayers do reduce the depth of the near-surface of the mineral contributing to the pattern. It is estimated that the LEED patterns shown here originate from the top 2 to 4 monolayers of the mineral.

Galena {001}. Galena (PbS) has the NaCl structure, with every Pb atom bonded to six S atoms, and every S atom bonded to six Pb atoms. If the surface in question does not relax laterally, one would expect a simple square surface unit cell (Fig. 18). If we arbitrarily assign a S atom to the origin, then they would occupy the 4 corners of the cell with one in the middle. Pb atoms would occupy a site mid-way along each edge of the cell. This arrangement is described by the two dimensional space group (i.e. plane group) $p4g$, and with S at 0,0 and Pb at 1/2,0, the special condition hk: $h + k = 2n$ limiting possible reflections applies. The LEED pattern of the galena {001} surface is shown in Figure 19 and agrees exactly with the pattern expected for the arrangement described above. The unit cell edge length measured from this pattern is 6.00 Å, in reasonable agreement with the bulk cell edge of 5.94 Å considering the 1-2% potential error in these LEED measurements. The surface structure is confirmed with an STM image (Fig. 20) showing both the S and Pb atoms in the positions expected.

Hematite {001}. The bulk structure of hematite (α-Fe_2O_3) can be most easily described as a slightly distorted, hexagonal close-packed array of oxygens with Fe filling two-thirds of the octahedral sites. The ideal {001} surface can be envisioned as a slightly distorted close-packed sheet of oxygens with Fe atoms evenly distributed in one-sixth of the available sites between the three oxygens that are closest to each other in the sheet (Fig. 21). The plane group describing this arrangement is $p31m$ which has no conditions restricting reflections. Again, the LEED pattern (Fig. 22) from this surface is exactly that expected for the simple termination of the bulk structure described above. The cell edge derived from this pattern is 5.07 Å, in good agreement with the known bulk a-cell edge of 5.04 Å. However, recent STM images of the hematite {001} surface (Eggleston, unpublished data) suggests that the top layer of oxygens may be slightly relaxed, resulting in a relatively undistorted close-packed layer.

As mentioned previously, surface structures can undergo structural transformations with temperature quite independently from the bulk. Kurtz and Henrich (1983), starting with an amorphous and non-stoichiometric Fe_2O_3 {001} surface, found that heating to 700°C in vacuum produces a stoichiometric 2 × 2 reconstruction. Heating to 820°C produces an additional reconstruction which has remained uncharacterized.

112

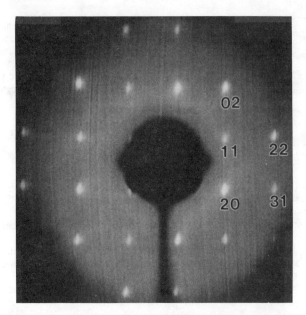

Figure 19. Indexed LEED pattern of the galena {001} cleavage surface taken with a primary beam voltage of 98 eV. The central shadow in this and other LEED patterns in this chapter is due to the miniature electron gun and holder which is positioned just above the sample. From Hochella et al. (1989).

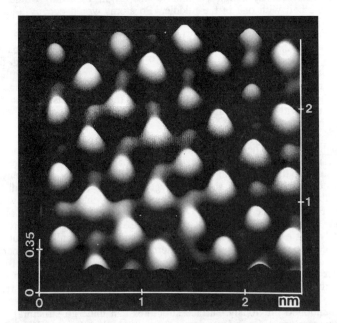

Figure 20. STM atomic-resolution image of a galena {001} surface taken at +200 mV sample bias and 1.8 nA tunneling current. The larger mounds are S sites, and the smaller mounds between them are Pb sites. This atomic arrangement is in agreement with a termination of the bulk structure as shown in Figure 18 and the LEED pattern shown in Figure 19. From Eggleston and Hochella (1990).

Rutile {110}. Rutile (TiO_2) is tetragonal with each Ti coordinated to six oxygens in a slightly distorted octahedron. One-half of the oxygen octahedra are vacant. Each oxygen is coordinated to three Ti atoms. The one good direction of cleavage is the {110}, and the nominal atomic structure on this surface is relatively complex. A model of this surface is shown in Figure 23, derived by assuming that this surface is generated by breaking the fewest number of Ti-O bonds and that the surface does not, at least initially, reconstruct. This is supported by the 1 × 1 LEED pattern collected on this surface by Henrich and Kurtz (1981). Upon annealing in vacuum, the pattern changes to 2 × 1, but the structural changes responsible for this have not yet been determined.

The model rutile {110} surface portrayed in Figure 23, showing two O vacancies discussed below, is interesting for several reasons. First, there are two near-surface Ti sites, one with its full coordination complement of six oxygens, and the other with one O missing. Also, the surface is not atomically flat in the usual sense due to the row of oxygens which complete the coordination shell along the row of Ti atoms beneath. An O vacancy along this row as shown in the figure results in the lowering of the coordination of two Ti atoms from six to five. Likewise, an O vacancy in the next layer down (also shown) reduces the coordination of two Ti atoms from five to four. In both cases, the cation-cation screening in the vacancy regions is reduced, and structural relaxation would be expected to occur in these areas.

Olivine {010}. Olivine (($Mg,Fe)_2SiO_4$) is orthorhombic with Mg and Fe filling octahedral sites among unlinked Si tetrahedra, making this a nesosilicate. The {010} cleavage plane is usually poorly developed, but it can often be recognized when breaking apart large crystals. The {010} cleavage plane cuts through a portion of the structure consisting of a sheet of Mg,Fe octahedra, with Si tetrahedra on either side pointing alternately towards and away from this plane. The structure of the {010} cleavage surface is probably complex. Nominally, the outermost surface would consist of undercoordinated Mg and Fe atoms within a loosely packed sheet of oxygens, with Si and O in the next layer down. However, on this cleavage plane and perpendicular to it, the atoms in this structure are evenly distributed. Therefore, one might expect, despite the lack of high symmetry, that this surface will stay relatively fixed after exposure. We have confirmed this with LEED which shows the 1 × 1 spot pattern expected for the unreconstructed surface (Fig. 24). The measured cell dimensions on this surface (a = 4.75, c = 6.03 Å) agree well with those measured from the bulk for this particular olivine (a = 4.76, c = 5.99 Å) (Tingle et al., 1990a).

Albite {010}. Albite ($NaAlSi_3O_8$) is a triclinic framework structure consisting of Si and Al tetrahedra (each tetrahedra is linked to four other tetrahedra). Sodium resides in large irregular cavities in the framework. The {010} cleavage plane is along a relatively open part of the structure with Na and O atoms on the surface and Si, Al and O atoms in the next layer down. The LEED pattern taken from this surface is shown in Figure 25. The pattern is consistent with the expected symmetry of the surface, i.e., the oblique plane group $p1$ (no systematic absences). However, as might be expected, the repeat distances measured from the pattern are close but do not match the equivalent values in the bulk. The pattern gives d(10) = 7.22 Å, d(01) = 6.51 Å, β^* = 60°. The nominal dimensions for bulk albite are d(10) = 7.28 Å, d(01) = 6.39 Å, and β^* = 63.5°. The discrepancy in d(01) and β^* are relatively small but significant. This suggests that the surface structure is slightly distorted from what would be expected for a simple termination of the bulk structure. This is not surprising considering the low symmetry of this structure and its low density of packing (openness) on this plane. In addition, it might also be expected that the topmost layers on the albite {010} are relaxed inward. However, although we do not have the data needed to determine the relaxation perpendicular to the surface, inward relaxation, if it exists at all, would be expected to be slight because of the presence of an adventitious overlayer.

114

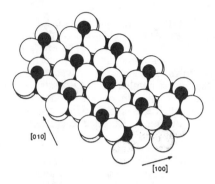

Figure 21. Ideal model of the top few layers of the hematite {001} surface. The solid and open circles are iron and oxygen, respectively. From Kurtz and Henrich (1983).

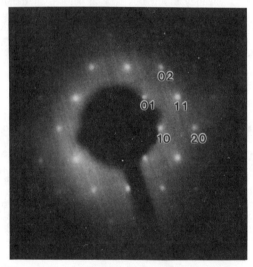

Figure 22. Indexed LEED pattern of the hematite {001} parting surface taken with a primary beam voltage of 90 eV. From Hochella et al. (1989).

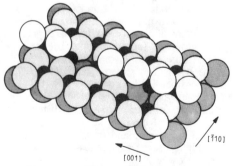

Figure 23. Model of the possible structure of the rutile {110} surface obtained by breaking the fewest number of bonds along this planar direction. Two types of oxygen vacancies are shown and discussed in the text. From Henrich (1985).

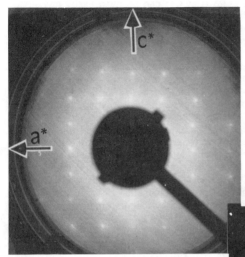

Figure 24 (left). LEED pattern of the {010} olivine cleavage surface collected at a primary beam energy of 117 eV. This is a rectangular net pattern without systematic extinctions and spacings which are consistent with an unreconstructed termination of the bulk structure. From Tingle et al. (1990).

Figure 25 (below). LEED pattern of the {010} albite cleavage surface collected at a primary beam energy of 87 eV. See text for details. From Hochella et al. (1990).

Figure 26 (below). LEED pattern of the {101} calcite cleavage surface collected with a primary beam energy of 97 eV. The rectangular pattern of reflections give repeat distances that are consistent with equivalent distances in the bulk, but the extra reflections (e.g., see arrow) indicate a surface 2 × 1 cell which may be the result of the rotation of surface carbonate groups away from their bulk positions. From Stipp and Hochella (1990).

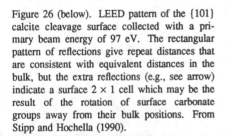

Calcite {101}. The structure of calcite ($CaCO_3$) can be most easily envisioned by taking the NaCl structure and replacing Na by Ca and Cl by CO_3 groups, and then compressing the cube diagonal to produce a rhombohedral cell. The perfect {101} cleavage surface has an orthogonal arrangement of equal numbers of Ca and CO_3 groups alternating with one another. The major spots of the LEED pattern (Fig. 26) collected from a {101} cleavage plane gives a rectangular surface unit cell with repeat distances of 4.97 and 8.11 Å. This is in excellent agreement with the equivalent distances in the bulk of 4.99 and 8.10 Å, suggesting that the {101} surface has the same structure as the equivalent plane in the bulk. However, there are extremely faint spots in the pattern (e.g., see arrow, Fig. 26) that correspond to a 2×1 surface cell. One possible explanation of the origin of these weak extra reflections is that the carbonate groups in the 4.99 Å direction of the cell are no longer all equivalent; due to rotations of the carbonate groups, every other one is equivalent. It should also be noted that, although the LEED pattern shown in Figure 26 was taken from a calcite surface exposed only briefly to air, the same LEED pattern was observed from calcite surfaces exposed to calcite undersaturated solutions (Stipp and Hochella, 1990). This suggests that, even under dissolution conditions, the near-surface of calcite remains ordered with the structure described above.

REACTIVITY OF SURFACES

The previous sections in this chapter have dealt with the composition, microtopography, and atomic structure of mineral surfaces. Of course the primary reason for studying these three factors is because of the thesis that the chemical reactivity of a surface fully depends on them. In this section, among other things, a number of examples that will demonstrate specifically how they affect reactivity will be presented. In as many cases as possible, examples where only one of these three factors is varied at a time will be presented so that its specific effect on reactivity can be exhibited. First, however, a few underlying principles of surface reactivity will be explored, including the two-dimensional phase approximation and the steps necessary for surface catalysis. At the end of this section, an example of a surface reaction which has been followed in great detail, atom by atom, will be presented so that the reader can see what is ultimately available in surface science today. Finally, it should be stated that because the amount of research to date in mineral surface science is much less than similar research in other disciplines, many of the examples presented are on non-mineral systems. However, the underlying principles are the same, and all of these studies are very much relevant to minerals.

Some general concepts concerning surface reaction

The two-dimensional phase approximation. When a species is instantaneously resident on a surface, it has a number of possible options. It can become tightly bound to a single site, it can diffuse into the bulk, it can desorb into the overlying medium, or it can hop to another site on the surface. The two-dimensional phase approximation is a model which states, maybe somewhat surprisingly, that of these four, the latter is the most likely to occur. The model presumes that the activation energy for surface diffusion is lower than any of the other possibilities, and that species at an interface have the opportunity to visit all of the various sites on that surface and stay in equilibrium with other surface-dwelling species. The model has actually been very useful in helping to explain many catalytic reactions (see below), and also crystal growth, evaporation, and certain sorption reactions. Recently, spectroscopic techniques which result in a direct measurement of diffusion coefficients of species across surfaces have been developed (e.g., Mak et al., 1986, and Deckert et al., 1989). These studies generally support the model, showing that the activation barrier for diffusion is generally less than that for desorption. A geologically relevant example can be found in the sorption of gold from solution onto a sulfide surface. Although gold complexes in solution can be reduced by electron transfer reactions at the sulfide surface (see

Bancroft and Hyland, this volume), the gold does not end up evenly distributed over the surface, but in individual gold spheres. There is now STM evidence (Eggleston and Hochella, unpublished data) that gold is highly mobile on these sulfide surfaces, suggesting that surface diffusion results in gold agglomeration.

Heterogeneous (surface) catalysis. Catalysis on surfaces is a topic which has only begun to be studied in the field of geochemistry (e.g., see Coyne and McKeever, 1989, and references therein). On the other hand, it has received enormous attention in applied physics, chemistry and chemical engineering, and materials science. There are a number of excellent books on the subject (e.g., see King and Woodruff, 1984; Bradley et al., 1989; Richardson, 1989). The basic principles and general requirements for heterogeneous catalysis have been defined, and the factors that will influence geochemical heterogeneous catalytic reactions will be very briefly reviewed here.

Step 1 in the heterogeneous catalytic process is that of selective adsorption of the reactant(s) that will be involved in the reaction. Adsorption of other species may block (poison) the active sites needed for the reaction to precede. Step 2 involves surface diffusion of the reactant(s) to the proper surface site for reaction. Step 3 involves the reaction of reactant(s) to product(s), usually through some transition state. In reality, the transition state usually involves the dissociation of reactants (for example, producing carbon and oxygen atoms from CO molecules) from which products can form. The reactant(s) must bind strongly enough to the reaction site(s) for bond activation to occur, but not so strongly as to restrict the product(s) from forming. Step 4 is the release (desorption) of the product(s) from the surface, allowing surface sites to be available for further reactions.

Note that for industrial applications, the catalytic process should be fast and efficient. For example, if Step 4 is sluggish, the catalytic surface will not be efficient as the reactive sites will be blocked by the very products that are formed. However, in geologic systems, the efficiency may not be critical. Any catalytic reaction could potentially be important. In this regard, our main goal as surface and interface-oriented geochemist should be to simply identify catalytic reactions that might occur in nature. This will depend on a great deal of laboratory work, as by their very nature, true catalytic reactions leave no record behind.

Examples of the effect of surface microtopography on reactivity

Carbon monoxide dissociation on Pt. Iwasawa et al. (1976) studied the reaction of CO on coordinatively unsaturated sites on the platinum (111) surface with and without the presence of surface steps. These experiments were performed by preparing the Pt surface in vacuum to assure its cleanliness, and then bleeding CO into the vacuum and exposing the surface for various lengths of time. The main results of this study are that CO will absorb but not dissociate on the Pt (111) surface, but the molecule will dissociate in the presence of steps on this surface. In this case, it is interesting to note that this process could be observed with XPS. Figure 27 shows the photoelectron spectral region of C1s electrons after different length exposures of CO on a stepped Pt(111) surface. The C peak present at the lowest exposure has a binding energy of 283.8 eV, indicative of a carbide, and therefore showing dissociative sorption. With increasing exposure, a new C peak appears at 286.7 eV, indicative of carbonyl (CO) groups. This peak represents associatively sorbed carbonyl ligands at terrace sites, as this peak is the only one that appears when CO is sorbed to flat (111) surfaces. Likewise, the intensity of the carbide peak at 286.7 eV increases as the step density on the Pt increases.

Ethylene decomposition on various metals. Not only can microtopography promote molecular dissociation not likely to occur at terrace sites as shown above, but it can also promote such reactions at lower temperatures. An excellent case in point can be found from the work of Somorjai et al. (1988). In this study, the breakdown of

118

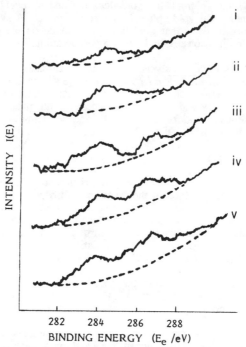

Figure 27. XPS spectra of the C1s region taken from a stepped Pt (111) surface after various lengths of exposure to CO in vacuum. (i) no exposure; (ii) 0.05 L CO (1 Langmuir, L, is an exposure of a gas at 10^{-6} torr pressure to a substrate for 1 sec); (iii) 0.2 L CO; (iv) 2.0 L CO; and (v) 100 L CO. The peak near 283.8 eV represents carbide (Pt-C) bonds; the peak near 286.7 eV represents carbonyl (Pt-CO) bonds. See text for further details. From Iwasawa et al. (1976).

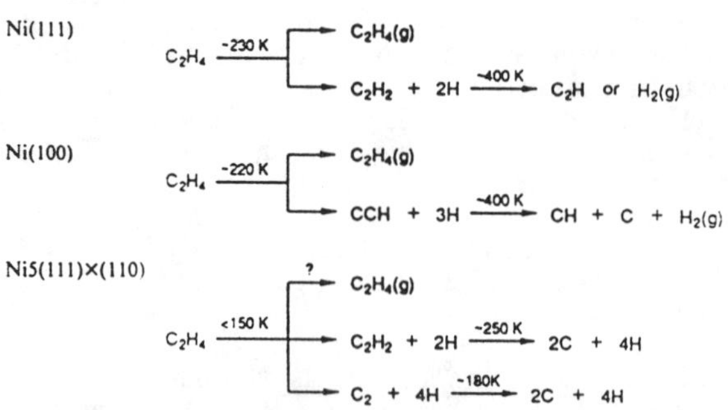

surface	ethylene thermal reaction pathways

Fe(111)

$$C_2H_4 \xrightarrow{\sim 250\ K} \begin{cases} C_2H_4(g) \\ 2C + 4H \end{cases}$$

Fe(110)

$$C_2H_4 \xrightarrow{\sim 270\ K} \begin{cases} \overset{?}{\longrightarrow} C_2H_4(g) \\ C_2H_2 + 2H \end{cases}$$

Ni(111)

$$C_2H_4 \xrightarrow{\sim 230\ K} \begin{cases} C_2H_4(g) \\ C_2H_2 + 2H \xrightarrow{\sim 400\ K} C_2H \text{ or } H_2(g) \end{cases}$$

Ni(100)

$$C_2H_4 \xrightarrow{\sim 220\ K} \begin{cases} C_2H_4(g) \\ CCH + 3H \xrightarrow{\sim 400\ K} CH + C + H_2(g) \end{cases}$$

Ni5(111)×(110)

$$C_2H_4 \xrightarrow{< 150\ K} \begin{cases} \overset{?}{\longrightarrow} C_2H_4(g) \\ C_2H_2 + 2H \xrightarrow{\sim 250\ K} 2C + 4H \\ C_2 + 4H \xrightarrow{\sim 180\ K} 2C + 4H \end{cases}$$

Figure 28. Decomposition temperatures and reaction pathways for ethylene on various transition-metal crystal faces. From Somorjai (1990).

ethylene in the presence of various metal surfaces in vacuum was studied. The results are shown in Figure 28. Although the breakdown schemes differ from metal to metal and surface to surface as shown, bond activation and bond breaking in the ethylene molecule occurs on the Fe (111) and (110) and Ni (111) and (100) surfaces between 220 and 270 K. However, a Ni (111) surface with (110) steps allows ethylene decomposition at only 150 K. More specifically, a stepped Ni (111) surface promotes the breakdown of ethylene (C_2H_4) to C_2H_2 + 2H at temperatures approximately 80 K below the same reaction on an unstepped Ni (111) surface. On the other hand, decomposition reactions of ethylene occur within 20 K of each other on Ni and Fe (111) flat surfaces. Therefore, at least in this case, the presence of steps appears to be more important in this decomposition reaction than performing the same reaction on the same crystallographic face of different metals.

Adsorption of gases on metals. There are several other vacuum-based studies which clearly point to the importance of step and kink sites in the sorption process. For example, Perdereau and Rhead (1971) used LEED to study the sorption of O and H_2S gases, among others, onto the Cu (100) surface with varying step densities. With O and H_2S exposed to this surface, the width and intensity of LEED spots representing stepped faces on the surface were modified before spots representing the terrace (100) face. From this, they concluded that sorption of these gases took place preferentially at steps and that the rate of sorption would increase with increased step density. Christmann and Ertl (1976) used LEED, AES, and high resolution electron energy loss spectroscopy (HREELS) to show, among other things, that H_2 was much more likely to interact with and attach to stepped Pt (111) surfaces than unstepped ones. They speculated that the approaching H_2 molecule to the Pt surface must be in a particular orientation for the sorption process to occur, and a step or kink sight may polarize the incoming molecule and bring it into proper alignment for reaction. Another possibility would be that the step or kink sight might allow a longer lifetime of the molecule in a weakly bonded physisorbed state, giving it more time to find the appropriate orientation for reaction and chemisorption.

Theoretical studies have also found evidence for the increased reactivity of sites associated with microtopography as opposed to terrace sites. For example, Jones et al. (1975) and Painter et al. (1975) used a multiple scattered wave technique on small metallic clusters to show that atoms at the edge of a step should have a variety of bonding orbitals available which are not present on the same atom in a terrace site. The larger number of bonding possibilities at edges as opposed to terrace sites may be enough to explain the increased reactivity of the former.

Dissolution of minerals. Reactions at the mineral-water interface are considerably more complex and difficult to study than the metal-gas systems discussed above. However, it is likely that at the atomic level, similar underlying and fundamental processes are operating that make microtopography just as important in the reactivity of mineral surfaces with solutions as it is in the simpler systems that are discussed above. Certainly on a theoretical basis, this is indeed the case. Dibble and Tiller (1981), in their theoretical treatment of nonequilibrium and interface-controlled rock-water interactions, have strongly emphasised the importance of surface roughness, specifically using the terrace, step, and kink surface model (Fig. 8). They have even suggested that reactions at atomic steps and kink sites may impose a controlling influence on certain mineral-water interface reactions.

Specifically concerning dissolution, it has been known for some time that fine particles have a higher solubility than course particles of the same material (e.g., Enustun and Turkevich, 1960, and references therein). This theory is usually couched in terms of a modified version of the Kelvin equation (e.g., see Adamson, 1982; Petrovich, 1981) which relates the solubility of a grain to its size and surface free energy as

$$\frac{S}{S_0} = \exp\left[\frac{2\,\gamma\,\overline{V}}{R\,T\,r}\right] \, , \qquad (10)$$

where S is the solubility of grains with inscribed radius r, S_0 is the bulk solubility, γ is the surface free energy, $\overline{V}$ is the molar volume, R is the gas constant, and T is the absolute temperature. As the radius of curvature of a grain decreases, the theory predicts that its solubility will increase exponentially as shown. This theory can also be applied to small radius of curvature features on surfaces. If the radius r is sufficiently small to describe a step or kink site, the prediction of increase solubility, this time for a specific site, is the same.

More recently developed theoretical models continue to emphasize the importance of microtopography of mineral surfaces in dissolution phenomena. A good example is the theory of Lasaga and Blum (1986) which uses elastic properties, dislocation characteristics, surface free energies, and Monte Carlo simulation methods to model mineral dissolution. Lasaga (this volume) presents a review of this and other theoretical treatments that are relevant to this subject. In addition, more and more experimental evidence for the importance of microtopography in mineral dissolution is accumulating (see Holdren and Speyer, 1985, 1987, for work with feldspars, and Brantley et al., 1986, for work with quartz). Finally, the recent work of Schott et al. (1989) has suggested that for calcite, more desorption reactions occur at steps and kinks than at defect sites.

[235]U sorption on sheet silicates. Lee and Jackson (1977) studied the surface charge density on the {001} surfaces of micaceous minerals using the [235]U fission particle track method. In the course of their study, they observed the strong adsorption of uranyl cations at "defects" on micaceous {001} surfaces, such as steps. Figure 29a shows an SEM image of fission particle tracks on a {001} biotite surface, and Figure 29b shows a light microscope image of tracks on a {001} muscovite surface. In both cases, it is apparent that uranyl complexes are preferentially attaching to steps, which in this specific case are the edges of micaceous sheets. Lee and Jackson observed that uranyl groups attached to these sites, as opposed to terraces sites, were much less likely to desorbed from washings in neutral salt solutions. They concluded that these step sites on mica surfaces were important in the retention of actinide elements in soils and rocks where sheet silicates are abundant.

Examples of the effect of surface composition on reactivity

Bonding modifiers on the surfaces of catalysts. It has been known for some time that slight modifications to the surface compositions of catalysts can greatly inhibit (poison) or enhance (promote) the reactivity of the surface. It is often the case that submonolayers of certain elements or molecules can be deposited onto the catalyst surface without significantly modifying the catalyst surface atomic structure or the microtopography present. Yet these compositional modifications can make vast differences in the catalyst performance. An excellent example of this principle is found in ammonia synthesis using pure iron as a catalyst. In this case, the rate limiting step in ammonia formation is the dissociation of N_2. The presence of alkalis on the Fe surface greatly increase the breakdown of N_2 and therefore promotes more efficient ammonia formation (Ertl et al., 1979). However, it has recently been determined that the presence of an alkali, particularly K, may play another important role (Strongin and Somorjai, 1988). When ammonia is formed on the Fe surface, the active site on which it formed will not be available for another conversion until the ammonia molecule desorbs. If this ammonia is sluggish in the desorption stage, the efficiency of the whole process will suffer, and failure to desorb would eventually result in a complete poisoning of the catalytic surface with its own product. Coadsorbed K weakens the bonding of ammonia on the Fe surface, reducing its residence time, and dramatically

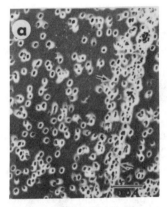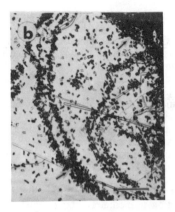

Figure 29. Fission particle tracks from the sorption of ^{235}U onto the {001} surface of sheet silicates. a) Tracks as viewed by SEM on biotite (bar = 0.04 mm). The arrows point to higher concentrations of ^{235}U sorption on steps. b) Tracks as viewed with a light microscope on muscovite (bar = 0.1 mm). The arrow points to increased sorption along a step. From Lee and Jackson (1977).

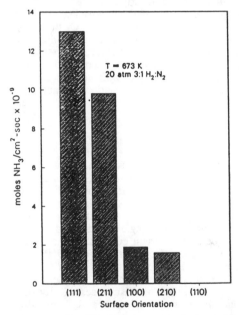

Figure 30. The rate of NH$_3$ synthesis on five crystallographic surfaces of iron. The rate of synthesis on the (110) surface is so low that it does not plot on this scale. From Strongin et al. (1987).

increasing the heterogeneous catalytic activity. Therefore, in this case, K has a dual role; it enhances the synthesis of ammonia by promoting the breakage of the N-N bond in N$_2$, and it weakens the bond strength of the product molecule promoting rapid desorption.

Dissolution rates across the plagioclase series. An excellent example of the dependence of surface reactivity with compositional change without variation in atomic structure can be found by measuring the dissolution rates of plagioclases. The coupled substitution of Na + Si for Ca + Al from albite to anorthite is essentially isostructural. The dissolution rates in strongly acidic aqueous solutions increases considerably towards the anorthite end. Data and discussion on these reactions are presented by Casey and Bunker (this volume).

Examples of the effect of surface atomic structure on reactivity

Catalytic reactions on single crystal metals. There are many examples in the catalysis literature of how the surface chemical reactivity of a material changes as a function of surface atomic structure. The most direct way to demonstrate this is to measure the catalytic activity on various crystallographic faces of monoatomic single crystals. The crystals are cut into flat thin disks to preferentially expose any desired face, and the surface of interest is sputter-cleaned in ultra-high vacuum, annealed, and examined with LEED to insure that the face has the appropriate structure. Precautions can be taken to assure that differences seen in reactivity from face to face are not due to variations in surface roughness (e.g., Strongin et al., 1987, and references therein). The specific example given here involves ammonia synthesis at 673 K on iron substrates using H_2 and N_2 as the starting gases at 20 atm pressure. This example is used both because it is a straight-forward demonstration of the principle of structural effects on reactivity, and because ammonia catalyzed synthesis has already been discussed in the section immediately above. For this synthesis, the (111) and (211) surfaces of Fe have been found to be approximately an order of magnitude more reactive than the (100) and (210) surfaces (Fig. 30) (Spencer et al., 1982; Strongin et al., 1987; Somorjai, 1990). The close-packed (110) face is the least active of all those studied, showing a reactivity two orders of magnitude less than the most active (111) and (211) surfaces. The reason for this tremendous variation in reactivity for a simple monoatomic material seems to be as follows. Bulk iron has a body-centered cubic packed structure, with each Fe atom coordinated to 8 other Fe's at the corners of a cube. Of the atomically flat faces tested, the most active ones in ammonia synthesis are those that have the highest density of 7-coordinated Fe atoms on the surface. All faces tested which did not have 7-coordinated Fe atoms were relatively unreactive. Theoretical calculations by Falicov and Somorjai (1985) have shown that highly coordinated metal atoms on terraces may show the greatest catalytic activity for certain reactions because they experience larger electronic charge fluctuations than less coordinated atoms at the surface.

Mineral and glass dissolution reactions. The reactivity of mineral and glass surfaces should be just as sensitive to surface atomic structure as the simple single crystal metal example given above. However, very few experiments have been done to actually quantify this reactivity dependence in any particular mineralogic case. Perhaps the best (or at least classic) examples come from some of the older mineral dissolution literature where etching rates were measured on different faces of the same crystal (e.g., see Parrish and Gordon, 1945, and references therein). Equivalently, it was often demonstrated that ground mineral spheres prepared from single crystals did not dissolve uniformly, and that this property could be used to orient euhedral crystals (Parrish and Gordon, 1945). A specific example is taken from Ernsberger (1960), who reports from his own and others work that HF will dissolve quartz 100 times faster on a {001} face than on prism faces perpendicular to this. He attributes this solely to the surface structural differences between these surfaces. On the prism faces, ideally silica tetrahedra with one dangling oxygen predominate, while on the basal faces, silica tetrahedra with two dangling oxygens are most common. This later configuration may be sterically more favorable for HF attack. In another example, Casey and Bunker (this volume) very nicely demonstrate dissolution rate dependence on structure with borate glasses containing varying amounts of trigonally and tetrahedrally coordinated boron.

It should be noted that some of the more recent work on dissolution anisotropy, most of which concerns non-mineral phases (e.g., Sangwal, 1987, and references therein; see also Liang and Readey, 1987), suggests that dissolution rate may vary, at least in part, because of differences in the microtopography on different crystallographic faces. Faces with higher densities of kink and edge sites would be expected to show higher dissolution rates. This shows have difficult it can be to clearly separate all of the factors that affect surface chemical reactions.

Sorption reactions on minerals. At this time, the role that the atomic structure of mineral surfaces plays in sorption reactions from overlying solutions can only be inferred. Several workers have shown strong indirect evidence for the existence of multiple structurally active sites for sorption on mineral surfaces (some of these sites have been characterized by direct spectroscopic means as, for example, by Hayes et al., 1987; see also Brown, this volume). For example, Hayes and Leckie (1986), studying the sorption of Pb on goethite, and Benjamin and Leckie (1981a,b), studying the sorption of Cd, Cu, Pb, and Zn on iron oxyhydroxides, present indirect evidence for the existence of several discrete sites for sorption on these surfaces, all with different reactivities for different sorbates. Loganathan and Burau (1973) and Guy et al. (1975) have come to similar conclusions in work involving the sorption of various cations into MnO_2 surfaces. The population of the various active sorption sites on mineral surfaces must ultimately depend on the atomic structure of the mineral and the face exposed to the solution. Different faces will expose different sites to solution, and an added complication is that, depending on conditions, the mineral surface may atomically reconstruct completely independently of the bulk. Much more work is needed to systematically identify and categorize these sites. Until this is done, the dependence of sorption characteristics on surface atomic structure for minerals can only be implied.

Surface reactivity observed atom by atom

All of the reactions which have been considered above, no matter how complex, can be reduced to the workings of individual atoms. The behavior of individual atoms at an interface is at the core of surface science, but until recently there was no way to directly observe such processes. Now, with the advent of tunneling microscopy and spectroscopy, this is at least possible with conducting and semiconducting materials. Perhaps the best example of how it has become possible to observe surface chemistry at the atomic level is given by Avouris and Wolkow (Wolkow and Avouris, 1988; Avouris and Wolkow, 1989; Avouris, 1990). This work involves the reaction of NH_3 with a silicon surface in vacuum. Again, we are forced to use this non-geologic example because a geologically relevant example does not yet exist. However, the enormous advance in the understanding of surface reactions as a result of these new techniques is well worth demonstrating on any material system. The applications to a number of important conducting and semiconducting minerals, including certain iron oxides, is clear.

The silicon surface used in these NH_3 sorption experiments is generated by heating the Si(111) surface to approximately $600^{\circ}C$. Near this temperature, the Si(111) surface reconstructs from a 1×1 structure to a complex 7×7 structure. As a result, there are only 19 Si dangling bonds in the 7×7 area, compared to 49 in the same area before the reconstruction. Therefore, the reconstruction, which persists after the sample is cooled, provides for a highly reduced surface free energy and much greater stability. The 7×7 structure, first purposed by Takayanagi et al. (1985), is shown in Figure 31. In this reconstruction, the upper layer of Si atoms are called adatoms, of which there are 12 per cell. Half are called corner adatoms because they surround the holes at the corners of the hexagonal cell, and the other half are called center adatoms as labeled in Figure 31. The adatoms have three sp^3 bonds to other Si atoms, and one partial backbond to a fourth Si directly beneath it (dashed bond, Fig. 31c). The next layer of Si atoms are called restatoms, of which there are 6 per cell and labeled A and B in Figure 31a and b. These Si atoms have three sp^3 bonds to neighboring Si atoms and one dangling bond (Fig. 31c).

Figure 32 shows STM images and STS spectra of the Si(111)-(7×7) both before (a) and after (b) exposure to NH_3. The STM images are taken at a sample bias of 800 mV, and therefore the STM is imaging unoccupied states just above the Fermi level associated with dangling bonds. Under these conditions, it has been shown that it is the adatoms that are preferentially imaged, resulting in the white dots in Figure 32

124

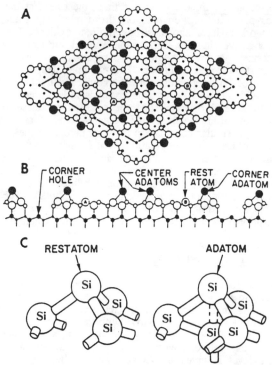

Figure 31. (A) Top view of the 7 × 7 reconstruction model of the Si (111) surface as proposed by Takayanagi et al. (1985). The open and solid circles, all representing Si atoms, are drawn smaller the deeper they are with respect to thè surface. The 7 × 7 cell is outlined by the solid lines. (B) Side view of the reconstruction along the left-right diagonal of the unit cell. The labeling and circle sizes correspond to the drawing in (A). Adatoms and restatoms are labeled. (C) Local bonding and configuration around adatom and restatom sites. The restatom has one dangling bond, while the adatom has a partial backbond to a silicon below (dashed bond). From Avouris (1990).

which can be directly compared with the arrangement of adatoms in the structure drawing in Figure 31. As an adatom reacts with NH_3 (actually with either NH_2 or H due to the dissociation of NH_3 on this surface), it's partial dangling bond is consumed, and the unoccupied state associated with it is removed. Therefore, in the STM images, reacted adatoms become dark under these imaging conditions. One can literally watch individual adatoms react with NH_3 in this way. One can also visually establish that center adatoms are approximately 4 times more reactive with NH_3 than corner adatoms. This would be much more difficult or impossible to establish with other surface sensitive spectroscopies.

The lower part of Figure 32 shows STS spectra on individual sites on the Si(111)-(7 × 7) surface. The x-axis is an energy scale on either side of the Fermi level (referenced to an energy of zero). Positive energy represents unoccupied electron levels above the Fermi energy (into the conduction band), while negative energy represents occupied levels below the Fermi energy (into the valence band). This is plotted against (dI/dV)/(I/V), where I is tunneling current and V is bias voltage. This quantity, differentiated conductance over conductance, has been shown to be a good approximation to the local density of states (LDOS) (Strocio et al., 1986). Curve A on the left of Figure 32 shows a LDOS spectrum of a restatom site before exposure to NH_3. The large peak just below the Fermi level are occupied states of the restatom dangling bond. On the right, spectrum A shows that this particular restatom has

125

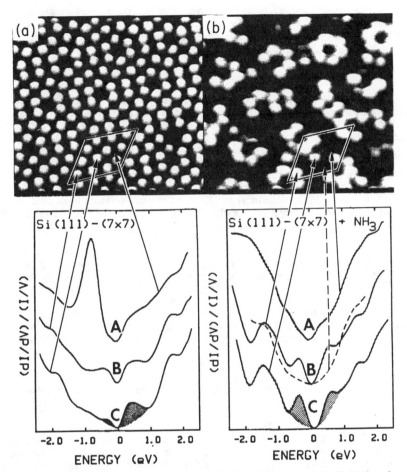

Figure 32. Atomic-scale STM images of an unreacted (a) and NH₃-exposed (b) Si (111) surface (top) along with STS spectra of various positions on these surfaces as indicated (bottom). In (a), the white spots in the STM image represent adatom sites. When an adatom reacts with NH₃, its tunneling characteristics change and the adatom site goes "dark" in the image (b). See text for additional details. From Avouris (1990).

reacted after exposure to NH₃ (the electrons in the dangling bond are gone, having now filled a bonding state much deeper in the valence band). Note that in this case, the STM image does not give evidence of the reaction of this restatom, but the STS spectrum clearly does. Spectra B and C (solid curves) are of adatoms before and after exposure, both of which have not reacted. However, one can see how the electronic structure of adatoms changes when Si atoms around them have reacted. The dashed spectrum on the right (under B) shows the LDOS spectrum of an adatom that has reacted. Just as in the restatom case, the electronic states due to the dangling bond are quenched.

In the final analysis, this example shows the ability of STM to "see" surface reactivity atom by atom, and of STS to follow the electronic nature of these surface reactions. As already stated, STM was used in this case to show that center adatoms are 4 times more reactive than corner adatoms. In addition, LDOS spectra showed that restatoms are more reactive than even center adatoms. Clearly, STM and STS can provide details of surface reactions that have never before been obtainable. Further development of these techniques, and application to mineral-water interfaces, will bring a new level of discovery to mineral surface geochemistry.

CONCLUSIONS

Since the 1950's, spectroscopic and microscopic observations, combined with indirect measurements, have dramatically improved our understanding of the characteristics of solid surfaces in a variety of environments, and of how surface properties influence chemical reactivity. Perhaps the most astonishing outcome of this revolution is that we no longer need to rely strictly on theoretical models of solid-surface reactivity, but rather we may observe directly surface phenomena using a combination of recently developed surface-sensitive instrumental techniques. Many examples of such work have been presented in this chapter. As these techniques continue to develop, the understanding of the geochemical reactivity of mineral surfaces in natural environments will advance. Applications of this information to more applied geochemical subfields should proliferate at a rapid pace.

Although most of the research in the general field of surface science to date has dealt with non-mineral inorganic materials, many of the resulting underlying principles are directly applicable to the study of mineral surface characteristics and reactivity. The following principles, formulated based on these principles as well as on the direct spectroscopic and microscopic observations of mineral surfaces, should serve as a fundamental framework for our present understanding of mineral surface reactivity. It should be emphasized that these principles are meant to be general guidelines; as usual, exceptions can be found, and some have been highlighted or eluded to above.

1. *Mineral surface compositions are not representative of the bulk and are laterally inhomogeneous.* Generally speaking, mineral surface compositional modification can occur via attachment (sorption), exchange, or detachment (desorption) reactions, or some combination of these. In natural environments, direct spectroscopic measurements show that this will happen to all but the most inert minerals to some degree under practically all conditions. Compositional modification may be restricted to just the top monolayer, or it may extent into the mineral, in some cases to considerable depths (thousands of Angstroms or more). In addition, sorption, exchange, and desorption modification of surface composition rarely happens uniformly over a mineral surface. Certain attaching species will prefer certain sorption sites, such as steps, kinks, or at a particular atomic site on a flat terrace. Likewise, certain desorption reactions will occur preferentially at certain sites, potentially resulting in a non-uniform leached layer.

2. *Mineral surface microtopography is complex.* Mineral surfaces are very rarely flat over large areas. Atomic and molecular-scale steps, undulations, and/or mounds generally exist even on "perfect" cleavage surfaces. Surface mircocracks and other defects are common. All of these features exist, to varying degrees, on crystal growth faces, on surfaces that have undergone dissolution, and on cleavage surfaces. Atomic and molecular-scale roughness of mineral surfaces is a key factor in mineral surface reactivity (see below).

3. *The atomic structure of the top few monolayers of a mineral is not representative of the bulk structure.* First considering only the atomic structure laterally across a surface, the structure will only be a simple termination of the bulk structure when it exhibits a close-packed arrangement as, for example, on the {001} surface of galena. Otherwise, surface atoms will shift laterally (i.e. relax) to obtain a lower energy arrangement. In the direction perpendicular to the surface, the spacing between the upper few atomic layers will decrease if the surface atoms are underbonded compared to like atoms in the bulk, and increase if the surface atoms are overbonded compared

to like atoms in the bulk. (In this case, under- and overbonding refers to the Pauling bond strength calculated on a central atom.) This expansion or contraction should rarely be in excess of 5 to 10%, and its effects diminish rapidly with each layer into the bulk. (These effects should rarely be observed more than 5 or so monolayers into the bulk.) Further, if enough thermal energy is added to the system (e.g., in a hydrothermal system), there is a possibility that the surface will atomically reconstruct, resulting in an entirely different atomic arrangement at the surface. Major reconstructions usually result in a much more stable arrangement that persists to lower temperatures.

4. *The reactivity of a mineral in aqueous solution depends ultimately on its surface composition, microtoprography, and atomic structure, all of which can be considered independent factors.* There are a number of examples in the surface science literature (some described in this chapter) where the reactivity of a surface is tracked as one of these factors is modified and the remaining two are held constant. Seemingly subtle changes in one of these factors can result in a dramatic change in reactivity. This reactivity dependence has an analogy in the general field of mineralogy, where the differences among the properties of polymorphs (same composition, different structure) and isomorphs (same structure, different composition) are well-known. One could also draw an analogy between the effects of bulk defects on bulk properties with surface microtoprography and its effects on surface reactivity. However, in general, surface microtoprography plays a larger role in the properties of a surface than bulk defects play in properties of the bulk.

5. *The reactivity of a mineral surface will generally increase as its atomic and molecular-scale roughness increases.* Surface roughness on this scale creates surface atoms which are less coordinated to other substrate atoms than they would be on terraces. These sites on the edges of steps, kinks and other discontinuous features have more flexibility in potential reaction (bonding) configurations, and they commonly have an increased affinity for species undergoing surface diffusion. These sites may be particularly conducive for promoting mineral surface catalysis reactions. In this reaction sequence, (1) reactant(s) attach, (2) molecular dissociation occurs, (3) recombination occurs producing a product or products, and (4) product(s) detach.

6. *Mineral surfaces are not static; they are dynamic systems sensitive to ambient conditions and local reactions.* For example, a sorbed surface species can (1) diffuse along the surface, (2) diffuse into the substrate, (3) detach from the surface, or (4) remain attached to (i.e., react with) a specific site. The potential barrier to surface diffusion may be (and often is) lower than those of the other three possibilities, and species at mineral-water interfaces can be highly mobile. The surface atomic structure, in response to the nearby presence of a sorbed species, can be modified on a local basis to best accommodate the sorbate and obtain the lowest energy configuration possible. In addition, as ambient conditions change, the entire surface atomic structure can change independently of the bulk as mentioned above. Pressure-temperature phase diagrams for mineral surfaces, if they existed, would be different from those known for the bulk. Even the microtopography and/or the composition of the surface often exists in a transient nature, as for example through processes involving dissolution and reprecipitation like Ostwald ripening (see Parks, this volume) or other "aging" effects (Eggleston et al., 1989). It is this dynamic nature of surfaces within the complex and highly variable chemistry of the Earth that allows them to play such a critical role in geochemical phenomena.

ACKNOWLEDGMENTS

I am very much indebted to the graduate students and post-docs with whom I have had the privilege to work with on a great number of surface science projects. To them, for their creativity, enthusiasm, scientific talents, and plain hard work, I dedicate this chapter. They are (in alphabetical order) Barbara Bakken, Alex Blum, Kris Butcher, Altaf Carim, Susan Cohen, Steve Didziulus, Carrick Eggleston, Pat Johnsson, Roland Hellmann, Jodi Junta, Heather Ponader, Susan Stipp, Tracy Tingle, and Bruce Tufts. In addition, Carrick Eggleston, Pat Johnsson, and Jodi Junta reviewed the original manuscript and provided many helpful suggestions, and Ray Browning obtained some of the high spatial resolution Auger maps. Financial support has been provided through the the the Center for Materials Research at Stanford, the Gas Research Institute, the National Science Foundation, the Petroleum Research Fund of the American Chemical Society, the United States Geological Survey, Newmont Metallurgical Services, and Chevron. Instrumental support has been provided by the Center for Materials Research and the Center for Integrated Systems at Stanford, the Physical Electronics Division of Perkin-Elmer, Charles Evans and Associates, and Digital Instruments, Inc.

REFERENCES

Abraham, F.F. and Batra, I.P. (1989) Theoretical interpretation of atomic force microscope images of graphite. Surface Sci., 209, L125-L132.

Adamson, A.W. (1982) Physical Chemistry of Surfaces (4th edition). Wiley, New York, 664 p.

Albrecht, T.R., and Quate, C.F. (1987) Atomic resolution imaging of a nonconductor by atomic force microscopy. J. Appl. Phys., 62, 2599-2602.

Albrecht, T.R., and Quate, C.F. (1988) Atomic resolution with the atomic force microscope on conductors and nonconductors. J. Vac. Sci. Tech., 6, 271-274.

Avouris, P. (1990) Atom-resolved surface chemistry using the scanning tunneling microscope. J. Phys. Chem., 94, 2246-2256.

Avouris, Ph., and Wolkow, R. (1989) Atom-resolved surface chemistry studied by scanning tunneling microscopy and spectroscopy. Phys. Rev. B, 39, 5091-5100.

Bakken, B.M., and Einaudi, M.T. (1986) Spatial and temporal relations between wall-rock alteration and gold mineralization, main pit, Carlin gold mine, Nevada, U.S.A. In: Proceedings of Gold '86, Int'l. Symp. on the Geology of Gold (A.J. Macdonald, ed.) Toronto, Canada, p. 388-403.

Bakken, B.M., Hochella, M.F., Jr., Marshall, A.F., and Turner, A.M. (1989) High-resolution microscopy of gold in unoxidized ore from the Carlin mine, Nevada. Econ. Geol., 84, 171-179.

Becker, C.H., and Gillen, K.T. (1984) Surface analysis by non-resonant multiphoton ionization of desorbed or sputtered species. Anal. Chem., 56, 1671-1674.

Benjamin, M.M., and Leckie, J.O. (1981a) Multiple-site adsorption of Cd, Cu, Zn, and Pb on amorphous iron oxyhydroxide. J. Colloid Interface Sci., 79, 209-221.

Benjamin, N.M., and Leckie, J.O. (1981b) Competitive adsorption of Cd, Cu, Zn, and Pb on amorphous iron oxyhydroxide. J. Colloid Interface Sci., 83, 410-419.

Berner, R.A., Holdren, G.R., Jr., and Schott, J. (1985) Surface layers on dissolving silicates. Geochim. Cosmochim. Acta, 49, 1657-1658.

Bethge, H., and Krohn, M. (1965) Wachstumsvorgang auf NaCl-kristallen nach Anlosung. In: Adsorption et Croissance Cristalline, C.N.R.S., 389-406.

Binnig, G., Quate, C.F., and Gerber, Ch. (1986) Atomic force microscope. Phys. Rev. Letters, 56, 930-933.

Binnig, G., Rohrer, H., Gerber, Ch., and Weibel E. (1982a) Tunneling through a controllable vacuum gap. Appl. Phys. Letters, 40, 178-180.

Binnig, G., Rohrer, H., Gerber, Ch., and Weibel E. (1982b) Surface studies by scanning tunneling microscopy. Phys. Rev. Letters, 49, 57-61.

Binnig, G., Rohrer, H., Gerber, Ch., and Weibel E. (1983) 7 × 7 reconstruction on Si(111) resolved in real space. Phys. Rev. Letters, 50, 120-123.

Binnig, G., Gerber, Ch., Stoll, E., Albrecht, T.R., and Quate, C.F. (1987) Atomic resolution with atomic force microscope. Surface Sci., 189/190, 1-6.

129

Blake, D.F. (1990) Scanning electron Microscopy. In: Instrumental Surface Analysis of Geologic Materials (D.L. Perry, ed.), VCH Publishers, New York, 11-44.

Bradley, S.A., Gattuso, M.J., and Bertolacini, R.J. (1989) Characterization and Catalyst Development. ACS Symposium Series 411, American Chemical Society, Washington, D.C., 451 p.

Brantley, S.L., Crane, S.R., Crerar, D.A., Hellmann, R., and Stallard, R. (1986) Dissolution at dislocation etch pits in quartz. Geochim. Cosmochim. Acta, 50, 2349-2361.

Brion D. (1980) Etude par spectroscopy de photoelectrons de la degradation superficielle de FeS_2, $CuFeS_2$, ZnS, and PbS a l'air dans l'eau. Appl. Surface Sci., 5, 133-152.

Browning, R., and Hochella, M.F., Jr. (1990) Auger electron spectroscopy and microscopy. In: Instrumental Surface Analysis of Geologic Materials (D.L. Perry, ed.), VCH Publishers, New York, 87-120.

Buckley, A.N., and Woods, R. (1984) An x-ray photoelectron spectroscopic study of the oxidation of galena. Appl. Surface Sci., 17, 401-414.

Casey, W.H., Westrich, H.R., and Arnold, G.W. (1988) Surface chemistry of labradorite feldspar reacted with aqueous solutions at $pH = 2$, 3, and 12. Geochim. Cosmochim. Acta, 52, 2795-2807.

Casey, W.H., Westrich, H.R., Arnold, G.W., and Banfield, J.F. (1989) The surface chemistry of dissolving labradorite feldspar. Geochim. Cosmochim. Acta, 53, 821-832.

Chou, L., and Wollast, R. (1985) Study of the weathering of albite at room temperature and pressure with a fluidized bed reactor. Geochim. Cosmochim. Acta, 49, 1659-1660.

Christmann, K., and Ertl, G. (1976) Interaction of hydrogen with Pt(111): The role of atomic steps. Surface Sci., 60, 365-384.

Chu, W.K., Mayer, J.W., and Nicolet, M.A. (1978) Backscattering Spectroscopy. Academic Press, New York.

Clarke, L.J. (1985) Surface Crystallography, an Introduction to Low Energy Electron Diffraction. Wiley, New York, 329 p.

Coyne, L.M., and McKeever, S.W.S. (1989) Spectroscopic characterization of minerals and their surfaces, an overview. In: Spectroscopic Characterization of Minerals and Their Surfaces (L.M. Coyne, S.W.S. McKeever, D.F. Blake, eds.), American Chemical Society, Symp. Ser. 415, 1-29.

Deckert A.A., Brand, J.L., Arena, M.V., and George, S.M. (1989) Surface diffusion of carbon monoxide on Ru(001) studied using laser-induced thermal desorption. Surface Sci., 208, 441-462.

Dibble, W.E., Jr., and Tiller, W.A. (1981) Non-equilibrium water/rock interactions - I. Model for interface-controlled reactions. Geochim. Cosmochim. Acta, 45, 79-92.

Eggleston, C.M., and Hochella, M.F., Jr. (1990) Scanning tunneling microscopy of sulfide surfaces. Geochim. Cosmochim. Acta, 54, 1511-1517.

Eggleston, C.M., Hochella, M.F., Jr., and Parks, G.A. (1989) Sample preparation and aging effects on the dissolution rate and surface composition of diopside. Geochim. Cosmochim. Acta, 53, 797-804.

Enustun, B.V., and Turkevich, J. (1960) Solubility of fine particles of strontium sulfate. J. Am. Chem. Soc., 82, 4502-4509.

Ernsberger, F.M. (1960) Structural effects in the chemical reactivity of silica and silicates. J. Phys. Chem. Solids, 13, 347-351.

Ertl, G., Weiss, M., and Lee, S.B. (1979) The role of potassium in the catalytic synthesis of ammonia. Chem. Phys. Lett., 60, 391-394.

Falicov, L., and Somorjai, G.A. (1985) Correlation between catalytic activity and bonding and coordination number of atoms and molecules on transition metal surfaces: Theory and experimental evidence. Proc. Nat. Acad. Sci. USA, 82, 2207-2211.

Goldstein, F.I., Newbury, D.E., Echlin, P., Joy, D.C., Fiori, C., and Lifshin, E. (1981) Scanning Electron Microscopy and X-ray Microanalysis. Plenum Press, New York, 673 p.

Guy, R., Chakrabarti, C., and Schramm, L. (1975) The application of a simple chemical model of natural waters to metal fixation in particulate matter. Canadian J. Chem., 53, 661-669.

Hagstrom, A.L., and Fahlman, A. (1978) The interaction between oxygen and the lead chalcogenides at room temperature studied by photoelectron spectroscopy. Appl. Surface Sci., 1, 455-470.

Hansma, P.K., Elings, V.B., Marti, O., and Bracker, C.E. (1988) Scanning tunneling microscopy and atomic force microscopy: Application to biology and technology. Science, 242, 209-216.

Hartman, H., Sposito, G., Yang, A., Manne, S., Gould, S.A.C., and Hansma, P.K. (1990) Molecular-scale imaging of clay mineral surfaces with the atomic force microscope. Clays and Clay Minerals, 38, 337-342.

Hayes, K., and Leckie, J.O. (1986) Mechanism of lead ion adsorption at the goethite-water interface.

130

In: Geochemical Processes at Mineral Surfaces (J.A. Davis and K.F. Hayes, eds.), Am. Chem. Soc., Washington, D.C., 114-141.

Hayes, K.F., Roe, A.L., Brown, G.E., Jr., Hodgson, K.O., Leckie, J.O., and Parks, G.A. (1987) In-situ X-ray absorption study of surface complexes: Selenium oxyanions on α-FeOOH. Science, 1987, 783-786.

Hellmann, R., Eggleston, C.M., Hochella, M.F., Jr., Crerar, D.A. (1990) Formation of leached layers on albite during dissolution under hydrothermal conditions. Geochim. Cosmochim. Acta, 54, 1267-1281

Henrich, V.E. (1985) The surfaces of metal oxides. Reports Progress Phys., 48, 1481-1541.

Henrich, V.E., and Kurtz, R.L. (1981) Surface electronic structure of TiO_2: Atomic geometry, ligand coordination, and the effect of adsorbed hydrogen. Phys. Rev. B, 23, 6280-6287.

Hochella, M.F., Jr. (1988) Auger electron and x-ray photoelectron spectroscopies. In: Spectroscopic Methods in Mineralogy and Geology (F.C. Hawthorne, ed.), Reviews in Mineralogy, 18, 573-637.

Hochella, M.F., Jr., and Carim, A.F. (1988) A reassessment of electron escape depths in silicon and thermally grown silicon dioxide thin films. Surface Sci., 197, L260-L268.

Hochella, M.F., Jr., Harris, D.W., and Turner A.M. (1986a) Scanning Auger microscopy as a high resolution microprobe for geologic materials. Am. Mineral., 71, 1247-1257.

Hochella, M.F., Jr., Turner, A.M., and Harris, D.W. (1986b) High resolution scanning microscopy of mineral surfaces. Scan. Electron Micros., 1986, 337-349.

Hochella, M.F., Jr., Ponader, H.B., Turner, A.M., and Harris, D.W. (1988a) The complexity of mineral dissolution as viewed by high resolution scanning Auger microscopy: Labradorite under hydrothermal conditions. Geochim. Cosmochim. Acta, 52, 385-394.

Hochella, M.F., Jr., Lindsay, J.R., Mossotti, V.G., and Eggleston, C.M. (1988b) Sputter depth profiling in mineral surface analysis. Am. Mineral., 73, 1449-1456.

Hochella, M.F., Jr., Eggleston, C.M., Elings, V.B., Parks, G.A., Brown, G.E., Jr., Wu, C.M., and Kjoller, K. (1989) Mineralogy in two dimensions: Scanning tunneling microscopy of semiconducting minerals with implications for geochemical reactivity. Am. Mineral., 74, 1235-1248.

Hochella, M.F., Jr., Eggleston, C.M., Elings, V.B., and Thompson, M.S. (1990) Atomic structure and morphology of the albite {010} surface: An atomic-force microscope and electron diffraction study. Am. Mineral., 75, 723-730.

Holdren, G.R., Jr., and Speyer, P.M. (1985) Reaction rate-surface area relationships during the early stages of weathering. I. Initial observations. Geochim. Cosmochim. Acta, 49, 675-681.

Holdren, G.R., Jr., and Speyer, P.M. (1987) Reaction rate-surface area relationships during the early stages of weathering. II. Data on eight additional feldspars. Geochim. Cosmochim. Acta, 51, 2311-2318.

Holland, B.W., Duke, C.B., and Paton, A. (1984) The atomic geometry of Si(100)-(2 × 1) revisited. Surface Sci., 140, L269-L278.

Hyland, M.M., and Bancroft, G.M. (1989) An XPS study of gold deposition at low temperature on sulfide minerals: Reducing agents. Geochim. Cosmochim. Acta, 53, 367-372.

Iwasawa, Y., Mason, R., Textor, M., and Somorjai, G.A. (1976) The reactions of carbon monoxide at coordinatively unsaturated sites on a platinum surface. Chem. Phys. Letters, 44, 468-470.

Jean, G.E., and Bancroft, G.M. (1985) An XPS and SEM study of gold deposition at low temperatures on sulfide mineral surfaces: Concentration of gold by adsorption/reduction. Geochim. Cosmochim. Acta, 49, 979-987.

Jones, R.O., Jennings, P.J., and Painter, G.S. (1975) Cluster calculations of the electronic structure of transition metal surfaces. Surface Sci., 53, 409-428.

King, D., and Woodruff, D.P. (1984) The Chemical Physics of Solid Surfaces and Heterogeneous Catalysis. Elsevier, New York.

Kirkpatrick, R.J. (1975) Crystal growth from the melt: A review. Am. Mineral., 60, 798-814.

Kirkpatrick, R.J. (1981) Kinetics of crystallization of igneous rocks. In: Kinetics of Geochemical Processes (A.C. Lasaga and R.J. Kirkpatrick, eds.), Reviews in Mineralogy, Vol. 8, 321-398.

Krohn, M., and Bethge, H. (1977) Step structures of real surfaces. In: Current Topics in Materials Science, Vol. 2 (E. Kaldis, ed.), North-Holland, Amsterdam, p. 141.

Kurtz, R.L., and Henrich, V.E. (1983) Geometric structure of the α-Fe_2O_3(001) surface: A LEED and XPS study. Surface Sci., 129, 345-354.

Lanford, W.A., Davis, K., Lamarche, P., Laursen, T., and Groleau, R.J. (1979) Hydration of soda-lime glass. J. Non-Cryst. Solids, 33, 249-266.

Lasaga, A.C., and Blum, A.E. (1986) Surface chemistry, etch pits and mineral-water reactions.

131

Geochim. Cosmochim. Acta, 50, 2363-2379.

Lee, S.Y., and Jackson, M.L. (1977) Surface charge density determination of micaceous minerals by ^{235}U fission particle track method. Clays and Clay Minerals, 25, 295-301.

Liang, D., and Readey, D.W. (1987) Dissolution kinetics of crystalline and amorphous silica in hydrofluoric-hydrochloric acid mixtures. J. Am. Ceram. Soc., 70, 570-577.

Loganathan, P., and Burau, R. (1973) Sorption of heavy metal ions by a hydrous manganese oxide. Geochim. Cosmochim. Acta, 37, 1277-1293.

Mak, C.H., Brand, J.L., Deckert, A.A., and George, S.M. (1986) Surface diffusion of hydrogen on Ru(001) studied using laser-induced thermal desorption. J. Chem. Phys., 85, 1676-1680.

Mathez, E.A. (1987) Carbonaceous matter in mantle xenoliths: Composition and relevance to the isotopes. Geochim. Cosmochim. Acta, 51, 2339-2347.

Metson, J. (1990) Secondary Ion Mass Spectrometry. In: Instrumental Surface Analysis of Geologic Materials (D.L. Perry, ed.), VCH Publishers, New York, 311-352.

Mogk, D.W. (1990) Application of Auger electron spectroscopy to studies of chemical weathering. In: Reviews of Geophysics (G. Sposito, ed.), American Geophysical Union, Washington, in press.

Mogk, D.W., and Locke, W.W. (1988) Application of Auger electron spectroscopy (AES) to naturally weathered hornblende. Geochim. Cosmochim. Acta, 52, 2537-2542.

Muir, I.J., Bancroft, G.M., and Nesbitt, H.W. (1989) Characteristics of altered labradorite surfaces by SIMS and XPS. Geochim. Cosmochim. Acta, 53, 1235-1241.

Nesbitt, H.W., and Muir, I.J. (1988) SIMS depth profiles of weathered plagioclase and processes affecting dissolved Al and Si in some acidic soil solutions. Nature, 334, 336-338.

Newbury, D.E., Joy, D.C., Echlin, P., Fiori, C.E., Goldstrin, J.I. (1985) Advanced Scanning Electron Microscopy and X-ray Microanalysis. Plenum Press, New York, 454 p.

Onuferko, J.H., Woodruff, D.P., and Holland, B.W. (1979) LEED structure analysis of the Ni{100}(2 × 2)C(p4g) structure; a case of adsorbate-induced substrate distortion. Surface Sci., 87, 357-374.

Painter, G.S., Jennings, P.J., and Jones, R.O. (1975) Electronic structure of stepped transition metal surfaces. J. Phys., C8, L199-L202.

Parrish, W., and Gordon, S.G. (1945) Orientation techniques for the manufacture of quartz oscillatorplates. Am. Mineral., 30, 296-325.

Perdereau, J., and Rhead, G. (1971) LEED studies of adsorption on vicinal copper surfaces. Surface Sci., 24, 555-571.

Perry, D.L., Taylor, J.A., and Wagner, C.D. (1990) X-ray induced photoelectron and Auger spectroscopy. In: Instrumental Surface Analysis of Geologic Materials (D.L. Perry, ed.), VCH Publishers, New York, 45-86.

Petit, J.-C., Della Mea, G., Dran J.-C., Schott, J., and Berner, R.A. (1987) Mechanism of diopside dissolution from hydrogen depth profiling. Nature, 325, 705-707.

Petrovich, R. (1981) Kinetics of dissolution of mechanically comminuted rock-forming oxides and silicates - II. Deformation and dissolution of oxides and silicates in the laboratory and at the Earth's surface. Geochim. Cosmochim. Acta, 45, 1675-1686.

Richardson, J.T. (1989) Principles of Catalyst Development. Plenum Press, New York, 288 p.

Sangwal, K. (1987) Etching of Crystals. North-Holland, Amsterdam, 497 p.

Schott, J., and Petit, J.-C. (1987) New evidence for the mechanisms of dissolution of silicate minerals. In: Aquatic Surface Chemistry (W. Stumm, ed.), Wiley Interscience, New York, p. 293-315.

Schott, J., Brantley, S., Crerar, D., Guy, C., Borcsik, M., and Willaime, C. (1989) Dissolution kinetics of strained calcite. Geochim. Cosmochim. Acta, 53, 373-382.

Schuhle, U., Pallix, J.B., and Becker, C.H. (1988) Sensitive mass spectrometry of molecular adsorbates by stimulated desorption and single photon ionization. J. Am. Chem. Soc., 110, 2323-2327.

Somorjai, G.A. (1981) Chemistry in Two Dimensions: Surfaces. Cornell University Press, Ithaca, New York, 575 p.

Somorjai, G.A. (1990) Modern concepts in surface science and heterogeneous catalysis. J. Phys. Chem., 94, 1013-1023.

Somorjai, G.A., and Van Hove, M.A. (1989) Adsorbate-induced restructuring of surfaces. Prog. Surface Sci., 30, 201-231.

Somorjai, G.A., Van Hove, M.A., and Bent, B.J. (1988) Organic monolayers on transition-metal surfaces. The catalytically important sites. J. Phys. Chem., 92, 973-978.

Spencer, N.D., Schoonmaker, R.C., and Somorjai, G.A. (1982) Iron single crystals as ammonia synthesis catalysts: Effects of surface structure on catalyst activity. J. Catalysis, 74, 129-135.

Stipp, S.L., and Hochella, M.F., Jr. (1990) Structure and bonding environments at the calcite surface as

132

observed with x-ray photoelectron spectroscopy (XPS) and low energy electron diffraction (LEED). Geochim. Cosmochim. Acta, in review.

Strocio, J.A., Feenstra, R.M., and Fein, A.P. (1986) Electronic structure of the Si(111) 2 × 1 surface by scanning-tunneling microscopy. Phys. Rev. Letters, 57, 2579-2582.

Strongin, D.R., and Somorjai, G.A. (1988) The effects of potassium on ammonia synthesis over iron single-crystal surfaces. J. Catalysis, 109, 51-60.

Strongin, D.R., Carrazza, J., Bare, S.R., and Somorjai, G.A. (1987) The importance of C_7 sites and surface roughness in the ammonia synthesis reaction over iron. J. Catalysis, 102, 213-215.

Sunagawa, I. (1961) Step heights of spirals on natural hematite crystals. Am. Mineral., 46, 1216-1226.

Sunagawa, I. (1984) Morphology of natural and synthetic diamond crystals. In: Materials Science of the Earth's Interior (I. Sunagawa, ed.), Terra Scientific Publishing, Tokyo, 303-330.

Sunagawa, I. (1987) Surface microtopography of crystal faces. In: Morphology of Crystals (I. Sunagawa, ed.), Terra Scientific Publishing, Tokyo, 323-365.

Sunagawa, I., and Koshino, Y. (1975) Growth spirals on kaolin group minerals. Am. Mineral., 60, 407-412.

Sunagawa, I., and Bennema, P. (1979) Modes of vibrations in step trains: Rhythmical bunching. J. Crystal Growth, 46, 451-457.

Takayanagi, K., Nanishiro, Y., Takahashi, M., Motoyoshi, H., and Yagi, K. (1985) Structural analysis of Si(111)-7 × 7 by UHV-transmission electron diffraction and microscopy. J. Vac. Sci. Technol. A, 3, 1502-1506.

Tiller, W.A. (1977) On the cross-pollination of crystallization ideas between metallurgy and geology. Phys. Chem. Minerals, 2, 125-151.

Tingle, T.N., and Fenn, P.M. (1984) Transport and concentration of molybdenum in granite molybdenite systems: Effects of fluorine and sulfur. Geology, 12, 156-158.

Tingle, T.N., Hochella, M.F., Jr., Becker, C.H., and Mulhotra, R. (1990a) Organic compounds on crack surfaces in olivine from San Carlos, Arizona and Hualalai Volcano, Hawaii. Geochim. Cosmochim. Acta, 54, 477-485.

Tingle, T.N., Mathez, E.A., and Hochella, M.F., Jr. (1990b) Carbonaceous matter in peridotites and basalts studied by XPS, SALI, and LEED. Geochim. Cosmochim. Acta, in review.

Tolansky, S. (1948) Multiple-beam Interferometry of Surfaces and Films. Clarendon Press, Oxford.

Tolansky, S. (1968) Microstructures of surfaces. American Elsevier Publishing, New York, 65 p.

Tolansky, S. (1973) An Introduction to Interferometry. Longman, London, 253 p.

Tossell, J.A., and Vaughan, D.J. (1987) Electronic structure and the chemical reactivity of the surface of galena. Canadian Mineral., 25, 381-392.

Tromp, R.M., Smeenk, R.G., Saris, R.W., and Chadi, D.J. (1983) Ion beam crystallography of silicon surfaces. II. Si(100) - (2 × 1). Surface Sci., 133, 137-158.

Van Hove, M.A., Weinberg W.H., and Chan C.-M. (1986) Low-energy electron diffraction: Experiment, theory and surface structure determination. Springer-Verlag, New York, 603 p.

Walker, R.J. and Fassett, J.D. (1986) Isotopic measurement of subnanogram quantities of rhenium and osmium by resonance ionization mass spectrometry. Anal. Chem., 58, 2923-2927.

Walker, R.J., Shirey, S.B., Hanson, G.N., Rajamani, V., and Horan, M.F. (1989) Re-Os, Rb-Sr, and O isotopic systematics of the Archean Kolar schist belt, Karnataka, India. Geochim. Cosmochim. Acta, 53, 3005-3013.

Wolkow, R., and Avouris, P. (1988) Atom-resolved surface chemistry using scanning tunneling microscopy. Phys. Rev. Letters, 60, 1049-1052.

White, A.F. and Yee, A. (1985) Aqueous oxidation-reduction kinetics associated with coupled electron-cation transfer from iron-containing silicates at 25 C. Geochim. Cosmochim. Acta, 49, 1263-1275.

Willis, R.D., Thonnard, N., and Cole, D.R. (1989) Resonance ionization spectroscopy and its potential application in geosciences. U.S. Geol. Survey Prof. Paper 1890, p. 117-128.

Zheng, N.J., Wilson, I.H., Knipping, U., Burt, D.M., Krinsley, D.H., and Tsong, I.S.T. (1988) Atomically resolved scanning tunneling microscopy images of dislocations. Phys. Rev. B, 38, 12780-12782.

CHAPTER 4   G. A. PARKS

SURFACE ENERGY AND ADSORPTION AT MINERAL-WATER INTERFACES: AN INTRODUCTION

INTRODUCTION

The properties of materials at or near surfaces or interfaces are likely to be different than the properties of the same material in the bulk. We expect this because the bonding of surface atoms is different—part of their coordination environment is either missing or different than it would be in bulk, so their potential energies must be different as well. We are interested in the thermodynamic implications of the special character of surfaces and interfaces, especially of surface excess free energy and adsorption. This chapter explores the relationship between surface area and volume or mass, surface excess quantities and some of the thermodynamic consequences of their existence, evidence for sorption, or the accumulation of minor and trace constituents at interfaces, and the nature of the complexes in which they are bound. The chapter closes with some brief speculations on the impact of surface chemistry and physics in some challenging and perhaps surprising fields.

PARTICLE SIZE, SHAPE, AND SURFACE AREA

A surface is the exterior boundary of a condensed phase in a vacuum or gaseous environment. An interface is the boundary between two condensed phases, e.g., a solid-liquid interface or a solid-solid grain boundary. We will often use the term 'surface' in a generic sense; the distinction between surfaces and interfaces should be clear in context.

Surface or interfacial area increases, relative to volume, as particle or grain size decreases and with increasing intricacy of particle shape. The area and volume of a particle of any shape are related to its "size" through shape factors, k_a and k_v, respectively; thus $A = k_a d^2$, and $V = k_v d^3$. The particle size, d, can be any measure of size; a diameter or closely analogous measure of size is convenient. The specific surface area per unit volume (A_v) or mass (A_s), and molar surface area (A_m) are related to size through another shape factor, $k_s = k_a/k_v$,

$$A_v = \frac{k_s}{d}; \quad A_s = \frac{k_s}{\rho d}; \quad A_m = \frac{k_s fw}{\rho d} \quad . \tag{1}$$

The quantities ρ and fw are the specific gravity and formula weight of the solid, respectively. The change in surface area, dA, with the number of moles of a key substance reacted, dn, determines the influence of surface energy on many reactions,

$$\frac{dA}{dn} = \frac{2k_s fw}{3\rho}\left(\frac{1}{d}\right) = \frac{2A_m}{3} = \frac{2A_s fw}{3} \quad . \tag{2}$$

Values of k_a, k_v, and k_s computed for some simple geometric forms, and empirical values for crushed quartz and some amorphous and biogenic silicas are collected in Table 1.

Surface area can be measured in many ways. The method in most common use today utilizes interpretation of the adsorption densities of gases and is known as the BET method, named for its originators, Brunauer, Emmett, and Teller (Adamson, 1982). The adsorbing gas is sometimes indicated by adding its chemical formula or name to the acronym, e.g., N_2-BET. Figure 1 contains typical BET surface area data for crushed quartz as a function of particle size and demonstrates that Equation 1 is obeyed remarkably

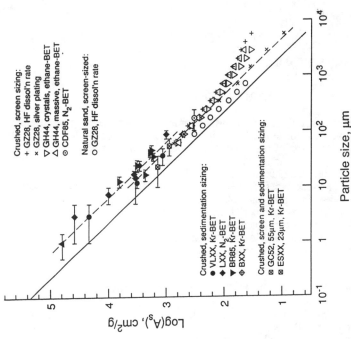

Crushed, screen sizing:
+ GZ28, HF dissol'n rate
× GZ28, silver plating
▽ GH44, crystals, ethane-BET
△ GH44, massive, ethane-BET
◎ CDF85, N₂-BET

Natural sand, screen-sized:
○ GZ28, HF dissol'n rate

Crushed, sedimentation sizing:
● VLXX, Kr-BET
♦ LXX, N₂-BET
▼ BR85, Kr-BET
◈ BXX, Kr-BET

Crushed, screen and sedimentation sizing:
⊠ GC52, 55μm, Kr-BET
⊠ ESXX, 23μm, Kr-BET

Particle size, μm

Log(A_s), cm²/g

Figure 1. Relationship between particle size d, and specific surface area A_s, for crushed quartz and a natural quartz sand with well rounded particles (Ottawa sand). All A_s were measured on size fractions, and the plotted d is an arithmetic average of limiting sizes. The solid line represents the theoretical surface area of spheres. The two sizing methods respond to different measures of "size", so they should yield different shape factors, as observed. The lower dashed line was fitted to the five smallest screen-sized fractions only, with fixed slope of –1. The upper dashed line was fitted to all sedimentation-sized materials. Sources: GZ28, Gross and Zimmerly, 1928; GH44, Gaudin and Hukki, 1944; CDF85, Cases et al., 1985; VL60, Van Lier et al., 1960; LI58, Li, 1958.

Table 1. Shape factors. Proportionality constants relating particle area, k_a, volume, k_v, and mass, k_s, to particle size for a variety of shapes.

Spheres and regular polyhedra. Particle "size" is measured by a diameter of edge length:

	size	k_a	k_v	k_s
sphere	diameter	π	$\pi/6$	6
cube	edge	6	1	6
tetrahedron	edge	1.7321	0.1179	14.691
octahedron	edge	3.4641	0.4714	7.348

Crushed quartz. Particle size is measured by screening or sedimentation.[1]

screening	---	---	---	12.3±1.5
sedimentation	---	---	---	17.3±5.5

Amorphous and biogenic silicas. Particles are aggregates or diatom tests. Size reported is the gross diameter. (Kent, 1983)

	diameter, μm	A_s, m²/g	k_s
BDH-silica[2]	28±1	227	≈14,000
Aerosil-II[3]	21±2	209±3	≈10,000
Diatoms:			
T. decipiens	20-55	258	>11,000
N. pelliculosa	5±2	36	≈1,300

1. derived from BET surface areas of crushed quartz size fractions prepared by either screening or sedimentation in air or water. Data are plotted in Figure 2.

2. a precipitated and acid washed amorphous silica with N₂-BET A_s ≡ 227 m²/g, resuspended in water to form aggregates averaging 28±1μm diam.

3. an amorphous silica with N₂-BET A_s ≡ 210m²/g, resuspended in water to form aggregates 18 to 23mm diam.

well, especially in the smaller size ranges. The near constancy of the slope of −1 in the log(A_s)-log(d) plot implies that particle shape does not change greatly with particle size. The apparent shape factor does increase in the largest size ranges suggesting either a less regular shape or accessible internal surface area unrelated to external particle size.

Other methods of measuring surface area yield different results. The method should be chosen to measure a surface area appropriate for each specific problem. Smaller or more polar probe molecules presumably penetrate smaller pores, cracks and re-entrant features than larger, less reactive molecules, so may yield higher estimates of surface area. In an early exploration of the apparent increase in shape factor with increasing size, Gross and Zimmerly (1928) measured the surface area of crushed quartz by two methods, the rate of dissolution in HF, and silver plating, on the assumption that the highly reactive HF would penetrate smaller pores and cracks of narrower aperture than would precipitated silver. Their results, illustrated in Figure 1, are consistent with the postulated internal area.

Many natural materials have high specific surface area as a result of complex shape. The silica in the cell walls of many diatoms occurs as 30-40 nm spherical aggregates of subunits as small as 12 nm (*Thalassiosira eccentrica* and *Navicula pelliculosa*; Schmid et al., 1981). If all of the cell wall were composed of these subunits, if they were spherical, and if all of the surface area of each were accessible, A_s should be about 227 m^2/g, in reasonable agreement with the BET area, 258 m^2/g observed for fresh, organic-free cell walls of *Thalassiosira decipiens* by Kamatani and Riley (1979).

THERMODYNAMICS

<u>Surface free energy</u>

Water spilled in a gravity-free environment does not flatten into sheets or break up into a fog of droplets; instead it spontaneously balls up, eventually taking spherical form. Crystals do not cleave spontaneously; cracks stop propagating and may heal when stress is relieved if the crack surfaces have not been contaminated in any way. In these two examples, if the experiment is performed reversibly, the only spontaneous change is a reduction of surface area. We must do work to increase surface area, so there must be a positive free energy of formation of surface, the surface free energy, γ. If G is the Gibbs free energy, γ is defined as

$$\gamma \equiv \left(\frac{\partial G}{\partial A}\right)_{P,T,n} . \tag{3}$$

The subscripts P, T, and n specify that pressure, temperature, and composition remain constant. The total free energy of a system thus comprises the sum of the molar free energies of formation of its constituents, i, $\sum_i \mu_i n_i$, plus the total excess free energy of formation of all surfaces and interfaces, j, $\sum_j \gamma_j A_j$. If a single anisotropic solid is present, then j identifies different crystal faces; if several solids or immiscible phases are present, j identifies all crystal faces of all solid-solid, solid-liquid, liquid-liquid interfaces.

<u>Gibbs' definition of surface excess properties: the dividing surface</u>

We need a definition of the extensive properties of surfaces. We will use composition as a vehicle to introduce the convention Gibbs adopted for this purpose (see Adamson, 1982). Studies of surface composition and of the properties of water near surfaces, for example, have demonstrated that the presence of a surface can perturb the properties of adjacent phases to a finite depth into each adjacent phase, i.e., surfaces have finite, but poorly defined, thickness. Adsorption, or segregation of minor constituents at the surface is common and an example will illustrate the point. Li and Kingery (1984) used

136

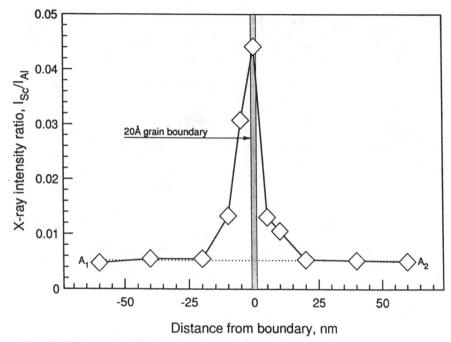

Figure 2. Relative concentration of scandium near a grain boundary in polycrystalline alumina containing 0.2 weight percent Sc_2O_3 as determined by scanning transmission electron microscopy (Li and Kingery, 1984). The x-ray intensity ratio, I_{Sc}/I_{Al}, was measured on a profile perpendicular to the grain boundary, which itself was parallel to the electron beam. The apparent grain boundary thickness was less than 20 Å. Electron beam diameter was less than 25 Å, but beam spreading within the sample reduces resolution. Correction for beam size and scattering would increase segregation in the grain boundary and reduce the width of the perturbed zone. No evidence of a second phase in the grain boundary near the analytical trace was observed, so the segregation probably represents true adsorption. The dotted line A_1-A_2 represents the bulk Sc concentration if it were not perturbed by the grain boundary.

Scanning Transmission Electron Microscopy (STEM) to detect segregation or sorption of several impurity elements at grain boundaries in polycrystalline Al_2O_3, with results summarized in Figure 2. In some cases this accumulation was caused by precipitation of a second phase in the grain boundary, but in others, there was convincing evidence that no second phase had formed, i.e. that the accumulation was a result of adsorption. In these cases, after correcting for lack of resolution caused by a finite electron beam-width (15-20 Å), they observed enrichments of Sc, for example, of as much as 40 times the bulk concentration. They found that the extent of grain boundary segregation increased as the difference in ionic size between the impurity ion and aluminum, increased, and assigned lattice misfit strain energy as the probable driving force for segregation.

To define a surface excess concentration or adsorption density rigorously, we must decide whether or not to recognize the finite thicknesses of surfaces. In view of the difficulty of defining surface thickness, Gibbs defined the surface for thermodynamic purposes as a mathematical plane or dividing surface of zero thickness near the physical surface, and surface properties as the net positive or negative excess in the vicinity of the surface over the magnitude of the same property in the bulk (Adamson, 1982). In Figure 2, for example, if the x-ray intensity profile were a true concentration profile, the adsorption density would be the net amount of Sc near the interface in excess over that which would be present if the bulk concentrations in each grain were constant up to the dividing surface, per unit area of interface. For a bar of material one square meter in cross sectional area, extending in length, x, from A_1 to A_2 to include the grain boundary, the adsorption density of Sc is:

$$\Gamma_{Sc} \equiv \int_{A_1}^{A_2}(c_x - c_b)dx \quad , \tag{4}$$

where c_x is the measured concentration of Sc at any point, x, in the concentration profile, and c_b is the bulk concentration of Sc, far from the interface, where it is not influenced by the presence of the grain boundary, and A_1 and A_2 are points at arbitrary positions within the bulk of the grains on each side of the boundary.

Other extensive properties of the surface are defined analogously.

Contribution of surfaces and interfaces to thermodynamic criteria of equilibrium

The Gibbs free energy of a system at equilibrium should not change if the system is perturbed slightly, i.e., dG = 0 at equilibrium. Processes or reactions for which dG is negative will proceed spontaneously in the direction of the proposed change. If dG is positive the process will proceed spontaneously in the direction opposite to that of the proposed change. Changes in surface area or energy, like other non-thermal forms of energy, can be related to dG by starting with the first law,

$$dE = \delta q + \Sigma \delta w \quad , \tag{5}$$

in which dE is the change in total energy of a system undergoing an infinitesimal change in state, δq is the heat absorbed by the system, and $\Sigma \delta w$ includes contributions of PV work, -PdV; surface work, γdA; and chemical work, μdn, done on the system or contributing to its energy content. If we limit ourselves to reversible processes and reactions, δq is related to the change in entropy accompanying the change in state,

$$\delta q = TdS \quad . \tag{6}$$

Finally, the Gibbs free energy and its total derivative are defined in terms of enthalpy, H, entropy, and energy, as follows:

$$G \equiv H - TdS = E + PV - TS \quad ; \tag{7}$$

$$dG = dE + PdV + VdP - TdS - SdT \quad . \tag{8}$$

Combining Equations 5, 6, and 8, after replacing dw with the three specific kinds of non-thermal work of most importance here, we obtain the useful relationship,

$$dG = -SdT + VdP + \sum_i \mu_i dn_i + \sum_j \gamma_j dA_j \quad , \tag{9}$$

for a system containing a variety of chemical species, i, and interfaces, j.

The Laplace (or Young-Laplace) Equation: Curved surfaces imply a pressure gradient

When an air bubble in water expands at constant temperature, the pressure inside the bubble must be larger than the pressure outside. As the bubble expands, the expanding air does work on the bubble wall. At the same time, the bubble wall does work on the water and the total amount of PV work done is $-(P_{in}-P_{out})dV$. As the bubble expands, however, its surface area increases, and the work required to do this is $\delta w_s = \gamma dA$. At equilibrium, if the process is conducted at constant overall pressure (both P_{in} and P_{out} held constant), the net work done to accomplish this small perturbation, must be zero, so

$$dG = \delta w = 0 = -(P_{in} - P_{out})dV_{in} + \gamma dA \quad ; \tag{10}$$

$$P_{in} - P_{out} = \gamma \frac{dA}{dV} \quad . \tag{11}$$

Equation 11 is the most general form of the LaPlace or Young-LaPlace equation. For spherical bubbles in a fluid, or for spherical drops of diameter, d, $dA/dV = 4/d$; for spherical soap bubbles, because there are two surfaces, $dA/dV = 8/d$; for non-spherical curved surfaces, if r_1 and r_2 are the principle radii of curvature, $dA/dV = (1/r_1 + 1/r_2)$ (Denbigh, 1971; Adamson, 1982).

Carmichael et al. (1974) used the LaPlace equation to estimate the effective pressure inside a strand of Pelèe's hair, i.e., a filament of molten basalt for which $r_1 \approx 10\mu m$ and $r_2 \approx \infty$, at 1200°C. If the surface free energy of the basalt at this temperature is about 350 mJ/m^2 (McBirney and Murase, 1971), and the external pressure is 1 bar, the internal pressure is 1.35 bars, perhaps enough to alter chemical equilibria slightly. The LaPlace equation also explains why the walls separating two bubbles in foams and low density pumices bulge toward the larger bubble and are flat only when the adjacent bubbles are of exactly the same size.

Tolman (1949; see also Adamson, 1982) suggested that the surface free energy itself should change with curvature, decreasing slightly when the radius of curvature approaches atomic dimensions. He proposed the relationship:

$$\gamma_d = \gamma \left(\frac{1}{1 + \dfrac{4\delta}{d}} \right) \quad , \tag{12}$$

in which γ_d is the surface or interfacial free energy appropriate for particle size d; γ is the surface free energy intrinsic to the material; and δ is a constant on the order of 1 Å. We will refer to this size dependence of γ as the Tolman effect.

The Kelvin effect: Equilibrium constants depend on A_s

The magnitudes of equilibrium constants change when significant changes in surface area accompany reaction. Evaporation of water from small droplets, for example, results in an increase in the specific surface area of the single liquid-vapor interface. Precipitation of gibbsite from aqueous solution increases the total amount of interfacial area at the {001}, {110}, and {100} crystal faces (Smith and Hem, 1972) and the alteration of biogenic silica involves loss of area in one interface and an increase in the area of at least one different interface as opal-A dissolves and opal-CT precipitates (Kastner et al., 1977). We can derive a general relationship between the equilibrium constants and the areas and interfacial free energies of interfaces by considering the free energy change accompanying reaction at constant pressure and temperature,

$$dG = \sum_i \mu_i dn_i + \sum_j \gamma_j dA_j \quad . \tag{13}$$

The reaction is written with the solid or liquid of most interest, identified with the letter and subscript ϕ, among the reactants. When dn_ϕ of ϕ has reacted, the amounts of each of the other species participating, dn_i, are related to dn_ϕ through the stoichiometric coefficients, f_i, which are positive for reaction products and negative for reactants, so that $dn_i = -f_i dn_\phi$, and

$$dG = -\sum_i f_i \mu_i dn_\phi + \sum_j \gamma_j dA_j \quad . \tag{14}$$

Because $\mu_i = \mu_i^\circ + RT ln\{i\}$, in which $\{i\}$ is the thermodynamic activity of i, Equation 14 can be expanded:

$$\frac{dG}{dn_\phi} = -\sum_i f_i \mu^\circ - \sum_i f_i RT ln\{i\} + \sum_j \gamma_j \frac{dA_j}{dn_\phi} \quad . \tag{15}$$

Because $\sum f_i \mu^\circ$ is the standard free energy change for the reaction, the first term can be replaced by $\Delta G^\circ = -RT lnK$, where K is the normal equilibrium constant, unperturbed by interfaces. The second term can be reformatted as $RT ln (\prod_i (i)^{f_i})$ or, since $\prod_i (i)^{f_i}$ has the form of an equilibrium constant, by $RT lnK^S$, where K^S is the equilibrium constant for the reaction as influenced by the interface. Making these substitutions and rearranging,

$$lnK^S = lnK + \frac{1}{RT} \sum_j \gamma_j \left(\frac{dA_j}{dn_\phi}\right) \quad . \tag{16}$$

This relationship between an equilibrium constant and changes in interfacial area that accompany reaction takes several named forms, as outlined in the following paragraphs.

The Kelvin equation and vapor pressures. For evaporation of a pure liquid, e.g.,

$$H_2O(l) = H_2O(g) \quad . \tag{17}$$

$K = [H_2O(g)]/[H_2O(l)] \cong P_0$, where P_0 is the saturation vapor pressure over a flat surface; and $K^S \cong P_S$, where P_S is the vapor pressure observed over small droplets. There is only one interface affected by the reaction, so, for spherical droplets, Equation 16 can be expressed in the form known as the Kelvin equation:

$$lnP_S = lnP_0 + \frac{\gamma_i}{RT} \left(\frac{4fw}{\rho d}\right) \quad . \tag{18}$$

The Kelvin equation requires that the equilibrium vapor pressure associated with small droplets be greater than the vapor pressure over a flat surface, where $dA/dn = 0$. The relative humidity in monodisperse water fogs, estimated with the Kelvin equation, using the 25°C surface tension of water (72.75 mJ/m^2; Adamson, 1982) is shown as a function of droplet size in Figure 3. Humidities observed in clouds and fogs can reach 110% (Hobbs and Deepak, 1981), implying an equilibrium droplet size of 0.02 μm. The minimum droplet sizes observed in significant numbers are closer to 0.5 μm (Jiusto, 1981), so, unless the size measurement is faulty, droplet size must be controlled by the rates of nucleation and growth rather than by equilibrium between the liquid and vapor. Because dA/dn is negative inside a cavity, the vapor pressure of water should be smaller than normal in a pore or crack, leading to capillary condensation. Pores in silica gels and zeolites have diameters of angstroms to tens of angstroms, thus the vapor pressure and activity of water in equilibrium with these materials should be much reduced. In a 10 Å pore, for example $P/P_0 = 0.12$ for water at 25°C.

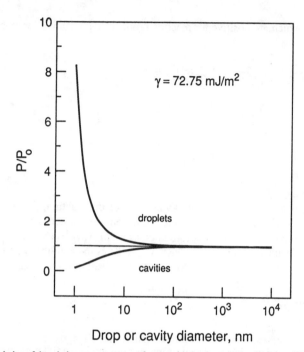

$\gamma = 72.75 \ mJ/m^2$

Drop or cavity diameter, nm

Figure 3. Variation of the relative vapor pressure of water with droplet and pore size as predicted by the Kelvin equation. Droplet sizes in fogs and clouds are larger than 500 nm (Hobbs and Deepak, 1981). Pore diameters in gels and zeolites can be smaller than 2 nm (Adamson, 1982). P is the vapor pressure in equilibrium with water in drops or lining cavities of any one size, and P_0 is the normal saturation vapor pressure over bulk water.

The Freundlich-Ostwald equation. When a mineral dissolves in water congruently, its surface area changes and, if dA/dn is large and positive, solubility is enhanced. Equation 16 applies, but because both solubility and dA/dn can be expressed in many different but equivalent ways, equations relating solubility to dA/dn or particle size take many forms and bear many names including Kelvin, Freundlich-Ostwald, and Gibbs-Thompson. Enüstün and Turkevich (1960), in a study of the solubility of fine particles of $SrSO_4$, found that the smallest particles in a polydisperse population determined solubility. Of course, high-area solids are more soluble than low area solids, and should dissolve and reprecipitate, or ripen, producing forms of lower A_S. Thus a polydisperse population cannot reach equilibrium with a solution. For this reason, measurements of the solubilities of high-surface area materials are not strictly equilibrium measurements and have meaning only if ripening is slow relative to the rate of dissolution. Another difficulty arises because we have assumed a single-valued γ when, in truth, several crystal faces are exposed during dissolution. These and other problems with the derivation and use of equations based on the Kelvin effect are discussed by Thompson (1987), Adamson (1982), and Enüstün and Turkevich (1960).

Dissolution of quartz and amorphous silicas provides a simple example. The process can be described by the reaction,

$$SiO_2(s) + H_2O(l) = H_4SiO_4(aq) \ , \tag{19}$$

in low salinity waters at pH < 9 where $H_4SiO_4(aq)$ is the predominant dissolved species. Ignoring non-ideality, the equilibrium constant and solubility, S, are identical for this reaction, i.e., $S \approx K \approx \{H_4SiO_4(aq)\} \approx \Sigma Si(dissolved)$. With these substitutions and expansion of dA/dn, the Freundlich-Ostwald equation can be expressed in terms of solibility and specific surface area as follows (Williams et al., 1985, after correction of a typographical error),

$$ln(S_s) = ln(S_n) + \frac{2\gamma fwA_s}{3RT} \quad ; \tag{20}$$

in which S_s is the surface-perturbed solubility, and S_n is the intrinsic solubility of a large particle or flat surface. For spheres of diameter, d, small enough that the Tolman effect (Eqn. 12) might come into play,

$$ln\left(\frac{S_s}{S_n}\right) = \frac{4\gamma fw}{\rho RT(d+4\delta)} = \frac{2\gamma fwA_s}{3RT}\left(\frac{A_{ref}}{(A_{ref} + A_s)}\right) \quad , \tag{21}$$

where δ and A_{ref} are constants with little, if any, physical meaning; $\delta \approx 1$ Å and A_{ref} is the specific surface area of spheres of diameter δ. These relationships allow estimation of S_n, and the solid-solution interfacial free energy, γ, from experimental measurements of equilibrium solubility as a function of specific surface area.

The Gibbs Equation: Adsorption reduces surface free energy

By allowing the composition of the surface to change, we can derive a relationship between adsorption density and surface free energy, the Gibbs Equation. So far, we have applied (Eqn. 9) to entire systems and bulk phases. As defined by Gibbs, surfaces and interfaces have free energy, heat content, entropy, and composition, just as do bulk phases; we will denote surface excess properties with the subscript s. For the surface alone, at constant temperature and pressure,

$$dG_s = \gamma dA + \sum_i \mu_{s,i} dn_{s,i} \quad . \tag{22}$$

At equilibrium, the chemical potential of each component is the same in every phase, including the "interphase" or interface, so a single μ_i represents species i throughout the entire system and we need not distinguish between surface and bulk chemical potentials.

By comparing this free energy change with the free energy change associated with the same differential change in state achieved by a process that allows γ and the μ_i to change as well as area and the n_i, we can derive a relationship between changes in adsorption density and changes in surface free energy. At any given temperature and pressure, the total free energy of formation or free energy content of the surface, G_s, is the sum of the normal contributions of each chemical component present in the surface region and the surface free energy. G_s and its total derivative are

$$G_s = \gamma A + \sum_i \mu_i n_{s,i} \quad , \tag{23}$$

$$dG_{s,tot} = \gamma dA + Ad\gamma + \sum_i \mu_i dn_{s,i} + \sum_i n_{s,i} d\mu_i \quad . \tag{24}$$

$dG_{s,tot}$ is the differential free energy associated with a change in state accomplished at constant P and T allowing γ, A, μ_i, and $n_{i,s}$ to change.

The Gibbs free energy is a state function, so the change in free energy associated with a particular process depends only on the difference between the initial and final states of the system, and not on the mechanism or route through which the change was achieved.

For this reason $dG_s = dG_{s,tot}$. Equating the two expressions for dG, defining $n_{s,i}/A$ as the Gibbs adsorption density, Γ_i, and rearranging, yields the classical Gibbs Equation:

$$d\gamma = -\sum_i \frac{n_{s,i}}{A} d\mu_i = -\sum_i \Gamma_i d\mu_i \quad . \tag{25}$$

The Gibbs adsorption equation requires that a decrease in surface free energy accompany positive adsorption of anything at the interface.

REACTIONS WITH WATER

The surfaces of many minerals, particularly oxides and silicates, hydroxylate upon exposure to water; i.e. they react with water, producing hydroxide functional groups. They also hydrate; molecular water adsorbs to the hydroxide groups, resulting in a multilayer three or more molecules thick that has properties different than bulk liquid water. The hydroxide groups behave as weak Brönsted acids or bases, capable of ionizing to yield a charged surface or of reacting with dissolved ions or molecules to produce surface coordination complexes in which the surface oxide or hydroxide serves as a ligand. We will review some of the evidence for this behavior and some of the consequences for adsorption and surface energy using oxide minerals, primarily quartz and amorphous silica, as examples.

Fracture surfaces

When quartz is fractured silicon-oxygen bonds are broken. The resulting surface contains "dangling bonds," some of which are unpaired electrons trapped on silicon atoms (Hochstrasser and Antonini, 1972). These surface sites are reactive and are destroyed quickly by adsorption of impurity molecules, even under ultra high vacuum conditions (Hochstrasser and Antonini, 1972; Carrière and Lang, 1977). The electric field gradient near a fresh fracture surface is likely to be large. Thermal annealing of silica glasses yields an anhydrous, reconstructed surface dominated by Si-OH (silanol) and/or Si-O-Si (siloxane) groups depending on sample history and annealing temperature (Snoeyenk and Weber, 1972; Hair, 1967). Yates and Sheppard (1957), using infrared (IR) spectroscopy, found that methane adsorbed on a silica glass outgassed at 300°C is distorted to an extent suggesting a local electrical field gradient $\nabla\Psi \approx 7 \times 10^4$ V/m.

Hydroxylation

In 1929 Gaudin (1929) observed that "it appears established, however, that H^+ and OH^- ions react with the mineral surfaces in some instances, and that they have much to do with adherence of {ionic surfactants} on mineral surfaces in other instances". Later, utilizing the crystal structure of quartz (Bragg and Gibbs, 1925; Gibbs, 1926) and bonding ideas of Pauling (1927), Gaudin and Rizo-Patròn (1942) offered an explanation, suggesting that water would dissociate upon reaction with fracture surfaces, the OH^- bonding to unsatisfied Si and the H^+ to unsatisfied O, producing ionizable surface hydroxide groups. We will refer to these functional groups generically as $\underline{S}OH$ sites, the "S" standing for a cation in the solid structure and the underscoring signifying a surface site. The $\underline{S}OH$ sites in SiO_2 and Fe_2O_3, for example, would be $\underline{Si}OH$ and $\underline{Fe}OH$, respectively.

High field gradients enhance dissociation of electrolytes. The high field gradient associated with surfaces may enhance dissociation of water and other weak acids and bases. In aqueous solutions, this is known as the Wien effect or the dissociation field effect (Harned and Owen, 1958; Moore, 1972). If the Wien effect is applicable to water molecules at the solid-vacuum interface, then the dissociation constant of water, K_w, on the silica surface should be reduced by many orders of magnitude relative to that of bulk water. In agreement with Gaudin, we would expect water to dissociate when it adsorbs onto silica surfaces, the H^+ and OH^- reacting with dangling bonds to produce bound hydroxide surface sites.

Infrared spectrophotometry provides a means of verifying hydroxylation and characterizing the surface in more detail. The stretching fundamentals of hydroxide groups in many molecules, including water, fall in the 2.5 to 3 μm region (MacDonald, 1958; Hair, 1967). Amorphous silicas dried and evacuated after prior equilibration with water have a single sharp band at about 3750 cm^{-1}; under the same conditions, quartz has two very sharp bands at 3627 and 3649 cm^{-1} (Gallei and Parks, 1972). The sharp band at 3750 cm^{-1} (or 3627 and 3649cm^{-1} for quartz) is attributed to isolated hydroxide groups bonded to silicon, i.e. $\underline{Si}OH$ or silanol groups, in recognition of the following observations: The $\underline{Si}OH$ groups are isolated, i.e. not hydrogen bonded to other species, because their frequencies are very close to those of the OH groups in isolated water molecules; they are on the surface because they change intensity rapidly in response to changes in humidity and because they change intensity and frequency rapidly upon exposure to D_2O (MacDonald, 1958). Anderson and Wickersheim (1964) show that heating to about 120°C removes all molecular water, leaving a surface dominated by silanol sites. Other work, summarized by Hair 1967) shows that heating to temperatures between 170 and 400°C causes reversible, partial dehydroxylation. Total dehydroxylation requires temperatures as high as 800°C. The dehydroxylated surface of quartz is hydrophobic and rehydroxylation is slow after such drastic treatment (Pashley and Kitchener, 1979). In contrast to quartz, hematite, γ-Al_2O_3, and ThO_2 dehydroxylate upon outgassing at temperatures as low as 100°C and the process is reversible, i.e., they rehydroxylate readily (McCafferty and Zettlemoyer, 1971).

The densities of $\underline{S}OH$ sites can be determined in several ways, including $^3H/H$ exchange, reaction with organic acids or bases in gas-solid systems, and acid-base titrations in aqueous suspensions (James and Parks, 1982). Measured site densities, in sites/nm^2, range from 3.5 (titration) to 11.4 (isotopic exchange) for quartz and from <1 to 22 for all oxides and all methods. These differences arise, at least in part, because some methods of estimation measure the acidity of the site relative to a selected reactant molecule either in the gas phase or in liquid water, while the isotopic exchange methods measure all protons with exchange rates rapid enough to respond in the time allowed. The method should be chosen to detect sites of the type expected to participate in the process or reaction of interest, but we know so little about the characteristics of functional groups on surfaces immersed in liquid water, that this is rarely possible; we need independent, perhaps spectroscopic, means of defining which sites respond to each method of measurement.

Hydration

Additional water adsorbs, first by hydrogen bonding to one or more silanol groups as individual molecules, then, as water vapor pressure is increased, by hydrogen bonding to other adsorbed water molecules, forming clusters and, at high humidities, continuous networks. Anderson and Wickersheim (1964) used overtones and combination bands involving the stretching and bending vibrations of OH in silanol groups and water molecules to investigate this process. The first overtone of the $\underline{Si}OH$ stretching frequency appears at 7326 cm^{-1}. Upon adsorption of small amounts of water the intensity of the 7326 band decreases and new bands appear. One of these is a new OH combination band involving an OH stretching frequency about 200 cm^{-1} lower than the silanol frequency. At the same time, several new bands identified with molecular water appear. These simultaneous changes suggest that the new bands correspond to individual water molecules hydrogen bonded to silanol groups. The binding energy estimated from the shift in silanol stretching frequency is 8 to 12 kJ/mole, a little smaller than the hydrogen bond energies in ice and water (Anderson, 1965). At higher humidities, more water adsorbs, some of it bonding to pre-adsorbed water molecules to form clusters of hydrogen bonded water, even before all of the isolated silanol sites are occupied (Anderson and Wickersheim, 1964). Apparently it is not possible to produce a true monolayer of adsorbed water molecules. A statistical monolayer incorporates enough water to cover the surface, but clustering results in parts of the surface remaining free of molecular water, while multilayer islands occupy other areas, in the classical BET multilayer adsorption mode (Adamson, 1982). Klier et al. (1973) came to very similar conclusions after a combined adsorption and spectroscopic study of partially hydrophobic silicas.

The properties of adsorbed water are different from those of bulk water, but the difference decreases as adsorption density or surface coverage increases. Many investigators have measured the adsorption density of water on quartz and silicas over wide ranges of relative humidity. These measurements require silicas of high A_s, and there is some risk of capillary condensation in the interstitial spaces in a powder at high humidity. Miyata (1968) measured Γ_{H_2O} at humidities up to 98% and corrected for capillary condensation to obtain the surface coverage data shown in Figure 4. He observed an adsorption density of about 170 μmoles/m^2 or 26 monolayers at $P/P_0 \approx 0.98$, for a typical sample of quartz. Pashley and Kitchener (1979), using ellipsometry, found a 150 nm film of water adsorbed on clean quartz at saturation ($P = P_0$). This film thickness corresponds to approximately 50 monolayers, and is a reasonable extrapolation from Miyata's result. Although this is a thick film, the first monolayer adsorbs at much lower fugacity of water (P/P_0) than the rest (Fig. 2), suggesting different bonding at low coverage. Fripiat et al. (1982) have observed that the heat of immersion of silica saturated with adsorbed water near P_0 approaches the surface energy of pure bulk water but is greater when statistical surface coverage drops below about three monolayers. They also cite NMR results suggesting a difference in microdynamic properties that vanishes when surface coverage exceeds three monolayers. Zettlemoyer et al. (1975) summarize evidence that the dielectric constant of water adsorbed on silica is very low in the first monolayer and increases very rapidly in the 2nd and 3d monolayers. These effects are not limited to silica; Cases and François (1982) used microcalorimetry to show that the surface field of kaolinite influences the properties of adsorbed water to a depth of about three statistical monolayers. Derjaguin and Churaev (1986) point out that water adsorbed in the interlayer region in swelling clays is highly structured and has a dielectric constant ranging from 3 to 4 when the thickness of the water layer is 5 to 6 Å and from 25 to 40 when the film thickness is 15 to 80 Å.

Water adsorbed on hematite behaves in much the same way as we have described for silica (McCafferty and Zettlemoyer, 1971). Water in the first monolayer is hydrogen bonded to two adjacent FeO sites, is immobile in comparison with liquid water, and has a very low dielectric constant, independent of adsorption density in the sub-monolayer range. The next few layers apparently develop a structure with a higher density of hydrogen bonding than liquid water, perhaps similar to ice. In this region, the dielectric constant depends strongly on adsorption density and rises from below 5 at one monolayer to about 30 at two monolayers.

Fripiat et al. (1965) and Anderson and Parks (1968) measured the electrical conductivity of silica gel as a function of water adsorption density and found that conductivity increases dramatically as adsorption density increases. The exponential dependence of conductivity on Γ_{H_2O}, and the polarization behavior, non-ohmic conductivity, and reduction in conductivity upon substitution of D_2O for H_2O all suggest that protons or hydronium ions are the dominant charge carriers when Γ_{H_2O} is high enough to provide water for solvation. Fripiat et al. (1982) argued that the H^+ are provided by dissociation of adsorbed water molecules; their conductivity data suggest approximately one percent dissociation, corresponding to $K_w \approx 10^{-4}$ ($K_w = 10^{-14}$ for H_2O(liq)) in agreement with the enhanced dissociation predicted by the Wien effect. Roco (1966), Ahmed (1967), and Parks(1968) found that positive charge transferred from silica to Al_2O_3 and MgO upon solid-solid contact in very low humidity environments where adsorbed molecular water is absent or scarce, which suggests that the are silanol groups are proton

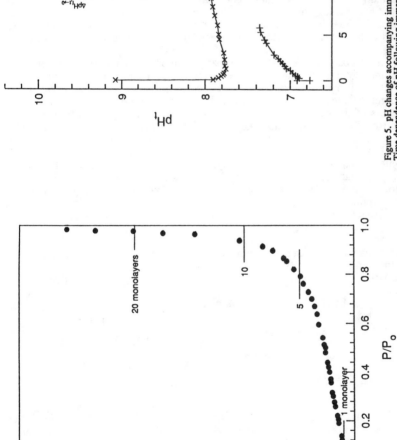

Figure 4. Adsorption of water vapor on quartz and the reduction in surface free energy or free energy of adsorption as estimated with the Gibbs equation (after Miyata, 1968). P is the water vapor pressure in equilibrium with the adsorption sample and P_0 is the normal saturation vapor pressure over bulk water.

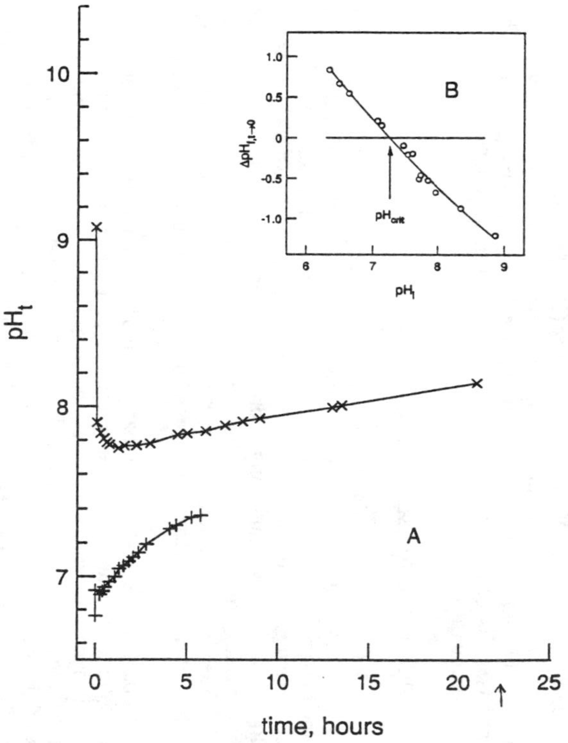

Figure 5. pH changes accompanying immersion of α-Al_2O_3 (data from Korpi, 1965). A: Time dependence of pH following immersion in water initially adjusted to pH_i. At pH_i = 9.1, pH drops rapidly to ≈7.8 presumably due to dissociation of H^+ from $AlOH$ surface sites on the solid, then rises slowly, presumably due to dissolution of the solid. Extrapolation of the slow change to zero time provides an estimate of the change in pH caused by surface reactions alone, $\Delta pH_{f,t\to 0}$. Immersion at pH_i = 6.7 results in a rise in pH; $\Delta pH_{f,t\to 0}$ is negative for all $pH_i > pH_{crit}$ and positive for all $pH_i < pH_{crit}$.

donors and that $\underline{Al}OH$ and $\underline{Mg}OH$ groups are proton acceptors (or OH^- donors). The relative acidities of the $\underline{M}OH$ sites on the three solids are qualitatively consistent with acidities of their bulk hydroxides. Anderson and Parks (1968) assumed dissociation of both silanol groups and adsorbed water in explanation of their conductivity observations. We can conclude from these observations that surface hydroxide groups behave as Brönsted acids or bases, dissociating to yield protons or hydroxide ions that are mobile in the oxide-water vapor interface.

Immersion, surface ionization, and electrified interfaces

There is a thick film of water adsorbed on clean oxide surfaces in equilibrium with water vapor at vapor pressures near saturation. Because the properties of water in the outermost adsorbed layers are indistinguishable from those of liquid water, the only free energy change accompanying total immersion of such a surface is a loss of surface area; the process is analogous to simply immersing a drop of water in a larger volume. The hydrogen (or hydroxide) ions that are mobile in two dimensions on oxide surfaces, however, should diffuse in three dimensions after immersion. If they do, then the surface should be left with a net electrical charge and the pH of the solution phase should change.

Numerous observations have shown that pH almost invariably does change when solids are immersed in liquid water. As seen in Figure 5, immersion of Al_2O_3 results in a very rapid change, much too rapid to be ascribed to dissolution, followed by a much slower change. Whereas the slow step probably includes contributions from dissolution and hydrolysis of dissolved species, and possibly diffusion of protons into the solid (Berube and de Bruyn, 1968; Casey et al., 1988)), the fast step can be attributed to adsorption or desorption. The direction and magnitude of the rapid change depend on the initial pH (pH_i) of the solution, and there is a critical pH (pH_{crit}), characteristic of each solid, at which no change occurs. Apparently, the solid is an H^+ donor or OH^- acceptor when $pH_i > pH_{crit}$, and an H^+ acceptor or OH^- donor when $pH_i < pH_{crit}$. If the surface is an H^+ donor, then either H^+ desorbs or OH^- adsorbs. We cannot distinguish between negative adsorption of hydrogen ion and positive adsorption of hydroxide, but, in any case, $\Gamma_H - \Gamma_{OH} < 0$ when $pH_i > pH_{crit}$ and $\Gamma_H - \Gamma_{OH} > 0$ when $pH_i < pH_{crit}$.

These changes are commonly interpreted in terms of the ionization of $\underline{S}OH$ functional groups in the surface (Parks, 1965; Sposito, 1984; Schindler and Stumm, 1987), thus for an Al_2O_3 surface, H^+ adsorbs by Reaction (26a) and desorbs by Reaction (27):

$$\underline{Al}OH + H^+ = \underline{Al}OH_2^+: \tag{26a}$$

$$\underline{Al}OH_2^+ = \underline{Al}OH + H^+ : \qquad K_{a1} = \frac{\{\underline{Al}OH\}\{H^+\}}{\{\underline{Al}OH_2^+\}} \tag{26b}$$

$$\underline{Al}OH = \underline{Al}O^- + H^+: \qquad K_{a2} = \frac{\{\underline{Al}O^-\}\{H^+\}}{\{\underline{Al}OH\}} \quad , \tag{27}$$

in which $\{i\}$ denotes the thermodynamic activity of species i. The activity coefficients for charged species adsorbed on charged surfaces should include the electrical potential energy of the ions in the surface field (Davis et al., 1978), but in most of the discussion to follow we will ignore the difference between activity and concentration. Surface complexation models utilize Reactions 26b and 27. An analogy between these reactions and dissociation reactions in solution has been drawn by Schindler and Stumm (1987), who find a rough correlation between surface acidity constants, K_{a1}, and the analogous aqueous acidity constants.

In terms of surface species, $\Gamma_H - \Gamma_{OH} = [\underline{Al}OH_2^+] - [\underline{Al}O^-]$. At pH_{crit}, $\Gamma_H - \Gamma_{OH}$ goes to zero, $[\underline{Al}OH_2^+] = [\underline{Al}O^-]$, and $pH_{crit} \approx -\log(K_{a1}K_{a2})/2$. $\Gamma_H - \Gamma_{OH}$ can be computed from the pH change associated with immersion or from acid-base titration of suspensions, corrected for dissolution and hydrolysis. Typical data are shown in Figure 6A.

Surface charge If $\Gamma_H - \Gamma_{OH} \neq 0$, then, unless something other than H^+ and OH^- adsorbs, the surface should have a net electrical charge,

$$\sigma_H = e(\Gamma_H - \Gamma_{OH}) \ , \tag{28}$$

in which e is the absolute value of the charge on an electron. The quantity σ_H, is the surface charge density from this source; because the proton is so small, and is assumed bonded directly to oxygen ions that are part of the structure of the solid surface, σ_H is considered to reside in the plane of the surface. The term σ_0 will be used to denote charge residing in the surface itself, regardless of origin. If there are no sources of surface charge other than adsorption of H^+ and OH^-, then $\sigma_0 = \sigma_H = 0$ when $\Gamma_H - \Gamma_{OH} = 0$, and pH_{crit} corresponds to zero surface charge. For this reason, pH_{crit} is called the Point of Zero Net Proton Charge, PZNPC (Sposito, 1984) or simply the point of zero charge, PZC or pH_{pzc} (Parks, 1975). The PZNPC plays an important role in the adsorption of weakly adsorbed ions, as we shall see.

The reality of surface charge and its pH dependence can be demonstrated by observing that suspended particles migrate in an imposed electrical field (electrophoresis) or that fluid flow in a packed column of particles generates an electrical potential gradient in the direction of flow (the streaming potential, Ψ_{sp}). Typical data are presented in Figures 6B and 7.

Origins of charge. Oxides and silicates apparently develop surface charge and potential through ionization of $\underline{S}OH$ functional groups, as we have described. The PZNPCs of simple oxides are roughly proportional to the ionic potential (z/r) of the cation (Parks, 1965). The PZNPCs of silicates and spinels are often approximately the average of the PZNPCs of the constituent oxides, weighted to reflect the bulk composition Parks (1967). Substitution of M^{3+} for Si^{4+} on tetrahedral sites, or of M^{2+} for Al^{3+} on octahedral sites, in clays leads to a net charge on the structural framework which imparts cation exchange capacity and can be interpreted as surface charge. The dangling bonds on the edges of clay mineral particles can hydroxylate, again producing ionizable surface functional groups. Parks (1967, 1975) and Yoon et al. (1979) have reviewed compositional and structural factors that affect the PZNPC.

The origins of surface charge are less clear for other solids. Silver sulfide (Freyberger, 1957, Iwasaki, 1957) develops surface charge and potential reversibly by adsorption of Ag^+ and/or S^{2-} ions, i.e. $\sigma_0 = \Gamma_{Ag^+} - \Gamma_{S^{2-}}$. A PZC (not a PZNPC, because surface charge is apparently not determined by adsorption of H^+) occurs at $pAg \approx 10$. We might expect water to react with the surface, producing $\underline{Ag}SH$ or $\underline{Ag}OH$ functional groups in analogy with the reactions of oxide surfaces, but surface charge is independent of {HS$^-$} and pH for $4.7 < pH < 9.2$, suggesting that hydroxylation is not important. ZnS (Moignard et al., 1977), on the other hand, has a pH-sensitive surface charge and an IEP in the range $4 < pH < 8$, depending on details of sample preparation. Adsorption of Cd^{2+} and H^+ are apparently both involved in determining surface charge on CdS (Park and Huang, 1987). Though expected for the same reasons given for oxides, spectroscopic evidence for $\underline{M}SH$ or $\underline{M}OH$ sites on sulfides is apparently lacking; H^+ may be involved only indirectly, through control of the activity of S^{2-} and HS^-.

Ionic solids such as barite, calcite, apatites, fluorite, and scheelite (Honig and Hengst, 1969; Healy and Fuerstenau, 1972; Parks, 1975) exhibit electrokinetic behavior

148

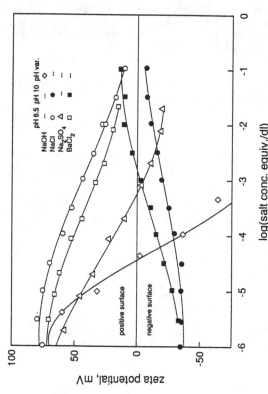

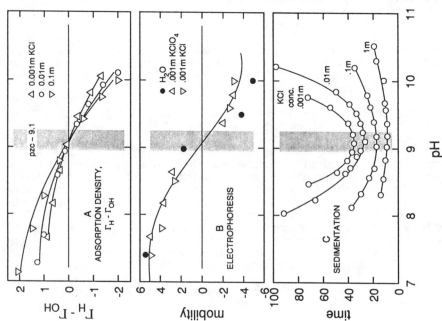

Figure 6 (left). Surface chemistry of α-Al_2O_3 in aqueous KCl solutions – H^+ adsorption density, electrophoretic mobility (μ), and settling of suspensions (data from Yopps and Fuerstenau, 1964). Plots show correct relative trends, but units other than pH are arbitrary. (A) $\Gamma_H - \Gamma_{OH}$ computed from acid/base titration. B. Electrical surface charge as reflected in μ; positive μ corresponds to positive particle charge. Surface charge, as measured by either μ or $\Gamma_H - \Gamma_{OH}$, is zero at pH 9.1 ± 0.15. This pH is both PZNPC and IEP. (C) Setting or sedimentation of suspensions of μm-sized alumina. "Time" measures relative time required for a standard suspension to clarify by sedimen-tation. Data for different KCl concentrations have been displaced vertically for clarity in plotting; all systems have about the same settling time when pH = IEP. Short settling times correspond to flocculated suspensions. $\Gamma_H - \Gamma_{OH}$ increase with increasing KCl concentration or ionic strength for all pH ≠ IEP, while surface charge decreases. All are insensitive to ionic strength when pH = IEP.

Figure 7 (above). Effects of electrolytes on the zeta potential, Ψ_Z, of corundum (Modi and Fuerstenau, 1957). Ψ_Z is derived from empirical measurements of streaming potential. Counterions are anions when the surface is positive, as indicated by positive Ψ_Z. Notice differences among OH^-, which causes charge reversal at low concentration, SO_4^{2-}, which causes charge reversal at higher concentration, and Cl^-, which does not cause charge reversal, even at high concentration. Chloride adsorbs in the diffuse layer, sulfate apparently forms a weak outer-sphere complex with surface sites, and H^+ and OH^- are "potential determining" in the sense that they are constituents of the hydroxylated surface and their influence on surface potential might obey the Nernst equation if a reversible alumina electrode could be made (Parks, 1975).

suggesting that surface charge develops through non-stoichiometric dissolution of the solid. At constant pH, the constituent ions of the solid play the charge-determining role that H^+ and OH^- play for oxides. Surface charge on some of these solids (e.g., calcite and apatite), at least, also is pH-dependent. This pH dependence may arise through control of the activities of charge-determining ions in the aqueous phase, through hydrolysis, but also might reflect reaction with water to form ionizable surface functional groups such as $\underline{Ca}CO_3H$, $\underline{Ca}OH$, and $\underline{Ca}PO_4H_x$. Again, spectroscopic evidence of the presence or absence of such groups is apparently lacking.

Electrified interfaces. We saw in Figure 6A that the absolute value of surface charge, measured as $\sigma_H = e(\Gamma_H - \Gamma_{OH})$, increases as the difference |pH – PZNPC| increases, as expected if ionization of surface hydroxide sites is the source of charge. Increasing ionic strength with a simple, non-reactive salt such as $NaNO_3$, however, reduces the charge and potential observed electrokinetically. This screening effect suggests that cations accumulate on or near negatively charged surfaces, and anions near positively charged surfaces, to minimize electrical potential energy and preserve electrical neutrality. These counterions must be reasonably labile, however, because if they were all bonded to the surface in non-labile complexes, then the charge would be zero, in contradiction to experiment. Apparently, the counterions accumulate near the surface in a manner allowing them to be partially stripped away, or the whole array polarized, by an external field or fluid motion. Guoy and Chapman (see Adamson, 1982) suggested that the screening ions were weakly adsorbed, attracted to the surface coulombically, in opposition to the randomizing influence of thermal kinetic energy, leading to a diffuse array of counterions, most concentrated close to the surface, and exponentially more dilute with increasing distance from the surface.

This model—a charged surface (at a distance $x = 0$) and a diffuse layer of counterions near the surface on the solution side of the interface $(x \geq 0)$—was originally proposed and expressed in mathematical terms by Guoy and Chapman in 1910-17 (see Adamson, 1982). The original model has been expanded and refined to recognize the finite sizes of counterions, and to distinguish among the locations of adsorbed ions on the basis of hydration and bond type and the variation of electrical properties of water with proximity to the surface, etc. Bockris and Reddy (1970) outline the history of these changes from an electrochemical point of view and Davis and Kent (this volume), Sposito (1984; this volume) and Schindler and Stumm (1987; this volume) compare models derived for use with oxides and other surfaces whose behavior is dominated by ionizable surface functional groups—surface complexation models. Shaw (1970) presents a particularly clear derivation of the equations governing the structure of electrified interfaces. All models of electrified interfaces assume two or more layers of charge, including, at least, bound surface charge and a diffuse layer of counterions. We will refer to any and all generically as Electrical Double Layer (EDL) models.

The EDL models provide an explanation for the effects of added electrolytes. When a particle moves in an electrical field or when fluid flows past a particle, an immobile hydrodynamic boundary layer of fluid moves with the particle, but most of the fluid slips past it. If the diffuse layer is thicker than the boundary layer, then the outermost counterions will be stripped away, resulting in charge separation at the "slipping plane". The particle is left with a net charge and a potential gradient is generated in the direction of relative motion. A rough estimate of the net charge within the boundary layer, σ_{ek}, comprising σ_H plus the fraction of counterion charge immobilized within the boundary layer, can be calculated from electrophoretic mobility measurements (Adamson, 1982). The potential is observable as the streaming potential in another family of electrokinetic experiments and can be related to the potential at the slipping plane or the outer limit of the hydrodynamic boundary layer, the zeta potential, ζ or Ψ_z (Adamson, 1982). The quantities σ_{ek} and Ψ_z change with changing pH, ionic strength, and the adsorption densities of solutes (Figs. 6B and 7). The surface is said to be isoelectric when electrokinetic charge and potential fall to zero, and the pH (for example) at which the surface is isoelectric can be designated an isoelectric point, IEP. The IEP and PZNPC are analogous in concept, but not necessarily quantitatively equal; the PZNPC corresponds to zero net protonic surface charge, $\sigma_H = 0$, whereas the IEP corresponds to zero σ_{ek}.

The stabilities of suspensions offer additional evidence of surface charge (Fig. 6C). As a consequence of the separation of bound surface charge from the diffuse layer, suspended particles of the same material repel because the like-charged diffuse layers interact at greater distance than do the physical surfaces of the particles themselves. At the IEP or at any pH at high ionic strength, the diffuse layer collapses and particles in suspension can approach closely, permitting van der Waals bonding between particles, resulting in agglomeration or flocculation followed by rapid sedimentation. At low ionic strength, the thick diffuse layer acts as a bumper, keeping the particles too far apart to permit agglomeration. For this reason, suspensions are more stable at low ionic strength, and less stable at high ionic strength, as illustrated in Figure 6C. The high ionic strengths of estuarine and marine waters agglomerate suspended sediments introduced by rivers, enhancing sedimentation (Stumm and Morgan, 1981).

SORPTION: REACTIONS WITH AQUEOUS SOLUTES

Solutes in natural waters are often taken up by the solids they contact (e.g., the minerals in an aquifer or the suspended solids in surface waters) without alteration of the solid. If the specific reaction responsible for uptake has not been identified, the process is called "sorption". If the sorbing species is an ion, it can be called an adion. Sorption of the adion can occur through true adsorption or surface complexation, through absorption or diffusion into the solid, or through surface precipitation (to form a single, distinct, adherent phase) or coprecipitation (to form an adherent mixed precipitate or solid solution with a second sorbate). In this section we will describe the influence of some key variables on sorption density, emphasizing characteristics that have proven helpful in identifying the uptake reaction or surface complex.

pH and ionic strength dependence.

Cations and anions adsorb with opposite pH dependence, each modified by ionic strength, hydrolysis, complex formation, and the ratio of total adsorbate present to total adsorbent surface area. These empirical modes of behavior are illustrated in Figures 8 and 9. In general, sorption of cations is weak at low pH and stronger at high pH; sorption of anions is weak at high pH and stronger at low pH. The pH at which 50% of the total adsorbate present is sorbed can be called an adsorption edge or pH_{ads}. Judging from the data in these figures and other information reviewed by Parks (1965, 1967, 1975), we can identify two extremes of behavior. For one class of adsorbate ions,

1. Adions sorb on surfaces of opposite sign; sorption is low at the PZNC and very low or negative when the charge of surface and adion are the same (Fig. 8).
2. The PZNPC and IEP are approximately equal to each other (for α-Al_2O_3 in $NaNO_3$ solutions, Figs. 6A,B).
3. σ_{ek} is reduced, but not reversed as ionic strength or the concentrations of counterions increases (Na^+ and Cl^- on α-Al_2O_3, Fig. 7).
4. Increasing ionic strength decreases sorption densities (Mg^{2+} on goethite, Fig. 8; SeO_4^{2-} on goethite, Fig. 9).

Another class of adsorbates behaves quite differently,

1. Sorption is insensitive to surface charge; adions may adsorb on surfaces of like charge or zero charge (Fig. 9).
2. The PZNPC and IEP, measured in the presence of the adsorbate, are different.

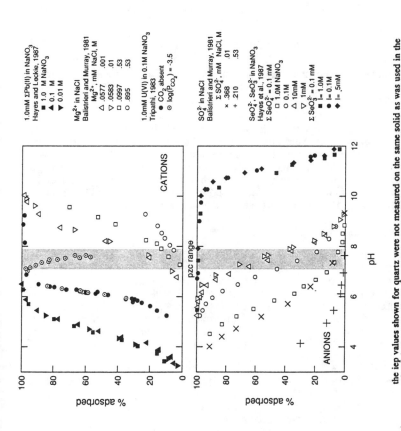

1.0mM ΣPb(II) in NaNO₃
Hayes and Leckie, 1987
- ■ 1.0 M NaNO₃
- ▲ 0.1 M
- ▼ 0.01 M

Mg²⁺ in NaCl
Balistrieri and Murray, 1981

	Mg²⁺, mM	NaCl, M
△	.0577	.001
▽	.0583	.01
□	.0997	.53
○	.895	.53

1.0mM U(VI) in 0.1M NaNO₃
Tripathi, 1983
- ● CO_2 absent
- ⊙ $\log(P_{CO_2}) = -3.5$

SO₄²⁻ in NaCl
Balistrieri and Murray, 1981

	Σ SO₄²⁻, mM	NaCl, M
×	.368	.01
+	.210	.53

SeO₄²⁻, SeO₃²⁻ in NaNO₃
Hayes et al., 1987
Σ SeO₄²⁻ = 0.1 mM
- □ 1.0M NaNO₃
- ○ 0.1M
- △ 10mM
- ▽ 1mM

Σ SeO₃²⁻ = 0.1 mM
- ■ I = 1.0M
- ● I = 0.1M
- ◆ I = .5mM

Figure 9 (right). Contrasting styles in pH and ionic strength dependence of adsorption. This figure is discussed in detail in the text.

the iep values shown for quartz were not measured on the same solid as was used in the adsorption measurements. (2) The IEP occurs at very low pH, where the high ionic strength of the acidic medium makes electrokinetic measurements difficult. (3) Absolute charge densities on quartz are very low because K_{a1} and K_{a2} are unusually far apart, resulting in a wide pH region of low charge, thus of low adsorption.

Figure 8 (left). Adsorption of indifferent electrolytes. Systems and experimental conditions are identified on the figure. The PZNPCs were measured by potentiometric acid/base titration; IEPs were measured by various authors (see Parks, 1965) by electrokinetic methods. Notice the approximately symmetrical behavior of anions and cations and the close correspondence between pzc and the onset of adsorption or the intersection of anion and cations curves. This correspondence is less pronounced for quartz for three reasons: (1) We expect PZNPC and IEP to coincide approximately for indifferent electrolytes, but

3. σ_{ek} can be reversed as the concentrations of adions increases (Ba^{2+} and SO_4^{2-} on γ-Al_2O_3, Fig. 7).

4. Increasing the ionic strength, or the concentration of an indifferent background electrolyte, has little effect on sorption densities of the adion (Pb^{2+}, UO_2^{2+}, and SeO_3^{2-}, Fig. 9).

Solutes that adsorb with the first set of characteristics are said to adsorb non-specifically, because the charge of the surface appears to be the principal factor controlling the process and sorption is relatively non-selective, i.e. the identity of the adion is relatively unimportant. Adsorption exhibiting the second set of characteristics is said to be specific or chemical because a specific or chemical bond with the surface is implied by the fact that positive adsorption occurs even when the adion and surface have the same charge so that adsorption densities would be negative if the attraction were simply coulombic.

Non-specific or physical adsorption

In systems with non-specific characteristics, negligible sorption at the PZNPC and coincidence of the PZNPC and IEP suggests that adsorption occurs predominantly through coulombic attraction to the oppositely charged surface and that the adions are located largely in the diffuse layer, forming, at best, weak outer-sphere complexes in which the adions' inner solvation sheath (or that of the surface, or both) remains intact (Sposito, 1984). Non-specific adsorption can be called physical adsorption in the sense that no true chemical bond is expected between the adsorbate and the surface. If a charged adion were bonded directly to the surface at the PZNPC, then σ_{ek} will be non-zero, and the IEP and PZNPC would differ.

Selenate ion adsorbs on goethite with nearly non-specific characteristics. X-ray absorption spectroscopy (XAS) has been used to demonstrate that the adsorbed ion is indistinguishable from the aqueous selenate ion and has no detectable iron atoms among second nearest neighbors, thus probably forms an outer-sphere surface complex (Hayes et al., 1987). Ions that adsorb in this non-specific mode tend to be those that do not hydrolyze significantly, and do not form strong complexes in solution with any constituent of the solid, e.g., Na^+, K^+, NO_3^-, and ClO_4^-.

Concentration dependence. Counterions accumulate in the diffuse layer near surfaces of opposite charge or potential. The distribution of ions in the diffuse layer is described by the Boltzmann equation,

$$c_{i,x} = c_{i,\infty} \exp\left(\frac{-z_i F \Psi_x}{RT}\right) \quad , \tag{29}$$

in which $c_{i,x}$ is the concentration of ionic species i at a distance x from the surface; Ψ_x is the electrical potential (including algebraic sign) at x relative to the bulk solution; $c_{i,\infty}$ is the concentration of i in the bulk solution; z_i is the ionic charge, including sign; F, R, and T are the faraday, the gas constant, and absolute temperature. Of course x cannot go to zero because ions have finite sizes.

Counterions in the diffuse layer are adsorbed in the sense that they represent a surface excess concentration which, integrating Equation 29, amounts to:

$$\Gamma_i = \int_{x=\infty}^{\delta} (c_{i,x} - c_{i,\infty}) dx = \int_{x=\infty}^{\delta} (c_{i,\infty})[\exp\left(\frac{-z_i F \Psi_x}{RT}\right) - 1]) dx \quad . \tag{30}$$

Cations should adsorb when the pH > PZNPC and the surface is negative; anions should adsorb when pH < PZNPC and σ_0 and Ψ_x are positive; neither should adsorb when pH = PZNPC where the surface is uncharged. The data in Figure 8 show this to be approximately true for ions whose solution chemistry suggests are unlikely to bond strongly to surface sites. For quartz, the minimum adsorption densities occur over a wide pH range, suggesting that pK_{a1} and pK_{a2} differ more than they do for anatase or goethite, and therefore that the surface is dominated by neutral $\underline{S}OH$ sites for pH < pK_{a2}. Li and de Bruyn (1966) have shown that the Gouy-Chapman EDL model describes the adsorption density of Na^+ well if experimental zeta potentials are used to approximate the potential in plane of closest approach of counterions. Equations 29 and 30 also require negative adsorption of indifferent co-ions, ions that have the same charge as the surface. Negative adsorption of anions by clays has been observed (see Sposito, 1984). In compact sediments where the diffuse layers of adjacent particles overlap, this results in effective exclusion of anions from pore waters and limited, somewhat selective permeability or membrane filtration in sedimentary rocks (see Drever, 1982).

The adsorption capacity for indifferent electrolytes is determined by σ_H. Adsorption of indifferent electrolytes can reduce charge to near zero to achieve electroneutrality, but, by definition, does not change the sign of σ_H. This behavior is illustrated in Figure 8. Na^+ and Cl^- adsorb indifferently on corundum, whereas adsorption of Ba^{2+} and SO_4^{2-} is apparently superequivalent, i.e., in excess of surface charge, so these ions are not indifferent.

Adsorption in the diffuse layer is relatively non-selective. Equations 29 and 30 suggest that the ratio of the adsorption densities of two indifferent ions of the same charge should be approximately the same as the ratio of their concentrations in bulk solution,

$$\frac{\Gamma_i}{\Gamma_j} \approx \frac{c_{i,x}}{c_{j,x}} \approx \frac{c_{i,\infty}}{c_{j,\infty}} \quad \text{or} \quad \frac{\Gamma_i c_{j,\infty}}{\Gamma_j c_{i,\infty}} \approx 1 \ , \tag{31}$$

i.e., the surface is an ion exchanger with a selectivity coefficient on the order of 1.0, and an exchange capacity that is near zero at the PZNPC, but increases as pH departs from the PZNPC. A positively charged surface should be an anion exchanger and a negatively charged surface should be a cation exchanger. The selectivity coefficient should not be exactly unity, in part because ions of different sizes approach to different distances (δ), so that the integration limits in Equation 30 differ among species. By accounting for ionic size and additional interaction energies, Bolt (1955) has shown that selectivity coefficients for indifferent ions of the same charge can range from 1 to 4. Referring to Figure 8, the decrease in $\Gamma_{Mg^{2+}}$ on goethite accompanying an increase in sea salt concentration is qualitatively consistent with the expected ion exchange behavior.

Specific or chemical adsorption

For adsorbates exhibiting the second set of characteristics, reversal of surface charge indicates that adsorption is superequivalent, i.e., that the adion can adsorb at densities larger than electrically equivalent to the surface charge density. Superequivalent adsorption requires a non-coulombic, attractive contribution to the adsorption bond energy, because ions adsorbing onto a surface of like charge are electrostatically repelled. This contribution has been called a specific adsorption potential or a chemical contribution to the free energy of adsorption. The existence of a specific adsorption potential suggests chemical bonding, thus that the adsorbate ion is close to the surface, and exchanges water for one or more surface functional groups among first-shell coordinating ligands to form an inner-sphere surface complex, (Sposito, 1984). Ions that adsorb specifically are often multivalent and tend to hydrolyze significantly. They tend to form relatively strong aqueous complexes with some constituent of the solid.

Co(II) adsorbed on TiO_2 and SiO_2 (James and Healy, 1972a,b); Pb(II) adsorbed on $\gamma\text{-}Al_2O_3$ (Hohl and Stumm, 1976), and goethite (Hayes and Leckie, 1987); U(VI) on goethite (Tripathi, 1983); and selenite ion on goethite (Hayes et al., 1987) all show specific adsorption characteristics in their adsorption and/or electrokinetic behavior. The close proximity to the surface required by direct bonding has been confirmed in a few cases. X-ray absorption spectroscopy (XAS, see Brown, this volume) of these systems (SeO_3^{2-}-goethite, Hayes et al., 1987; Pb(II)-oxides, Roe et al., 1990; Chisholm-Brause et al., 1990; Co(II)-oxides, Chisholm-Brause et al., 1990; O'Day et al., 1990; Co(II)-kaolinite, O'Day et al., 1989; 1990; U(VI)-goethite, Combes, 1988) prove that the complex formed is an inner sphere complex, in the sense that the distance between the central atom of the adion and the metal atom of the adsorbent solid is too small to accommodate intervening water molecules bound to either surface or adion.

For cations, there is a rough correlation between pH_{ads} and the onset of hydrolysis (Fig. 10; James, 1971; James and Healy, 1972a), the solubility product of the adion hydroxide, or the electronegativity of the adion (Kinniburgh et al., 1976). Schindler and Stumm (1987) pointed out an analogy between proposed surface complexation reactions:

$$\underline{S}OH + M^{z+} = \underline{S}OM^{(z-1)+} + H^+ \tag{32}$$

or

$$2\underline{S}OH + M^{z+} = (\underline{S}O)_2M^{(z-2)+} + 2H^+ \ , \tag{33}$$

and aqueous hydrolysis:

$$HOH + M^{z+} = HOM^{(z-1)+} + H^+ \tag{34}$$

or

$$2HOH + M^{z+} = (HO)_2M^{(z-2)+} + 2H^+ \ . \tag{35}$$

By limiting their analyses to low adsorption densities, a single adsorbent, and a constant ionic medium, thus removing some of the ambiguity in definition and evaluation of activity coefficients for both aqueous and surface species, Schindler and Stumm (1987) were able to demonstrate an excellent correlation between the stability constants for the adsorption of several cations in a constant capacitance model (models for electrified interfaces are described in subsequent papers in this volume), and the stability constants for analogous aqueous hydrolysis reactions. These authors (Schindler and Stumm; 1987) also found a rough correlation between stability constants for anion surface complexation Reaction 36 and the analogous aqueous Reaction 37, e.g.,

$$\underline{Fe}OH + H_2A(aq) = \underline{Fe}AH + H_2O \ , \tag{36}$$

$$FeOH^{2+}(aq) + H_2A = FeAH^{2+}(aq) + H_2O \ . \tag{37}$$

Concentration dependence. If a single reaction is responsible for adsorption, we can derive an adsorption isotherm, or an explicit relationship between adsorption density and concentration in solution. Using Reaction 32, for example,

$$K_{32} = \frac{\{\underline{S}OM^{(z-1)+}\}\{H^+\}}{\{\underline{S}OH\}\{M^{z+}\}} \approx \frac{\Gamma_{Mz+}\{H^+\}}{\Gamma_{\underline{S}OH}\{M^{z+}\}} \ , \tag{38}$$

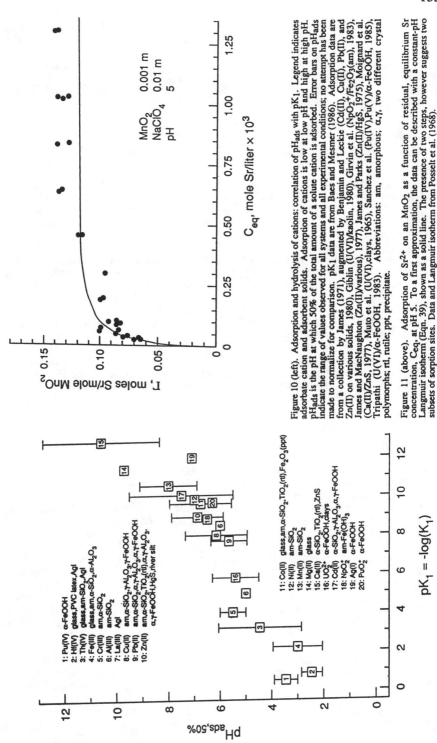

Figure 10 (left). Adsorption and hydrolysis of cations: correlation of pH_{ads} with pK_1. Legend indicates adsorbate cation and adsorbent solids. Adsorption of cations is low at low pH and high at high pH. pH_{ads} is the pH at which 50% of the total amount of a solute cation is adsorbed. Error bars on pH_{ads} indicate the range of values observed for all systems and all experimental conditions; no attempt has been made to normalize for comparison. pK_1 data are from Baes and Mesmer (1986). Adsorption data are from a collection by James (1971), augmented by Benjamin and Leckie (Cd(II), Cu(II), Pb(II), and Zn(II) on various solids, 1980), Giblin (U(VI)/kaolin, 1980), Girvin et al. (NpO_2^+/Fe_2O_3(am), 1983), James and MacNaughton (Zn(II)/various), 1977), James and Parks (Zn(II)/HgS, 1975), Moignard et al. (Ca(II)/ZnS, 1977), Muto et al. (U(VI),clays, 1965), Sanchez et al. (Pu(IV),Pu(V)/α-FeOOH, 1985), Tripathi (U(VI)/α-FeOOH, 1983). Abbreviations: am, amorphous; α,γ, two different crystal polymorphs; rtl, rutile; ppt, precipitate.

Figure 11 (above). Adsorption of Sr^{2+} on an MnO_2 as a function of residual, equilibrium Sr concentration, C_{eq}, at pH 5. To a first approximation, the data can be described with a constant-pH Langmuir isotherm (Eqn. 39), shown as a solid line. The presence of two steps, however suggests two subsets of sorption sites. Data and Langmuir isotherm from Posselt et al. (1968).

in which we have approximated the activities of surface species, $\{i\}$, by their surface concentrations, Γ_i. The subscript 32 simply identifies the equilibrium constant with Reaction 32. The activities of the ions involved in the reaction can be modified to account for the free energy involved in transferring an ion from the bulk solution to the surface, but we will ignore that for now (see chapters by Davis and Kent, Sposito, and Schindler in this volume). In a system in which only H^+ and M^{z+} are adsorbing, the sum, $\Gamma_{M^{z+}} + \Gamma_{\underline{S}OH}$, should be approximately constant, denoted arbitrarily by $\Gamma_{max,h}$, and

$$K_{32} \approx \frac{\{H^+\}\Gamma_{M^{z+}}}{(\Gamma_{max,h}-\Gamma_{M^{z+}})\{M^{z+}\}} \quad \text{or} \quad \Gamma_{M^{z+}} \approx \frac{K_{32}\Gamma_{max,h}\{M^{z+}\}/\{H^+\}}{(1 + K_{32}\{M^{z+}\}/\{H^+\})} \quad . \tag{39}$$

Equation 39 has the form of the Langmuir isotherm written for a single type of adsorption site with no interaction among sites (Adamson, 1982). When $\{M^{z+}\}/\{H^+\}$ is very small, the denominator approaches unity and $\Gamma_{M^{z+}}$ should increase linearly with increasing $\{M^{z+}\}/\{H^+\}$. When $\{M^{z+}\}/\{H^+\}$ is large, the concentration terms cancel and $\Gamma_{M^{z+}}$ approaches $\Gamma_{max,h}$.

Some experimental data appear to approximate Langmuir behavior, as shown in Figure 11, for the adsorption of Sr^{2+} on MnO_2, but there are usually deviations from the ideal isotherm (see, e.g., Kitchener, 1965). Compositional complexity in the solid and surface heterogeneity can be expected to introduce a variety of reactive sites and binding energies. Each subset of sites might be described with its own equilibrium constant (analogous to K_{32}) and adsorption capacity ($\Gamma_{max,h}$). Two subsets of sites with distinctly different binding constants and site densities might result in two distinct steps in the overall isotherm, resembling that for Sr^{2+} on MnO_2 (Fig. 11). Different sites may be selective for specific adions (Benjamin and Leckie, 1981). If there are many sites whose densities and binding energies cover a broad spectrum, as might be expected on a complex silicate or in a mixture of minerals, such as a real soil or sediment, adsorption densities may increase continuously with increasing adion concentration. The purely empirical Freundlich isotherm,

$$\Gamma_i = a \, [i]^n \quad , \tag{40}$$

in which a and n are empirical constants, often describes the concentration dependence of adsorption well in such systems. Kinniburgh and Jackson (1982) found it necessary to use three separate Freundlich isotherms to account for the adsorption of Ca^{2+} and $Zn(II)$ on a hydrous oxide of $Fe(III)$, over five orders of magnitude of adion concentration, suggesting that this adsorbent is complex indeed, or that adsorption behavior is complicated by other phenomena, such as adion-adion interactions, as well as by site heterogeneity.

Interactions among adsorbed species can be repulsive or attractive. If the adsorption complex is electrically charged, then the surface will accumulate charge as adsorption density increases, resulting in an increasingly repulsive potential, reduced binding energy, and reduced adsorption. Coulombic interactions are taken into account either empirically (Stroes-Gascoyne et al., 1986) or explicitly through models of the electrified interface (see Davis and Kent, this volume). Repulsive interactions can be expected to flatten the adsorption isotherm and reduce adsorption capacity. If the structure of the adsorbing species is incompatible with the structure of bound water at the surface, then entropy can be reduced by grouping adsorbate ions or molecules, and adsorbate ions or molecules may be attracted to each other by this "hydrophobic bonding" (Israelachvilli, 1985), probably leading to clustering or multilayer adsorption. Surface active solutes like the long-chain carboxylic acids and amines cluster upon adsorption forming "hemi-micelles" (Fuerstenau and Raghavan, 1978), or two dimensional analogues of the three-dimensional micelles so important in the solution chemistry of detergents (Adamson,

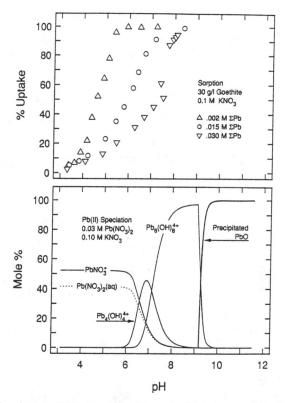

Figure 12. Sorption of aqueous Pb(II) onto goethite. The sorption data in the upper panel are taken from Roe et al. (1990). The lower panel shows the distribution of Pb(II) among dissolved hydroxo complexes and precipitated PbO, as estimated with the program HYDRAQL and thermodynamic data from Baes and Mesmer (1986). The precipitation curve would be shifted to higher pH if a true hydroxide formed instead of stable PbO. There is no direct correlation between the uptake curves and any single dissolved species. If the onset of adsorption near pH_{ads} at low ΣPb is attributed to sorption of hydroxo complexes, then the surface apparently increases their stability. If the abrupt increase in uptake at pH ≈ 7.5 is attributed to sorption of multinuclear species from solution or to precipitation of an oxide or hydroxide, then the surface apparently alters the stabilities of these species as well.

1982). Attractive interactions can be expected to increase binding energy with increasing adsorption density, leading to concave-upward portions of isotherms. Kitchener (1965) and Parks (1975) have reviewed many of these interactions and their influences on the shapes of isotherms, emphasizing heteropolar organic adsorbates.

Hydrolysis, polymerization, and precipitation. Hydrolysis, polymerization, and precipitation also modify the shapes of sorption isotherms. Metal ions (M^{z+}) in solution hydrolyze to produce mononuclear, $M(OH)_n^{z-n}(aq)$, and polymeric or multinuclear $M_p(OH)_q^{pz-q}(aq)$ hydroxo complexes. The number of coordinating hydroxide ions increases with increasing pH; the number of M^{z+} in the multinuclear complexes increases with increasing $\Sigma M(aq)$. James and Healy (1972b), in recognition of the correlation between pH_{ads} and hydrolysis constants and reversals of charge accompanying cation sorption, argue that either: (a) all of the aquo- and hydroxo- complexes of hydrolyzable cations should be expected to adsorb, most strongly for complexes of least charge density; or (b) if M^{z+} adsorbs as such, then a parallel set of hydrolysis reactions occurs on the surface. As pH and $\Sigma M(aq)$ approach saturation with respect to $M(OH)_z$, James and Healy would expect surface polymerization and precipitation. Farley et al. (1985) come to a similar conclusion, that sorption and precipitation are part of a continuous process, reasoning that each adsorbed M^{z+} becomes a sorption site, and that surface sites and sorbed

158

ions can react to form surface coprecipitates. We have already observed the correlation between pH_{ads} and mononuclear hydrolysis. There is sometimes also an obvious correlation between the appearance of multinuclear hydroxo complexes and sharp increases in sorption with increasing pH, yielding an uptake curve that mimics a precipitation curve, but is shifted to lower pH (Fig. 12). This suggests surface precipitation, but spectroscopic evidence for the presence of a precipitate remains ambiguous (Brown, this volume). We will return to surface precipitation during discussion of surface charge reversal.

Competition and synergism between adsorbate and complexing ligands. Complexing ligands can inhibit or enhance adsorption densities and change pH dependence (Benjamin and Leckie, 1982; Davis and Leckie 1978a; Schindler, this volume). Ligands which themselves adsorb weakly or non-specifically, such as Cl^- or CO_3^{2-} on goethite, can be expected to inhibit adsorption of metals with which they form aqueous complexes, by reducing the activity of the adsorbing species and binding the metal ion in complexes which adsorb less strongly than the free M^{z+} or its hydroxo complexes. Hg(II), for example, sorbs onto hydrous Fe(III) oxides from chloride solutions in which $HgCl_2(aq)$ predominates. In this case sorption is apparently a ligand exchange process, substituting $\underline{S}O^-$ for Cl^-; Kinniburgh and Jackson (1978) proposed this mechanism upon observing adsorption of Hg(II) without accompanying adsorption of Cl^-. At high total chloride concentrations, Cl^- competes successfully with $\underline{S}O^-$ in the exchange reaction, inhibiting sorption of Hg(II) (Avotins, 1975; MacNaughton and James, 1974). Chloride and SO_4^{2-} inhibit adsorption of Cd(II) on am-$Fe(OH)_3$, quartz, lepidocrocite, and γ-Al_2O_3 (Benjamin and Leckie, 1982) and carbonate inhibits adsorption of U(VI) on goethite (Langmuir, 1978; Tripathi, 1983) and kaolinite (Giblin, 1980). In contrast, hydrolysis (or hydroxo complex formation) enhances sorption of cations (James and Healy, 1972a,b) and $S_2O_3^{2-}$ enhances adsorption of Ag(I) on am-$Fe(OH)_3$ (the prefix "am" is used to identify an amorphous material) at low pH, while inhibiting adsorption at high pH (Davis and Leckie, 1978a). This type of synergism is discussed in detail by Schindler (this volume).

Multivalent inorganic cations can enhance adsorption of organic anions such as carboxylic acids and inhibit adsorption of organic cations such as amines; multivalent inorganic anions may enhance adsorption of amines and inhibit adsorption of acids. These effects might arise from alteration of surface charge through preadsorption by the inorganic modifier or alternatively by formation of an aqueous complex between the inorganic ion and the organic solute and its subsequent adsorption. Most efforts to characterize these ternary surface complexes have resorted to drying the sample, thus introducing the possibility that species hydrated while immersed are altered during drying. Direct adsorption measurements and characterization studies are scarce; the generalizations quoted here are inferred from electrokinetic and froth flotation observations (Fuerstenau and Raghavan, 1978; Rea and Parks, 1990). The presence of long-chain organic surfactants and natural organic materials can enhance or inhibit adsorption of non-ionic and nonpolar organic solutes. Penetration of nonionic pesticides into soils, for example, can be retarded by low concentrations of surfactants and enhanced by high concentrations (Smith and Bayer, 1967; Huggenberger et al., 1973). Retardation is attributed to co-adsorption of the pesticide with the surfactant, and enhancement to solubilization in surfactant micelles. In the presence of added surfactants, sorption of PCBs by goethite exhibits the same pH dependence as does sorption of the surfactants alone, suggesting coadsorption (Ainsworth et al., 1985; Ainsworth, pers. comm.). In acidic solutions, for example, Na-dodecyl sulfate, an anionic surfactant, enhances sorption of Aroclor 1242 (one of the nonionic PCBs), and a long-chain amine (a cationic surfactant) inhibits sorption of Aroclor. In alkaline solutions sorption of the surfactants and their influence on sorption of Aroclor are reversed.

Identifying sorption reactions: Clues to the composition and structure of surface complexes

Quantitative interpretation or modelling of adsorption should start with a physically realistic representation of the reaction or reactions responsible. We understand the reactants fairly well, the aqueous species containing the adsorbate, and, to a lesser extent, the surface functional groups, but we know very little about the surface complex that is the product of the adsorption reaction. Spectroscopic methods are essential in the effort to determine the composition and structure of adsorption complexes (Sposito, 1984; Motschi, 1987). There are, however, clues that can be interpreted independently of spectroscopy and of electrical double or triple layer models in the stoichiometry of the adsorption reaction, in the pH and ionic strength dependence of adsorption, and in the effects of adsorption on surface charge.

In aqueous systems we usually have little or no information about sorption sites or surface complexes. Identification of sorption reactions is largely a trial and error process, guided by solution chemistry and a conceptual model of the solid-water interface. As a first approximation, we might suppose that the most abundant species in solution reacts with the most abundant appropriate surface species, with or without accompanying ligand exchange, cation exchange, or hydrolysis. The implications of the resulting models for reaction stoichiometry and surface charge can be tested against experimental observation.

Figure 13 presents (B) the pH dependence of adsorption of Cd(II) onto TiO_2, and (D) corresponding changes in zeta potential, as measured by James et al. (1981). The figure also presents (A) one possible, approximate distribution of surface sites on TiO_2 in the absence of a sorbing solute, and (C) the distribution of Cd among aqueous species in solution in the absence of a sorbing solid. Over most of the range of interest, $4 < pH < 8$, the predominant species in solution is Cd^{2+}, and on the surface it is the unionized TiOH site. We can conceive of several adsorption reactions among these species, e.g., reactions of one Cd ion with a single site producing mononuclear, monodentate complexes:

$$Cd^{2+} + \underline{Ti}OH = [\underline{Ti}O^-Cd^{2+}]^+ + H^+ , \tag{41a}$$

$$Cd^{2+} + \underline{Ti}OH + H_2O = [\underline{Ti}O^-CdOH^+]^0 + 2H^+ , \tag{41b}$$

and reactions with two adjacent sites producing mononuclear, bidentate complexes:

$$Cd^{2+} + 2\underline{Ti}OH = [(\underline{Ti}O^-)_2Cd^{2+}]^0 + 2H^+ , \tag{41c}$$

$$Cd^{2+} + 2\underline{Ti}OH + H_2O = [(\underline{Ti}O^-)_2CdOH^+]^- + 3H^+ . \tag{41d}$$

Above pH 8, adsorption of $CdOH^+$or $Cd(OH)_2(aq)$ onto the TiO^- site might be considered. James et al. (1975) examined the adsorption isotherms implied by several reactions producing mononuclear sorption complexes, concluding that they are nearly indistinguishable on the basis of the concentration and pH dependence of adsorption density alone.

Surface charge. Electrokinetic measurements reflect the net surface charge, σ_{ek}, and can be used to detect changes in surface charge accompanying sorption. Changes in surface charge can be used to test proposed sorption reactions; e.g., if σ_{ek} increases, the sorption complex must be more positive than the surface site with which the sorbing species (the adion, Ad^z) reacted. In Figure 13D, for example, sorption of Cd(II) results in a positive σ_{ek} at the IEP. Among Reactions 41, only 41a produces a positively charged surface complex, so if these reactions actually represent the only possibilities, then Reaction 41a must be included in the description of sorption.

The typical pH-dependent influence of hydrolyzable cations on electrokinetic properties is illustrated in Figures 13D and 14 for the Cd(II)-TiO_2 and Co(II)-SiO_2 sorption systems, respectively. There are two extremes of behavior:

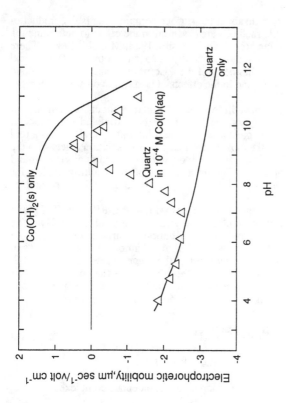

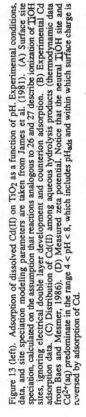

Figure 13 (left). Adsorption of dissolved Cd(II) on TiO₂ as a function of pH. Experimental conditions, data, and site speciation modeling parameters are taken from James et al. (1981). (A) Surface site speciation, calculated on the assumption that reactions analogous to 26 and 27 describe ionization of ⎯TiOH sites, ignoring electrical double layer development and counterion adsorption. (B) Experimental Cd adsorption data. (C) Distribution of Cd(II) among aqueous hydrolysis products (thermodynamic data from Baes and Mesmer, 1986). (D) Measured zeta potential. Notice that the neutral ⎯TiOH site and Cd²⁺(aq) predominate in the range 4 < pH < 8, which includes pH_ads and within which surface charge is reversed by adsorption of Cd.

Figure 14 (above). Effect of Co(II) on the surface charge of quartz, as revealed by electrophoretic mobility. Surface charge is positive when mobility is positive, negative when mobility is negative. Notice that Co(II) has little effect on the quartz at low pH, but reverses charge and causes the quartz to approximate the behavior of Co(OH)₂ at high pH.

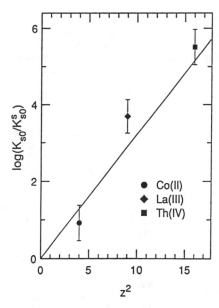

Figure 15. Solubility constants for surface precipitation of hydroxides and hydrous oxides, Kss0, relative to those for the same solids in bulk aqueous solution, K_{s0}. The solids are increasingly less soluble at the interface than in bulk with increasing cationic charge. The K^s_{s0} were estimated from the pH and SM(aq) concentration at which an adion reversed surface charge on a host adsorbent, as revealed by electrokinetic methods. Error bars reflect reproducibility and scatter in experimental data. The solid line predicts $\log(K_{s0}/K^s_{s0})$ on the basis of an electrostatic model. After James and Healy (1972b).

1. At low pH and low adion concentrations, ΣAd, σ_{ek} may be more positive than the adsorbate-free surface, but its pH dependence roughly parallels that of the host solid (Fig. 13D, pH $\leq$ IEP; Fig. 14, pH $\leq$ 6). .

2. At high pH and high ΣAd, σ_{ek} roughly parallels that of the solid hydroxide, $Ad(OH)_z$, but is usually less positive than the hydroxide surface (Fig. 14, pH > 9).

The low-pH behavior suggests that much of the solid surface is free of adsorbates. The overall positive shift requires generation of at least a small number of positive sorption complexes. The alternatives are not limited to those produced by Reactions 41; the required positive charge could be attributed as well to a few small, highly charged multinuclear adsorbate complexes, or to an adherent precipitate covering a small fraction of the surface, or even to adherence of a few small particles of a positively charged precipitate originally formed in solution.

The high-pH, high ΣAd behavior so closely imitates the behavior of bulk $Ad(OH)_z(s)$ that the surface must be completely dominated by something having properties similar to $Ad(OH)_z$. This might be near 100% occupancy of all reactive surface sites by amphoteric hydroxo Ad complexes, but is more commonly assumed to be a coherent, adsorbed hydroxy-polymer, or an adherent second phase or surface precipitate (James and Healy, 1972b; James et al., 1977, 1981; Sposito, 1984). Several types of spectroscopic evidence for clustering and "hydroxide-like" sorption complexes which are not well ordered hydroxide precipitates have been reported (Brown, 1990).

Dramatic surface charge reversals in the intermediate pH-ΣAd region have been explained alternatively by sorption of highly charged, polymeric species (Matijevic; 1967; Healy et al., 1968), partial-to-multilayer masking of the surface by precipitates (James and Healy, 1972b), and heterocoagulation of a hydroxide (precipitated in the solution phase) onto the surface of the solid (Fuerstenau, 1970; James et al, 1975). James and Healy

(1972b) discuss this entire spectrum of behavior in detail, ultimately deciding in favor of a surface precipitation model in which the solubility product of the precipitate is reduced significantly by the low dielectric constant of water near the surface. Their estimates of the difference between normal and surface-modified solubility products for Co(II), La(III), and Th(IV) are summarized in Figure 15. Should this interpretation prove correct, there could be far-reaching implications for dissolution and precipitation in sediments and soils, where much of the water present is bound at surfaces, owing to the abundant surface area and high solid-liquid ratios. Sposito (1984) pointed out that the experimental observation that a solution phase is undersaturated with respect to precipitation of a single, pure phase is insufficient evidence that precipitation has not happened because the solution may well be saturated with respect to coprecipitation with another species. If James and Healy are correct, alteration of the solubility products of even pure, single phases at interfaces provides another mechanism for precipitation from apparently subsaturated solutions.

Proton stoichiometry. When a solute adsorbs, something invariably comes off the surface in exchange. Loganathan and Burau (1973), for example, found that H^+ and Mn^{2+} were released in various proportions upon adsorption of transition metals, Ca^{2+}, or Na^+ onto δ-MnO_2. Methods of determining the molar ratio of hydrogen released to metal adsorbed, n_h/n_m, have been reviewed by Perona and Leckie (1985) and Honeyman and Leckie (1986). After correction for changes in solution chemistry, any net release of H^+ must be involved in the adsorption reaction or in re-equilibration among surface sites. The adsorption reaction might involve H^+ in at least three obvious ways: (1) M^{z+}/H^+ exchange in the diffuse layer; (2) M^{z+} exchange for H^+ bound in surface functional groups such as $\underline{S}OH$ or $\underline{S}OH_2^+$, exemplified by Reactions 41a and 41c; and (3) hydrolysis of the adsorbed cation during adsorption, as in Reactions 41b and 41d. Integer values for n_h/n_m might imply single, simple reactions such as 32, 33, 41. Non-integer proton release implies exchange in the diffuse layer, or, for specifically adsorbed species, either (a) that at least two reactions are responsible for the net adsorption observed, or (b) that the adsorption complex includes more than one solute cation on a single site. For Cd(II) on TiO_2, $n_h/n_m \leq 2$, so, if specific adsorption is assumed, no single reaction among 41a-d is sufficient to account for the observed proton stoichiometry; 41a must be accompanied by at least one of the others, but there are other alternatives.

Experimentally, the molar ratio of hydrogen released to metal adsorbed, n_h/n_m can be as low as 0.05 at one extreme, to nearly 3 (for Zn^{2+} on amorphous $Fe(OH)_3$ at pH > 6.5 (Perona and Leckie (1985)). The cations least likely to adsorb specifically, and thus most likely to adsorb predominantly in the diffuse layer (e.g., the alkaline earths ions) adsorb with $n_h/n_m < 0.5$ (on an MnO_2 in 0.1M NaCl at 4.5 < pH < 7.5 (Murray, 1975)). A low value of n_h/n_m suggests cation exchange in the diffuse layer, exchange with pre-adsorbed cations (Na^+ in this example) instead of H^+, or possibly adsorption of a multinuclear hydroxo complex present in bulk solution prior to the adsorption event (if the multinuclear complex were formed during adsorption, then more protons would be released). The alkaline earths do not form multinuclear complexes under these conditions, so the first two alternatives seem most likely. For ions more likely to adsorb specifically, such as the transition metals, n_h/n_m is between 0.8 and 2. Murray (1975) observed 0.8 < n_h/n_m < 1.3 for transition metal ions on α-FeOOH in 0.1M NaCl;, others, usually working at lower ionic strength, find 1 < n_h/n_m < 2 (Perona and Leckie, 1985; Honeyman and Leckie, 1986).

The last analysis. Solution chemistry, especially the stoichiometry, pH, and concentration dependence of the sorption process, together with changes in surface charge and many other observations and clues give us a great deal of help in characterizing the surface complexes resulting from sorption and identifying sorption reactions. Invariably, however, several alternative sets of reactions will satisfy all observations. Davis and Leckie (1978b) and Hohl and Stumm (1976), for example interpreted the same data for adsorption of Pb(II) on γ-Al_2O_3 with equal success, the former using Reactions 41a and 41b, and the latter using 41a and 41c. In the last analysis, we need models of the electrified interface capable of accounting for electrophoretic charge or potential measurements, and independent means of characterizing surface complexes *in situ*, without altering the local environment of the complex on a molecular scale. Subsequent chapters in this book will address these issues in detail. Using a triple layer model for the EDL, James et al. (1981) found that a single surface complexation reaction (41b) alone was sufficient to account for the general pH-dependence of the adsorption of Cd(II) on TiO_2 and its effect on σ_{ek}, but no independent evidence for the reality of this reaction has been reported.

SURFACE AND INTERFACIAL FREE ENERGIES OF QUARTZ

The Gibbs adsorption equation (Eqn. 25) requires that changes in interfacial free energy accompany adsorption. We can gain some insight into the magnitude of this effect by tracing the surface energy of quartz from fracture to immersion in a saline solution.

Fracture surface energy. The only experimental data we have on the free energy of pristine quartz surfaces, γ_{vac}, are derived from attempts to measure the work required to initiate non-catastrophic fracture. If the experiment can be performed reversibly, the only change in state produced by fracture is an increase in surface area. The Griffith equation (Lawn and Wilshaw, 1975),

$$\sigma_{crit}^2 = \frac{2\gamma E}{\pi c} \quad , \tag{42}$$

relates the stress, σ_{crit}, needed to re-start propagation of a pre-existing crack of length, c, to the surface free energy appropriate for the environment of fracture, γ, and Young's modulus, E, of the solid. Many attempts have been made to estimate γ by this means (Atkinson, 1984; Parks, 1984); results for quartz range from 410 mJ/m² for the {1011} face, as measured by Brace and Walsh (1962) to 11,500 mJ/m² for the {1011} face as measured by Hartley and Wilshaw (1973). All fracture measurements suffer from the strong possibility that energy is dissipated as light, heat, sound, electrical discharge, multiple fracture, and plastic deformation. No measurements have been made in vacuum, so most are subject to the strong possibility of contamination by adsorption of at least water. In short, the lowest estimates of γ are undoubtedly too low, and the highest, too high. Obviously, we need new measurements or an independent means of deciding which among the fracture measurements are closest to the truth.

An estimate of the quartz-water interfacial free energy can be derived from measurements of solubility as a function of particle size. Figure 16 presents the solubility of quartz as a function of A_s, extracted from dissolution rate measurements by Stöber (1967). These data have been used to estimate γ for quartz in an $NaCl-NaHCO_3$ brine at pH $\approx$ 8.4, by application of the Freundlich-Ostwald-Tolman Equation (21) with results summarized in Table 2 (see also Parks, 1984) together with similar results for amorphous silicas. The difference between γ_{brine} and γ_{vac} includes hydroxylation, adsorption of molecular water, immersion, and adsorption of adions.

Water vapor adsorbs onto oxide surfaces in two stages, as we have seen: dissociative adsorption, forming a hydroxylated surface with energy, γ_{hydrox}, and, at $P/P_0 \approx 1$, a thick film of hydrogen bonded molecular water with energy $\gamma_{v,sat}$. No one, apparently, has measured the free energy of the hydroxylation process, but we can estimate $\Delta\gamma_{v,sat} = \gamma_{v,sat} - \gamma_{hydrox}$, from a water vapor adsorption isotherm. An isotherm representing adsorption of water onto quartz, outgassed at low temperature to produce a silanol surface, is presented in Figure 4. The apparent adsorption density includes water that is condensed at particle-particle contacts (capillary condensation; Miyata, 1968). If we express the chemical potential of water in terms of its vapor pressure, the Gibbs equation (25) can be integrated from $P/P_0 = 0$ to the maximum vapor pressure achieved in the experiment to estimate the free energy of adsorption of water, i.e., $\Delta\gamma_{v,sat}$. For the data

164

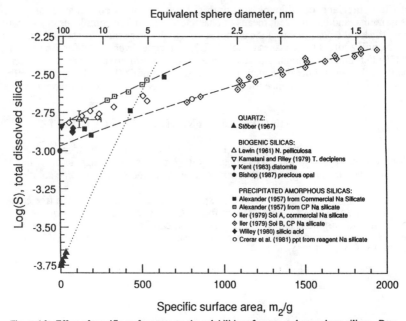

Figure 16. Effect of specific surface area on the solubilities of quartz and amorphous silicas. Data represent different solution compositions, methods of estimating surface area, and varying approximations of equilibrium as well as different materials. The lines represent the Freundlich-Ostwald equation, adapted to allow variation of γ with particle size, fitted to selected experimental data and extrapolated far beyond the range of fitted data. The uppermost dashed line utilizes only Alexander's (1957) data for CP Na Silicate sols; the lower dashed line utilizes only Iler's (1979) sol B; the dotted line utilizes only Stöber's (1967) data. Notice that, for particle sizes smaller than about 5nm, quartz may be more soluble, or less stable, than amorphous silicas.

Table 2. Intrinsic solubilities and interfacial free energies of silicas in aqueous solutions.

	T°C	pH	$\log(S_n)$ molar	γ_0 mJ/g	Notes	Source
Linear regression:						
quartz	25	8.4	-3.75±.01	351±23	a	Parks (1984)
am-SiO$_2$	25	2.2	-2.89±.02	42±2	b	Iler (1979)
am-SiO$_2$	25	2-8	-2.82±.01	32±3	c	Alexander (1957)
Tolman regression:						
quartz	25	8.4	-3.75±.01	353±23	a	Parks (1984)
am-SiO$_2$	25	2.2	-2.97±.02	61±3	b	Iler (1979)
am-SiO$_2$	25	2-8	-2.83±.01	33±3	c	Alexander (1957)
Independent measurements:						
am-SiO$_2$	50	4-9	nd	70-115	d	Weres et al. (1980)
am-SiO$_2$	75-105	4.5-6.5	nd	45	e	Makrides et al. (1980)
am-SiO$_2$	rt	nd	nd	50	f	Kendall et al. (1987)

a. Estimated from A_s-dependence of solubility of quartz in an NaHCO$_3$ (0.0119 m)-NaCl brine (0.1539 m) (Stöber, 1967). b. Estimated from the particle size-dependence of the solubility of an amorphous SiO$_2$ prepared by polymerization of H$_4$SiO$_4$ at pH 8. Size estimated by computation from unreported A_s determined by NaOH titration. c. Estimated from the A_s-dependence of solubility of an am-SiO$_2$ prepared by acidification of CP Na$_2$SiO$_3$. A_s determined by NaOH titration. d. Estimated from nucleation rate of am-SiO$_2$ in .05 to .5M NaCl brines at pH from 4 to 9 . e. Estimated from the nucleation rate of am-SiO$_2$ in 0.06 to 0.45M NaCl, KCl, and MgCl$_2$ brines. f. Estimated from the elastic modulus of compacted am-SiO$_2$ powder, immersed in water.

presented in Figure 4, $\Delta\gamma$ is 89.4 mJ/m^2; other investigators have obtained estimates between 65 and 244 mJ/m^2, without correction for capillary condensation (Parks, 1984).

Because the properties of thick layers of adsorbed water are indistinguishable from those of bulk water, $\gamma_{v,sat}$ should be close to γ_{H_2O} ($\approx$ 73 mJ/m^2; Adamson, 1982) and the only change expected upon immersion of a silica surface pre-equilibrated with water vapor at saturation is the loss of surface area—the exterior surface area of the water-jacketed material. Upon immersion, then, $\Delta\gamma_{imm} \approx -73$ mJ/m^2.

Once immersed in a water solution, quartz develops surface charge and adsorbs an atmosphere of counterions, thus further reducing its surface energy. This change, $\Delta\gamma_{adion}$, can be estimated from the adion adsorption isotherm. Li (1955) measured the adsorption density of Na$^+$ on quartz, for pH > IEP, as a function of pH and NaCl concentration (shown, in part, in Fig. 8), and used the Gibbs adsorption Equation (25) to estimate $\Delta\gamma$. His results are summarized in Figure 17. The solubility data used to estimate γ_{brine} was obtained in a solution 0.166 molar in Na$^+$ at pH $\approx$ 8.4. Judging from Figure 17, this combination of pH and Na$^+$ concentration should result in $\Delta\gamma_{adion} \approx -10$ to -20 mJ/m^2.

All of these estimates are summarized in Table 3. We still cannot determine γ_{vac} by computation from γ_{brine}, but apparently $480 < \gamma_{vac} + \Delta\gamma_{hydrox} < 713$. Because $\Delta\gamma_{hydrox}$ is undoubtedly greater than $\Delta\gamma_{v,sat}$, γ_{vac} is undoubtedly greater than 550 mJ/m^2 (assuming the minimum value of $\Delta\gamma_{v,sat}$ as a minimum estimate of $\Delta\gamma_{hydrox}$).

APPLICATIONS

There are many applications and motivations for the study of surface and interfacial phenomena in the earth sciences. Partitioning processes, and sorption in particular, play important roles in determining the distributions of minor and trace elements among minerals and water, wherever water and solids come in contact. Quantitative understanding of sorption processes is essential to development of robust thermodynamic and kinetic models. Robust models, in turn, are needed in efforts to use simulation in the interpretation of geochemical history or to forecast future consequences of present-day changes. Several of the other papers in this volume deal with experimental means of identifying sorption processes and with development of simulation models, therefore, in this section, we will outline three areas in which surface chemistry may have an important, but not quite so obvious role. Each of the three areas, fracture propagation, Ostwald ripening and Ostwald's step rule, and ultralow frequency electromagnetic radiation accompanying earthquakes, is a complex topic in itself, and the discussion to be offered is purposely biased toward an explanation involving surface and interfacial geochemistry.

Fracture and crack propagation

Water and many solutes reduce the fracture strengths of rocks and minerals, and increase crack propagation velocities, effects that may arise from adsorption of water or ions at the crack tip. Molecular models of this hydrolytic or chemomechanical weakening are emerging (Michalske and Bunker, 1984; Wiederhorn and Fuller, 1989), but much older, macroscopic interpretations based on the influence of interfacial free energy are still instructive. The subject deserves attention because the weakening effects are large, and have lead some investigators to the conclusion that water-rock interactions may be involved in large-scale inelastic deformation in the upper crust (Kirby, 1984).

Hartley and Wilshaw (1973) used an indentation method to estimate the critical load, P_c, needed to produce fracture in quartz. Their results demonstrate a clear increase in P_c when dry N$_2$ is substituted for laboratory air in the experimental environment. We cannot derive useful estimates of γ from P_c, because, although P_c is proportional to γ, the proportionality constant is unknown. We can, however, compare the ratio $P_{c,air}/P_{c,N_2}$,

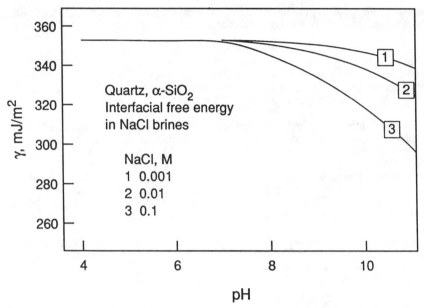

Figure 17. Changes in the interfacial free energy of quartz caused by adsorption of Na^+ as a function of pH and $\Sigma Na(aq)$. Li (1955) estimated $\Delta\gamma_{Na}$, by integration of experimental Na^+ adsorption isotherms, using the Gibbs equation (Eqn. 25).

Table 3. Average Surface and Interfacial Free Energies of Quartz. Quantities in parentheses are computed estimates.

	Surface State	$\bar{\gamma}$ mJ/m2	$\Delta\bar{\gamma}$ mJ/m^2	Source
γ_{fract}	fracture	550 to 11,500	--	fracture measurements
$\Delta\gamma_{hydrox}$		}	(>65)	estimate
$\gamma_{silanol}$	silanol	(478 to 713)		
$\Delta\gamma_{v,sat}$		}	65 to 244	$\Gamma_{H2O(P/Po\rightarrow1)}$, Eqn.(25)
$\gamma_{v,sat}$	adsorbed water	(413 to 469)		
$\Delta\gamma_{imm}$		}	73	measured $\gamma_{H2O(liq)}$
γ_{imm}	liquid water	(340 to 376)		
$\Delta\gamma_{adion}$		}	10 to 20	Γ_{Na^+}, Eqn. (25)
γ_{brine}	brine	353±23		solubility measurements, Eqn. (21)

with $\gamma_{air}/\gamma_{N_2}$. If we assume that exposure to the dry N_2 atmosphere is effective in desorbing molecular water from the quartz surface, leaving a silanol surface; and also assume, on the other hand, that the laboratory air has a humidity high enough to produce a surface nearly saturated with adsorbed molecular water, then $\gamma_{air}/\gamma_{N_2} \approx \gamma_{v,sat}/\gamma_{silanol} > 0.66$ to 0.86 (Table 3). Because $P_{c,air}/P_{c,N_2}$ lies between 0.3 and 0.6 (for the {100} and {001} crystal faces, respectively), this comparison, although highly speculative, suggests that interfacial free energies are sufficient to predict the directions and orders of magnitude of change in fracture properties accompanying adsorption.

Changes in the composition of water in the local environment modify fracture properties as well. Millimolar concentrations of, e.g., $Al(NO_3)_3$ or the cationic surfactant dodecyl tetraammonium chloride, and other surfactants can cause a 30 to 40% reduction in the fracture strength and order of magnitude increases in drilling rates in quartz (Ishido and Mizutani, 1980; Dunning et al., 1984).

Fracture propagation velocities are far more sensitive to environmental conditions than fracture strength. Fracture velocities in quartz under constant stress intensity, for example, increase from 10^{-7} m/s in vacuum ($P/P_0 \approx 10^{-7}$) to over 10^{+3} m/s when immersed in liquid water ($P/P_0 = 1$). Any proton-containing molecule small enough to diffuse to the crack tip can be expected to increase fracture velocity, perhaps because the proton adsorbs at a bridging oxygen, weakening the Si-O bonds and reducing the activation energy of the fracture process (Michalske and Bunker, 1984). Changes in the composition of ambient water also change fracture velocities. Again at constant stress intensity, fracture velocities in quartz are low in 2M HCl, an order of magnitude faster in pure water, and another order of magnitude faster still in 2M NaOH (Atkinson and Meredith, 1981). In models of fracture propagation viewing the process as a chemical reaction (Lawn and Wilshaw, 1975), interfacial free energy is thought to contribute to the activation energy, but the relationship is not simple. The changes in γ caused by adsorption of the solutes in these systems would be expected to result in velocity changes of the same sign as are observed. The effects of solutes change with stress intensity however, approaching zero as stress intensity is reduced toward the critical fracture stress, so interfacial energies are not the sole factors controlling velocity.

Interpretations of hydrolytic weakening in terms of interfacial energies appear to have some value, but are incomplete. Several alternative conceptual models have been offered to explain of the effects of chemical environment on fracture strength and velocity. These models invoke migration of charged defects toward the crack tip in response to surface potentials, coulombic repulsion between the electrical double layers on opposing crack surfaces, alteration of the activation energies for rupture of chemical bonds by adsorption of water, protons, or solutes at stressed bonds near the crack tip, and mechanical wedging by adsorbed molecules (Freiman, 1984; Ishido and Nishizawa, 1984; Wiederhorn, 1974a,b; Wiederhorn and Fuller, 1989). As models of the fracture process continue to evolve, we will need to know more about the energies involved in water and proton adsorption, at free and stressed surface sites, about the structures and energy contents of adsorbed water or solutes, and about the mechanisms and rates of surface diffusion.

<u>Ostwald ripening and the Ostwald step rule</u>

Minerals formed by low temperature precipitation from water and the shells or skeletons of microfauna often have extraordinarily complex microstructures and shapes, which are reflected in very high specific surface area in comparison with gross particle size (Table 1). The energy stored in this very large surface area is one of the driving forces for diagenesis (Williams et al., 1985) and development of texture and differentiation in metamorphic rocks (Thompson, 1987). The solubilities of amorphous silicas and quartz are compared in Figure 16. A precipitated am-SiO_2 with $A_s = 500$ m^2/g is more soluble than another am-SiO_2 with smaller A_s, so silica should dissolve from the high area material and re-precipitate on (or as) material of lower area. The growth in particle size and simplification of shape resulting from this dissolution and re-precipitation is known as Ostwald ripening (Adamson, 1982; Morse and Casey, 1988). It is observed in nature and in laboratory alteration of diatom tests (Kastner et al., 1977), for example. Ostwald ripening is slow because the driving force is small. The molar free energy difference between an am-SiO_2 silica with $A_S = 500$ m^2/g and another with negligible surface area, for example, is only about -1.2 kJ.

Thermodynamics can be very useful in predicting the outcome of a chemical process, but the prediction is very often wrong at low temperatures. Pyrite, for example, is the stable iron sulfide in reducing sediments, yet the first Fe(II) sulfides to precipitate are mackinawite, a tetragonal Fe_{1+S}, and greigite, cubic Fe_3S_4 (Berner, 1971). Silica usually precipitates as an amorphous, hydrous SiO_2, yet the stable phase is quartz. These are examples of the "Ostwald Step Rule," which states that a reaction that can produce any of several polymorphs should be expected to yield the least stable among them, not the most stable (see Morse and Casey, 1988). The fundamental explanation of this apparently perverse behavior is kinetic and has been discussed in detail by Morse and Casey (1988). The difference in interfacial free energies among polymorphs at the particle size (or A_S) of initial precipitation, provides a simple, qualitative explanation (Schindler, 1967).

The free energies of formation of amorphous materials are less negative, and their water solubilities are higher, than those of their crystalline counterparts, as inspection of compilations of thermodynamic data will show. Their rates of dissolution and precipitation are also higher than those of the crystalline materials (Stumm and Morgan, 1981). The high concentrations of structural defects and disorder responsible for these differences should also lead to lower interfacial energies, if only because the density of bonds in the surface should be lower in the amorphous material. The correlation of high solubility with low surface energy predicted by these speculations may be born out for silica, as suggested by the data in Figure 4. Schindler (1967) made similar observations for oxides and hydroxides of Zn(II) and Cu(II) and used them in an explanation of the Ostwald step rule. Again referring to Figure 16, amorphous silicas are more soluble than quartz in micrometer size ranges, and solutions saturated with respect to am-SiO_2 are thus supersaturated with respect to quartz. Quartz would be expected to precipitate, but unless particles of quartz are already present, precipitation must begin with nucleation of very small particles. In particle sizes ranges below about 5 nm (or $A_S > 500$ m^2/g), quartz is more soluble than am-SiO_2 (if the long extrapolation in Fig. 16 is correct), so am-SiO_2 should precipitate, not quartz. If, then, the rate of dissolution of am-SiO_2 is larger than the rates of nucleation and growth of quartz, am-SiO_2 is the expected product of homogeneous precipitation, whereas precipitation in the presence of seed particles larger than the 5 nm critical size should bypass the step rule and allow quartz to grow directly.

Earthquake prediction.

After the dust and anxiety settled, one of the most exciting things to emerge from the October 17, 1989 Loma Prieta (California) earthquake was the discovery of ultralow frequency (ULF; 0.01 to 10 Hz) electromagnetic radiation recorded near the epicenter for a few days before and after the event (ULF background noise was being recorded for an unrelated purpose, see Fraser-Smith et al., 1990). This is exciting to everyone who lives in earthquake country because it may eventually lead to an early warning system; but it should be exciting to interface geochemists for another reason. The electromagnetic radiation may originate in changes in streaming potential caused by fluctuations in groundwater velocity accompanying pre-fracture deformation of the crust. The reality of streaming potentials in natural aquifers is not the key question; streaming potentials have been observed between geothermal wells in Japan (Ishido et al., 1983). The questions of concern are the magnitudes and rates of change of streaming potential that can be expected, and the degree of attenuation of the signal as it travels through kilometers of water-saturated rock toward the surface. If streaming potentials contribute to the ULF signals, then if we learn enough about the surface chemistry involved we should be able to identify variables that will affect the magnitude and frequency dependence of the signal. With this understanding and the cooperation of hydrologists, geophysicists, and electrical engineers (who may understand the coupling between deformation and flow, and the relationship between the abundance and composition of groundwater and signal attenuation), it might be possible to develop a quantitative model and experimental tests of the phenomenon. There is a great deal of surface chemistry and physics involved, ranging from the influence of surface-induced order in bound water on viscosity to the pressure dependence of the character of adion-surface bonds and the role that will play in determining the proximity of counterions to the surface, hence the geometry of the diffuse layer and the magnitude of the streaming potential. We are not going to solve this problem here, but the topic provides a fascinating motivation for study and research in surface and interfacial geochemistry.

ACKNOWLEDGMENTS

Several generations of students and TAs in my course "Surfaces and Interfaces" have suffered through my attempts to collect, organize, and communicate the information and ideas presented here and suggested I write them down. Pat Johnsson, Carrick Eggleston, and Rebecca Rea studied early drafts of the paper, insisted on clarification and documentation in many important places, then helped with tedious proofreading. Gordon Brown, Mike Hochella, and Jonathan Stebbins have frequently asked my own favorite questions, "Why is that important?" and "How do you know that?" during discussion of some of the topics covered in this paper. Rob James and Jim Leckie often steered me toward answers. I am much indebted to all of these colleagues and I hope that some of their wisdom has survived. I am also grateful for financial support provided by the National Science Foundation, through grant EAR-8805440, and by the McGee Fund of the School of Earth Sciences at Stanford.

REFERENCES

Adamson, A.W. (1982) Physical Chemistry of Surfaces, 4th ed., John Wiley and Sons, New York, 664 pp.

Ahmed, A. (1967) Contact electrification of oxides in humid atmospheres. In: Parks, G.A., ed., Research in Mineral Processing, Progress Report 67-1, Dept. Mineral Engineering, Stanford University, Stanford, California, 69-77.

Ahmed, A. (1971) The Sorption of Aqueous Nickel (II) Species on Alumina and Quartz. PhD dissertation, Stanford University, Stanford, California, 113 pp.

Ainsworth, C.C., Chou, S.F.J., and Griffin, R.A. (1985) Sorption of Aroclor 1242 by earth materials in the presence of surfactants. 24th Hanford Life Sciences Symp., Oct. 1985, Richland, Washington (abstract).

Alexander, G.B. (1957) The effect of particle size on the solubility of amorphous silica in water. J. Phys. Chem. 61, 1563-1564.

Anderson, J.H. (1965) Calorimetric vs. infrared measures of adsorption bond strengths on silica. Surface Sci. 3, 290-291.

Anderson, J.H., and Parks, G.A. (1968) The electrical conductivity of silica gel in the presence of adsorbed water. J. Phys. Chem. 72, 3662-3668.

Anderson, J.H., and Wickersheim, K.A. (1964) Near infrared characterization of water and hydroxyl groups on silica surfaces. Surface Sci. 2, 252-260.

Atkinson, B.K. (1984) Subcritical crack growth in geological materials. J. Geophys. Res. 89, 4077-4114.

Atkinson, B.K., and Meredith, P.G. (1981) Stress corrosion cracking of quartz: a note on the influence of chemical environment. Tectonophysics 77, T1-T11

Avotins, P. (1975) Adsorption and Coprecipitation Studies of Mercury on Hydrous Iron Oxide. PhD dissertation, Stanford University, Stanford, California, 124 pp.

Baes, C.F., Jr., and Mesmer, R.E. (1986) The Hydrolysis of Cations. Robert E. Krieger Publ. Co., Malabar, Florida, 489 pp.

170

Balistrieri, L., and Murray, J.W. (1981) The surface chemistry of goethite (α-FeOOH) in major ion seawater. Am. J. Sci. 281, 788-806.

Benjamin, M.M., and Leckie, J.O. (1980) Adsorption of metals at oxide interfaces: Effects of the concentrations of adsorbate and competing metals. In: Baker, R.A., ed., Contaminants and Sediments, Vol. 2, Ann Arbor Science Publishers, Ann Arbor, Michigan, 305-321.

Benjamin, M.M., and Leckie, J.O. (1981) Competitive adsorption of Cd, Cu, Zn, and Pb on amorphous iron oxyhydroxide. J. Colloid Interface Sci. 83, 410-419.

Benjamin, M.M., and Leckie, J.O. (1982) Effects of complexation by Cl, SO_4, and S_2O_3 on adsorption behavior of Cd on oxide surfaces. Envir. Sci. Technol. 16, 162-170.

Berner, R.A. (1971) Principles of Chemical Sedimentology. McGraw Hill, New York, 240 pp.

Berube, Y.G., and de Bruyn, P.L. (1968) Adsorption at the rutile-solution interface, II. Model of the electrochemical double layer. J. Colloid Interface Sci. 28, 92.

Blake, P., and Ralston, J. (1985) Controlled methylation of quartz particles. Colloids Surfaces 15, 101-118.

Bockris, J.O'M., and Reddy, A.K.N. (1970) Modern Electrochemistry, Vol. 2. Plenum, New York, 1432 pp.

Bolt, G.H. (1955) Analysis of the validity of the Gouy-Chapman theory of the electric double layer. J. Colloid Sci. 10, 206-218.

Brace, W.F., and Walsh, J.B. (1962) Some direct measurements of the surface energy of quartz and orthoclase. Am. Mineral. 47, 1111-1122.

Bragg, W., and Gibbs, R.E. (1925) The structure of alpha and beta quartz. Proc. Roy. Soc. (London) A109, 405-427.

Brown, G.E. (1990) Spectroscopic studies of chemisorption reactions mechanisms at oxide-water interfaces. This volume.

Carmichael, I.S., Turner, F.J., and Verhoogen, J. (1974) Igneous Petrology. McGraw-Hill Book Co., San Francisco, California, 739 pp.

Carrière, B, and Lang, B. (1977) A study of the charging and dissociation of SiO_2 surfaces by AES. Surface Sci. 64, 209-223.

Cases, J., Doerler, N., and François, M. (1985) Influence de differents types de broyage fin sur les mineraux: I. Cas du quartz. XV Int'l Congress on Mineral Processing, Vol. 1, Editions GEDIM, Cannes, France, 169-179.

Cases, J., and François, M. (1982) Etude des propriétés thermodynamiques de l'eau au voisinage des interfaces. Agronomie 2, 931-938.

Casey, W.H., Westrich, H.R., and Arnold, G.W. (1988) Surface chemistry of labradorite feldspar reacted with aqueous solutions at pH = 2, 3, and 12. Geochim. Cosmochim. Acta 52, 2795-2807.

Chisholm-Brause, C.J., Hayes, K.F., Roe, A.L., Brown, G.E., Jr., and Parks, G.A. (1990) Sorption mechanisms of Pb(II) at the γ-Al_2O_3/water interface. Geochim. Cosmochim. Acta 54, (in press).

Chisholm-Brause, Parks, G.A., and Brown, G.E., Jr. (1990) EXAFS study of changes in Co(II) sorption complexes on γ-Al_2O_3 with increasing adsorption densities. Submitted for publication in Proceeding of XAFS VI, 6th Int'l Conf. on X-Ray Absorption Fine Structure, York, England, August, 1990.

Combes, J-M. (1988) Evolution de la Structure Locale des Polymeres et Gels Ferriques Lors de la Cristallisation des Oxydes de Fer. PhD dissertation, Universitè Pierre et Marie Curie, Paris, 194 pp.

Crerar, D.A., Axtmann, E.V., and Axtmann, R.C. (1981) Growth and ripening of silica polymers in aqueous solutions. Geochim. Cosmochim. Acta 45, 1259-1266.

Davis, J.A., and Leckie, J.O. (1978a) Effect of complexing ligands on trace metal uptake by hydrous oxides. Envir. Sci. Technol. 12, 1309-1315.

Davis, J.A., and Leckie, J.O. (1978b) Surface ionization and complexation at the oxide/water interface. II. Surface properties of amorphous iron oxyhydroxide and adsorption of metal ions. J. Colloid Interface Sci. 67, 90-107.

Davis, J.A., James, R.O., and Leckie, J.O. (1978) Surface ionization and complexation at the oxide/water interface. I. Computation of electrical double layer properties in simple electrolytes. J. Colloid Interface Sci. 63, 480-499.

Denbigh, K.G. (1971) Principles of Chemical Equilibrium, 3rd. ed. Cambridge University Press, Cambridge, England.

Derjaguin, B.V., and Churaev, N.V. (1986) Properties of water layers adjacent to interfaces. In: Croxton, C. A., ed., Fluid Interfacial Phenomena. John Wiley and Sons, Ltd., London, 663-738.

Drever, J.I. (1988) The Geochemistry of Natural Waters. Prentice Hall, Englewood Cliffs, New Jersey, 437 pp.

171

Dunning, J.D., Petrovski, D., Schuyler, J., and Owens, A. (1984) The effects of aqueous chemical environments on crack propagation in quartz. J. Geophys. Res. 89, 4115-4123.

Enüstün, B.V., and Turkevich, J. (1960) Solubility of fine particles of strontium sulfate. J. Phys. Chem. 82, 4503-4509.

Every, R.L., Wade, W.A., and Hackerman, N. (1961) Free energy of adsorption. I. The influence of substrate structure in the SiO_2-H_2O, SiO_2-n-hexane, and SiO_2-Ch_3OH systems. J. Phys. Chem. 65, 25-29.

Farley, K.J., Dzombak, D.A., and Morel, F.M.M. (1985) A surface precipitation model for sorption of cations on metal oxides. J. Colloid Interface Sci. 106, 226-242.

Fraser-Smith, A.C., Bernardi, A., McGill, P.R., Ladd, M.E., Helliwell, R.A., and Villard, O.G., Jr. (1990) Low-frequency magnetic field measurements near the epicenter of the M_S 7.1 Loma Prieta Earthquake. Accepted by Geophys. Res. Lett.

Freiman, S.W. (1984) Effects of chemical environments on slow crack growth in glasses and ceramics. J. Geophys. Res. 89, 4072-4076.

Freyberger, W.L., and de Bruyn, P.L. (1957) The electrochemical double layer on silver sulfide. J. Phys. Chem. 61, 586-592.

Fripiat, J., Cases, J, François, M., and Letellier, M. (1982) Thermodynamic and microdynamic behavior of water in clay suspensions and gels. J. Colloid Interface Sci. 89, 378-400.

Fuerstenau, D.W. (1970) Interfacial processes in mineral/water systems. Pure Appl. Chem. 24, 135-164.

Fuerstenau, D.W., and Raghavan, S. (1978) The surface and crystal chemistry of silicate minerals and their flotation behavior. Freiberger Forschungshefte, VEB Deutshcer Verlag für Grundstoffindustrie, A593, 75-108.

Gallei, E., and Parks, G.A. (1972) Evidence for surface hydroxyl groups in attenuated total reflectance spectra of crystalline quartz. J. Colloid Interface Sci. 38, 650-651.

Gaudin, A.M., and Rizo-Patron, A. (1942) The mechanism of activation in flotation. Am. Inst. Mining Metal. Engin., Tech. Pub. 1453, 1-9.

Gaudin, A.M., and Hukki, R.T. (1944) Principles of comminution: Size and surface distribution. Am. Inst. Mining Metal. Engin., Tech. Pub. 1779, 1-17.

Gibbs, R.E. (1926) Structure of alpha quartz. Proc. Roy. Soc. A110, 443-455.

Giblin, A.M. (1980) The role of clay adsorption in genesis of uranium ores. In: Ferguson, J., and Goleby, A.B., eds., Uranium in the Pine Creek Geosyncline. Int'l Atomic Energy Agency, Vienna, 521-529.

Girvin, D.C., Ames, L.L., Schwab, A.P., and McGarrah, J.E. (1983) Neptunium adsorption on synthetic amorphous iron oxyhydroxide. Report BNL-SA-11229, Battelle Northwest Laboratories, Richland, Washington.

Gross, J., and Zimmerley, S.R. (1928) Crushing and grinding, II. The relation of measured surface of crushed quartz to sieve sizes. Am. Inst. Mining Metal. Eng., Tech. Pub. 126, 1-8.

Hahn, H.H., and Stumm, W. (1968) Coagulation by Al(III). The role of adsorption of hydrolyzed aluminum in the kinetics of coagulation. In: Weber, W.J., Jr., and Matijevic, E., eds., Adsorption From Aqueous Solution, Advances in Chem. Series, 79, Am. Chem. Soc., Washington, D.C., 91-111.

Hair, M.L. (1967) Infrared Spectroscopy in Surface Chemistry. Marcel Dekker, New York.

Harned, H.S., and Owen, B.B. (1958) The Physical Chemistry of Electrolytic Solutions, 3rd ed. Reinhold Publishing Corp., New York, 803 pp.

Hartley, N.E.W., and Wilshaw, T.R. (1973) Deformation and fracture of synthetic a-quartz. J. Mat. Sci. 8, 265-278.

Hayes, K.F., and Leckie, J.O. (1987) Modeling ionic strength effects on cation adsorption at hydrous oxide/solution interfaces. J. Colloid Interface Sci. 115, 564-572.

Hayes, K.F., Roe, A.L., Brown, G.E., Jr., Hodgson, K.O., Leckie, J.O., and Parks, G.A. (1987) In situ x-ray absorption study of surface complexes at oxide/water interfaces: selenium oxyanions on α-FeOOH. Science 238, 783-786.

Healy, T.W., James, R.O., and Cooper, R. (1968) The adsorption of aqueous Co(II) at the silica-water interface. In: Weber, W.J., Jr., and Matijevic, E., eds., Adsorption from Solution, Advances in Chem Series., 79. Am. Chem. Soc., Washington, D.C., 62-73.

Healy, T.W., and Fuerstenau, D.W. (1972) Principles of mineral flotation. In: Lemlich, R., ed., Adsorptive Bubble Separation Techniques. Academic Press, New York, 91-131

Hiemenz, P.C. (1977) Principles of Colloid and Surface Chemistry. Marcel Dekker, New York, 516 pp.

172

Hobbs, P.V., and Deepak, A. (1981) Clouds, Their Formation, Optical Properties, and Effects. Academic Press, San Francisco, 497 pp.

Hochstrasser, G., and Antonini, J.F. (1972) Surface states of pristine silica surfaces. 1. ESR studies of E'$_s$ dangling bonds and of CO_2^- adsorbed radicals. Surface Sci. 32, 644-664.

Hohl, H., and Stumm, W. (1976) Interaction of Pb^{2+} with hydrous γ-Al_2O_3. J. Colloid Interface Sci. 55, 281-288.

Honeyman, B.D., and Leckie, J.O. (1986) Macroscopic partitioning coefficients for metal ion adsorption. In: Hayes, K.F. and Davis, J.A., Geochemical Processes at Mineral Surfaces. ACS Symposium Series, Vol. 323, Am. Chem. Soc., Washington, D.C., 162-190.

Honig, E.P., and Hengst, J.H.Th. (1969) Points of zero charge of inorganic precipitates. J. Colloid Interf. Sci. 29, 510-520.

Huggenberger, F., Letey, J., and Farmer, W.J. (1973) Effect of two nonionic surfactants on adsorption and mobility of selected pesticides in a soil system. Proc. Soil Sci. Soc. Am. 37, 215-219.

Iler, R.K. (1979) The Chemistry of Silica: Solubility, Polymerization, Colloid, and Surface Properties, and Biochemistry. John Wiley and Sons, New York, 866 pp.

Ishido, T., and Mizutani, H. (1980) Relationship between fracture strength of rocks and ζ-potential. Tectonophysics 67, 13-23.

Ishido, T., Mizutani, H., and Baba, K. (1983) Streaming potential observations, using geothermal wells and in-situ electrokinetic coupling coefficients under high temperature. Tectonophysics 91, 89-104.

Ishido, T., and Nishizawa, O. (1984) Effects of ζ-ptotential on microcrack growth in rock under relatively low uniaxial compression. J. Geophys. Res. 89, 4153-4159.

Israelachvili, J.N. (1985) Intermolecular and Surface Forces. Academic Press, New York.

Iwasaki, I. (1957) Electrochemical Studies of Adsorption on Silver Sulfide Surfaces. PhD dissertation, Massachusetts Institute of Technology, Cambridge, Massachusetts, 115 pp.

James, R.O. (1971) The Adsorption of Hydrolyzable Metal Ions at the Oxide/Water Interface, PhD dissertation, University of Melbourne, Parkville, Australia, 286 pp.

James, R.O., and Healy, T.W. (1972a) Adsorption of hydrolyzable metal ions at the oxide-water interface. I. Co(II) adsorption on SiO_2 and TiO_2 as model systems. J. Colloid Interface Sci. 40, 42-52.

James, R.O., and Healy, T.W. (1972b) Adsorption of hydrolysable metal ions at the oxide-water interface. II. Charge reversal of SiO_2 and TiO_2 colloids by adsorbed Co(II), La(III), and Th(IV) as model systems. J. Colloid Interface Sci. 40, 53-64.

James R.O. and MacNaughton, M.G. (1977) The adsorption of aqueous heavy metals on inorganic minerals. Geochim. Cosmochim. Acta 41, 1549-1555.

James, R.O., and Parks, G.A. (1975) Adsorption of zinc(II) at the cinnabar (HgS)/H_2O interface. Am. Inst. Chem. Eng. Symp. Ser. 71, #150, 157-64.

James, R.O., and Parks, G.A. (1982) Characterization of aqueous colloids by their electrical double layer and intrinsic surface chemical properties. Surface and Colloid Sci. 12, 119-216.

James, R.O., Stiglich, P.J., and Healy, T.W. (1975) Analysis of models in adsorption of metal ions at the oxide/water interface. Disc. Farad. Soc. 59, 142-156.

James, R.O., Stiglich, P.J., and Healy, T.W. (1981) The TiO_2/aqueous electrolyte system—Applications of colloid models and model colloids. In: Tewari, P.H., ed., Adsorption from Aqueous Solutions. Plenum, New York, 19-40.

James, R.O., Wiese, G. R., and Healy, T.W. (1977) Charge reversal coagulation of colloidal dispersions by hydrolyzable metal ions. J. Colloid Interface Sci. 59, 381-385.

Jiusto, J.E. (1981) Fog structure. In: Hobbs, P.V., and Deepak, A., eds., Clouds, Their Formation, Optical Properties, and Effects. Academic Press, San Francisco, California, 187-239.

Kamatani, A., and Riley, J.P. (1979) Rate of dissolution of diatom silica walls in seawater. Marine Biology 55, 29-35.

Kastner, M., Keene, J.B., and Gieskes, J.M. (1977) Diagenesis of siliceous oozes—I. Chemical controls on the rate of opal-A to opal-CT transformation—an experimental study. Geochim. Cosmochim. Acta 41, 1041-1059.

Kendall, K., Alford, N.McN., and Birchall, J.D. (1987) A new method for measuring the surface energy of solids. Nature 325, 794-796.

Kent, D.B. (1983) On the Surface Chemical Properties of Synthetic and Biogenic Amorphous Silica. PhD dissertation, University of California, San Diego, 420 pp.

Kinniburgh, D.G., and Jackson, M.L. (1978) Adsorption of mercury(II) by iron hydrous oxide gel. Soil Sci. Soc. Am. J. 42, 45-47.

Kinniburgh, D.G., and Jackson, M.L. (1982) Concentration and pH dependence of calcium and zinc adsorption on iron hydrous oxide gel. Soil Sci. Soc. Am. J. 40, 796-799.

Kinniburgh, D.G., Jackson, M.L., and Syers, J.K. (1976) Adsorption of alkaline earth, transition, and heavy metal cations by hydrous oxide gels of iron and aluminum. Soil Sci. Soc. Am. J. 46, 56-61.

Kirby, S. H. (1984) Introduction and digest to the special issue on chemical effects of water on the deformation and strength of rocks. J. Geophys. Res. 89, 3991-3995.

Kitchener, J.A. (1965) Mechanism of adsorption from aqueous solutions: some basic problems. J. Photog. Sci. 13, 152-159.

Klier, K., Shen, J.H., and Zettlemoyer, A.C. (1973) Water on silica and silicate surfaces. I. Partially hydrophobic silicas. J. Phys. Chem. 77, 1458-1465.

Korpi, G.K. (1965) Electrokinetic and Ion Exchange Properties of Aluminum Oxide and Hydroxides. PhD dissertation, Stanford University, Stanford, California, 132 pp.

Langmuir, D. (1978) Uranium solution-mineral equilibria at low temperatures with applications to sedimentary ore deposits. Geochim. Cosmochim. Acta 42, 547-569.

Lawn, B.R., and Wilshaw, T.R. (1975) Fracture of Brittle Solids. Cambridge, New York, 204 pp.

Lewin, J.C. (1961) The dissolution of silica from diatom walls. Geochim. Cosmochim. Acta 21, 182-198.

Li, C-W, and Kingery, W.D. (1984) Solute segregation at grain boundaries in polycrystalline Al_2O_3. In: Kingery, W.D., ed., Structure and Properties of MgO and Al_2O_3 Ceramics, Am. Ceram. Soc., Columbus, Ohio, 368-378.

Li, H.C. (1955) The Adsorption of Inorganic Ions on Quartz. MS thesis, Dept. Metallurgy, Massachusetts Institute of Technology, Cambridge, Massachusetts, 45 pp.

Li, H.C. (1958) Adsorption of Organic and Inorganic Ions on Quartz. PhD dissertation, Department of Metallurgy, Massachusetts Institute of Technology, Cambridge, Massachusetts, 118 pp.

Li, H.C. and de Bruyn, P.L. (1966) Electrokinetic and adsorption studies on quartz. Surface Sci. 5, 203-220.

Loganathan, P., and Burau, R.G. (1973) Sorption of heavy metal ions by a hydrous manganese oxide. Geochim. Cosmochim. Acta 37, 1277-1293.

MacDonald, R.S. (1958) Surface functionality of amorphous silica by infrared spectroscopy. J. Phys. Chem. 62, 1168-1176.

McCafferty, E., and Zettlemoyer, A.C. (1971) Adsorption of water vapour on a-Fe_2O_3. Disc. Farad. Soc. 52, 239-255.

McBirney, A.R., and Murase, T. (1971) Factors governing the formation of pyroclastic rocks. Bull. Volc. 34, 372-384.

MacNaughton, M.G., and James, R.O. (1974) Adsorption of aqueous mercury complexes at the oxide/water interface. J. Colloid Interface Sci. 47, 431-440.

Makrides, A.C., Turner, M., and Slaughter, J. (1980) Condensation of silica from supersaturated silicic acid solutions. J. Colloid Interface Sci. 73, 345-367.

Matijevic, E. (1961) Detection of metal ion hydrolysis by coagulation. III. Aluminum. J. Phys. Chem. 65, 826-830.

Matijevic, E. (1967) Charge reversal of lyophobic colloids. In: Faust, S.D. and Hunter, J.V., eds., Principles and Applications of Water Chemistry. John Wiley and Sons, New York, 328-369.

Matijevic, E. (1973) Colloid stability and complex chemistry. J. Colloid Interface Sci. 43, 217-245.

Michalske, T.A., and Bunker, B.C. (1984) Slow fracture model based on strained silicate structures. J. Appl. Phys. 56, 2686-2693.

Miyata, K. (1968) Free energy of adsorption of water vapor on quartz. Nippon Kagaku Zasshi 89, 346-349.

Modi, H. J., and Fuerstenau, D.W. (1957) Streaming potential studies on corundum in aqueous solutions of inorganic electrolytes. J. Phys. Chem. 61, 640-643.

Moignard, M. S., James, R.O., and Healy, T.W. (1977) Adsorption of calcium at the zinc sulphide-water interface. Aust. J. Chem. 30, 733-740.

Moore, W.J. (1972) Physical Chemistry. Prentice-Hall, Englewood Cliffs, New Jersey, 977 pp.

Morse, J.W., and Casey, W.H. (1988) Ostwald processes and mineral paragenesis in sediments. Am. J. Sci. 288, 537-560.

Motschi, H. (1987) Aspects of the molecular structure in surface complexes; Spectroscopic investigations. In: Stumm, W., ed., Aquatic Surface Chemistry. Wiley-Interscience, New York, 111-125.

Murray, J.W. (1975) The interaction of metal ions at the manganese dioxide-solution interface. Geochim. Cosmochim. Acta 39, 505-519.

174

Muto, T., Hirono, S., and Kurata, H. (1965) Some aspects of fixation of uranium from natural waters. Kozan Chishitsu, 15, 287-298.

O'Day, P.A., Brown, G.E., Jr., and Parks, G.A. (1990) EXAFS study of changes in Co(II) sorption complexes on kaolinite and quartz surfaces. Submitted for publication in Proceedings of XAFS VI, 6th Int'l Conference on X-Ray Absorption Fine Structure, York, England, August, 1990.

O'Day, P.A., C.J. Chisholm-Brause, G.E. Brown, Jr., and G.A. Parks (1989) Synchrotron-based XAS study of Co(II) sorption complexes at mineral/water interfaces: Effect of mineral surfaces (abstr.), 28th Int'l Geological Congress, 2-534.

O'Melia, C.R., and Stumm, W. (1967) Aggregation of silica dispersions by iron(III). J. Colloid Interface Sci. 23, 437-447.

Park, S.W., and Huang, C.P. (1987) The surface acidity of hydrous CdS(s). J. Colloid Interface Sci. 117, 431-441.

Parks, G.A. (1965) The isoelectric points of solid oxides, solid hydroxides, and aqueous hydroxo complex systems. Chem. Rev. 65, 177-198.

Parks, G.A. (1967) Surface chemistry of oxides in aqueous systems. In: Stumm, W., ed., Equilibrium Concepts in Aqueous Systems. Adv. in Chem. Ser. 67, Am. Chem. Soc., Washington, D.C., 121-160.

Parks, G.A. (1968) Contact Electrification of Oxides in Humid Atmospheres. Terminal Report: NSF Grant GP 172, Department of Mineral Engineering, Stanford University, Stanford, California, 23 pp.

Parks, G.A. (1975) Adsorption in the marine environment. In: Riley, J.P., and Skirrow, G., eds., Marine Geochemistry, 2nd ed., Vol. I. Academic Press, New York, 241-308.

Parks, G.A. (1984) Surface and interfacial free energies of quartz. J. Geophys. Res. 89, 3997-4008.

Pashley, R.M., and Kitchener, J.A. (1979) Surface forces in adsorbed layers of water on quartz. J. Colloid Interface Sci. 71, 491-500.

Pauling, L. (1927) The sizes of ions and the structure of ionic crystals. J. Am. Chem. Soc. 49, 765-790.

Perona, M.J. and Leckie, J.O. (1985) Proton stoichiometry for the adsorption of cations on oxide surfaces. J. Colloid Surf. Sci. 106, 64-69.

Posselt, H.S., Anderson, F.J., and Weber, W.J., Jr. (1968) Cation adsorption on colloidal hydrous manganese dioxide. Envir. Sci. Technol. 2, 1087-1093.

Rea, R.L., and Parks, G.A. (1990) Numerical simulation of coadsorption of ionic surfactants with inorganic ions on quartz. In: Melchior, D.C., and Bassett, R.L., eds., Chemical Modelling in Aqueous Systems II. ACS Symposium Series 416, Am. Chem. Soc., Washington, D.C., 260-271.

Roco, J.N. (1966) Contact electrification of oxides in humid atmospheres. In: Parks, G.A., ed., Research in Mineral Processing, Progress Report 66-1, Department of Mineral Engineering, Stanford University, Stanford, California, 94305, 16-26.

Roe, A.L., Hayes, K.F., Chisholm-Brause, C.J., Brown, G.E., Jr., and Parks, G.A. (1990) X-ray absorption study of lead complexes at α-FeOOH/water interfaces. Langmuir (in press).

Sanchez, A.L., Murray, J.W., and Sibley, T.H. (1985) The adsorption of plutonium IV and V on goethite. Geochim. Cosmochim. Acta 49, 2297-2307.

Schindler, P.W. (9167) Heterogeneous equilibria involving oxides, hydroxides, carbonates, and hydroxide carbonates. In: Stumm, W., ed. Equilibrium Concepts in Natural Water Systems. Am. Chem. Soc., Washington, D.C., 196-221.

Schindler, P.W., and Stumm, W. (1987) The surface chemistry of oxides, hydroxides, and oxide minerals. In: Stumm, W., ed., Aquatic Surface Chemistry. Wiley-Interscience, New York, 83-110.

Schmid, A-M.M., Borowitzka, M.A., and Volcani, B.E. (1981) Morphogenesis and biochemistry of diatom cell walls. In Kiermayer, O., ed., Cell Biology Monographs, Vol. 8, Cytomorphogenesis in Plants. Springer-Verlag, New York, 63-97.

Shaw, D.J. (1970) Introduction to colloid and Surface Chemistry, 2nd ed., Butterworths, London, 236 pp.

Smith, L.W., and Bayer, D.E. (1967) Soil adsorption of diuron as influenced by surfactants. Soil Sci. 103, 328-330.

Smith, R.W., and Hem, J.D. (1972) Effect of aging on aluminum hydroxide complexes in dilute aqueous solution. U.S. Geol. Survey Water Supply Paper 1827-D, 51 pp.

Snoeyink, V.L., and Weber, W.J. (1972) Surface functional groups on carbon and silica. Progress in Surface Membrane Sci. 5, 63-119.

Sposito, G. (1984) The Surface Chemistry of Soils. Oxford, New York, 234 pp.

Stöber, W. (1967) Formation of silicic acid in aqueous suspensions of different silica modifications. In: Stumm, W., ed. Equilibrium Concepts in Natural Water Systems. Am. Chem. Soc., Washington, D.C., 161-181.

Stroes-Gascoyne, S., Kramer, J.R., and Snodgrass, W.J. (1986) A new model describing the adsorption of copper on MnO_2. Envir. Sci. Technol. 20, 1047-1050.

Stumm, W., and Morgan, J. J. (1981) Aquatic Chemistry. John Wiley and Sons, New York, 780 pp.

Thompson, J.B. (1987) A simple thermodynamic model for grain interfaces. In: Helgeson, H.C., Chemical Transport in Metasomatic Processes. D. Reidel, Dordrecht, The Netherlands, 169-188.

Tolman, R.C. (1949) The effect of droplet size on surface tension. J. Chem. Phys. 17, 333-337.

Tripathi, V.S. (1983) Uranium (VI) Transport Modeling: Geochemical Data and Submodels. PhD dissertation, Stanford University, Stanford, California., 297 pp.

Van Lier, J.A., De Bruyn, P.L., and Overbeek, J.Th.G. (1960) The solubility of quartz. J. Phys. Chem. 64, 1675-1682.

Weres, O., Yee, A, and Tsao, L. (1980) Kinetics of Silica Polymerization. Tech. Rept. LBL-7033, Earth Sciences Division, Lawrence Berkeley Lab., University of California, Berkeley, California, 256 pp.

Whalen, J.W. (1961) Thermodynamic properties of water adsorbed on quartz. J. Phys. Chem. 65, 1676-1681.

Wiederhorn, S.M. (1974a) Subcritical crack growth in ceramics. In: Bradt, R.C., Hasselman, D.P.H., and Lange, F.F., eds., Fracture Mechanics of Ceramics, Vol. 2, Microstructure, Materials, and Applications. Plenum Press, New York, 613-646.

Wiederhorn, S. M. (1974a) Subcritical crack growth in ceramics. In: Bradt, R.C., Hasselman, D.P.H., and Lange, F.F., eds., Fracture Mechanics of Ceramics, Vol. 4, Crack Growth and Microstructure. Plenum Press, New York, 549-580.

Wiederhorn, S.M., and Fuller, E.R., Jr. (1989) Effect of surface forces on subcritical crack growth in glass. J. Am. Ceram. Soc. 72, 248-251.

Williams, L.A., Parks, G.A., and Crerar, D.A. (1985) Silica diagenesis, I. Solubility controls. J. Sed. Pet. 55, 301-311.

Willey, J.D. (1980) Effects of aging on silica solubility: a laboratory study. Geochim. Cosmochim. Acta 44, 573-578.

Yates, D.J.C., and Sheppard, N. (1957) Studies of physically adsorbed gases by means of infra-red spectroscopy. In: Schulman, J.H., ed., Solid/Gas Interface, Vol. II of Proc. 2nd Int'l Congress on Surface Activity. Butterworths, London, 27-34.

Yoon, R.H., Salman, T., and Donnay, G. (1979) Predicting points of zero charge of oxides and hydroxides. J. Colloid Interface Sci. 70, 483-493.

Yopps, J.A., and Fuerstenau, D.W. (1964) The zero point of charge of alpha alumina. J. Colloid Sci. 19, 61-71.

Zettlemoyer, A.C., Micale, F.J., and Klier, K. (1975) Adsorption of water on well characterized solid surfaces. In: Franks, F., ed., Water, A Comprehensive Treatise, Vol. 5. Plenum, New York, 249-291.

CHAPTER 5   J. A. DAVIS AND D. B. KENT

SURFACE COMPLEXATION MODELING
IN AQUEOUS GEOCHEMISTRY

INTRODUCTION

The quantitative description of *adsorption* as a purely macroscopic phenonmenon is achieved through the concept of *relative surface excess* (see chapter by Sposito, this volume). For accumulation of solute i at the mineral-water interface, the relative surface excess, Γ_i, or adsorption density of solute i, may be defined as:

$$\Gamma_i = \frac{n_i}{\mathcal{A}} \quad , \tag{1}$$

where n_i is the moles of surface excess of solute i per unit mass of the mineral phase, and $\mathcal{A}$ is the specific surface area of the mineral phase. This definition assumes that solute i does not enter the structure of the mineral phase. The term, *sorption*, is used in a more general sense to include processes such as surface precipitation or diffusion of solutes into porous materials (Sposito, 1986).

While Equation 1 appears simple, thermodynamic definition and measurement of the quantities n_i and $\mathcal{A}$, especially for soils, sediments, and rock surfaces, are complex topics. The surfaces of minerals are *loci* of a large suite of chemical reactions, including adsorption, ion exchange, electron transfer reactions, precipitation, dissolution, solid solution formation, hydrolysis, and polymerization (Davis and Hayes, 1986; Stumm, 1987). A thorough, mechanistic understanding of adsorption-desorption reactions is needed to understand their role as preliminary steps in various geochemical processes (Stone, 1986; Davis et al.,1987; and see chapter by Stumm, this volume). Soil scientists and analytical chemists have recognized for many decades that mineral surfaces and precipitates can adsorb a large number of ions and molecules (Bolt, 1982). However, the interest in adsorption processes now extends to many other fields, including geochemistry, hydrogeology, chemical oceanography, aquatic toxicology, water and wastewater treatment, and chemical, metallurgical, and mining engineering. As a result, there has been an explosive growth in experimental data describing the adsorption of ions on mineral surfaces in the last 30 years. Simultaneously, there have been significant advances in understanding the fundamental mechanisms involved in adsorption-desorption equilibria.

Experimental sorption data are described by various empirical means, including partition coefficients, iostherm equations, and conditional binding constants (Honeyman and Leckie, 1986). Because these relationships and empirical parameters are highly dependent on the chemical composition of aqueous solutions, empirical approaches to describing adsorption equilibria in aqueous geochemistry are considered unsatisfactory by many scientists (Davis et al., 1990; Honeyman and Santschi, 1988; Kent et al., 1986). Instead, it is envisioned that *surface complexation theory*, which describes adsorption in terms of chemical reactions between surface functional groups and dissolved chemical species, can be coupled with aqueous speciation models to describe adsorption equilibria within a general geochemical framework. While this approach has been adopted in computer models developed by experimental aquatic chemists (e.g., MINEQL, MINTEQ and HYDRAQL; Westall et al., 1976; Brown and Allison, 1987; Papelis et al., 1988), it has not been included in models used widely within the geochemical community (e.g., PHREEQE and EQ3/EQ6; Parkhurst et al., 1980; Wolery, 1983). A recent modification of the PHREEQE code that includes surface complexation equilibria has been published by Brown et al. (1990).

One of the reasons for the lack of acceptance of the surface complexation approach by geochemical modelers has been the development of several competing models, each with its own set of thermodynamic data and other model parameters. In addition to this problem, most of the apparent equilibrium constants for adsorption equilibria that are published in the literature are not self-consistent. Hopefully, this situation will soon improve. Dzombak and Morel (1990) recently published an excellent treatise on one of the surface complexation models (the diffuse double layer model) that included a critically reviewed and self-consistent thermodynamic database for use of the model with ferrihydrite. The extensive data sets in the literature at present should make it possible to compile databases for other mineral phases and each of the surface complexation models. In this chapter we argue that each of the models has advantages and disadvantages when evaluated in terms of model applicability to various geochemical environments. Thus, the existence of several adaptations of surface complexation theory should be viewed optimistically, since it provides modelers with several ways to approach the extremely complex interactions of natural systems.

Extensive experience has shown that the each of the models is extremely successful in simulating adsorption data from laboratory experiments with well-characterized synthetic minerals (Dzombak and Morel, 1987). Guidelines for the use of surface complexation models in geochemical applications are slowly developing (Fuller and Davis, 1987; Charlet and Sposito, 1987, 1989; Zachara et al., 1989b; Payne and Waite, 1990). However, application of the modeling approach to natural systems is considerably more difficult than when applied in simple mineral-water systems (Luoma and Davis, 1983; Bolt and van Riemsdijk, 1987; McCarthy and Zachara, 1989). In particular, there are severe difficulties associated with describing the reactive functional groups of soil, sediment, and rock surfaces and the electrical double layer properties of mixtures of mineral phases and their surface coatings. A realistic appraisal of these problems can lead to a pessimistic view of the future of surface complexation modeling in aqueous geochemistry (e.g., Bolt and van Riemsdijk, 1987).

In this chapter we review fundamental aspects of electrical double layer theory and the adsorption of inorganic ions at mineral-water interfaces. An important prerequisite for applying surface complexation theory is a detailed knowledge of surface functional groups, surface area, and the porosity of adsorbing mineral phases. Accordingly, we have discussed these topics in some detail. Both surface complexation and empirical approaches to adsorption modeling in natural systems are reviewed. The surface complexation models are compared in terms of their capabilities to simulate data and applicability to natural systems. Practical issues involving parameter estimation and experimental techniques applied to soils and sediments are discussed. Finally, guidelines for the application of surface complexation theory to natural systems are formulated based on the current state of knowledge.

SURFACE FUNCTIONAL GROUPS

Surface functional groups and mineral types

In surface complexation theory, adsorption is described in terms of a set of complex formation reactions between dissolved solutes and surface functional groups (Sposito, 1984). The free energies of the reactions can be divided into chemical and electrostatic contributions for modeling purposes. Surface functional groups influence both terms. The nature of the surface functional groups controls the *stoichiometry* of the adsorption reaction, and hence the variation in adsorption with solution chemistry. Surface functional groups also influence the electrical properites of the interface, and their density controls the adsorption capacity. Accordingly, the definition of *surface functional groups* represents the most basic concept of surface complexation theory.

A variety of materials are of interest as adsorbents in the field of aqueous geochemistry. For the purposes of this discussion, these materials are classified in terms of the nature of the surface functional groups that they possess (Kent et al., 1986). Hydrous oxide minerals and natural organic particulate matter possess proton-bearing surface functional groups. Consequently, adsorption onto these solid phases is pH dependent. Aluminosilicate minerals are divided into those that possess a permanent structural charge and those that do not.

Aluminosilicate minerals without permanent charge have proton-bearing surface functional groups, hence they are discussed along with hydrous oxides. Minerals with permanent structural charge, such as clay minerals, micas, zeolites and most Mn oxides, have ion-bearing *exchange* sites in addition to proton-bearing surface functional groups. The surface functional groups of *salt-type* minerals bear the cation or anion of the salt, e.g., Ca^{2+} or CO_3^{2-} on the calcite surface. Sulfide minerals are potentially important in many reducing environments. These minerals possess both proton-bearing and salt-type surface functional groups; however, metal ion sorption by sulfides may be controlled primarily by surface precipitation reactions.

Oxides and aluminosilicates without permanent charge

Surface hydroxyl groups constitute the complexation sites on oxide and aluminosilicate minerals that do not possess a fixed charge (James and Parks, 1982; Bolt and van Riemsdijk, 1987). Various types of surface hydroxyl groups are illustrated in Figure 1. M and O moieties present at the surface suffer from an imbalance of chemical forces relative to the bulk. This imbalance is satisfied by "chemisorbing" water to form surface hydroxyl groups (see Parks, this volume). Hydrogen bonding between surface hydroxyl groups and either water vapor or liquid water gives rise to layers of physically adsorbed water (Fig. 1).

Types of surface hydroxyl groups. Analyses of the crystal structures of oxide and aluminosilicate minerals indicate that the different types of surface hydroxyls have different reactivities (James and Parks, 1982; Sposito, 1984). Goethite (α-FeOOH), for example, has four types of surface hydroxyls whose reactivities depend upon the coordination environment of the O in the FeOH group (Fig. 2). The FeOH sites are designated A-, B-, or C-type sites, depending on whether the O is coordinated with 1, 3, or 2 adjacent Fe(III) ions. A fourth type of site, designated a Lewis-acid site, results from chemisorption of a water molecule on a "bare" Fe(III) ion. Sposito (1984) argues that only A-type sites are basic, i.e., can from a surface complex with H+, while both A-type and Lewis acid sites can release a proton. B- and C-type sites are considered unreactive. Thus, A-type sites can act as either a proton acceptor or a proton donor (i.e., are *amphoteric*). The water coordinated with Lewis-acid sites may act only as a proton donor site (i.e., an acidic site). Aluminosilicates posses both *aluminol* (≡ AlOH) and *silanol* (≡ SiOH) groups. Thus kaolinite has three types of surface hydroxyl groups: aluminol, silanol, and water adsorbed to Lewis acid sites (Fig. 3).

Spectroscopic studies have confirmed the presence of different types of surface hydroxyl groups on oxide minerals. IR (Infra-red) spectra of silica and aluminosilicate catalysts show bands corresponding to two different types of surface OH groups (James and Parks, 1982). Oxides of Al, Fe, Ti, and other metals show more bands, indicating the presence of several types of surface OH groups (James and Parks, 1982). Parfitt and Rochester (1976) discuss the relationship between IR spectra of molecules adsorbed at surface OH groups and the acidity of the surface OH groups. Application of Si-29 cross-polarization magic angle spinning NMR spectroscopy (CP MAS NMR) has confirmed the presence of surface hydroxyl groups with different chemical environments on amorphous silica (am-SiO_2). Cross-polarization is a method by which the spin from the ^1H (proton) spin system is transfered to nearby ^{29}Si centers (Pines et al., 1973). This technique eliminates spectral contributions from framework ^{29}Si centers and reduces complications arising from the presence of adsorbed water, which hamper the interpretation of spectral contributions of surface OH groups. ^{29}Si CP MAS spectra of am-SiO_2 samples show three peaks that correspond to O_4Si, $O_3Si(OH)$, and $O_2Si(OH)_2$ (Maciel and Sindorf, 1980; Sindorf and Maciel, 1981, 1983).

Density of surface hydroxyl groups. James and Parks (1982) offer a thorough review of the various methods used to quantify the density of surface hydroxyl groups. Analysis of the densities of groups that outcrop on the various crystal and cleavage faces of a mineral are possible if the crystal structure and habit are known (James and Parks, 1982; Sposito, 1984; Davis and Hem, 1989). This type of analysis yields insight as to the reactivity of most surface functional groups but does not take into account surface defects, which can be

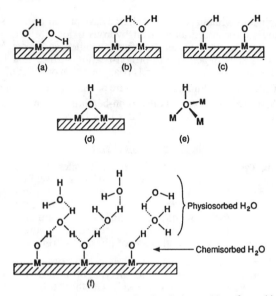

Figure 1. Different types of surface hydroxyl groups found on hydrous oxide surfaces: (a) Geminal hydroxyl groups, (b) vicinal groups, H-bonded, (c) isolated hydroxyls, (d) doubly coordinated hydroxyl, (e) triply coordinated hydroxyl, (f) idealized relationship between surface hydroxyls (= chemisorbed H_2O) and two layers of physiosorbed H_2O.

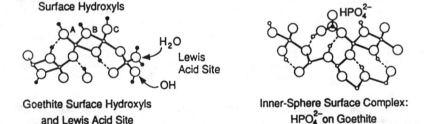

Goethite Surface Hydroxyls and Lewis Acid Site

Inner-Sphere Surface Complex: HPO_4^{2-} on Goethite

Figure 2. (a) Types of surface hydroxyl groups on goethite. Type A, B, and C groups are singly, triply, and doubly coordinated to Fe(III) ions (one Fe-O bond not represented for type B and C groups), and a Lewis acid site. (b) Phosphate adsorbed onto a Type A site. Reprinted from Sposito (1984), The Surface Chemistry of Soils, Oxford University Press.

Kaolinite Surface Hydroxyls

Outer-Sphere Surface Complex: $Na(H_2O)_6^{2+}$ on Kaolinite

Figure 3. (a) Surface hydroxyl groups on kaolinite. Besides the OH groups on the basal plane, there are aluminol groups, Lewis acid sites (at which H_2O is adsorbed), and silanol groups, all associated with ruptured bonds along the edges of the crystallites. (b) Outer sphere surface complex between Na^+ and singly ionized H_2O (at a Lewis acid site) and ionized silanols. Reprinted from Sposito (1984), The Surface Chemistry of Soils, Oxford University Press.

important sites for reactions at mineral-water interfaces. The tritium exchange method has been applied to a large number of oxide minerals to determine the density of surface hydroxyl groups (Berube and de Bruyn, 1968; Yates and Healy, 1976; Yates et al., 1980). A mineral is equilibrated in a solution containing tritiated water, dried to remove physically adsorbed tritiated water, and reequilibrated with water to assay for tritium rapidly released by the surface, which is assumed to be present as ≡MeOH groups. Another method involves drying the mineral to remove physically adsorbed water, reacting the mineral with an -OH labile compound, such as a CH_3-Mg-I, and determining the amount of reagent consumed (Boehm, 1971). Other methods that have been applied include themogravimetric analysis and water vapor adsorption. Attempts have been made to combine IR spectroscopy with either isotopic exchange, thermogravimetric anlysis, or reaction with methylating agents, to distinguish between the different types of surface hydroxyls (see James and Parks, 1982).

Ranges for the total densities of surface functional groups for several oxide minerals are collected in Table 1. Except for ferrihydrite, the ranges of total site densities were obtained using the methods described in the previous paragraph. For the crystalline solids, the breadth of the ranges reflects not only differences in the techniques used to obtain total site densities but also actual differences in the solids. It is important to consider that solid surfaces retain a partial memory of the history of the sample. Processes such as nucleation and crystal growth, grinding, heating, evacuating, and exposure to solutions containing stongly adsorbing solutes can leave their mark at the surfaces of solids. Synthetic minerals used in adsorption studies are often pretreated in an attempt to erase as much of this memory as possible. To the extent that this memory cannot be erased, one must expect variations in important surface properties between samples of different origin and history.

For ferrihydrite, (poorly crystalline hydrous ferric oxide), *microporosity* can be another source of variation in total site density. Micropores are pores of molecular dimensions (less than 2 nm; see section on "Surface Area and Porosity", below). Natural and synthetic ferrihydrites typically are formed by aqueous polymerization reactions at meteoric temperatures (Flynn, 1984; Schwertmann, 1988). Microporosity arises from incomplete cross-linking of the coalesced hydroxy polymer strands that comprise the primarly particles of the sample. Microporosity in ferrihydrite can extend well into the bulk of the solid (Yates, 1975; Cornejo, 1987). Because of this, the total density of surface sites for ferrihydrite is usually normalized to the Fe content rather than the surface area (Davis and Leckie, 1978; Dzomback and Morel, 1990). The site densities for ferrihydrite in Table 1 come from maximum extents of adsorption of various solutes as well as from tritium exchange experiments (see Dzomback and Morel, 1990) for an exhaustive review). Likewise, the variations reported for the total site density of synthetic amorphous silica (am-SiO_2) primarily reflect variations in the extent of microporosity between samples (Yates and Healy, 1976; Iler, 1979). Microporosity occurs in many natural materials, including biogenic amorphous silica, weathered aluminosilicate minerals and allophanes (Kent, 1983; and see chapter by Casey and Brunker, this volume).

Densities of proton donor and proton acceptor sites for goethite, rutile, gibbsite, and kaolinite, calculated from crystallographic considerations, are presented on the right hand side of Table 1. Clearly these sites represent a subset of the array of all types of surface hydroxyl groups. The density of proton donor sites exceeds that for proton acceptor sites because groups such as ≡MOH_2, which result from attachment of water at Lewis acid sites, cannot accept protons (Sposito, 1984). Since silica sols do not acquire a detectable positive charge in aqueous suspensions, even at low pH (Iler, 1979), it has been suggested that silica may lack proton acceptor groups. This hypothesis is supported by observations that selenite ions do not adsorb on amorphous silica at pH values as low as 5 (Anderson and Benjamin, 1990). Evidence to the contrary can be found in the study of Dalas and Koutsoukos (1990), who have recently shown that orthophosphate ions adsorb on silica at pH 2.

Table 1. Density of Surface Functional Groups on Oxide and Hydrous Oxide Minerals

Mineral Phase	Range of Site Densities (sites/nm²)	Reference	Proton Acceptor Groups (sites/nm²)	Proton Donor Groups (sites/nm²)	Reference
α-FeOOH	2.6-16.8	James and Parks (1982) Sposito (1984)	4.4	6.7	Sposito (1984)
α – Fe₂O₃	5-22	James and Parks (1982)	-	-	-
Ferrihydrite	0.1-0.9 moles per mole of Fe	Dzombak and Morel (1990)	-	-	-
TiO₂ (rutile)	12.2	James and Parks (1982)	2.6	4.2	James and Parks (1982)
TiO₂ (rutile and anatase)	2-12	James and Parks (1982)	-	-	-
α-Al(OH)₃	2-12	Davis and Hem (1989)	2.8	5.6	Sposito (1984)
γ– Al₂O₃	6-9	Davis and Hem (1989)	-	-	-
SiO₂ (am)	4.5-12	James and Parks (1982)	0	all	James and Parks (1982)
Kaolinite	1.3-3.4	Fripiat (1964)	0.35	1.0	Sposito (1984)

Table 2. Estimates of Site Densities for α-FeOOH from Adsorption Measurements

Ion	Adsorption maximum (sites/nm²)	Method	Reference
OH⁻/H⁺	4.0	a	f
OH⁻	2.6	a	g
F⁻	5.2	b	h
F⁻	7.3	c	f
SeO₃²⁻	1.5	d	i
PO₄³⁻	0.8	d	i
Oxalate	2.3	b	j
Pb²⁺	2.6	b	k
Pb²⁺	7.0	e	k

Methods:
    a) Acid-base titration
    b) Adsorption isotherms
    c) Excess F at pH 5.5
    d) Adsorption isotherm at pH$_{zpc}$
    e) Fit with model

References:
    f) Sigg and Stumm (1981)
    g) Balistrieri and Murray (1981)
    h) Hingston et al. (1968)
    i) Hansmann and Anderson (1985)
    j) Parfitt et al. (1977)
    k) Hayes (1987)

Site density determined by adsorption. It is instructive to compare the site densities assembled in Table 1 to the maximum densities of adsorbed solutes. The maximum extent to which a solute adsorbs onto a mineral is controlled by the affinity of the adsorbing species for the surface sites as mitigated by steric and electrostatic effects. Thus, maximum site occupancy by an adsorbing species is favored when the species forms strong surface complexes that are not bulky and do not lead to an accumulation of charge. Adsorption is sometimes normalized per unit of surface area, in which case it is referred to as an *adsorption density*, Γ. "Maximum" adsorption densities (Γ_{max}) for various solutes adsorbing onto goethite are presented in Table 2. For F^- and oxalate, the Γ_{max} values are derived from plateaus in plots of Γ versus the aqueous concentration of the adsorbing solute, called *adsorption isotherms*. The isotherms were measured at the pH of maximum adsorption (see section on "Adsorption of anions", below). The Γ_{max} for F^- greatly exceeds that for oxalate. Hingston and coworkers found that F^- adsorption densities greatly exceeded those for other anions studied, including Cl^-, selenite, arsenate, phosphate, and silicate. F^- adsorption onto goethite is consistent with a two-step ligand exchange pathway (Sposito, 1984; Sigg and Stumm; 1981):

$$\equiv FeOH + H^+ \leftrightarrow \ \equiv FeOH_2^+ \ , \tag{2}$$

$$\equiv FeOH_2^+ + F^- \leftrightarrow \ \equiv FeF + H_2O \ , \tag{3}$$

which gives the net reaction:

$$\equiv FeOH + H^+ + F^- \leftrightarrow \ \equiv FeF + H_2O \ , \tag{4}$$

where $\equiv FeOH_2^+$ and $\equiv FeOH^o$ represent positively charged and uncharged surface sites, respectively, on the oxide surface, and $\equiv FeF$ is a fluoride ion bound by ligand exchange to a surface iron atom. Extensive adsorption of F^- is promoted by the strength of the resulting Fe-F bond, small size of F^-, its strong nucleophilic character, and the lack of increase in surface charge resulting from F^- adsorption. The Γ_{max} for F^- lends support to the density of reactive surface sites calculated by Sposito (Table 1) as compared to the higher estimates represented by the upper end of the range of values in the second column of Table 1. The fact that the Γ_{max} for F^- exceeds the density of proton acceptor sites suggests that it may react with Lewis acid sites in addition to proton acceptor sites.

Hansmann and Anderson (1985) determined Γ_{max} values from selenite and phosphate adsorption isotherms on goethite, obtained at the pH of the isoelectric point (pH_{IEP}, see section on "The Electrified Mineral-Water Interface", below). The relatively low Γ_{max} values obtained indicate that for selenite and phosphate adsortion at the pH_{IEP}, steric factors limit the maximum amount adsorbed. The Γ_{max} for OH^- and H^+ were derived from acid-base titrations at high ionic strength. Goethite titration curves continue to rise at higher pH values, so it is not surprising that this value is much lower than the density of proton donor sites calculated by Sposito (Table 1). Two Γ_{max} were derived from the Pb^{2+} adsorption data of Hayes (1987). Hayes performed several experiments in which there was an excess in moles of Pb^{2+} in comparison to the moles of surface sites in the suspension. The lower value for the site density (2.6 sites/nm^2) represents the largest density of adsorbed Pb^{2+} measured experimentally. The higher value (7 sites/nm^2) was derived from an optimization of the fit between model simulations and experimental data collected over a wide range of conditions. The maximum molar ratio of Pb^{2+} to surface sites in Hayes' experiments was about 2, based on the 7 sites/nm^2 value for the total functional group density. The general agreement between the adsorption densities shown in Table 2 with the densities of proton donor and acceptor sites in Table 1 lends support to the hypothesis that only a portion of the hydroxyl groups present at oxide surfaces are active in surface complexation reactions.

Chang et al. (1987) measured a Γ_{Zn} of 5 sites/nm^2 for a TiO_2 sample by decreasing the solid-water ratio to low values (2 mg/liter) in solutions containing 1 μM Zn. This site density was somewhat lower than that estimated by tritium exchange (12.2 sites/nm^2; James and Parks, 1982). The maximum Zn/surface site molar ratio was only about 0.4, based on 12.2 sites/nm^2, suggesting that Γ_{max} for Zn^{2+} was possibly not reached.

Organic matter

A variety of functional groups are present in organic compounds that polymerize to form the humic substances commonly found in soils and sediments, e.g., carboxyl, carbonyl, amino, imidazole, phenylhydroxyl, and sulfhydryl groups (Sposito, 1984; Hayes and Swift, 1978). Dissolved natural organic material typically contains 6 to 12 millimoles/g of weakly acidic functional groups, primarily carboxylic and phenolic groups (Perdue et al., 1980). The stabilities of complexes between these groups and protons range from weak to very strong, and thus, a considerable range of functional group reactivity with dissolved constitutents in water can be expected. Organic material may contribute important surface functional groups for metal complexation (Sigg, 1987; Davis, 1984), either in the form of live organisms, detrital organic particulates, or organic coatings adsorbed on mineral surfaces (Davis, 1982; Tipping, 1981).

The importance of organic matter on the chemical properties of oxide surfaces is most simply evidenced by the negative charge observed on particles suspended in natural waters that would normally be positively charged at neutral pH values (Davis and Gloor, 1981; Hunter, 1980). Synthetic alumina particles suspended in a filtered lakewater with a dissolved organic carbon concentration of 2.3 mg/l attained a high surface coverage of adsorbed organic matter within a few minutes (Davis, 1982). Luoma and Davis (1983) reviewed the literature on the quantities of functional groups of organic particulate material in sediments and estimated a total functional group density of about 0.001 moles/g, based on the results of metal complexation studies.

The bulk of particulate matter settling in the open oceans is produced by biological processes (Whitfield and Turner, 1987), and a large percentage of suspended particulate matter in lakes is composed of biological material (Sigg, 1987). The capacities of these materials for binding metal ions is not well known. Morel and Hudson (1985) estimated that a representative phytoplankton cell contains 10^7 to 10^8 high-affinity sites for metal coordination. Sigg (1987) estimated the average dry weight of a diatom cell in Lake Zurich as 560 picograms, and applying this weight to the above site concentration yields a high-affinity site density of 0.03 to 0.3 $\mu moles/g$. Fisher et al. (1983) concluded that the adsorption of transuranic elements on algal cells was similar, regardless of whether the cells were viable or dead.

Minerals with permanent structural charge

Phyllosilicate minerals. Clay minerals are important adsorbents in many systems of interest in aqueous geochemistry (Sposito, 1984; Bolt and van Riemsdijk, 1987). Kaolinite and minerals in the smectite, vermiculite, and illitic mica groups are especially important because they often occur as extremely small particles with high surface areas and they are widespread. In addition to surface hydroxyl groups, these minerals have rings of siloxane groups, $\equiv Si_2O$, which occur on the basal planes and interlayer regions that dominate the surface area of these minerals. In many phyllosilicate minerals these groups are not hydroxylated because the coordination environments of these bridging oxygen ions are satisfied by the two Si(IV) ions with which they are coordinated. The importance of the siloxane rings as surface functional groups is enhanced by the magnitude of the permanent charge of the crystal lattice of clay minerals, which is due to isomorphic cationic substitutions.

Most clay minerals are comprised of composite sheets consisting of layers of $M^{z+}O_6$ octahedra and SiO_4 tetrahedra (Figs. 4a,b). In the mineral kaolinite, each Al(III)-containing octahedral layer is linked to a Si(IV)-containing tetrahedral layer (Fig. 4c). In the smectite, vermiculite, and illitic mica groups, each octahedral layer is sandwiched in between two tetrahedral layers (Fig. 4d). Linkage of SiO_4 tetrahedra to form the tetrahedral layer gives rise to hexagonal cavities bounded by siloxane groups (Fig. 4a). Joining of the tetrahedral and octahedral layers causes distortion of these cavaties from hexagonal to ditrigonal symmetry (Fig. 5). Substitution of divalent cations for trivalent cations in the octahedral layer and trivalent cations for Si(IV) in the tetrahedral layers gives rise to the permanent structural charge on the composite layer that is compensated by complexation of mono- or divalent cations at the ditrigonal cavities between the sheets.

Kaolinite. Kaolinite has five types of surface functional groups: ditrigonal siloxane cavities on the face of tetrahedral sheets (Fig. 4c), aluminols on the face of octahedral sheets (Fig. 4c), silanols and aluminols exposed at the edge of the sheets (Fig. 3), and Lewis-acid sites at the edge (Fig. 3). The oxygen ions of the face aluminols are coordinated with two Al ions, hence are considered unreactive (Figs. 4b, c). The degree of ionic substitution in kaolinite is very low, less than 0.01 ion per unit cell (Sposito, 1984). The resulting low permanent charge renders the ditrigonal cavities along the tetrahedral sides of the sheets unreactive. The principal surface complexation sites are therefore the silanols, aluminols, and Lewis acid sites located along the edge of the sheets (Fig. 3). All three site types are proton donor groups, which can form complexes with metal ions. Only the aluminols are proton acceptor groups (Table 1), and these groups can complex anions.

Smectites, vermiculites, and illitic micas. These minerals all have significant permanent charges resulting from isomorphic cation substitutions. The permanent charge in the smectite group results from substitution of divalent cations (e.g., Mg^{2+} and Fe^{2+}) for trivalent cations in the octahedral sheet (e.g., Al^{3+} and Fe^{3+}). The degree of substitution in smectites imparts a permanent charge of 1-2 μequiv./m^2, where the denominator refers to the total sheet area (Bolt and van Riemsdijk, 1987). The charge resulting from the substitution of a single cation in the octahedral layer is distributed among the 10 surface oxygens of the 4 tetrahedra linked to the site (Sposito, 1984). The diffuse nature of the charge distribution in the ditrigonal cavities leads to the formation of outer sphere complexes between the exchangeable cation and the ditrigonal cavity. An outer sphere complex is one in which there is at least one layer of solvent between the complexing ion and the functional group. This is illustrated for the smectite mineral montmorillonite in Figure 6. The permanent charge in illitic micas results largely from the substitution of Al or Fe(III) for Si(IV) in the tetrahedral layer. A larger degree of substitution leads to a higher permanent charge, viz., about 3 μequiv./m^2 (Bolt and van Riemsdijk, 1987). The resulting negative charge is much less diffuse. Substitution in the tetrahedral layer leads to distribution of the resulting negative charge over the 3 surface oxygens of the tetrahedron, hence the charge on the ditrigonal cavity is much less diffuse. Higher permanent charge and higher density of charge on ditrigonal cavities favor the formation of inner sphere complexes between interlayer cations and ditrigonal cavities, i.e., complexes lacking solvent molecules between cation and site (Fig. 6). The ionic radius of K^+ corresponds almost exactly to that of the ditrigonal cavity, which enhances the strength of the resulting surface complex. Vermiculites resemble illitic micas in the mode of substitution, but often show lower degrees of substitution, and hence, lower permanent charges. As a result, both inner sphere and outer sphere complexes may occur in vermiculites.

Sposito (1984) has discussed methods of quantifying ditrigonal siloxane cavities. In order to form complexes with solutes, ditrigonal siloxane cavities need to be accessible to cations in solution. Clay minerals that form outer sphere complexes between interlayer cations and ditrigonal cavities fulfill this criterion. Thus, the interlayer cations of smectite minerals are exchangeable. The nature of the complexes formed in the interlayer region suggest that ditrigonal cavities in the interlayer region will react primarily with the major cations in solution, i.e., those present at the highest concentration.

186

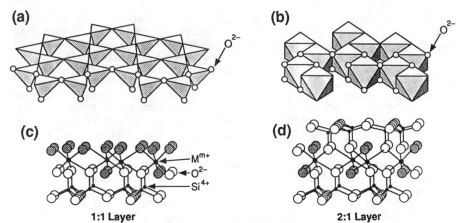

(a) O^{2-}

(b) O^{2-}

(c)
M^{m+}
O^{2-}
Si^{4+}

(d)

1:1 Layer   **2:1 Layer**

Figure 4. Structural elements of phyllosilicate minerals. (a) SiO_4 tetrahedra linked to form the tetrahedral sheet. (b) MO_6 octahedra linked to form the octahedral sheet. Note that only two of the three possible sites in the octahedral sheet shown are occupied, hence this is a *dioctahedral* sheet. (c) Tetrahedral and octahedral sheets linked to form a 1:1 layer structure (e.g., kaolinite). Shaded circles represent OH groups. (d) Tetrahedral and octahedral sheets linked to form a 2:1 layer structure. Reprinted from Sposito (1984), The Surface Chemistry of Soils, Oxford University Press.

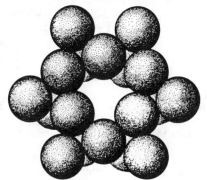

Figure 5. The siloxane ditrigonal cavity. Reprinted from Sposito (1984), The Surface Chemistry of Soils, Oxford University Press.

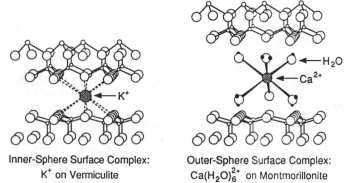

K^+ H_2O Ca^{2+}

Inner-Sphere Surface Complex: **Outer-Sphere Surface Complex:**
K^+ on Vermiculite $Ca(H_2O)_6^{2+}$ on Montmorillonite

Figure 6. Surface complexes between cations and siloxane ditrigonal cavities on 2:1 phyllosilicates, shown in exploded view. Outer sphere complex between Ca^{2+} and ditrigonal siloxane cavities in the smectite mineral montmorillonite and inner sphere complex between K^+ and ditrigonal siloxane cavities in illite or vermiculite. Reprinted from Sposito (1984), The Surface Chemistry of Soils, Oxford University Press.

The relative importance of ditrigonal cavities and surface hydroxyl groups depends on the system under investigation. Binding at ditrigonal cavities is predominantly electrostatic and the majority of ditrigonal cavities are located in the interlayer region. Consequently, these are the sites at which the major cations in natural waters bind. Anion binding on clay minerals occurs predominantly at surface hydroxyl groups along the *edges* of clay crystallites (Honeyman, 1984; Bar-Yosef and Meek, 1987; Zachara et al., 1989b; Neal et al., 1987a,b). Intuitively, one would expect that hydrolyzable cations present at trace concentrations would react primarily with edge sites. Electrostatic and hydration forces dominate the interaction between ditrigonal cavities and cations hence these sites should be populated by the dominant multivalent cations in the solution (e.g., for many natural water systems, Ca^{2+}). The proton stoichiometry of Cd^{2+} adsorption onto montmorillonite is much lower than that onto pure oxides, but increases sharply with increasing pH (Honeyman, 1984). This observation suggests that Cd^{2+} binds to both surface hydroxyls, releasing protons, and ditrigonal cavities, releasing exchangeable cations, *and* that the relative importance of surface hydroxyls increases with increasing pH. This is consistent with the results of Ziper et al. (1988) on adsorption of Cd^{2+} onto a variety of phyllosilicates at low pH. The site density of the edges of montmorillonite is believed to be similar to that of kaolinite (Bar-Yosef and Meek, 1987).

Manganese oxides. Hydrous Mn oxides that are important in natural systems are complex minerals characterized by poor crystallinity, structural defects, domain intergrowths, cation vacancies, and solid solutions (Burns and Burns, 1979; Sposito, 1984). These minerals consist of chains of $Mn(IV)O_6$ octahedra linked to form tunnels or sheets (Burns and Burns, 1979; Turner and Buseck, 1981; Turner et al., 1982). $Mn(IV)$ vacancies and substituion of $Mn(II)$ and $Mn(III)$ for $Mn(IV)$ in the structure give rise to a permanent structural charge that is compensated by loosely bound cations that occupy a range of interstitial positions. These cations are exchangeble, as are some of the $Mn(II)$ ions in the structure. Clearly, Mn oxides can exhibit a large variety of types of adsorption sites. Considering the complexity of Mn oxide minerals, it is interesting to note the success of Balistrieri and Murray (1982) in applying a surface complexation model with homogeneous sites to describe the adsorption properties of a synthetic δ-MnO_2 in major ion sea water. Studies of adsorption onto Mn oxides confirm that these minerals have large capacities for metal ion adsorption because of their high surface area (James and MacNaughton, 1977; Dempsey and Singer, 1980; McKenzie, 1980; Catts and Langmuir, 1986; Zasoski and Burau, 1988). Zasoski and Burau (1988) observed a site density of 2.2 sites/nm^2 at Γ_{max} for Zn^{2+} and Cd^{2+}, similar to the values for goethite reported in Table 2.

Carbonate minerals

Interfaces between aqueous solutions and carbonate minerals are dynamic. Dissolution and precipitation reactions driven by differences in particle sizes (Ostwald ripening), surface dislocations, and, in some cases, phase transformations lead to a persistent flux of cations and carbonate ions across the interface (e.g., Lahann and Siebert, 1982). The interface consists of an array of sites that form surface complexes with cations and anions (herein designated $\equiv S_c$ and $\equiv S_a$, respectively). For salt-type minerals, these cation and anion sites are generally considered to be completely occupied with adsorbed cations, anions, or their hydrolysis products (Parks, 1975). The relative concentrations of cation surface complexes (e.g., $\equiv S_c-Ca^{2+}$ and anion (carbonate) surface complexes (e.g., $\equiv S_a-CO_3^{2-}$, $\equiv S_a-HCO_3^-$), varies with the relative concentrations of Ca^{2+} and CO_3^{2-} in solution (Somasundaran and Agar, 1967; Parks, 1975).

Studies of the interaction of Cd(II) (Davis et al., 1987) and Zn(II) (Zachara et al., 1988, 1989a) with calcite have illustrated the complexity of sorption behavior on carbonate minerals. Both of these cations form sparingly soluble carbonates. Cation uptake occurs in two steps. A relatively rapid step, which reaches completion within one day, is followed by a slow step, whereby the uptake rate appears to be constant over a long period of time (at least several days). The rapid step results from sorption onto a hydrated $CaCO_3$ layer. For Cd, the slow step results from incorporation into calcite as a $(Cd,Ca)CO_3$ solid solution during recrystallization. The adsorption step is consistent with the exchange of Cd or Zn with Ca in the hydrated $CaCO_3$ layer. Reactions with different types of sites or with different stoichiometries may also occur (Zachara et al., 1988), as has been observed with cation adsorption onto hydrous oxides. The importance of the slow step depends on the properties of the solid, adsorbing solute, and the solution chemistry. The calcite sample used by Zachara et al. had a much slower recrystallization rate than that used by Davis et al. The ionic radius and relative ease of dehydration of Cd favored incorporation into a $(Cd,Ca)CO_3$ solid solution. The ionic radius and relative difficulty of dehydration of Zn helped retard the rate of solid solution formation. Inhibition of the recrystallization rate by solutes such as Mg^{2+} also affect the rate of solid solution formation (Davis et al., 1987). Other cations appear to behave similarly (e.g., Mn^{2+}, McBride, 1979; see also references in Davis et al, 1987, and Zachara et al., 1988). Interpretation of the results of many sorption studies is difficult because the experiments were conducted in solutions supersaturated with respect to metal carbonate phases.

For cation adsorption on calcite, it appears of interest to determine the density of $\equiv S_c-Ca^{2+}$ groups. By analogy, for anion adsorption on carbonates it is of interest to determine the density of $\equiv S_a-CO_3^{2-}$ and $\equiv S_a-HCO_3^-$ groups. The density of Ca^{2+} (and CO_3^{2-}) lattice positions exposed on the $(10\overline{1}4)$ cleavage face of calcite is 5 sites/nm² (8.3 µmole/m², Moeller and Sastri, 1974). ^{45}Ca isotopic exchange studies with the calcite surface have shown that the density of exchangeable Ca is consistent with this value, but decreases with decreasing excess of dissolved Ca^{2+} over CO_3^{2-} or increasing Mg^{2+} concentrations (Moeller and Sastri, 1974; Davis et al., 1987). Under similar solution conditions, isotherms of Γ_{Zn} versus Zn^{2+} concentration reached a plateau at about 0.6 µmole/m² (0.33 sites/nm²; Zachara et al., 1988). Surface irregularities such as dislocations and etch pits are potentially important sites for adsorption but their role has not been investigated.

Wersin et al. (1989) have studied the adsorption of Mn^{2+} on the surface of siderite ($FeCO_3$). Like the investigations of the calcite surface, the results exhibited complex kinetics, which were interpreted in terms of an adsorption step followed by incorporation of Mn^{2+} as a $(Mn,Fe)CO_3$ solid solution at the surface. The co-precipitation hypothesis was supported by electron spin resonance spectroscopy. Adsorption of Mn^{2+} was proportional to dissolved Mn^{2+} until the surface carbonate sites were filled (9.3 atoms Mn^{2+}/nm^2). The site density of siderite (11 sites/nm²) estimated from crystallographic data (Lippmann, 1980) was in good agreement with the observed adsorption maximum.

Knowledge of the solution chemistry is an essential aspect of interpreting sorption behavior on carbonates. Chemical equilibrium cannot be assumed because long periods of time are required to attain equilibrium in some systems (especially those in which CO_2 gas must be exchanged between the gas and solution phases, Somasundaran and Agar, 1967), and equilibrium will not be achieved without considerable effort expended to pretreat the solid phase (Plummer and Busenberg, 1982). Crushing carbonate minerals should be avoided, since it can lead to production of high energy surfaces that can produce irreversible effects (e.g., Goujon and Mutaftshiev, 1976).

Sulfide minerals

Chemical interaction with sulfide minerals is likely to have a major impact on the transport and fate of trace metals in sulfidic environments (Berner, 1981; Raiswell and Plant, 1980; Emerson et al, 1983, see chapter by Bancroft and Hyland, this volume). The iron sulfide minerals, mackinawite, greigite, and pyrite (Berner, 1964; 1970), are especially important because of their widespread occurrence (e.g., Goldhaber and Kaplan, 1974; Raiswell and Plant, 1980). Removal of metal ions from solution by iron sulfide minerals can be extensive (Caletka et al., 1975; Framson and Leckie, 1978; Brown et al., 1979).

Systematic studies of sorption onto sulfide minerals are lacking. Previous investigations have identified two pathways by which metal ions are taken up by sulfide minerals: the metathetical reaction and adsorption. These pathways are discussed below. Experimental investigations have been hampered by the ease of oxidation of surface layers of sulfide minerals; strict anaerobic conditions must be maintained (Wolf et al.,1977; Framson and Leckie, 1978; Park and Huang, 1987).

The metathetical reaction between sulfide minerals and dissolved metals has been widely documented (Gaudin et al., 1957, 1959; Phillips and Kraus, 1963, 1965):

$$A_xS(s) + yB^{n+} = B_yS(s) + xA^{m+} \quad ,$$ (5)

where $x=2/m$ and $y=2/n$. The reaction is driven by the lower solubility of $B_yS(s)$ in comparison to $A_xS(s)$. Replacement of the first two or three layers of $A_xS(s)$ proceeds rapidly, during which the reaction can be reversed by, for example, adding a complexing agent that has a much higher affinity for B^{n+} than A^{m+}. Extensive replacement of $A_xS(s)$ by $B_yS(s)$ can occur over short periods of time at room temperature. The metathetical pathway could be extremely important in natural systems because mackinawite is more soluble than the sulfides of most transition metals (Caletka et al., 1975; Framson and Leckie, 1978).

Adsorption onto sulfide minerals has been studied in systems where the dissolved metal forms a sulfide that is more soluble than the adsorbent (Gaudin and Charles, 1953; Iwasaki and deBruyn, 1958; James and Parks, 1975; James and MacNaughton, 1977; Wolf et al., 1977; Park and Huang, 1987; Park and Huang, 1989). The results support the hypothesis that the operative surface functional groups are $\equiv$MOH and $\equiv$SH (sulfhydryl or thiol) groups. The availability of both oxygen and sulfur donor ligands adds an interesting dimension to the adsorption properties of sulfide minerals. The relative abundance of these groups is controlled by the concentration of M^{n+}, $[M^{n+}]$. At high $[M^{n+}]$, $\equiv$MOH groups are in excess, whereas at low $[M^{n+}]$, $\equiv$SH groups dominate (Iwasaki and deBruyn, 1958; Park and Huang, 1987). As with hydrous oxides, the crystal structure of the particular sulfide mineral will dictate the types of groups that are available, their reactivities, and their densities. Some authors suggest that the sulfhydryl groups are inactive on some sulfide minerals, thus the surface chemistry is controlled by $\equiv$MOH groups (e.g., Horzempa and Helz, 1979; Williams and Labib, 1985). Oxidation of surface layers could explain this behavior (Park and Huang, 1987). A complete understanding of the surface chemistry of sulfide minerals awaits the results of systematic studies of cation and anion adsorption over a wide range of solution conditions.

SURFACE AREA AND POROSITY

In this section, we will discuss methods for determining $\mathcal{A}$, the *specific surface area*, which is the amount of reactive surface area available for adsorbing solutes per unit weight of the material. The SI units for $\mathcal{A}$ are m^2/kg or cm^2/g, but herein we will use m^2/g for convenience. Knowledge of the amount of reactive surface area enables normalization of solute adsorption data to surface area, and is required for applying electrical double layer models (see section on "The Electrified Mineral-water Interface"). More importantly, it allows an estimation of the quantity of surface functional groups per unit mass of solids, if the group density per unit area is already known.

Physical Methods

The surface area of a solid phase is related to the size and morphology of the particles that make up the sample For example, a sample that consists of uniform spherical particles has a *geometric surface area* (A'), in m^2/g, given by:

$$A' = \frac{6 \times 10^{-4}}{\rho d} \quad ,$$ (6)

where ρ is the density (g/cm^3) and d is the diameter (cm). Formulae for obtaining A' from particle size distributions and for other particle geometries can be found in Gregg and Sing (1982). Particle size and morphology of solids of interest to adsorption studies are usually determined using electron microscopy (EM).

Natural and synthetic materials often have much larger $\mathcal{A}$ values than the geometrical surface areas determined from particle size and morphology (White and Peterson, 1990). Crystalline solids often possess surface roughness caused by fractures and defects on cleavage and crystal faces. Materials with high surface areas usually consist of very small *primary* particles linked together to form larger *secondary* particles. The secondary particles are called *aggregates* if the primary particles are loosely held together and *agglomerates* if they are rigidly held together (Gregg and Sing, 1982; note that this terminology is not universal). Surface functional groups within the aggregates or agglomerates are accessible to solutes through the contiguous pore structure within the secondary particles.

Discussion of the characterization of pore structures of materials is beyond the scope of this chapter (see Gregg and Sing, 1982). It is useful, however, to introduce the standard operational definitions of pore sizes. Pores with diameters less than 2 nm, i.e., pores of molecular dimensions, are called *micropores*. Pores with diameters between 2 and 50 nm are called *mesopores*; capillary forces are important for pores in this size range. Pores with diameters in excess of 50 nm are called *macropores*. Macropores grade into *voids*, i.e., interparticle space. Porosity primarily affects the *rates* of adsorption and desorption reactions. Microporosity, however, is an important consideration in the determination of $\mathcal{A}$ and the density of adsorption sites. Amorphous and poorly crystalline hydrous oxides of Fe, Al, and Si that result from low-temperature polymerization reactions in solution usually possess extensive microporosity. Meso- and macroporosity in natural materials can result from agglomeration of colloidal-sized particles during polymerization, precipitation reactions, and low-temperature weathering reactions.

The $\mathcal{A}$ of phyllosilicates can by calculated from the unit cell dimensions and the chemical composition (van Olphen, 1977; Sposito, 1984). These calculations are most reliable for montmorillonite because (1) the unit cell dimensions are accessible from X-ray diffraction data, (2) the relationship between chemical composition and structure of the *quasicrystal* (the result of stacking phyllosilicate layers along the crystallographic c axis) is fairly well known (van Olphen, 1977), (3) the region between phyllosilicate layers of unit cell thickness is occupied by hydrated cations and accessible to solutes, and (4) the lateral dimensions of crystals often greatly exceed their thicknesses (Sposito, 1984). Crystallographic $\mathcal{A}$'s of montmorillonites lie in the range 600-800 m^2/g, most of which is accounted for by the interlayer region (van Olphen, 1977).

Clearly these physical methods are of limited application. It is therefore necessary to measure $\mathcal{A}$. Because we are interested in determining $\mathcal{A}$ on a molecular level, most methods involve determining the extent of adsorption of either a gas or solute and relating the amount adsorbed to the area occupied. Aspects of the most common used methods are described in the following sections and in Appendix A.

Gas adsorption

Adsorption isotherms on minerals. The most widely used non-polar adsorptives are N_2 and Kr, although Ar is used occasionally. Isotherms are collected by measuring the amount of gas adsorbed at the boiling temperature of liquid N_2 at atmospheric pressure (viz., 77 °K or -196 °C) as a function of the *relative pressure, p/p^0*, where p is the partial pressure of the adsorptive and p^0 is its equilibrium vapor pressure. The isotherm for N_2 on goethite shown in Figure 7 illustrates some basic features of gas adsorption isotherms. The goethite was synthesized in our laboratory using the method of Atkinson et al. (1972), rinsed with base, acid, and water, and then freeze-dried. The isotherm rises rapidly at low pressures, bends over near p/p^0 0.05, which gives rise to a "knee" in the isotherm, increases linearly over a short range of relative pressures (about 0.1 to 0.25 for this sample), has positive curvature above the linear region (above p/p^0 of 0.25 for this sample), and rises rapidly as p^0 is approached. For nonporous materials, the rapid rise at low relative pressure is due to build

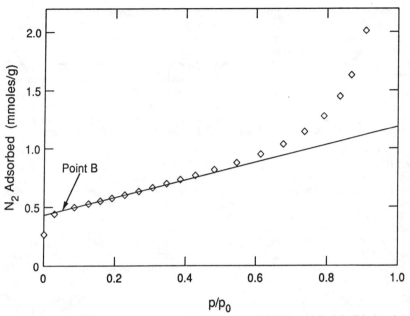

Figure 7. Adsorption isotherm for N_2 gas, at 77.4°K, on synthetic α-FeOOH (goethite). Point B is the point at which the line extending through the linear portion of the isotherm in the multilayer region departs from the isotherm. Point B corresponds to 0.48 μmoles/g for this sample. The sample was outgassed at 110°C for 36 hours, after which the pressure was below 10^{-4} torr.

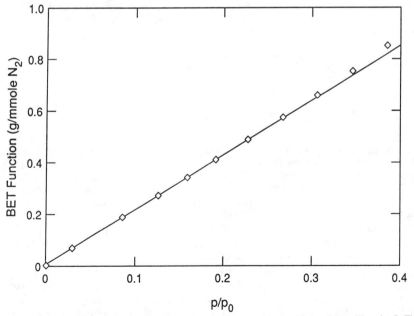

Figure 8. BET plot from the data in Figure 7. Plotted along the ordinate is the right hand side of Equation 7. The regression line of the data across the domain 0.08 to 0.23 is shown. Note that the intercept is near zero.

up of a monolayer of the adsorptive. Above the knee, multilayer adsorption occurs. The rapid increase at high relative pressure is due to the onset of condensation of liquid N_2 on the solid. Monolayer coverage is thought to correspond to "point B' (Fig. 7), which is where the line through the linear portion of the isotherm diverges from the isotherm on the low pressure side (Gregg and Sing, 1982). A is usually determined either by applying the BET model to the data in the multilayer region or by comparing the isotherm to that of a nonporous material with known A. Mesoporous materials exhibit enhanced adsorption above relative pressures of about 0.4, which results from capillary condensation of liquid N_2 in mesopores. Microporous materials exhibit enhanced adsorption at extremely low relative pressures, which results in a sharp knee and flattening of the linear portion of the isotherm.

BET analysis. The most popular method for deriving A from gas adsorption isotherms is the BET method (Brunauer, et al., 1938). The BET model, whose derivation is presented elsewhere (Gregg and Sing, 1982), is based on generalization of the Langmuir isotherm to an inifinite number of layers of adsorptive. Several critical assumptions are embodied in the model. First, the number of adsorbed layers approaches infinity as p approaches p^0. This is a good assumption except for samples with a high density of very small pores. Second, all adsorption sites are identical, which is to say that adsorption occurs uniformly across the surface rather than in clumps. Third, for all layers except the first, the heat of adsorption equals the molar heat of condensation. Fourth, for all layers except the first, the tendency for a molecule to adsorb and desorb is independent of how many layers underlie its adsorption site. The BET equation, in linear form, is

$$\frac{p/p^0}{n(1-p/p^0)} = \frac{1}{C_{BET}n_1} + \frac{(C_{BET}-1)}{C_{BET}n_1}\frac{p}{p^0} \quad , \tag{7}$$

where n is the amount adsorbed (e.g., moles/g), n_1 is the monolayer capacity (same dimensions as n), and C_{BET} is a dimensionless parameter related to the difference between the heat of adsorption onto the first layer and the heat of condensation, which should equal the heat of adsorption onto all layers except the first. A linear regression of the left hand side of the BET equation against p/p^0 yields a slope m and intercept b from which n_1 and C_{BET} are computed according to:

$$n_1 = \frac{1}{m+b} \qquad\qquad C_{BET} = \frac{m}{b}+1 \quad .$$

A is calculated from n_1:

$$A = n_1 a_m L \quad , \tag{8}$$

where a_m is the average area of an adsorbate molecule in a completed monolayer and L is a conversion factor, which equals 6.02×10^5 for a_m in nm^2 and n_1 in moles per unit mass of solid. According to the original paper, the range of relative pressures over which the BET equation is valid is 0.05 to 0.30. Many solids, however, have more restricted BET ranges. A BET plot derived from the data shown in Figure 7 is shown in Figure 8; the regression line has also been plotted. The BET range for this sample is between 0.04 and 0.25. The n_1 and C_{BET} are 472 µmole/g and 305, respectively. Note that n_1 calculated from the BET equation is in good agreement with the estimated position of point B (Fig. 7). The A calculated from n_1 using Equation 8 is 46.0 m^2/g, based on the widely accepted value of 0.162 nm^2 for a_m for N_2 (Gregg and Sing, 1982).

Problems caused by microporosity. The assumptions of the BET model appear to be valid for a wide variety of solids, especially for N_2 as the adsorptive (Gregg and Sing, 1984). Major problems arise for microporous materials, however. Most adsorptives have a very high affinity for micropores, which results in clumping of the adsorptive in and around micropores (Gregg and Sing, 1982). This violates the assumptions of the BET model. The C_{BET} provides a means of assessing the validity of the BET model. The C_{BET} is related to the difference between the heat of adsorption on the surface of the solid and the heat of adsorption onto any of the layers except the first (which is assumed to be the heat of condensation). Materials with extensive microporosity have high C_{BET} values. Extremely low values of C_{BET} imply that multilayer adsorption occurs before monolayer coverage nears completion; the knee of the isotherm is diminished for such samples. Gregg and Sing (1982) suggest that BET theory is valid for materials with C_{BET} values in the range 50 to 150. Our experience suggests that, for N_2 as the adsorptive, samples with C_{BET} values as high as 475 showed no evidence of microporosity; microporous samples had C_{BET} values in excess of 700. BET plots for hydrous oxides tend to intercepts that are near zero (Fig. 8). The intercept is critical in calculating C_{BET}. Our experience indicates that it is important to have at least four points in the BET range to obtain a reliable value for C_{BET}. Negative values for the intercept yield negative C_{BET} values, which are meaningless. Usually, negative intercepts result from including points that are outside the linear BET range.

Low surface area materials. N_2 cannot be used with low A samples (i.e., below about 1 m^2/g) because of the non-ideal gas behavior of N_2 or buoyancy problems. Kr adsorption is usually used for these types of samples because it is an ideal gas under the conditions required for isotherm building. However, it has a number of disadvantages as an adsorptive (see Gregg and Sing, 1982). First, -196 °C is below the triple point so that there is disagreement over whether to use the vapor pressure of solid Kr or of the super cooled liquid for p^0. Usually, the vapor pressure of the super-cooled liquid is used (viz., about 2.5 torr). Second, BET plots of Kr adsorption isotherms often have smaller linear ranges than those for N_2. Third, Kr is *larger* and more polarizable than N_2. Consequently, a_m values reported for Kr (0.202 ±0.026 nm^2; McClellan and Harnsberger, 1967) vary over a wider range than those for N_2, and a_m for Kr has a much greater dependency on the nature of the solid than does the a_m of N_2. Gregg and Sing (1982) recommend that, if working with a solid for which a_m has not already been determined, it should be determined on a sample for which A is known (e.g., from N_2 adsorption); otherwise an uncertainty of ± 20% should be assigned to A. Another consequence of the polarizability is that step-like isotherms are much more common for Kr than for N_2. In some cases, step-like isotherms are due to extreme regularity of the surface of the sample (Gregg and Sing, 1982). In other cases, they occur on highly disturbed samples.

Partial isotherms for N_2 and Kr adsorption onto a highly disturbed calcite sample are shown in Figure 9. The sample had been ground, rinsed briefly with dilute HCl and then with distilled water, and dried. The Kr adsorption isotherm exhibits a step near a relative pressure of 0.15. The high pressure branch of the isotherm is parallel to the isotherm for N_2, indicating that multilayer adsorption does not occur on this sample until the relative pressure exceeds 0.15. The high pressure arms of the isotherms for N_2 and Kr do not coincide because Kr has a larger a_m than N_2. A BET plot of the N_2 data yields an A of 4.5 m^2/g, but the C_{BET} is 776, indicating clumping of N_2 on regions of the surface. BET analysis yields an n_1 value of 38.2. Disagreement between the n_1 value from BET analysis (38.2 μmoles/g) and that from point B (50.4 μmoles/g; Fig. 9) also suggests problems with the BET A. Experience with other calcite samples, both synthetic and commercial, shows that many calcites have similar steps in their Kr adsorption isotherms. Aging the calcites using the method of Plummer and Busenberg (1982) greatly reduces the magnitude of the step. Synthetic aragonites tested yield "normal" isotherms: linear BET ranges from 0.05 to 0.2, C_{BET} values around 60, and n_1 values from the BET equation in good agreement with point B.

Evaluation of microporosity. Comparison plots (t- and α_s-plots, see Appendix A) facilitate identification of micro- and mesoporosity; an estimate of the micorpore volume can also be obtained. t-plots for our goethite sample and a commercial amorphous SiO_2 sample (BDH precipitated silica) are presented in Figure 10. For goethite, the linear branch extrapolates through zero, indicating the lack of microporosity. The small deviations from linearity along the high pressure branch indicate the presence of a small degree of mesoporosity in the sample. The slope of the linear branch yields a value for A of 47.1 m^2/g, which agrees quite well with the BET A (Table 3). For BDH precipitated am-SiO_2, the large intercept of the extrapolated linear branch of the t-plot indicates extensive microporosity.

194

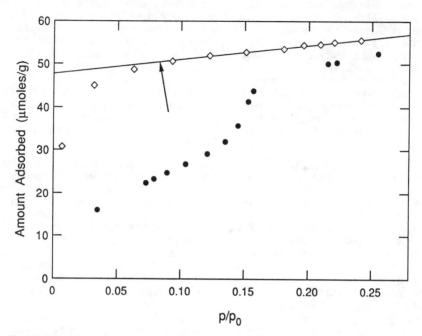

Figure 9. Partial adsorption isotherms for N_2 (open diamonds) and Kr (filled circles) at 77.4°K, on a highly disturbed calcite powder (see text). Point B is shown for the N_2 isotherm. The sample was outgassed at room temperature for 2 days (N_2 isotherm) or overnight at 70°C (Kr isotherm), but the results were independent of outgassing conditions.

Table 3. Specific Surface Areas by BET and t-plot Methods

Solid Phase	Outgassing Temperature (°C)	BET Area (m^2/g)	C_{BET}	BET Range (p/p^o)	t-plot area (m^2/g)	t-plot V_μ (mm^3/g)
α-FeOOH	110	46.0	305	0.03-0.23	47.1	0
SiO_2 (am) BDH precipitated	115	292	1070	0.01-0.17	53.4	102
Linde $\alpha - Al_2O_3$	60, 120	15.8	475	0.02-0.30	16.6	0
Linde $\alpha - Al_2O_3$	180	13.0	1450	0.05-0.20	10.0	1.35
Linde $\alpha - Al_2O_3$; Treated (see text)	100	16.0	275	0.03-0.24	16.0	0
Linde $\alpha - Al_2O_3$; Treated (see text)	195	16.3	467	0.01-0.21	15.6	0

An estimate of the micropore volume can be obtained from the intercept (Table 3; see Appendix A). Yates and Healy (1976) present an α_s plot for a silica sample from the same source, which also reveals the presence of micropores. The slope of the linear branch yields an $\mathcal{A}$ of 53.4 m^2/g. BET analysis yields an $\mathcal{A}$ of 292 m^2/g and a C_{BET} of 1070 (Table 3). The disagreement between t-plot and BET $\mathcal{A}$, the high C_{BET}, and the large intercept of the t-plot all corroborate that this am-SiO_2 sample is highly microporous.

The $\mathcal{A}$ of nonmicroporous materials does not depend on outgassing conditions as long as they are sufficient to desorb physically adsorbed water. The $\mathcal{A}$ of microporous materials, however, may vary with outgassing conditions. An example of this is shown in Table 3 for a synthetic corundum sample (i.e., α-Al_2O_3). The untreated sample shows no evidence of microporosity after being outgassed at 60 or 120°C. After outgassing at 180°C, evidence of microporosity is revealed by both BET and t-plot analyses. Treating the sample by washing with dilute acid, dilute base, water, and freeze-drying removes the microporous fraction (Table 3). Similar observations have been made for other oxides and alumino-silicates (e.g., Aldcroft et al., 1968; Yates, 1975; Gregg and Sing, 1982), although some apparently microporous materials do not show this effect (e.g., Pyman and Posner, 1978). Outgassing at temperatures above 200°C, however, may lead to a reduction in density of surface functional groups for some hydrous oxides (e.g., Aldcroft et al, 1968; Bye and Howard, 1971; Iler, 1979).

Surface area of clay minerals and soils. The evaluation of the $\mathcal{A}$ of clay minerals by gas adsorption is confounded by the microporosity of the interlayer regions. Murray and Quirk (1990) discuss the assumptions necessary and empirical methods for obtaining $\mathcal{A}$ for various clay minerals from adsorption and desorption isotherms of N_2.

Empirical methods based on the retention of polar organic compounds have been applied to measuring $\mathcal{A}$. The adsorptives include glycerol (1,2,3-trihydroxypropane; van Olphen, 1970), ethylene glycol (EG, 1,2-dihydroxyethane; Dyal and Henricks, 1950; Rawson, 1969), and ethylene glycol monoethyl ether (EGME, 2-ethoxyethanol; Bower and Goertzen, 1959; Carter et al., 1965; Eltantawy and Arnold, 1973). Each of these adsorptives can penetrate the interlayer region of expandable clays. Retention rather than adsorption of the molecules is measured because adsorption from the vapor phase is extremely slow (Hajek and Dixon, 1966). $\mathcal{A}$ is calculated from a calibration curve of surface area versus sample weight obtained from a reference material with known $\mathcal{A}$. A montmorillonite for which $\mathcal{A}$ can be calculated from crystallographic considerations is a convenient reference material for samples dominated by expandable clays. EGME has the distinct advantage that it volatizes much more rapidly than the other adsorptives. Desorption isotherms for glycerol on kaolinite, montmorillonite, and vermiculite samples indicate that adsorption is quasi-Langmuirian for the expandable clayes but multilayer adsorption appears to occur on kaolinite (Hajeck and Dixon, 1966). Pyman and Posner (1978) found that $\mathcal{A}$ values for hydroxypolymers of Fe(III), Al, and Si were much greater than those determined from water and N_2 adsorption if montmorillonite was used as the reference material. Agreement was much better if hydrous ferric oxide, whose $\mathcal{A}$ had been determined by water and N_2 adsorption, was used instead. Their results provide additional evidence that multilayer adsorption occurs on the external surfaces of minerals. The nature of interaction between these adsorptives and microporous materials is unknown. Steric considerations would suggest that they would be excluded from microporous regions that cannot expand to accomodate these bulky molecules. These methods appear to be most useful for determining $\mathcal{A}$ of phyllosilicate minerals whose $\mathcal{A}$ is dominated by the interlayer region.

EGME retention has been extended to determining the $\mathcal{A}$ of clay-dominated soils (Heilman et al., 1965; Cihacek and Bremner, 1979; Ratner-Zohar et al., 1983). It is difficult to achieve good precision for samples with $\mathcal{A}$ below about 50 m^2/g (Ball et al., 1990). Considering the results for single minerals, an accurate measurement of $\mathcal{A}$ requires an understanding of whether the reactive surface area is dominated by interlayer regions of phyllosilicate minerals or the external surfaces of hydrous oxides.

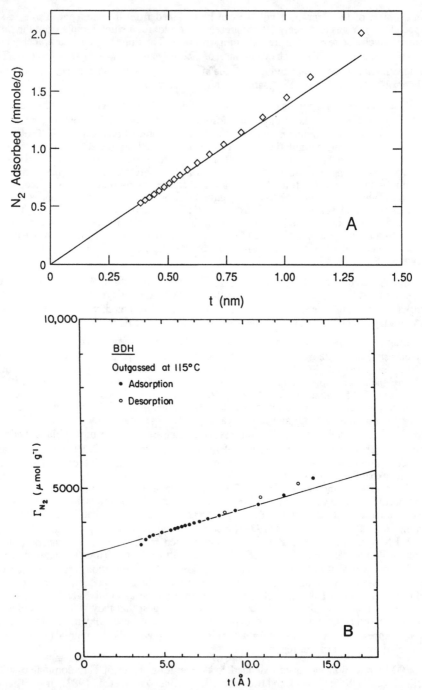

Figure 10. Plots of N_2 adsorbed versus the thickness of the statistical multilayer (t plots, see Appendix A). (a) Goethite sample from Figure 7. The zero intercept indicates lack of microporosity and the deviation from the regression line at high relative pressure (i.e., high t values) indicates the presence of mesoporosity. (b) A microporous synthetic amorphous silica sample (BDH precipitated silica), which was outgassed at 115°C overnight to a pressure of less than 10^{-4} torr.

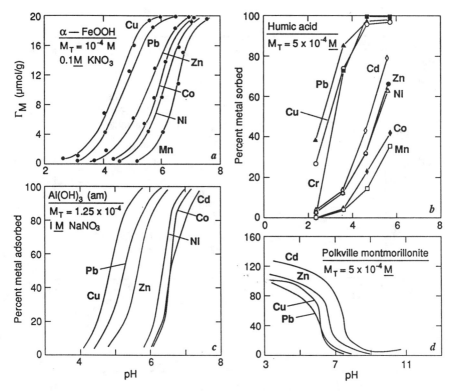

Figure 11. The effect of pH on the adsorption of metal cations by constituents of soils and sediments. (a) Adsorption on goethite from McKenzie (1980); (b) Adsorption on humic acids, from Kerndorf and Schnitzer (1980); (c) Adsorption on freshly precipitated aluminum hydroxide, from Kinniburgh et al. (1975); (d) Adsorption on montmorillonite (*note that metal concentrations [μmoles/dm3] remaining in solution are plotted*), from Farrah and Pickering (1977). Reprinted from Sposito (1984), The Surface Chemistry of Soils, Oxford University Press.

ADSORPTION OF IONS AT HYDROUS OXIDE SURFACES IN WATER

The following material presents a brief review of some important characteristics of the adsorption of ionic solutes on oxide minerals commonly found in soils, sediments, aquifers and other geological formations. Much of the material comes from the comprehensive reviews by Dzombak and Morel (1987) and Hayes (1987). The surface chemistry of hydrous aluminum oxides was reviewed recently by Davis and Hem (1989).

Adsorption of cations

Metal cations that form strong complexes with OH⁻ in water also bind strongly to hydrous oxides (Dzombak and Morel, 1987). Figure 11 illustrates the pH dependence of cation adsorption on various soil components with oxygen-containing functional groups as a function of pH. For most transition metal cations, adsorption typically increases from near zero to nearly complete (>99%) removal from water within a narrow pH range (Benjamin and Leckie, 1981a). It is known that cation adsorption on oxides is generally accompanied by the release of protons (or adsorption of OH⁻) by the surface (James and Healy, 1972a; Hohl and Stumm, 1976; Kinniburgh, 1983), but the actual net stoichiometry of the adsorption process is still poorly understood (Hayes and Leckie, 1986; Fokkink et al., 1987). Cations and anions adsorb onto oxide surfaces in response to both chemical and electrostatic forces (James and Healy, 1972b). However, transition metal cations can be strongly adsorbed

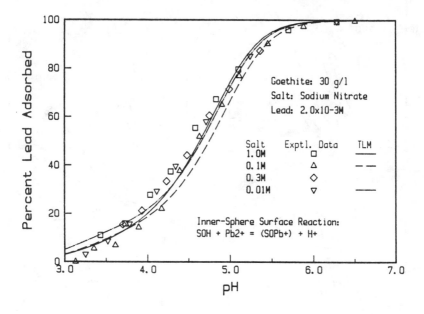

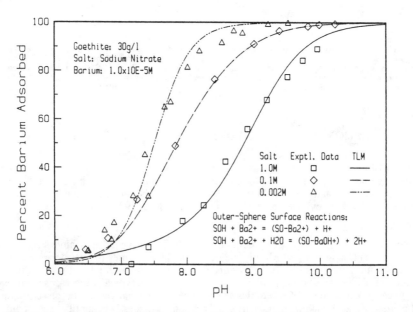

Figure 12. The effect of pH and NaNO₃ concentration on dilute Pb²⁺ and Ba²⁺ adsorption by goethite. Points denote experimental data. Curves denote triple layer model simulations (section "Triple layer model"). (a) Model simulations with Pb²⁺ coordinated as an inner-sphere complex; (b) Model simulations with Ba²⁺ coordinated as an outer-sphere complex. Reprinted from Hayes (1987), Equilibrium, spectroscopic, and kinetic studies of ion adsorption at the oxide/aqueous interface. Ph.D., Stanford University.

against electrostatic repulsion (Davis and Leckie, 1978a; Hohl and Stumm, 1976), demonstrating that it is the coordination chemistry of metal ion reactions with surface hydroxyls that is most significant (Stumm et al., 1976; Schindler, 1981). These ions are soft acids (Pearson, 1968) that tend to react with the least basic hydroxyl groups on the surface (Hayes, 1987). Adsorption of these cations by hydrous iron oxides (Fig. 12a) has been found to be nearly independent of ionic strength (Swallow et al., 1980; Hayes and Leckie, 1986), although adsorption on other hydrous oxides (quartz, titania) and some clay minerals does exhibit some ionic strength dependence (Vuceta, 1976; Chang et al., 1987, Hirsch et al., 1989). Weakly adsorbing cations (hard acids) like the alkaline earth ions (Ca^{2+}, Mg^{2+}, Ba^{2+}) do not generally form covalent bonds, and adsorption of these ions (Fig. 12b) has been shown to be influenced by electrostatic forces, via its dependence on ionic strength (Kinniburgh et al., 1975; Kent and Kastner, 1985; Hayes, 1987; Cowan et al., 1990). An emerging area of research is the investigation of cation adsorption by calorimetry (Fokkink, 1987) and as a function of temperature (Johnson, 1990; Bruemmer et al, 1988). Generally it has been found that cation adsorption increases with increasing temperature, but further studies of this type are needed.

Adsorption of anions

Anion adsorption on oxide surfaces is also dependent on pH, but in contrast to cations, adsorption is generally greater at lower pH values and decreases with increasing pH (Hayes et al., 1988; Zachara et al., 1987; Balistrieri and Chao, 1987; Sigg and Stumm, 1981; Hingston, 1981). Figure 13 shows the adsorption of some anions on ferrihydrite, (i.e., poorly crystalline hydrous ferric oxide) as a function of pH, oxidation state, and ionic strength. Anion adsorption on oxides is usually accompanied by an uptake of protons by the surface (or release of OH⁻) (Hingston et al., 1972; Sigg and Stumm, 1981; Davis, 1982). The importance of the oxidation state in the adsorptive reactivity of anions is illustrated in Figures 13b and 13c. Some anions, e.g., phosphate, selenite (Fig. 13c), are strongly adsorbed in a manner that is independent of ionic strength (Hayes et al., 1988; Barrow et al., 1980; Ryden et al., 1977; Hingston et al., 1968). Like strongly bound cations, it is believed that this is caused by the formation of coordinative complexes at the surface. Adsorption of weakly bound anions, e.g., chromate, sulfate, selenate (Fig. 13c), is more dependent on ionic strength (Hayes et al., 1988; Rai et al., 1986; Davis and Leckie, 1980), suggesting a greater dependence on electrostatic energy contributions to the free energy of adsorption. The results of EXAFS spectroscopic studies indicate that selenite binds to goethite via an inner-sphere complex like structure (c) in Figure 14; structures (a) and (b) could be eliminated based on the EXAFS analysis (Hayes et al., 1987). The coordinative complex formation involves a two-step ligand exchange reaction in which a protonated surface hydroxyl is exchanged for the adsorbing anion. On the other hand, selenate does not lose its primary hydration sheath during adsorption, and it binds as an outer-sphere complex (Hayes, 1987). This interpretation of the differences between strongly and weakly adsorbed anions is supported by a unique kinetic study (Yates and Healy, 1975) and infrared adsorption results (Cornell and Schindler, 1980; Sposito, 1984). The chapter by Brown (this volume) contains a more detailed review of surface spectroscopic investigations. Investigations of anion adsorption by calorimetry have been published by Machesky et al. (1989) and Zeltner et al. (1986). It was postulated by these authors that the free energies of anion adsorption on hydrous oxides are dominated by large favorable entropies, with enthalpic contributions of minor importance.

Surface site heterogeneity and competitive adsorption of ions

As was noted in Figure 11, adsorption of cations on hydrous oxides increases rapidly over a narrow pH range, sometimes referred to as an *adsorption edge*. It is typically observed that an adsorption edge will shift to greater pH values in such experiments when the molar ratio of aqueous metal ion/surface site concentrations is increased (Kurbatov et al., 1951; Benjamin and Leckie, 1981a; Dzombak and Morel, 1986). In order for the adsorption edge to remain within a fixed pH range, the metal ion adsorption must be proportional to the metal ion concentration. The shift of the adsorption edge can be caused by surface site saturation when the metal ion to site concentration ratio is high, but such shifts are observed in cation adsorption experiments even when the site concentrations are in excess. Benjamin and Leckie (1981a) and Kinniburgh et al. (1983) interpreted the phenomenon as evidence

200

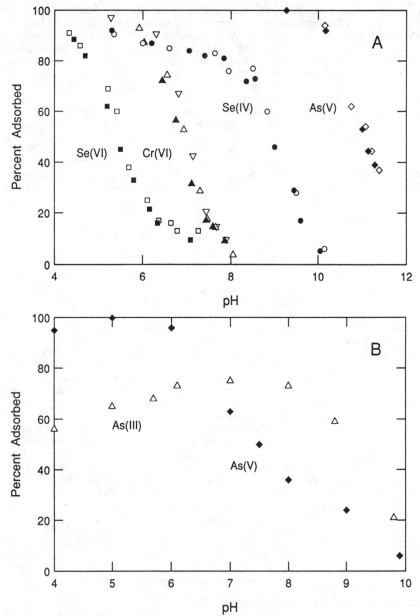

Figure 13. The effect of pH, ionic strength, oxidation state, and trace anion concentrations on the adsorption of oxyanions by ferrihydrite.

(a) Data for various anions adsorbed on 0.001 M Fe (as ferrihydrite) suspended in 0.1M NaNO$_3$:
 Se(VI): open squares, 0.2 μM Se; filled squares, 10 μM Se; data from avis and Leckie (1980).
 Se(IV): open circles, 0.5 μM Se; filled circles, 5.0 μM Se; data from Leckie et al. (1980).
 Cr(VI): open upward triangles, 0.12 μM Cr; open downward triangles, 1.0 μM Cr; filled triangles, 5.0 μM; data from Honeyman (1984).
 As(V): open diamonds, 0.5 μM As; filled diamonds, 5.0 μM As; data from Leckie et al. (1980).

(b) Adsorption of As(V) (filled diamonds) and As(III) (open triangles) on 42 μM Fe (as ferrihydrite), total As = 1.33 μM in each experiment; data from Pierce and Moore (1980, 1982).

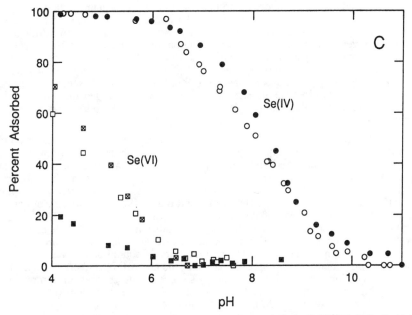

(c) Effect of pH and NaNO₃ concentration on Se(VI) and Se(IV) adsorption by 0.001 M Fe (as ferrihydrite), total Se = 100 μM in each experiment; data from Hayes et al. (1988).
Se(VI): x-box, 0.013 M NaNO₃; open squares, 0.1 M NaNO₃; filled squares, 1.0 M NaNO₃.
Se(IV): open circles, 0.013 M NaNO₃; filled circles, 1.0 M NaNO₃.

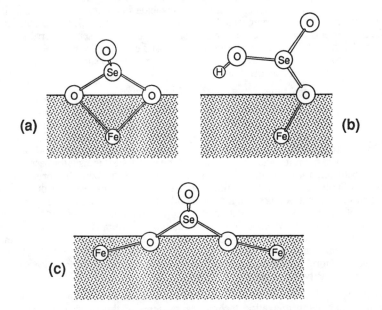

Figure 14. Possible molecular structures for selenite anion, SeO_3^{2-}, coordinated with Fe atoms at the goethite surface. It is possible to distinguish among these model structures for surface complexes using Extended X-ray Absorption Fine Structure (EXAFS) spectroscopy, see Hayes et al. (1987). Reprinted from Hayes (1987), Equilibrium, spectroscopic, and kinetic studies of ion adsorption at the oxide/aqueous interface. Ph.D., Stanford University.

of the heterogeneity of surface hydroxyl groups, as was mentioned in the section on "Types of surface hydroxyl groups". This results in a lack of adherence of the adsorption data to the Langmuir isotherm (see section on "Empirical adsorption models," below). Although it is likely that several types (or a continous distribution) of nonequivalent surface hydroxyl groups exist on oxide surfaces (van Riemsdijk et al., 1986; 1987), adsorption data can usually be modeled with an assumption of only two nonequivalent sites (Loganathan and Burau, 1973; Kinniburgh, 1986; Dzombak and Morel, 1990). In two-site models, these are usually referred to as high and low energy sites or strong and weak complexation sites. Anion adsorption is usually modeled with a single site model, with some exceptions, e.g., phosphate (Sposito, 1982).

Site heterogeneity is also revealed in competitive adsorption experiments. Competitive effects among metal ions can be quite specific and concentration-dependent (Benjamin and Leckie, 1980; 1981b; Balistrieri and Murray, 1982; Zasoski and Burau, 1988). A small proportion of functional groups on hydrous oxides bind metal ions more strongly than others, and competition among metal ions for these high-energy sites occurs at very low adsorption densities (Benjamin and Leckie, 1981b; Zasoski and Burau, 1988). The specific effects are exhibited as different degrees of competition, e.g., Zn^{2+} having more effect on Cd^{2+} adsorption than Cu^{2+} or Pb^{2+}, which could not be explained in terms of the free energies of adsorption. Benjamin and Leckie (1981b) proposed that each metal ion was preferentially adsorbed at different high-energy sites that represent small fractions of the total site density. At high adsorption densities, the specific nature of the competitive effects is diminished, since most metal ions are bound to low energy sites. Cd^{2+} adsorption onto iron oxides is reduced by high concentrations of weakly-adsorbing alkaline earth cations (Cowan et al., 1990; Balistrieri and Murray, 1982), and Zn^{2+} competes with Ca^{2+} for adsorption sites on Mn oxides (Dempsey and Singer, 1980). Studies of strongly-bound anions also exhibit specific effects at low adsorption density (Hingston et al., 1971). Weakly-binding anions typically do not show these effects in competition experiments (Davis and Leckie, 1980; Leckie et al., 1984; Zachara et al., 1987). Competition among adsorbing anions may be important in regulating the concentration of phosphate in natural systems, especially the competitive interactions of organic acids and silicate (Davis, 1982; Sigg and Stumm, 1981). Competitive effects between cations and anions appear to be absent on ferrihydrite (Benjamin and Bloom, 1981; Benjamin, 1983). Synergistic effects have been observed in these types of experiments, and these effects have been attributed to surface precipitation or the formation of ternary complexes.

<u>Kinetics of sorption reactions</u>

The mechanisms that control the rates of adsorption reactions are poorly understood and require further study. In general, sorption of inorganic ions on mineral surfaces is a two-step process consisting of a short period of rapid initial uptake of adsorbate followed by a slower process(es) (van Riemsdijk and Lyklema, 1980; Dzombak and Morel, 1986; Davis et al., 1987; Fuller and Davis, 1987). The rapid step is usually assumed to be a diffusion-controlled adsorption reaction that takes minutes to reach equilibrium when mass transport in the bulk solution is not limiting. The slow step has been attributed to various processes, such as surface precipitation or solid solution formation, micropore diffusion, formation of aggregates via coagulation, or structural rearrangment of surface species, and these processes can take weeks to months to attain equilibrium. The relative importance of the second step appears to increase with increasing ionic strength or with an increasing molar ratio of adsorbate/surface site concentrations (Dzombak and Morel, 1990). The study by Dzombak and Morel (1986) demonstrated clearly that the rate-limiting process can change as adsorption density increases.

Recently the kinetics of the fast adsorption processes have been studied by pressure and temperature jump relaxation techniques; reviews of this work have been given by Yasunaga and Ikeda (1986) and Hayes (1987). Much of the following is taken from Hayes (1987). The kinetics of proton and hydroxide ion adsorption on hydrous oxides exhibit a single relaxation process (Ashida et al., 1980), which can be attributed to the rate-limiting desorption step. The relative acidities of the surface hydroxyl groups on iron oxides was found to be related to the magnitude of the desorption rate constant (Astumian et al., 1981).

Although the adsorption of protons is fast, proton transfer reactions can continue for weeks or months (Onoda and de Bruyn, 1966; Berube and de Bruyn, 1968) due to structural rearrangements at the surface (Yates, 1975) or micropore diffusion (Kent, 1983). Sasaki et al. (1983) studied the binding of weak binding electrolyte anions (e.g., ClO_4^-) and found that the adsorption was a two-step process, with the first step being a fast bulk diffusion step that was followed by a slower surface reaction, involving surface diffusion and ion pair formation. The relaxation time of the slower process observed was of the order of microseconds. Mehr et al. (1989) have shown that hysteresis occurs in calorimetric measurements of the enthalpy of proton adsorption by TiO_2 during acid-base titrations. The hysteresis is due to the slow release of weakly bound cations present as ion pairs at the surface, and the degree of hysteresis varies considerably among univalent cations (Mehr et al., 1990).

The mechanisms of adsorption and desorption of stongly-binding cations and anions have also been investigated by these techniques. Double relaxations are typically observed in these experiments (Mikami et al., 1983a; 1983b; Hachiya et al., 1984a,b), but the interpretation of the mechanisms that contribute to these relaxations is not as straightforward (Hayes, 1987). Hachiya et al. (1984a,b) postulated that the faster relaxation process was due to an adsorption/desorption reaction involving the release of a water molecule and a proton. The data were consistent with the rate-limiting step involving the removal of water from the hydration sheath of the metal ion to form an inner-sphere complex. The slower process was attributed to parallel adsorption/desorption reactions at heterogeneous sites with different adsorption energies. However, Hayes (1987) has shown that the slow relaxation process observed during the adsorption of Pb^{2+} is dependent on the ionic strength, suggesting that the slow process may be due to the adsorption/desorption of metal ions at surface hydroxyl groups that have formed an ion pair with a weakly binding anion.

Rates of adsorption onto aggergated and agglomerated solids are, in some cases, controlled by diffusion through the pore structure (Hansmann and Anderson, 1985; Kent, 1983; Smit, 1981; Smit et al., 1978). The rate of adsorption depends on the nature of the pore structure. For macroporous and mesoporous materials, the rate is proportional to the concentration of adsorbate in solution, the inverse square of the radius of the agglomerates, and the radius of the pores that comprise the structure (Kent, 1983; Helfferich, 1965). For Na^+ and OH^- adsorption on porous silicas, equilibrium was achieved in minutes to a few hours (Kent, 1983). Adsorption onto microporous materials can be much slower (Kent, 1983; Helfferich, 1965). The rate of cation adsorption is controlled primarily by the *acidity* of the functional groups in the microporous region (Helfferich, 1965; Smit et al., 1978; Smit, 1981; Kent, 1983). This results from the fact that adsorption involves interdiffusion of cations and protons in the microporous region; only free protons can diffuse and their concentration is controlled by the acidity of the functional groups (Helfferich, 1965). For example, adsorption of Na^+ onto microporous silica required several hours to reach equilibrium by this mechanism (Kent, 1983).

Reversibility of sorption processes. Hysteresis between adsorption and desorption of strongly-bound ions is frequently observed, with desorption taking more time than adsorption to attain equilibrium. Padmanabham (1983a,b) investigated the desorption of various transition metal cations from geothite and found that the desorption rate and extent of desorption was dependent on pH and the time of reaction allowed before initiating desorption. The effects were attributed to the formation of bidentate surface complexes at higher pH values and longer reaction times; such complexes would be expected to have greater activation energies for dissociation. Similar results have been obtained for anions like phosphate (Hingston, 1981). In some cases, the extent of reversibility has been used to estimate relative adsorption affinities of ions (Hayes, 1987, and references therein); however, this correlation does not generally apply because other processes may cause hysteresis. For example, when porous materials or aggregated particles are used as adsorbents, ions may diffuse into pore or void spaces and then diffuse out slowly after solution conditions are changed (Fuller and Davis, 1989). The formation of solid solutions or surface precipitates on mineral surfaces may also cause the slow release of sorbed ions (Davis et al., 1987).

Effect of solution speciation on ion adsorption

Complexation of metal ions by ligands can significantly alter their adsorption by mineral surfaces (Bourg and Schindler, 1978; Davis and Leckie, 1978b; Vuceta and Morgan, 1978; Schindler, 1981). Chloride and sulfate complexes of Cd(II) are weakly adsorbed by ferrihydrite in comparison to Cd^{2+} (Benjamin and Leckie, 1982), and metal-EDTA and metal-fulvate complexes are generally not adsorbed by the surfaces of silica, manganese oxides, calcite, or aluminosilicate minerals (van den Berg, 1982; Bowers and Huang, 1986; Davis, 1984; Davis et al., 1987; Hunter et al., 1988). In these cases, the mineral surface sites and dissolved ligands compete thermodynamically for coordination of metal ions, and the net adsorption of the metal ion at equilibrium can be estimated from straightforward equilibrium calculations (Benjamin and Leckie, 1982; Fuller and Davis, 1987). However, relatively basic oxides, e.g., hydrous aluminum oxide, may adsorb metal-EDTA, metal-fulvate, or metal-thiosulfate complexes in acidic solutions, resulting in a complicated pattern of increasing metal adsorption at low pH and decreasing metal adsorption at high pH (Bowers and Huang, 1986; Davis, 1984; Davis and Leckie, 1978b). Such complexes are called ternary surface complexes and their formation is reviewed in the chapter by Schindler (this volume). The pH dependence of ferrihydrite solubility can also introduce complex interactive effects in these systems (Bowers and Huang, 1987). Thus, it is difficult to generalize about the adsorption of complexes that form in solution between cations and ligands. For modeling, one hopes in general that adsorption equilibria can be formulated based on the activities of free (hydrated) cations and anions. This is often the case, but such an assumption must always be verified by experiment or other evidence.

Adsorption of hydrophobic molecules.

Detailed consideration of this topic is outside the scope of this chapter, and the reader is referred to the reviews of Karickhoff (1984) and Curtis et al. (1986) for more detailed information. Sorption of hydrophobic compounds can be considered as a solvent extraction process in which the hydrophobic solute partitions into organic particulate matter to avoid molecular interactions with the solvent, water. The process is driven by the incompatability of nonpolar compounds with water, not by the attraction of these compounds to the organic sorbent (Westall, 1987). Previous research has demonstrated that sorption of hydrophobic compounds by soils and sediments can be estimated with a small number of parameters, such as the solute's octanol/water partition coefficient and the mass fraction of organic carbon found in the sediment. Curtis et al. (1986) have considered the conditions under which mineral surfaces may influence the sorption of hydrophobic solutes on low-organic sandy aquifer materials. Westall (1987) reviewed the important molecular interactions involved in the sorption of hydrophobic compounds with ionizable or ionic moieties.

THE ELECTRIFIED MINERAL-WATER INTERFACE

The creation of an interface between mineral and liquid phases induces fundamental dissymmetry in the molecular environment of the interfacial region. The net effect on the molecular constituents of the two phases is a structural reorganization at the interface that reflects a compromise among competing interactions originating in the bulk phases (Sposito, 1984). These perturbations of the molecular environment invariably lead to a separation of electrical charge and the establishment of an electrical potential relative to the bulk solution phase.

Definitions of mineral surface charge

Surface charge can be classified into three types: (1) *permananent* structural charge, (2) *coordinative* surface charge, and (3) *dissociated* surface charge. Permanent structural charge is associated with the charge due to isomorphic substitutions in minerals, such as that due to substitution of Al^{3+} for Si^{4+} in tetrahedral sites of the crystal lattice of phyllosilicate minerals. Although in principle the permanent structural charge may be positive or negative charge, it is almost always negative among minerals commonly found in soils and sediments. Methods of estimating this component of charge in soils and sediments are given in Sposito (1984). The coordinative surface charge is the charge associated with the reactions of *potential-determining ions* with surface functional groups. For oxides, such reactions include the adsorption of H^+ or OH^- by the surface, but also include coordination reactions of other ions with surface functional groups. The coordinative surface charge may be negative or positive, and as will be shown below, may be subdivided into other subcategories in electrical double layer models.

The charge on particles is usually expressed as a *surface density*, σ_p, in units of charge per unit area ($C\ m^{-2}$). The net particle surface charge is defined as the sum of the surface densities of permanent structural charge, σ_s, and coordinative surface charge, σ_o, i.e.,

$$\sigma_p = \sigma_s + \sigma_o \quad . \tag{9}$$

In general, this sum will not be equal to zero, and to preserve electroneutrality, a counterion charge must accumulate near the particle surface. The counterion charge may accumulate as dissociated charge, σ_d, a diffuse atmosphere of counterions fully dissociated from the surface, or the counterion charge may take the form of a compact layer of bound counterions in addition to the diffuse atmosphere. In this case, the portion of the counterion charge that is present only as dissociated charge in the diffuse atmosphere is referred to as σ_d. The surface, compact, and diffuse layer charges are referred to collectively as an electrical double layer (EDL).

Classical electrical double layer models

The separation of charges in the EDL results in an electrical potential difference across the particle-water interface. Gouy (1910) and Chapman (1913) derived equations to describe the distribution of counterions in a diffuse swarm formed at a charged planar surface. The distribution of charge and potential within the EDL in the Gouy-Chapman theory are given by solutions of the Poisson-Boltzmann equation derived for a planar double layer. Detailed derivations and discussions of this equation are given in Bolt (1982) and Sposito (1984). In the model, all counterion charge is present as dissociated charge, σ_d, and the electroneutrality condition is given by:

$$\sigma_p + \sigma_d = \sigma_o + \sigma_s + \sigma_d = 0 \quad , \tag{10}$$

and σ_d, for a symmetical electrolyte with ions of charge z at 25°C, is derived from the Poisson-Boltzmann equation as:

$$\sigma_d = -0.1174\sqrt{I}\sinh\frac{ze\psi_o}{2kT} \quad , \tag{11}$$

where ψ_o is called the electrical potential at the surface. In the case of an asymmetrical electrolyte, a different charge-potential relationship is involved (Hunter, 1987). The calculated potential decays exponentially with distance from the surface (Fig. 15).

The Gouy-Chapman theory was in poor agreement with measurements made on charged mercury electrodes, because the predictions of electrical capacity greatly exceeded experimental observations. The theory was refined by Stern (1924) and Grahame (1947) to recognize the limitations imposed on electrical capacity by the finite size of ions and the likelihood that counterions may only approach the surface within some finite distance, probably close to the ionic radii of anions and the hydrated radii of cations. These authors introduced the concept of "specific adsorption", by which ions could bond chemically to the surface, and proposed that specifically adsorbed ions were located close to the surface, in essentially the same plane as the closest counterions (Fig. 15). With these modifications, Grahame (1947) was able to calculate EDL charges and electrical potentials that were in reasonable agreement with measured values for the mercury surface. In addition, the model provided a good description of the EDL properties of reversible electrodes (such as AgI and Ag_2S) at low ionic strength (Hunter, 1987). In the Stern-Grahame EDL model, the particle surface charge is now balanced by the charge in the Stern layer (referred to as the β plane charge) plus that of the dissociated charge, i.e.,

$$\sigma_p + \sigma_\beta + \sigma_d = 0 \quad . \tag{12}$$

206

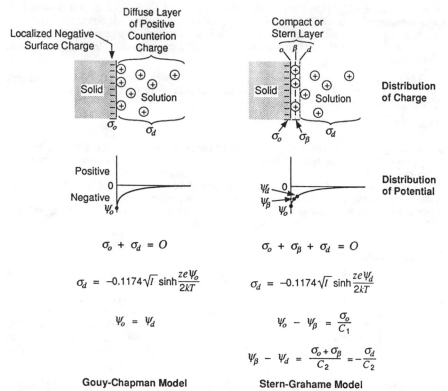

Figure 15. Schematic drawings of the electrical double layer in the classical Gouy-Chapman and Stern-Grahame models. Charge-potential relationships assumed for each model are shown below the diagrams. The relationships shown assume that σ_s equals zero. Reprinted from James and Parks (1982), in Surface and Colloid Science, v. 12, Plenum Press.

The Stern-Grahame model of the EDL also assumes that the electrical potential decays linearly with distance between the charged planes, leading to the following charge-potential relationships:

$$\psi_o - \psi_\beta = \frac{\sigma_p}{C_1} \quad , \tag{13}$$

$$\psi_\beta - \psi_d = \frac{\sigma_p + \sigma_\beta}{C_2} = -\frac{\sigma_d}{C_2} \quad , \tag{14}$$

where C_1 and C_2 are the integral capacitances of the interfacial layers. The Gouy-Chapman charge-potential relationship (Eqn. 11) is used to describe the decay of electrical potential away from the d-plane toward bulk solution (Fig. 15).

The Electrical Double Layer at Oxide Surfaces

The Nernst equation and proton surface charge. In applying the model to the reversible AgI electrode, it was noted that coordinative surface charge is acquired by nonstoichiometric transfer of the potential-determining ions, Ag^+ and I^-, from the solution phase to the electrode surface. As explained in more detail by James and Parks (1982), the coordinative surface charge can then be defined in terms of the adsorption densities of the potential-determining ions:

$$\sigma_o = F(\Gamma_{Ag^+} - \Gamma_{I^-}) \quad , \tag{15}$$

where F is the Faraday constant. The surface charge was shown to vary as a smooth function of the solution concentration of Ag^+, $[Ag^+]$. At one unique value of $[Ag^+]$, the surface charge equals zero, because Γ_{Ag} equals Γ_I, and this condition was called the *point-of-zero charge* (or PZC). Differences in the surface potential, ψ_o, as a function of potential-determining ions can then be expressed via a modified form of the Nernst equation (James and Parks, 1982), i.e.,

$$\psi_o = \frac{2.3RT}{F}(pAg_{pzc} - pAg) \quad , \tag{16}$$

where pAg is the negative log of $[Ag^+]$.

The success of the Stern-Grahame model in describing these EDL properties on reversible electrodes led other research groups to use this approach in modeling the surface of oxides, with H^+ and OH^- as the potential-determining ions. Parks and de Bruyn (1962) studied the formation of surface charge on hematite (α-Fe_2O_3) particles suspended in KNO_3 solutions by acid-base titration as a function of ionic strength. They argued that coordinative surface charge developed via proton transfer reactions of surface hydroxyl groups that formed when the virgin surface became hydrated (Fig. 1).

$$\equiv FeOH^o \leftrightarrow \equiv FeO^- + H^+ \quad , \tag{17}$$

$$\equiv FeOH^o + H^+ \leftrightarrow \equiv FeOH_2^+ \quad , \tag{18}$$

where $\equiv FeOH_2^+$, $\equiv FeOH^o$, and $\equiv FeO^-$ represent positively charged, uncharged, and negatively charged hydroxyl groups, respectively, on the oxide surface. To distinguish this type of coordinative surface charge from other types, we shall define the term, *proton surface charge*, or σ_H, i.e.,

$$\sigma_H = \Gamma_{H^+} - \Gamma_{OH^-} \quad . \tag{19}$$

The most common method of measuring the proton surface charge is by potentiometric acid-base titration of a mineral suspension in solutions of variable ionic strength (Bolt, 1957; Yates and Healy, 1980; James and Parks, 1982; Dzombak and Morel, 1990). If $CO_2(g)$ and all other acids and bases (other than the mineral surface) are absent, the net consumption of H^+ or OH^- can be calculated with the expression:

$$\Gamma_{H^+} - \Gamma_{OH^-} = (c_A - c_B - [H^+] + \frac{[OH^-]}{\mathcal{A}W}) \quad , \tag{20}$$

where c_A and c_B are the molar concentrations of added acid and base, $\mathcal{A}$ is the specific surface area per unit weight of the mineral, and W is the weight of mineral in suspension per unit weight of water. The molar concentrations of H^+ and OH^- are calculated from the pH measurements. The mineral must be highly insoluble, so that the uptake or release of H^+ or OH^- by dissolved constituents of the solid are negligible (Parker et al., 1979). Proton surface charge may also be calculated by subtracting the titration curve of the background electrolyte (in the absence of the mineral) from the titration curve of the mineral suspension (Fig. 16) to yield the excess acid, q, defined as:

$$q = c_A - c_B - [H^+] + [OH^-] \quad . \tag{21}$$

Excess acid can then be plotted as a function of pH as shown in Figure 16b. The value of the proton surface charge is arbitrary until a value for zero charge is established for the given system. As will be discussed below, the pH at which the proton surface charge is assigned a value of zero is usually determined by a unique intersection point of q for a family of titration curves as a function of ionic strength (Fig. 17).

In the model of Parks and de Bruyn (1962), the proton surface charge forms via the reactions in Equations 17 and 18, and is defined as:

208

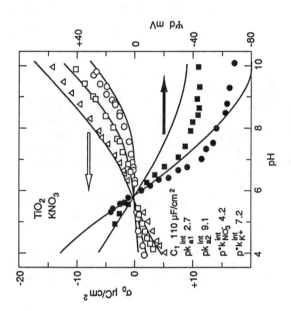

Figure 16 (left). Different methods of presenting data for acid-base titrations of oxide surface functional groups. (a) raw data (moles of protons in the titrant versus pH) for titration of a rutile (TiO$_2$) suspension in 0.01 M KNO$_3$ solution (data from Yates, 1975). (b) Proton charge density (σ_H), in units of μC/cm^2 as a function of pH. σ_H is computed from the excess acid, q (defined in Eqn. 21), which is derived from the raw data in part (a). Reprinted from Dzombak and Morel (1990), Surface Complexation Modeling: Hydrous Ferric Oxide, John Wiley & Sons.

Figure 17 (above). Proton charge density and zeta potential of rutile (TiO$_2$) in KNO$_3$ solutions as a function of pH and electrolyte concentration at 25°C. Data are taken from Yates(1975) and Wiese and Healy (1975a). Points denote experimental data: circles, 0.001 M; squares, 0.01 M; triangles, 0.1 M. Solid lines represent triple layer model (section "Triple layer model") simulations of the data using the model parameters shown in the figure. Reprinted from James and Parks (1982), in Surface and Colloid Science, v. 12, Plenum Press.

$$\sigma_H = [\equiv FeOH_2^+] - [\equiv FeO^-] \quad . \tag{22}$$

Parks and de Bruyn (1962) observed a unique pH at which σ_H was equal to zero at all ionic strengths, analogous to the behavior of the AgI electrode.

Despite this analogy, Li and de Bruyn (1966) and Hunter and Wright (1971) found that the Nernst equation was invalid in describing the surface potential of oxides in conjunction with the Stern-Grahame model. The lack of applicability of the Nernst equation to oxides, as opposed to reversible electrodes, has now been examined in great detail in many studies (Levine and Smith, 1971; Lyklema, 1971; Healy and White, 1978; Bousse and Meindl, 1986), with most authors concluding that oxides do not behave as reversible electrodes because (1) H^+ is not a constituent of the oxide lattice and the ionization of surface hydroxyl groups (Eqns. 17 and 18) is controlled by chemical reactions, and (2) a wide variety of ions may participate in coordination reactions with surface hydroxyl groups and thus become potential-determining ions (Dzombak and Morel, 1990).

The zero surface charge condition. The unique pH at which σ_H equals zero has been referred to as the *point-of-zero charge*, or pH$_{PZC}$, by most authors. Other authors have referred to the unique intersection point of a family of titration curves at different ionic strengths as the *point-of-zero-salt-effect*, or pH$_{PZSE}$ (Parker et al., 1979; Pyman et al., 1979; Sposito, 1984). As mentioned above, many ions may participate in coordination reactions with surface hydroxyl groups. Thus, the coordinative surface charge, σ_o, can be divided into two subgroups, the proton surface charge, σ_H, and the coordinative complex surface charge, σ_{CC}, i.e.,

$$\sigma_o = \sigma_H + \sigma_{CC} \quad . \tag{23}$$

The technique for determining σ_H measures a mass balance of moles of protons and hydroxide ions that are "bound" in some manner by the mineral surface (Eqn. 19), and as such it is not actually a measurement of surface charge. The pH$_{PZC}$ has been defined as a pH value at which σ_H is zero. However, according to Equation 23, it is not necessary for the coordinative surface charge, σ_o, to be equal to zero when σ_H is zero. Thus, the pH$_{PZC}$ may occur at a pH value at which σ_o is not equal to zero, and under these conditions, the term point-of-zero charge, no longer seems appropriate. Therefore, we shall refer to the pH value at which the proton surface charge, σ_H, equals zero as the *point-of-zero-net-proton charge*, or pH$_{PZNPC}$.

The observation above led to the definition of the *pristine point-of-zero charge* (pH$_{PPZC}$) of oxides (Bolt and van Riemsdijk, 1982). When an oxide surface is suspended in a solution in which H^+ and OH^- are the only potential-determining ions, then σ_{CC} equals zero, and $\sigma_o = \sigma_H$. The pH$_{PPZC}$ distinguishes the pH$_{PZNPC}$ of such a system from that in which coordination reactions of other ions take place. Surface coordination reactions of strongly adsorbing ions may shift the pH$_{PZNPC}$ of an oxide to a new value. For example, in a system containing colloidal oxide particles suspended in a $Pb(NO_3)_2$ solution, the proton surface charge could be defined as in Equation 19. Since Pb^{2+} may also form coordinative complexes with the surface, the coordinative surface charge could be defined as follows:

$$\sigma_o = \Gamma_{H^+} + 2\Gamma_{Pb^{2+}} - \Gamma_{OH^-} \quad . \tag{24}$$

A specific pH value may exist in this system (defined by a particular Pb^{2+} concentration) at which the proton surface charge is zero, thus meeting the definition of pH$_{PZNPC}$. However, this pH$_{PZNPC}$ will differ from the one found in a system in which Pb^{2+} is absent (Fig. 18), because the adsorption of H^+ is influenced by the adsorption of Pb^{2+} (Hohl and Stumm, 1976). The value of the pH$_{PPZC}$ is indicative of the intrinsic acidity of the mineral surface alone in its reaction with pure water. A compilation of these values is given in Table 4. The more general term, pH$_{PZNPC}$, is indicative of a specific system defined by the mineral phase and solution composition. Little is known about the temperature dependence of σ_H and the pH$_{PZNPC}$; comprehensive studies on this topic have only recently been completed (Fokkink et al., 1989).

210

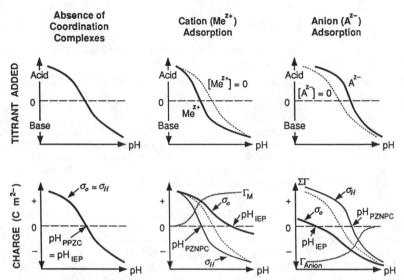

Figure 18. The coordinative surface charge (σ_o) on an oxide surface is established by proton transfer reactions and specific adsorption of anions and cations. Upper graphs show the way in which cation and anion adsorption affect acid-base titration curves. Lower graphs illustrate the effect of cation and anion adsorption on the coordinative surface charge (σ_o) and the proton surface charge (σ_H). Note the separation of the isoelectric point, pH_{IEP}, and point of zero net proton charge, pH_{PZNPC}, when cations or anions are specifically adsorbed. Modified from Hohl et al. (1980).

Table 4. Estimates of the pH_{PPZC} for various minerals

Mineral	pH_{PPZC}	Reference
$\gamma - Al_2O_3$	8.5	a
Anatase (TiO_2)	5.8	b
Birnessite ($\delta - MnO_2$)	2.2	c
Calcite ($CaCO_3$)	9.5	d
Corundum ($\alpha - Al_2O_3$)	9.1	e
Goethite ($\alpha - FeOOH$)	7.3	f
Hematite ($\alpha - Fe_2O_3$)	8.5	g
Hydroxyapatite ($Ca_5(PO_4)_3OH$)	7.6	h
Magnetite ($\alpha - Fe_3O_4$)	6.6	i
Rutile (TiO_2)	5.8	j
Quartz ($\alpha - SiO_2$)	2.9	e

References:
a) Huang and Stumm (1973)
b) Berube and de Bruyn (1968)
c) Balistrieri and Murray (1982)
d) Parks (1975)
e) Sposito (1984)
f) Atkinson et al. (1967)
g) Breeuwsma and Lyklema (1973)
h) Bell et al. (1973)
i) Tewari and McLean (1972)
j) Yates (1975)

Zeta potential. The electrical potential difference across a portion of the mineral-water interface can be estimated from electrokinetic methods such as electrophoresis or streaming potential measurements (Hunter, 1981; James, 1979). For electrophoresis, data are reported in terms of the electrophoretic mobility, the average, steady-state velocity of charged particles moving in response to an applied, constant electric field. Water near the particle surfaces has a greater viscosity than bulk water, and thus, some water molecules move along with the particles in the electric field. The zeta potential correponds to the electrical potential at the effective shear (or slipping) plane between the moving and stationary phases. The zeta potential can be calculated from the electrophoretic mobility for particles of simple geometry (Hunter, 1981; James and Parks, 1982, and references therein). The Gouy-Chapman theory can then be used to estimate the dissociated counterion charge in the diffuse atmosphere that is located on the solution side of the slipping plane.

The location of the slipping plane is not known with certainty. It is known that the distribution of dissociated charge on the solution side of the slipping plane is well described by Gouy-Chapman theory (Li and de Bruyn, 1966). It has been proposed that the approximation be made that the slipping plane lies near the distance of closest approach of dissociated counterion charge (Hunter, 1987; Lyklema, 1977). Thus, all dissociated counterion charge would be located outside the slipping plane and the zeta potential could be used to estimate σ_d. Probably the most important measurement that can be made by electrophoresis is that of the isoelectric point, or pH_{IEP}. For oxides, it is usually found that a suspension of colloidal particles in a solution of defined composition will have a unique pH value (or values) at which the electrophoretic mobility is zero. These pH values are referred to as isoelectric points (Breeuwsma and Lyklema, 1973), and the measurement defines conditions under which particles have no dissociated counterion charge, i.e., $\sigma_d = 0$. In oxide systems in which H^+ and OH^- are the only potential-determining ions, the pH at which there is zero counterion charge (the pH_{IEP}) is also the pristine point-of-zero charge (pH_{PPZC}), and,

$$pH_{IEP} = pH_{PPZC} = pH_{PZNPC} \quad . \tag{25}$$

However, as was the case with the pH_{PZNPC}, surface coordination reactions of strongly adsorbing ions shift the pH_{IEP} of an oxide to a new value (Fig. 18). For example, in a system containing colloidal oxide particles suspended in a $Pb(NO_3)_2$ solution, the coordinative surface charge is defined by Equation 24. At the pH_{IEP} of the system (defined by a particular Pb^{2+} concentration), the dissociated surface charge, σ_d, is equal to zero. Assuming that σ_s is zero, then by Equation 10, the coordinative surface charge, σ_o, must also be equal to zero (assumes the Gouy-Chapman model; in the Stern-Grahame model this condition also requires that σ_β be equal to zero). In this case, the zero value for coordinative surface charge represents a balance of positive and negative charges contributed by all ions coordinated at the surface plane. However, the pH_{IEP} in this system will differ from one in a system in which Pb^{2+} is absent, because of the influence of Γ_{Pb} on Equation 24 (Fig. 18).

In the Gouy-Chapman and Stern-Grahame models, electrical charge and potential are assumed uniform in any particular plane. It is recognized, however, that the physical discreteness-of-charge on the surface sites means that the actual surface potential cannot be equated with ψ_o, because the actual potential is expected to be much larger in the vicinity of the sites (Healy and White, 1978). Instead, it is assumed that the surface potential includes another term, the micropotential (Levine and Smith, 1971), which is insensitive to the values of surface charge or ionic strength, and thus, can be included within the values of acidity constants. The discreteness-of-charge is known to have a significant effect on the predictions of electrical potential at the beginning of the diffuse layer, Ψ_d (Hunter, 1987; Healy and White, 1978).

Interpretation of electrophoretic mobility measurements is complicated by the complex relationships between mobility and zeta potential that exist for complex particle geometries and the dependence of the relationships on ionic strength. Because of this, some authors have questioned the usefulness of zeta potential measurements in understanding EDL properties (e.g., Dzombak and Morel, 1987; 1990; Westall and Hohl, 1980). While it is true that the absolute value of the zeta potential will generally not be known unambiguously, measurement of the pH_{IEP} is not subject to these errors, and this measurement provides

212

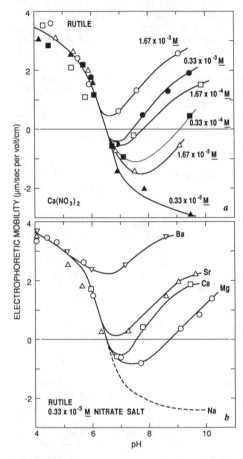

Figure 19. Dependence of the electrophoretic mobility of rutile particles on pH and electrolyte concentration in solutions containing alkaline earth cations. (a) Mobility at various concentrations of $Ca(NO_3)_2$; (b) Mobility at an electrolyte concentration of 0.330 mM in various alkaline earth nitrate solutions. Reprinted from Fuerstenau et al. (1981), in Adsorption from Aqueous Solutions, Plenum Press.

interesting additional information about the EDL that can be included in tests of EDL models. For example, Fuerstenau et al. (1981) and James et al. (1981) have investigated detailed effects of cation adsorption on the electrophoretic mobility of rutile suspensions in solutions with variable concentrations of metal and alkaline earth ions (Fig. 19). Wiese and Healy (1975a,b) have shown that measurement of the zeta potential is extremely effective in estimating the coagulation rates of colloidal suspensions. The establishment of pH_{IEP} is extremely important in many applications of EDL models, in particular for the flotation of minerals (Hornsby and Leja, 1982; de Bruyn and Agar, 1962) and in the optimization of drinking water and wastewater treatment by oxide precipitates (Dempsey et al., 1988; Benjamin et al., 1982; Sorg, 1979; 1978).

<u>Early developments of surface coordination theory</u>

As noted by Dzombak and Morel (1990), while the differences between the EDL properties of oxides and classical reversible electrodes were being discovered, other groups were emphasizing the importance of chemical reactions in describing the adsorption of ions by oxide surfaces. Kurbatov et al. (1951) studied the adsorption of cobalt ions by ferric hydroxide and considered the adsorption equilibria to be controlled by chemical reactions with specific surface sites. These equilibria were described in terms of mass law equations involving the exchange of protons for metal ions. Other groups (Ahrland et al., 1960; Dugger et al., 1964) applied this approach to describe the adsorption of many metal ions on the surface of silica. Long-range electrostatic interactions were not evaluated in these studies, but Dugger et al. noted that the equilibrium "constants" derived from the mass law expressions were only apparent constants that needed correction for the activities of the surface species. These developments were an important step toward the surface complexation models developed several years later.

Parks and de Bruyn (1962) noted that the titration procedure and analysis of the pH_{PZNPC} for hematite was essentially the same as that used commonly in characterizing proteins (Tanford, 1961), and that proton surface charge formation as a function of pH and ionic strength on the oxide was similar to that observed for colloidal proteins. The electrical charge on molecular and colloidal proteins arises from the ionization of carboxylate and amine functional groups, e.g.,

$$\equiv COOH \leftrightarrow \ \equiv COO^- + H^+ \ , \tag{26}$$

$$\equiv NH_2 + H^+ \leftrightarrow \ \equiv NH_3^+ \ . \tag{27}$$

Unlike monomeric solutes in water, however, the ionization constants for the reactions of Equations 26 and 27 are not invariant with solution composition, but are observed to depend on surface charge (Steinhardt and Reynolds, 1969).

An important development of this research, which was to be applied later to hydrous oxide minerals, was that charge and electrical potential developed as a result of chemical reactions at specific surface sites. As explained by James and Parks (1982), what distinguishes the protein surface reactions from analogous reactions in homogeneous solution is the variable electrostatic energy of interaction caused by the variable charge on the surface. An invariant acidity constant results when the activity of H^+ at the surface, $\{H^+\}_s$, is used in the equilibrium expression, for example, of Equation 26, i.e.,

$$K_{Eq.26}^{intr} = \frac{\{\equiv COO^-\}\{H^+\}_s}{\{\equiv COOH\}} \ , \tag{28}$$

and $\{H^+\}_s$ is related to $\{H^+\}_{aq}$ through a function of surface charge or potential. A considerable degree of success has been achieved in modeling the pH and ionic strength dependence of surface charge of colloidal proteins by applying a coulombic correction factor, derived from either Gouy-Chapman or Stern-Grahame theory, to the mass law equations for Equations 26 and 27 (Tanford, 1961; Steinhardt and Reynolds, 1969; King, 1965).

MODELS FOR ADSORPTION-DESORPTION EQUILIBRIUM

Two types of models for describing the equilibria of adsorption-desorption reactions at mineral surfaces can be distinguished: (1) empirical partitioning relationships, and (2) conceptual models for surface complexation that use the formalism of ion association reactions in solution as a representation of surface reactions (Hayes, 1987). Because of the complexity of natural systems, the empirical approach has been widely used in describing the partitioning of solutes between mineral and water phases in geochemical applications, especially in transport models and engineering applications. Surface complexation models, on the other hand, have been used primarily by aquatic scientists interested in developing a thermodynamic understanding of the coordinative properties of mineral surface ligand groups via laboratory investigations. It must be emphasized that the adsorption models discussed here are *equilibrium* models, and an assumption of adsorptive equilibrium should be justified before they are applied to laboratory or field observations. Attempts to apply the models to natural systems and typical problems encountered are discussed in the next section.

Empirical adsorption models

Distribution coefficients. Adsorption is often described in terms of equations or partitioning relationships that relate the activity of a solute in water to the amount of the solute adsorbed at constant temperature. The simplest of these expressions is the distribution coefficient derived from the association reaction:

$$J_{aq} \leftrightarrow J_{ads} \, , \tag{29}$$

where J is an adsorbing solute, J_{aq} represents solute J dissolved in the aqueous phase, and J_{ads} is solute J adsorbed. The amount of solute adsorbed is frequently expressed as an adsorption density parameter, Γ, in units of adsorbed solute per unit area or weight of adsorbent. The distribution coefficient, K_d, is usually defined as:

$$K_d = \Gamma_J / [J_{aq}] \, . \tag{30}$$

Usually $[J_{aq}]$ represents the concentration of all dissolved species of J, although a distribution coefficient for a specific species has been defined (Tessier et al., 1989). For inorganic ions, the distribution coefficient is highly dependent on the conditions under which it is measured, e.g., pH, background salt composition, concentration of dissolved carbonate, concentrations of competing adsorbates, etc. The calculation usually assumes that all aqueous species of solute J (including all complexes) have equal affinity for the surface and that all exposed surfaces of the adsorbent (even a complex mineral assemblage) have equal affinity for J. The value of K_d is often very strongly influenced by pH. Thus, the distribution coefficient has little value in predicting the response of solute adsorptive behavior that may result from changes in either the aqueous or mineralogical composition of a system. Applications of the distribution coefficient to natural systems are discussed in the section on "Applications in Aqueous Geochemistry".

Langmuir isotherm. Langmuir (1918) derived an equation to describe adsorption by considering the reaction:

$$\equiv S + J \leftrightarrow \equiv SJ \, , \tag{31}$$

where J is an adsorbing solute, $\equiv S$ is an adsorptive surface site, and $\equiv SJ$ represents the species of adsorbed solute J. Assuming that all surface sites have the same affinity for solute J, a mass law equation can be written for the equation above as follows:

$$K_L = \frac{\Gamma_J}{[J_{aq}]\Gamma_S} \, , \tag{32}$$

where Γ_S is the surface density of uncomplexed adsorptive sites and K_L is the conditional Langmuir equilibrium constant. Making the further assumption that the density of all adsorptive sites on the surface, S_T, is fixed and that J is the only adsorbing solute, the mass law can be combined with the mass balance for surface sites:

$$S_T = \Gamma_S + \Gamma_J \, , \tag{33}$$

to yield the expression commonly known as the Langmuir isotherm:

$$\Gamma_J = S_T \left[\frac{K_L[J_{aq}]}{1 + K_L[J_{aq}]} \right] \, . \tag{34}$$

The generalized shape of the Langmuir isotherm on a log-log plot is shown in Figure 20, where the x-axis represents $[J_{aq}]$ (shown as C). A useful method of evaluating whether data are consistent with the Langmuir isotherm is to plot the distribution coefficient, K_d, as a function of Γ_J (Veith and Sposito, 1977a). Multiplying both sides of the Langmuir isotherm by $\frac{1}{[J_{aq}]} + K_L$, and solving for K_d yields the following linearized form of the Langmuir equation:

$$K_d = S_T K_L - K_L \Gamma_J \, . \tag{35}$$

Graphical procedures can then be utilized to estimate the values of S_T and K_L if the Langmuir equation is applicable to experimental data. However, as discussed by Kinniburgh (1986), other linearizations of the Langmuir isotherm can be made, and each of the transformations has deficiencies with respect to parameter determination by linear regression, even when data are properly weighted for error. The parameters are best determined by nonlinear regression, which also allows testing and comparison with description by other isotherms (Kinniburgh, 1986).

The adherence of data to an adsorption isotherm provides no evidence as to the actual mechanism for the association of a solute with a mineral phase (Sposito, 1986). It has been shown that special cases of precipitation reactions may also exhibit data that conform to the Langmuir isotherm (Veith and Sposito, 1977b). Like the distribution coefficient, K_L is dependent on strictly constant solution conditions and is very sensitive to changes in pH. Thus, it cannot be applied in a straightforward manner under highly variable conditions.

Freundlich and other isotherms. For inorganic ions, it is frequently observed that a plot of K_d versus Γ_J results in a curve that is convex to the Γ_J axis instead of the straight line expected from the homogeneous site Langmuir expression. In these cases, the data must be fitted to a multiple-site Langmuir expression or a generalized exponential isotherm (Kinniburgh, 1986), such as the Freundlich isotherm (Fig. 20). Isotherms for cation adsorption on hydrous oxides typically exhibit slopes of less than one on a log-log plot of adsorption density versus pH (Benjamin and Leckie, 1981a; Kinniburgh and Jackson, 1982; Honeyman and Leckie, 1986). The two-site Langmuir isotherm assumes that there are two types of surface sites that may participate in adsorption reactions, and this isotherm is frequently suitable for describing adsorption data on mineral surfaces with heterogeneous sites (Sposito, 1982; Sposito, 1984; Kinniburgh, 1986; Dzombak and Morel, 1990).

Other isotherms that exhibit a wide range of applicability to minerals are the Toth and the modified Dubinin-Radushkevich. One of the frequently mentioned drawbacks of empirical expressions is that they are applicable only to a specific set of conditions. In particular, the dependence of adsorption equilibria on pH is significant, and an empirical expression that is valid only for a single pH value would usually be of limited use in modeling the migration of inorganic contaminants in a groundwater system (Honeyman and Leckie, 1986). However, Kinniburgh (1986) has shown that generalized, pH-independent versions of the two-site Langmuir, Toth, and modified Dubinin-Radushkevich isotherms can be derived and successfully applied to problems such as phosphate adsorption by soils or zinc adsorption by ferrihydrite. If linked with geochemical aqueous speciation models, such generalized empirical expressions for adsorption could prove as useful for certain practical applications as the more elegant (but data intensive) surface complexation models. The migration of hydrophobic organic contaminants may be satisfactorily accomplished with distribution coefficients or isotherm relationships (McCarthy and Zachara, 1989).

General partitioning equation. Because adsorption of many inorganic ions is highly dependent on the concentration of H^+, it is important that an adsorption model be capable of predicting adsorptive behavior as a function of pH. The empirical relationships that relate adsorption density to adsorbate concentration are valid only for constant conditions, and thus, these models are generally insufficient for modeling in environmental geochemistry. Honeyman and Leckie (1986) have considered a modified form of the distribution coefficient which describes the adsorption of ion J in terms of the macroscopic observations of proton (or hydroxyl) exchange:

$$\equiv S + J \leftrightarrow \ \equiv SJ + \chi H^+ \quad , \tag{36}$$

and,

$$K_{part} = \frac{\Gamma_J [H^+]^\chi}{\Gamma_S [J]} \quad , \tag{37}$$

216

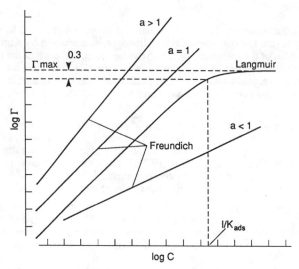

Figure 20. Characteristic shapes of Langmuir and Freundlich adsorption isotherms. Reprinted from Morel (1983), Principles of Aquatic Chemistry, John Wiley & Sons.

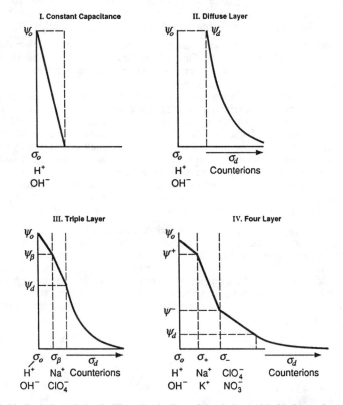

Figure 21. Idealized schematic drawing illustrating the decay of electrical potential with distance from the surface for the constant capacitance, diffuse double layer, triple layer, and four-layer models. Diagram assumes conditions typical of an oxide surface bathed in a simple electrolyte (e.g., NaCl) at a pH value different than the pH_{PPZC}.

where χ is the apparent ratio of moles of protons released per mole of solute J adsorbed. Since the relationship described in Equation 37 corresponds to macroscopic observations of all surface reactions involving H^+, no information is provided about the source of protons in the adsorption process other than the generic relationship between adsorption and changes in the activity of H^+. In particular, it does not reveal stoichiometric information about microscopic adsorption reactions at the surface (Honeyman and Leckie, 1986), although some investigators continue to relate the macroscopic parameter to the stoichiometry of microscopic reactions (Belzile and Tessier, 1990; Tessier et al., 1989). The objective in using such a model should be to "calibrate" the values of χ and K_{part} for a range of system compositions of interest to specific applications. Since the speciation of J is usually not considered, the modeling approach is only useful for systems of constant composition. Despite the conditional nature of such constants, the approach has been useful in examining specific issues of adsorptive behavior (Balistrieri and Murray, 1983). Applications of this approach are discussed in the section on "Applications to Aqueous Geochemistry" (below).

Surface complexation models

An alternative to the empirical modeling approaches are the surface complexation models (SCM), which extend the ion-association model of aqueous solution chemistry to include chemical species on surfaces (Hayes, 1987). The concepts used in modeling the ionization of protein surface groups (see section on "Early developments of surface complexation theory," above) were extended by other research groups, particularly those of Schindler and Stumm (reviewed below), to the consideration of adsorption reactions on hydrous oxides. These concepts form the basic tenets of all surface complexation models:

1. The surface is composed of specific functional groups that react with dissolved solutes to form surface complexes (coordinative complexes or ion pairs) in a manner analogous to complexation reactions in homogeneous solution.

2. The equilibria of surface complexation and ionization reactions can be described via mass law equations, with correction factors applied for variable electrostatic energy using EDL theory.

3. Surface charge and surface electrical potential are treated as necessary consequences of chemical reactions of the surface functional groups. Unlike the Gouy-Chapman and Stern-Grahame models for polarizable and reversible electrodes, the specific chemical interactions of protein (and oxide) functional groups dominate the EDL properties; the electric field and electrostatic effects are secondary factors that result from the surface coordination reactions themselves.

4. The apparent binding constants determined for the mass law equations are empirical parameters related to thermodynamic constants via the rational activity coefficients of the surface species (Sposito, 1983).

A number of different surface complexation models have been proposed in the last two decades. The models are distinguished by differences in their respective molecular hypotheses. Each model assumes a particular interfacial structure (Fig. 21), resulting in the consideration of various kinds of surface reactions and electrostatic correction factors to mass law equations. While the models differ in their consideration of interfacial structure, all the models reduce to a set of simultaneous equations that can be solved numerically (Dzombak and Morel, 1987). These equations include: (1) mass law equations for all surface reactions under consideration, (2) a mole balance equation for surface sites, (3) an equation for computation of surface charge, and (4) a set of equations representing the constraints imposed by the model of interfacial structure. The models are illustrated below as applied to the surfaces of hydrous oxides, but the treatment could be extended in a straightforward manner to other minerals that develop amphoteric surface charge. Some common features of the models are presented first before explaining the differences that distinguish the models.

Properties of solvent water at the interface. Water molecules near mineral surfaces have distinct properties (Mulla, 1986). In surface complexation models it is assumed that the dielectric constant of the solvent remains constant at the value of the bulk electrolyte solution up to the outer adsorption plane of the model. Mulla's theoretical calculations suggest that the dielectric constant should be much lower close to a mineral surface, and the predictions suggest that the value decays gradually between the surface and bulk water. Pashley and others (Pashley and Quirk, 1989; Pashley and Israelachvili, 1984a; 1984b) have shown that oscillatory forces of hydration exist when two mica plates are forced together. The forces are a function of the separation distance, and the oscillations indicate density variations and the layering of water on the order of molecular dimensions. Regardless of the interfacial model used, the low dielectric constant of solvent water molecules in the interfacial region should enhance the formation of uncharged surface complexes (Davis and Leckie, 1979). Because of this, the hydrolysis of adsorbed metal ions and protolysis of adsorbed anions at oxide surfaces may occur more readily and at different pH values than occurs in bulk water.

Surface acidity of hydrous oxides. The acidity of a surface hydroxyl group depends on numerous factors, e.g., acidity and coordination number of the metal atom to which it is bonded, electrostatic field strength and induction effects of the mineral, structural ordering of water molecules in the vicinity of the surface, and the local surface structure (face, edge, dislocation, etc.) (Huang, 1981; Westall, 1986; also see chapter by Parks, this volume). Strong acid-base titrations of colloidal oxide suspensions in univalent electrolyte solutions exhibit no clear inflection points, indicating that surface acidity is a function of the degree of surface protonation. The existence of various types of surface functional groups was reviewed earlier, and the titration results are consistent with a distribution of surface site acidities (van Riemsdijk et al., 1986). Nonetheless, to simplify the modeling, the variation in surface acidity with surface charge is usually treated as a consequence of electrostatic perturbations on identical diprotic surface sites, rather than considering heterogeneity among surface sites (Hayes, 1987).

Following the observations of Parks and de Bruyn (1962) and the approach used for modeling polyelectrolytes (Tanford, 1961; King, 1965), a series of papers appeared applying this approach to oxide surfaces (Schindler and Kamber, 1968; Stumm et al., 1970; Schindler and Gamsjager, 1972). Two mass law equations for hydroxyl ionization reactions were written to describe the amphoteric behavior of oxide surfaces (a generalization of Eqns. 17 and 18):

$$\equiv SOH_2^+ \leftrightarrow \ \equiv SOH + H^+ \ , \tag{38}$$

$$\equiv SOH \leftrightarrow \ \equiv SO^- + H^+ \ . \tag{39}$$

Following the theoretical arguments of Chan et al. (1975), the chemical potential of species i is written as follows:

$$\mu_i = \mu_i^o + kT \ln C_i + kT \ln \gamma_i \ , \tag{40}$$

where the terms containing $kT \ln_i$ represent the concentration-dependent part of the free energy of interaction of species i with its environment. All concentration-independent terms are included in the constant, μ_i^o. This equation defines the activity coefficient, γ_i. Thus, the thermodynamic equilibrium constant for Equation 38, defined for standard state conditions, K_{a1}^o, is given by:

$$K_{a1}^o = \frac{[\equiv SOH] \ [H^+]}{[\equiv SOH_2^+]} \frac{\gamma_{SOH} \gamma_{H^+}}{\gamma_{SOH_2^+}} \ , \tag{41}$$

where the terms in square brackets are concentrations and the term, γ_i, is the activity coefficient of species i.

To apply the surface ionization model, Schindler and Stumm defined conditional equilibrium constants (called "intrinsic" constants) in the following manner:

$$K^{intr} = \frac{x_{SOH}[H^+]}{x_{SOH_2^+}} , \qquad (42)$$

where x_i refers to the mole fraction of species i among all surface sites. As explained in detail by Sposito (1983), in calculating K^{intr}, the authors assumed a Standard State for the species $\equiv SOH_2^+$ wherein $\equiv SOH_2^+$ exists in a chargeless environment. The conditional constant for Equation 42 derived in this manner differs from the true thermodynamic constant by the mean interaction energy per adsorbed proton. Most definitions of equilibrium constants in surface complexation models have used analogous unconventional definitions of the Standard State for surface species.

Following this approach, operational acidity quotients were derived that varied smoothly with surface charge, and by extrapolating the values to zero surface charge (Huang and Stumm, 1973; Huang 1981), conditional acidity constants were obtained that are dependent on ionic strength (Davis et al., 1978). The graphical procedures described in these papers assumed a complete dominance of positive or negative sites on either side of the pH_{PPZC}, but this assumption has been shown to result in systematic errors in estimating the apparent acidity constants (Dzombak and Morel, 1987). While the diprotic acid representation of the oxide surface has been widely accepted, other models for surface acidity have been used. Westall (1986) and van Riemsdijk et al. (1986) have described surface ionization in terms of one acidity constant. Healy and White (1978) and James and Parks (1982) have considered zwitterionic surfaces containing two separate acidic and basic monoprotic surface groups.

Surface coordination reactions. Following Sposito (1984), the general reactions that describe surface coordination are as follows:

$$a(\equiv SOH) + pM^{m+} + qL^{l-} + xH^+ + yOH^- \;\leftrightarrow\; \equiv (SO)_a M_p (OH)_y H_x L_q^\delta + aH^+ \qquad (43)$$

$$b(\equiv SOH) + qL^{l-} + xH^+ \;\leftrightarrow\; \equiv S_b H_x L_q^\zeta + bOH^- , \qquad (44)$$

where $\delta = pm + x - a - ql - y$ and $\zeta = x + b - ql$ are valences of the surface complexes formed (hence are *whole numbers*). These valences contribute to the coordinative surface charge, σ_o. At the present stage of model evolution, all surface complexation models treat the adsorption of strongly-bound ions in accordance with Equations 43 and 44. Differences between the models are based on the interfacial structure of the EDL and the manner in which weakly-bound ions are treated. Mass law equations are defined for each coordination reaction (following the approach taken in Eqn. 41), but each model applies different terms for electrostatic correction factors that are consistent with the interfacial structure of the model.

The Constant Capacitance Model (CCM). The original development of the model can be found in Schindler and Kamber (1968) and Hohl and Stumm (1976), and it has been reviewed by Sposito (1984) and Schindler and Stumm (1987). The CCM is a special case of the diffuse double layer (DDL) model (reviewed below), applicable in theory only to systems at high, constant ionic strength. At high ionic strength, the EDL, consisting of the coordinative surface charge and dissociated counterion charge, can be approximated as a parallel plate capacitor (Fig. 21). The molecular hypotheses of this model are as follows (Hayes, 1987):

1. Amphoteric surface hydroxyl groups form ionized surface sites as described in Equations 38 and 39. Two apparent equilibrium constants, K_+^{CCM} and K_-^{CCM} describe these reactions, but the values are valid only for a particular ionic strength (Davis et al., 1978).

2. Only one plane in the interfacial region is considered: a surface plane for adsorption of H^+, OH^-, and all specifically adsorbed solutes. Only inner-sphere complexes are formed, via the surface coordination reactions (Eqns. 43 and 44). Certain ions, e.g., Na^+, K^+, Cl^-, NO_3^-, are assumed to be inert with respect to the surface.

3. The charge-potential relationship used in the model is $\sigma_o = C^{ccm}\Psi_o$, where C^{ccm} is the capacitance of the mineral-water interface.

4. The constant ionic medium Reference State is used for aqueous species; a zero charge Reference State is used for surface species.

In summary, for fitting acid-base titration data in a mineral bathed in a simple 1:1 electrolyte, the model has three adjustable parameters (K_+^{CCM}, K_-^{CCM}, and C^{ccm}) for each ionic strength.

The CCM has been widely applied in modeling the adsorption of dilute ions by hydrous oxides (Schindler and Stumm, 1987). Schindler et al. (1976) described the adsorption of transition metal cations by silica, and Hohl and Stumm (1976) applied the model to describe Pb^{2+} adsorption by γ-Al_2O_3. Adsorption of strongly binding anions has also been modeled with the CCM (Stumm et al., 1980; Sigg and Stumm, 1981; Goldberg and Sposito, 1984a,b; Goldberg, 1985).

The Diffuse Double Layer Model (DDLM). In the DDLM, introduced by Stumm et al. (1970) and Huang and Stumm (1973), all ions are adsorbed as coordination complexes within the surface plane, except the dissociated counterions present in the diffuse layer (Fig. 21). The model accounts for the ionic strength effects on ion adsorption through the explicit dependence of the diffuse-layer charge, σ_d, on ionic strength (Eqn. 11). A finite number of sites is designated, thus limiting σ_o to reasonable values, unlike the original Gouy-Chapman theory. Like the CCM, the model describes surface reactions in terms of amphoteric hydroxyl groups that form ionized sites and only inner-sphere complexes are formed in coordination reactions. However, the molecular hypotheses of the DDLM differ in the following ways:

1. There are two planes in the interfacial region: (1) a surface plane for adsorption of H^+, OH^-, and all specifically adsorbed solutes, and (2) a diffuse layer plane, representing the closest distance of approach for all counterions.

2. The Gouy-Chapman theory is applied for the charge-potential relationship (Eqn. 11) in the diffuse layer. The electrical potential at the beginning of the diffuse layer (the d-plane) is equal to the surface potential (see Fig. 21). Unlike the CCM, this model should be applicable at variable ionic strengths. Westall and Hohl (1980) have shown proton surface charge of rutile as a function of pH is well described at $I < 0.1$, in the absence of specifically adsorbing ions.

3. The infinite dilution Reference State is used for aqueous species; a zero surface charge Reference State is used for surface species.

For fitting acid-base titration data in a mineral bathed in a simple 1:1 electrolyte, the model has two adjustable parameters, K_+^{DDL} and K_-^{DDL}, that are applied uniformly to variable (but low) ionic strength solutions.

Until recently, the DDLM was not widely applied in modeling the adsorption of dilute ions from solution. Its first application was by Huang and Stumm (1973), who modeled the adsorption of alkaline earth cations by γ-Al_2O_3. The model was used by Harding and Healy (1985a,b) and Dzombak and Morel (1986) to model cadmium adsorption on amphoteric polymeric latex particles and ferrihydrite. The excellent treatise recently published by Dzombak and Morel (1990) describes an improved version of the DDLM in detail, provides a reviewed database for its use with ferrihydrite, and compares the model characteristics with other surface complexation models. The improved model uses the two-site Langmuir approach with strong and weak coordinative sites to simulate site heterogeneity and the lack of proportionality between metal ion adsorption density and aqueous metal ion concentrations. Example calculations using the DDLM of Dzombak and Morel (1990) to describe the adsorption of Cr(VI) and Zn^{2+} by ferrihydrite are shown in Figures 22 and 23.

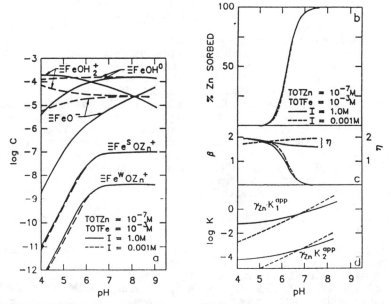

Figure 22. Variations of surface parameters in the two-site DDLM of Dzombak and Morel (1990) as a function of pH and ionic strength in the presence of low adsorbing cation (Zn^{2+}) concentration. (a) Concentrations of surface species, (b) pH adsorption edge for Zn^{2+}, (c) slope of $\log \Gamma_{Zn}$ versus pH (β) and stoichiometry of proton release (η), (d) apparent adsorption constants, i.e., conditional equilibrium constants corrected for electrostatic terms. Reprinted from Dzombak and Morel (1990), Surface Complexation Modeling: Hydrous Ferric Oxide, John Wiley & Sons.

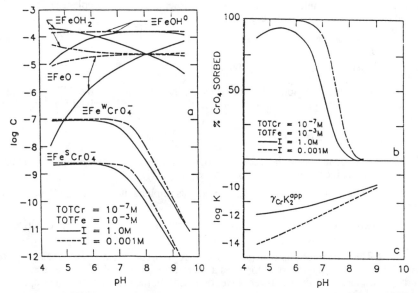

Figure 23. Variations of surface parameters in the two-site DDLM of Dzombak and Morel (1990) as a function of pH and ionic strength in the presence of low adsorbing anion concentration, [Cr(VI)]. (a) Concentrations of surface species, (b) pH adsorption edge for Cr(VI), (c) apparent adsorption constants, i.e., conditional equilibrium constants corrected for electrostatic terms. Reprinted from Dzombak and Morel (1990), Surface Complexation Modeling: Hydrous Ferric Oxide, John Wiley & Sons.

Triple Layer Model (TLM). Because the CCM and DDLM discussed above have only one adsorption plane in the interfacial structure, they are limited in their ability to distinguish between weakly and strongly bound ions (Hayes, 1987). Stern (1924) recognized this limitation and proposed the first three-plane model for the EDL (Westall, 1986). In this model, there was a surface plane, for adsorption of potential-determining ions, and a second adsorption plane for weakly bound counterions (the inner Helmholtz plane (IHP), also called the β plane). The capacitance of the layer between the IHP and the plane of closest approach of dissociated counterions (the outer Helmholtz plane, or OHP) was ignored (Westall, 1986). A significant problem in applying the DDLM to hydrous oxides was that the dissociated counterion charge was usually a small fraction of the proton surface charge, suggesting either specific adsorption of counterions or a microporous "gel" layer (Yates et al., 1974). Yates and others subsequently discounted the possibility of a gel layer for most crystalline minerals based on experimental evidence (Yates and Healy, 1976; Yates et al., 1980; Smit et al., 1978a; 1978b; Smit and Holten, 1980). The three-plane model with specific counterion binding was proposed for the oxide-water interface by Yates et al. (1974), but in this model the capacitance between the IHP and OHP was evaluated based on classical EDL studies, resulting in a decrease in potential between the β and d planes (Fig. 21). Quantitative application of the TLM to EDL data for hydrous oxides was subsequently developed by Davis et al. (1978).

A large body of literature now supports the hypothesis that essentially all electrolyte ions form complexes with surface hydroxyls on oxides (Smit el al., 1978a,b; Smit and Holten, 1980; Foissy et al., 1982; Sprycha, 1983; Sprycha, 1984; Sprycha and Szczypa, 1984; Jafferzic-Renault et al., 1986; Sprycha, 1989a,b; Sprycha et al., 1989; Mehr et al., 1990). It is likely that weakly adsorbed ions, e.g., alkali cations, alkaline earth cations, halides, have at least one layer of water separating them from surface oxygen or metal atoms, i.e., they form ion-pairs or outer-sphere complexes (Hayes, 1987; Hayes et al, 1987). These conclusions are also supported by reaction kinetic studies (see section on "Kinetics of sorption reactions," above).

In the EDL of the TLM applied by Davis et al. (1978), counterion binding in the β layer is incorporated directly into the model structure, allowing ion-pair complexes to form with charged surface hydroxyl groups, e.g.,

$$\equiv SOH + Na^+ + H_2O \leftrightarrow \ \equiv SO^-(H_2O)Na^+ + H^+ \ , \tag{45}$$

$$\equiv SOH + NO_3^- + H^+ \leftrightarrow \ \equiv SOH_2^+NO_3^- \ . \tag{46}$$

The counterions present as ion-pair complexes are included in σ_β in the TLM. Assuming that there is no permanant structural charge, the charge balance is derived from Equations 9 and 12, i.e.,

$$\sigma_o + \sigma_\beta + \sigma_d = 0 \ . \tag{47}$$

This allows an explanation of the increase in σ_o as a function of ionic strength as a simple consequence of an increase in counterion binding (Fig. 17).

The molecular hypotheses of the TLM model as implemented by Hayes (1987) are as follows:

1. Amphoteric surface hydroxyl groups form ionized surface sites as described in Equations 38 and 39.

2. There are three planes in the interfacial region: (1) a surface plane for adsorption of H^+, OH^-, and strongly-adsorbed ions, (2) a near-surface plane (the β-plane) for weakly adsorbed ions, and (3) a diffuse layer plane, representing the closest distance of approach of dissociated charge.

3. The Stern-Grahame interfacial model is applied for the charge-potential relationships for the two regions between the three layers (Eqns. 13 and 14). The Gouy-Chapman theory is applied for the relationship in the diffuse layer (Eqn. 11).

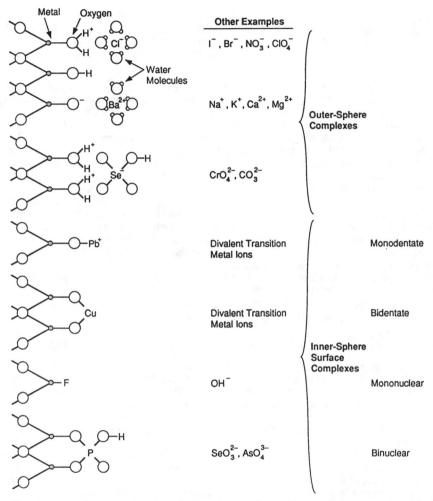

Figure 24. Schematic representation of coordinative surface complexes and ion pairs formed between inorganic ions and hydroxyl groups of an oxide surface in the triple layer model. Reprinted from Hayes (1987), Equilibrium, spectroscopic, and kinetic studies of ion adsorption at the oxide/aqueous interface. Ph.D., Stanford University.

4. Different Reference States have been applied in the model. These are discussed below.

In fitting acid-base titration data, the model has five adjustable parameters, K_+^{TLM}, K_-^{TLM}, C^{TLM}, and the apparent equilibrium constants for the ion-pair formation reactions (Eqns. 45 and 46).

Although the DDLM also predicts an increase in σ_o with ionic strength, it does so via an increase in σ_d and its dependence on ionic strength. In the triple layer model, σ_d is evaluated as the dissociated counterion charge only (not in ion-pair complexes), making it possible to estimate the electrical potential at the OHP. In the model simulations of Davis et al. (1978), these estimates compared favorably with experimental measurements of the zeta potential (Fig. 17), although no effort has been made to optimize TLM performance in this regard. The value of capacitance between the β plane and the OHP is usually given the constant value of 0.2 F/m²; however, this value could be adjusted to achieve better agreement with zeta potential measurements. The TLM allows the introduction of specificity in describing the development of proton surface charge in different electrolyte solutions. For example, the charge of rutile in KNO_3 solutions is quite different than that found in $LiNO_3$ solutions (Yates and Healy, 1980). No attempt has been made to explain this phenomena with the DDLM.

Use of the TLM to model the adsorption of dilute cations and anions by hydrous oxides was first reported by Davis and Leckie (1978a, 1980). Since then the model has been widely applied (e.g., Balistrieri and Murray, 1982; Hsi and Langmuir, 1985; Catts and Langmuir, 1986; LaFlamme and Murray, 1987; Zachara et al., 1987; Hunter et al., 1988; Zachara et al, 1989b; Payne and Waite, 1990); earlier reviews were published by James and Parks (1982) and Sposito (1984). In the application of the model by Davis and Leckie, only charge from ionized surface hydroxyls contributed to the coordinative surface charge; all specifically-adsorbed ions were placed in the β plane as outer-sphere complexes. This implementation was largely based on the expectation that adsorption of cations was somewhat dependent on ionic strength, as had been reported by Vuceta (1976) for Cu^{2+} and Pb^{2+} adsorption on quartz. Hayes and co-workers (Hayes and Leckie, 1986; 1987; Hayes et al., 1988) recently completed a comprehensive study of the ionic strength dependence of cation and anion coordination by the goethite surface. The results demonstrate that the TLM overestimates the effects of ionic strength when cations or anions that are strongly-bound on goethite are assumed to occupy positions in the β plane of the EDL (Hayes and Leckie, 1986; Hayes et al., 1988). For such strongly-bound ions, the TLM simulates the experimental behavior more accurately (Fig. 12a) when the adsorbing ions are placed in the surface plane, thus contributing directly to the coordinative surface charge, σ_o. Weakly-bound ions, like Ca^{2+} or SO_4^{2-}, are significantly affected by ionic strength; in this case adsorption is best described as an outer-sphere complex in the TLM (Fig. 12b). Figure 24 illustrates the various types of surface species proposed in the more recent implementation of the TLM.

The original implementation and subsequent applications of the TLM (except Hayes and co-workers) used the infinite dilution Reference State for aqueous species and the zero charge Reference State for surface species (Davis et al., 1978; Sposito, 1983). The activity coefficients of surface species were assumed to be equal, following the arguments of Chan et al. (1975), which results in cancellation of these terms in evaluating surface stability constants (Hayes and Leckie, 1986, 1987). Hayes and co-workers adopted a different approach in defining activity coefficients. The Standard State for both solution and surface species was defined as 1 mole/liter at zero surface charge and no ionic interaction. The Reference State was chosen for all species as infinite dilution relative to the aqueous phase and zero surface charge (Hayes, 1987). Equation 40 was then redefined such that the activity coefficient term is replaced by an electrochemical potential, representing the free energy required to bring a species from the reference state potential to a given potential, Φ. A possible problem with the thermodynamic development may arise from the manner in which the electrochemical potential term is defined, such that uncharged surface species, e.g., $\equiv FeOH$, $\equiv FeF$, behave ideally and require no activity correction. A critical review of the thermodynamic arguments of Hayes (1987) has not yet been published.

Generalization of the implementation of the TLM used by Hayes to minerals other than hydrous iron oxides seems likely; however, it should be noted that adsorption of strongly-bound cations on quartz, titania, and soils does exhibit some ionic strength dependence (Vuceta, 1976; Shuman, 1986; Chang et al., 1987). James and Parks (1982) developed an application of the TLM to clay minerals. An interesting application of the TLM was recently made by Zachara et al. (1990) in a study of aminonaphthalene and quinoline adsorption by silica. Adsorption of these organic compounds was modeled using outer-sphere surface species, and the simulations agreed well with experimental behavior as a function of ionic strength.

Four layer models. Bowden et al. (1980) introduced a four-plane model to allow strongly-bound ions to be placed closer to the surface than the β plane of the TLM without being in the surface plane of adsorption. Although the model can simulate experimental data quite well, we have not considered it here because the model is based on empirical constraint equations rather than mass action equilibrium principles (Hayes, 1987; Sposito, 1984).

The four-layer model of Bousse and Meindl (1986) introduced a second plane for the adsorption of outer-sphere complexes (Fig. 21). As has been noted by Sposito (1984), it is frequently observed that the capacitance of the interface at pH values less than the pH_{PPZC} (positive surface charge) is less than that observed when the surface is negatively charged. It is likely that outer-sphere cation complexes can approach the surface plane more closely than outer-sphere anion complexes, and this may explain the observed differences in interfacial capacitance (Davis and Leckie, 1979). The four-layer model of Bousse and Meindl (1986) can describe proton surface charge in this manner and may provide a more accurate representation of EDL structure.

The non-electrostatic surface complexation model. The simplest SCM approach is to ignore the electrical double layer by excluding electrostatic terms from the mass law equations for surface equilibria (Kurbatov et al., 1951; Ahrland et al., 1960; Dugger et al., 1964). When this approach is utilized in equilibrium models, surface functional groups are treated computationally in exactly the same manner as dissolved ligands (for cation adsorption) or complexing metal ions (for anion adsorption). This is justified by the observation that for moderately or strongly adsorbing ions, the chemical contribution to the free energy of adsorption dominates over the electrostatic contribution (see sections on "Adsorption of cations and anions", above). To apply the model, integral stoichiometries for reactions between the adsorbing species and surface sites are proposed and apparent equilibrium constants for the reactions are derived by fitting experimental data. James and Parks (1975) used this approach to describe Zn^{2+} adsorption on cinnabar (HgS) with the following reaction:

$$Zn^{2+} + \ \equiv SH \leftrightarrow \ \equiv S-Zn + H^+ \ , \tag{48}$$

where $\equiv SH$ represents a functional group on the cinnabar surface. By including $\equiv S-Zn$ as a species in an equilibrium computation, they were able to describe Zn^{2+} adsorption over the entire range of pH and Zn^{2+} concentrations investigated. Davis et al. (1987) used an analogous approach in describing Cd^{2+} adsorption by calcite.

As described above, this approach appears the same as the partitioning equation approach (Eqn. 37) with χ equal to one. However, the conceptual basis of the SCM approach allows the description of competitive effects from other ions in the solution, by computing their adsorption simultaneously. Thus, new constants need not be derived for each solution composition if the adsorption of all aqueous species is adequately described. The SCM approach without electrostatic correction has not been used frequently to simulate ion adsorption in systems with well-characterized mineral phases. However, interest in this approach has recently been renewed because of the complexities involved in applying the SCM to natural materials (see section on "Applications to Aqueous Geochemistry", below). For example, Cowan et al. (1990) applied the non-electrostatic model to describe Cd-Ca adsorptive competition on the ferrihydrite surface. Interestingly, the performance of the non-electrostatic model for this data set appeared to be as good as the TLM. Krupka et al. (1988) used this modeling approach to develop a small database for the adsorption of H^+, Ca^{2+}, Cd^{2+}, $CrOH^{2+}$, Zn^{2+}, SO_4^{2-}, CrO_4^{2-} on ferrihydrite from experimental data of other investigators. The goodness-of-fit of model simulations to experimental data was poor in comparison to DDLM and TLM simulations of the same data sets published in the literature. On the other hand, it should be noted that the simple modeling approach was adopted in order to increase the computational efficiency of the FASTCHEM solute transport computer code (Krupka et al., 1988). It remains to be seen if the increased efficiency of this approach is worth the potential loss of accuracy in adsorption simulations.

Proton stoichiometry in surface complexation reactions

That protons are released by cation adsorption and consumed by anion adsorption is well documented and is consistent with the coordination reactions shown in Equations 43 and 44. Nonintegral proton exchange stoichiometries per ion adsorbed are typically observed (Fokkink et al., 1987; Kinniburgh, 1983), and some authors have attempted to relate these values directly to microscopic reaction stoichiometry. However, as discussed in detail by others (Hayes and Leckie, 1986; Hayes, 1987; Dzombak and Morel, 1990), the proton release observed in experimental systems is the net result of adsorption reactions plus the adjustment of surface ionization and ion pair reactions to a new equilibrium condition (Fig. 18). Thus, calculated values of net proton release/uptake are dependent on the interfacial model chosen for the EDL and should be performed in a self-consistent model-dependent analysis. This has rarely been done. Hayes (1987) showed that the simulated net proton release using the TLM was considerably different for inner-sphere lead complexes than outer-sphere lead complexes. The experimental measurements of proton released supported the inner-sphere complex formation of Pb^{2+} on the goethite surface (in agreement with spectroscopic and kinetic measurements) with a release of only one proton during the adsorption reaction, even though the predicted net release of protons was closer to two protons per adsorbed lead ion. Similar differences between microscopic reaction stoichiometry and macroscopic observations of proton release are predicted with the DDLM, and an excellent detailed analysis is given by Dzombak and Morel (1990). As shown in Figure 22, the stoichiometry of proton release is a function of pH and ionic strength, and it has noted by Honeyman and Leckie (1986) that the proton coefficient is dependent on adsorption density. Comprehensive studies of net proton exchange stoichiometry need to be conducted in the future to allow further refinement of surface complexation models and constraints on adsorption reaction stoichiometry.

Parameter estimation

Objective determination of parameters for surface complexation models is best accomplished with the use of numerical optimization techniques (Dzombak and Morel, 1990; Hayes et al., 1990). It must be emphasized that the values of parameters obtained are dependent on the model chosen for fitting experimental data. The model-dependent parameters have frequently been compared in an inappropriate manner in the literature, e.g., in comparing surface acidity (Huang, 1981). Despite the objective nature of such numerical techniques, however, the parameter estimation procedure also has inherent limitations, due to (1) the covariance of estimated parameters (Westall and Hohl, 1980; Hayes et al., 1990) and (2) non-unique solutions during optimization, i.e., the solution has been shown to be dependent on the initial conditions assumed for variables (Koopal et al., 1987). The practice of not reporting goodness-of-fit parameters in comparisons of different models of interfacial structure, reaction stoichiometries or surface speciation (e.g., Cowan et al., 1990) seems inappropriate, since this parameter should be compared with the lower limit of the parameter established by a rigorous error analysis. Although the fits of calculated curves can sometimes be optimized beyond the variation expected from errors involved in data collection, comparisons of model performance are only valid above this lower limit for the goodness-of-fit parameter (Koopal et al., 1987). Bousse and Meindl (1986) concluded that acid-base titration data were only useful for an evaluation of ion-pair formation constants; surface ionization (acidity) constants should be determined from measurements of surface potential or from zeta potentials (Sprycha, 1989a,b).

Recent studies have shown that the ability of surface complexation models to fit adsorption data is relatively insensitive to the value of the site density used (Kent et al., 1986; Hayes et al., 1990). Clearly, the absolute value of the binding constants that describe the adsorption reactions are dependent on the choice of the site density. However, Hayes et al. (1990) showed that the ability to fit experimental data over a wide range of conditions (with a consistent set of model parameters) is independent of the choice of the site density over two orders of magnitude. This is true as long as the molar ratio of adsorbate to surface sites is small, i.e., there is an excess of surface sites over adsorbate in the system. When the adsorbing solute is present in excess, the ability to fit adsorption data becomes more sensitive to the value of surface site density used (Hayes, 1987). Under these conditions, the site density can be derived by optimizing the fit between model simulations and experimental data.

Because the goodness-of-fit between model simulations and experimental data for *dilute* ion adsorption is insensitive to the value of functional group density, it has been argued that the site density can be used as an additional fitting parameter. Three comments can be made about this modeling strategem:

1. Surface complexation theory treats surface functional groups in the same fashion as dissolved ligands (or metals) in an equilibrium speciation framework. It would be unacceptable to compute metal complexation by chloride ions in solution by co-varying the chloride concentration and chloride complexation constants within the model. The geochemist requires an accurate analysis of water composition before computing aqueous speciation at equilibrium. Ideally, surface complexation modeling should be performed with an accurate determination of surface site density, which can be made experimentally or estimated from previous studies.

2. The fit of model simulations to experimental data is dependent on the value of site density chosen at high ion adsorption densities. While many modeling applications will involve only dilute solutes, problems involving high adsorption densities will also need to be solved, e.g., near contamination sources, or in estuaries. If the site density is used as a fitting parameter and derived from experiments with dilute ions in a swamping 1:1 electrolyte, its value may no longer be optimal when applied under variable conditions. Thus, a general view of applicability should be considered in deriving model parameters.

3. Other model parameters (especially binding constants) are dependent on the value chosen for the density of surface functional groups (Luoma and Davis, 1983). In addition to being dependent on the interfacial model, the apparent binding constants reported in the literature are usually not self-consistent, because varying values for the site densities of minerals have been applied to derive the constants. For example, the binding constant reported for Pb^{2+} complexation with goethite surface hydroxyls by Hayes (1987), i.e.,

$$\equiv FeOH + Pb^{2+} \leftrightarrow \equiv FeOPb^+ + H^+ \qquad \log K = 2.3 \quad ,$$

is self-consistent with a goethite site density of 7 sites/nm². That is to say, this apparent binding constant should not be applied elsewhere with a different value for goethite site density. A self-consistent thermodynamic database for surface complexation models can only be developed when values for the density of functional groups of minerals have been established and accepted by the general scientific community. Dzombak and Morel (1990) recognized this in developing their database for ferrihydrite and the DDLM. These authors rederived apparent binding constants from original data sets in the literature, using a single value for the surface functional group density.

In addition to the considerations above, a parsimonious modeling approach is needed in order to extend surface complexation theory to applications in natural systems. In complex mixtures of minerals it is often difficult to quantify the numbers of surface functional groups that are present from various minerals. *We recommend that binding constants for strongly-binding solutes be derived with a site density of 2.31 sites/nm²* (3.84 μmoles/m²) *for all minerals*; this recommendation is based on the review in the section on "Surface Functional Groups" (above). While the actual Γ_{max} may vary from 1 to 7 sites/nm², it is important that one value be selected to encourage the development of a self-consistent thermodynamic database that can be applied easily to soils and sediments (see section on "Applications to Aqueous Geochemistry," below). To encourage unanimity within the field, we have chosen the particular value of 2.31 sites/nm² because it is consistent with the value chosen by Dzombak and Morel (1990) to describe ferrihydrite (0.205 moles per mole of Fe), assuming a specific surface area of 600 m²/g of $Fe_2O_3 \cdot H_2O$. The value recommended closely approximates the site densities found by adsorption on various minerals, including goethite (Table 2), manganese oxides (Zasoski and Burau, 1988), and the edge sites of clay minerals (Motta and Miranda, 1989; Bar-Yosef and Meek, 1987; Inskeep and Baham, 1983).

Eventually this site density may be subdivided into high and low energy site densities for specific minerals, as was done by Dzombak and Morel (1990); however, that is not possible for most minerals without further research investigations.

Comparison of the performance of surface complexation models

Only a few examples of useful and objective model comparisons in the literature can be cited. Westall and Hohl (1980) examined the ability of various models to simulate proton surface charge and found that all models could perform this task well. Morel et al. (1981) compared the models in describing adsorption of Pb^{2+} by γ-Al_2O_3, and found that all the models tested performed well for a single pH adsorption edge. These conclusions of the equality of model performance have been used by some authors to argue that differences between the models are insignificant. With the evolution of implementation of the TLM made by Hayes and co-workers (Hayes and Leckie, 1986, 1987; Hayes et al., 1988), differences between the surface complexation models in describing dilute cation and anion adsorption have diminished even further. As has been noted by Hunter (1987) and Hayes (1987), the models differ primarily in their predictions of diffuse layer potential and the ionic strength dependence of weakly adsorbed solutes. However, the differences between model simulations are not easily demonstrated with small data sets. For example, it would have been a more significant test of model performance if Westall and Hohl (1980) had compared the abilities of models to describe the proton surface charge of rutile in KNO_3 and $LiNO_3$ solutions, which are significantly different (Yates, 1975).

From a geochemical point of view, the primary distinction among the models, at their present stage of evolution, is in the way that interactions of major ions in natural waters with mineral surfaces are treated, e.g., the surface reactions of Na^+, Ca^{2+}, Mg^{2+}, Cl^-, HCO_3^-, and SO_4^{2-}. Each of the models has a distinct set of assumptions for the interfacial structure of the EDL. Thus, differences in model simulations would be expected to be most obvious as the major ion composition of water is varied.

For example, differences in model simulations can be exhibited by either varying ionic strength or in competitive adsorption experiments. Figure 25 shows the effect of sulfate concentration on adsorption of Cr(VI) by ferrihydrite and the ability of the TLM and DDLM to simulate the data. Only outer-sphere species were assumed for sulfate and chromate in the TLM, using the modeling approach of Davis and Leckie (1980). The ionic strength dependence of Cr(VI) adsorption on ferrihydrite (Music et al., 1986; Rai et al., 1986) suggests that it may form an outer-sphere complex. For the DDLM, the constants of Dzombak and Morel (1990) were used without any attempt at further optimization of the fit. Only slight differences between the model performances can be seen, and it can be concluded that either model performs well in describing the competition of sulfate and chromate adsorption. It has already been stated that ions like Na^+, K^+, and Cl^- do not affect the adsorption of strongly-bound ions (Figs. 12a and 13c). Dempsey and Singer (1980) showed that Ca^{2+} had little effect on Zn^{2+} adsorption by ferrihydrite. Both the TLM and DDLM predict the observed negligible effects on trace cation adsorption. Cowan et al. (1990) illustrated that Ca^{2+} decreases Cd^{2+} adsorption on ferrihydrite, but only limited success was achieved in simulating the data with the TLM. A combination of inner-sphere and outer-sphere complexes for Ca^{2+} was used in the modeling. Interestingly, a comparable degree of success was achieved with the non-electrostatic surface complexation model. We have found that neither the TLM or DDLM was able to simulate well the data of Hayes (1987) for Ba^{2+} adsorption by ferrihydrite as a function of ionic strength (calculations not shown). Hayes (1987) was able to simulate the data with the TLM (curves shown in Fig. 12b), but only after significant manipulation of the ion pair formation constants of Na^+ and NO_3^-. This appears to be the first set of *dilute* cation adsorption data ever collected that would require the four-layer model (Fig. 21) for an accurate description of all experimental data with a single set of model parameters. The justification for such a model stems from the need to place outer-sphere cation complexes closer to the surface than outer-sphere anion complexes (see section on ""Four layer models," above).

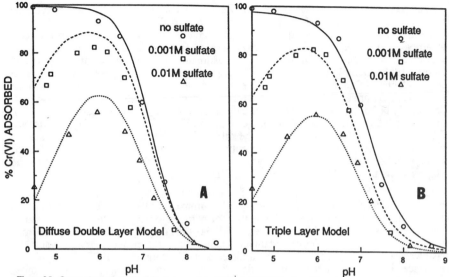

Figure 25. Comparison of model performances of the TLM and DDLM in describing the competitive adsorption of Cr(VI) and sulfate anions by 0.001 M Fe (as ferrihydrite) in 0.1 M NaNO₃ solution.

(a) DDLM parameters taken from Dzombak and Morel (1990).

(b) TLM: log $K_{CrO_4^{2-}}$, 10.1; log $K_{HCrO_4^-}$, 17.6; log $K_{SO_4^{2-}}$, 9.4; log $K_{HSO_4^-}$, 15.05; other model parameters taken from Davis and Leckie (1980), aqueous solution complex formation constants taken from Dzombak and Morel (1990).

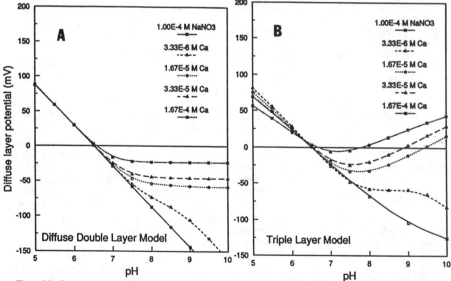

Figure 26. Comparison of model performances of the TLM and DDLM in describing the diffuse layer potential of rutile (TiO₂) suspensions in solutions of varying concentrations of Ca(NO₃)₂. Rutile properties in both models: $W = 100$ mg/dm³; $A = 26$ m²/gm; $S_T = 4.96$ μmoles/m²

(a) DDLM constants at I=0: log K_+, 5.5; log K_-, -7.5; log $K_{Ca^{2+}}$, -4.5.

(b) TLM parameters at I=0: C_1, 1.05 F/m²; log K_+, 4.4; log K_-, -8.6; log K_{Na^+}, -6.9; log $K_{NO_3^-}$, 6.1; log K_{CaOH^+}, -11.2.

As was mentioned in the section on "Zeta potential" (above), accurate conversion between measurements of electrophoretic mobility and diffuse layer potential is made difficult by complex relationships involving particle geometry and viscoelectric effects (Dzombak and Morel, 1990; Hunter, 1987). It is generally acknowledged that the diffuse layer potential, Ψ_d, in the DDLM greatly overestimates measured values of zeta potential (Hunter, 1987). Dzombak and Morel (1987) argue that this problem can be overcome with the introduction of an empirical function describing the location of the shear plane in the DDL, and this approach was demonstrated for rutile suspended in KNO_3 solutions. However, it can easily be shown that such an empirical approach would be unsuccessful in $Ca(NO_3)_2$ solutions, which cause a reversal of charge (Fig. 19). Figure 26 illustrates the performance of the DDLM and TLM in simulating the diffuse layer potential of rutile suspended in calcium nitrate solutions of various concentrations. Ca^{2+} binding constants were derived from adsorption data. The TLM clearly performs better than the DDLM at simulating the pH_{IEP} as a function of pH (compare Fig. 26a with Fig. 19a).

It is widely recognized that the rate of coagulation of colloidal particles is dependent on surface charge (Hunter, 1987; Stumm, 1977; Stumm et al., 1970). Effective coagulation is of considerable significance in the settling of particles during waste treatment and in natural systems (Farley and Morel, 1986; O'Melia, 1987). Colloidal particles of hydrous aluminum and iron oxides are important adsorbents in the treatment of drinking water and wastewaters (Dempsey et al., 1988; Benjamin et al., 1982; Sorg, 1978, 1979). Manipulation of surface charge is of interest in the modeling of various engineering and industrial operations, such as ore processing and mineral flotation, secondary oil recovery from geological deposits, and the manufacture of ceramics, pigments, and other industrial products (Hunter, 1987; Hornsby and Leja, 1982; Fuerstenau, 1976; de Bruyn and Agar, 1962). Generally, the onset of rapid coagulation occurs when the absolute value of the diffuse layer potential decreases below about 14 mV (Wiese and Healy, 1975a,b). Thus, the prediction of diffuse layer potential, and more importantly, the pH_{IEP} is of considerable importance in these applications. One of the foremost objectives in the development of the TLM (Yates et al., 1974; Davis et al., 1978) was to create a model with more realistic simulations of σ_d and Ψ_d. The simulations shown in Figure 26 and the arguments of Hunter (1987) and James et al. (1981) support the view that the TLM better serves these objectives.

The primary objective of many applications of surface complexation models in aqueous geochemistry is to simulate the adsorption and transport of ions in natural systems. With the possible exception of the non-electrostatic model (Krupka et al., 1988), it can be stated that all surface complexation models simulate ion adsorption data adequately in simple mineral-water systems. It has frequently been argued that the DDLM and CCM models should be applied instead of the TLM, because of their simpler interfacial models. As noted by Dzombak and Morel (1987), the TLM in fact contains fewer model parameters than the CCM when applications involving variable solution compositions are considered. The few parameters required by the DDLM and its successful performance in complex solutions with single mineral phases (Fig. 25) are appealing features. The availability of the self-consistent, validated database and two-site model for ferrihydrite compiled by Dzombak and Morel (1990) is also an important development that will undoubtedly increase usage of the DDLM. However, the additional model parameters required by the TLM (two electrolyte binding constants in 1:1 electrolyte solutions and the inner layer capacitance) are of little consequence when incorporated in numerical computer algorithms, and electrolyte binding constants have now been determined for many hydrous oxides (Charlet and Sposito, 1987; James and Parks, 1982). Many difficult problems need to be addressed in order to apply surface complexation theory to soils and sediments. As will be shown in the next section, it is the ease of applicability that is most important in choosing a model for geochemical applications.

APPLICATIONS IN AQUEOUS GEOCHEMISTRY

Numerous field studies have demonstrated that sorption processes are important in natural systems. Sorption has been shown to be important for Zn and V in pristine rivers (Shiller and Boyle, 1985, 1987) and for Ni, Cu, Zn, Pb, and As in rivers contaminated by acid mine drainage and other industrial activities (Mouvet and Bourg, 1983; Johnson, 1986; Fuller and Davis, 1989). In rivers, sorption processes can control the dissolved concentrations of solutes, give rise to high concentrations of toxic metals in the bed load (Johnson, 1986; Fuller and Davis, 1989), and affect the discharge rates of solutes into estuarine and coastal marine systems (Shiller and Boyle, 1985; 1987). In lakes and the oceans, sorption processes are important in the flux of many elements from the water column to bed sediments (Belzile and Tessier, 1990; Tessier et al., 1989; Honeyman and Santchi, 1988; Sigg et al., 1987). Sorption may regulate the availability and toxicity of trace elements to phytoplankton (Kuwabara et al., 1986; Morel and Hudson, 1985), and the uptake of sorbed metals by filter feeding and burrowing organisms is a potential environmental risk in surface waters (Luoma and Davis, 1983). In soils and aquifers, sorption processes can retard the transport of solutes (Cherry et al., 1984; Lowson et al., 1986; Jacobs et al., 1988; Davis et al., 1990) and give rise to the formation of secondary repositories, i.e., regions where contaminants are accumulated downgradient from a source, from which they can be released in response to a change in the geochemical environment. Sorption processes have long been utilized in metallurgical processes (Robins et al., 1988; Bryson and te Reile, 1986; de Bruyn and Agar, 1962) and are important in the treatment of wastewaters and mining tailings (Khoe and Sinclair, 1990; Leckie et al., 1985; Benjamin et al., 1982).

In this section we discuss the problems typically encountered in the application of surface complexation theory to natural systems. Perhaps the most significant problem lies in the identification and quantification of surface functional groups in heterogeneous mixtures of mineral phases. The reactivity of functional groups among the minerals present may vary significantly, and the sorption of an ion may be controlled by interaction with the surface of a particular mineral that is present in minor or trace quantities or represents only a small fraction of the total surface area. Another problem is the definition of electrical properties and determination of model parameters for the electrical double layer. Interacting double layers of heterogeneous particles and the formation of organic coatings cause significant changes in the electrical properties of mineral-water interfaces; the techniques used in simple mineral-water systems to define surface charge densities cannot be applied in these complex systems. The determination of binding constants in laboratory experiments is complicated by variable solution compositions that may evolve from the dissolution of mineral phases and the desorption of ions. An accurate assessment of solution speciation is required to determine the binding constants and to apply them in natural systems. Guidelines for the resolution of some of these problems are presented.

Surface area and functional groups of soils and sediments

The solid phases involved in adsorption processes in natural systems are typically *composite* materials, that is, they consist of mixtures of various minerals and organic debris with a wide range of intrinsic chemical properties. Weathering reactions, diagenetic processes, and interactions with bacteria give rise to the leaching of surface layers of minerals (Casey and Brunker, this volume) and the deposition of extremely fine-grained, poorly crystalline minerals (Benson and Teague, 1982; Fritz and Mohr, 1984). These processes also lead to the formation of organic and hydroxypolymer (i.e., amorphous Al or Fe hydrous oxide) coatings on mineral grains (Jenne, 1977; Lion et al., 1982; Davis, 1984). Some important differences between well-characterized mineral phases typically used in laboratory studies and the composite materials of natural systems are illustrated in Figure 27. In many cases, the surface chemical properties of natural materials are dominated by secondary minerals and coatings, which usually constitute only a minor fraction of the whole sample (Jackson and Inch, 1989; Fuller and Davis, 1987; Davis, 1984; Fordham and Norrish, 1979). These surficial deposits are at least partly responsible for the differences observed between mineral dissolution rates in the laboratory and in natural systems (Velbel, 1986; White and Peterson, 1990).

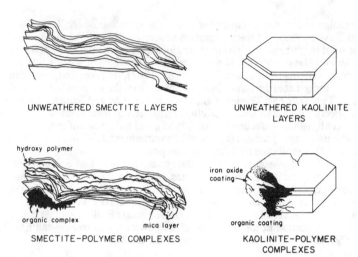

UNWEATHERED SMECTITE LAYERS

UNWEATHERED KAOLINITE
LAYERS

SMECTITE-POLYMER COMPLEXES

KAOLINITE-POLYMER
COMPLEXES

Figure 27. The unweathered smectite and kaolinite particles represent the idealized particles used in laboratory studies of adsorption reactions. Natural weathering processes can alter the surfaces of minerals, e.g., causing illitization of some smectite layers and the acquisition of coatings of organic matter or poorly crystalline Fe and Al hydroxypolymers. Typical surfaces of natural materials are comprised of mixtures of functional groups with variable chemical properties. Reprinted from Sposito (1984), The Surface Chemistry of Soils, Oxford University Press.

The method selected for characterizing the surface area depends on the nature of the composite material and the types of interactions that give rise to the adsorptive process under investigation. For example, for samples abundant in smectites, the role of interlayer ditrigonal cavities in the adsorptive process needs to be assessed. Processes such as anion exclusion and the adsorption of major solution cations or various organic solutes respond to both external and interlayer sites on smectites. In this case, a method that evaluates interlayer as well as external surface area, such as EGME retention, is appropriate. Adsorption of anions and possibly trace cations, however, occurs primarily at external surface sites of clays (see section on "Smectites, vermiculites, and illitic micas," above). Thus, a method such as N_2 or Kr adsorption may provide a better estimate of *reactive* surface area. If the reactive surface area is dominated by potentially microporous components, such as allophane or hydroxypolymer coatings, measurements of $\mathcal{A}$ must be interpreted with caution. Analysis of gas adsorption isotherms using t- or α_s-plots is recommended for such samples, because these analyses yield estimates of the external surface area and micropore volume (Pyman and Posner, 1978; see section on "Surface Area and Porosity," above).

The solid phase-water ratios (W) in geochemical applications vary from a few $\mu g/dm^3$ in the deep oceans to several kg/dm^3 in sediments and aquifers. Given typical values for site density, this variation yields a range of 10^{-10} to approximately 0.5 Molar in the concentration of surface functional groups available for complexation with ions (Honeyman and Santschi, 1988). Ideally, the density of each type of surface functional group would be determined for surface complexation modeling. In practice, this information is usually not known for composite materials, and simplified approaches must be adopted. Balistrieri and Murray (1983) used tritium exchange (see section on "Density of surface hydroxyls," above) to estimate the total density of surface sites on some marine sediments. Later, they noted that the site densities estimated by tritium exchange agreed well with those from much simpler cation exchange measurements (Balistrieri and Murray, 1984), but the generality of this result has not been tested. As was discussed in the section on "Surface Functional Groups," these methods may overestimate the densities of *reactive* surface sites. Another method that has been used to estimate site densities of natural materials is acid-base titration (Mouvet and Bourg, 1983; Goncalves et al., 1987), but this approach is complicated by the possible dissolution of mineral phases or organic matter during the titration. Several investigators have estimated site densities directly from adsorption data, applying the Langmuir isotherm equation. Linearization of the Langmuir isotherm (Eqn. 35) has been used to obtain Γ_{max} values for phosphate adsorption on soils (Goldberg and Sposito, 1986), Na^+, K^+, Cl^-, and NO_3^- adsorption on soils (Charlet and Sposito, 1987), and $Cr(VI)$ adsorption on subsurface earth materials (Zachara et al., 1989b). Determination of the density of surface functional groups by adsorption experiments, however, may be complicated by the low solubilities of many solutes and the release of ions from solid phases during batch experiments (Luoma and Davis, 1983). If the solubility is exceeded well before the surface sites are fully complexed, then the method cannot accurately estimate the site density.

An assessment of the predominant adsorbent(s) in a mineral assemblage often facilitates the estimate of site density for modeling purposes. For example, if hydroxypolymer coatings and poorly crystalline precipitates of hydrous iron oxides are assumed to dominate the adsorptive reactions of interest, then the site density for modeling purposes can be estimated from the abundance of these mineral phases in a sample of composite material (Payne and Waite, 1990; Belzile and Tessier, 1990; Tessier et al., 1989). Fuller and Davis (1987) identified incorporation into calcite in a sandy aquifer material as the most important sorptive process for Cd^{2+}. The site density for modeling was estimated from the weight abundance of $CaCO_3$ in the composite material and the density of $\equiv S_c$ sites on pure calcite.

If it cannot be shown that a specific mineral component dominates the adsorptive interactions of a composite material, a more general approach is needed. We believe the best approach is to multiply the measurement of *reactive* surface area by an average or typical value of site density for the major minerals present (Luoma and Davis, 1983). Based on the arguments given in the section on "Parameter estimation," we recommend a surface site density of 2.31 sites/nm^2 (3.84 μmoles/m^2) for general modeling of bulk composite materials. When applying apparent binding constants from the literature, it may be necessary to rederive the constants from the original data using a consistent value for the site density. Composite materials that have large abundances of minerals with permanent structural charge (e.g., smectites) require special consideration, both in terms of the definition and measurement of reactive surface area for the solute of interest and in the choice of surface site density.

Observations of sorption phenomena in complex mineral-water systems

The composite nature of mineral assemblages in soils and sediments and the variable aqueous speciation of solutes confront the researcher interested in describing sorption equilibria with a problem of considerable complexity. Experimental investigations in simple mineral-water systems have shown that there are three critical elements necessary for describing sorption phenomena (Kent et al., 1986). First, the range of aqueous speciation of the sorbing solute must be determined. Second, various chemical and physical properties of the composite materials are characterized. Third, sorption of the solute is determined experimentally for a relevent range of geochemical conditions. In the following paragraphs, we will illustrate that, in many cases, sorption of ions by soils and sediments can be understood with reference to the adsorption phenomena observed in simple mineral-water systems.

Effect of aqueous composition. Characterizing the speciation of solutes is central to understanding their adsorption from aqueous solutions. Identification of trends in adsorption with variations in solution composition should be established over the range of chemical conditions appropriate for the specific application. Important chemical variables include pH, ionic strength, concentration of the adsorbate, solid-water ratio, and concentrations of other solutes that either decrease or enhance the adsorption of the solute under investigation. In laboratory studies, aqueous speciation should be computed from a knowledge of the system composition and equilibrium constants for the array of complexes that can form. Monitoring the solution composition during laboratory experiments with soils and sediments is important, because elements such as Ca, Fe, Mn, Al, and Si and organic compounds can be released slowly by desorption and dissolution reactions, thereby affecting the experimental data (Goldberg and Glaubig, 1986a; Neal et al., 1987a; Davis et al., 1990). Accumulation of dissolved carbonate species can occur at high pH and have similar effects (Hsi and Langmuir, 1985; Zachara et al., 1989b). In natural systems, thorough chemical analyses must be performed to determine the identity and concentrations of adsorbing species and potential complexing ligands, especially fulvic acids. Depending upon the nature of the application, it may be necessary to examine the influence of spatial or temporal variability in the composition of solutions or composite materials. The oxidation state of elements must also be considered, if variable. For example, Se(IV) forms strong coordinative complexes with the surface hydroxyl groups of ferrihydrite, while Se(VI) forms only weak complexes (Fig. 13). Similar observations have been made in experiments with soils (Neal et al., 1987a,b; Neal and Sposito, 1989).

The effects of solution composition on ion adsorption in simple mineral-water systems have general applicability to natural adsorbents. The importance of ionic strength on the adsorption of weakly-bound ions has been demonstrated for natural materials (Mayer and Schinck, 1981; Neal et al., 1987b; Charlet and Sposito, 1989). Unlike the findings for iron oxides, the adsorption of a strongly-bound cation, Zn^{2+}, on soils was found to depend on ionic strength (Shuman, 1986). The formation of dissolved complexes affects the adsorption of metal ions on sediments, soils, and suspended particles in the presence of natural organic material (Lion et al, 1982; Mouvet and Bourg, 1983; Davis, 1984). Studies with soils and suspended particulate matter from rivers have shown that proton dissociation reactions must be considered in characterizing the sorption of oxyanions and other weak acids (Goldberg and Sposito, 1984a,b; Goldberg and Glaubig, 1986; Shiller and Boyle, 1987; Zachara et al., 1989b).

Competitive effects need to be considered in the design of laboratory experiments and in modeling. Zachara et al. (1989b) showed that sulfate and carbonate anions compete with Cr(VI) for surface sites on subsurface soils. Neal et al. (1987b) found that Se(IV) adsorption was not affected by a large excess of sulfate, but was reduced significantly by equimolar concentrations of phosphate. Unusual synergistic effects were also observed, e.g., the enhancement of Se(IV) adsorption on soils in the presence of Ca^{2+}. Such synergistic effects of variable solution composition are difficult to predict *a priori*; these effects need to be identified in the course of characterizing adsorption trends with solution composition for specific applications.

Identification of dominant sorptive mineral components in composite materials. Various indirect methods have been developed to identify specific mineral components that dominate the sorption of ions on soil and sediment samples. Such studies may suggest dominant sorptive mechanisms and guide the modeling of sorption. Lion et al. (1982) studied Cd^{2+} and Pb^{2+} sorption on organic-rich sediments from San Francisco Bay (California). Chemical treatment targeted at removing organic matter had no effect on Pb^{2+} sorption behavior but reduced the sorption of Cd^{2+}. Amending the organic content of the treated sediments by adding humic material restored the Cd^{2+} sorption capacity. On the other hand, chemical treatment designed to leach Fe and Mn hydrous oxides had no effect on Cd^{2+} sorption, but reduced the extent of Pb^{2+} sorption. Based on a similar approach, Zachara et al. (1989b) suggested that Al-substituted goethite dominated the Cr(VI) sorption behavior of several subsurface soils comprised of clay minerals and hydrous Fe oxides. Aggett and Roberts (1986) observed a strong correlation of Fe and As released from As-contaminated sediments when leached with solutions containing EDTA, suggesting an association of these elements in the sediment.

The results of studies that utilize *selective extractions* must be interpreted with caution, because the chemical treatments employed are not completely selective (Lion et al., 1982; Gruebel et al., 1988). Attempts have been made to quantify the amount of trace elements associated with clay minerals, $CaCO_3$, Fe and Mn hydrous oxides, and organic matter using *sequential selective extractions* (Tessier et al., 1985). However, because the chemical treatments are not perfectly selective, and because they may allow trace elements to redistribute among other mineral phases or cause unintended redox reactions, such quantifications must be viewed with skepticism (Gruebel et al., 1988, and references therein). Such methods should be supported by isotopic confirmation that trace elements are fully desorbed during each extraction step (Belzile et al., 1989). In addition, multiple techniques and strategies should be applied to provide other evidence of particular element-mineral associations. For example, Fordham and Norrish (1979) used manual separation of minerals, autoradiography, and elemental analysis by electron microscope techniques to demonstrate a close association of As and Fe in soil samples. Lowson et al. (1986) and Waite and Payne (1990) demonstrated that the $^{234}U/^{238}U$ ratios in Tamm's oxalate extracts of subsurface profiles near uranium ore bodies were equal to the ratio of the isotopes found in the groundwater, while the U isotopic ratio of the bulk composite material was quite different. The authors concluded that dissolved U(VI) in the groundwater was in equilibrium with U(VI) adsorbed by poorly crystalline iron oxides in the composite material. Fuller and Davis (1987) showed that adsorption and coprecipitation on calcite dominated Cd^{2+} reactions with a calcareous aquifer material comprised of quartz, feldspars, calcite, and trace quantities of Fe and Mn oxides. Three methods were used in the study: selective extraction, manual separation of component minerals after sorption experiments employing a radioisotope of Cd^{2+}, and kinetic studies that determined that the rates of Cd^{2+} and Ca^{2+} sorption on the aquifer material were similar to those for pure calcite.

Trace element-mineral correlations do not prove that a mineral dominates the adsorption of a trace element in a composite material. The trace element may have been coprecipitated within a mineral and then released by dissolution of the mineral during the chemical extraction. Even when it can be demonstrated that a solute is probably at adsorptive equilibrium with a particular phase (Lowson et al., 1986), it does not mean that phase alone *dominates* adsorption by the bulk composite material. Sophisticated geochemical techniques should be applied to composite materials to provide the best understanding possible of their surface composition and quantities of surficial deposits. Hopefully, scanning Auger microscopy or other techniques can be developed for the purpose of estimating the types and quantities of surface functional groups of composite materials (see chapter on surface composition by Hochella, this volume).

Interactive effects of mineral phases. Honeyman (1984) compared adsorption measurements in experiments with binary mechanical mixtures of titania, alumina, ferrihydrite, and Na-montmorillonite to that of the individual minerals. Qualitatively, adsorption edges in mixtures were broader than those on single components, reflecting the presence of two or more sets of surface functional groups with different binding intensities. In most cases, adsorption in binary mixtures could not be predicted quantitatively by summing the adsorptive characteristics of the single components. Similar conclusions follow from the work of Altmann (1984) on Cu adsorption in binary mixtures of humic acid with geothite, rutile, or corundum. This phenomenon results from non-linear interactions between unlike particles that affect the stoichiometry and binding intensity of adsorption reactions. Simplified theoretical treatments of interacting electrical double layers of dissimilar surfaces suggest that substantially different surface potentials result from the interaction (Prieve and Ruckenstien, 1978). In addition, the work of Anderson and Benjamin (1990) gives evidence of chemical factors involved in changing surface chemical properties, based on studies of the adsorption of Ag^+, Zn^{2+}, Cd^{2+}, phosphate and Se(IV) on binary mixtures of Fe, Al and Si oxides. In mixtures containing Al oxides, surfaces of another mineral phase became enriched with Al, significantly altering surface chemical properties. Wiese and Healy (1975b) made similar observations in mixtures of titania and alumina.

Few conclusions can be drawn from this work at the present time for application to adsorption studies with soils and sediments. Adsorption on mixtures of minerals is likely to be somewhat less sharply dependent on pH than observed in experiments with single minerals, because of the heterogeneous surface functional groups available for complexation with ions. The relative importance of this effect is still the subject of some debate (Dzombak and Morel, 1990; Honeyman and Santschi, 1988), but the effects of adsorbed Al and Si on the surface chemistry of ferrihydrite and goethite seem significant (Anderson and Benjamin, 1990; Zachara et al., 1989b; Zachara et al., 1987). Component minerals in the assemblage will not necessarily have the same surface chemical properties in mineral assemblages that they possess in monomineralic systems. Thus, unless evidence can be given to the contrary, adsorption properties of composite materials should probably be characterized as a whole.

Special problems in sorption experiments with natural composite materials. Various physical and chemical properties of the mineral assemblage need to be considered in designing sorption experiments with soils and sediments. Most studies of sorption are conducted with batch reactors, which may allow the concentration of dissolved constituents to increase with time. A careful consideration of dissolution, precipitation and redox reactions of mineral phases over the range of pH and electrolyte concentrations used in the sorption experiments is required. At high pH, dissolution of soluble silicates can cause precipitation of metal silicates (Kent and Kastner, 1985). At low pH, dissolution of aluminum oxides can give rise to precipitation of phosphate (Veith and Sposito, 1977b; Goldberg and Sposito, 1984b). Carbonate minerals and ferrihydrite may have significant recrystallization rates, thus enhancing the formation of solid solutions at mineral surfaces via surface precipitation (Wersin et al., 1989; Davis et al., 1987; Dzombak and Morel, 1986). Characterization of the redox chemistry of the mineral phases is especially important in studies of the sorption of redox sensitive elements (White and Hochella, 1989; Eary and Rai, 1989; Rai et al., 1988).

Charlet and Sposito (1987, 1989) discussed the importance of pretreating samples to convert them to a form where all sites are populated by a particular cation or anion that can be later displaced by an adsorbing solute of interest. While the concept is of significance, it was not made clear how their procedure (saturation with $LiClO_4$) would displace strongly-adsorbed ions such as Al^{3+}. In some cases, the mineral assemblage may be contaminated with the solute whose adsorption behavior is to be determined. In these cases, it is necessary to pretreat the sample to remove the contaminant (Mouvet and Bourg, 1983) or account for the release of the contaminant in describing adsorption behavior (Goldberg and Glaubig, 1986b; Neal and Sposito, 1989). Isotopic exchange techniques can be effectively employed for this purpose (Davis et al., 1990; Payne and Waite, 1990).

A variety of sorptive mechanisms may be operative during experiments with soils and sediments. The mesoporous and microporous nature of reactive components in some materials can result in slow attainment of adsorption equilibrium (Fuller and Davis, 1989). When a strongly-bound ion diffuses into aggregated or agglomerated materials, its effective molecular diffusion coefficient may be orders of magnitude smaller than its diffusion coefficient in water (Berner, 1980), resulting in very slow equilibration. Quantification and comparison of the rates of sorption processes in laboratory and natural systems is not straightforward. Empirical rate laws have been proposed from laboratory studies (Honeyman and Santschi, 1988; Sparks, 1985), but these must be applied with caution since they have limited applicability outside the conditions under which they were determined (Helfferich, 1962). Rimstidt and Dove (1986) and Sparks (1985) discuss the importance of reactor design on the determination of rates for application to natural systems. In particular, it must be recognized that the rates of sorption processes may be limited by mass transfer in natural systems (Stollenwerk and Kipp, 1990), whereas that is often not the case in batch studies of simple mineral-water systems.

The electrical double layer of soils and sediments

Bolt and van Riemsdijk (1987) discuss the problems associated with the determination of the pH_{PZNPC} of composite materials. Acid-base titrations of soils and sediments are not expected to yield useful microscopic information as is the case in simple mineral-water systems. Kinetic effects and dissolution reactions during such titrations are expected to confound the interpretation of data; this includes the dissolution of organic matter throughout the entire pH range of titrations and dissolution of Al oxides and secondary minerals at low and high pH values (Parker et al., 1979). For didactic purposes, Bolt and van Riemsdijk (1987) simulated the titration curves that would result from combining various mineral types in mixtures, assuming no interaction. The results illustrate that the common intersection point found in acid-base titrations of simple mineral-water systems (Fig. 17) cannot be expected for many natural materials.

The determination of electrical double layer properties of composite materials was approached in a different manner by Charlet and Sposito (1987, 1989). They determined and modeled the properties of an oxisol consisting of kaolinite, gibbsite, and hematite. After a pretreatment to achieve saturation with $LiClO_4$, the adsorption of H^+ (or OH^-) and electrolyte ions was measured directly in 1:1 electrolytes (NaCl and KNO_3). The univalent electrolyte ions were assumed to be adsorbed either as outer-sphere complexes (thus contributing to σ_β in the triple layer model, TLM) or dissociated counterions (σ_d). The *point-of-zero-net-charge*, or pH_{PZNC}, is defined as the pH at which $\sigma_\beta + \sigma_d = 0$ (Sposito, 1984). The pH_{PZNC} was determined at several ionic strengths from the pH at which the adsorption densities of the electrolyte ions were equivalent. Because of the lack of minerals with permanent structural charge, $\sigma_s = 0$, and from Equation 47, the coordinative surface charge, σ_o, must also be zero at the pH_{PZNC}. The authors assumed that inner-sphere complexes were removed by the pretreatment, and hence, $\sigma_{CC} = 0$. It follows from Equation 23 that $\sigma_o = \sigma_H$, and thus, $\sigma_H = 0$ at the pH_{PZNC} given these assumptions. Thus, the pH_{PZNPC} must equal the pH_{PZNC} under these conditions.

In Charlet and Sposito's experiments, the adsorption of H^+ (or OH^-) and electrolyte ions were determined simultaneously, and curves for the variation of proton surface charge (σ_H) with pH and ionic strength could be determined once the values of the pH_{PZNPC} (= pH_{PZNC}) were known. For both the 1:1 and 2:1 electrolytes ($Ca(ClO_4)_2$, $Mg(ClO_4)_2$, and Li_2SO_4), significant adsorption of cations occurred at pH < pH_{PZNPC} and of anions at pH > pH_{PZNPC}, which attests to the heterogeneous nature of the adsorbent. For all ions except Ca^{2+}, increasing ionic strength increased the extent of adsorption; in most cases the effect was greater at pH > pH_{PZNPC} for cations and at pH < pH_{PZNPC} for anions. The authors applied the triple layer model (TLM) to the data, since this model is consistent with the observations of ion pair formation for univalent electrolyte ions at the pH_{PZNPC}. The authors used one "average amphoteric site" instead of attempting to consider the wide array of possible surface functional groups. Site density was determined by applying Equation 35 to the data. A good simulation of both the adsorption and surface charge data required that a combination of inner-sphere and outer-sphere complexes be used for the divalent ions.

The weak, but specific, interactions of electrolyte ions that are allowed within the interfacial structure of the TLM were useful for simulating the data of Charlet and Sposito (1987, 1989). It appears unlikely that simulations with the DDLM would have been as satisfactory in describing these data. Adsorption of weakly binding cations or anions occurs under conditions where the net charge on the composite material is positive or negative, respectively (Charlet and Sposito, 1987, 1989). This is consistent with the conclusion of Bousse and Meindl (1986) that proton surface charge data are only useful for the determination of ion pair formation constants. Surface acidity constants should be evaluated from measurements of surface electrical potentials or zeta potentials (Sprycha, 1989a,b). Given the interactive effects of minerals on surface potentials (Prieve and Ruckenstein, 1978), it is unlikely that the surface acidity constants can be meaningfully evaluated for composite materials. This means that the DDLM may be relatively ineffective in describing the electrical double layer properties of composite materials. Although it has not yet been tried, the TLM could be effectively used without surface ionization reactions (Eqns. 38 and 39), if estimates of the diffuse layer potential are not needed. This would effectively eliminate two of the five adjustable parameters in the interfacial model. This can be done because the proton surface charge and surface potential in the model are primarily controlled by the ion pair formation reactions (Eqns. 45 and 46).

Use of empirical adsorption models for soils and sediments

Distribution coefficients. Until the early 1980's, attempts to characterize sorption processes on natural materials focused on generating empirical parameters like distribution coefficients (K_d) or sorption ratios (R_d; e.g., Higgo and Rees, 1986; Kent et al., 1986; Duursma and Gross, 1971). The distribution coefficient (Eqn. 30) is usually determined in batch reactors by suspending a solid phase in a solution with a known concentration of solute and observing the amount of solute removed after a specified time (e.g., Higgo and Rees, 1986; Balistrieri and Murray, 1986; Nyffeller et al., 1984; Li et al., 1984). Although some authors have used a K_d value normalized for reactive surface area (e.g., Balistrieri and

Murray, 1984) or aqueous speciation (Tessier et al., 1989), in most cases, very little characterization of the solution speciation or chemical properties of the composite materials is performed. The approach involves collecting K_d values over the range of conditions encountered in the systems under investigation.

K_d values vary over several orders of magnitude with solution composition even in simple mineral-water systems. The pH is a primary variable (Tessier et al., 1989); cation and anion adsorption onto oxide minerals vary extensively with pH (Figs. 11-13). Variable concentrations of complexing ligands can cause large changes in K_d values for some metal ions (e.g., Hsi and Langmuir, 1985; Leckie and Tripathi, 1985; Tripathi 1983; Means et al., 1978). Changes in oxidation state of an adsorbing solute greatly influence K_d values, as has been shown for Pu sorption on marine sediments (Higgo and Rees, 1986). K_d values are affected by the concentrations of ions that compete for adsorption sites (Davis et al, 1990; Balistrieri and Chao, 1987) and other interactions between sorption sites and solution components (Honeyman and Santchi, 1988; Kent et al., 1986). The dependence on solution composition limits the usefulness of K_d values in systems of variable composition, such as estuaries or contaminated surface waters and groundwaters (Davis et al., 1990; Kent et al., 1986; Cederberg et al., 1985; Reardon, 1981).

Modeling based on the partitioning equation. The partitioning equation (Eqns. 36 and 37) has been applied to describe sorption processes in surface waters (Tessier et al., 1989; 1985; Honeyman and Santschi, 1988; Balistrieri and Murray, 1983). There are three parameters in the model: total site density (S_T), the apparent proton coefficient (χ), and the partitioning coefficient (K_{part}). S_T can be estimated by methods described earlier. K_{part} and χ, however, are empirical parameters. Even in simple mineral-water systems containing an oxide, one dilute solute (J), and a fixed background electrolyte, both parameters are functions of pH and adsorption density (Γ_J). These dependencies result from using a single reaction to describe the entire suite of surface ionization and complexation reactions (Honeyman and Leckie, 1986). Two methods have been used to evaluate K_{part} and χ: Kurbatov plots (Kurbatov et al., 1951) and isotherm subtraction (Perona and Leckie, 1985). Kurbatov plots are based on transforming Equation 37 to:

$$\log \frac{\Gamma_J}{\Gamma_S[J]} = \chi \cdot pH + \log K_{part} \quad . \tag{49}$$

K_{part} and χ are computed from the intercept and slope of a regression of the left hand side of Equation 48 against pH. Honeyman and Leckie (1986) have shown that Kurbatov plots are of limited applicability because the dependencies of K_{part} and χ on pH and Γ_J are not considered. The findings are significant because of a renewed interest in applying the Kurbatov method to natural systems (Belzile and Tessier, 1990; Tessier et al., 1989; 1985; Johnson, 1986; Davies-Colley et al., 1984). The better approach, isotherm subtraction, determines χ graphically using the equation:

$$\chi = \left(\frac{\Delta \log[J]}{\Delta pH} \right)_{\Gamma_J} \quad . \tag{50}$$

χ is evaluated as a function of pH and Γ_J; K_{part} can then be determined (Eqn. 37). Ignoring the variation in χ and K_{part} with pH and adsorption density can lead to large errors in the predicted adsorption of ions (Honeyman and Santschi, 1988). Adsorption data should be collected experimentally over the entire range of solution compositions and conditions of interest before this modeling approach is applied.

Johnson (1986) applied the partitioning approach in a study of Cu^{2+} and Zn^{2+} adsorption onto suspended particles in a river contaminated by acid-mine drainage. Adsorption was estimated in field samples, from determinations of dissolved and particulate Cu, Zn, and Fe, and in laboratory experiments using mixtures of acid-mine waters (pH 3) and sea water (pH 7.5). Concentrations of major and minor solutes were used to compute the speciation of Cu^{2+} and Zn^{2+}. Reactive surface area was assumed to be proportional to particulate Fe concentration. Good agreement was obtained between adsorption determined in the experimental mixtures and those in the field samples over a wide range of solution conditions (pH 3 to 7.5; water composition ranging from river water to sea water).

The partitioning approach is most appropriate for systems where the concentrations of the components of the background electrolyte are nearly invariant, such as sea water (Balistrieri and Murray, 1983) or some lakes (Tessier et al., 1989). Typically, the speciation of the dissolved solute is ignored (Honeyman and Santschi, 1988), although Tessier et al. (1989) considered the *inorganic* speciation of Zn in their analysis. Inclusion of speciation in the model enhances its applicability, especially in systems where changes in speciation occur as a function of pH.

Use of surface complexation models with soils and sediments

The use of empirical modeling approaches has been rationalized by the complexities associated with understanding sorption mechanisms in natural systems (Higgo and Rees, 1986), and until recently, only these approaches were used for applications in aqueous geochemistry. For example, when adsorption has been considered within solute transport models, distribution coefficients or isotherm equations were typically used to describe adsorption equilibria (Bencala, 1984; Grove and Stollenwerk, 1987). The advancement of surface complexation models, however, has greatly enhanced the understanding of adsorption processes, and an extension of this approach to natural systems is now beginning. The consideration of specific site-binding models that include aqueous speciation have been included in some recent solute transport models (Cederberg et al., 1985; Lewis et al., 1987; Krupka et al., 1988).

The surface complexation modeling (SCM) approach has advantages over empirical ones, even if employed in a semi-empirical manner. The advantages arise from a consideration of aqueous speciation and the sorptive reactivity of all aqueous species. As in the empirical approaches, adsorption must still be determined over the range of solution compositions expected in a field application, but the smaller number of model parameters can be incorporated within geochemical equilibrium computer codes to facilitate *interpolation* of adsorption behavior in natural systems. Clearly, *extrapolation* of adsorption behavior is not appropriate.

Modeling with the non-electrostatic SCM. $Ko\beta$ (1988) used the SCM approach without EDL correction in modeling U(VI) adsorption in groundwater systems, using an average surface site for the composite aquifer materials, with site density estimated from the cation exchange capacity. Adsorption was described with a single equation, i.e.,

$$\equiv SOH + UO_2^{2+} \leftrightarrow \ \equiv SOUO_2^+ + H^+ \ . \tag{51}$$

Mouvet and Bourg (1983) carried this simplified SCM approach a step further by considering two types of surface functional groups for suspended particulate matter in a contaminated river system. The authors compared laboratory determinations of Zn^{2+}, Cu^{2+} and Cd^{2+} sorption on suspended particles collected from the river with field measurements. Estimates of adsorption in the field were made by measuring dissolved and particulate concentrations of the metals. The concentrations of major dissolved cations and anions, as well as humic acid, were determined. Aqueous speciation computations accounted for complexation of the metals with carbonate, sulfate, chloride and humic acid ligands. Surface site density and apparent binding constants were adjusted to fit the laboratory data. Good agreement between predicted and measured adsorption for Zn^{2+} was obtained without further manipulation. Successful simulations of Cu^{2+} adsorption, however, required adjustment of the complex formation constant of the dissolved Cu-humate complex and introducing an additional equation for formation of a ternary complex involving Cu, humate, and surface hydroxyl groups. A similar modeling approach was used by Davis (1984), who found that the Cu complexation by adsorbed fulvate ligands had essentially the same stability constants (and pH dependence) as Cu-fulvate complexes in solution.

Fuller and Davis (1987) described the sorption of Cd^{2+} by a calcareous aquifer material suspended in artificial groundwater, with or without added EDTA, using the SCM approach without EDL correction. In the absence of EDTA, Cd^{2+} sorbed extensively on the aquifer material. Cd-EDTA complexes, however, did not sorb. Adsorption, at the early stage of the sorption process, was described by the reaction:

$$Cd^{2+} + \equiv S_c \leftrightarrow \equiv S-Cd^{2+} , \tag{52}$$

where $\equiv S_c$ denotes a calcite surface site. Sorption of Cd^{2+} varied with pH due to dissociation of Cd-EDTA complexes. Inclusion of a pH-independent binding constant for Equation 52 in a computation of Cd speciation in the presence of the aquifer material, artificial groundwater, and EDTA gave good agreement with the adsorption data throughout the pH range studied. Site density was estimated from calcite content and the calcite surface site density. The binding constant found for the composite material using Equation 52 had the same value as that found for Cd^{2+} adsorption on pure calcite (Davis et al., 1987).

Modeling with electrical double layer corrections. The constant capacitance model (CCM) has been applied by Goldberg and coworkers (Goldberg and Sposito, 1984b; Goldberg and Glaubig, 1986a) to simulate sorption of phosphate and borate on soils. For the phosphate sorption study, published investigations of phosphate sorption onto non-calcareous, non-allophanic soils were modeled. $\mathcal{A}$ and total site density from the original studies were used when reported, otherwise they were estimated from phosphate sorption data. Average values for K_+^{CCM}, K_-^{CCM}, and C^{CCM} had previously been determined for a variety of Fe and Al oxides over a range of ionic strengths (Goldberg and Sposito, 1984a). These values were used for modeling phosphate sorption on the soils. Following the same methods used to model phosphate adsorption onto hydrous oxides, apparent binding constants for three surface complexation reactions were derived by fitting experimental data for each soil (44 in all), i.e.,

$$\equiv SOH + H_3PO_4 \leftrightarrow \equiv SOPO_3H_2 + H_2O \tag{53}$$

$$\equiv SOH + H_3PO_4 \leftrightarrow \equiv SOPO_3H^- + H_2O + H^+ \tag{54}$$

$$\equiv SOH + H_3PO_4 \leftrightarrow \equiv SOPO_3^{2-} + H_2O + 2H^+ . \tag{55}$$

Values for the apparent binding constants obtained for each soil were averaged to obtain a single value for each of the three binding constants. Model fits to the data were quite good. The lack of sensitivity of the values of the phosphate binding constants to the values of K_+^{CCM} and K_-^{CCM} had previously been established (Goldberg and Sposito, 1984a) and is consistent with the hypothesis that the electrostatic contribution to the free energy of adsorption for strongly-binding ions is small. Thus, it is possible that similar success could have been achieved with the non-electrostatic SCM.

A similar approach was used to simulate borate adsorption onto a variety of soils (Goldberg and Glaubig, 1986a). A single binding constant was used with stoichiometry analogous to that of Equation 53. For some soils, it was necessary to adjust K_+^{CCM} and K_-^{CCM} to optimize the fit. This is consistent with the hypothesis that, as a weakly adsorbing solute, the binding of borate is more sensitive to the electrostatic terms. Considering the variety of soil compositions and electrolyte compositions, it is encouraging that good model simulations could be obtained with a single type of binding site.

SCM modeling to dominant adsorptive components of composite materials. Some investigators have applied the SCM approach to a specific mineral phase of composite materials to simulate ion adsorption on soils or sediments. Payne and Waite (1990) applied this approach to describe the adsorption of U(VI) on ferrihydrite present in two subsurface samples derived from cores within the vicinity of U ore bodies. One sample (Koongarra) was a highly weathered schist composed of quartz, kaolinite, vermiculite, goethite, hematite, and aluminum oxides. The other (Ranger) contained smectite, kaolinite, quartz, mica, goethite, hematite, and anatase. Adsorptive equilibrium of U(VI) with ferrihydrite was suggested by equal ^{234}U/^{238}U ratios in the groundwaters and oxalate extracts of the composite materials, while the isotopic ratio of the bulk samples were quite different. Dominance of the adsorptive reactions by ferrihydrite was assumed. The authors applied the TLM, using the binding constants and EDL parameters derived in the study of Hsi and Langmuir (1985) without further manipulation. In that study, it was found that the adsorption of two U(VI) carbonate complexes, i.e., $UO_2(CO_3)_2^{2-}$, $UO_2(CO_3)_4^{4-}$, had to be considered to fit model simulations with experimental data. Differences in the aqueous conditions in the study by Payne and Waite resulted in some solution compositions in which $UO_2(CO_3)^0$ was the predominant dissolved species of U(VI). To describe their data fully, it was necessary to add an additional adsorption reaction for this species. The binding constant for this species was derived with data from the Ranger subsurface sample. Interestingly, the same set of binding constants described U(VI) adsorption by the Koongarra sample, even though the ferrihydrite abundances in these samples were substantially different. It is surprising (and encouraging) that surface complexation modeling could be applied to these composite materials with binding constants derived from experiments with synthetic ferrihydrite. It suggests that the limitations caused by the interactive effects of mineral phases (Anderson and Benjamin, 1990; Altmann, 1984; Honeyman, 1984) may not always be significant.

The most sophisticated application of surface complexation modeling to natural materials has been performed by Zachara et al. (1989b). Adsorption of Cr(VI) on three subsurface soils was modeled using the TLM, based on a companion study of Cr(VI) adsorption onto Al-substituted goethite (Ainsworth et al., 1989). The soils were primarily composed of clay minerals, micas, and hydrous oxides, with very low organic matter and carbonate contents. Cr(VI) adsorption edges were obtained over modest ranges of ionic strength, soil-water ratios (W), Cr(VI) concentration, and concentrations of competing anions (carbonate and sulfate). Modeling focused on the role of hydrous Fe and Al oxides, which were thought to dominate the adsorptive interactions of Cr(VI) with the subsurface soils. Adsorption on Al-substituted goethites was modeled with a two site model whereby Cr(VI) was allowed to adsorb on both $\equiv$ FeOH and $\equiv$ AlOH functional groups. Each group had its own set of surface ionization, electrolyte binding, and Cr(VI) binding constants. Cr(VI) adsorption on each site was modeled with two reactions, i.e.,

$$\equiv FeOH + CrO_4^{2-} + H^+ \leftrightarrow \quad \equiv FeOH_2^+ - CrO_4^{2-} \tag{56}$$

$$\equiv FeOH + CrO_4^{2-} + 2H^+ \leftrightarrow \quad \equiv FeOH_2^+ - HCrO_4^- \tag{57}$$

$$\equiv AlOH + CrO_4^{2-} + H^+ \leftrightarrow \quad \equiv AlOH_2^+ - CrO_4^{2-} \tag{58}$$

$$\equiv AlOH + CrO_4^{2-} + 2H^+ \leftrightarrow \quad \equiv AlOH_2^+ - HCrO_4^- \;. \tag{59}$$

Surface ionization and electrolyte binding constants were taken from previous investigations on pure Fe and Al hydrous oxides in the same background electrolyte ($NaNO_3$). Apparent binding constants for Cr(VI) adsorption on $\equiv$ AlOH and $\equiv$ FeOH sites were obtained by modeling adsorption data onto corundum and goethite, respectively. With the exception of Cr(VI) binding constants on $\equiv$ AlOH sites, all the same parameters were used to model adsorption onto subsurface soils. The functional group density was varied to optimize the fit to Cr(VI) adsorption data at one Cr(VI) concentration and soil-water ratio (W). The model was then applied to other data sets by scaling the site density to the reactive surface area of Fe and Al oxides. The binding constants of Cr(VI) with $\equiv$ AlOH were adjusted also. Simulated Cr(VI) adsorption agreed well with the experimental data over most of the range of pH, ionic strength, and solids concentrations studied. Deviations were confined to low pH, where the model simulations underestimated Cr(VI) adsorption because the site density was too small. Competition with sulfate and carbonate was modeled using previously derived binding constants for these ions on ferrihydrite (Zachara et al., 1987); site density was again used to optimize the fit. The model simulations matched the data closely in the pH range 6 to 9. At lower pH values, actual Cr(VI) and sulfate adsorption exceeded the site density used in the model. The results demonstrate the ability of a heterogeneous, two-site surface complexation model to broaden the range of model applicability. In addition, these studies illustrate how one can optimize model simulations to fit adsorption data in ranges of solution composition that are of highest priority for specific applications.

Guidelines for surface complexation modeling with natural composite materials

A review of the studies that have applied surface complexation theory to the complex mixtures of phases present in natural materials suggests that the most important aspect of the work is detailed characterization of the solid phases and their surface composition. The best model simulations of experimental data or predictions of adsorption coupled with transport will probably be performed when the solid phases that dominate adsorptive interactions are known. Identifying the dominant adsorptive phases facilitates evaluating the stoichiometry of the adsorption process (see section on "Surface Functional Groups," above). It is well accepted in aqueous geochemistry that a thorough analysis of water composition is required to compute aqueous speciation as part of solute transport or geochemical flowpath modeling. By analogy, in surface complexation theory, surface functional groups are the reactants with ions that determine surface speciation, and a thorough understanding of the concentration (surface density) and types of functional groups is needed to calculate the effects of adsorption equilibria on aqueous composition. Therefore, studies involving the adsorption of ions by composite materials should first focus on the surface composition of the bulk material and techniques that can be applied to identify particular components of the solid phases that may dominate adsorptive interactions. The significance of interactive effects of mineral phases is not well understood; in two cases involving strongly-binding ions (Fuller and Davis, 1987; Payne and Waite, 1990), the effects appeared minimal. On the other hand, the effects were more substantial in the study of weakly-binding Cr(VI) conducted by Zachara et al. (1989b). Thus, the question of the applicability of binding constants found in simple mineral-water systems to mineral components of composite materials needs further research. In addition, systems in which organic functional groups dominate surface complexation are still poorly understood and deserve special attention, because of their likely significance in many natural systems. In the absence of evidence of adsorptive dominance by a particular component, the bulk composite material can be modeled as a whole. In this case, particular attention needs to be paid to a determination of the *reactive* surface area for the adsorptive interactions of particular solutes.

Given the complexity of natural materials, it can be argued that the simplest surface complexation model should be applied for modeling purposes. The non-electrostatic surface complexation model is clearly the simplest; this is especially true given the difficulties involved in the determination of double layer model parameters for natural materials. However, this model has not been widely used and its performance and general applicability cannot be evaluated. Simulations with the non-electrostatic model in simple mineral-water systems exhibited less pH dependence than experimental adsorption data (Krupka et al., 1988). On the other hand, the pH dependence of ion adsorption on heterogeneous materials has been shown to be less dependent than found in simple mineral-water systems. The non-electrostatic model would be best applied to strongly-binding solutes, i.e., those solutes whose adsorptive interactions exhibit little ionic strength dependence. The approach could be enhanced by inclusion of a two-site Langmuir model to allow for site heterogeneity and consistency with observed Freundlich isotherms for metal ion adsorption.

Ideally, an electrical double layer model would be included in applications of surface complexation theory. For weakly-binding solutes, electrostatic correction factors to the mass law equations for adsorption equilibria appear necessary. However, an assessment of surface electrical potentials and other double layer properties of soils and sediments is extremely complex. For modeling bulk composite materials, the flexibility of the TLM model appears preferable because the proton surface charge is defined primarily by the adsorption of electrolyte ions, which can be determined experimentally (Charlet and Sposito, 1987; 1989). However, if a particular mineral phase dominates adsorptive interactions, then either the DDLM or TLM may be applied if binding constants from the literature are available.

A convenient modeling approach for seawater uses apparent binding constants that incorporate pH effects, ionic strength activity coefficients, and electrostatic corrections within the constant (Dzombak and Morel, 1987; Luoma and Davis, 1983; Balistrieri and Murray, 1983). If the speciation of dilute ions is to be considered (e.g., complexes with organic molecules), it can be argued that the constant capacitance model (CCM) is most appropriate for simulating dilute ion adsorption in marine systems. The constant ionic medium Reference State is used in this model, and thus, binding constants can be determined in laboratory experiments in artificial seawater without need for further activity corrections. However, for model simulations of trace element scavenging in the deep oceans, it may be necessary to derive binding constants from laboratory experiments conducted under unusual conditions, e.g., at very low solid-water ratios (Honeyman and Santschi, 1988; Chang et al., 1987) or with free metal concentrations buffered at low values (Hunter et al., 1988).

For some applications, it may be necessary to examine the spatial variability of the surface chemical properties of composite materials. Neal et al. (1987a,b) examined the role of adsorption processes on the mobility of Se(IV) in alluvial soils in the San Joaquin Valley (California), an area affected severly by Se contamination from agricultural drain waters. Soil samples from a variety of environments in the San Joaquin Valley were examined. In the acidic pH range, Se(IV) adsorption correlated with the amounts of readily soluble Fe, Mn, and Al, suggesting the importance of oxide and hydroxypolymer coatings in the adsorptive interactions. No differences were observed between various soils in the alkaline pH range. In contrast, Zachara et al. (1989b) showed that Cr(VI) adsorption on subsurface materials increased with increasing Fe and Al oxide content and decreasing soil pH. A sample with a soil pH of about 10.6 did not adsorb Cr(VI) even at low pH. Our research group is conducting an investigation of the transport of Zn, Ni, Cr, and Se in oxic and suboxic zones of a shallow, sewage-contaminated, sand and gravel aquifer (e.g., Davis et al., 1989; 1990; Kent et al., 1989). Under identical aqueous conditions, Cr(VI) adsorption differs on subsurface materials collected from oxic recharge and suboxic, sewage-contaminated zones. These trends correlate with differences in the abundance of Mn oxides and poorly-crystalline Fe oxides in the different zones.

CONCLUDING REMARKS

Surface complexation theory has become increasingly popular for describing ion adsorption in simple mineral-water systems. The attractive feature of the theory is that it adopts the formalism of ion association reactions in solution as a representation of adsorption reactions at the mineral-water interface. The nature of surface functional groups is of primary importance because it determines which ions are typically exchanged for an adsorbing ion, e.g., H^+ for cations on hydrous oxides, Ca^{2+} for cations on calcite, Na^+ and H^+ for cations on montmorillonite. This characteristic, in turn, governs the ways in which adsorption on various minerals varies as a function of solution composition. Ion adsorption on oxides is highly dependent on the pH, but in addition, the ionic strength of solutions has now been demonstrated as an important variable that characterizes adsorption. Those ions whose adsorption is independent of ionic strength are strongly-bound ions that form inner-sphere coordination complexes with surface functional groups. Like complex formation in aqueous solution, the free energies of adsorption of these ions is dominated by covalent bonding rather than electrostatic attraction. Conversely, those ions that exhibit a dependence of adsorption on ionic strength are weakly-bound ions that form outer-sphere complexes with surface functional groups. Adsorption of these ions is more dependent on electrostatic attraction, which is revealed by the relationship between ionic strength and electrical potentials in the EDL. The major ions in natural waters are typically weakly-bound ions, whereas trace elements and many inorganic contaminants are strongly-bound ions. Although the effects of many variables on adsorption equilibria are now well documented, the stoichiometry of H^+ in surface complexation reactions and the temperature dependence of adsorption equilibria are still poorly understood.

A governing paradigm for the application of surface complexation theory to natural systems has not yet been fully developed. Two approaches have been used: (1) treatment of composite materials as an integrated whole with adsorption described as complexation with average surface functional groups or (2) consideration of a specific mineral surface in composite materials that is proposed to dominate adsorptive interactions. More experience is needed with each of these approaches before either can be accepted or rejected as expedient methods. Application of the second approach appears preferable, but this will require the development of more sophisticated techniques for characterizing surface composition and a greater understanding of how particle-particle interactions in heterogeneous systems affect adsorption behavior. Ideally, the results of such techniques could be interpreted in terms of the quantities of various types of surface functional groups. Based on experience with simple systems, it can be concluded that models with at least two types of sites (high and low energy) will probably need to be applied, because of the general observations of Freundlich isotherms for strongly-bound cations.

Considerable research effort has been expended by various groups to demonstrate the superior performance of particular interfacial models. Evolution of model implementations has narrowed their differences. Experience with simulating experimental data suggests that comparisons of model performance in simple mineral-water systems is a secondary consideration in evaluating model applicability to natural systems. Each of the models appears to have specific strengths and weaknesses when evaluated in terms of applicability to various geochemical environments. Given the complexities involved in describing the electrical double layer properties of soils and sediments, it is not surprising that interest in the non-electrostatic surface complexation model has been revived. This simplified description of surface complexation equilibria can be more easily incorporated in computationally intensive megamodels, e.g., solute transport models that are coupled with geochemical equilibrium speciation models. As a first approximation, it is certainly preferable that adsorption reactions be considered in this manner rather than lumped into empirical parameters as has been done previously.

Many of the experimental techniques that are used to characterize adsorption and electrical double layer properties in simple mineral-water systems cannot be applied in a straightforward manner to soils and sediments. For example, the methods commonly used to estimate surface functional group densities (tritium exchange, acid-base titration, and adsorption isotherms) may suffer from experimental artifacts when applied to natural materials. Based on a review of the important adsorbing mineral phases and arguments for a parsimonious modeling approach, a value of 3.84 μmoles/m^2 for total site density is recommended for future modeling applications. Because the fit of model simulations to experimental data is relatively insensitive to the value chosen, it is more important that a universal value be adopted for modeling than it is that accurate site densities be used for each mineral surface in a composite sample. The acceptance of a general value for the site density will aid the intercomparison of future studies and allow the development of a self-consistent database of apparent binding constants for specimen minerals. It is emphasized that the values of binding constants in surface complexation models are dependent on the value chosen for total site density, as well as the interfacial model. To apply this value for total site density, it is necessary to measure the reactive surface area. A review of literature on this subject suggests that gas adsorption techniques yield the most useful information when applied carefully, including estimates of external surface area and an evaluation of the extent of microporosity.

The experimental techniques employed to determine apparent binding constants may also be subject to artifacts when applied to natural materials. Complex and unpredictable interactive effects have been observed even in simple mineral-water systems as the solution composition was varied. Precipitation reactions, slow kinetics, and organic materials may confound the interpretation of adsorptive equilibrium in laboratory experiments conducted with soils and sediments. Adsorption may be only the first of a series of reactions that remove an ion from water. Subsequent reactions, e.g., redox reactions and surface precipitation, may occur on time scales only slightly longer than adsorption, thereby invalidating an assumed equality of sorption and adsorption in the experiment. Whenever possible, adsorptive equilibrium should be established by reversibility studies that approach the proposed equilibrium condition in several ways (adsorption and desorption by dilution, pH changes, etc.). In addition, isotopic exchange techniques can be applied effectively to demonstrate the maintenance of a constant equilibrium condition as a function of time.

ACKNOWLEDGEMENTS

The senior author was on sabbatical at the Australian Nuclear Science and Technology Organisation (Lucas Heights, Australia) during preparation of the manuscript and is grateful to ANSTO and T. David Waite for providing facilities, support, a stimulating environment, and an excellent facsimile machine for communication with his co-author. Comments by Linda Anderson, David Waite, Sam Luoma, Art White, and Chris Fuller were helpful in improving the comprehensibility of the manuscript. The authors' enthusiasm for the subject stems largely from interactions with many colleagues and mentors, in particular, K. F. Hayes, J. O. Leckie, G. A. Parks, and W. Stumm.

LIST OF TERMS AND SYMBOLS

a_m = cross sectional are of adsorbed gas molecule [m^2]

$\mathcal{A}$ = specific surface area [m^2/g]

A' = geometric surface area [m^2/g]

C_{BET} = parameter from BET gas adsorption analysis

CCM = constant capacitance model

C^{model} = interfacial capacitance of the near surface region for a particular SCM

DDLM = diffuse double layer model

e = unit of electronic charge [Coulombs]

EDL = electrical double layer

F = Faraday constant

$\{H^+\}_s$ = activity of the hydrated proton at the surface

I = ionic strength

IHP = Inner Helmholtz plane (= Stern or β plane)

$[J]$ = aqueous concentration of solute or species J [moles/dm^3]

k = Boltzmann constant

K_d = distribution coefficient [dm^3/g]

K_L = conditional equilibrium constant evaluated from Langmuir isotherm [dm^3/mole]

K_{part} = conditional equilibrium constant of the partitioning mass law equation [Eqn. 37]

K_+^{model} = conditional equilibrium constant for protonation of a surface hydroxyl group for a particular SCM

K_-^{model} = conditional equilibrium constant for proton release from a surface hydroxyl group for a particular

n = amount of adsorbed gas [moles/g]

n_1 = monolayer capacity for gas adsorptives [moles/g]

n_i = surface excess (adsorption) of solute i per unit mass of adsorbent [moles/g]

OHP = Outer Helmholtz plane (= diffuse layer plane)

p = partial vapor pressure [torr]

p^o = equilibrium vapor pressure [torr]

pH_{PZNC} = pH at which $\sigma_\beta + \sigma_d = 0$

pH_{PZNPC} = pH at which $\sigma_H = 0$

pH_{IEP} = pH at which $\sigma_d = 0$

pH_{PPZC} = pH value when $\sigma_\beta = 0$ *and* $\sigma_H = 0$

pH_{PZSE} = pH at which $\frac{\delta\sigma_H}{\delta I} = 0$

246

q = concentration of excess acid in acid-base titrations [moles/dm^3]

R = gas constant

SCM = surface complexation model(ing)

S_T = surface density of all surface functional groups [moles/m^2]

TLM = triple layer model

t = thickness of statistical multilayer (see Appendix)

T = temperature

W = mass concentration of particles per unit volume of aqueous solution [g/dm^3]

z = sign and magnitude of ionic charge

$\equiv S$ = general symbol for a surface functional group

$\equiv$MeOH = surface hydroxyl group coordinated to metal or metalloid atom, Me

$\equiv S_c$ = surface functional group for cation complexation on carbonate mineral

α_s = normalized extent of gas adsorption (see Appendix)

γ_i = activity coefficient of species i

Γ_i = adsorption density of solute i [moles/m^2]

Γ_S = surface density of uncomplexed surface functional groups [moles/m^2]

Γ_{max} = maximum adsorption density [moles/m^2]

μ_i = chemical potential of species i in water

μ_i^o = chemical potential of species i in a chosen standard state

ψ_o = averaged electrical potential in the surface plane [volts]

ψ_β = electrical potential at the Stern or β plane [volts]

ψ_d = electrical potential at the diffuse layer plane [volts]

σ_p = surface density of net particle charge [Coulombs/m^2]

σ_s = surface density of permanent structural charge [Coulombs/m^2]

σ_o = surface density of coordinative charge (ions adsorbed in the surface plane) [Coulombs/m^2]

σ_{CC} = surface density of coordinative complex charge (ions other than H$^+$ or OH$^-$ adsorbed in the surface plane) [Coulombs/m^2]

σ_β = surface density of charge in the Stern or β plane (outer-sphere complexes) [Coulombs/m^2]

σ_d = surface density of dissociated counterion charge [Coulombs/m^2]

σ_H = surface density of net proton charge [Coulombs/m^2]

θ_J = solid phase concentration of solute or species J per unit weight [moles/g]

χ = apparent (macroscopic) quantity of protons released in moles per mole of adsorbate

Appendix A. DETAILS OF SURFACE AREA MEASUREMENT

Gas Adsorption

Sample drying. In order to apply gas adsorption methods to determining $\mathcal{A}$, a sample must be removed from aqueous suspension and dried. Simply allowing the water to evaporate is unsatisfactory for materials that have hydrophilic goups at the surface or consist of small particles. As water evaporates, *menisci* form between particles and the surface tension of water draws particles together as the *menisci* shrink. This leads to agglomeration of the particles. Agglomeration introduces porosity, reduces the surface area due to formation of isolated pores, and can create microporous zones near particle contacts (e.g., Tyler et al., 1969; Rousseaux and Warkentin, 1976). Organic solvents such as acetone are sometimes added to decrease the surface tension of water and speed up drying. Freeze-drying, whereby the aqueous supension is frozen and the ice is removed by sublimation under vacuum, minimizes agglomeration. For example, amorphous silica samples consisting of 2 to 80 μm particles (volume-averaged particle sizes ranged from 5 to 30 μm) exhibited minimal agglomeration upon freeze-drying and no increase in microporosity (Kent, 1983). Precautions such as maximizing the exposed area of ice and minimizing its thickness produced the best results; these precautions minimized the amount of melting that occurred during the freeze-drying process. Colloidal particles are difficult to remove from aqueous suspension without causing some agglomeration of the sample. Whenever it is necessary to remove a sample from aqueous suspension to determine $\mathcal{A}$, the results must be examined critically for artifacts introduced by the drying process.

t- and α_s plots. $\mathcal{A}$ can also be obtained by comparing the isotherm for a sample to that for a reference solid of known surface area. If the reference material is nonporous, information on the existence of microporosity and mesoporosity can also be obtained. Empirical isotherms for N_2 adsorption on nonporous silica and alumina are available in the literature (Gregg and Sing, 1982; Payne and Sing, 1969). The quantity adsorbed is plotted as a function of either t, the thickness of the statistical multilayer, or α_s, the quantity adsorbed at p/p^0 divided by that adsorbed at $p/p^0 = 0.4$. The multilayer thickness is calculated from:

$$t = \frac{n}{n_m} t_1 \quad , \qquad\qquad (A1)$$

where t_1 is the average thickness of an adsorbed monolayer, (0.345 nm for N_2). t or α_s is obtained from the isotherm for the reference material (Gregg and Sing, 1982). It is advisable to use data for relative pressures *above* 0.35 since the shape of the isotherm below relative pressures of 0.35 is strongly affected by the nature of the solid (e.g., as reflected in the magnitude of the C_{BET}; Gregg and Sing, 1982; Lecloux et al., 1979, 1986). If the sample is nonmesoporous and nonmicroporous, the t or α_s plot yields a straight line through the origin. With microporous samples, the slope of the linear branch yields an estimate of the *external* surface area. The $\mathcal{A}$ is calculated from the slope of the linear branch of the t plot using

$$\mathcal{A} = a_m t_1 L m_t \quad , \qquad\qquad (A2)$$

or,

$$\mathcal{A} = 3.45 x 10^5 m_t \quad , \qquad\qquad (A3)$$

for $\mathcal{A}$ in m^2/g and N_2 as the adsorptive, where m_t is the slope of the t-plot. Similar equations can be derived to calculate $\mathcal{A}$ from α_s plots (see Gregg and Sing, 1982). Samples with mesopores show deviations from linearity along the high pressure branch of the t or α_s plot. For microporous samples, the linear branch does not extrapolate through the origin. An estimate of the micropore volume can be otained from the y intercept by multiplying by the molar volume of liquid N_2 at -196°C. This estimate should be used for comparative purposes only (Gregg and Sing, 1982). Comparison plots thus offer several advantages: the existence of porosity can be determined, $\mathcal{A}$ can be determined for samples with high C_{BET} values, and, for α_s plots, $\mathcal{A}$ can be determined for adsorptives other than N_2 without need of evaluating a_m (Gregg and Sing, 1982).

Adsorption from solution

Various methods for determining $\mathcal{A}$ based on adsorption from aqueous solution have been proposed. There are methods based on the measurement of negative adsorption (co-ion exclusion from the electrical double layer), the determination of adsorption isotherms for organic cations, and empirical correlations between titration data and $\mathcal{A}$. These methods share the obvious advantage over gas adsorption methods that the sample does not have to be removed from suspension and dried. On the other hand, there are serious limitations to their applicability to a wide variety of natural materials.

The negative adsorption method has been used to determine $\mathcal{A}$ of a variety of hydrosols (James and Parks, 1982; Sposito, 1984). It is based on the concept that the surface charge is compensated by both positive adsorption of counterions (ions of opposite charge to the surface) and negative adsorption (exclusion) of co-ions (ions of like charge to the surface). Diffuse double theory yields a relationship between the exclusion volume (see Sposito, 1984) or excess concentration of co-ion in solution (see van den Hul and Lyklema, 1968) and $C^{-1/2}$, where C is the concentration of electrolyte. The measurement must be made at low ionic strength because the fraction of the surface charge compensated by co-ion exclusion decreases with increasing ionic strength. At low ionic strength, the thickness of the electrical double layer is large; porosity on a smaller scale than the double layer thickness will not be detected by this method (van den Hul and Lyklema, 1968). For porous materials, the method probes the geometrical $\mathcal{A}$ of the aggregates or agglomerates rather than the $\mathcal{A}$ on the scale of surface functional groups. For natural materials that consist of mixtures of minerals with different surface properties, the theoretical basis of the method may be invalid.

Adsorption of N-cetylpyridinium bromide (CPB) and methylene blue, both of which are monovalent organic cations, have been used to determine the $\mathcal{A}$ of clay minerals. CPB, which consists of a cetyl group ($CH_3(CH_2)_{14}CH_2$-) attached to the N of a pyridine ring, forms complexes with negatively charged sites on basal surfaces of phyllosilicates and in interlayer regions of expandable clay minerals. CPB exhibits Langmuir-type adsorption (see section on "Langmuir isotherm"); the maximum adsorption density value is used to calculate $\mathcal{A}$ (Greenland and Quirk, 1964; Sposito, 1984). The a_m value has been determined from X-ray diffraction studies to be 0.27 nm^2 on basal surfaces and 0.54 nm^2 on interlayer sites. Free soil organic matter sorbs CPB strongly and therefore must be removed prior to the analysis. CPB has a low affinity for neutral surface functional groups on hydrous oxides and aluminosilicates without fixed charge (Greenland and Quirk, 1964; Aomine and Otsuba, 1968). The methylene blue adsorption method is based on the same principal and has similar limitations to the CPB method (Hang and Brindley, 1970; van Olphen, 1977).

REFERENCES

Aggett, J. and Roberts, L. S. (1986) Insight into the mechanism of accumulation of arsenate and phosphate in hydro lake sediments by measuring the rate of dissolution with ethylenediaminetetraacetic acid. Environ. Sci. Tech. 20, 183-186.

Ahrland, S., Grenthe, I., and Noren, B. (1960) The ion exchange properties of silica gel. I. The sorption of Na$^+$, Ca^{2+}, Ba^{2+}, UO$_2^{2+}$, Gd^{3+}, Zr(IV) + Nb, U(IV), and Pu(IV). Acta Chem. Scand. 14, 1059-1076.

Ainsworth, C. C., Girvin, D. C., Zachara, J. M., and Smith, S. C. (1989) Chromate adsorption on goethite: Effects of aluminum sustitution. Soil Sci. Soc. Am. J. 53, 411-418.

Aldcroft, D., Bye, G. C., Robinson, J. G., and Sing, K. S. W. (1968) Surface chemistry of the calcination of gelatinous and crystalline aluminium hydroxides. J. Appl. Chem. 18, 301-306.

Altmann, S. A. (1984) Copper binding in heterogeneous, multicomponent aqueous systems: Mathematical and experimental modeling. Ph.D. dissertation, Stanford University, Stanford, Calif.

Anderson, P. R. and Benjamin, M. M. (1990) Surface and bulk characteristics of binary oxide suspensions. Environ. Sci. Technol. 24, 692-698.

Anderson, M. A., Tejedor-Tejedor, M. I., and Stanforth, R. R. (1985) Influence of aggregation on the uptake kinetics of phosphate on goethite. Environ. Sci. Technol. 19, 632-637.

Aomine, S. and Otsuka, H. (1968) Surface of soil allophanic clays. Trans. 9th Inter. Congr. Soil Sci. 1, 731-737.

Ashida, M., Sasaki, M., Hachiya, K. and Yasunaga, T. (1980) Kinetics of adsorption-desorption of OH$^-$ at TiO$_2$-H$_2$O interface by means of pressure-jump technique. J. Colloid Interface Sci. 74, 572-574.

Astumian, R. D., Sasaki, M., Yasunaga, T. and Schelly, Z. A. (1981) Proton adsorption-desorption kinetics on iron oxide in aqueous suspensions, using the pressure-jump method. J. Phys. Chem. 85, 3832-3835.

Atkinson, R. J., Posner, A.M., and Quirk, J. P. (1972) Kinetics of heterogeneous isotopic exchange of phosphate at the α-FeOOH aqueous solution interface. J. Inorg. Nucl. Chem. 34, 2201-2211.

Balistrieri, L. S., Brewer, P. G., and Murray, J. W. (1981) Scavenging residence times of trace metals and surface chemistry of sinking particles in the deep ocean. Deep-Sea Res. 28A, 101-121.

Balistrieri, L. S. and Chao, T. T. (1987) Selenium adsorption by goethite. Soil Sci. Soc. Am. J. 51, 1145-1151.

Balistrieri, L. S. and Murray, J. W. (1981) The surface chemistry of goethite in major ion seawater. Am. Jour. Sci. 281, 788-806.

Balistrieri, L. S. and Murray, J. W. (1982) The surface chemistry of δ-MnO$_2$ in major ion sea water. Geochim. Cosmochim. Acta 46, 1041-1052.

249

Balistrieri, L. S. and Murray, J. W. (1983) Metal-solid interactions in the marine environment: Estimating apparent equilibrium binding constants. Geochim. Cosmochim. Acta 47, 1091-1098.

Balistrieri, L. S. and Murray, J. W. (1984) Marine scavenging: Trace metal adsorption by interfacial sediment from MANOP Site H. Geochim. Cosmochim. Acta 48, 921-929.

Balistrieri, L. S. and Murray, J. W. (1986) The surface chemistry of sediments from the Panama Basin: The influence of Mn oxides on metal adsorption. Geochim. Cosmochim. Acta 50, 2235-2243.

Ball, W. P., Buehler, C., Harmon, T. C., MacKay, D. M., and Roberts, P. V. (1990) Characterization of a sandy aquifer material at the grain scale. J. Contam. Hydrol., in press.

Barrow, N. J., Bowden, J. W. Posner, A. M. and Quirk, J. P. (1980) An objective method for fitting models of ion adsorption on variable charge surfaces. Austral. J. Soil Research 18, 37-47.

Bar-Yosef, B. and Meek, D. (1987) Selenium adsorption by kaolinite and montmorillonite. Soil Sci. 144, 11-19.

Bell, L. C., Posner, A. M. and Quirk, J. P. (1973) The point of zero charge of hydroxyapatite and fluorapatite in aqueous solutions. J. Colloid Interface Sci. 42, 250-261.

Belzile, N., Lecomte, P. and Tessier, A. (1989) Testing readsorption of trace elements during partial chemical extraction of bottom sediments. Environ. Sci. Tech. 23, 1015-1020.

Belzile, N. and Tessier, A. (1990) Interactions between arsenic and iron oxyhydroxides in lacustrine sediments. Geochim. Cosmochim. Acta 54, 103-110.

Bencala, K. E. (1984) Interactions of solutes and streambed sediment: Part 2. A dynamic analysis of coupled hydrologic and chemical processes that determine solute transport. Water Resourc. Res. 20, 1804-1814.

Benjamin, M. M. (1983) Adsorption and surface precipitation of metals on amorphous iron oxyhydroxide. Environ. Sci. Technol. 17, 686-692.

Benjamin, M. M. and Bloom, N. S. (1981) Effects of strong binding adsorbates on adsorption of trace metals on amorphous iron oxyhydroxide. In: Adsorption from Aqueous Solutions, P. H. Tewari (ed.) Plenum, New York, p. 41-60.

Benjamin, M. M., Hayes, K. F. and Leckie, J. O. (1982) Removal of toxic metals from power-generation waste streams by adsorption and coprecipitation. J. Water Poll. Control Fed. 54, 1472-1481.

Benjamin, M. M. and Leckie, J. O. (1980) Adsorption of metals at oxide interfaces: Effects of the concentration of adsorbate and competing metals. In: Contaminants and Sediments, R. A. Baker (ed.), Ann Arbor Science, Ann Arbor, MI, p. 305-322.

Benjamin, M. M. and Leckie, J. O. (1981a) Multiple-site adsorption of Cd, Cu, Zn, and Pb on amorphous iron oxyhydroxide. J. Colloid Interface Sci. 79, 209-221.

Benjamin, M. M. and Leckie, J. O. (1981b) Competitive adsorption of Cd, Cu, Zn, and Pb on amorphous iron oxyhydroxide. J. Colloid Interface Sci. 83, 410-419.

Benjamin, M. M. and Leckie, J. O. (1982) Effects of complexation by Cl, SO_4, and S_2O_3 on the adsorption behavior of cadmium on oxide surfaces. Environ. Sci. Tech. 16, 162-170.

Benson, L. V. and Teague, L. S. (1982) Diagenesis of basalts from the Pasco Basin, Washington I. Distribution and composition of the secondary mineral phases. J. Sed. Petrol. 52, 595-613.

Berner, R. A. (1964) Iron sulfides formed from aqueous solution at low temperatures and atmospheric pressure. J. Geol. 72, 293-306.

Berner, R. A. (1970) Sedimentary pyrite formation. Am. J. Sci. 268, 1-23.

Berner, R. A. (1980) Early Diagenesis. Princeton Univ. Press, Princeton, NJ.

Berner, R. A. (1981) A new geochemical classification of sedimentary environments. J. Sed. Petrol. 51, 359-365.

Berube, Y. G. and de Bruyn, P. L. (1968) Adsorption at the rutile-solution interface I. Thermodynamic and experimental study. J. Colloid Interface Sci. 27, 305-318.

Boehm, P. (1971) Acidic and basic properties of hydroxylated metal oxide surfaces. Disc. Farad. Soc. 52, 264-275.

Bolland, M. D. A., Posner, A. M., and Quirk, J. P. (1976) Surface charge on kaolinite in aqueous suspension. Austral. J. Soil Res. 14, 197-216.

Bolland, M. D. A., Posner, A. M., and Quirk, J. P. (1977) Zinc adsorption by goethite in the absence and presence of phosphate. Aust. J. Soil Res. 15, 279-286.

Bolt, G. H. (1957) Determination of the charge density of silica sols. J. Phys. Chem. 61, 1166-1169.

Bolt, G. H. (editor) (1982) Soil Chemistry, Vol. B: Physico-Chemical Models., Elsevier, Amsterdam.

Bolt, G. H. and van Riemsdijdk, W. H. (1982) Ion adsorption on inorganic variable charge constituents. In: Soil Chemistry, Vol. B, G. H. Bolt (ed.), Elsevier, Amsterdam, p. 459-504.

Bolt, G. H. and van Riemsdijdk, W. H. (1987) Surface chemical processes in soil. In: Aquatic Surface Chemistry, W. Stumm (ed.), Wiley, New York, p. 127-164.

Bourg, A. C. M. and Schindler, P. W. (1978) Ternary surface complexes 1. Complex formation in the system silica-Cu(II)-ethylenediamine. Chimia 32, 166-168.

Bousse, L. and Meindl, J. D. (1986) The importance of Ψ_o/pH characteristics in the theory of the oxide/-electrolyte interface. In: Geochemical Processes at Mineral Surfaces, J. A. Davis and K. F. Hayes (eds.), ACS Symp. Ser. 323, Am. Chem. Soc., Washington, D. C., p. 79-98.

Bower, C. A. and Goertzen, J. O. (1959) Surface area of solids and clays by an equilibrium ethylene glycol method. Soil Sci. 87, 289-292.

Bowers, A. R. and Huang, C. P. (1986) Adsorption characteristics of metal-EDTA complexes onto hydrous oxides. J. Colloid Interface Sci. 110, 575-590.

Bowers, A. R. and Huang, C. P. (1987) Role of Fe(III) in metal complex adsorption by hydrous solids. Water Res. 21, 757-764.

250

Breeuwsma, A. and Lyklema, J. (1973) Physical and chemical adsorption of ions in the electrical double layer on hematite (α-Fe_2O_3). J. Colloid Interface Sci. 43, 437-448.

Brown, D. S. and Allison, J. D. (1987) MINTEQA1, Equilibrium metal speciation model: A user's manual. EPA/600/3-87/012, U. S. Environmental Protection Agency, Athens, GA.

Brown, J. R., Bancroft, G. M., Fyfe, W. S., and McLean, R. A. N. (1979) Mercury removal from water by iron sulfide minerals. Electron spectroscopy for chemical analysis (ESCA) study. Environ. Sci. Technol. 13, 1142-1144.

Brown, P. L., Haworth, A., Sharland, S. M., and Tweed, C. J. (1990) HARPHRQ: An extended version of the geochemical code PHREEQE, NIREX Safety Studies Rept. NSS-R.188, Harwell Laboratory, Oxfordshire, UK.

Bruemmer, G. W., Gerth, J. and Tiller, T. G. (1988) Reaction kinetics of the adsorption and desorption of nickel, zinc, and cadmium by goethite. I. Adsorption and diffusion of metals. J. Soil Sci. 39, 37-52.

Brunauer, S., Emmett, P. H., and Teller, E. (1938) Adsorption of gases in multimolecular layers. J. Phys. Chem. 60, 309-319.

Bryson, A. W. and te Reile, W. A. M. (1986) Factors that affect the kinetics of nucleation and growth and the purity of goethite precipitates produced from sulphate solutions. In: Iron Control in Hydrometallurgy, J. E. Dutrizac and A. J. Monhemius (eds.), John Wiley, New York, p. 377-390.

Burns, R. G. and Burns, V. M. (1979) Manganese oxides. In: Marine Minerals, R. G. Burns (ed.), Rev. Mineralogy Ser. 6, Mineral. Soc. Am. p. 1-46.

Bye, G. C. and Howard, C. R. (1971) An examination by nitrogen adsorption of the thermal decomposition of pure and silica-doped goethite. J. Applied Chem. Biotechnol. 21, 324-329.

Caletka, R., Tympl, M., and Kotas, P. (1975) Sorption properties of iron (II) sulphide prepared by the sol-gel method. J. Chromatogr. 111, 93-104.

Carter, D. L., Heilman, M. D., and Gonzalez, C. L. (1965) Ethylene glycol monoethyl ether for determining surface area of silicate minerals. Soil Sci. 100, 356-360.

Catts, J. G. and Langmuir, D. (1986) Adsorption of Cu, Pb, and Zn onto birnessite (δ-MnO_2). J. Appl. Geochem. 1, 255-264.

Cederberg, G. A., Street, R. L., and Leckie, J. O. (1985) A groundwater mass transport and equilibrium chemistry model for multicomponent systems. Water Resourc. Res. 21, 1095-1104.

Chan, D. Perram, J. W., White, L. R. and Healy, T. W. (1975) Regulation of surface potential at amphoteric surfaces during particle-particle interaction. J. Chem. Soc. Faraday Trans. I, 71, 1046-1057.

Chang, C. C. Y., Davis, J. A. and Kuwabara, J. S. (1987) A study of metal ion adsorption at low suspended solid concentrations. Estuar. Coast. Shelf Sci. 24, 419-424.

Chapman, D. L. (1913) A contribution to the theory of electrocapillarity. Philos. Mag., 6 (25), 475-481.

Charlet, L. and Sposito, G. (1987) Monovalent ion adsorption by an oxisol. Soil Sci. Soc. Am. J. 51, 1155-1160.

Charlet, L. and Sposito, G. (1989) Bivalent ion adsorption by an oxisol. Soil Sci. Soc. Am. J. 53, 691-695.

Cihacek, L. J. and Bremner, J. M. (1979) A simplified ethylene glycol monoethyl ether procedure for assessment of soil surface area. Soil Sci. Soc. Am. J. 43, 821-822.

Cherry, J. A., Gillham, R. W., and Barber, J. F. (1984) Contaminants in groundwater: Chemical processes. In: Groundwater Contamination. National Academy Press, Washington D. C., p. 46-64.

Cornejo, J. (1987) Porosity evaluation of thermally treated ferric oxide gel. J. Colloid Interface Sci. 115, 260-265.

Cornell, R. M. and Schindler, P. W. (1980) Infrared study of the adsorption of hydroxycarboxylic acids on α-FeOOH and amorphous Fe(III) hydroxide. Coll. Polym. Sci. 258, 1171-1175.

Cowan, C. E., Zachara, J. M. and Resch, C. T. (1990) Cadmium adsorption on iron oxides in the presence of alkaline earth elements. Environ. Sci. Tech., submitted for publication.

Curtis, G. P., Reinhard, M. and Roberts, P. V. (1986) Sorption of hydrophobic organic compounds by sediments. In: Geochemical Processes at Mineral Surfaces, J. A. Davis and K. F. Hayes, ACS Symp. Ser. 323, Am. Chem. Soc., Washington, D. C., p. 191-216.

Dalas, E. and Koutsoukos, P. G. (1990) Phosphate adsorption at the porous glass/water and SiO_2 interfaces. J. Colloid Interface Sci. 134, 299-304.

Davies-Colley, R. J., Nelson, P. O. and Williamson, K. J. (1984) Copper and cadmium uptake by estuarine sedimentary phases. Environ. Sci. Tech. 18, 491-499.

Davis, J. A. (1982) Adsorption of natural dissolved organic matter at the oxide/water interface. Geochim. Cosmochim. Acta 46, 2381-2393.

Davis, J. A. (1984) Complexation of trace metals by adsorbed natural organic matter. Geochim. Cosmochim. Acta 48, 679-691.

Davis, J. A., Fuller, C. C., and Cook, A. D. (1987) Mechanisms of trace metal sorption by calcite: adsorption of Cd^{2+} and subsequent solid solution formation. Geochim. Cosmochim. Acta 51, 1477-1490.

Davis, J. A. and Gloor, R. (1981) Adsorption of dissolved organics in lakewater by aluminum oxide: Effect of molecular weight. Environ. Sci. Tech. 15, 1223-1229.

Davis, J. A. and Hayes, K. F. (1986) Geochemical processes at mineral surfaces: An overview. In: Geochemical Processes at Mineral Surfaces, J. A. Davis and K. F. Hayes, ACS Symp. Ser. 323, Am. Chem. Soc., Washington, D. C., p. 2-18.

Davis, J. A. and Hem, J. D. (1989) The surface chemistry of aluminum oxides and hydroxides. In: The Environmental Chemistry of Aluminum, G. A. Sposito (ed). CRC Press, Boca Raton, FL., p. 185-219.

Davis, J. A., James, R. O., and Leckie, J. O. (1978) Surface ionization and complexation at the oxide/water interface. I. Computation of electrical double layer properties in simple electrolytes. J. Colloid Interface Sci. 63, 480-499.

Davis, J. A., Kent, D. B., and Rea, B. A. (1989) Field and laboratory studies of coupled flow and chemical reactions in the ground-water environment. In: G. E. Mallard and S. E. Ragone (eds.), Water-Resources Investigations Rept. 88-4220, U. S. Geological Survey, p. 189-196.

Davis, J. A., Kent, D. B., Rea, B. A., Maest, A. S. and Garabedian, S. P. (1990) Influence of redox environment and aqueous speciation on metal transport in groundwater: Preliminary results of tracer injection studies. In: H. Allen, D. Brown, and E. M. Perdue (eds.), Metals in Groundwater, Lewis Publishers, Chelsea, MI, in press.

Davis, J. A. and Leckie, J. O. (1978a) Surface ionization and complexation at the oxide/water interface. II. Surface properties of amorphous iron oxyhydroxide and adsorption of metal ions. J. Colloid Interface Sci. 67, 90-107.

Davis, J. A. and Leckie, J. O. (1978b) Effect of adsorbed complexing ligands on trace metal uptake by hydrous oxides. Environ. Sci. Tech. 12, 1309-1315.

Davis, J. A. and Leckie, J. O. (1979) Speciation of adsorbed ions at the oxide/aqueous interface. In: Chemical Modeling in Aqueous Systems (ed. E. A. Jenne), ACS Symp. Ser. 93, Amer. Chem. Soc., Washington, D.C., p. 299-317.

Davis, J. A. and Leckie, J. O. (1980) Surface ionization and complexation at the oxide/water interface. III. Adsorption of anions. J. Colloid Interface Sci. 74, 32-43.

de Bruyn, P. L. and Agar, G. E. (1962) Surface chemistry of flotation. In: Froth Flotation, D. W. Fuerstenau (ed.), Amer. Inst. Mining Petrol. Eng., New York.

Dempsey, B. A., Davis, J. A. and Singer, P. C. (1988) A review of solid-solution interactions and implications for the control of trace inorganic materials in water treatment. J. Am. Water Works Assoc. 80, 56-64.

Dempsey, B. A. and Singer, P. C. (1980) The effects of calcium on the adsorption of zinc by $MnO_x(s)$ and $Fe(OH)_3$(am). In: Contaminants and Sediments, Vol. 2, R. A. Baker (ed.), Ann Arbor Science, Ann Arbor, MI., p. 333-352.

Dugger, D. L., Stanton, J. H., Irby, B. N., McConnell, B. L., Cummings, W. W. and Maatman, R. W. (1964) The exchange of twenty metal ions with the weakly acidic silanol group of silica gels. J. Phys. Chem. 68, 757-760.

Duursma, E. K. and Gross, M. G. (1971) Marine sediments and radioactivity. In: Radioactivity in the Marine Environment, National Academy of Sciences, USA, Chap. 6.

Dyal, R. S. and Hendricks, S. B. (1950) Total surface of clays in polar liquids as a characteristic index. Soil Sci. 69, 421-432.

Dzombak, D. A. and Morel, F. M. M. (1986) Sorption of cadmium on hydrous ferric oxide at high sorbate/sorbent ratios: Equilibrium, kinetics, and modeling. J. Colloid Interface Sci. 112, 588-598.

Dzombak, D. A. and Morel, F. M. M. (1987) Adsorption of inorganic pollutants in aquatic systems. J. Hydraul. Eng. 113, 430-475.

Dzombak, D. A. and Morel, F. M. M. (1990) Surface Complexation Modeling: Hydrous Ferric Oxide, John Wiley, New York.

Eary, L. E. and Rai, D. (1989) Kinetics of chromate reduction by ferrous ions derived from hematite and biotite at 25°C. Am. J. Sci. 289, 180-213.

Eltantawy, I. M. and Arnold, P. W. (1973) Reappraisal of ethylene glycol mono-ethyl ether (EGME) method for surface area estimations of clays. J. Soil Sci. 24, 232-238.

Emerson, S., Jacobs, L. J., and Tebo, B. (1983) The behavior of trace metals in marine anoxic waters: Solubilities at the oxygen-hydrogen sulfide interface. In: Trace Metals in Sea Water, C. S. Wong, E. Boyle, K. W. Bruland, J. D. Burton, and E. D. Goldberg (eds.), NATO Conference Series IV: 9, 579-608, Plenum, New York.

Farley, K. J. and Morel, F. M. M. (1986) The role of coagulation in the kinetics of sedimentation. Environ. Sci. Tech. 20, 187-195.

Farrah, H. and Pickering, W. F. (1977) Influence of clay-solute interactions on aqueous heavy metal ion levels. Water, Air and Soil Poll. 8, 189-197.

Fisher, N. S., Bjerregaard, P. and Fowler, S. W. (1983) Interactions of marine plankton with transuranic elements. 1. Biokinetics of neptunium, plutonium, americium, and californium in phytoplankton. Limnol. Oceanogr. 28, 432-447.

Flynn, C. M. (1984) Hydrolysis of inorganic Fe(III) salts. Chem. Rev. 84, 31-41.

Foissy, A., M'Pandou, A. and Lamarche, J. M. (1982) Surface and diffuse layer charge at the TiO_2-electrolyte interface. Colloids Surfaces 5, 363-368.

Fokkink, L. G. J. (1987) Ion adsorption on oxides. Ph.D. thesis, Landbouw University, Wageningen, Netherlands.

Fokkink, L. G. J., de Keizer, A. and Lyklema, J. (1987) Specific ion adsorption on oxides: Surface charge adjustment and proton stoichiometry. J. Colloid Interface Sci. 118, 454-462.

Fokkink, L. G. J., de Keizer, A. and Lyklema, J. (1989) Temperature dependence of the electrical double layer on oxides: Rutile and hematite. J. Colloid Interface Sci. 127, 116-131.

Fordham, A. W. and Norrish, K. (1979) Arsenate-73 uptake by components of several acidic soils and its implications for phosphate retention. Aust. J. Soil Res. 17, 307-316.

Framson, P. E., and Leckie, J. O. (1978) Limits of coprecipitation of cadmium and ferrous sulfides. Environ. Sci. Technol. 12, 465-469.

Fripiat, J. J. (1964) Surface properties of aluminosilicates. Clays Clay Min., Proc. 12th Natl. Conf. Clays Clay Min., p. 327-358.

Fritz, S. J. and Mohr, D. W. (1984) Chemical alteration in the micro-weathering environment within a spheroidally-weathered anorthosite boulder. Geochim. Cosmochim. Acta 48, 2527-2535.

252

Fuerstenau, D. W. (1976) Flotation - A. M. Gaudin Memorial Volume, Vols. 1 and 2, Am. Inst. Mining Metal. Petrol. Eng., New York.

Fuerstenau, D. W., Manmohan, D. and Raghavan, S. (1981) The adsorption of alkaline-earth metal ions at the rutile/aqueous interface. In: Adsorption from Aqueous Solutions, P. H. Tewari (ed.), Plenum Press, New York, p. 93-117.

Fuller, C. C. and Davis, J. A. (1987) Processes and kinetics of Cd^{2+} sorption by a calcareous aquifer sand. Geochim. Cosmochim. Acta 51, 1491-1502.

Fuller, C. C. and Davis, J. A. (1989) Influence of coupling of sorption and photosynthetic processes on trace element cycles in natural waters. Nature, 340, 52-54.

Gaudin, A. M. and Charles, W. D. (1953) Adsorption of calcium and sodium ions on pyrite. Trans. AIME 196, 195-200.

Gaudin, A. M., Fuerstenau, D. W., and Turkanis, M. M. (1957) Activation and deactivation of sphalerite with Ag and CN ions. Trans. AIME 208, 65-69.

Gaudin, A. M., Fuerstenau, D. W. and Mao, G. W. (1959) Activation and deactivation studies with copper on sphalerite. Trans. AIME 214, 430-436.

Goldberg, S. (1985) Chemical modeling of anion competition on goethite using the constant capacitance model. Soil Sci. Soc. Am. J. 49, 851-856.

Goldberg, S. and Glaubig, R. A. (1986a) Boron adsorption on California soils. Soil Sci. Soc. Am. J. 50, 1173-1176.

Goldberg, S. and Glaubig, R. A. (1986b) Boron adsorption and silicon release by the clay minerals kaolinite, montmorillonite, and illite. Soil Sci. Soc. Am. J. 50, 1442-1448.

Goldberg, S. and Sposito, G. (1984a) A chemical model of phosphate adsorption by soils. I. Reference oxide minerals. Soil Sci. Soc. Am. J. 48, 772-778.

Goldberg, S. and Sposito, G. (1984b) A chemical model of phosphate adsorption by soils. II. Noncalcareous soils. Soil Sci. Soc. Am. J. 48, 779-783.

Goldhaber, M. B. and Kaplan, I. R. (1974) The sulfur cycle. In: The Sea, Vol. V., E. D. Goldberg (ed.) Wiley-Interscience, New York, N. Y., p. 569-655.

Goncalves, M. d. L. S., Sigg, L., Reutlinger, M. and Stumm, W. (1987) Metal ion binding by biological surfaces: Voltammetric assessment in the presence of bacteria. Sci. Tot. Environ. 60, 105-119.

Goujon, G. and Mutaftshiev, B. (1976) On the crystallinity and the stoichiometry of the calcite surface. J. Colloid Interface Sci. 57, 148-161.

Gouy, G. (1910) Sur la constitution de la charge electrique a la surface d'un electolyte. J. Phys. 9, 457-468.

Grahame, D. C. (1947) The electrical double layer and the theory of electrocapillarity. Chem. Rev. 41, 441-501.

Greenland, D. J. and Quirk, J. P. (1964) Determination of the total specific surface areas of soils by adsorption of cetyl pyridinium bromide. J. Soil Sci. 15, 178-191.

Gregg, S. J. and Sing, K. S. W. (1982) Adsorption, Surface Area and Porosity. Academic Press, London.

Grove, D. B. and Stollenwerk, K. G. (1987) Chemical reactions simulated by groundwater quality models. Water Res. Bull. 23, 601-615.

Gruebel, K. A., Davis, J. A., and Leckie, J. O. (1988) The feasibility of using sequential extraction techniques for arsenic and selenium in soils and sediments. Soil Sci. Soc. Am. J. 52, 390-397.

Gschwend, P. M. and Wu, S. Ch. (1985) On the constancy of sediment-water partition coefficients of hydrophobic organic pollutants. Environ. Sci. Technol. 19, 90-96.

Hachiya, K., Sasaki, M., Saruta, Y., Mikami, N. and Yasunaga, T. (1984a) Static and kinetic studies of adsorption-desorption of metal ions on a $\gamma-Al_2O_3$ surface. I. Static study of adsorption-desorption. J. Phys. Chem. 88, 23-27.

Hachiya, K., Sasaki, M., Ikeda, T. Mikami, N. and Yasunaga, T. (1984b) Static and kinetic studies of adsorption-desorption of metal ions on a $\gamma-Al_2O_3$ surface. II. Kinetic study by means of pressure-jump technique. J. Phys. Chem. 88, 27-31.

Hajek, B. F. and Dixon, J. B. (1966) Desorption of glycerol from clays as a function of glycol vapor pressure. Soil Sci. Soc. Am. Proc. 30, 30-34.

Hang, P. T. and Brindley, G. W. (1970) Methylene blue absorption by clay minerals. Determination of surface areas and cation exchange capacities. Clays Clay Minerals 18, 203-212.

Hansmann, P. D. and Anderson, M. A. (1985) Using electrophoresis in modeling sulfate, selenite, and phosphate adsorption onto goethite. Environ. Sci. Technol. 19, 544-551.

Harding, I. H. and Healy, T. W. (1985a) Electrical double layer properties of amphoteric polymer latex colloids. J. Colloid Interface Sci. 107, 382-397.

Harding, I. H. and Healy, T. W. (1985b) Adsorption of aqueous cadmium(II) on amphoteric latex colloids. II. Isoelectric point effects. J. Colloid Interface Sci. 107, 371-381.

Hayes, K. F. (1987) Equilibrium, spectroscopic, and kinetic studies of ion adsorption at the oxide/aqueous interface. Ph.D. thesis, Stanford University, Stanford, CA.

Hayes, K. F. and Leckie, J. O. (1986) Mechanisms of lead ion adsorption at the goethite-water interface. In: Geochemical Processes at Mineral Surfaces, J. A. Davis and K. F. Hayes, ACS Symp. Ser. 323, Am. Chem. Soc., Washington, D. C., p. 114-141.

Hayes, K. F. and Leckie, J. O. (1987) Modeling ionic strength effects on cation adsorption at hydrous oxide/solution interfaces. J. Colloid Interface Sci. 115, 564-572.

Hayes, K. F., Papelis, C., and Leckie, J. O. (1988) Modeling ionic strength effects on anion adsorption at hydrous oxide/solution interfaces. J. Colloid Interface Sci. 78, 717-726.

Hayes, K. F., Redden, G., Ela, W., and Leckie, J. O. (1990) Surface complexation models: An evaluation of model parameter estimation using FITEQL and oxide mineral titration data. J. Colloid Interface Sci., in press.

Hayes, K. F., Roe, A. L., Brown, G. E., Hodgson, K. O., Leckie, J. O. and Parks, G. A. (1987) In-situ X-ray absorption study of surface complexes: Selenium oxyanions on α-FeOOH. Science 238, 783-786.

Hayes, M. H. B. and Swift, R. S. (1978) The chemistry of soil organic colloids. In: The Chemistry of Soil Constituents, D. J. Greenland and M. H. B. Hayes (eds.), John Wiley & Sons, New York, p. 179-320.

Healy, T. W. and White, L. R. (1978) Ionizable surface group models of aqueous interfaces. Advan. Colloid Interface Sci. 9, 303-345.

Heilman, M. D., Carter, D. L., and Gonzalez, C. L. (1966) The ethylene glycol monoethyl ether (EGME) technique for determining soil-surface area. Soil Sci. 100, 409-413.

Helfferich, F. (1962) Ion Exchange. McGraw-Hill, New York.

Helfferich, F. (1965) Ion exchange kinetics. V. Ion exchange accompanied by reaction. J. Phys. Chem. 69, 1178-1187.

Higgo, J. J. and Rees, L. V. C. (1986) Adsorption of actinides by marine sediments: effect of the sediment/seawater ratio on the measured distribution ratio. Environ. Sci. Technol. 20, 483-490.

Hingston, F. J. (1981) A review of anion adsorption. In: Adsorption of Inorganics at Solid-Liquid Interfaces, M. A. Anderson and A. J. Rubin (eds.), Ann Arbor Science, Ann Arbor, Mich., p. 51-90.

Hingston, E. J., Atkinson, R. J., Posner, A. M., and Quirk, J. P. (1967) Specific adsorption of anions. Nature 215, 1459-1461.

Hingston, F. J., Atkinson, R. J., Posner, A. M., and Quirk, J. P. (1968) Specific adsorption of anions on goethite. 9th Int. Cong. Soil Sci. Trans. 1, 669-678.

Hingston, F. J., Posner, A. M. and Quirk, J. P. (1971) Competitive adsorption of negatively charged ligands on oxide surfaces. Disc. Farad. Soc. 52, 334-342.

Hingston, F. J., Posner, A. M. and Quirk, J. P. (1972) Anion adsorption by goethite and gibbsite. I. The role of the proton in determining adsorption envelopes. J. Soil Sci. 23, 177-192.

Hirsch, D., Nir, S., and Banin, A. (1989) Prediction of cadmium complexation in solution and adsorption to montmorillonite. Soil Sci. Soc. Am. J. 53, 716-721.

Hohl, H., Sigg, L. and Stumm, W. (1980) Characterization of surface chemical properties of oxides in natural waters. In: Particulates in Water, M. C. Kavanaugh and J. O. Leckie (eds.), ACS Adv. Chem. Ser. 189, Am. Chem. Soc., Washington, D. C., p. 1-31.

Hohl, H. and Stumm, W. (1976) Interaction of Pb^{2+} with hydrous α-Al_2O_3. J. Colloid Interface Sci. 55, 281-288.

Honeyman, B. D. (1984) Cation and anion adsorption at the oxide/solution interface in systems containing binary mixtures of adsorbents: An investigation of the concept of adsorptive additivity. Ph.D. thesis, Stanford University, Stanford, CA.

Honeyman, B. D. and Leckie, J. O. (1986) Macroscopic partitioning coefficients for metal ion adsorption: Proton stoichiometry at variable pH and adsorption density. In: Geochemical Processes at Mineral Surfaces, J. A. Davis and K. F. Hayes, ACS Symp. Ser. 323, Am. Chem. Soc., Washington, D. C., p. 162-190.

Honeyman, B. D. and Santschi, P. H. (1988) Metals in aquatic systems. Environ. Sci. Tech. 22, 862-871.

Hornsby, D. and Leja, J. (1982) Selective flotation and its surface chemical characteristics. Surf. Colloid Sci., Vol. 12, E. Matijevic (ed.), p. 217-313.

Horzempa, L. M. and Helz, G. R. (1979) Controls on the stability of sulfide sols: colloidal covellite as an example. Geochim. Cosmochim. Act. 43, 1645-1650.

Hsi, C. D. and Langmuir, D. (1985) Adsorption of uranyl onto ferric oxyhydroxides: Application of the surface complexation site-binding model. Geochim. Cosmochim. Acta 49, 1931-1941.

Huang, C. P. (1981) The surface acidity of hydrous solids. In: Adsorption of Inorganics at Solid-Liquid Interfaces, M. A. Anderson and A. J. Rubin (eds.), Ann Arbor Science, Ann Arbor, MI, p. 183-217.

Huang, C. P. and Stumm, W. (1973) Specific adsorption of cations on hydrous α-Al_2O_3. J. Colloid Interface Sci. 22, 231-259.

Hunter, K. A. (1980) Microelectrophoretic properties of natural surface-active organic matter in coastal seawater. Limnol. Oceanogr. 25, 807-823.

Hunter, K. A., Hawke, D. J. and Choo, L. K. (1988) Equilibrium adsorption of thorium by metal oxides in marine electrolytes. Geochim. Cosmochim. Acta 52, 627-636.

Hunter, R. J. (1981) Zeta Potential in Colloid Science. Academic Press, New York.

Hunter, R. J. (1987) Foundations of Colloid Science. Oxford University Press, Oxford.

Hunter, R. J. and Wright, H. J. L. (1971) The dependence of electrokinetic potential on concentration of electrolyte. J. Colloid Interface Sci. 37, 564-580.

Iler, R. K. (1979) The Chemistry of Silica. John Wiley, New York.

Inskeep, W. P. and Baham, J. (1983) Adsorption of Cd(II) and Cu(II) by montmorillonite at low surface coverage. Soil Sci. Soc. Am. J. 47, 660-665.

Iwasaki, I. and deBruyn P. L. (1958) The electrochemical double layer on silver sulfide at pH 4.7. I. In: the absence of specific adsorption. J. Phys. Chem. 62, 594-599.

Jackson, R. E. and Inch, K. J. (1989) The in-situ adsorption of ^{90}Sr in a sand aquifer at the Chalk River Nuclear Laboratories. J. Contam. Hydrol. 4, 27-50.

Jacobs, L. A., von Gunten, H. R., Keil, R. and Kuslys, M. (1988) Geochemical changes along a river-groundwater infiltration flow path: Glattfelden, Switzerland. Geochim. Cosmochim. Acta 52, 2693-2706.

254

Jafferzic-Renault, N., Pichat, P., Foissy, A. and Mercier, R. (1986) Effect of deposited Pt particles on the surface charge of TiO_2 aqueous suspensions by potentiometry, electrophoresis, and labeled ion adsorption. J. Phys. Chem. 90, 2733-2738.

James, A. M. (1979) Electrophoresis of particles in suspension. Surface and Colloid Science, vol. 11, E. Matijevic (ed.), Plenum Press, New York, p. 121-185.

James, R. O. and Healy, T. W. (1972a) Adsorption of hydrolyzable metal ions at the oxide-water interface. I. Co(II) adsorption on SiO_2 and TiO_2 as model systems. J. Colloid Interface Sci. 40, 42-52.

James, R. O. and Healy, T. W. (1972b) Adsorption of hydrolyzable metal ions at the oxide-water interface. III. Thermodynamic model of adsorption. J. Colloid Interface Sci. 40, 65-81.

James, R. O. and MacNaughton, M. G. (1977) The adsorption of aqueous heavy metals on inorganic materials. Geochim. Cosmochim. Acta 41, 1549-1555.

James, R. O. and Parks, G. A. (1975) Adsorption of zinc(II) at the cinnabar (HgS)/H_2O interface. Am. Inst. Chem. Eng. Symp. Ser. (150) 71, 157-164.

James, R. O. and Parks, G. A. (1982) Characterization of aqueous colloids by their electrical double layer and intrinsic surface chemical properties. Surface and Colloid Science, Vol. 12, E. Matijevic (ed.), p. 119-216.

James, R. O., Stiglich, P. J. and Healy, T. W. (1981) The TiO_2/aqueous electrolyte system: Applications of colloid models and model colloids. In: Adsorption from Aqueous Solution, P. H. Tewari (ed.), Plenum Press, New York, p. 19-40.

Jenne, E. A. (1977) Trace element sorption by sediments and soils -- sites and processes. In: Symposium on Molybdenum in the Environment, Vol. 2, W. Chappel and K. Peterson (eds.), Marcel-Dekker, New York, Pgs. 425-553.

Johnson, C. A. (1986) The regulation of trace element concentrations in river and estuarine waters contaminated with acid mine drainage: The adsorption of Cu and Zn on amorphous Fe oxyhydroxide. Geochim. Cosmochim. Acta 50, 2433-2438.

Johnson, B. B. (1990) Effect of pH, temperature, and concentration on the adsorption of cadmium on goethite. Environ. Sci. Tech. 24, 112-118.

Karickhoff, S. W. (1984) Organic pollutant sorption in aquatic systems. J. Hydraul. Eng. 110, 707-735.

Kent, D. B. (1983) On the surface chemical properties of synthetic and biogenic silica. Ph.D. thesis, University of California at San Diego, San Diego, CA.

Kent, D. B., Davis, J. A., Maest, A. S., and Rea, B. A. (1989) Field and laboratory studies of the transport of reactive solutes in groundwater. In: Water-Rock Interactions, WRI-6, D. L. Miles (ed.), p. 381-383.

Kent, D. B. and Kastner, M. (1985) Mg^{2+} removal in the system Mg^{2+}-amorphous SiO_2-H_2O by adsorption and Mg-hydroxysilicate precipitation. Geochim. Cosmochim. Acta 49, 1123-1136.

Kent, D. B., Tripathi, V. S., Ball, N. B., and Leckie, J. O. (1986) Surface-complexation modeling of radionuclide adsorption in sub-surface environments. Stanford Civil Engineering Tech. Rept. #294, Stanford, CA; also NUREG Rept. CR-4897, SAND 86-7175 (1988).

Kerndorf, H. and Schnitzer, M. (1980) Sorption of metals on humic acid. Geochim. Cosmochim. Acta 44, 1701-1708.

Khoe, G. H. and Sinclair, G. (1990) Chemical modeling of the neutralization process for acid uranium mill tailings. Proc., Hydrometallurgy and Aqueous Processing Symp., 1991 AIME Annual Mtg., New Orleans, LA., EPD Congress 91, in press.

King, E. J. (1965) Acid-Base Equilibria. Pergamon Press, Oxford.

Kinniburgh, D. G. (1983) The H^+/M^{2+} exchange stoichiometry of calcium and zinc adsorption by ferrihydrite, J. Soil Sci. 34, 759-768.

Kinniburgh, D. G. (1986) General purpose adsorption isotherms. Environ. Sci. Tech. 20, 895-904.

Kinniburgh, D. G., Barker, J. A. and Whitfield, M. (1983) A comparison of some simple isotherms for describing divalent cation adsorption by ferrihydrite. J. Colloid Interface Sci. 95, 370-384.

Kinniburgh, D. G. and Jackson, M. L. (1982) Concentration and pH dependence of calcium and zinc adsorption by iron hydrous oxide gel. Soil Sci. Soc. Am. J. 46, 56-61.

Kinniburgh, D. G., Jackson, M. L. and Syers, J. K. (1976) Adsorption of alkaline earth, transition, and heavy metal cations by hydrous oxide gels of iron and aluminum. Soil Sci. Soc. Am. J. 40, 796-799.

Kinniburgh, D. G., Syers, J. K. and Jackson, M. L. (1975) Specific adsorption of trace amounts of calcium and strontium by hydrous oxides of iron and aluminum. Soil Sci. Soc. Am. J. 39, 464-470.

Koopal, L. K., van Riemsdijk, W. H. and Roffey, M. G. (1987) Surface ionization and complexation models: A comparison of methods for determining model parameters. J. Colloid Interface Sci. 118, 117-136.

Koβ, V. (1988) Modeling of uranium(VI) sorption and speciation in a natural sediment-groundwater system. Radiochim. Acta 44/45, 403-406.

Krupka, K. M., Erikson, R. L., Mattigod, S. V., Schramke, J. A. and Cowan, C. E. (1988) Thermochemical data used by the FASTCHEM package. Electric Power Research Institute (EPRI) Rept. EA-5872, Palo Alto, Ca.

Kurbatov, M. H., Wood, G. B. and Kurbatov, J. D. (1951) Isothermal adsorption of cobalt from dilute solutions. J. Phys. Chem. 55, 1170-1182.

Kuwabara, J. S., Davis, J. A. and Chang, C. C. Y. (1986) Algal growth response to particle-bound ortho-phosphate and iron. Limnol. Oceanogr. 31, 503-511.

LaFlamme, B. D. and Murray, J. W. (1987) Solid/solution interaction: The effect of carbonate alkalinity on adsorbed thorium. Geochim. Cosmochim. Acta 51, 243-250.

Lahann, R. W. and Siebert, R. M. (1982) A kinetic model for distribution coefficients and application to Mg-calcites. Geochim. Cosmochim. Acta 46, 2229-2237.

Langmuir, I. (1918) The adsorption of gases on plane surfaces of glass, mica, and platinum. J. Am. Chem. Soc. 40, 1361-1403.

Leckie, J. O., Appleton, A. R., Ball, N. B., Hayes, K. F., and Honeyman, B. D. (1984) Adsorptive removal of trace elements from fly-ash pond effluents onto iron oxyhydroxide. Electric Power Research Institute (EPRI) Rept. RP-910-1, Palo Alto, CA.

Leckie, J. O., Benjamin, M. M., Hayes, K. F., Kaufmann, G. and Altmann, S. (1980) Adsorption/coprecipitation of trace elements from water with iron oxyhydroxide. EPRI Rept. CS-1513, Electric Power Research Institute, Palo Alto, CA.

Leckie, J. O., Merrill, D. T., and Chow, W. (1985) Trace element removal from power plant wastestreams by adsorption/coprecipitation with amophous iron oxyhydroxide. In: Separation of Heavy Metals and other Trace Contaminants, P. W. Peters and B. M. Kim (eds.), AIChE Symp. Ser. #243, Vol. 81, p. 28-42.

Leckie, J. O. and Tripathi, V. S. (1985) Effect of geochemical parameters on the distribution coefficient. Proc. 5th Inter. Conf. Heavy Metals in the Environ. Vol. 2, p. 369-371.

Lecloux, A. and Pirard, J. P. (1979) The importance of standard isotherms in the analysis of adsorption isotherms for determining the porous texture of solids. J. Colloid Interface Sci. 70, 265-281.

Lecloux, A. J., Bronckart, J., Noville, F., Dodet, C., Marchot, P., Pirard, J. P. (1986) Study of the texture of monodisperse silica sphere samples in the nanometer size range. Colloids and Surfaces 19, 350-374.

Levine, S. and Smith, A. L. (1971) Theory of the differential capacity of the oxide/aqueous interface. Disc. Faraday Soc. 52, 290-301.

Lewis, F. M., Voss, C. I. and Rubin, J. (1987) Solute transport with equilibrium aqueous complexation and either sorption or ion exchange: Simulation methodology and applications. J. Hydrol. 90, 81-115.

Li, Y-H., Burkhardy, L., and Teraoka, H. (1984) Desorption and coagulation of trace elements during estuarine mixing. Geochim. Cosmochim Acta 48, 1879-1884.

Li, H. C. and de Bruyn, P. L. (1966) Electrokinetic and adsorption studies on quartz. Surf. Sci. 5, 203-220.

Lion, L. W., Altmann, R. S., and Leckie, J. O. (1982) Trace-metal adsorption characteristics of estuarine particulate matter: Evaluation of contributions of Fe/Mn oxide and organic surface coatings. Environ. Sci. Technol. 16, 660-666.

Lippmann, F. (1980) Phase diagrams depicting aqueous solubility of binary mineral systems. N. Jahreb. Miner. Abh. 139, 1-25.

Loganathan, P. and Burau, R. G. (1973) Sorption of heavy metals by a hydrous manganese oxide. Geochim. Cosmochim. Acta 37, 1277-1293.

Lowson, R. T., Short, S. A., Davey, B. G. and Gray, D. J. (1986) $^{234}U/^{238}U$ and $^{230}Th/^{234}U$ activity ratios in mineral phases of a lateritic weathered zone. Geochim. Cosmochim. Acta 50, 1697-1702.

Luoma, S. N. and Davis, J. A. (1983) Requirements for modeling trace metal partitioning in oxidized estuarine sediments. Marine Chem. 12, 159-181.

Lyklema, J. (1971) The electrical double layer on oxides. Croat. Chem. Acta 43, 249-260.

Lyklema, J. (1977) Water at interfaces: A colloid-chemical approach. J. Colloid Interface Sci. 58, 242-250.

Machesky, M. L. (1990) Influence of temperature on ion adsorption by hydrous metal oxides. In: Chemical Modeling of Aqueous Systems II. J. C. Melchoir and R. L. Bassett (eds.), ACS Symp. Ser. 416, Amer. Chem. Soc., Washington, D.C., p. 282-292.

Machesky, M. L., Bischoff, B. L. and Anderson, M. A. (1989) Calorimetric investigation of anion adsorption onto geothite. Environ. Sci. Tech. 23, 580-587.

Maciel, G. E. and Sindorf, D. W. (1980) Silicon-29 nuclear magnetic resonance study of the surface of silica gel by cross polarization and magic-angle spinning. J. Am. Chem. Soc. 102, 7606-7607.

Mayer, L. M., and Schinck, L. L. (1981) Removal of hexavalent chromium from estuarine waters by model subtrates and natural sediments. Environ. Sci. Technol. 15, 1482-1484.

McBride, M. B. (1979) Chemisorption and precipitation of Mn^{2+} at $CaCO_3$ surfaces. Soil Sci. Soc. Am. J. 43, 693-698.

McCarthy, J. F. and Zachara, J. M. (1989) Subsurface transport of contaminants. Environ. Sci. Tech. 23, 496-502.

McClellan, A. L. and Harnsberger, H. F. (1967) Cross-sectional areas of molecules adsorbed on solid surfaces, J. Colloid Interface Sci. 23, 577-599.

McKenzie, R. M. (1980) The adsorption of lead and other heavy metals on oxides of manganese and iron. Aust. J. Soil Res. 18, 61-73.

Means, J. L., Crerar, D. A., and Duguid, J. O. (1978) Migration of radioactive wastes: Radionuclide mobilization by complexing agents. Science 200, 1479-1481.

Mehr, S. R., Eatough, D. J., Hanson, L. D., Lewis, E. A. and Davis, J. A. (1989) Calorimetry of heterogeneous systems: H^+ binding to TiO_2 in NaCl. Thermochim. Acta 154, 129-143.

Mehr, S. R., Eatough, D. J., Hanson, L. D., Lewis, E. A. and Davis, J. A. (1990) Calorimetric studies of Li^+, Na^+, K^+, $(CH_3)_4N^+$, and Ca^{2+} interactions with TiO_2 in water. J. Colloid Interface Sci., submitted.

Mikami, N., Sasaki, M., Hachiya, K., Astumian, R. D., Ikeda, T. and Yasunaga, T. (1983a) Kinetics of the adsorption-desorption of phosphate on the γ-Al_2O_3 surface using the pressure-jump technique. J. Phys. Chem. 87, 1454-1458.

Mikami, N., Sasaki, M., Kikuchi, T. and Yasunaga, T. (1983b) Kinetics of the adsorption-desorption of chromate on the γ-Al_2O_3 surface using the pressure-jump technique. J. Phys. Chem. 87, 5245-5248.

Moeller, P. and Sastri, C. S. (1974) Estimation of the number of surface layers of calcite involved in Ca-^{45}Ca isotopic exchange with solution. Zeit. Physik. Chem. Neue. Folg. 89, 80-87.

Morel, F. M. M. (1983) Principles of Aquatic Chemistry, Wiley, New York.

256

Morel, F. M. M. and Hudson, R. J. M. (1985) The geobiological cycle of trace elements in aquatic systems: Redfield revisited. In: Chemical Processes in Lakes, W. Stumm (ed.), Wiley, New York, p. 251-281.

Morel, F. M. M., Yeasted, J. G. and Westall, J. C. (1981) Adsorption models: A mathematical analysis in the framework of general equilibrium calculations. In: Adsorption of Inorganics at Solid-Liquid Interfaces, M. A. Anderson and A. J. Rubin (eds.), Ann Arbor Science, Ann Arbor, MI, p. 263-294.

Motta, M. M. and Miranda, C. F. (1989) Molybdate adsorption on kaolinite, montmorillonite, and illite: Constant capacitance modeling. Soil Sci. Soc. Am. J. 53, 380-385.

Mouvet, C. and Bourg, A. C. M. (1983) Speciation (including adsorbed species) of copper, lead, and zinc in the Meuse River. Water Res. 17, 641-649.

Mulla, D. J. (1986) Current methods and limitations of simulating liquid water near hydrophobic mineral surfaces. In: Geochemical Processes at Mineral Surfaces, J. A. Davis and K. F. Hayes, ACS Symp. Ser. 323, Am. Chem. Soc., Washington, D.C., p. 20-36.

Murray, R. S. and Quirk, J. P. (1990) Surface area of clays. Langmuir 6, 122-124.

Music, S., Ristic, M. and Tonkovic, M. (1986) Sorption of Cr(VI) on hydrous iron oxides. Z. Wasser Abwasser. Forsch. 19, 186-196.

Neal, R. H. and Sposito, G. (1989) Selenate adsorption on alluvial soils. Soil Sci. Soc. Am. J. 53, 70-74.

Neal, R. H., Sposito, G., Holtzclaw, K. M., and Traina, S. J. (1987a) Selenite adsorption on alluvial soils: I. Soil composition and pH effects. Soil Sci. Soc. Am. J. 51, 1161-1165.

Neal, R. H., Sposito, G., Holtzclaw, K. M., and Traina, S. J. (1987b) Selenite adsorption on alluvial soils: II. Solution composition effects. Soil Sci. Soc. Am. J. 51, 1165-1169.

Nyffeller, U. P., Li, Y-H., and Santschi, P. H. (1984) A kinetic approach to describe trace-element distributions between particles and solution in natural aquatic systems. Geochim. Cosmochim. Acta 48, 1513-1522.

O'Melia, C. R. (1987) Particle-particle interactions. In: Aquatic Surface Chemistry, W. Stumm (ed.), John Wiley, New York, p. 385-404.

Onoda, G. Y. and de Bruyn, P. L. (1966) Proton adsorption at the ferric oxide/aqueous solution interface. I. A kinetic study of adsorption. Surf. Sci. 4, 48-63.

Padmanabham, M. (1983a) Adsorption-desorption behavior of copper(II) at the goethite-solution interface. Aust. J. Soil Res. 21, 309-320.

Padmanabham, M. (1983b) Comparative study of the adsorption-desorption behavior of copper(II), zinc(II), cobalt(II), and lead(II) at the goethite-solution interface. Aust. J. Soil Res. 21, 515-525.

Papelis, C., Hayes, K. F. and Leckie, J. O. (1988) HYDRAQL: A program for the computation of chemical equilibrium composition of aqueous batch systems including surface complexation modeling of ion adsorption at the oxide/solution interface. Tech. Rept. 306, Dept. of Civil Eng. Stanford University, Stanford, CA.

Parfitt, G. D. and Rochester, C. H. (1976) Surface characterization: Chemical. In: Characterization of Powder Surfaces, G. D. Parfitt and K. S. W. Sing (eds.), Academic Press, New York, p. 57-105.

Parfitt, R. L., Farmer, V. C., and Russell, J. D. (1977) Adsorption on hydrous oxides 1. Oxalate and benzoate on goethite. J. Soil Sci. 28, 29-39.

Park, S. W. and Huang, C. P. (1987) The surface acidity of hydrous CdS(s). J. Colloid Interface Sci. 117, 431-441.

Park, S. W. and Huang, C. P. (1989) The adsorption characteristics of some heavy metal ions onto hydrous CdS(s) surface. J. Colloid Interface Sci. 128, 245-257.

Parker, J. C., Zelazny, W., Sampath, S., and Harris, W. G. (1979) Critical evaluation of the extension of the zero point charge (ZPC) theory to soil systems. Soil Sci. Soc. Am. J. 43, 668-674.

Parkhurst, D. L., Thorstenson, D. C. and Plummer, L. N. (1980) PHREEQE: A computer program for geo-chemical calculations. U. S. Geol. Surv. Water Resources Invest. PB81-167801.

Parks, G. A. (1975) Adsorption in the marine environment. In: Chemical Oceanography, 2nd ed., Vol. 1. J. P. Riley and G. Skirrow (eds.), Academic Press, San Francisco, CA, p. 241-308.

Parks, G. A. and de Bruyn, P. L. (1962) The zero point of charge of oxides. J. Phys. Chem. 66, 967-973.

Pashley, R. M. and Israelachvili, J. N. (1984a) DLVO and hydration forces between mica surfaces in Mg^{2+}, Ca^{2+}, Sr^{2+} and Ba^{2+} chloride solutions. J. Colloid Interface Sci. 97, 446-455.

Pashley, R. M. and Israelachvili, J. N. (1984b) Molecular layering of water in thin films between mica surfaces and its relation to hydration forces. J. Colloid Interface Sci. 101, 511-523.

Pashley, R. M. and Quirk, J. P. (1989) Ion exchange and interparticle forces between clay surfaces. Soil Sci. Soc. Am. J. 53, 1660-1667.

Payne, P. A. and Sing, K. S. W. (1969) Standard data for the adsorption of nitrogen at -196°C on non-porous alumina. Chem. Ind. (1969), 918-919.

Payne, T. E. and Waite, T. D. (1990) Surface complexation modeling of uranium sorption data obtained by isotopic exchange techniques. Radiochim. Acta, in press.

Pearson, R. G. (1968) Hard and soft acids and bases, HSAB: Part I. J. Chem. Education 45, 581-587.

Perdue, E. M., Reuter, J. H. and Ghosal, M. (1980) The operational nature of acidic functional group analyses and its impact on mathematical descriptions of acid-base equilibria in humic substances. Geochim. Cosmochim. Acta 44, 1841-1850.

Perona, M. J. and Leckie, J. O. (1985) Proton stoichiometry for the adsorption of cations on oxide surfaces. J. Colloid Interface Sci. 106, 64-69.

Phillips, H. O. and Kraus, K. A. (1963) Adsorption on inorganic materials. V. Reaction of cadmium sulfide with copper (II), mercury (II), and silver (I). J. Amer. Chem. Soc. 85, 487-488.

Phillips, H. O. and Kraus, K. A. (1965) Adsorption on inorganic materials. VI. Reaction of insoluble sulfides with metal ions in aqueous media. J. Chromatogr. 17, 549-557.

Pierce, M. L. and Moore, C. B. (1980) Adsorption of arsenite on amorphous iron hydroxide from dilute aqueous solution. Environ. Sci. Tech. 14, 214-216.

Pierce, M. L. and Moore, C. B. (1982) Adsorption of arsenite and arsenate on amorphous iron hydroxide. Water Res. 16, 1247-1253.

Pines, A., Gibby, M. G., and Waugh, J. S. (1973) Proton-enhanced NMR of dilute spins in solids. J. Chem. Phys. 59, 569-590.

Plummer, L. N. and Busenberg, E. (1982) The solubilities of calcite, aragonite and vaterite in CO_2-H_2O solutions between 0° and 90°C, and an evaluation of the aqueous model for the system $CaCO_3$-CO_2-H_2O. Geochim. Cosmochim. Acta 46, 1011-1040.

Posselt, H. S., Anderson, F. J. and Weber, W. J. (1968) Cation sorption on colloidal hydrous manganese dioxide. Environ. Sci. Tech. 2, 1087-1093.

Prieve, D. C. and Ruckenstein, E. (1978) The double-layer interaction between dissimilar ionizable surfaces and its effect on the rate of deposition. J. Colloid Interface Sci. 63, 317-329.

Pyman, M. A. and Posner, A. M. (1978) The surface area of amorphous mixed oxides and their relation to potentiometric titration. J. Colloid Interface Sci. 66, 85-94.

Pyman, M. A., Bowden, J. W and Posner, A. M. (1979) The movement of titration curves in the presence of specific adsorption. Aust. J. Soil Res. 17, 191-??.

Rai, D., Zachara, J. M., Eary, L. E., Girvin, D. C., Moore, D. A. and Schmidt, R. L. (1986) Geochemical behavior of chromium species. Elec. Power Res. Inst. (EPRI) Report EA-4544, Palo Alto, CA.

Rai, D., Zachara, J. M., Eary, L. E., Ainsworth, C. C., Amonette, J. E., Cowan, C. E., Szelmeczka, R. W., Schmidt, R. L., Girvin, D. C., and Smith, S. C. (1988) Chromium reactions in geologic materials. Electric Power Res. Inst. (EPRI) Rept. EA-5741, Palo Alto, CA.

Raiswell, R. and Plant, J. (1980) The incorporation of trace elements into pyrite during diagenesis of black shales, Yorkshire, England. Econ. Geol. 75, 684-699.

Ratner-Zohar, Y., Banin, A., and Chen, Y. (1983) Oven drying as a pretreatment for surface-area determinations of soils and clays. Soil Sci. Soc. Am. J. 47, 1056-1058.

Rawson, R. A. G. (1969) A rapid method for determining the surface area of aluminosilicates from the adsorption dynamics of ethylene glycol vapour. J. Soil Sci. 20, 325-335.

Reardon, E. J. (1981) K_d's - Can they be used to describe reversible ion sorption reactions in contaminant migration? Ground Water, 19, 279-286.

Relayea, J. F., Serne, R. J., and Rai, D. (1980) Methods for determining radionuclide retardation factors. Pacific Northwest Laboratories Rept. PNL-3349, Richland, WA.

Rimstidt, J. D. and Dove, P. M. (1986) Mineral/solution reaction rates in a mixed flow reactor: wollastonite hydrolysis. Geochim. Cosmochim. Acta 50, 2509-2516.

Robins, R. G., Huang, J. C. Y., Nishimura, T. and Khoe, G. H. (1988) The adsorption of arsenate ion by ferric hydroxide. Proceedings, Arsenic Metallurgy Symp., AIME Annual Mtg., Phoenix, AZ, p. 99-112.

Ryden, J. C., McLaughlin, J. R., and Syers, J. K. (1977) Mechanisms of phosphate adsorption by soils and hydrous ferric oxide gel. J. Soil Sci. 28, 72-92.

Rousseaux, J. M. and Warenkentin, B. P. (1976) Surface properties and forces holding water in allophane soils. Soil Sci. Soc. Am. J. 40, 444-451.

Sasaki, M., Moriya, M. Yasunaga, T., and Astumian, R. D. (1983) A kinetic study of ion-pair formation on the surface of α-FeOOH in aqueous suspensions using the electric field pulse technique. J. Phys. Chem. 87, 1449-1453.

Schindler, P. W. (1981) Surface complexes at oxide-water interfaces. In: Adsorption of Inorganics at Solid-Liquid Interfaces, M. A. Anderson and A. J. Rubin (eds.), Ann Arbor Science, Ann Arbor, MI, p. 1-49.

Schindler, P. W., Furst, B., Dick, R. and Wolf, P. U. (1976) Ligand properties of surface silanol groups. I. Surface complex formation with Fe^{3+}, Cu^{2+}, Cd^{2+}, and Pb^{2+}. J. Colloid Interface Sci. 55, 469-475.

Schindler, P. W. and Gamsjager, H. (1972) Acid-base reactions of the TiO_2 (anatase)-water interface and the point of zero charge of TiO_2 suspensions. Kolloid Z. u. Z. Polymere 250, 759-763.

Schindler, P. W. and Kamber, H. R. (1968) Die aciditat von silanolgruppen. Helv. Chim. Acta 51, 1781-1786.

Schindler, P. W. and Stumm, W. (1987) The surface chemistry of oxides, hydroxides, and oxide minerals. In: Aquatic Surface Chemistry, W. Stumm (ed.), John Wiley, New York, p. 83-110.

Schwertmann, U. (1988) Occurrence and formation of iron oxides in various pedoenvironments. In: Iron in Soils and Clay Minerals, J. W. Stucki, B. A. Goodman, and U. Schwertmann (eds.), NATO ASI Ser. C, vol. 217, D. Reidel, Dordrecht, Netherlands, p. 267-308.

Sears, G. W. (1956) Determination of specific surface area of colloidal silica by titration with sodium hydroxide. Anal. Chem. 28, 1981-1983.

Shiller, A. M. and Boyle, E. (1985) Dissolved Zn in rivers. Nature 317, 49-52.

Shiller, A. M. and Boyle, E. (1987) Dissolved vanadium in rivers and estuaries. Earth Planet. Sci. Lett. 86, 214-224.

Shuman, L. M. (1986) Effect of ionic strength and anions on zinc adsorption by two soils. Soil Sci. Soc. Am. J. 50, 1438-1442.

Sigg, L. (1987) Surface chemical aspects of the distribution and fate of metal ions in lakes. In: Aquatic Surface Chemistry, W. Stumm (ed.), John Wiley, New York, p. 319-350.

Sigg, L. and Stumm, W. (1981) The interaction of anions and weak acids with the hydrous goethite (α-FeOOH) surface. Colloids and Surfaces 2, 101-117.

Sigg, L., Sturm, M. and Kistler, D. (1987) Vertical transport of heavy metals by settling particles in Lake Zurich. Limnol. Oceanogr. 32, 112-130.

258

Sindorf, D. W. and Maciel, G. E. (1981) ^{29}Si CP/MAS NMR studies of methylchlorosilane reactions on silica gel. J. Am. Chem. Soc. 103, 4263-4265.

Sindorf, D. W. and Maciel, G. E. (1983) ^{29}Si NMR study of dehydrated/rehydrated silica gel using cross polarization and magic-angle spinning. J. Am. Chem. Soc. 105, 1487-1493.

Smit, W. (1981) Note on tritium exchange studies on microporous silica. J. Colloid Interface Sci. 84, 272-273.

Smit, W. and Holten, C. L. M. (1980) Zeta potential and radiotracer adsorption measurements on EFG γ-Al$_2$O$_3$ single crystals in NaBr solutions. J. Colloid Interface Sci. 78, 1-14.

Smit, W., Holten, C. L. M., Stein, H. N., de Goeij, J. J. M., and Theelen, H. M. J. (1978a) A radiotracer determination of the adsorption of sodium ion in the compact part of the double layer of vitreous silica. J. Colloid Interface Sci. 63, 120-128.

Smit, W., Holten, C. L. M., Stein, H. N., de Goeij, J. J. M., and Theelen, H. M. J. (1978b) A radiotracer determination of the sorption of sodium ions by microporous silica films. J. Colloid Interface Sci. 67, 397-407.

Somasundaran, P. and Agar, G. E. (1967) The zero point of charge of calcite. J. Colloid Interface Sci. 24, 433-440.

Sorg, T. J., Csandy, M. and Logsdon, G. S. (1978) Treatment technology to meet the interim primary drinking water regulations for inorganics: Part 3. J. Am. Water Works Assoc. 70, 680-691.

Sorg, T. J. (1979) Treatment technology to meet the interim primary drinking water regulations for inorganics: Part 4. J. Am. Water Works Assoc. 71, 454-466.

Sparks, D. L. (1985) Kinetics of ionic reactions in clay minerals and soils. Adv. Agronomy, Vol. 38, N. C. Brady (ed.), Academic, New York, p. 231-266.

Sposito, G. (1982) On the use of the Langmuir equation in the interpretation of "adsorption" phenomena: II. The "two-surface" Langmuir equation. Soil Sci. Soc. Am. J. 46, 1147-1152.

Sposito, G. (1983) On the surface complexation model of the oxide-aqueous solution interface. J. Colloid Interface Sci. 91, 329-340.

Sposito, G. (1984) The Surface Chemistry of Soils. Oxford University Press, New York.

Sposito, G. (1986) On distinguishing adsorption from surface precipitation. In: Geochemical Processes at Mineral Surfaces, J. A. Davis and K. F. Hayes, ACS Symp. Ser. 323, Am. Chem. Soc., Washington, D.C., p. 217-228.

Sprycha, R. (1983) Attempt to estimate σ_β charge components on oxides from anion and cation adsorption measurements. J. Colloid Interface Sci. 96, 551-554.

Sprycha, R. and Szczypa, J. (1984) Estimation of surface ionization constants from electrokinetic data. J. Colloid Interface Sci. 102, 288-291.

Sprycha, R. (1989a) Electrical double layer at alumina/electrolyte interface. I. Surface charge and zeta potential. J. Colloid Interface Sci. 127, 1-11.

Sprycha, R. (1989b) Electrical double layer at alumina/electrolyte interface. II. Adsorption of supporting electrolyte ions. J. Colloid Interface Sci. 127, 12-25.

Sprycha, R. Kosmulski, M. and Szczypa, J. (1989) Ionic components of charge on oxides. J. Colloid Interface Sci. 128, 88-95.

Stanton, J. and Maatman, R. W. (1963) The reaction between aqueous uranyl ion and the surface of silica gel. J. Colloid Sci. 18, 132-146.

Steinhardt, J. and Reynolds, J. A. (1969) Multiple Equilibria in Proteins. Academic Press, London.

Stern, O. (1924) Zur theory der electrolytischen doppelschicht. Z. Electrochem. 30, 508-516.

Stollenwerk, K. G. and Kipp, K. L. (1990) Simulation of molybdate transport with different rate-controlled mechanisms. In: Chemical Modeling in Aqueous Systems II, D.C. Melchior and R. L. Bassett (eds.), ACS Symp. Ser. #416, Am. Chem. Soc., Washington, D.C., p. 243-257.

Stone, A. T. (1986) Adsorption of organic reductants and subsequent electron transfer on metal oxide surfaces. In: Geochemical Processes at Mineral Surfaces, J. A. Davis and K. F. Hayes, ACS Symp. Ser. 323, Am. Chem. Soc., Washington, D.C., p. 446-461.

Stumm, W. (1987) Aquatic Surface Chemistry, John Wiley, New York.

Stumm, W. (1977) Chemical interaction in particle separation. Environ. Sci. Tech. 11, 1066-1070.

Stumm, W., Hohl, H., and Dalang, F. (1976) Interaction of metal ions with hydrous oxides. Croat. Chem. Acta 48, 491-504.

Stumm, W. Huang, C. P. and Jenkins, S. R. (1970) Specific chemical interactions affecting the stability of dispersed systems. Croat. Chem. Acta 42, 223-244.

Stumm, W. Kummert, R. and Sigg, L. (1980) A ligand exchange model for the adsorption of inorganic and organic ligands at hydrous oxide interfaces. Croat. Chem. Acta 53, 291-312.

Stumm, W. and Morgan, J. J. (1981) Aquatic Chemistry, 2nd Edition. John Wiley. New York.

Swallow, K. C., Hume, D. N., and Morel, F. M. M. (1980) Sorption of copper and lead by hydrous ferric oxide. Environ. Sci. Tech. 14, 1326-1331.

Tanford, C. (1961) Physical Chemistry of Macromolecules. John Wiley, New York, p. 526-586.

Tessier, A. Rapin, F. and Carignan, R. (1985) Trace metals in oxic lake sediments: Possible adsorption onto iron oxyhydroxides. Geochim. Cosmochim. Acta 49, 183-194.

Tessier, A., Carignan, R., Dubreuil, B. and Rapin, F. (1989) Partitioning of zinc between the water column and the oxic sediments in lakes. Geochim. Cosmochim. Acta 53, 1511-1522.

Tewari, P. H. and McLean, A. W. (1972) Temperature dependence of point of zero charge of alumina and magnetite. J. Colloid Interface Sci. 40, 267-272.

Tipping, E. (1981) The adsorption of aquatic humic substances by iron oxides. Geochim. Cosmochim. Acta 45, 191-199.

259

Tripathi, V. S. (1983) Uranium transport modeling: goechemical data and sub-models. Ph.D. Dissertation, Stanford University, Stanford, CA.

Turner, S. and Buseck, P. R. (1981) Todorokites: A new family of naturally occurring manganese oxides. Science 212, 1024-1027.

Turner, S., Siegel, M. D., and Buseck, P. R. (1982) Structural features of todorokite intergrowths in mangnese nodules. Nature 296, 841-842.

van den Berg, C. M. G. (1982) Determination of copper complexation with natural organic ligands in seawater by equilibration with MnO$_2$. II. Experimental procedures and application to seawater. Marine Chem. 11, 323-342.

van den Hul, H. J. and Lyklema, J. (1968) Determination of specific surface areas of dispersed materials. Comparison of the negative adsorption method with some other methods. J. Am. Chem. Soc. 90, 3010-3015.

van Olphen, H. (1970) Determination of surface areas of clays - evaluation of methods. In: Surface Area Determination, D. H. Everett and R. H. Ottewill (eds.), Butterworths, London, p. 255-268.

van Olphen, H. (1977) An Introduction to Clay Colloid Chemistry, 2nd Ed. John Wiley, New York.

van Riemsdijk, W. H., Bolt, G. H., Koopal, L. K. and Blaakmeer, J. (1986) Electrolyte adsorption on heterogeneous surfaces: Adsorption models. J. Colloid Interface Sci. 109, 219-228.

van Riemsdijk, W. H., DeWit, J. C. M., Koopal, L. K. and Bolt, G. H. (1987) Metal ion adsorption on heterogeneous surfaces: Adsorption models. J. Colloid Interface Sci. 116, 511-522.

van Riemsdijk, W. H. and Lyklema, J. (1980) The reaction of phosphate with aluminum hydroxide in relation with phosphate bonding in soils. Colloids Surfaces 1, 33-44.

Veith, J. A. and Sposito, G. (1977a) On the use of the Langmuir equation in the interpretation of "adsorption" phenomena. Soil Sci. Soc. Am. J. 41, 697-702.

Veith, J. A. and Sposito, G. (1977b) Reactions of aluminosilicates, aluminum hydrous oxides, and aluminum oxide with o-phosphate: The formation of X-ray amorphous analogs of variscite and montebarasite. Soil Sci. Soc. Am. J. 41, 870-876.

Velbel, M. A. (1986) Influence of surface area, surface characteristics, and solution composition on feldspar weathering rates. In: Geochemical Processes at Mineral Surfaces, J. A. Davis and K. F. Hayes, ACS Symp. Ser. 323, Am. Chem. Soc., Washington, D.C., p. 615-634.

Vuceta, J. (1976) Adsorption of Pb(II) and Cu(II) on α-quartz from aqueous solutions: Influence of pH, ionic strength, and complexing ligands. Ph.D. thesis, California Institute of Technology, Pasadena, CA.

Vuceta, J. and Morgan, J. J. (1978) Chemical modeling of trace metals in fresh waters: Role of complexation and adsorption. Environ. Sci. Technol. 12, 1302-1308.

Waite, T. D. and Payne, T. E. (1990) Uranium transport in the sub-surface environment: Koongarra - A case study. In: Metals in Groundwater, H. Allen, D. Brown and E. M. Perdue (eds.), Lewis Publishers, Chelsea, MI, in press.

Wersin, P., Charlet, L., Karthein, R. and Stumm, W. (1989) From adsorption to precipitation: Sorption of Mn^{2+} on FeCO$_3$(s). Geochim. Cosmochim. Acta 53, 2787-2796.

Westall, J. C. (1987) Adsorption mechanisms in aquatic surface chemistry. In: Aquatic Surface Chemistry, W. Stumm (ed.), Wiley, New York, p. 3-32.

Westall, J. C. (1986) Chemical and electrostatic models for reactions at the oxide-solution interface. In: Geochemical Processes at Mineral Surfaces, J. A. Davis and K. F. Hayes, ACS Symp. Ser. 323, Am. Chem. Soc., Washington, D.C., p. 54-78.

Westall, J. C. and Hohl, H. (1980) A comparison of electrostatic models for the oxide/solution interface. Advan. Colloid Interface Sci. 12, 265-294.

Westall, J. C., Zachary, J. L. and Morel, F. M. M. (1976) MINEQL: A computer program for the calculation of chemical equilibrium composition of aqueous systems. Tech. Note 18, Dept. of Civil Eng., Mass. Inst. Tech., Cambridge, MA.

White, A. F. and Hochella, M. F. (1989) Electron transfer mechansims associated with the surface dissolution and oxidation of magnetite and ilmenite. In: D. Miles (ed.), Water-Rock Interaction WRI-6. Balkema, Rotterdam, The Netherlands, pgs. 765-768.

White, A. F. and Peterson, M. L. (1990) Role of reactive-surface area characterization in geochemical kinetic models. In: D. C. Melchior and R. L. Bassett (eds.), ACS Symp. Ser. #416, Am. Chem. Soc., Washington, D.C., p. 461-475.

Whitfield, M. and Turner, D. R. (1987) The role of particles in regulating the composition of seawater. In: Aquatic Surface Chemistry, W. Stumm (ed.), John Wiley, New York, p. 457-494.

Wiese, G. R. and Healy, T. W. (1975a) Coagulation and electrokinetic behavior of TiO$_2$ and Al$_2$O$_3$ colloidal dispersions. J. Colloid Interface Sci. 51, 427-433.

Wiese, G. R. and Healy, T. W. (1975b) Solubility effects in Al$_2$O$_3$ and TiO$_2$ colloidal dispersions. J. Colloid Interface Sci. 52, 452-458.

Williams, R. and Labib, M. E. (1985) Zinc sulfide surface chemistry: An electrokinetic study. J. Colloid Interface Sci. 106, 251-254.

Wolery, T. J. (1983) EQ3NR, A computer program for geochemical aqueous speciation-solubility calculations. User's guide and documentation. Rept. UCRL-53414, Lawrence Livermore Laboratory, Livermore, CA.

Wolf, P. U., Schindler, P. W., Berthou, H., and Jorgensen, C. K. (1977) X-ray induced photoelectron spectrometric evidence for heavy metal adsorption by molybdenum disulfide from aqueous solution. Chimia 31, 223-225.

260

Yasunaga, T. and Ikeda, T. (1986) Adsorption-desorption kinetics at the metal-oxide/solution interface studied by relaxation methods. In: Geochemical Processes at Mineral Surfaces, J. A. Davis and K. F. Hayes, ACS Symp. Ser. 323, Am. Chem. Soc., Washington, D.C., p. 230-253.

Yates, D. E. (1975) The structure of the oxide/aqueous electrolyte interface. Ph.D. dissertation, Univ. Melbourne, Melbourne, Australia.

Yates, D. E. and Healy, T. W. (1980) Titanium dioxide-electrolyte interface 2. Surface charge (titration) studies. J. Chem. Soc. Faraday Trans. I, 76, 9-18.

Yates, D. E. and Healy, T. W. (1976) The structure of the silica/electrolyte interface. J. Colloid Interface Sci. 55, 9-19.

Yates, D. E. and Healy, T. W. (1975) Mechanism of anion adsorption at the ferric and chromic oxide/water interfaces. J. Colloid Interface Sci. 52, 222-228.

Yates, D. E., James, R. O., and Healy, T. W. (1980) Titanium dioxide-electrolyte interface 1. Gas adsorption and tritium exchange studies. J. Chem. Soc. Faraday Trans. I, 76, 1-8.

Yates, D. E., Levine, S. and Healy, T. W. (1974) Site-binding model of the electrical double layer at the oxide/water interface. J. Chem. Soc. Faraday Trans. I, 70, 1807-1818.

Zachara, J. M., Girvin, D.C., Schmidt, R. L. and Resch, C. T. (1987) Chromate adsorption on amorphous iron oxyhydroxide in the presence of major groundwater ions. Environ. Sci. Tech. 21, 589-594.

Zachara, J. M., Kittrick, J. A., and Harsh, J. B. (1988) The mechanism of Zn^{2+} adsorption on calcite. Geochim. Cosmochim. Acta 52, 2281-2291.

Zachara, J. M., Kittrick, J. A., and Harsh, J. B. (1989a) Solubility and surface spectroscopy of zinc precipitates on calcite. Geochim. Cosmochim. Acta 53, 9-19.

Zachara, J. M., Ainsworth, C. C., Cowan, C. E. and Resch, C. T. (1989b) Adsorption of chromate by subsurface soil horizons. Soil Sci. Soc. Am. J. 53, 418-428.

Zachara, J. M., Ainsworth, C. C., Cowan, C. E. and Schmidt, R. L. (1990) Sorption of aminonaphthalene and quinoline on amorphous silica. Environ. Sci. Tech. 24, 118-126.

Zasoski, R. J. and Burau, R. G. (1988) Sorption and sorptive interaction of cadmium and zinc on hydrous manganese oxide. Soil Sci. Soc. Am. J. 52, 81-87.

Zeltner, W. A, Yost, E. C., Machesky, M. L., Tejedor-Tejedor, M. I. and Anderson, M. A. (1986) Characterization of anion adsorption on goethite using titration calorimetry and CIR-FTIR. In: Geochemical Processes at Mineral Surfaces, J. A. Davis and K. F. Hayes, ACS Symp. Ser. 323, Am. Chem. Soc., Washington, D.C., p. 142-161.

Ziper, C., Komarneni, S., and Baker, D. E. (1988) Specific cadmium sorption in relation to the crystal chemistry of clay minerals. Soil Sci. Soc. Am. J. 52, 49-53.

MOLECULAR MODELS OF ION ADSORPTION
ON MINERAL SURFACES

INTRODUCTION

From the perspective of molecular theory, adsorption phenomena represent accumulations of matter at interfaces between two phases (usually a liquid and a solid), such that a three-dimensional molecular arrangement does not develop (Sposito, 1986). This concept thus precludes surface precipitates, even if their structures are highly constrained by the surface on which they form, and it permits only the influence of nearest-neighbor adsorbate layers on a given layer in multilayer adsorption processes. Two-dimensional solid solutions are not excluded, nor are disordered molecular structures extending broadly in two dimensions on a regular or even uniform surface. [See Hochella (this volume) and Lasaga (this volume) for a discussion of mineral surface structures.] Adsorption can be either a positive or a negative accumulation of matter, corresponding to the attraction or the repulsion, respectively, of an adsorptive by an adsorbent. Repulsive interactions lead to the exclusion of an adsorptive from an interfacial region (e.g., the exclusion of mobile ions from the neighborhood of a surface of like charge).

The molecular modeling of adsorption processes, properly a branch of statistical mechanics, is a development in physical chemistry during the present century (Fowler and Guggenheim, 1949; Hill, 1960). Therefore, its foundation and basic conceptualizations are not dissimilar in spirit from those of, say, the molecular theory of liquids. An excellent discussion of the precepts and applications of statistical mechanics has been given by Hill (1960) in a classic textbook. Sposito (1983, 1984) has described the use of statistical mechanics for describing adsorption by natural colloids, such as clay minerals and metal oxides. In this latter context, the concern of the present chapter, molecular approaches can be classified on the basis of their *chemical constraints* and their *molecular hypotheses*. Chemical constraints appear in the form of equations that develop from the laws of mass and charge balance and of$chemical equilibrium or kinetics. These equations express the general requirements that a molecular model must meet in order to reflect fundamental chemical behavior in an adsorption phenomenon as perceived through conventional *macroscopic* experiments. They do not come from the model itself, but instead are imposed on it from the data of the adsorption process. Examples include the conservation of surface charge and the law of mass action as represented through an equilibrium constant for a chemical reaction (Sposito, 1984).

Molecular hypotheses, on the other hand, are constraints based (ideally) on the results of molecular-level theory and experiment, notably quantum chemistry and surface spectroscopy. Often, however, they are in fact derived from the (over-) interpretation of the macroscopic experiments that provide the basis for the chemical constraints, but this is an approach fraught with pitfalls (Johnston and Sposito, 1987). Molecular hypotheses involve both qualitative and quantitative statements about the surface species and mechanisms that contribute to an adsorption process. But traditional adsorption experiments are notoriously insensitive to surface mechanisms and speciation, with the unfortunate result that molecular models having intrinsically different mechanistic or speciation concepts each can do justice to the quantitative description of the same adsorption phenomenon. Molecular-level data, particularly those from non-invasive, *in situ* spectroscopic methods, are the only direct source of information about surface speciation (Johnston and Sposito, 1986; Davis and Hayes, 1986; Sposito, 1989). Some of these methods are discussed by Brown (this volume) and Bancroft and Hyland (this volume).

The purpose of this chapter is to introduce the molecular modeling of adsorption via consideration of two classic statistical mechanical models: the Modified Gouy-Chapman theory of mobile species and the Bragg-Williams model of site-bound species. The discussion is intended both to inform the modern interpretation of adsorption data and to provide a basis for understanding popular chemical models of ion adsorption by natural colloids. The plan of the chapter is to describe the models in careful terms, without excessive mathematical detail, then to discuss briefly their properties and typical applications as encountered in the literature of aquatic surface chemistry. Davis and Kent (this volume) give additional discussion of the applications and experimental context for adsorption models.

DIFFUSE DOUBLE LAYER MODELS

Modified Gouy-Chapman theory

The statistical mechanics of diffuse ion swarms near charged surfaces has been reviewed comprehensively by Carnie and Torrie (1984). They point out that a full theoretical treatment of an aqueous solution/charged surface interface at the molecular level is not yet possible, but that progress has been made with model approaches in which the solvent (water) molecular coordinates are averaged to leave consideration only to ionic coordinates. In one of the simplest of these model approaches, termed *Modified Gouy-Chapman (MGC) theory*, the ions are assumed to be hard spheres of uniform diameter into which are embedded point charges. For a strong electrolyte solution, this picture constitutes the simple Primitive Model. Ion polarizability and solvation are neglected, except insofar as they contribute to ion size. Modified Gouy-Chapman theory then extends this simplification to the charged surface, which is assumed to be perfectly impenetrable and polarizable. Generalization to permit differing cation and anion sizes for the electrolyte is direct (Valleau and Torrie, 1982), but will not be discussed in the present chapter.

The basic physical postulates of MGC theory for a planar interface can be summarized as follows:

(a) The charged surface is a uniform, infinite plane characterized by a charge density σ_0 (expressed conventionally in coulombs per square meter).

(b) The charged species in aqueous solution are hard-sphere ions, of diameter d_0, which are dissociated completely from the planar surface. These ions interact among themselves and with the surface through the coulomb force.

(c) The water in the system is a uniform continuum liquid, characterized by the relative permittivity (dielectric constant), D. Nonlinear dielectric behavior is neglected.

(d) The mean electrostatic potential, $\psi(x)$, is assumed to be proportional to $W_i(x)$, the average energy required to bring an ion i from a point at infinity to a point at x measured positively outward from the charged surface. Since $W_i(x)$ actually includes the effects of both noncoulombic interactions (e.g., short-range repulsive forces that determine ion size) and fluctuations of the true electrostatic potential about its mean value, $\psi(x)$, it follows that these effects and, therefore, ion-ion correlations, are neglected in the assumption:

$$W_i(x) = Z_i F \psi(x) \quad , \tag{1}$$

where Z_i is the valence of ion i, including the sign of the valence, and $F = 96,485$ C $\bullet$ mol$_c^{-1}$ is the Faraday constant.

The mathematical content of MGC theory stems from the molecular hypothesis that connects Equation (1) with a statistical mechanics expression involving the surface charge density. This expression is the Poisson-Boltzman equation (Carnie and Torrie, 1984), which, for a planar charged surface can be written in the form:

$$\frac{du}{dx} = \begin{cases} -\beta\sigma_o/2F & 0 < x < d_o/2 \\ \left\{ \beta\Sigma_i c_{\infty i} \left[\exp\left(-Z_i u(x)\right) - \exp\left(-Z_i u_\infty\right) \right] \right\}^{1/2} & x > d_o/2 \end{cases} \tag{2}$$

where $u = \psi(x)/\psi_D$; $\psi_D = RT/F = 25.69$ mV at 298.15 K; $\beta = 2F^2/\varepsilon_o DRT = 1.084 \times 10^{16}$ m $\bullet$ mol$_c^{-1}$ at 298.15 K; ε_o is the permittivity of vacuum; $c_{\infty i}$ is the "bulk" concentration of an ion in the electrolyte "far" from the charged surface; and u_∞ is the value of u where du/dx vanishes. If the colloidal platelets modeled by Equation (2) are separated widely in suspension, $u_\infty \equiv 0$; otherwise u_∞ is the value of u at a point midway between opposing planar surfaces of nearest-neighbor platelets. The sum in Equation (2) is over each species of ion in solution.

The nonlinear ordinary differential equation above is subject to a constraint, viz., that the electric field intensity (strictly, the electric displacement) and, therefore, du/dx, be continuous at the distance of closest approach of the hard-sphere ions to the charged surface (Carnie and Torrie, 1984). This constraint may be expressed in the form:

$$\sigma_o = -sgn(u) \left\{ (4F^2/\beta) \Sigma_i c_{\infty i} \left[\exp\left(-Z_i u_{d_o/2}\right) - \exp\left(-Z_i u_\infty\right) \right] \right\}^{1/2}, \tag{3}$$

where

$$sgn(u) = \begin{cases} +1 & u > 0 \\ -1 & u < 0 \end{cases} \tag{4}$$

and $u_{d_o/2}$ is $u(x)$ at $x = d_o/2$. Equation (4) is simply the result of equating both parts of the right side of Equation (2) at $x = d_o/2$. If σ_0 is independent of electrolyte effects, the values of $u_{d_o/2}$ and u_∞ in general will depend on the bulk ion concentration $c_{\infty i}$. The value of $u(x)$ at the charged surface, u_0, is related to $u_{d_o/2}$ by solving Equation (2) for $0 < x < d_o/2$ and imposing the boundary conditions, $u(0) = u_0$ and $u(d_o/2) = u_{d_o/2}$ (Carnie and Torrie, 1984):

$$u_{d_o/2} = u_o - (\beta\sigma_o/4F)\, d_o \ . \tag{5}$$

Equations (3) and (5) imply that u_0 also will depend on $c_{\infty i}$, in general.

The Poisson-Boltzman equation can be solved analytically in three important special cases: symmetric (i.e., Z:Z), 2:1, and 1:2 electrolyte solutions (Grahame, 1953). The resulting mean electrostatic potential functions are summarized in Table 1 for the case $u_\infty = 0$ (Sposito, 1984). When $Z = 1$ (1:1 electrolyte solution), the complete solution of Equation (2) for this case is:

$$u(x) = \begin{cases} u_o - (\beta\sigma_o/2F)\, x & 0 < x < d_o/2 \\ 4\, \tanh^{-1}\left[a\, \exp\left(-\kappa\,(x - d_o/2)\right) \right] & d_o/2 < x \end{cases} \tag{6}$$

where $a = \tanh(u_{d_o/2})$ and $\kappa = (\beta c_\infty)^{1/2}$, c_∞ being the bulk 1:1 electrolyte concentration.

Accuracy of MGC theory

Carnie and Torrie (1984) have compared a variety of predictions based on Equation (6) with the results of accurate Monte Carlo calculations for a system of hard-sphere ions immersed in a continuum dielectric fluid that contacts a planar charged surface. Some of their findings are reproduced in Figure 1, which shows the local concentration of cations (c_+) or anions (c_-) in a 1:1 electrolyte,

$$c_\pm(x)/c_\infty = \exp\left[\mp u(x) \right] \ , \tag{7}$$

plotted against distance from a charged surface. The value of c_∞ was set at 100 mol • m^{-3} and σ_0 was given two values that bracket the normal range of surface charge density for the ubiquitous clay minerals known as smectites, viz., -0.087 to -0.208 C • m^{-2} (Sposito, 1984). These colloidal platelet minerals are especially suitable for characterization by Equation (2). The graphs in Figure 1 demonstrate that, for $c_\infty = 100$ mol • m^{-3}, MGC theory is indeed an accurate model of a 1:1 electrolyte solution comprising hard-sphere ions and a continuum solvent near a uniformly-charged plane. Above this bulk concentration, the model fails because it cannot describe the effects of ion-ion correlations [neglected when Eqn. (1) is imposed] on the local ion concentration (Carnie and Torrie, 1984). Even at lower bulk concentrations, MGC theory overestimates $|u(x)|$ and $u_{d_o/2}$ for $|\sigma_0| > 0.1$ C • m^{-2} (Carnie and Torrie, 1984). Thus it may be concluded that Equation (2) is reasonably, but not perfectly, accurate for the chemical system it is intended to model if $c_\infty < 100$ mol • m^{-3}. On the other hand, MGC theory is completely inaccurate for electrolyte solutions containing *bivalent* ions at concentrations even as low as 5 mol • m^{-3} (Carnie and Torrie, 1984; Sposito, 1984).

Table 1. Analytical solutions of the Poisson-Boltzmann equation for single-electrolyte solutions.

Electrolyte Solution	Mean electrostatic potential[a]
Symmetric	$u(x) = (4/Z)\tanh^{-1}(ae^{-Z\kappa x'})$ with $a = \tanh(Zy_o/4)$
$2:1\,(\sigma_o < 0)$	$u(x) = \ln\left[\frac{3}{2}\tanh^2\left(\frac{\sqrt{3}}{2}\kappa x' + b\right) - \frac{1}{2}\right]$
	with $b = \tanh^{-1}\left\{[1 + 2\exp(y_o)]^{1/2}/\sqrt{3}\right\}$
$2:1\,(\sigma_o > 0)$	$u(x) = \ln\left\{1 + \frac{6b'\exp(\sqrt{3}\kappa x')}{[b'\exp(\sqrt{3}\kappa x') - 1]^2}\right\}$
	with $b' = \dfrac{(1 + 2e^{y_o})^{1/2} + \sqrt{3}}{(1 + 2e^{y_o})^{1/2} - \sqrt{3}}$
$1:2\,(\sigma_o < 0)$	$u(x) = -\ln\left\{1 + \frac{6c\exp(\sqrt{3}\kappa' x')}{[c\exp(\sqrt{3}\kappa' x') - 1]^2}\right\}$
	with $c = \dfrac{(1 + 2e^{-y_o})^{1/2} + \sqrt{3}}{(1 + 2e^{-y_o})^{1/2} - \sqrt{3}}$
$1:2\,(\sigma_o > 0)$	$u(x) = -\ln\left[\frac{3}{2}\tanh^2\left(\frac{\sqrt{3}}{2}\kappa' x' + c'\right) - \frac{1}{2}\right]$
	with $c' = \tanh^{-1}\left\{[1 + 2\exp(-y_o)]^{1/2}/\sqrt{3}\right\}$

[a] $\kappa = (\beta c_{\infty+})^{1/2}$ with $c_{\infty+}$ = cation concentration in bulk solution

$\kappa' = (\beta c_{\infty-})^{1/2}$ with $c_{\infty-}$ = anion concentration in bulk solution

$x' = x - d_o/2$, $x > 0$

$y_o = \psi(d_o/2)/\psi_D$

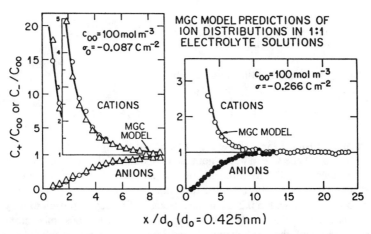

$$x/d_o\ (d_o = 0.425\,\text{nm})$$

Figure 1. Predictions of the local concentrations of cations or anions near a planar charged surface by Modified Gouy-Chapman theory (solid curves) compared with Monte Carlo calculations (data points) of the ion concentrations.

In this case, ion-ion correlations cannot be neglected except at extremely low concentrations that obviate any practical application of the model to natural colloids.

Counterion condensation

Counterion condensation is a fundamental property of the electrostatic interaction between a charged colloid or macromolecule and the ions it attracts in the diffuse-ion swarm (Manning, 1979). Let $F(x)$ be the fraction of colloid surface charge neutralized by diffuse-swarm counterions occupying the region between the surface and the point at x. In general, $F(x)$ will depend on both the surface charge density and the bulk electrolyte concentration. Counterion condensation exists if $F(x)$ does not vanish as the bulk electrolyte concentration goes to zero in a limiting sense (Zimm and Le Bret, 1983). This means that there remains a finite fraction of excess counterions in the vicinity of the charged surface of a colloid even when the solution in which it is suspended becomes infinitely dilute. The electrostatic attraction is thus strong enough to prevent the counterions from moving away into the bulk solution even when the latter decreases to zero in concentration.

Zimm and Le Bret (1983) have shown that counterion condensation *never* occurs for spherical colloids; that it occurs at surface charge densities (in $mol_c \bullet m^{-1}$) larger in magnitude than $8\pi/Z\beta$ for cylindrical colloids, where Z is the valence of the counterion; and that it occurs at *any* non-zero surface charge density for planar surfaces. In the context of the MGC theory applied to 1:1 electrolytes, and for $\sigma_0 < 0$:

$$F(x) = \left(-F/\sigma_0\right) \int_{d_0/2}^{x} \left[c(x') - c_\infty \right] dx'$$

$$= \left(-Fc_\infty/\sigma_0\right) \int_{u_{d_0/2}}^{u(x)} \frac{[\exp(-u) - 1]\, du}{du/dx}$$

$$= - \left(2Fc_\infty/\kappa\sigma_0\right) \left[\exp\left(\frac{-u(x)}{2}\right) - \exp\left(-\frac{u_{d_0/2}}{2}\right) \right] \tag{8}$$

for the case $u_\infty = 0$. The limiting value of $F(x)$ as $c_\infty \downarrow 0$ is:

$$\lim_{c_\infty \downarrow 0} F(x) \equiv F_0(x) = 1 - \left\{ 1 - \left[\beta\sigma_0(x - d_0/2)/4F \right] \right\}^{-1} \tag{9}$$

(cf. Zimm and Le Bret, 1983). Table 2 lists values of the right side of Equation (9) for $\sigma_0 = -0.087$ and -0.208 C $\bullet$ m^{-2}, the nominal endpoints of the range of surface charge densities for smectites. The most remarkable feature of Table 2 is the fact that between one-third and one-half of the monovalent cations that neutralize surface charge are predicted to be located within 0.2 nm of the distance of closest approach to the charged surface. Since $d_0/2 \approx 0.2$ nm as well for typical monovalent ions, this means that a significant portion of the neutralizing diffuse-layer cations reside within about one molecular diameter of the charged surface. Indeed, three-fourths of the neutralizing cations reside within 1.0 nm of the surface. These results (which do not change if $F(x)$

Table 2. Values of $F_0(x)$ for two representative values of σ_0.

$F_0(x)$		$x - d_0/2$
$\sigma_0 = -0.087$ C $\bullet$ m^{-2}	$\sigma_0 = -0.208$ C $\bullet$ m^{-2}	nm
0.196	0.369	0.1
0.328	0.539	0.2
0.550	0.745	0.5
0.710	0.854	1.0
0.786	0.898	1.5

is evaluated with Equation (8) for any $c_\infty < 0.5 \bullet$ mol m^{-3}) imply that a monovalent diffuse-ion swarm near a planar colloidal surface is compressed significantly, even in dilute solution. This behavior is already apparent in Figure 1, where the local cation concentration rises sharply for small x/d_0.

SURFACE COORDINATION MODELS

Types of surface coordination

Reactive molecular units that protrude from a solid adsorbent surface into an aqueous solution are termed *surface functional groups* (Sposito, 1984). In the case of solid organic adsorbents, surface functional groups are necessarily organic molecular units, but in general they can be either organic or inorganic, and they can have any molecular structural arrangement. Because of the variety of natural colloid compositions, a broad spectrum of surface functional group reactivity occurs. Superimposed on this variability is that created by the wide range of stereochemical and surface charge distribution characteristics possible in a heterogenous solid matrix.

Metal oxides found in natural colloids react with water to create solvated metal cations at the colloid/aqueous solution interface (Schindler and Sposito, 1990). This combination of a metal cation and a water molecule at an interface is a *Lewis acid site*, with the metal cation identified as the Lewis acid. Lewis acid sites exist, for example, on the edge surfaces of gibbsite (γ-Al(OH)$_3$) and goethite (α-FeOOH), as well as the edge surfaces of clay minerals like kaolinite (Al$_4$Si$_4$O$_{10}$(OH)$_8$). These surface functional groups are very reactive, because a positively-charged water molecule is quite unstable and can be either deprotonated or exchanged readily for an anion in aqueous solution. The inorganic surface functional group of greatest abundance and reactivity in clay-sized colloids is the hydroxyl group exposed on the outer periphery of a mineral. This kind of OH group also is found on metal oxides, oxyhydroxides, and hydroxides, and on clay minerals and amorphous silicates, like allophane. [See Figs. 2 and 3 in Davis and Kent (this volume).]

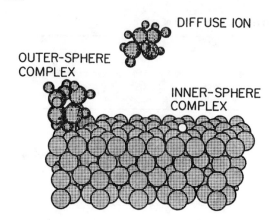

DIFFUSE ION

OUTER-SPHERE
COMPLEX

INNER-SPHERE
COMPLEX

Figure 2. The three modes of adsorption of a monovalent cations by the siloxane surface of a 2:1 layer silicate. From Sposito (1989).

The plane of oxygen atoms on the cleavage surface of a 2:1 clay mineral is called a *siloxane surface* (Sposito, 1984). This plane is characterized by a distorted hexagonal symmetry among the constituent oxygen ions, and the functional group associated with the siloxane surface is a roughly hexagonal cavity formed by the bases of six corner-sharing silica tetrahedra. This cavity has a diameter of about 0.26 nm and is bordered by six sets of "lone-pair" orbitals emanating from the surrounding ring of oxygen ions. [See Fig. 5 in Davis and Kent (this volume).]

The reactivity of the siloxane cavity depends on the nature of the electronic charge distribution in the layer silicate structure. If there are no neighboring cation substitutions to create local deficits of positive charge in the underlying structure, the siloxane cavity will function as a very mild electron donor that can bind only neutral, dipolar molecules, such as water molecules. If isomorphic substitution of Al^{3+} by Fe^{2+} or Mg^{2+} occurs deep in the layer, the excess negative charge makes it possible for the siloxane cavity to form reasonably strong complexes with cations as well as dipolar molecules. If isomorphic substitution of Si^{4+} by Al^{3+} occurs in the silica tetrahedra, the excess negative charge is located much nearer to the surface oxygen ions, and much stronger complexes with cations and molecules become possible because of this greater localization.

The complexes formed between surface functional groups and constituents of an aqueous solution can be classified analogously to the complexes that form entirely among aqueous species (Sposito, 1984; Stumm, 1986, 1987). If no water molecule is interposed between the surface functional group and the ion or molecule it binds, the complex is termed *inner-sphere*. If at least one water molecule is interposed between the functional group and the bound ion or molecule, the complex is *outer-sphere*. As a general rule, outer-sphere surface complexes involve electrostatic bonding mechanisms and, therefore, are less stable than inner-sphere surface complexes. These types of complex are illustrated in Figure 2 for the case of a monovalent cation adsorbed by a 2:1 clay mineral. Comprehensive discussions of surface complex formation can be found in the books by Sposito (1984) and edited by Stumm (1987). Methods of investigating these complexes are described in the book edited by Davis and Hayes (1986).

The Bragg-Williams approximation

From the point of view of statistical mechanics, the molecular theory of surface complexes is a special case of a multispecies lattice model. One imagines that the surface complexes are molecular species immobilized (over the time scale for the diffusive motion of an ion in aqueous solution) on an array of sites that represent surface functional groups. If there are two different surface species on the sites (e.g., a protonated and an unprotonated surface hydroxyl group, or a free siloxane cavity and one that has complexed K^+), and if each site has z nearest neighbors, then any distribution of the two species (call them A and B) over the sites must satisfy the conditions (Hill, 1960):

$$zN_A = 2N_{AA} + N_{AB} \quad , \quad \text{and} \tag{10a}$$

$$zN_B = 2N_{BB} + N_{AB} \quad , \tag{10b}$$

where N_A is the total number of A species, N_{AA} is the number of nearest-neighbor A pairs (with analogous definitions for N_B and N_{BB}) and N_{AB} is the number of nearest-neighbor AB pairs. It is assumed that each species binds just to one site. If the array of sites is not regular, z can be interpreted as the average number of nearest neighbors of a site.

The chemical properties of the distribution of the two species on the sites can be calculated with standard expressions in statistical mechanics once the "canonical partition function" is known (Hill, 1960). For a two-species lattice model, this function has the form (Hill, 1960):

$$Q = q_A^{N_A} \, q_B^{N_B} \sum_{N_{AB}} g\left(N_A, \, M, \, N_{AB}\right) \exp\left(-W/k_B T\right) \, , \tag{11}$$

where q is the canonical partition function for a single species on a single site, k_B is the Boltzmann constant, T is absolute temperature, and

$$W \equiv N_{AA} \, \varepsilon_{AA} + N_{BB} \, \varepsilon_{BB} + N_{AB} \, \varepsilon_{AB} \tag{12}$$

is the total energy of interaction between species on nearest-neighbor sites. In Equation (11), the function $g(N_A, M, N_{AB})$ is the number of ways that N_A species can be distributed on M total sites such that N_{AB} nearest-neighbor AB pairs occur. This function is subject to the constraint (Hill, 1960):

$$\sum_{N_{AB}} g\left(N_A, \, M, \, N_{AB}\right) = M! \, / N_A! \, N_B! \tag{13}$$

where ! refers to the factorial function and the right side is the total number of ways that N_A indistinguishable species of type A can be distributed among $M = N_A + N_B$ sites (Feller, 1968). The sum in Equations (11) and (13) is over all possible numbers of nearest-neighbor AB pairs that can be placed on the sites.

The interaction energy W is assumed to depend only on the respective pair interaction energies ε, which can be positive, negative, or zero. Thus Equation (11) can be interpreted as an expression of the relative likelihood of a given choice of N_A species A molecules and N_B species B molecules being found on an array of M sites with emphasis on nearest-neighbor interactions only. If there were no interactions, $W = 0$ and Equation (11) would reduce to the relative probability that there is independent occupancy of N_A sites by species A and N_B sites by species B. [Each q is interpreted in statistical mechanics as the relative probability of a single site containing a single species (Hill, 1960).] The function $g(N_A, M, N_{AB})$ appears when the site occupancy is influenced by nearest-neighbor interactions, with $\exp(-W/k_BT)$ being the appropriate "Boltzmann factor" for weighting each combinatorial term. Of principal concern in the Boltzmann factor is the relative importance of AA and BB interactions versus AB interactions. On defining

$$\varepsilon \equiv \varepsilon_{AA} + \varepsilon_{BB} - 2\varepsilon_{AB} \tag{14}$$

(Hill, 1960) and making use of Equations (10), one can transform Equation (11) into the expression:

$$Q = \left[q_A \exp\left(-z\varepsilon_{AA}/2k_BT\right) \right]^{N_A} \left[q_B \exp\left(-z\varepsilon_{BB}/2k_BT\right) \right]^{N_B}$$
$$\times \sum_{N_{AB}} g\left(N_A, M, N_{AB}\right) \exp\left(N_{AB}\varepsilon/2k_BT\right) . \tag{15}$$

Equation (15) exposes the significance of ε as an energy difference that determines the "statistical mechanical weighting" of each possible N_{AB} that contributes to Q.

The evaluation of the g-factor in Equation (15) is a formidable task to perform in general (Hill, 1960), so model assumptions are always made in order to facilitate the calculation of Q. One of the simplest is the *Bragg-Williams approximation*, which can be shown to be equivalent to the well known van der Waals model of liquids (Hill, 1960). It consists of replacing N_{AB} in Equation (15) by its average value as computed for a *random* distribution of the two species over the M sites, viz.,

$$N_{AB}^{BW} \equiv zN_A N_B/M . \tag{16}$$

In this approximation, all N_{AB} values are set equal to the total number of nearest neighbors of A species (zN_A) times the fraction of total sites occupied by B species (N_B/M). This result would be strictly true only if $\varepsilon = 0$, such that no advantage accrued to any particular nearest-neighbor association. In the Bragg-Williams approximation, one assumes that, even when $\varepsilon \neq 0$, the fraction of B species overall still determines the short-range ordering into AB pairs. Alternatively, Equation (16) is equivalent to the mathematical operation of letting z become infinite while ε goes to zero in such a way that the product $z\varepsilon$ remains finite (Fowler and Guggenheim, 1949; Sposito, 1983). This "van der Waals limit" is interpreted physically as the result of placing each surface species in an average potential energy field determined collectively by *all* of its neighbors on other surface sites (Sposito, 1984).

The combination of Equations (13), (15), and (16) produces the Bragg-Williams partition function:

$$Q_{BW} = \left[q_A \exp\left(-z\varepsilon_{AA}/2k_BT\right) \right]^{N_A} \left[q_B \exp\left(-z\varepsilon_{BB}/2k_BT\right) \right]^{N_B}$$

$$x \left(M!/N_A! \, N_B! \right) \left[\exp\left(z\varepsilon/2k_BT\right) \right]^{N_AN_B/M} . \tag{17}$$

All of the present generation of chemical models of surface complexation can be derived from Equation (17). Generalizations of Q_{BW} for an arbitrary number of surface species, each of which can bind to an arbitrary number of sites, have been derived by Sposito (1983, 1984).

Surface complexation equilibria

Chemical equilibrium conditions for surface species described by Equation (17) can be derived readily by application of the standard relation (Hill, 1960; Sposito, 1983):

$$\mu_A - \mu_B = -k_BT \left(\frac{\partial \ln Q_{BW}}{\partial N_A} \right) , \tag{18}$$

where μ is a chemical potential and the derivative of Q_{BW} is calculated under the reaction mass-balance constraint $dN_A = -dN_B$ (i.e., the reaction $A \leftrightarrow B$). The result is:

$$\mu_A - \mu_B = k_BT \ln \left[q_B \exp\left(-z\varepsilon_{BB}/2k_BT\right)/q_A \exp\left(-z\varepsilon_{AA}/2k_BT\right) \right]$$

$$+ k_BT \ln\left(x_A/x_B\right) + \left(x_A - x_B\right) z\varepsilon/2$$

$$= k_BT \ln\left(q_B \, x_A/q_A \, x_B\right) + \left(x_A z\varepsilon\right) + \left(\varepsilon_{AB} - \varepsilon_{BB}\right) , \tag{19}$$

where $x = N/M$ is a mole fraction, and $x_A = 1 - x_B$ along with Equation (14) has been used to derive the second step. Chemical equilibrium between the surface species and the cognate aqueous species exists if (Sposito, 1983):

$$\mu_A - \mu_B = \mu\left[A_{aq}\right] - \mu\left[B_{aq}\right]$$

$$= \mu_A^\circ - \mu_B^\circ + k_BT \ln\left(a_A/a_B\right) , \tag{20}$$

where $\mu[\ ]$ is the chemical potential of an aqueous species whose standard state chemical potential is μ° and whose activity is a. Given the conditional equilibrium constant for the 1:1 exchange reaction, $A \leftrightarrow B$:

$$^cK \equiv x_A a_B / x_B a_A , \tag{21}$$

it follows from Equations (19) and (20) that

$$^cK = (K_A/K_B) \exp(-z\varepsilon x_A/k_BT) \tag{22}$$

in the Bragg-Williams approximation, where

$$K \equiv q \, exp(\mu°/k_B T) \tag{23}$$

for either species. [The difference between ε_{AB} and ε_{BB} has been ignored, consistent with the "van der Waals limit" (Sposito, 1983)]. Sposito (1983) has derived the generalization of Equation (22) when ^cK contains arbitrary powers of the mole fractions and activities caused by an exchange reaction stoichiometry between A and B that differs from 1:1.

APPLICATIONS

Proton adsorption

Proton adsorption by a surface hydroxyl group (e.g., aluminol groups on the edge surfaces of gibbsite or kaolinite) can be described by the generic chemical reaction (Schindler and Stumm, 1987):

$$SOH(s) + H^+(aq) = SOH_2^+(s) \quad , \tag{24}$$

where SOH(s) represents 1 mol of surface hydroxyl groups and SOH_2^+(s) represents 1 mol of protonated surface hydroxyl groups. In principle, SOH_2^+(s) includes bound protons as diffuse-swarm ions, outer-sphere complexes, and inner-sphere complexes. Particular molecular models of Equation (24) then will differ depending on how SOH_2^+(s) is interpreted; i.e., which surface species of adsorbed H^+ are assumed to exist. Experiments designed to determine precisely the surface speciation of protons on hydroxylated solids are not common, with the result that competing molecular models of the protonation reaction for the same adsorption data often have been applied (Sposito, 1984).

One of these models is the *Constant Capacitance Model* of Schindler and Stumm (Schindler and Kamber, 1968; Stumm et al., 1970; Schindler and Stumm, 1987). On this model, SOH(s) is interpreted as an "average" surface hydroxyl group (which is *not* the same as assuming the hydroxylated surface is homogeneous) and SOH_2^+(s) is interpreted as an inner-sphere complex. Equation (21) is specialized to the form,

$$^cK \equiv 1/K_{a1}^s = x_{SOH_2^+} \left[H^+ \right] / x_{SOH} \quad , \tag{25}$$

with the omission of any aqueous species relating to SOH because of the form of Equation (24). [The conditional equilibrium constant K_{a1}^s actually refers to the reverse of the reaction in Equation (24).] The constant ionic medium reference state is used for aqueous-species activity coefficients, so a_{H^+} is represented by the molar proton concentration, $[H^+]$ (Sposito, 1984). The mole fractions and aqueous proton concentration in K_{a1}^s are measured conventionally by potentiometric titration experiments (Schindler and Gamsjäger, 1972; Stumm et al., 1976).

The Constant Capacitance Model of proton adsorption is a special case of Equation (22). It can be obtained by making the identifications (Sposito, 1983, 1984; Schindler and Stumm, 1987):

$$K_B/K_A \equiv K_{a1}^s(\text{int}) \quad , \quad \text{and} \tag{26a}$$

$$z\varepsilon/k_BT \equiv F^2SOH_T/CRT \quad , \tag{26b}$$

where C is a model differential capacitance density [F • m^{-2}] at the mineral surface and SOH_T is the total concentration of protonatable surface hydroxyls [mol • m^{-2}]. The combination of Equations (22) and (26) produces the result:

$$K_{a1}^s = K_{a1}^s(int) \; \exp\left(F^2SOH_T \; x_{SOH_2^+}/CRT\right) \quad . \tag{27}$$

Equation (27) also can be developed from the idea that, in the "van der Waals limit," the maximum surface proton charge density, $F\ SOH_T$, is proportional to a "surface electrostatic potential," $z\varepsilon$, that acts uniformly on all protonated SOH (Sposito, 1984):

$$\sigma_{max} \equiv F\ SOH_T = C\ z\varepsilon \quad , \tag{28}$$

with the capacitance density C linking the surface charge density on the left to the electrostatic potential on the right. The "intrinsic" equilibrium constant, $K_{a1}^s(\text{int})$, is the result of "correcting" the conditional constant, K_{a1}^s, for electrostatic potential effects:

$$K_{a1}^s(int) = K_{a1}^s \; \exp\left(-F^2SOH_T \; x_{SOH_2^+}/CRT\right) \quad , \tag{29}$$

where the exponential factor then is interpreted as a ratio of activity coefficients for the surface species, SOH_2^+ and SOH (Sposito, 1983, 1984). Semilogarithmic graphs of K_{a1}^s against $x_{SOH_2^+}$ can be used to evaluate $K_{a1}^s(\text{int})$ and SOH_T/C, as indicated in Figure 3 for γ-Al_2O_3 (Hohl and Stumm, 1976). Typical deviations from the strict linearity in these graphs required by Equation (27) reflect the fact that the Bragg-Williams approximation of random site occupation is most accurate when $x_{SOH_2^+} \ll 1$.

Wieland et al. (1988) (see also Stumm and Wollast, 1990) have extended the chemical hypotheses underlying Equation (27) to the phenomenon of proton-catalyzed dissolution of metal oxides. They point out that the combination of Equations (25) and (27) produces an adsorption isotherm equation (the so-called Frumkin-Fowler-Guggenheim isotherm) for protons that agrees well with experimental data on a variety of metal oxides. In the spirit of this equation, i.e., the Bragg-Williams approximation, Wieland et al. (1988) postulate that the steady-state rate of metal oxide dissolution catalyzed by proton adsorption should be proportional to the surface fraction of "active" sites, S_a, and the probability, P_n, that a metal center on the surface has n protonated hydroxyl groups as nearest neighbors. These protonated OH groups destabilize the bonding of the metal to the bulk mineral and thereby promote metal detachment. Thus

$$R_H = k_H S_a P_n \quad , \tag{30}$$

where k_H is a rate constant and R_H is the steady-state rate of dissolution. If the protons are complexed *randomly* on the mineral surface as assumed in the Bragg-Williams approximation, then (Wannier, 1966):

274

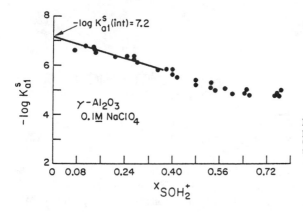

Figure 3. Semilogarithmic plot of proton titration data for γ-Al$_2$O$_3$ (Hohl and Stumm, 1976) according to Equation 29.

$$P_n = \frac{m!}{n!(m-n)!} \, x_{SOH_2^+}^n \left(1 - x_{SOH_2^+}\right)^{m-n} \qquad (n \leq m) \quad, \qquad (31)$$

where m is the maximum number of protonatable nearest-neighbor OH groups. Equation (31) gives P_n as the product of two factors: the number of ways that n protons can be placed randomly on m sites and the probability that n sites are protonated (randomly) while $m - n$ are not. Under the condition $x_{SOH_2^+} \ll 1$, Equation (31) reduces to the approximation,

$$P_n \approx \frac{m!}{n!(m-n)!} \, x_{SOH_2^+}^n \,, \qquad (32)$$

and Equation (30) becomes

$$R_H \approx k_H \, S_a \, (m!/n!(m-n)!) \, x_{SOH_2^+}^n \quad. \qquad (33)$$

Equation (33) has been found (Stumm and Wollast, 1990) to provide good agreement with experiment when n is identified with the oxidation state of the metal center (e.g., n = 3 for Al(III)).

Metal cation adsorption

The adsorption of alkali metal cations by mineral surfaces is widely believed to involve outer-sphere surface complexes and the diffuse ion swarm as principal mechanisms (Sposito, 1984). For a hydroxylated surface, the chemical reaction underlying adsorption can be expressed

$$SOH(s) + M^+(aq) = SOM(s) + H^+(aq) \quad, \qquad (34)$$

where M is an alkali metal and SOM(s) includes both outer-sphere complexes and diffuse-layer cations. A prototypical situation in which the reaction above occurs is the adsorption of M^+ by a metal oxide from a solution containing the electrolyte MClO$_4$. The *Triple Layer Model* (Davis et al., 1978; Hayes and Leckie, 1987) offers a molecular description of this situation based on a combination of the Bragg-Williams approximation for surface complexes and Modified Gouy-Chapman theory for the diffuse-ion swarm (Sposito, 1983; 1984). The basic postulates of the model are discussed by Davis et al. (1978), James and Parks (1982), and Sposito (1984). The present application will consider outer-sphere surface complexes, but the model can describe inner-sphere complexes as well (Hayes and Leckie, 1987).

A conditional equilibrium constant for the reaction in Equation (34), applied *only* to the surface-complex portion of SOM(s), can be written as a special case of Equation (21) with A ≡ SOM and B ≡ SOH (Davis and Leckie, 1978):

$$^{c}K \equiv {}^{*}Q_M = x_{SOM} \, a_H / x_{SOH} \, a_M \, , \tag{35}$$

where the infinite dilution reference state now is used for aqueous species activity coefficients. The Bragg-Williams model of $^{*}Q_M$ then follows from Equation (22):

$$^{*}Q_M = {}^{*}K_M^{int} \exp\left(-e^2 SOH_T \, x_M / C_1 k_B T\right) \, , \tag{36}$$

where e is the protonic charge and C_1 is a model capacitance density assigned to the molecular-scale region between the point of closest approach of M^+ and the adsorbent surface (i.e., the domain $0 < x < d_0/2$ in Modified Gouy-Chapman theory). Similar to Equations (26), the identifications

$$K_A / K_B \equiv {}^{*}K_M^{int} \tag{37a}$$

$$z\varepsilon \equiv e^2 SOH_T / C_1 \tag{37b}$$

are made to convert Equation (22) into Equation (36). The chemical interpretation of Equation (37b) is analogous to that of Equation (26b): $e \, SOH_T = C_1 z\varepsilon$, with $z\varepsilon$ regarded as an average electrostatic potential acting on the outer-sphere complexes (Sposito, 1983). The relation between $^{*}K_M^{int}$ and $^{*}Q_M$ is also analogous to that discussed for $K_{a1}^{s}(int)$ and K_{a1}^{s} in the Constant Capacitance Model.

In general, the negative charge on a hydroxylated surface whose ionization is generated by the deprotonation of SOH(s) will be balanced by M^+ in both outer-sphere complexes and in the diffuse-ion swarm. The balancing charge density thus can be expressed mathematically as the sum of $e \, SOH_T \, x_M$ and a term given by Equation (3) (divided by the suspension density of metal oxide colloids). For sufficiently large values of c_∞, the contribution of Equation (3) will be very small and the negative charge on the surface is balanced effectively by the outer-sphere complexes of M^+ alone. Under this condition, $e \, SOH_T \, x_M$ in Equation (36) can be replaced by the net surface proton charge, σ_H, and a semilogarithmic graph of $^{*}Q_M$ against σ_H can be used to evaluate $^{*}K_M^{int}$ and C_1. This exercise is illustrated in Figure 4 for TiO_2 adsorbing Li^+ from a solution of $LiNO_3$. The y-intercept of 7.2 is $-\log {}^{*}K_{Li}^{int}$ and the slope of 25 leads to $C_1 = 1.3$ F $\cdot$ m^{-2} (Sposito, 1984). The x-axis in Figure 4 actually is the same as x_M; the linearity observed at low values of this parameter indicates once again the accuracy of the assumption of random siting of surface complexes when $x_M << 1$.

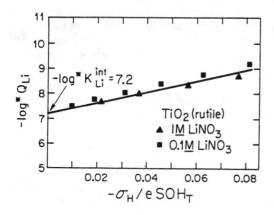

Figure 4. Semilogarithmic plot of proton tritration data for rutile (Davis et al., 1978) according to Equation 36.

Coion exclusion

If the sole mechanism of ion adsorption is via the diffuse-ion swarm, the anions in an electrolyte solution in which negatively-charged colloids are suspended will, in general, be excluded from a portion of the suspension volume near the colloid surface (Sposito, 1984). If q is the experimentally-measured (negative) surface excess of these anions [mol • kg^{-1}] resulting from exclusion and c_∞ is their bulk concentration in a 1:1 electrolyte, then

$$V_{ex} \equiv -q/c_\infty \tag{38}$$

is the operational definition of V_{ex}, the exclusion volume per unit mass of clay mineral in the suspension.

Modified Gouy-Chapman theory provides a model of a parameter closely related to the exclusion volume, viz., d_{ex}, the "exclusion distance." For a 1:1 electrolyte (Sposito, 1984):

$$d_{ex} \equiv \int_{d_o/2}^{\infty} \left[1 - \left(c_-(x)/c_\infty \right) \right] dx + d_o/2$$

$$= (2/\kappa) \left[1 - \exp\left(\frac{u_{d_o/2}}{2} \right) \right] + d_o/2 \ . \tag{39}$$

Since the term in brackets under the integral sign is the local fraction of anion depletion relative to the bulk concentration, d_{ex} is an average distance over which anions are excluded. A similar equation (with the sign of u reversed) applies to cation exclusion by a positively-charged colloid.

In the case of platelet colloids like smectites, the relation between the experimental variable V_{ex} and the theoretical variable d_{ex} is provided by the expression (Sposito, 1984):

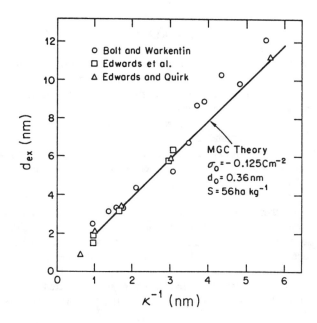

Figure 5. Predictions of the exclusion distance (d_{ex}) by Modified Gouy-Chapman theory (solid line) compared with experimental measurements.

$$V_{ex} = Sd_{ex} \quad . \tag{40}$$

where S is the specific surface area of the colloids. Equations (39) and (40) imply that a graph of V_{ex} vs. ($1/\kappa$) will yield a straight line whose slope is proportional to S. Alternatively, if S is known, a graph of V_{ex}/S against $1/\kappa$ can be treated as an experimental measurement of d_{ex}.

Tests of Equations (39) and (40) are complicated by the fact that surface complexes with counterions may form to reduce the net particle charge, such that the parameter S in Equation (40) is only the portion of the geometric surface area which repels coion, not the entire geometric surface area (Sposito, 1984). Nonetheless, a large variety of viscosity, photon scattering, and neutron scattering data indicate that, for the 2:1 clay mineral Na-montmorillonite, $S = 56 \times 10^4$ $m^2 \cdot kg^{-1}$. This clay mineral is one of the most important in the smectite group. A fairly stringent test of Equation (40) is provided once this value is inserted, since there are no adjustable parameters in Equation (39) and V_{ex} comes directly from experiment. Figure 5 shows a test based on three sets of published values of V_{ex} as a function of $1/\kappa$ (Bolt and Warkentin, 1958; Edwards and Quirk, 1962; Edwards et al., 1965) for Na-montmorillonite suspensions. The value of σ_0 for the montmorillonite samples was near -0.125 C $\cdot m^{-2}$. The data points are values of the measured V_{ex} divided by $S \equiv 56 \times 10^4$ $m^2 \cdot kg^{-1}$. The straight line through the data points (actually a curve whose changing slope is too subtle to perceive in the range of the $1/\kappa$ shown) is Equation (39). The agreement between experiment and theory — without adjustable parameters — is as good as the data warrant.

CONCLUDING REMARKS

The intent of the present chapter has been to show that two well-established statistical mechanical models, Modified Gouy-Chapman theory and the Bragg-Williams approximation for a lattice fluid, suffice to account for the leading molecular approaches to describe adsorption by natural colloids like clay minerals and metal oxides. The simplicity of these two models and their remarkable accuracy in the quantitative treatment of ion adsorption phenomena (Sposito, 1984, 1989; Schindler and Stumm, 1987) suggest that the adsorption experiments themselves are not particularly sensitive to molecular complexity. "Mean field" theories, then, are quite adequate to provide both descriptive and predictive capability at present.

It is well to remember that natural colloids in soils and sediments are highly heterogeneous mixtures of inorganic and organic materials. The more well-characterized constituents of these mixtures derive from aluminosilicates, metal oxides, and humus, and together they produce a complicated solid surface to react with an equally complicated natural aqueous environment, as stressed by Davis and Kent (this volume). This complexity makes it all the more imperative to establish the unified perspective on natural surface reactions that comes from the point of view of coordination chemistry.

ACKNOWLEDGMENTS

The preparation of this chapter was supported in part by NSF grant No. EAR-8915291. Gratitude is expressed to Dr. Douglas B. Kent for his helpful review, to Ms. Joan Van Horn for her excellent typing of the manuscript, and to Mr. Frank Murillo for preparation of the figures.

REFERENCES

Bolt, G. H. and Warkentin, B. P. (1958) The negative adsorption of anions by clay suspensions. Kolloid-Z. 156, 41-46.

Carnie, S. L. and Torrie, G. M. (1984) The statistical mechanics of the electrical double layer. Advan. Chem. Phys. 56, 141-253.

Davis, J. A. and Hayes, K. F., eds. (1986) Geochemical Processes at Mineral Surfaces. Am. Chem. Soc., Washington, D.C.

Davis, J. A., James, R. O. and Leckie, J. O. (1978) Surface ionization and complexation at the oxide/water interface. I. Computation of electrical double layer properties in simple electrolytes. J. Colloid Interface Sci. 63, 480-499.

Edwards, D. G. and Quirk, J. P. (1962) Repulsion of chloride by montmorillonite. J. Colloid Sci. 17, 872-882.

Edwards, D. G., Posner, A. M. and Quirk, J. P. (1965) Repulsion of chloride ions by negatively charged clay surfaces. Part 2. Monovalent cation montmorillonites. Trans. Faraday Soc. 61, 2816-2819.

Feller, W. (1968) An Introduction to Probability Theory and Its Applications. Vol. I. John Wiley, New York.

Fowler, R. A. and Guggenheim, E. A. (1949) Statistical Thermodynamics. Cambridge Univ. Press, London.

Grahame, D. C. (1953) Diffuse double layer theory for electrolytes of unsymmetrical valence types. J. Chem. Phys. 21, 1054-1060.

Hayes, K. F. and Leckie, J. O. (1987) Modeling ionic strength effects on cation adsorption at hydrous oxide/solution interfaces. J. Colloid Interface Sci. 115, 564-572.

Hill, T. L. (1960) An Introduction to Statistical Thermodynamics. Addison-Wesley, Reading, Massáchusetts.

Hohl, H. and Stumm, W. (1976) Interaction of Pb^{2+} with hydrous γ-Al_2O_3. J. Colloid Interface Sci. 55, 281-288.

James, R. O. and G. A. Parks. (1982) Characterization of aqueous colloids by their electrical double-layer and intrinsic surface chemical properties. Surface Colloid Sci. 12, 119-214.

Johnston, C. T. and Sposito, G. (1987) Disorder and early sorrow: Progress in the chemical speciation of soil surfaces. In: Future Developments in Soil Science Research, L. L. Boersma, ed., Soil Sci. Soc. of America, Madison, Wisconsin, 89-99.

Manning, G. S. (1979) Counterion binding in polyelectrolyte theory. Acc. Chem. Res. 12, 443-449.

Schindler, P. W. and Kamber, H. R. (1968) Die Acidität von Silanogruppen. Helv. Chim. Acta 51, 1781-1786.

Schindler, P. W. and Gamsjäger, H. (1972) Acid-base reactions of the TiO_2 (anatase)-water interface and the point of zero charge of TiO_2 suspensions. Kolloid Z. Polymere 250, 759-763.

Schindler, P. W. and Stumm, W. (1987) The surface chemistry of oxides, hydroxides, and oxide minerals. In: Aquatic Surface Chemistry, W. Stumm, ed., John Wiley, New York, 83-110.

Schindler, P. W. and Sposito, G. (1990) Surface complexation. In: Interactions at the Soil Colloid-Solution Interface, G. H. Bolt, M. F. De Boodt, M. B. McBride, and M. H. B. Hayes, eds., D. Reidel, Dordrecht, The Netherlands (in press).

Sposito, G. (1983) On the surface complexation model of the oxide-aqueous solution interface. J. Colloid Interface Sci. 91, 329-340.

Sposito, G. (1984) The Surface Chemistry of Soils. Oxford Univ. Press, New York.

Sposito, G. (1986) Distinguishing adsorption from surface precipitation. In: Geochemical Processes at Mineral Surfaces, J. A. Davis and K. F. Hayes, eds., Am. Chem. Soc., Washington, D.C., 217-228.

Sposito, G. (1989) Surface reactions in natural aqueous colloidal systems. Chimia 43, 169-176.

Stumm, W. (1986) Coordinative interactions between soil solids and water — An aquatic chemist's point of view. Geoderma 38, 19-30.

Stumm, W., ed. (1987) Aquatic Surface Chemistry. John Wiley, New York.

Stumm, W. and Wollast, R. (1990) Coordination chemistry of weathering: Kinetics of the surface-controlled dissolution of oxide minerals. Rev. Geophys. 28, 53-69.

Stumm, W., Huang, C. P. and Jenkins, S. R. (1970) Specific chemical interaction affecting the stability of dispersed systems. Croat. Chem. Acta 53, 291-312.

Stumm, W., Hohl, H. and Dalang, F. (1976) Interaction of metal ions with hydrous oxide surfaces. Croat. Chem. Acta 48, 491-504.

Valleau, J. P. and Torrie, G. M. (1982) The electrical double layer. III. Modified Gouy-Chapman theory with unequal ion sizes. J. Chem. Phys. 76, 4623-4630.

Wannier, G. H. (1966) Statistical Physics. John Wiley, New York.

Wieland, E., Wehrli, B. and Stumm, W. (1988) The coordination chemistry of weathering: III. A generalization on the dissolution rates of minerals. Geochim. Cosmochim. Acta 52, 1969-1981.

Zimm, B. H. and Le Bret, M. (1983) Counter-ion condensation and system dimensionality. J. Biomolecular Struc. & Dynamics 1, 461-471.

Co-adsorption of Metal Ions and Organic Ligands:
Formation of Ternary Surface Complexes

INTRODUCTION

This chapter deals with the co-adsorption of metals ions and (preponderantly) organic ligands. It thus requires the reader to be familiar with the principles that govern the adsorption of metal ions and of organic molecules. It is the aim of these introductory paragraphs to briefly discuss the relevant adsorption modes. Because adsorption of metal ions is already discussed in other chapters in this volume, the emphasis of this discussion will be placed on the adsorption of organic compounds.

Adsorption of dissolved species from aqueous solution at the mineral-water interface is based on forces acting between surface, solute and solvent. It is thus convenient to discuss adsorption modes according to the nature of these forces. These forces are themselves dependent on the properties of solutes and surfaces that are involved in the adsorption process. For the purpose of this chapter, it will be convenient to classify both solutes and surfaces according to the following criteria:
—Polarity
—Charge
—Lewis acidity
—Lewis basicity

Surface polarity controls the interaction with water, a polar solvent. Polar solutes show a high solubility; polar surfaces are wettable. Both polar solutes and polar surfaces are thus called hydrophilic. In contrast, nonpolar substances are but sparingly soluble and nonpolar surfaces show little affinity for water. They are thus called hydrophobic.

Charges give rise to strong electrostatic interactions. They are acquired by attachment or release of ions. In many cases the attached or released ions are hydrogen ions and the acquisition of a charge is controlled by the pH of the solution. Hence, surface "ionization" is often based on protonation or deprotonation of surface hydroxyls. Surface charging can also result from binding or release of metal ions and anions.

Lewis acidity is a property that results from the presence of energetically low lying unoccupied atomic or molecular orbitals. Uncoordinated ("naked") metal ions are Lewis acids. Thus, in aqueous solution, they coordinate with water molecules. Dry mineral surfaces exhibit coordinatively unsaturated metal ions that display Lewis acidity. In humid environments, these surface metal ions coordinate with water molecules or, more frequently, with surface hydroxyls.

Lewis bases contain non-bonding occupied atomic or molecular orbitals. A pair of electrons occupying such an orbital is also called a "lone pair." Atoms carrying one or more lone pairs are called "ligand atoms." Molecules or ions that carry ligand atoms are termed "ligands." The number of ligand atoms carried by a given ligand is reflected by the specifications monodentate, bidentate etc. Among the most prominent Lewis bases, we note the water molecule, the hydroxyde ion and the numerous anions originating from deprotonation of organic acids.

The above mentioned properties give rise to strong interactions. The reaction of Lewis acids with Lewis bases leads to formation of stable complexes or coordination compounds with strong covalent bonds. Ion-ion and ion-dipole interaction (both attraction and repulsion) are of similar strength whereas dipole-dipole interactions are somewhat weaker. In addition to these strong interactions we have to consider the van der Waals interactions. These comparatively weak short range forces increase with increasing molecular mass.

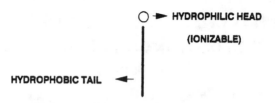

AMPHIPATHIC MOLECULE

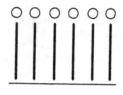

HYDROPHOBIC SURFACE

Figure 1. Mode of adsorption of amphipatic molecules on a hydrophobic surface. Upon adsorption the surface becomes hydrophilic.

Last is the important class of hydrophobic interactions. They are based on the fact that with nonpolar solutes in an aqueous environment both the solvent-solvent and the solute-solute interactions are stronger than the solute-solvent interaction. This leads to an expulsion of the nonpolar solute by the aqueous phase.

Adsorption of organic compounds

Based on the foregoing considerations, the adsorption of organic compounds can be described by three fundamental adsorption modes:
- Hydrophobic expulsion
- Electrostatic attraction
- Surface complexation.

Hydrophobic expulsion. Hydrophobic expulsion assisted by van der Waals inter-action is the dominant mode for the adsorption of hydrophobic and amphipatic compounds on nonpolar surfaces. Amphipatic compounds contain both polar (hydrophilic) and non-polar (hydrophobic) moieties. This class thus comprises important surfactants such as dodecylammonium-salts, dodecylsulfonates and fatty acids, all of them widely used as collectors in froth flotation. Ulrich et al. (1988) have investigated the adsorption of fatty acids on the nonpolar mercury surface. The adsorption mode is depicted in Figure 1.

The hydrophobic tails (hydrocarbon chains) are oriented towards the nonpolar surface, thus exposing the polar heads towards the aqueous solution. The adsorption data were found to follow the Frumkin equation,

$$\frac{\Theta}{1 - \Theta} = B \ c \ \exp(2\alpha\Theta) \ , \tag{1}$$

where Θ is the fraction of the mercury surface that is covered by the fatty acids, c is their concentration in the aqueous solution, B is the adsorption constant, and a is the interaction coefficient discussed below. B is related with the Gibbs free energy of adsorption by

$$DG_{ads} = - RT \ln B \ . \tag{2}$$

The exponential factor reflects the lateral interaction of adsorbed molecules. The Gibbs free energy of adsorption was found to linearly decrease with the length of the hydrophobic tail:

$$DG_{ads} = 7 - 31 \ n_c \ (kJ \ mol^{-1}) \ , \tag{3}$$

where n_c is the number of carbon atoms of the fatty acid. The interaction coefficient a was found to depend on the pH of the solution. At low pH values where the adsorbed fatty acids are not ionized, a was found to be a positive number that increases with increasing chain length. This indicates the presence of attractive lateral interaction based on van der Waals forces. At higher pH values where the fatty acids become ionized, α is a negative number and indicates the electrostatic repulsion between the charged hydrophilic groups.

Although this adsorption mode is not important for pure (hydrophilic) mineral surfaces, it becomes important for natural mineral mixtures such as aquifers and lake sediments containing 0.1% to 5% of organic carbon. Let's designate the hydrophobic molecule by R. Its adsorption from the aqueous solution to the mixture of minerals is then described by the equation

$$R_{aq} \leftrightarrow R_{(ads)} \ , \tag{4}$$

and the adsorption/desorption equilibrium is given by the distribution coefficient K_D:

$$K_D = [R_{(ads)}]/[R_{aq}] \ . \tag{5}$$

It has been shown (Karikhoff et al., 1979; Schwarzenbach and Westall, 1981) that K_D is just dependent on one property of the solute, e.g., its water/octanol distribution coefficient K_{ow} and one property of the adsorbing solid, e.g., f_{oc}, the fraction of the solid that consists of organic carbon:

$$\log K_D = a \log K_{ow} + \log f_{oc} + b \ , \tag{6}$$

where a and b are empirical coefficients that vary but slightly with the set of hydrophobic compounds considered (Lyman, 1982).

Note on K_{ow}: The partition between water and an organic solvent has frequently been used as an indicator of the hydrophobicity of a given solute. Although several organic liquid-water partition coefficients could be used, the most widely used is K_{ow}. This dimension-less parameter is defined as

K_{ow} = Concentration in octanol / Concentration in water

K_{ow} is slightly temperature dependent with most data being reported for 298 K.

Electrostatic attraction. Electrostatic attraction is the principal mode of adsorption of ionizable organics on polar ionizable surfaces. It occurs if surface and solute carry opposite charges. Ionization is based on protonation and deprotonation processes that are dependent on the pH of the solution. We shall first briefly discuss the fundamental aspects of surface ionization. Protonation and deprotonation of the surface hydroxyl are based on the equilibria

284

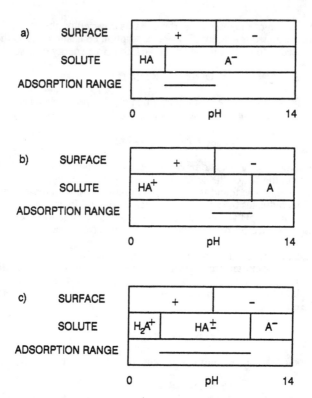

Figure 2. Interaction of ionizable solutes with charged interfaces. Schematic presentation of the charge distributions for the case of (a) fatty acids, (b) ammonium salts, (c) amino acids.

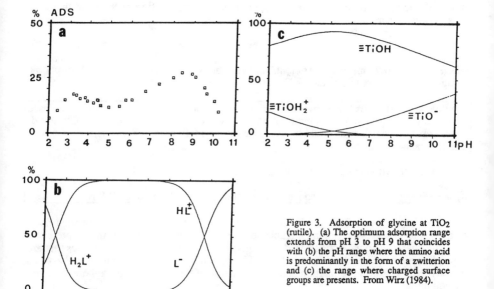

Figure 3. Adsorption of glycine at TiO$_2$ (rutile). (a) The optimum adsorption range extends from pH 3 to pH 9 that coincides with (b) the pH range where the amino acid is predominantly in the form of a zwitterion and (c) the range where charged surface groups are presents. From Wirz (1984).

$$\equiv S\text{-}OH_2^+ = \equiv S\text{-}OH + H^+ \; ; \; Ka_1^s = [\equiv S\text{-}OH][H^+]/[\equiv S\text{-}OH_2^+] \; ; \tag{7}$$

$$\equiv S\text{-}OH = \equiv S\text{-}O^- + H^+ \; ; \; Ka_2^s = [\equiv S\text{-}O^-][H^+]/[\equiv S\text{-}OH] \; , \tag{8}$$

where $\equiv S\text{-}OH$ is the surface hydroxyl group and [i] is the concentration of i in mole per dm^3 of solution. Ka_1^s and Ka_2^s are conditional equilibrium constants. A more rigorous formulation of surface equilibria will be given in the next section. It can be seen from Equations (7) and (8) that the pH of zero point of charge, i.e., the pH where $[\equiv S\text{-}OH_2^+]$ and $[\equiv S\text{-}O^-]$ become equal, is given by

$$pH_{pzc} = \frac{1}{2}(pKa_1^s + pKa_2^s) \tag{9}$$

For pH $<$ pH_{pzc}, we find that $[\equiv S\text{-}OH_2^+] > [\equiv S\text{-}O^-]$. This means that the surface as a whole is positively charged but that there are still $\equiv S\text{-}O^-$ groups present, although their concentration will steadily decrease with decreasing pH. The same applies for the alkaline region where $S\text{-}OH_2^+$ groups may exist at pH values well above the pH_{pzc}.

The occurrence of charges on ionizable solutes depends on the pKa value of the dissolved acids and the pH of the solution. A schematic presentation of the charges of both surfaces and solutes are given in Figure 2. Because electrostatic attraction requires opposite charges, the pH range of optimum adsorption lies between the pH_{pzc} of the adsorbing mineral and the pH that equals the pKa of the dissolved acid. For amino acids, this range of optimum adsorption is expected to lie between pH = pKa_1 and pH = pKa_2. The accuracy of this prediction depends on whether the attraction of a dissolved ion requires an overall oppositely charged surface or just the presence of negatively charged groups. If the adsorbed ion is part of a diffuse layer (as assumed by the Gouy Chapman model), it is the overall charge of the surface that governs the adsorption behavior in terms of the above predictions. For the case where the adsorbed ions form ion pairs with oppositely charged surface hydroxyls, a considerable extension of this range will be observed.

The foregoing general considerations may be illustrated by taking the adsorption of glycine on TiO_2 (rutile) as an example (Wirz, 1984). Figure 3c shows the distribution of charged and uncharged surface groups as a function of pH calculated from $pKa_1^s = 3.77$ and $pKa_2^s = 6.62$ (298 K, I = 1 M $NaClO_4$) under consideration of corrections arising from surface potentials (as discussed in the next section). Figure 3b depicts the distribution of the three glycine species calculated from $pKa_1 = 2.54$ and $pKa_2 = 9.74$, respectively. As seen from Figure 3a, the region of optimum adsorption actually lies between pH 2.5 and pH 9.7, but in addition there is noticeable adsorption beyond these limits.

Surface complexation. Adsorption of dissolved ligands can be based on surface complexation. The surface hydroxyls ($\equiv S\text{-}OH$) that coordinate the coordinatively undersaturated surface metal ions ($\equiv S$) can be exchanged against the dissolved ligand L^- (the negative charge assigned to the dissolved ligand implies that it was formed by deprotonation of an organic acid HL) such that L^- is directly bond to the surface metal $\equiv S$,

$$\equiv S\text{-}OH + L^- = \equiv S\text{-}L + OH^- . \tag{10}$$

Combining Equation (10) with

$$H^+ + OH^- = H_2O \tag{11}$$

results in

$$\equiv S\text{-}OH + H^+ + L^- = \equiv S\text{-}L + H_2O \ . \tag{12}$$

Equation (12) can be written in alternative forms, i.e.,

$$\equiv S\text{-}OH + HL = \equiv S\text{-}L + H_2O \ , \tag{12a}$$

and

$$\equiv S\text{-}OH_2^+ + L^- = \equiv S\text{-}OH_2^+ \ L^- \ . \tag{12b}$$

Equation (12a) describes the formation of a surface complex whereas Equation (12b) stands for the electrostatic attraction discussed in the previous paragraph. Stoichiometry does not permit one to discriminate between these two adsorption modes. Moreover, the changes in Gibbs free energy associated with either (12a) or (12b) are quite similar. Spectroscopic observations (Wirz, 1984; Motschi, 1987) seem to indicate that for mono-carboxylic acids, electrostatic attraction (12b) is more likely than surface complexation (12a). On the other hand, studies on the kinetics of the adsorption of acetate on silica-alumina surfaces (Ikeda et al., 1982) have shown that the rate determining step consists of

$$\equiv Al\text{-}OH_2^+ + Ac^- = \equiv AlAc + H_2O \ .$$

For bidentate (chelating) ligands such as salicylate, evidence from spectroscopy (Motschi, 1987) and from dissolution kinetics (Furrer and Stumm, 1986; Stumm and Furrer, 1987) unequivocally supports surface complexation.

<u>Adsorption of metal ions</u>

Adsorption of metal ions can be based on electrostatic attraction or on surface complexation. We shall briefly discuss the two important adsorption modes placing some emphasis on experimental criteria that permit us to discriminate between the two basic mechanisms.

<u>Electrostatic attraction</u> is to be expected for pH values above the pH_{pzc} of the adsorbing mineral; an upper value is imposed by the formation of negatively charged hydroxo complexes of the dissolved metal (Fig. 4). Again we have to discriminate between (i) situations where the adsorbed metal ion is part of the diffuse double layer and (ii) arrangements where adsorption is based on ion pairing with deprotonated surface hydroxyls. In the latter case, adsorption can be observed at pH values well below the pH_{pzc}. The pertinent reaction can then be described by

$$\equiv S\text{-}O^- + M(H_2O)_6^{z+} = \equiv S\text{-}O^- (H_2O)M(H_2O)_5^{z+} \ . \tag{13}$$

Please note, that the metal ion is still coordinated with six water molecules and that there is no direct link between the metal ion and the deprotonated surface hxdroxyl group. The ion pair formed in Reaction (13) is therefore also designated as "outer sphere" complex.

Electrostatic attraction is the prevailing adsorption mode for metal ions with little Lewis acidity such as the alkali ions and the heavier members of the alkaline earth metals ions (Sr^{2+} and Ba^{2+}).

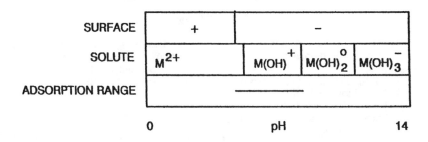

Figure 4. Electrostatic interaction of hydrolyzable metal ions with charged interfaces.

Surface complexation is based on reaction of deprotonated surface hydroxyls and metal ions:

$$\equiv S\text{-}O^- + M(H_2O)_6^{z+} = \equiv S\text{-}O\text{-}M(H_2O)_5^{(z-1)+} + H_2O \quad . \tag{13a}$$

It should be emphasized that the metal ion is now directly bound to the ligand atom of the surface ligand forming, what is called an inner sphere complex.

How to distinguish between inner sphere and outer sphere complexes. Stoichiometry will not be of any help because the water molecule released in Reaction (13a) is hard to be detected in an aqueous environment. Thermodynamics does not provide a direct tool for discrimination because (i) the equilibrium constants of the Reactions (13) and (13a) are formally identical and (ii) the Gibbs free energies of formation of inner and outer sphere complexes are not fundamentally different. Thermodynamics may indirectly assist in discriminating between the two metal adsorption modes because ion pair formation will depend on the charge density of the adsorbed metal ion. The observation (Huang and Stumm, 1973) that the equilibrium constants for the adsorption of the alkali earth ions Mg^{2+}, C^{2+}, Sr^{2+} and Ba^{2+} on alumina decrease with increasing ionic radius is in accordance with electrostatic attraction. On the other hand data on the adsorption of some divalent ions on "hydrous ferric oxide" (Dzomback and Morel, 1990) clearly show that the stability constants of the surface complexes formed follow the series

$$Ni^{2+} (83) < Zn^{2+} (86) < Cu^{2+} (87) < Pb^{2+} (133)$$

(numbers in parentheses: ionic radii in picometer).

They tend to increase with decreasing charge density indicating surface complexation. The most important difference between inner sphere and outer sphere complexation lies in the fact that inner sphere complexation affects the orbital systems of the coordinated metal ions. Spectroscopic methods are therefore the most sensitive for detecting inner sphere complexes. There is ample evidence from spectroscopy that adsorption of most of the metal ions (except the foregoing mentioned alkali and alkaline earth metal ions) form inner sphere surface complexes.

288

Co-adsorption of metal ions and organic ligands

Studies on the adsorption of metal ions at mineral surfaces have been, in most cases, carried out in the presence of "innocent" ligands such as NO_3^- and ClO_4^- that do not form complexes with the dissolved metal ion nor participate in surface reactions. Similarly, adsorption of organic ligands has usually been investigated in the presence of alkali ions. In natural aquatic systems as well as in the course of technical processes such as flotation adsorption of metal ions, organic ligands occur in the presence of each other. Under such conditions, both metal ions and ligands can still adsorb independently. In addition we have to look for possible mechanisms for co-adsorption.

We shall first consider the possible reactions of a metal-ligand complex ML with the mineral surface. As a starting point, we shall assume L to be a bidentate ligand (i.e., a ligand that carries two ligand atoms), and the metal ion is assumed to have a coordination number of six. Hence in the complex ML, two sites in the coordination shell of the metal ion are occupied by the ligand L and four sites are still occupied by water molecules. These four sites are thus available for surface complexation:

$$\equiv S\text{-}OH + M(H_2O)_4 L^{z+} = \equiv S\text{-}O\text{-}M(H_2O)_3 L^{(z-1)+} + H^+ \ . \tag{14}$$

The species formed by complexing ML with the surface hydroxyls is a metal ion that carries (besides the ubiquitous water molecules) two different sets of ligands. Such species that are known to form in solution are called "mixed" or "ternary" complexes. We shall adapt the later term and designate the species schematically depicted by (I) as a type A ternary surface complex.

I. $$\equiv S\text{-}O\text{-}M\text{-}L$$

Next we shall assume L to be a tetradentate ligand. If the geometrical arrangements of the ligand atoms permit the simultaneous coordination of four sites in the metal ions coordination shell, we have a situation similar to the one already described. The complex still has two coordinated water molecules that can be exchanged for surface ligands. If the geometrical arrangements of the ligand atoms permit the simultaneous coordination of just two of the four ligand atoms, we have an modified situation: Two of the ligand atoms may replace surface hydroxyls attaching thus the ligand to the surface while the remaining ligand atoms may coordinate with the metal ion:

$$2 \equiv S\text{-}OH + M(H_2O)_2 L^{z+} = (\equiv S)_2\text{-}L\text{-}M(H_2O)_2^{(z+2)+} + 2\ OH^- \ . \tag{14a}$$

The so formed surface species (symbolized by the scheme II) is a type B ternary surface complex.

II: $$= S\text{-}L\text{-}M$$

For sake of completeness, it remains to consider the case of L being a hexadentate ligand such as EDTA with all ligand atoms available for simultaneous coordination with the metal ion. In this case the coordination shell of the metal ion is completely filled by the ligand atoms of L, and ML can not react with deprotonated S-OH groups except for the case where the S-O⁻ groups can successfully compete with the ligand atoms of L. The chelate effect that greatly stabilizes ML prevents, however, such a competition and we expect that ML is not adsorbed by surface complexation. This expectation is supported by experimental evidence.

As mentioned in the foregoing sections, the adsorption of metal ions is generally favored by high pH values, whereas low pH values promote adsorption of ligands. Thus, Ag⁺-ions typically adsorb on iron(III)hydroxide at pH values above 7, whereas a ligand such as thiosulfate ($S_2O_3^{2-}$) adsorbs in the acidic region. Hence, when Davis and Leckie (1978) found that in the presence of thiosulfate Ag⁺-ions are adsorbed at low pH, they concluded that type B ternary surface complexes must be formed. Bourg and Schindler (1978), in a study on the effect of ethylenediamine (en) on the adsorption of copper(II) on silica presented the first evidence for the formation of type A ternary surface complexes: This argumentation was based on the observation that the amount of adsorbed copper(II) was considerably higher than that calculated on the basis of just competition for copper(II) of the surface ligand Si-O⁻ and the dissolved ethylenediamine, respectively. In order to account for the experimentally observed adsorption data, they had to assume the formation of two type A ternary surface complexes.

Subsequent studies, especially on the effect of dissolved ligands upon the adsorption of metal ions, gave further evidence for the preferential formation of type A complexes (Basak et al., 1990). In most cases, the thermodynamic stabilities of these ternary species were not particularly high. Ternary surface complexes of considerable stability were found with organic molecules that belong to the class of π-acceptor ligands. The result from adsorption studies was supported by EPR measurements (von Zelewsky and Bemtgen, 1982). Further support came from ENDOR (electron nuclear double resonance) and ESEEM (electron spin echo envelope modulation) spectroscopy (Rudin and Motschi, 1984; Motschi, 1987).

THERMODYNAMIC STABILITY

The objectives of this section are
— to introduce the conventions adapted for the description of the thermodynamic stability of surface complexes,
— to outline the experimental methods established for evaluating the stability constants of ternary surface complexes, and
— to discuss the factors that effect the stabilities of ternary surface complexes.
Although it is tacitly assumed that the surface species discussed in this section are inner sphere complexes, it should be remembered that most of the conclusions apply also to outer sphere complexes.

Conditional and intrinsic constants

Let's consider the simple equilibrium

$$\equiv\text{S-OH} + M^{z+} = \equiv\text{S-OM}^{(z-1)+} + H^+ \tag{15}$$

The intrinsic microscopic constant (King, 1965) is defined by the equation

$$K^s_{(int)} = \frac{a_{\equiv SOM} \, a_H}{a_{\equiv SOH} \, a_M} \tag{16}$$

where a_i denotes the activity of the surface species i. For convenience this activity of a surface species can be expressed as the product of the surface concentration $\{i\}$ (in mole per kg of adsorbing solid) and the surface activity coefficient y_i (in kg mole^{-1}):

$$K^s_{(int)} = \frac{\{\equiv SOM\} \, y_{\equiv SOM} \, a_H}{\{\equiv SOH\} \, y_{\equiv SOH} \, a_M} \tag{16a}$$

It is possible to measure the surface concentrations $\{SOH\}$ and $\{SOM\}$. It may sometimes be convenient to express the concentrations of the surface species in the same units as the concentrations of the solutes (mole dm^{-3}). This conversion can be performed with the aid of equations such as

$$[\equiv SOH] = \frac{A}{V} \{\equiv SOH\}$$

where A is the mass of the suspended solid mineral (kg) and V is the volume of the solution (dm^3). It is, unfortunately, in most cases impossible to experimentally evaluate the surface activities a_H and a_M, respectively. For the case that surface and solution are at electrochemical equilibrium, the electrochemical potentials of H^+ and M^{z+} are equal in both phases, so that

$$\mu^*_{(H,aq)} = \mu^*_{(H,surf)} \; ; \; \mu^*_{(M,aq)} = \mu^*_{(M,surf)} \tag{17}$$

The electrochemical potential of a species i in a phase j is given by

$$\mu^*_{(i,j)} = \mu^{*0}_{(i,j)} + RT \ln a_{(i,j)} + z_i F y_{(j)} \tag{18}$$

where $y_{(j)}$ is the potential at the phase j and z_i is the charge number of the species i and μ^{*0} is the standard potential. In the presence of an ionic medium of constant ionic strength, the activities can, without any loss of rigor, be replaced by concentrations. By choosing the standard states to be the same in both phases

$$\mu^{*0}_{(H,aq)} = \mu^{*0}_{(H,surf)} \; ; \; \mu^{*0}_{(M,aq)} = \mu^{*0}_{(M,surf)} \tag{19}$$

and by assigning a zero potential to the aqueous solution we obtain from (17) and (18)

$$a_H = [H^+] \exp(-Fy/RT) \; ; \; a_M = [M^{z+}] \exp(-zFy/RT) \tag{20}$$

Combining Equations (16) and (20) results in

$$K^s_{(int)} = \frac{\{\equiv SOM\}[H^+]}{\{\equiv SOH\}[M^{z+}]} \frac{y_{\equiv SOM}}{y_{\equiv SOH}} \exp((z-1)Fy/RT) = K^s \Gamma \tag{21}$$

or

$$K^s = K^s_{(int)} (\Gamma)^{-1} \tag{22}$$

where

$$K^S = \frac{\{\equiv SOM\}[H^+]}{\{\equiv SOH\}[M^{z+}]} \tag{23}$$

is the experimentally accessible conditional stability constant and

$$\Gamma = \frac{y_{SOM}}{y_{SOH}} \exp((z-1)Fy/RT) \tag{24}$$

collects terms that contribute to deviation from ideality.

Evaluating intrinsic stability constants

The previous section leaves us with the problem of evaluating intrinsic stability constants from the experimentally available conditional constants. Any solution of this problem requires consideration of the physical nature of the factors that are responsible for non-ideal behavior. These factors are (Schindler and Stumm, 1987):

The above mentioned surface potentials that originate from the adsorption of charged species. (Note that this chapter uses the concept of potential determining ions in an extended sense as to include every kind of adsorbed ions.) For the simple case that only one charged species is adsorbed, the surface potential will vanish at zero surface coverage. For the case of simultaneous adsorption of several species (i.e., for co-adsorption of metal ions and ligands) the surface potential will again vanish at zero surface coverage or at zero charge conditions where the charges of the various adsorbed species compensate each other.

Lateral interaction (both attraction and repulsion) are based on coulombic and van der Waals forces. Again these lateral interactions will vanish at zero surface coverage.

Surface heterogeneity, i.e., the presence of $\equiv$S-OH groups with different binding properties, are in turn reflected by different $K^S_{(int)}$ values. At low surface coverage, adsorption will be limited to reactions involving surface hydroxyls with maximum $K^S_{(int)}$ values.

With charged species, there is a strong overlap of the first two factors (Sposito, 1983), and the above separation of Γ into terms comprising activity coefficients and surface potentials, respectively, is not unambiguous. On the basis of the foregoing consideration, we will outline the methods used for evaluating intrinsic stability constants.

Studies at low surface coverage. The above discussions indicate that the influence of both surface potentials and lateral interactions decrease with decreasing surface coverage so that the difference between K^S and $K^S_{(int)}$ diminishes and may eventually escape experimental observation. Hence data on trace adsorption of Fe(III), Pb(II), Cu(II), and Cd(II) on amorphous silica (Schindler et al., 1976) extending over a considerable pH range could reasonably well be modelled with the aid of conditional constants.

Extrapolation techniques. In favorable cases where adsorption can be described by just one reaction leading to a single surface complex i, the intrinsic constant is available from extrapolation of K^S values obtained at different surface coverages (surface charges) to zero coverage (zero charge) conditions. This extrapolation is greatly assisted in cases where K^S and $K^S_{(int)}$ are related by a Frumkin type equation

$$K^S = K^S_{(int)} \exp (\alpha\{i\}) \tag{25}$$

$$\log K^S = \log K^S_{(int)} + \frac{\alpha}{\ln (10)} \{i\} \tag{26}$$

and $\log K^S$ is a linear function of the concentration of the surface complex i. The extrapolation procedure can (although with considerable loss of reliability) also be applied in cases where $\log K^S$ is not a linear function of $\{i\}$. Details of the extrapolation technique are given by Schindler and Stumm (1987).

Double layer techniques. By double layer techniques, we will designate the various methods used to evaluate $K^S_{(int)}$ from Equation (21) using different models (i.e., the Helmholtz model, the Gouy-Chapman model, the Stern model and the triple layer model) of the electrified water-mineral interface. For a detailed discussion the reader is referred to Chapter 6 of this volume. The different approaches have in common that they neglect lateral interactions assuming, therefore, the surface activity coefficient to be unity. They are furthermore presumed to be equivalent in fitting experimental data (Westall and Hohl, 1980; Morel et al., 1981) although this presumption is supported by a rather limited experimental basis. They differ, however, in their perception on how the Gibbs free energy of adsorption

$$DG_{ads} = - RT \ln(10) \log K^S \tag{27}$$

is split into

$$DG_{int} = - RT \ln(10) \log K^S_{(int)} \tag{28}$$

and

$$DG_{coul} = zF \tag{29}$$

Therefore they produce somewhat different values of K^S and sometimes even different surface species. Despite these complications, the use of double layer technique is strongly indicated for adsorption processes involving several simultaneous reactions. Westall's (1982) FITEQL V 2.0 is an almost ideal computer program for evaluating intrinsic stability constant of simultaneously occurring surface reactions.

In comparing the results of the three methods, it can be presumed that they may fundamentally produce similar results for homogeneous surfaces with identical $\equiv$S-OH groups. For heterogeneous surfaces, both studies at low surface coverage and extrapolation techniques will result in $K^S_{(int)}$ values of the thermodynamically most active sites, whereas double layer techniques will presumably produce some average values.

Stability constants of ternary surface complexes

Definitions. Let's consider the formation of a type A ternary surface complex

$$m \equiv S\text{-}OH + M^{z+} + n L = (\equiv S\text{-}O)_m ML_n^{(z-m)+} + m H^+ \tag{30}$$

According to Equation (21) its stability can by described by

$$K^s_{(int)} = \frac{\{(\equiv SO)_m ML_n^{(z-m)+}\}[H^+]^m}{\{\equiv SOH\}^m [M^{z+}][L]^n} \Gamma \tag{31}$$

It is, however, convenient to combine Equation (30) with Equation (32)

$$ML_n^{z+} = M^{z+} + nL \; ; \; K = (\beta_n)^{-1} \tag{32}$$

where

$$\beta_n = [ML_n^{z+}]/[M^{z+}][L]^n \tag{33}$$

is the stability constant of the dissolved complex ML_n. The combination results in

$$m \equiv S\text{-}OH + ML_n^{z+} = (\equiv S\text{-}O)_m ML_n^{(z-m)+} + m \, H^+ \tag{34}$$

$$K^s_{m(int)}(ML_n) = \frac{\{(\equiv SO)_m ML_n^{(z-m)+}\}[H^+]^m}{\{\equiv SOH\}^m [ML_n^{z+}]} \Gamma(ML_n) \tag{35}$$

The stability constant of the ternary surface complex as defined by Equation (35) can now be compared with the stability constant of the corresponding binary complex

$$m \equiv S\text{-}OH + M^{z+} = (\equiv S\text{-}O)_m M^{z+} \tag{36}$$

$$K^s_{m(int)}(M) = \frac{\{(\equiv SO)_m M^{z+}\}[H^+]^m}{\{\equiv SOH\}^m [M^{z+}]} \Gamma(M) \tag{37}$$

As in solution chemistry, it is convenient to make this comparison by defining the relative stability of the type A ternary surface complex by the ratio

$$R_{m,n} = \frac{K^s_{m(int)}(ML_n)}{K^s_{m(int)}(M)} \tag{38}$$

where the indices m and n represent the numbers of (deprotonated) surface hydroxyls and (mono- or polydentate) ligands respectively. A similar expression can be defined for type B ternary surface complexes.

Predictions from statistics. For statistical reasons, ternary complexes are less stable than binary complexes. We shall illustrate this fact taking binary and ternary surface complexes of copper(II) as examples (Fig. 5). Cu(II) complexes are subject to Jahn-Teller deformation transforming the O_h symmetry into a D_{4h} symmetry.

294

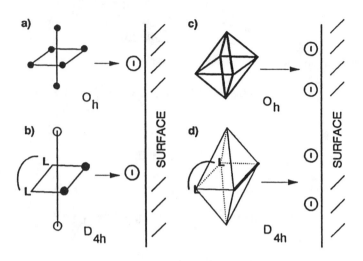

Figure 5. Statistical stabilities of ternary surface complexes. (a) The relative probability of an octahedrally coordinated metal ion to become attached to a surface is equal to the number (6) of available corners. (b) Upon coordination of a rigid bidentate ligand (each ligand atom is symbolized by L), the Jahn-Teller deformation of Cu(II) complexes is likely to become static. Two equatorial corners (full circles) are left for coordination with a surface site. The axial corners (open circles) are too distant to provide surface coordination. (c) The relative probability to coordinate with two surface sites is equal to the number (12) of available edges. (d) After coordination of a rigid bidentate ligand there is just one edge left over.

The Jahn-Teller theorem states that any nonlinear molecular system in a degenerate electronic state will be unstable and will undergo some kind of distortion that will lower its symmetry and split the degenerate states. The electronic state of the Cu^{2+} ion placed in the center of an octahedron of ligands is a degenerate E_g state. According to the Jahn-Teller theorem the octahedron must be disturbed in some way, for instance by stretching the octahedron along one of its axes. For details the reader is referred to standard textbooks such as J.E. Huheey, Inorganic Chemistry, Third Edition, Harper & Row, New York, 1983.

We will assume that the Jahn-Teller deformation is dynamic for the aquaion $Cu(H_2O)_6^{2+}$ and that oscillation of the coordinated water molecules along the axes of the coordination polyhedron is fast as compared with ligand exchange reactions. It is reasonable to assume that coordination of bidentate or polydentate ligands will lead to a static deformation. Moreover, it has been shown by McBride et al.(1984) that tetragonally deformed Cu(II) coordinates to surface hydroxyls at gibbsite by using the equatorial positions only.

Let's first compare the (statistical) stabilities of the surface complexes ≡S-OCu+ and ≡S-OCu-L+ for the case that L is a bidentate ligand. In a statistical sense, we can say that the relative stability of the surface complex is equal to the number of available positions in

TABLE 1. Stability constants of type A ternary surface complexes
          (298.2 K)

System	$R_{1,1}$	$R_{1,2}$	$R_{2,1}$	Ref.
SiO_2-Cu(II)-en	2	0	4.2 E-2	a
SiO_2-Cu(II)-gly	0.38	0	8.9 E-4	b
SiO_2-Cu(II)-ox	0.13	0	5.1 E-3	b
SiO_2-Cu(II)-bipy	4.9 E2	4.57 E3	0	b
SiO_2-Zn(II)-bipy	-	-	11.5	c
SiO_2-Mg(II)-gly	0.68	0	0	b,d
SiO_2-Mg(II)-ala	1	0	0	b,d
TiO_2-Mg(II)-gly	0.94	0	0	b,d
TiO_2-Co(II)-gly	1.15	0	4.6 E-3	b,d

a) Bourg and Schindler (1978), b) Basak et al. (1990),
c) Schindler and Vayloyan (unpublished), d) Gisler (1980)

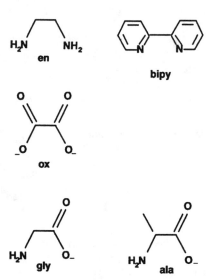

Figure 6. Ligands found to form ternary Cu(II) complexes on silica. en: ethylenediamine, ox: oxalate, gly: glycine, ala: alanine, bipy: 2,2' bipyridine.

the coordination shell of the Cu(II) species. Hence, referring to Figure 5, the relative stabilities of $\equiv$S-OCu$^+$ and $\equiv$S-OCuL$^+$ are 6 and 2, respectively, and we obtain

$$R_{1,1(stat)} = \frac{K^s_{(int)}(CuL)}{K^s_{(int)}(Cu)} = \frac{2}{6} \tag{39}$$

For metals not undergoing Jahn-Teller deformation we would from similar argumentation expect a value of $R_{1,1(stat)} = 4/6$.

Next we will compare the relative stabilities of $(\equiv SO)_2Cu^0$ and $(\equiv SO)_2CuL^0$. Because the coordination to the surface requires one edge of the coordination polyhedron, the relative stabilities of the two complexes are 12 and 1, respectively, and

$$R_{2,1(stat)} = \frac{K^s_{2(int)}(CuL)}{K^s_{2(int)}(Cu)} = \frac{1}{12} \tag{40}$$

For non-Jahn-Teller deformed metals we obtain $R_{2,1(stat)} = 5/12$.

Finally we would expect $R_{1,2,(stat)}$ and $R_{2,2(stat)}$ to be zero for copper complexes coordinated with two bidentate ligands because the ligands L will coordinate using the equatorial sites.

Experimental results. Although there is ample evidence for the formation of ternary surface complexes, published values of stability constants are scarce. Bourg and Schindler (1978), Gisler (1980) and Basak et al. (1990) obtained stability constants of type A ternary surface complexes from studies on the effect of dissolved bidentate ligands (Fig. 6) on the adsorption of metal ions on amorphous silica and on TiO$_2$ (rutile). The experimental approach shall be outlined by taking the system Cu^{2+}-SiO$_{2(am)}$-glycine as an example (Fig. 7; Basak et al., 1990). At 298.2 K Cu^{2+} was found to adsorb on silica from 1 M KNO$_3$ solutions in the range $5 < pH < 7$ by forming two binary surface complexes

$$\equiv Si\text{-}OH + Cu^{2+} = \equiv Si\text{-}OCu^+ + H^+ \qquad \log K^s_{1(int)} = -5.22 \tag{41}$$

$$2\equiv SiOH + Cu^{2+} = (\equiv Si\text{-}O)_2Cu^0 + 2H^+ \qquad \log K^s_{2(int).} = -9.24 \tag{42}$$

Addition of glycine inhibits adsorption especially at high pH values. This inhibition is based of the formation of complexes with the dissolved ligand HL

$$Cu^{2+} + HL = CuL^+ + H^+ \qquad \log {}^*K1 = -1.44 \tag{43}$$

$$Cu^{2+} + 2\,HL = CuL_2^0 + 2\,H^+ \qquad \log {}^*\beta2 = -4.38 \tag{44}$$

and thus by competition of dissolved ligands with surface ligands

$$\equiv Si\text{-}OCu^+ + HL = CuL^+ + \equiv Si\text{-}OH \qquad \log K = 3.76 \tag{45}$$

$$\equiv Si\text{-}OCu^+ + 2HL = CuL_2^0 + \equiv Si\text{-}OH + H^+ \qquad \log K = 0.48 \tag{46}$$

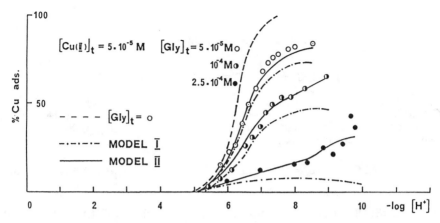

Figure 7. Adsorption of Cu(II) at silica in the absence and presence of glycine. The dashed curve (model I) was calculated with the aid of the equations (38)-(45). The solid lines were computed with a model that includes the formation of two ternary surface complexes. From Basak et al. (1990).

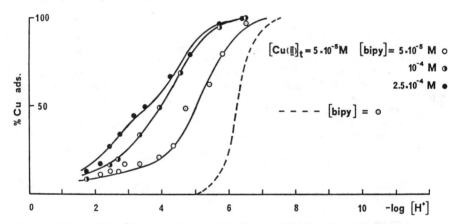

Figure 8. Effect of 2,2'-bipyridine on the adsorption of Cu(II) on amorphous silica. From Schindler and Stumm (1987).

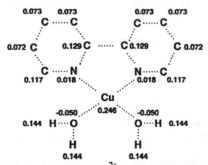

Figure 9. Tentative charge distribution in $Cu(bipy)(H_2O)_4^{2+}$ (Ludwig, 1990). The calculations were performed with the aid of the program ICONC (Calzaferri et al., 1989) that is based on the program ICON8 by Howell et al. (1978). The calculated figure should be considered as an illustration of the general nature of the charge distribution. The axial water molecules and the hydrogen atoms of the bipy ligand are omitted for simplicity.

$$(\equiv\text{Si-O})_2\text{Cu}^0 + \text{HL} + \text{H}^+ = \text{CuL}^+ + 2 \equiv\text{Si-OH} \qquad \log K = 7.75 \qquad (47)$$

$$(\equiv\text{Si-O})_2\text{Cu}^0 + 2 \text{HL} = \text{CuL}_2^0 + 2 \equiv\text{Si-OH} \qquad \log K = 4.86 \qquad (48)$$

Calculations based on Equilibria (41)-(48) (model I in Fig. 7) underestimate the extent of adsorption. The best fit could be obtained by a model (model II in Fig. 7) that assumes the formation of two ternary surface complexes

$$\equiv\text{Si-OH} + \text{CuL}^+ = \equiv\text{Si-OCuL}^0 + \text{H}^+ \qquad \log K_1^s(\text{CuL}) = -5.64 \qquad (49)$$

$$2 \equiv\text{Si-OH} + \text{CuL}^+ = (\equiv\text{Si-O})_2\text{CuL}^- + 2 \text{H}^+ \qquad \log K_2^s(\text{CuL}) = -12.29 \qquad (50)$$

The so far obtained stability constants of type A ternary surface complexes are collected in Table 1. Stability constants of type B ternary surface complexes have so far not been reported.

Charge effects. In this and the following paragraph we will discuss some factors that seem to control the stabilities of type A ternary surface complexes. The data (Table 1) indicate that the stabilities of complexes $\equiv$S-O-M-L (as expressed by $R_{1,1}$) are partially controlled by statistics. In addition, charge effects may be involved and may even play a major role (Schindler and Stumm, 1987). The negatively charged silica surface seems to exhibit a slight preference for positively charged complexes as seen from the sequence $R_{1,1}(\text{Cu-en}^{2+}) > R_{1,1}(\text{Cu-gly}^+) > R_{1,1}(\text{Cu-ox}^0)$. Among the species $(\equiv\text{SO})_2\text{ML}$, the stability of the complex $(\equiv\text{Si-O})_2\text{Cu(en)}^0$ is within the scope of statistical expectations. The negatively charged species $(\equiv\text{SO})_2\text{M(II)gly}^-$, $(\equiv\text{SO})_2\text{M(II)ala}^-$ and $(\equiv\text{SO})_2\text{M(II)ox}^{2-}$ are, however, rather unstable and in some cases not detectable. This is in agreement with the earlier qualitative observations of Farrah and Pickering (1976). It can thus tentatively be concluded that on negatively charged mineral surfaces, formation of positively charged surface complexes is (compared to statistical prediction) slightly favoured, whereas the formation of negatively charged species is inhibited.

Ternary surface complexes with π-acceptor ligands. As seen from Table 1, a striking effect is exhibited by 2,2' bipyridine. The presence of this ligand shifts the pH range of the adsorption of Cu(II) at silica by several pH units from the neutral into the acidic region (Fig. 8). In addition to the complexes recorded in Table 1, two more surface species are formed without proton removal

$$\equiv\text{Si-OH} + \text{Cubipy}^{2+} = \equiv\text{Si-OHCubipy}^{2+} \qquad \log K_{(\text{int})}^s = 0.8 \qquad (51)$$

$$\equiv\text{Si-OH} + \text{Cu(bipy)}_2^{2+} = \equiv\text{Si-OHCu(bipy)}_2^{2+} \qquad \log K_{(\text{int})}^s = 1.63 \qquad (52)$$

In these equations, the number of involved silanol groups is uncertain. A similar analogous adduct is formed with Zn(II) (Schindler and Vayloyan, unpublished) where

$$\equiv\text{Si-OH} + \text{Zn(bipy)}_2^{2+} = \equiv\text{Si-OHZn(bipy)}_2^{2+} \qquad \log K_{(\text{int})}^s = 1.67 \qquad (53)$$

A similar promotion of the adsorption of Cu(II) at silica has been observed for other π-acceptor ligands, i.e., 1,10 phenantroline and for 2,2',6',2" terpyridine (von Zelewsky and Bemptgen, 1982). In addition, it is known that 2,2' bipyridine stabilizes ternary complexes preferentially with oxygen ligands in solution (Griesser and Sigel, 1970).

π-acceptor ligands contain energetically low positioned empty π orbitals. It was therefore presumed by Griesser and Sigel (1970) that the π-acceptor ligands draw electron density from the d-orbitals of the metal ion, therefore enhancing thus the positive charge of the Cu(II) center. Extended-Hueckel-MO calculations (Fig. 9) do not support this hypothesis. Schindler and Stumm (1987) mentioned the possibility of a hydrophobic interaction of the coordinated ligand with the silica surface. On the other hand, adsorption of free (uncoordinated) 2,2' bipyridine on silica has so far not been observed.

SPECTROSCOPY

It was previously mentioned that formation of inner sphere complexes affects the orbital energies of the coordinated metal ions. This leads to a modification of the electron spectra. Unfortunately, it turns out, that changes induced by the replacement of one or two water molecules by deprotonated surface hydroxyls are marginal and not easily detectable by reflectance spectroscopy.

In this section we shall present information on ternary surface complexes gathered from spectroscopic analysis. The bulk of the informations originates from EPR (electron paramagnetic resonance) spectroscopy and related magnetic resonance methods. For details of the available techniques, the reader is referred to the competent review by Motschi (1987).

Methods

EPR spectroscopy can easily distinguish between freely tumbling complexes in solution and immobilized species. In aqueous solution with a rotational correlation time on the order of 10^{-10} s, the anisotropic contributions from the g and hyperfine coupling constants are averaged and an isotropic spectrum results. An anisotropic spectrum occurs upon immobilization. In the presence of adsorbing surfaces, the occurrence of an anisotropic spectrum is usually ascribed to adsorption by surface complexation. The extent of adsorption can even be evaluated from the relative contributions of isotropic and anisotropic spectra (von Zelewsky and Bemtgen, 1982). Moreover, the static EPR parameters ($g_{\parallel}$ and $A_{\parallel}$) are sensitive to changes in the coordinative environment of the adsorbed metal ion. Some precaution is indicated for the case of porous adsorbents. Bassetti et al. (1979) have demonstrated that the mobility of $Cu(H_2O)_6^{2+}$ can be reduced by narrow pores to an extent as to produce an anisotropic spectrum.

Although EPR is a highly valuable tool for identification of ternary surface complexes, it should not be overlooked that the obtained information is qualitative in nature. It seems that the evaluation of the number of surface ligands involved is especially difficult.

ENDOR (electron nuclear double resonance) measurements permit the rough determination of bond distances (for cases where the interaction between the magnetic moments of the electron and the nucleus are of a dipolar nature). The advantage of the higher resolution goes at expense of a reduced sensitivity.

The reader is referred to Chapter 7 for additional information on these spectroscopic techniques.

Results

Perhaps the first report of the formation of ternary surface complexes goes back to Takimoto and Miura (1972). From EPR spectra of Cubipy^{2+} and Cu(bipy)$_2^{2+}$ adsorbed on alumina and 50% silica-alumina, they found that Cubipy^{2+} forms two surface complexes, $\equiv$S-OCubipy$^+$ and $(\equiv$S-O)$_2$Cubipy^0. With Cu(bipy)$_2^{2+}$, the surface species cis-Cu(bipy)$_2(\equiv$SO-)$_2$ was identified.

Important contributions to our knowledge of ternary surface complexes have been provided by M.B.McBride. In a study on the adsorption of Cu(II) on microcrystalline gibbsite, McBride et al. (1984) first concluded (from the observed values of g$_\parallel$ and A$_\parallel$) that adsorbed Cu(II) is equatorially coordinated with two water molecules and presumably two deprotonated surface hydroxyls. After exposure to NH$_3$ vapor, the adsorbed Cu(II) remained firmly bound to the surface. From the change in the EPR spectrum, they suggested the formation of a ternary surface complex $(=$Al-O)$_x$Cu(NH$_3$)$^{(4-\bar{x})+}$ (x: 1 or 2). In a subsequent study (McBride, 1985), it was found that in the presence of glycine, Cu(II) adsorbs on gibbsite and boehmite to form ternary surface complexes. Three different species were identified (Fig. 10,a-c). More recent work (McBride, 1987) gives evidence for the formation of type A ternary surface complexes in the systems boehmite-VO^{2+}-phosphate and allophane-VO^{2+}-phosphate.

Von Zelewsky and Bemtgen (1982) investigated the formation of ternary Cu(II) surface complexes with π-acceptor ligands on silica. At a bipy to Cu ratio of 1, surface complex ($\equiv$SiO)Cubipy was identified. At bipy/Cu = 2, a species of the composition ($\equiv$SiO)$_x$Cu(bipy)$_2$ was found. In agreement with the earlier work of Takimoto and Miura (1972), they concluded that the observed values of the EPR parameters suggest the formation of cis-Cu(bipy)$_2(\equiv$Si-O)$_2^0$ (Fig. 10d).

Rudin and Motschi (1984) investigated the structure of ternary Cu(II) complexes on δ-Al$_2$O$_3$ with both EPR and ENDOR. The proton ENDOR study of the binary complex $(=$Al-O)$_2$Cu^0 revealed the presence of both equatorial and axial water molecules. Additional insight was gained from nitrogen ENDOR. The measured hyperfine tensors were found to be typical for nitrogen coordination in the molecular plane of the Cu(II) complex (Fig. 10e,f).

THE ROLE OF TERNARY SURFACE COMPLEXES IN NATURE AND TECHNOLOGY

It is the aim of this section to present some case studies that may elucidate the possible role of ternary surface complexes in both nature and technology. The presentation is not intended to be exhaustive. Important topics such as heterogeneous catalysis and heterogeneous photochemistry for solar energy conversion are not covered.

Effect of organic ligands upon the fate of trace metals in aquatic environments

The fate of trace metals in natural aquatic systems as well as in soils is largely controlled by adsorption/desorption processes involving both inorganic and organic particulate matter (Schindler 1975a,b, 1984, 1989; Sposito, 1984; Sigg, 1987). Transformations such as

$$M(\text{adsorbed}) \leftrightarrow M(\text{particulate}) \qquad (54)$$

are of fundamental relevance for the mobility of the involved metal ions in the particular reservoir. In aquatic systems, adsorbed metals are subject to sedimentation. In soils, desorption leads to mobilization and to transfer from the soil to the ground water system.

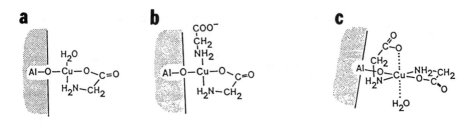

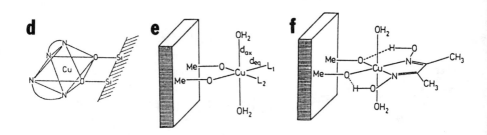

Figure 10. Proposed structures of ternary surface complexes with some bidentate ligands. (a)-(c) Ternary Cu(II) surface complexes with glycine (McBride, 1985), (d) cis-Cu(bipy)$_2$(Si-O)$_2$ (von Zelewsky and Bemtgen, 1982), (e) and (f) Cu(bipy)(H$_2$O)$_2$(Al-O)$_2$Cu(oxime)(H$_2$O)$_2$(Al-O)$_2$. From Rudin and Motschi (1984).

Assuming reaction (54) to be at equilibrium, the ratio of concentrations of particulate and dissolved metal can be described by the distribution coefficient

$$K_D = \{M\}/[M] \qquad (dm^3/kg) ,$$ (55)

where

$\{M\}$ = Concentration of particulate metal (mole per kg of particulate matter)

$[M]$ = Concentration of dissolved metal (mole dm^{-3})

K_D as defined by Equation (55) is a very important parameter in the convection-dispersion equation that describes the transport of adsorbable solutes in the soil column as well as in steady state models for trace metals in lakes and oceans (Schindler, 1975b).

Obviously $\{M\}$ is the sum of the concentrations of the metal bearing surface complexes. Thus in the absence of dissolved ligands we have

$$\{M\} = \sum_{m}^{m} \{\equiv SO\}_m M^{(z-m)+} = \sum^{m} K^s_{m(int)}(\Gamma)^{-1}\{\equiv SOH\}^m[M^{z+}][H^+]^{-m}$$ (56)

Since $[M^{z+}] = [M]$, one obtains

$$K_D = \sum^{m} K^s_{m(int)}(\Gamma)\text{-}1\{\equiv SOH\}^m[H^+]^{-m}$$ (57)

In the presence of a dissolved ligand L we have to consider both dissolved and adsorbed complexes:

$$\{M\} = \sum^{m}\{\equiv SO\}_m M^{(z-m)+} + \sum\sum^{m\ n}\{\equiv SO\}_m ML_n^{(z-m)+} \tag{58}$$

$$[M] = [M^{z+}] + \sum^{n}[ML_n^{z+}] \tag{59}$$

From Equations (33),(35), and (38) one obtains

$$K_D = \frac{F1 + F2}{1 + \sum \beta_n [L]^n} \tag{60}$$

where

$$F1 = \sum^{m} K_{m(int)}^s (\Gamma)^{-1}\{\equiv SOH\}^m [H^+]^{-1}$$

$$F2 = \sum\sum^{m\ n} K_{m(int)}^s R_{m,n} (\Gamma(ML_n))^{-1}\{\equiv SOH\}_m \beta_n [L]^n [H^+]^{-m}$$

Neglecting the coefficients Γ Equation (57) simplifies to

$$K_D = \frac{\sum^{m} K_{m(int)}^s \{\equiv SOH\}^m [H^+]^{-1}(1+\sum^{n} R_n^m \beta_n [L]^n)}{1 + \sum \beta_n [L]^n} \tag{61}$$

For the usual case that the values of $R_{n,m}$ are smaller than unity, the presence of dissolved ligands will thus diminish K_D and enhance the mobility of the metal. It must be emphasized that Equation (61) does not take into account the formation of type B ternary complexes with multidentate ligands. Nature provides us with multidentate ligands in the form of fulvic acids and humic acids. These substances are known to adsorb at oxide and clay surfaces and to strongly bind metal ions. The nature of the ternary complexes formed with clay and humic substances will be discussed in the next paragraph.

The structure of the clay-organic interface

In most mineral soils, practically all of the humic materials occurs in association with clay minerals (Stevenson, 1985). On the other hand, most of the inorganic particulate matter in lake water and sea water is covered with organic coatings consisting of humic substances. At the usual soil pH, both the clay surface and the humic substances are negatively charged. Electrostatic interaction is thus not in favor of the observed associations. Their stability may thus be based on ligand exchange involving polydentate groups of the humic material and/or of the formation of ternary surface complexes. The proposed schematic structure of the clay-organic interface (Fig. 11; Stevenson and Ardakani, 1972) exhibits the characteristic features of type A ternary surface complexes. Metal ions M (Al^{3+}, Fe^{3+}) are coordinated to both the clay surface and the organic polyelectrolyte. It

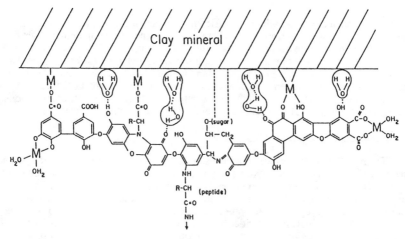

Figure 11. Schematic structure of the clay-humic acid complex. From Stevenson and Ardakani (1972).

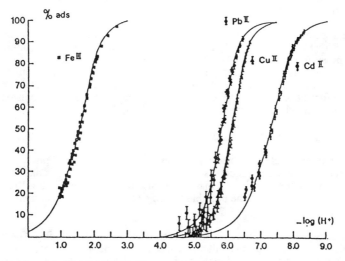

Figure 12. Adsorption of Fe(III), Pb(II), Cu(II) and Cd(II) on amorphous silica. From Schindler et al. (1976).

Table 2. pH ranges of 90% recovery of quartz in froth flotation with 1.E-4 M sulfonate and 1.E-4 M salt solution. Adapted from Fuerstenau et al. (1985).

Metal salt	pH range
Fe(III)	2.9 - 3.8
Al(III)	3.8 - 8.4
Pb(II)	6.5 - 12.0
Mn(II)	8.5 - 9.4
Mg(II)	10.9 - 11.7
Ca(II)	> 12

elucidates the mechanisms on which the efficiency of the various extractants used to isolate humic and fulvic substances from soils is based. EDTA and other polydentate ligands such as diphosphate and citrate remove the bridging Al(III) and Fe(III) centers. Elevated pH values assist the separation of clay and humic fraction (a) by ligand exchange to replace the anchoring polydentate groups and (b) by enhancing the negative charge of the humic material and thus weakening the ternary complexes. It is seen from Figure 11 that the functional -OH and -COOH groups exposed to the soil solution can coordinate further metal ions and metal-ligand complexes. This leads to apparently unlimited multilayer structures.

Ternary surface complexes in froth flotation

The role of ternary surface complexes in froth flotation shall be illustrated by taking quartz activation as an example. Quartz activation is usually performed with anionic collectors such as sulfonates and carboxylates (fatty acids). Anionic flotation is, however, dependent on the presence of metal ions. Each of the tested metal ions is observed to cover a specific pH range for optimum activation (Table 2). It is seen from Figure 12 that the pH range of optimum activation coincides with the pH range of adsorption on SiO_2. This coincidence strongly suggests that the adsorption of the collector anions is based on the formation of type A ternary surface complexes. A numerical simulation which accounts for the enhanced adsorption of anionic surfactants on quartz in the presence of Ca^{2+} on the basis of formation of type A ternary surface complexes was recently presented by Rea and Parks (1990).

The nature of the chemical bond between activator and collector anions has been investigated by Motschi and McEvoy (1985). In a study on the adsorption of linear alkylbenzene-sulfonates on δ-Al_2O_3 and TiO_2, they found that the presence of surface complexed metal ions such as Cu^{2+} promotes adsorption. The EPR spectrum of adsorbed Cu^{2+} was, however, not changed by the adsorbed sulfonate. This means that interaction is based on the formation of outer sphere complexes (ion pairs). This conclusion does not necessarily apply to other collectors.

Ternary surface complexes in heterogeneous redox reactions

This paragraph is concerned with heterogeneous redox reactions. This class of reactions includes important processes such as the corrosion of metals and the oxidative and reductive dissolution of minerals that will be discussed Chapter 10. It is the purpose of this paragraph to demonstrate the possible key role of ternary surface complexes in these reactions.

There are two general mechanisms for redox reactions in solutions. The first mechanism is called "outer sphere" and will not be discussed. The second "inner sphere" mechanism involves the formation of binuclear mixed complexes (Taube, 1970). For instance, the reduction on the non-labile $CoCl(NH_3)_5^{2+}$ by the labile $Cr(H_2O)_6^{2+}$ results in the formation of labile $Co(H_2O)_6^{2+}$ and non-labile $CrCl(H_2O)_5^{2+}$. The quantitative production of $CrCl(H_2O)_5^{2+}$ implies that the electron transfer from Cr(II) to Co(III) and the chloride transfer from Co(III) to Cr(II) are not independent of each other. It is generally accepted that an intermediate species

$$(H_2O)_5Cr-Cl-Co(NH_3)_5$$

is formed.

Zabin and Taube (1965) have studied the reductive dissolution of Fe_2O_3 by $Cr(H_2O)_6^{2+}$. In the presence of ClO_4^- the reaction was found to proceed slowly. Addition of Cl^- lead to an increase of the reaction rate by several orders of magnitude, the oxidized Cr(III) being present as $CrCl(H_2O)_5^{2+}$. This suggests a mechanism (Schindler, 1985) that involves the formation of a type B ternary surface complex

$$\equiv Fe\text{-}OH + H^+ + Cl^- = \equiv Fe\text{-}Cl + H_2O$$

$$\equiv Fe\text{-}Cl + Cr^{2+} = \equiv Fe\text{-}Cl\text{-}Cr^{2+}$$

followed by the electron transfer

$$\begin{array}{cc} III & II \\ \equiv Fe\text{-}Cl\text{-}Cr^{2+} \\ \downarrow \\ II & III \\ \equiv Fe\text{-}Cl\text{-}Cr^{2+} \end{array}$$

and the hydrolysis

$$\downarrow$$

$$\begin{array}{cc} II & III \\ \equiv Fe\text{-}OH_2 + CrCl^{2+} \end{array}$$

The Fe(II) centers are then rapidly dissolved by acid. Recent work on the reductive dissolution of magnetite (Baumgartner et al., 1983), goethite (Cornell and Schindler, 1987) and hematite by Stumm and his students (Suter et al., 1988, Banwart et al., 1989) has revealed the key role of the complexes

$$\begin{array}{cc} III & II \\ \equiv Fe\text{-}oxalate\text{-}Fe^+ \\ \downarrow \\ II & III \\ \equiv Fe\text{-}oxalate\text{-}Fe^+ \end{array}$$

in the reductive dissolution of iron minerals by oxalic acid (see Chapter 10).

ACKNOWLEDGMENTS

The bulk of this chapter was written during a stay at Stanford University. The author is greatly indebted to Professor George Parks for his generous hospitality and for valuable and inspiring discussions. Research grants from the Swiss National Foundation have supported the research on surface complexation.

REFERENCES

Banwart, S., Davies, S. and Stumm, W. (1989) The role of oxalate in accelerating the reductive dissolution of hematite (α-Fe$_2$O$_3$) by ascorbate. Colloids & Surfaces 39, 303-310.

Basak, M., Bourg, A.C.M., Cornell, R.M., Gisler, A., Schindler, P.W., Stettler, E. and Trusch, B. (1990) The effect of dissolved ligands upon the adsorption of metal ions at oxide water interfaces (to be published).

Bassetti, V., Burlamacchi, L. and Martini, G. (1979) Use of paramagnetic probes for the study of liquid adsorbed on porous supports. Cu(II) in water solution. J. Amer. Chem. Soc. 101, 5471-5477.

Baumgartner, E., Blesa, M.A., Marinovitch, H.A. and Maroto, A.J.G. (1983) Heterogeneous electron transfer pathways in dissolution of magnetite in oxalic acid. Inorg Chem. 22, 2224-2226.

Bourg, A.C.M. and Schindler, P.W. (1978) Ternary surface complexes I. Complex formation in the system silica-Cu(II)-ethylenediamine. Chimia 32, 166-168.

306

Calzaferri, G., Forss, L., Hugentobler, Th. and Kamber, I. (1989) EHMO-Calculalations: ICONC and INPUTC, Internal Report, Univ. of Bern, Bern, Switzerland.

Cornell, R.M. and Schindler, P.W. (1987) Photochemical dissolution of goethite in acid/oxalate solutions. Clays Clay Minerals, 35 347-352.

Davis, J.A. and Leckie, J.O. (1978) Effect of adsorbed complexing ligands on trace metal uptake by hydrous oxides. Environ. Sci. Technol. 12, 1309-1315.

Dzombak, D.A. and Morel, F.M.M (1990) Surface Complexation Modelling, John Wiley and Sons, New York.

Farrah, H. and Pickering, W.F. (1976) The sorption of copper species by clays I. Kaolinite. Aust. J. Chem. 29, 1167-1176.

Fuerstenau, M.C., Miller, J.D. and Kuhn, M.C. (1985) Chemistry of Flotation. Am. Inst. Mining, Metallurgical and Petroleum Engineers, New York, p. 115-118.

Furrer, G. and Stumm, W. (1986) The coordination chemistry of weathering: I.Dissolution kinetics of δAl_2O_3 and BeO. Geochim. Cosmochim. Acta 50, 1847-1860.

Gisler, A. (1980) Die Adsorption von Aminosäuren an Grenzflächen Oxid-Wasser, Ph.D. dissertation, Univ. of Bern, Bern, Switzerland.

Griesser, R. and Sigel, H. (1970) Ternary complexes in solution. VIII. Complex formation between the copper(II)-2,2'-bipyridyl 1:1 complex and ligands containing oxygen and/or nitrogen donor atoms. Inorg. Chem. 9, 1238-1243.

Howell, J., Rossi, A., Wallace, K., Haraki, K. and Hoffmann, R. (1978) ICON8 quantum chemistry program performing extended Hückel calculations. QCPE No. 344.

Huang, C.P. and Stumm, W. (1973) Specific adsorption of cations on hydrous Al_2O_3. J. Colloid Interface Sci. 43, 409-420.

Ikeda, T., Sasaki M., Hachiya, K., Astumian, R.D., Yasunaga, T. and Schelly, Z.A. (1982) Adsorption-desoprtion kinetics of acetic acid on silica-alumina particles in aqueous suspensions, using the pressure-jump relaxation method. J. Phys. Chem. 86, 3861-3866.

Karickhoff, S.W., Brown, D.S. and Scott, T.A. (1979) Sorption of hydrophobic pollutants on natural sediments. Water Res. 13, 241-248.

Ludwig, Ch. (1990) Personal communication.

Lyman, W.J. (1982) Adsorption coefficients for soils and sediments. In: Lymann, W.J., Reehl, W.F. and Rosenblatt, D.H., eds., Handbook of Chemical Property Estimation Methods. McGraw-Hill, New York.

McBride, M.C., Fraser, A.R. and McHardy, W.J. (1984) Cu^{2+} interaction with microcristalline gibbsite. Evidence for oriented chemisorbed copper ions. Clays Clay Minerals 32, 12-18.

McBride, M.C. (1985) Influence of glycine on Cu^{2+} adsorption by microcristalline gibbsite and boehmite. Clays Clay Minerals, 33 397-402.

McBride, M.C. (1987) Ternary VO^{2+}-ligand-surface complexes on boehmite and noncrystalline aluminosilicates. J. Coll. Interface Sci. 120, 419-429.

Morel, F.M.M., Westall, J.C. and Yeasted, J.G. (1981) Adsorption models: A mathematical analysis in the framework of general equilibrium analysis. In: Anderson, M.A. and Rubin, A.J., eds., Adsorption of Inorganics at Solid-Liquid Interfaces. Ann Arbor Science, Ann Arbor, Michigan.

Motschi, H. (1987) Aspects of the molecular structure in surface complexes; spectroscopic investigations. In: Stumm, W., ed., Aquatic Surface Chemistry. John Wiley and Sons, New York.

Motschi, H. and McEvoy, J. (1985) Influence of metal/adsorbate interactions on the adsorption of linear alkylbenzenesulfonates to hydrous surfaces. Naturwissenschaften 72, 654-655.

Rea, R.L. and Parks, G.A. (1990) Numerical simulation of coadsorption of ionic surfactants with inorganic ions on quartz. In: Melchior, D.C. and Basset, R.L., eds., Chemical Modeling of Aqueous Systems II. ACS Symposium Series 416, Am. Chem. Soc., Washington, D.C.

Rudin, M. and Motschi, H. (1984) A molecular model for the structure of copper complexes on hydrous oxide surfaces: An ENDOR study of ternary Cu(II) complexes on δ-alumina. J. Coll. Interface Sci. 98, 385-393.

Schindler, P.W. (1975a) Removal of trace metals from the oceans: a zero order model. Thalassia Jugosl. 11, 101-111.

Schindler, P.W. (1975b) The ragulation of trace metal concentrations in natural water systems: A chemical approach. Proc. 1st Speciality Symposium on Atmospheric Contributions to the Chemistry of Lake Waters, Int'l Assoc. Great Lakes Res., p. 132

Schindler, P.W. (1989) The regulation of heavy metal concentrations in natural aquatic systems. Proc. 7th Int'l Conf. on Heavy Metals in the Environments, Geneva, Sept. 1989, Vol. 2, p. 210-216.

Schindler, P.W. and Stumm, W. (1987) The surface chemistry of oxides, hydroxides, and oxide minerals. In: Stumm, W., ed., Aquatic Surface Chemistry. John Wiley and Sons, New York

Schindler, P.W., Fürst, B., Dick, R. and Wolf, P.U. (1976) Ligand properties of surface silanol groups. I. Surface complex formation with Fe^{3+}, Cu^{2+}, Cd^{2+} and Pb^{2+}. J. Coll. Interface Sci. 55, 469-475.

Schindler, P.W. and Vayloyan, A. (1990) Unpublished results.

Schwarzenbach, R.P. and Westall, J. (1981) Transport of nonpolar organic compounds from surface water to groundwater. Laboratory sorption studies. Environ. Sci. Technol. 15, 1360-1367.

Sigg, L. (1987) Surface chemical aspects of the distribution and the fate of metal ions in lakes. In: Stumm, W., ed., Aquatic Surface Chemistry. John Wiley and Sons, New York

Sposito, G. (1983) On the surface complexation model of the oxide-aqueous solution interface. J. Coll. Interface Sci. 91, 329-340.

Sposito, G. (1984) The Surface Chemistry of Soils. Oxford Univ. Press, New York.

Stevenson, F.J. and Ardakani, M.S. (1972) Organic matter reactions involving micronutrints in soils. In: Mortvedt, J.J., Giordano, P.M. and Lindsay, W.L., eds., Micronutrients in Agriculture. Soil Sci. Soc., Madison Wisconsin.

Stumm, W. and Furrer, G. (1987) The dissolution of oxides and aluminum silicates; examples of surface-coordination controlled kinetics. In: Stumm, W., ed., Aquatic Surface Chemistry. John Wiley and Sons, New York

Suter, D., Siffert, C., Sulzberger, B. and Stumm, W. (1988) Catalytic dissolution of iron(III)(hydr)oxides by oxalic acid in the presence of Fe(II). Naturwissenschaften 75, 571.

Takimoto, K. and Miura, M. (1972) Studies on the basic sites of the silica-alumina surface by means of the adsorption of copper(II) complexes with 2,2'-bipyridine. Bull. Chem. Soc. Japan 45, 653.

Taube, H. (1970) Electron Transfer Reactions of Complex Ions in Solution. Academic Press, New York.

Ulrich, H.J., Stumm, W. and Cosovic, B. (1987) Adsorption of fatty acids on aquatic interfaces. Comparison between two model surfaces: The mercury electrode and δ-Al_2O_3 colloids. Environ. Sci. Technol. 22, 37-41.

von Zelewsky, A. and Bemtgen, J.M. (1982) Formation of ternary copper(II) complexes at the surface of silica gel as studies by esr spectroscopy. Inorg. Chem. 21, 1771-1777.

Westall, J.C. (1982) FITEQL, A computer program for determination of chemical equilibrium constants from experimental data, Version 2.0, Report 82-02, Dept. of Chemistry, Oregon State Univ., Corvallis, Oregon.

Westall, J.C. and Hohl, H. (1980) A comparison of electrostatic models for the oxide/solution interface, Adv. J. Coll. Interface Sci. 12, 265-294.

Wirz, U. (1984) Die Adsorption proteinogener Aminosäuren an TiO_2 (Anatas). Ph.D. dissertation, Univ. of Bern, Bern, Switzerland.

Zabin, B.A. and Taube, H. (1964) The reactions of metal oxides with aquated chromium(II) ion. Inorg. Chem. 3, 963-968.

CHAPTER 8

GORDON E. BROWN, JR.

SPECTROSCOPIC STUDIES OF CHEMISORPTION REACTION MECHANISMS AT OXIDE-WATER INTERFACES

INTRODUCTION AND OVERVIEW

The chemisorption of adatoms and admolecules at mineral/water interfaces is an important microscopic process affecting many of the macroscopic geochemical processes that occur in the Earth's crust (for overview see Davis and Hayes, 1986). Chemisorption reactions at solid/liquid and solid/gas interfaces are also important in affecting the processes occurring at the surfaces of many technologically-important materials, such as insulator oxides and metals used in heterogeneous catalysis and semiconductors used in integrated circuits (e.g., Williams and McGovern, 1984; Lundqvist, 1984; Kiselev and Krylov, 1985). Systematic study of sorption processes at solid/gas interfaces began in the early 1900's with the pioneering work of Irving Langmuir (Langmuir, 1916; Suits, 1962), which led to recognition that solid surfaces play an important role in most chemical reactions involving solids. Interest in the field of surface science has literally exploded during the past several decades, driven in large part by the need to understand the molecular-level mechanisms of catalyzed surface reactions and how surface structure and chemisorption affect surface electronic processes in thin-film semiconductors and integrated circuits. Extension of the concepts and methodology of surface science to problems in mineralogy and geochemistry began in the 1970's with the first applications of x-ray photoelectron spectroscopy to the study of dissolution phenomena of silicates (e.g., Petrovic et al., 1976). Although the field of interface geochemistry is still in its infancy, it is already leading to changes in the way we think about many geochemical and mineralogical processes. Continuing studies in this rapidly growing field will undoubtedly lead to a more fundamental understanding of the physics and chemistry of mineral/water interfacial phenomena than we presently have and to new insights about how the hydrosphere interacts with the Earth's crustal rocks and how chemical species partition between minerals and aqueous fluids.

In this chapter we will discuss the molecular-level concept of sorption reactions at solid/water interfaces and the fact that molecular-level information is required for an adequate description of sorption processes. In order to gain some insight as to how the structure of a solid surface differs from that of the bulk solid and how adsorbates may perturb surface structure, we will review some of the general observations derived from surface crystallographic studies of metal and oxide surfaces. We will also examine some of the surface-sensitive spectroscopic methods that are used to obtain information on the structure, bonding, and composition of chemisorbed species at solid/liquid interfaces as well as selected results from recent studies using these methods. Our major focus will be on a few methods that can directly probe the structure of sorbates at solid/water interfaces, particularly synchrotron-based, x-ray absorption spectroscopy. We will then summarize some of the more recent results from in situ Extended X-ray Absorption Fine Structure (EXAFS) and Surface EXAFS (SEXAFS) spectroscopic measurements on sorbed species, in an attempt to lay a foundation for understanding the molecular mechanisms of chemisorption of metal ions and molecules at mineral/water interfaces. One of our primary aims is to give the reader a glimpse of the enormous literature in the field of surface science that deals with the structure and bonding of chemisorbed species and the growing number of structural probes that are now being used to study these species and the surfaces they sorb on. Most of this literature is concerned with the structure and electronic properties of single-crystal metal and oxide surfaces in ultra-high vacuum and gas-phase sorption reactions at these surfaces. Although it may not seem directly relevant to studies of sorption reactions at mineral/water interfaces, most of which have been done on high surface area, powdered minerals, this body of information contains much that we should be aware of. Many parallels can be drawn between the structure and chemical reactivity of

310

mineral surfaces in contact with liquid H_2O and that of metal, semiconductor, or oxide surfaces in contact with vacuum or a gas phase in terms of common methods and concepts.

Chemisorption versus physisorption

Early in the 20th century, many scientists thought that sorption at the solid/gas interface involved the attraction of gas-phase atoms toward the solid due to long-range attractive forces (van der Waals interaction), which is now referred to as *physisorption*. It was thought that gas-phase atoms became more concentrated toward the surface, much like retention of the Earth's atmosphere by the gravitational field, but with little interaction with the atoms of the solid substrate (Zangwill, 1988). This thinking was changed significantly by Langmuir's work on adsorption-induced changes in the work function of metals (e.g., Cs sorbed on W; Langmuir and Taylor, 1932) which investigated the possibility that strong, short-range forces can exist between adsorbates and the surface of a solid. Langmuir considered the atomic arrangement at the solid surface as defining a specific density of potential sorption sites to which gas-phase atoms or molecules might bind through the formation of a chemical bond to the surface. This process, referred to as *chemisorption*, is illustrated in Figure 1 using the simple potential energy model introduced by Lennard-Jones (1932) for gas-phase sorption. These plots show one-dimensional models of potential energy changes as a gas molecule, AB, or the dissociated gas atoms, A+B, approach a solid surface, shown schematically as the hachured area at zero separation. The dashed curve in each represents the physisorption interaction (van der Waals or London dispersion forces) between AB and a solid surface. The dash-dot curves in Figures 1a and 1b represent the change in potential energy when A and B form strong chemisorption bonds to the solid surface due to the deep energy minimum at small surface-adatom separation. Figure 1a illustrates the process of dissociative chemisorption, i.e., the AB molecule spontaneously dissociates at the separation marked by the crossing point of the two curves, which occurs below the zero of energy; the individual atoms form chemisorption bonds with surface atoms. Molecular physisorption occurs at low temperatures when the crossing point of the two curves occurs above the zero of energy (Fig. 1b). At higher temperatures the dissociated species would form because thermal

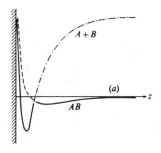

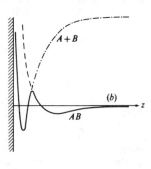

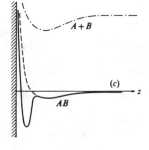

Figure 1. Schematic diagrams of potential energy vs. separation for the sorption of the gas-phase molecule AB on a solid substrate illustrating (a) dissociative chemisorption, (b) molecular physisorption, and (c) molecular chemisorption. From Zangwill (1988).

activation overcomes the potential energy barrier separating the molecular physisorption state and the dissociated chemisorption state. Here the physisorbed state is a precursor to the chemisorbed state, although this may not always be the case. Finally, Figure 1c illustrates molecular chemisorption, which occurs when the energy curve of the dissociated state (dash-dot curve) never crosses the curve of the associated state (dashed curve). However, a deeper minimum occurs in the potential energy curve of the molecular state at small separations resulting in the formation of a strong chemisorption bond between AB and the solid surface. A more quantitative view of the nature of these bonding forces is dealt with in the chapter by Lasaga (this volume), in the book edited by Rhodin and Ertl (1979), in the review by Van Hove et al. (1989), and in numerous papers that attempt to model chemisorption interactions on solid surfaces by molecular orbital calculations (e.g., Ban et al., 1987; Hoffman, 1988; Kobayashi and Yamaguchi, 1989; Pisani et al., 1989; Hong et al., 1990).

The need for molecular-level information about sorption mechanisms

Most of the information we have about sorption reactions of cations or anions at mineral/water interfaces comes from macroscopic measurements of sorption behavior, primarily solubility and kinetic measurements (Sposito, 1986; Parks, this volume) and from fitting these data using one of several surface complexation models (see Haworth, 1990 for a recent review; Davis and Kent, this volume). However, Sposito (1986) pointed out that "the adherence of experimental sorption data to an adsorption isotherm provides no evidence as to the actual mechanism of a sorption process" and concluded that "...the inherently macroscopic, indirect nature of the data produced by such measurements limits their applicability to determine sorption mechanisms in a fundamental way." Similar conclusions were also reached by Westall and Hohl (1980). Certain classes of surface-sensitive spectroscopy experiments can provide more direct structural information needed to define sorption mechanisms at mineral/water interfaces.

The general term *sorption* as used in this volume refers to any of several possible mechanisms by which a chemical species may partition from an aqueous solution to a solid surface in contact with it. Many aqueous surface chemists distinguish among precipitation, adsorption, and absorption as possible sorption mechanisms (Sposito, 1986). Because these mechanisms will be examined in some detail below through spectroscopic means, it is important to define operationally what we mean by these terms. We begin with Sposito's (1986) definitions of these terms but modify them as needed to reflect what can or cannot be observed by surface-sensitive spectroscopic or scattering measurements. *Precipitation* means the formation from solution of a solid phase of different composition and structure than the solid substrate that exhibits a three-dimensional structure, although one not necessarily exhibiting long-range periodicity. We deviate somewhat from Sposito's definition of a precipitate as "a primitive molecular unit (a complex) that repeats itself in three dimensions" because it could imply the formation of a phase exhibiting long-range, repetitive order as in a crystal. Precipitate phases exhibiting only short-range order (i.e., x-ray amorphous phases) or intermediate degrees of order between x-ray amorphous and crystalline are also possible. We also exclude the precipitation of a phase with the same composition and structure as the solid (i.e., crystal growth) because the spectroscopic methods used to study sorbed species cannot distinguish such precipitates from the original solid substrate. *Adsorption* implies the accumulation of matter at the interface between an aqueous solution phase and a solid adsorbent without the development of a three-dimensional molecular arrangement. This definition implies the formation of a two-dimensional molecular arrangement on the surface, although one could also envision linear or branched molecules attached to the solid surface by only one bond in the extreme case, with their long axis at an angle other than 0° to the surface. The terms *specific adsorption* and *nonspecific adsorption* are sometimes used to denote chemisorption and physisorption, respectively (see Parks, this volume for more detail). *Absorption* means the incorporation of an aqueous chemical species into a solid phase by diffusion or some other means, such as dissolution of the solid followed by reprecipitation of the solid with the formerly aqueous chemical species included as part of a solid solution or solid inclusion.

Figure 2 illustrates three possible adsorption mechanisms at the oxide/water interface as well as sorption through formation of a surface precipitate. Formation of an *outer-sphere* adsorption complex (non-specific adsorption) occurs when a positively or negatively charged complex retains its waters of hydration and is hydrogen bonded or attracted to the surface through long-range coulombic forces. Formation of an *inner-sphere* adsorption complex (specific adsorption) occurs through loss of waters of hydration and direct, short-range bonding of the complex to surface oxygens. Inner-sphere complexes may bond to one, two, or more surface oxygens, resulting in monodentate, bidentate, or multidentate molecular configurations. Due to the difference in strength of hydrogen bonds or long-range coulombic forces and short-range electrostatic or covalent bonds, inner-sphere complexes are typically more strongly bound to the surface than outer-sphere complexes. These different adsorption mechanisms can be expressed by the following reactions in which surface hydroxide sites are represented by $\underline{S}OH$, where $\underline{S}$ is a metal atom at the surface, and a hexa-hydrated, divalent aqueous metal ion is represented by $M^{2+}(aq)$:

$$\underline{S}OH + M^{2+}(aq) = \underline{S}OHM^{2+}(aq) \qquad \text{(outer-sphere complex)} \qquad (1)$$

$$\underline{S}OH + M^{2+}(aq) = \underline{S}OM^+ + H^+ \qquad \text{(monodentate, inner-sphere complex)} \quad (2)$$

$$2\,\underline{S}OH + M^{2+}(aq) = \underline{S}O_2M^0 + 2\,H^+ \qquad \text{(bidentate, inner-sphere complex)} . \qquad (3)$$

Because these reactions release 0, 1, and 2 protons, respectively, careful monitoring of pH changes should allow the experimentalist to distinguish among them. However, the form of such reactions is model dependent, and therefore one cannot unambiguously define the molecular-level sorption mechanism by this means. For example, Hohl and Stumm (1976) measured the sorption of aqueous Pb(II) on γ-Al_2O_3 as a function of pH. They used a constant capacitance surface complexation model to simulate their experimental data, which indicated that a non-interger number of protons between one and two were released during the sorption process. Moreover, they described the sorption process as being specific, from which we can infer that Pb sorbs as inner-sphere complexes. Thus some combination of Reactions (2) and (3) could be used to describe their results. Davis and Leckie (1978) used

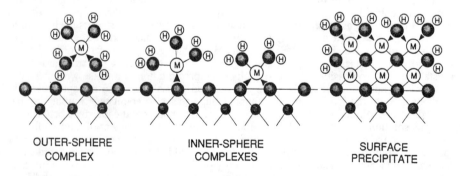

OUTER-SPHERE   INNER-SPHERE   SURFACE
COMPLEX   COMPLEXES   PRECIPITATE

Figure 2. Definition of possible sorption complexes at the solid/water interface, which is represented by the horizontal line. The solid substrate is below the line and the solution is above the line. The circles labeled M represent sorbed metal atoms in various types of sorption complexes. The larger shaded spheres in the substrate and surrounding the metal in the solution phase are oxygens. The smaller dark spheres in the substrate are metal ions, as are the spheres labeled M in the sorption complexes and surface precipitate.

a triple-layer surface complexation model to fit Hohl and Stumm's experimental data. Davis and Leckie's reaction replaces the bidentate complex formed in Reaction (2) with a monodentate complex as shown in Reaction (4):

$$\underline{A}lOH + Pb^{2+}(aq) + H_2O = \underline{A}lOPbOH + 2H^+ . \tag{4}$$

The second proton required to account for the non-interger number of protons released is provided by hydrolysis of the Pb^{2+}. Both modeling approaches fit the sorption data equally well and thus illustrate the conclusions of Sposito (1986) and Westall and Hohl (1980) that solution chemistry and surface complexation modeling do not uniquely define sorption mechanisms.

The structure of "clean" and "real" surfaces

Though we know a great deal about the average crystal structure of almost all of the 3000 or so mineral species from detailed single-crystal x-ray diffraction studies, far less is known about the structures of their surfaces or surface regions (the top few atomic layers adjacent to the solid/vacuum, solid/gas, solid/liquid, or solid/solid interface). We suspect that the surface region of a solid at such an interface may have a structure different from the bulk because a surface can be thought of as a major defect in the solid state. Assuming structural differences, the electronic and vibrational properties of the surface may differ considerably from the bulk properties, resulting in unique physical and chemical properties in the surface region. Surface structure, composition, and bonding together with solution conditions (pH, type of ion, solution species concentrations) control the reactivity at solid/water interfaces, so information about the atomic arrangement at the surface, particularly at reactive sites, is necessary for a complete description of chemisorption reactions at oxide/water interfaces.

Surface scientists commonly distinguish between "clean" and "real" surfaces. The term "clean" refers to a surface that is free of adsorbates or impurities, whereas a "real" or "dirty" surface has atoms or molecules chemisorbed to it and/or may contain impurity atoms. Mineral surfaces are likely to be hydroxylated or hydrated if in contact with air or water. Those containing cations or anions that can undergo reduction or oxidation may also have a different stoichiometry than the bulk mineral. Because such adsorbate layers or impurities affect the chemical, mechanical, and electronic properties of surfaces, it is necessary to study clean surfaces as well as real surfaces if one wishes to understand the intrinsic differences in surface structure and bonding and chemisorptive behavior among different solids. Clean surfaces can only be obtained by cleaning or preparing a real sample in an ultra-high vacuum system ($< 10^{-9}$ torr) by heat treatment, ion bombardment, cleavage, crushing, or some other appropriate means such as chemical processing. Unfortunately, each of these cleaning or preparation methods has some drawback. For example, high-temperature annealing and ion bombardment may lead to reconstruction of the surface and/or a change in its composition. Cleavage, when possible, can result in the release of dissolved gas necessitating heat treatment of the specimen before cleavage, which can cause redistribution of defects and impurities. Dry crushing is often accompanied by the aggregation of particles, which changes the surface area of the sample and affects the interpretation of certain experimental results (Kiselev and Krylov, 1985). Thus obtaining clean surfaces of a material with reproducible structure, composition and defect population is not a trivial task. Determining the structure of such a surface is even more formidable.

Clean and real surfaces may undergo relaxation or reconstruction relative to the bulk structure during cleavage, ion bombardment, high temperature annealing, or some other type of treatment. *Relaxation* implies that the surface atoms adopt new equilibrium positions that change the outermost interlayer spacing. Such relaxation will result in changes in the lengths of bonds between the first and second layer in the surface region (in some cases extending down to four or five layers) and in certain bond angles, but it does not result in changes in the number of nearest neighbors around a cation or anion. If there are no changes of in-plane atomic positions within the outermost layer, the two-dimensional surface unit cell will be the same as for the ideal surface structure, which is obtained by projection of the bulk unit cell to that surface. *Reconstruction* implies that the surface atoms move to new equilibrium positions such that the number of nearest neighbors, bond distances in and at angles to the surface, bond angles, and rotational symmetry in the surface region change relative to the bulk structure. Unreconstructed surfaces are referred to as (1x1) surfaces to reflect the fact that the surface unit cell is the same as the bulk unit cell projected onto that surface, whereas reconstructed surfaces can have quite different surface unit cells relative to the bulk (e.g., the Si (111) (7x7) surface, the rutile (100) (1x7) surface). This type of surface unit cell nomenclature is discussed in Somorjai (1981) and Hochella (this volume).

Before considering clean versus real oxide surfaces, let us examine some of the general conclusions concerning metal surfaces reached by Van Hove (1979), Somorjai (1981, 1990), Ohtani et al. (1986), Somerjai and Van Hove (1989), and Van Hove et al. (1989) based on Low Energy Electron Diffraction (LEED) and other structural studies of clean and adsorbate-covered metal surfaces.

(1) The surface unit cells of many clean metal surfaces generally have been found to be the same as that in the bulk metal (i.e., unreconstructed), and the spacing between the first and second layers (or "z-spacing") usually differs from the bulk value by 5 percent or less. There are exceptions to this limit, sometimes involving changes in z-spacing of 10 to 15 percent.

(2) Bond length contractions between the first and second surface layers occur on clean metal surfaces. Expansion is also theoretically possible, but has not yet been proven conclusively using surface crystallographic methods. The physical origin of this contraction can be rationalized by considering the fact that with fewer neighbors, the two-body repulsion energy is smaller, allowing greater atomic overlap and, therefore, shorter bond lengths. Alternatively, one can think of the electrons belonging to the severed bonds at the surface being partly shifted to the remaining bonds, which results in increased bond orders and shorter bond lengths, or of the surface electron density attempting to form a smooth surface, resulting in a redistribution of charge which produces electrostatic forces that draw the surface atoms toward the substrate.

(3) Atomic adsorption on metals takes place preferentially (not always) in sites of high symmetry and high coordination to metal atoms.

(4) Adatom-metal bond lengths tend to agree best with bulk compound values, though with substantial scatter, part of which may be real and related to valence, bond order, coverage, etc.

(5) Adsorption tends to cause expansion of underlying metal-metal bonds in cases where these bonds are already shortened on the clean surface.

Recent SEXAFS studies of metal surfaces (clean and real) have shown that bond length changes at a surface, relative to the bulk, are typically within the limits of ±0.15 Å (Citrin, 1987). The chapter by Hochella (this volume) should also be consulted for generalizations concerning the surface structures of selected oxides.

Should we assume that the surface structure of insulator oxides is the same as the bulk structure or should we assume that the surface structure relaxes or reconstructs relative to the bulk? How does cleavage of an oxide single crystal perturb the two new surfaces relative to the bulk? How does the structure of a hydroxylated or hydrated oxide surface differ from that of a clean surface? These questions will become important when we attempt to interpret the results of spectroscopic studies of chemisorbed species at solid/water or solid/gas interfaces, so we will consider them briefly here. Insights to these questions for the most studied oxide surfaces are provided by several reviews (e.g., Heinrich, 1983, 1985) and by Hochella (this volume). In addition, crystal chemical reasoning can give us some qualitative answers to these questions, although we must keep in mind the fact that crystal chemical principles were derived from studies of bulk structures, thus may not be entirely applicable to surface structures. However, the principles of bonding, as embodied in some of the simple rules of crystal chemistry, should be generally applicable to most forms of matter, including surfaces.

Consider, for example, the (100) surface of MgO produced by cleavage in a perfect vacuum (Fig. 3). This surface is often used as a support for epitaxial growth of thin films because of its ability to induce such growth for a large number of metals over a wide range of temperatures. Ideally, each of the Mg and O atoms is five coordinated at the (100) surface, whereas they are both six coordinated in the bulk. Among the factors affecting cation-oxygen bond lengths, $d(M-O)$, in oxides is the coordination number of the cation (Pauling, 1960). As the coordination number is reduced, $d(M-O)$ should decrease in response to an increase in mean bond strength s, which was defined by Pauling as the nominal valence of the cation (z) divided by its coordination number (v). The Pauling s value for Mg-O bonds in the cleavage surface is 0.4 valence units (v.u.) and the sum of Pauling bond strengths at a five-coordinated surface oxygen is 1.93 v.u. (4 x 0.4 v.u. + 0.33 v.u.). This value is less than the 2.0 v.u. required by Pauling's electrostatic valence rule for oxides (Pauling, 1960) and indicates that the oxygen is slightly underbonded, which means it has a slight excess of electron density or negative charge. This simple calculation assumes that all individual Mg-O bonds have the same bond strength. However, that should not be true if the bonds differ in length. Brown and Shannon (1973) developed an empirical bond-strength-bond-length relationship ($s_{B-S} = \exp [(r_0-r)/B]$, where r is the observed or predicted bond length and r_0 (the length of a bond of unit valence, 1.693 for MgO) and B (0.37) are fitted constants with which one can quantitatively adjust the strength of individual cation-anion bonds depending on bond length. This empirical relationship has since been refined by Brown (Brown, 1977; 1981; 1987; Brown and Altermatt, 1985) and can be used to predict the changes in bond length needed to produce bond strengths that satisfy Pauling's electrostatic valence rule. With this relationship we can predict qualitative changes in bond length at the MgO (100) cleavage surface relative to the bulk. For bulk MgO $d(Mg-O)$ is 2.10 Å, and the four Mg-O in-plane bonds to a given five-coordinated surface oxygen are predicted to contract to 2.02 Å (a 3.8% reduction) in order for Pauling's valence rule to be satisfied, assuming that the Mg-O bond perpendicular to the surface remains unchanged at 2.10 Å. Although consistent with Pauling's second rule, this predicted change of in-plane bond length is likely to be an overestimate because it would produce an outermost layer incommensurate with the second layer in the surface region of MgO. An alternative calculation can be made in which the in-plane Mg-O bonds are assumed to remain at the bulk value and the bonds perpendicular to the surface relax inward or outward. Simple chemical intuition would suggest that the relaxation should be inward, and a bond-strength-bond-length calculation like the one above predicts a contraction of ~0.27 Å (a 12.9% reduction). It is likely that relaxation at the surface of MgO would involve both of these mechanisms, with the second being the more dominant.

Although these simple predictions are chemically and structurally plausible, are they consistent with interpretations of experimental data on MgO (100) cleavage surfaces? The answer is yes or no, depending on which results one believes. The earliest LEED studies of the clean MgO (100) surface (Kinniburgh, 1975, 1976) were unable to detect contraction of the outermost interlayer spacing and observed no changes of the in-plane bond lengths. A study of the same surface based on reflection high-energy electron diffraction (RHEED) (Gotoh et al., 1981) interpreted anomalous enhancement of Kikuchi patterns after heating to 573 K and then cooling to room temperature as due to an outward motion of the surface oxygens upon annealing. The magnitude of the surface rumpling was estimated to be about 6% of the atomic spacing (corresponding to an increase in length of the Mg-O bonds

316

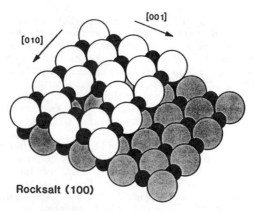

Rocksalt (100)

Figure 3. Model of the MgO(100) (1x1) cleavage surface. The small black spheres represent Mg and the large shaded or white spheres represent oxygens. The shaded oxygens are in the plane below the outer-most layer. Three types of defects are shown including (1) an oxygen-vacancy defect, resulting in four-coordinated Mg atoms, (2) a Mg-vacancy defect, resulting in four-coordinated oxygens, and (3) an edge step, with both Mg and O four-coordinated. These types of defect sites on the clean surface should be highly reactive on exposure to air, a selected gas, water, or electrolyte solutions. From Heinrich (1983).

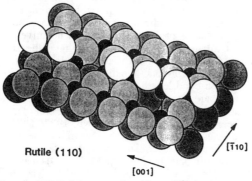

Rutile (110)

Figure 4. Model of the rutile (110) (1x1) fracture surface. The small black spheres represent Ti atoms, and the large spheres of different shading represent oxygens. Three layers of oxygens are shown, with the degree of shading increasing with increasing depth from the surface. Two types of O-vacancy defects are shown, resulting in two five-coordinated Ti atoms in one case and three five-coordinated oxygens in the second case. From Heinrich (1983).

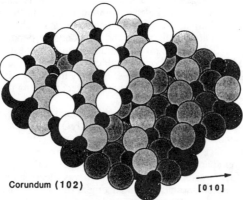

Corundum (102)

Figure 5. Model of the corundum (and hematite) (102) (1x1) fracture surfaces. The black spheres represent Al (corundum) or Fe (hematite), and the large spheres represent oxygens. The degree of shading increases with increasing depth from the surface. One type of O-vacancy defect is shown as an edge step. Both types of defects result in lower atom coordinations than on a perfect surface or in the bulk. From Heinrich (1983).

perpendicular to the (100) surface by ~0.13 Å), and it was assumed that the cleaved surface did not exhibit any rumpling before annealing. More recent LEED work by Urano et al. (1983) found no significant differences in the MgO (100) surface structure after vacuum cleaving, after annealing in vacuum at 573 K, and after annealing in O_2; their data are consistent with a model in which there is no surface rumpling and relaxation is limited to an outward displacement of the top layer of atoms by $\leq 2.5\%$ [≤ 0.05 Å increase in d(Mg-O) perpendicular to the surface], which is similar to the changes predicted by Welton-Cook and Berndt (1982) based on LEED and by the simulation study of Mackrodt (1988). These conclusions are at variance with recent ion scattering (Nakamatsu et al., 1988) and surface electron energy loss fine structure (SEELFS) measurements (Santoni et al., 1988), which were interpreted as indicating an inward rumpling of 15% and an in-plane relaxation of 0.3% in air-cleaved MgO (100) (cf. predictions above). All of these studies suggest that the vacuum-cleaved (100) surface of MgO does not undergo significant reconstruction, although significant relaxation may occur. However, the lack of a consensus concerning the structure of the MgO (100) cleavage surface based on different types of experiments (see Duriez et al., 1990) underscores the difficulty of surface structure determination, even for the simplest structure types.

Experimental studies of the structure and electronic properties of clean and real surfaces of more complex single-crystal oxides are reviewed by Heinrich (1983, 1985), Tsukada et al. (1983), and Hochella (this volume). Up to now, relatively few single crystal oxide surfaces have received much attention in the surface physics or chemistry communities, so far less is known about them than about single crystal metal and semiconductor surfaces. However, this situation is beginning to change because of the extensive use of oxides as supports for small-particle metal catalysts and the need to understand how the surface region of the oxide affects the metal's catalytic properties (e.g., Hall, 1986; Gates, 1986) and because geochemists now realize that mineral surfaces play an important role in many geochemical processes. Studies of oxide surfaces will also be stimulated undoubtedly by the use of high T_c superconductor oxide thin films in electronic devices. Among the oxide surfaces studied most thoroughly by surface crystallography and spectroscopy methods, those of greatest importance to geochemists and mineralogists are the rutile (TiO_2) (110), corundum (α-Al_2O_3) (102), and hematite (α-Fe_2O_3) (102) surfaces which are shown in Figures 4 and 5. Although rutile does not exhibit good cleavage, the (110) surface of rutile is relatively stable and non-polar (i.e., neutral charge), and breakage along this surface involves severing the smallest number of Ti-O bonds. Corundum and hematite do not cleave, although hematite does exhibit {001} and {101} parting due to twinning. Isostructural α-Ti_2O_3 and α-V_2O_3 do cleave along (102), so this surface has been studied most commonly. It is slightly corrugated like the (110) surface of rutile (Figs. 4 and 5). LEED studies of these surfaces before annealing show (1x1) LEED patterns, indicating no reconstruction and little if any relaxation of in-plane bonds (Heinrich, 1985). Little is known about contraction or expansion of the outermost interplanar spacing of these surfaces because this knowledge requires interpretation of LEED intensities using dynamical theory, which has not been attempted to my knowledge. What little we do know about these clean surfaces is that they show little if any in-plane relaxation in their unannealed states. Annealed (110) surfaces of rutile can exhibit either (1x1) or (2x1) LEED patterns, indicating that reconstruction can take place. Similarly, the (102) surfaces of α-Fe_2O_3 (Lad and Heinrich, 1988) and α-Al_2O_3 (Gignac et al., 1985) are found to undergo (1x2) reconstructions after Ar^+ ion bombardment followed by annealing in 10^{-10} torr O_2 at 900°C. Annealing the α-Fe_2O_3 (102) and (001) surfaces at a higher partial pressure of O_2 (10^{-6} torr) results in surface structures and stoichiometries approaching the "ideal" truncation of bulk α-Fe_2O_3 (Lad and Heinrich, 1988). The details of these reconstructed surfaces, however, are unclear.

One might predict that exposure of a clean (100) cleavage surface of an oxide such as MgO to O_2 would result in adsorption of oxygen and completion of the coordination sphere around the five-coordinated surface Mg ions based on reasoning that the fully

318

coordinated cation sites in the bulk structure should have a lower energy than the partially coordinated surface sites. However, experimental studies have shown that the nearly perfect (100) cleavage surface of MgO as well as the (110) fracture surface of TiO_2 are virtually inert to O_2 (Heinrich, 1979, 1983; Heiland and Lüth, 1984). No changes were observed in their XPS or UPS spectra after exposure to even thousands of Langmuirs (1 L = 10⁻⁶ torr) of O_2. Heinrich (1985) suggested that one reason for this inertness to O_2 may be related to the fact that a ligand field is required to stabilize the O^{2-} electronic configuration. In contrast to O_2, H_2O reacts readily with most oxide surfaces, in some cases undergoing dissociation to produce surface OH^- groups, often in monolayer amounts (Hair, 1967; Kiselev and Krylov, 1985, Heinrich, 1985; Schindler and Stumm, 1987). Such hydroxylated surfaces are stable in vacuum and require heating to several hundred °C to completely dehydroxylate the surface. Consider, for example, the results of an infrared investigation of the thermal decomposition of $Mg(OH)_2$ and subsequent rehydration of the oxide (Anderson et al., 1965; Razouk and Mikhail, 1958). Quoting from Hair's (1967) description of these results:

"The magnesium hydroxide initially exhibits a broad absorption band between 3650 and 3700 cm⁻¹ that is due to the antisymmetrical OH stretching vibration of the lattice hydroxide and a small band at 3770 cm⁻¹ that is a combination band also characteristic of the bulk material. Thermal decomposition at 300°C in vacuo causes these bands to be gradually removed, and new bands are produced at higher frequencies. A band at 3710 cm⁻¹ is first observed, and this in turn slowly disappears, leaving a sharp peak at 3752 cm⁻¹ together with a broad band at 3610 cm⁻¹. In conformity with the interpretation of similar bands observed in the spectrum of silica and other oxides, the sharp peak at 3752 cm⁻¹ is attributed to a free surface hydroxyl group and the band at 3610 cm⁻¹ is attributed to a hydrogen-bonded hydroxyl group. After a 2-hr heat treatment at 300°C, molecular water is no longer present on the surface, and this is evidenced by the absence of a bending vibration in the 1600-1650 cm⁻¹ region of the spectrum. Further heating causes the bands at 3610 and 3652 cm⁻¹ to disappear slowly until, at 750°C, only the free surface hydroxyl grouping can be seen. Further heating to 900°C causes the onset of the sintering process, and the (bands due to) hydroxyl groups become broadened and shifted to 3725 cm⁻¹. ....Readsorption of water, at less than monolayer coverage, onto a completely dehydrated surface causes replacement of the 3752 cm⁻¹ band and formation of a broad hydroxyl band at 3550 cm⁻¹. This latter frequency is to be compared with the value of 3610 cm⁻¹ observed in the original sample. Above a surface coverage of 0.11 molecules/Å² (9 H_2O/100 Å²), molecular water becomes adsorbed on the surface and all the hydroxyl bands are broadened and shifted to lower frequencies. The 3725 cm⁻¹ band, for instance, appears at 3675 cm⁻¹, a shift of 77 cm⁻¹. This shift is much less than that observed for H_2O on silica (Δv = 200 cm⁻¹) and is interpreted in terms of a more basic surface hydroxyl group."

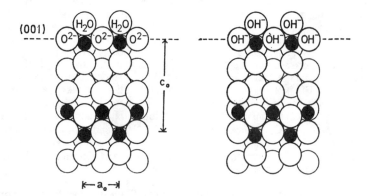

Figure 6. Schematic (010) projection of the anatase structure showing a cross-section of the (001) surface and possible sites for surface hydroxide groups and water molecules. The small dark spheres represent Ti and the large white spheres represent oxygen. From Boehm (1971).

These observations suggest that the surface of an oxide like MgO can be populated by both hydroxide groups and water molecules in air, and that substrates of different bond character can have a significant effect on adsorbate bonding. Additional IR and NMR observations of hydrated oxide surfaces indicate that the hydroxide cover of oxides is nonuniform and that hydroxide groups are arranged in various configurations (Kiselev and Krylov, 1985). These general ideas are illustrated in Figure 6, which shows possible configurations of H_2O molecules and OH⁻ groups on the (001) surface of anatase in air (Boehm, 1971). This schematic drawing also illustrates two types of OH⁻ groups: those bound to one Ti^{4+} ion and those bound to two Ti^{4+} ions. The latter type should be strongly polarized by the cations, which should reduce the strength of the O-H bond, resulting in an acidic character. The former type should have a stronger basic character and should be exchangeable for other anions. This simple model helps explain the amphoteric Bronsted acid/base character of surface hydroxide functional groups (see Parks, this volume, for further discussion).

Using a simple bond-strength-bond-length model calculation like we did above, one finds that the addition of OH⁻ or H_2O to a five-coordinated Mg in the (100) surface of MgO should satisfy the local valence balance at Mg. Addition of a proton to the five-coordinated surface oxygens should also help satisfy local charge-balance requirements at the oxygen. Although this type of reasoning may be simplistic, it suggests that hydration and/or hydroxylation of an oxide cleavage or fracture surface should result in local charge balance in the surface region of the oxide and in bond lengths much more similar to those in the bulk structure than on a clean oxide cleavage surface. Based on this reasoning, we will assume that the oxide structure at oxide/water interfaces can be considered near-perfect terminations of the bulk structure for purposes of simplifying our interpretation of spectroscopic data.

Another question that should be asked is what effect, if any, does the solid substrate have on the type of sorption complex formed at a surface. This question is addressed in the chapter by Hochella (this volume) which presents evidence that even a partial monolayer of sorbate can dramatically modify the substrate structure. One example of this effect at a solid/water interface comes from a study of the partitioning of aqueous Ni(II) on α-alumina and silica surfaces (Ahmed, 1971). This work indicates that Ni precipitates on α-alumina as the more crystalline, catalytically inactive form of Ni(OH)₂, whereas at higher pH it precipitates on silica as the active form of Ni(OH)₂. The surface structure of α-alumina is similar to that of Ni(OH)₂ (inactive), whereas that of silica differs from the structure of the active form of the precipitate. Ahmed interpreted these results as suggesting that the surface structure of the sorbent plays an important role in determining the structure and properties of the sorbate. It is not surprising that the structure of the surface should play some role in determining the structure of specifically adsorbed complexes at solid/liquid, solid/gas, or solid/vacuum interfaces. However, we don't have much direct knowledge of why surfaces might induce incommensurate or commensurate ordered versus disordered arrangements of adsorbing species. Further insights to this question depend on detailed spectroscopic studies of adsorption of a given species at the same surface coverage on structurally and chemically different surfaces (see Brown et al., 1989, and below).

There is the need for much additional structural data on clean and real oxide surfaces before we can fully answer the questions posed above. In the following sections we will examine some of the methods that can be used to provide these data and will review some of the more recent data on sorption complexes at mineral/water interfaces.

OVERVIEW OF STRUCTURAL METHODS PROVIDING MOLECULAR-LEVEL INFORMATION ABOUT CHEMISORBED SPECIES

Defining the average structure of a surface or of a chemisorbed species on a surface, particularly those at solid/liquid interfaces, is an inherently more difficult problem than defining the structure of bulk crystalline solids. This is true in part because of the non-periodic nature of some sorption complexes on surfaces, which renders x-ray diffraction methods relatively useless, and because of the low volume of the surface region or the low concentration of sorbates. In addition, x-rays or neutrons are highly penetrating, thus not particularly sensitive to surface regions unless a glancing-angle technique is employed. In contrast, electrons have relatively short mean free paths in solids (of the order of 5-10 Å in the electron kinetic energy range of 20 to ~1000 eV), although the attenuation lengths may be greater depending on the solid (see Hochella and Carim, 1988). Thus a variety of spectroscopic methods, which depend on electron scattering, emission, or absorption, have been developed to probe surface regions in solids (see Woodruff and Delchar, 1986; Van Hove et al., 1986; Zangwill, 1988). Other methods employing incident photon and ion beams have been adapted or specifically developed to characterize surface structure and/or adsorption complexes (see Vanselow and Howe, 1984; Van Hove and Tong, 1985). Synchrotron-based scattering and spectroscopic methods are particularly useful for studying surfaces and chemisorbed species because of the extremely high intensity of this radiation and its wavelength tunability (see Bassett and Brown, 1990, for discussion of synchrotron radiation and its applications in the earth sciences).

A far from exhaustive survey of the current literature showed that over 100 different surface-sensitive methods have been used to obtain information on the structure, composition, and/or bonding of surfaces and chemisorbed species (Tables 1 and 2) (see also Van Hove et al., 1989). Many of these methods, particularly those employing incident or detected electrons (e.g., XPS, UPS, LEED, EELS, SEELFS), require that the sample be placed in an ultra-high vacuum environment due to significant air absorption of the incident or emitted radiation. Thus they cannot be used to study sorption complexes at solid/water interfaces, in situ. When these methods are used in an "ex situ" fashion, the required drying and ultra-high vacuum have the potential for modifying or destroying sorption complexes formed at solid/water inter-faces. Few methods are capable of providing structural information on sorption complexes at solid/liquid interfaces in situ; most involve incident and emitted photons (see Tables 1 and 2; techniques marked by an asterisk in Table 2 have been used to characterize solid/liquid interfaces). Synchrotron-based x-ray absorption spectroscopy is one of the most important of these "in situ" methods and the one that has been used most in structural studies of sorption complexes at mineral/water interfaces; however, like other spectroscopic methods listed in Table 2, XAS also has some limitations for studying sorption complexes at solid/water interfaces. One of these limitations is that the element of interest in the sorption complex must not be present in the bulk solid or fluid phase in amounts larger than ~10 percent of the total element concentration. However, this situation can be achieved relatively easily by careful selection of systems and solution conditions. Another limitation of XAS for these types of studies is the fact that sorbing cations or anions of atomic number less than ~22 (Ti) cannot be studied due to the strong absorption by water of x-rays with energies less than ~5 keV. Unfortunately, this limitation precludes the study of sorbing elements like Na (E_{K-edge} = 1071 eV), Al (1560 eV), or Si (1839 eV) using XAS. The sample requirements for XAS study of adsorption mechanisms at the mineral/water interface will be discussed more thoroughly in the section on XAS applications.

Because we focus primarily on XAS results in our discussion below, we will examine this method in considerably more detail than other commonly-used in situ methods (EPR, NMR, Raman, FTIR, and Mössbauer spectroscopies). Numerous reviews of these methods exist in the literature, so they need not be covered here except to point out some of their unique attributes and limitations as well as a few of their applications. Goodman (1986) presents a relatively thorough review through 1985 of the applications of many of these non-synchrotron methods to adsorption on aluminosilicate minerals.

X-ray Absorption Spectroscopy (XAS)

X-ray absorption spectroscopy (XAS) was developed as a quantitative, short-range structural probe in the 1970's following the pioneering work of Sayers, Stern, and Lytle

Table 1. Summary of surface-sensitive methods listed by acronym and arranged according to the incident particle and detected particle (adapted from Rhodin and Gadzuk, 1979). The acronyms are defined in Table 2, although not all acronyms defined in Table 2 are listed in the Table 1.

INCIDENT PARTICLE	DETECTED PARTICLE					
	ELECTRON	PHOTON	NEUTRAL	ION	PHONON	E/H FIELD
ELECTRON	AEPS AEM AES DAPS EELS HEED HREELS IS LEED SEE RHEED SEM SLEEP STEM STM TEM	APS BIS CL CIS EM SXAPS SXES	ESDN SDMM	ESDI		
PHOTON	AEAPS AEM AES PEM PES SEE SEXAFS UPS XEM XES XPS	ATR ELL ENDOR ESR EXAFS GLAEXAFS IRAS IRS LS MOSS NEXAFS NMR ReflEXAFS SRS XANES XRD	LMP PD	LMP PD		
NEUTRAL	SEE	NIRS	MBRS MBSS $\sigma(p)$		ΔH_{ADS}	
ION	IMXA INS SEE	GDOS IIRS IIXS	ISD SDMM	GDMS IMMA ISD ISS RBS SIIMS SIMS		
PHONON	TE	ES TL	FD	SI	ASW	
	FEES FEM ITS	EL	FDM FDS	FIM FIM-APS FIS		CPD MS SC

Table 2. Definition of acronyms for different experimental methods used to characterize surfaces and chemisorbed species on surfaces. Those methods marked by an asterisk have been used to study sorption complexes at solid/liquid interfaces.

Acronym	Definition	Acronym	Definition
AEM-	Auger-Electron Microscopy	EXAFS-*	Extended X-ray Absorption Fine Structure
AEPS-	Auger-Electron Appearance Potential Spectroscopy	EYEXAFS-	Electron Yield Extended X-ray Absorption Fine Structure
AES-	Auger-Electron Spectroscopy	FD-	Flash Desorption
AFM-*	Atomic Force Microscope	FDM-	Field-Desorption Microscopy
APS-	Appearance Potential Spectroscopy	FDS-	Field Desorption Spectroscopy
ASW-	Acoustic Surface-Wave Measurements	FEES-	Field-Electron Energy Spectroscopy
ATR-	Attenuated Total Reflectance	FEM-	Field Emission Microscopy
BIS-	Bremsstrahlung Isochromat Spectroscopy	FIM-	Field Ion Microscopy
CEMS-*	Conversion-Electron Mössbauer Spectroscopy	FIM-APS-	Field Ion Microscope-Atom Probe Spectroscopy
CIS-	Characteristic Isochromat Spectroscopy	FTIR-*	Fourier Transform Infrared Spectroscopy
CL-	Cathodoluminescence (CL)	FYEXAFS-*	Fluorescence-Yield Extended X-ray Absorption Fine Structure
COL-*	Colorimetry (IR, Visible, UV, X-ray, and γ-ray Absorption Spectroscopy)	GDMS-	Glow-Discharge Optical Spectroscopy
		GDOS-	Glow-Discharge Optical Spectroscopy
CPD-	Contact Potential Difference (Work Function Measurements)	GLAEXAFS-*	Glancing Angle Extended X-ray Absorption Fine Structure
DAPS-	Disappearance-Potential Spectroscopy	HEED-	High Energy Electron Diffraction
ΔHABS-	Heat of Absorption Measurement	HEIS-	High-Energy Ion Scattering
EELS-	Electron Energy Loss Spectroscopy	HREELS-	High Resolution Electron Energy Loss Spectroscopy
EHAS-	Elastic Helium Atom Scattering		
EL-	Electroluminescence	IETS-	Inelastic Electron Tunneling Spectroscopy
ELL-	Ellipsometry	IIRS-	Ion-Impact Radiation Spectroscopy
EMA-	Electron Microprobe Analysis	IIXS-	Ion-Induced X-ray Spectroscopy
EMOSS-*	Emission Mössbauer Spectroscopy	ILS-	Ionization Loss Spectroscopy
ENDOR-*	Electron-Nuclear Double Resonance	IMMA-	Ion Microprobe Mass Analysis
ES-	Emission Spectroscopy	IMXA-	Ion Microprobe X-ray Analysis
ESDI-	Electron-Stimulated Desorption of Ions	INS-	Ion-Neutralization Spectroscopy
ESDN-	Electron-Stimulated Desorption of Neutrals	IPS-	Inverse Photoemission Spectroscopy
ESEEM-*	Electron Spin-Echo Envelope Modulation	IRAS-*	Infrared Absorption Spectroscopy
ESR or EPR-*	Electron Spin or Paramagnetic Resonance	IRS-*	Internal Reflectance Spectroscopy
		IS-	Ionization Spectroscopy

Table 2, continued.

Abbreviation	Name
ISD-	Ion-Stimulated Desorption
ISN-	Inelastic Scattering of Neutrons
ISS-	Ion-Scattering Spectroscopy
ITS-	Inelastic Tunneling Spectroscopy
LEED-	Low-Energy Electron Diffraction
LEIS-	Low-Energy Ion Scattering
LMP-	Laser Microprobe
LS-	Light Scattering
MBRS-	Molecular Beam Reactive Scattering
MBSS-	Molecular-Beam Surface Scattering
MOSS- *	Mössbauer Spectroscopy
MS-	Magnetic Saturation
NEXAFS- *	Near-Edge X-ray Absorption Fine Structrure
NIRS-	Neutral Impact Radiation Spectroscopy
NMR- *	Nuclear Magnetic Resonance
NS-	Neutron Scattering
PAS- *	Photoacoustic Spectroscopy
PD-	Photodesorption
PEM-	Photoelectron Microscopy
PES-	Photoelectron Spectroscopy
RBS- *	Rutherford Backscattering Spectroscopy
ReflEXAFS- *	Reflection Extended X-ray Absorption Fine Structure
RHEED-	Reflection High-Energy Electron Diffraction
σ(p)- *	Adsorption Isotherm Measurements
SC- *	Surface Capacitance
SDMM-	Scanning Desorption Molecule Microscope
SEE-	Secondary-Electron Emission
SEELFS-	Surface Extended Electron Energy-Loss Fine Structure
SEM-	Scanning Electron Microscopy
SERS- *	Surface-Enhanced Raman Spectroscopy
SEXAFS-	Surface Extended X-ray Absorption Fine Structure
SI-	Surface Ionization
SIIMS-	Secondary-Ion Imaging Mass Spectroscopy
SIMS-	Secondary-Ion Mass Spectroscopy
SIS- *	Surface Infrared Spectroscopy
SLEEP-	Scanning Low-Energy Electron Probe
SLEEP-	Scanning Low-Energy Electron Probe
SRS- *	Surface Reflectance Spectroscopy
STEM-	Scanning Transmission Electron Microscopy
STM-	Scanning Tunneling Microscopy
STS-	Scanning Tunneling Spectroscopy
SXAPS-	Soft X-ray Appearance Appearance Potential Spectroscopy
SXES-	Soft X-ray Emission Spectroscopy
TE-	Thermionic Emission
TEM-	Transmission Electron Microscopy
TL-	Thermoluminescence
UPS-	Ultraviolet Photoelectron or Photoemission Spectroscopy
XANES- *	X-ray Absorption Near Edge Structure
XEM-	Exoelectron Microscopy
XEM-	Exoelectron Microscopy
XES-	Exoelectron Spectroscopy
XES-	Exoelectron Spectroscopy
XPS-	X-ray Photoelectron or Photoemission Spectroscopy
XRD- *	X-ray Diffraction
XRSW- *	X-ray Standing Wave

324

(Sayers et al., 1970; 1971). They developed an adequate single-scattering theory for the extended fine structure beyond an x-ray absorption edge and showed that Fourier transformation of this fine structure yielded structural information. XAS has since been applied widely in most fields of science and engineering to problems requiring an element-specific, short-range structural probe. A number of recent reviews thoroughly discuss the theoretical and experimental details of XAS as well as a variety of applications (see e.g., Stern and Heald, 1983; Teo, 1986; Wong, 1986; Koningsberger and Prins, 1988; Brown et al., 1988; Gurman, 1989). The discussion that follows draws from Brown and Parks (1989) and covers only the highlights of XAS. In order to demonstrate how EXAFS spectroscopy is used to determine the average structure of a sorption complex at a mineral/water interface, an analysis of a hypothetical EXAFS spectrum of Pb^{2+} adsorbed on γ-Al_2O_3 is presented.

XAS is commonly subdivided into two methods known as EXAFS (Extended X-ray Absorption Fine Structure) spectroscopy and XANES (X-ray Absorption Near Edge Structure) or NEXAFS (Near-Edge X-ray Absorption Fine Structure) spectroscopy, which arise from different physical processes and provide complementary information. Modifications of the basic EXAFS data collection technique have led to surface-sensitive methods, including SEXAFS (Stöhr, 1988) and ReflEXAFS (Bosio et al., 1986; Barrett et al., 1990). The key features of XAS methods can be summarized as follows:

(1) Synchrotron-based XAS can be used to study most elements in solid, liquid, or gaseous states at concentrations ranging from parts per million to the pure element. The high intensity of synchrotron radiation allows the study of very small (μg) or dilute (millimolar, mM) samples and experimental conditions of high or low temperature or pressure and controlled atmospheres, including the presence of fluids such as water.

(2) XAS is an element specific, bulk method giving information about the average local structural and compositional environment of the absorbing atom. It can be used to study compositionally complex materials such as natural minerals.

(3) XAS is a local structural probe which "sees" only the two or three closest shells of neighbors around an absorbing atom (less than about 6 Å) due to the short electron mean free path in most materials. The sample size, on the other hand, is determined by the size of the x-ray beam, which ranges from about 2 x 20 mm (uncollimated) to less than 1 x 1 mm (when collimated).

(4) For many systems, EXAFS analysis is capable of yielding average distances accurate to ±0.02 Å and average coordination numbers to ±10-20%, assuming that systematic errors have been minimized in the experiment and data analysis and that static and thermal disorder are small.

(5) The time required for photon absorption is about 10^{-16} seconds, compared to about 10^{-12} to 10^{-14} seconds for interatomic vibrations. Thus XAS averages over all local distances around an absorbing atom.

(6) Although presently qualitative relative to EXAFS analysis, XANES or NEXAFS spectroscopy can provide information about the oxidation state of an atom and the symmetry and bonding of its local environment. By comparison of the XANES spectra of unknown and model compounds, the similarity of an atom's local environment in the two compounds can be tested.

Production of x-ray absorption spectra. Most XAS experiments consist of exposing a sample to a monochromatic beam of x-rays, which is scanned through a range of energies below and above the absorption edge of the element of interest. The very intense x-ray continuum from synchrotron radiation sources is most convenient for this purpose. Various types of processes occur when x-rays interact with matter, including x-ray scattering (both elastic and inelastic), production of optical phonons, production of photoelectrons and Auger electrons, production of fluorescence x-ray photons, and positron-electron pair production. In the x-ray energy range (0.5 to 100 keV), photo-electron production is the dominant process and is the primary cause of x-ray attenuation by matter. Depending on the absorber and the energy of incident x-rays, the photoelectrons can be from any atomic shell (K, L_I, L_{II}, L_{III}, etc.).

As mentioned above, the sample can be in almost any form (solid, liquid, suspension, or gas) with the element of interest at concentrations ranging from millimolar to 100%. SEXAFS measurements require a solid sample because they are carried out under ultra-high vacuum. Sample size can range from a cm^3 to as small as a few μm^3, depending on element concentration and x-ray flux. The major constraint on the lower limits of element concentration and sample size is achieving a spectral signal-to-noise ratio adequate for extracting structural data. The ranges quoted above are typical in our experi-ence. A solid sample is typically powdered and loaded into a mylar-windowed aluminum, stainless steel, or Teflon sample holder. Fluid samples or suspensions are contained in sealed mylar envelopes.

In an XAS experiment, the incident (I_0) and transmitted (I_1) x-ray intensities are recorded at sequential incident x-ray energies (E) to produce an absorption spectrum, which

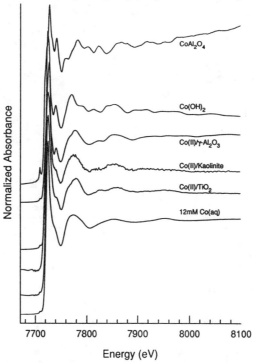

Figure 7. K-edge transmission x-ray absorption spectra of Co in (a) $CoAl_2O_4$ and (b) $Co(OH)_2$, and Co K-edge fluorescence-yield x-ray absorption spectra for samples of (c) Co(II) sorbed at the γ-Al_2O_3/water interface, (d) Co(II) sorbed at the kaolinite/water interface, (e) Co(II) sorbed at the TiO_2 (rutile)/water interface, and (f) 12 mM $Co(NO_3)_2$ solution. The XANES region extends from about 10 eV below to about 50 eV above the main absorption edge, whereas the EXAFS region extends from about 50 eV to ≥ 1000 eV above the absorption edge. Pre-edge features due to 1s to 3d bound-state electronic transitions are clearly visible in the spectrum of $CoAl_2O_4$ in which the Co^{2+} occupies a tetrahedral coordination environment. This feature is much weaker in the other spectra, which is indicative that Co^{2+} in these samples occupies a centrosymmetric site such as an octahedron.

326

(a)

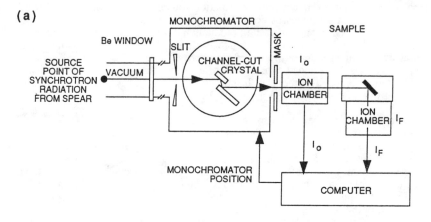

(b)

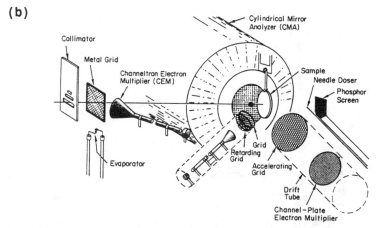

Figure 8. Schematic diagram showing the experimental arrangements for (a) fluorescence x-ray absorption measurements (after Wong, 1986) and (b) electron- or ion-yield SEXAFS measurements (from Stöhr et al., 1988). In the SEXAFS experiment, electrons are detected with a cylindrical mirror analyzer (CMA) for elastic Auger electron yield measurements and with a two grid retarding detector for total-yield and partial Auger yield studies. Ions are detected and mass analyzed with the time-of-flight detector opposite the CMA.

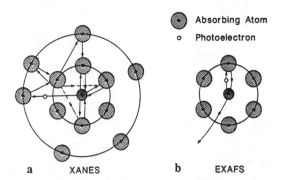

Figure 9. Electron scattering processes leading to EXAFS (a) and XANES (b). After Brown et al. (1988).

is typically plotted as ln (I_0/I_1) vs E (in eV). The relationship between these intensities and the linear absorption coefficient μ of a sample of thickness x (in cm) is ln $(I_0/I_1) = \mu x$. Alternatively, x-ray fluorescence from the sample, I_f, can be measured and ratioed with I_0 as I_f/I_0, which is proportional to μ for dilute samples. Fluorescence x-ray emission arises from de-excitation of the core hole generated by absorption of an incident x-ray photon. Thus I_f is directly proportional to the number of absorption events for elements at low concentrations (in practice less than 1 wt %). The main advantage of the fluorescence-yield method relative to the transmission method is its greater sensitivity to elements at low concentrations in highly absorbing matrices. This is true because of the effect of matrix absorption on transmitted x-rays and the fact that x-ray fluorescence yield can be measured from the surface of the sample that incident x-rays strike. Large solid angle detectors (Lytle et al., 1984) and appropriate filters (to reduce interference caused by Compton and elastic scattering) are necessary in x-ray fluorescence measurements because the process leading to x-ray fluorescence is relatively inefficient. Representative Co K-edge transmission and fluorescence-yield EXAFS spectra for a series of Co compounds are shown in Figure 7 (above). When the element of interest has an edge energy of less than 2000 eV (e.g., Al and Si) or when the sample is placed in an ultra-high vacuum environment for SEXAFS measurements, electron detection techniques become more feasible than fluorescence yield. Electron yield generally requires a vacuum environment for the sample because of the extremely short transmission path of electrons in air or light gases. Figure 8a contains a schematic drawing of the experimental arrangements for fluorescence XAS measurements; Figure 8b is a typical experimental arrangement for electron- and ion-yield SEXAFS studies.

When the energy of the incident x-ray beam (hv) is less than the binding energy (E_b) of a core electron on the absorbing element, little significant absorption takes place. However, when hv $\approx E_b$, electronic transitions to unoccupied, bound energy levels occur, contributing to the main absorption edge and producing features below the main edge referred to as the pre-edge spectrum (Fig. 7). As hv increases beyond E_b, the electron can be ejected to continuum (unbound) levels and remains in the vicinity of the absorber for a short time with excess kinetic energy. In the energy region extending from just above to about 50 eV above E_b, the electron is multiply scattered among neighboring atoms (Fig. 9a), which produces the XANES (X-ray Absorption Near-Edge Structure). When hv is about 50 to 1000 eV above E_b, the electron is ejected from the absorber, is singly scattered from first- or second-neighbor atoms back to the absorber, then leaves the vicinity of the absorber (Fig. 9b). The EXAFS (Extended X-ray Absorption Fine Structure) is due to interference between outgoing and backscattered photoelectron waves which modulates the atomic absorption coefficient (Fig. 10). Single-scattering processes are restricted to maximum path lengths of about 6 Å in solids and liquids. Thus the structural information obtained from EXAFS spectroscopy is limited to the local environment (first-, second-, and

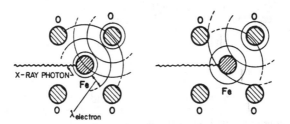

Figure 10. Schematic drawing showing the interference between outgoing and backscattered photoelectron waves produced by photoelectric absorption of a hypothetical Fe atom surrounded by four oxygen atoms. The diagram on the left shows backscattered waves at a position and energy where constructive interference occurs, resulting in amplification of EXAFS oscillations, whereas the diagram on the right shows a position at greater energy where destructive interference occurs, resulting in reduction in the amplitude of EXAFS oscillations. From Brown et al. (1988).

perhaps third-nearest neighbors) of the absorber. Multiple-scattering processes can involve greater path lengths, thus XANES spectra can provide information on more distant neighbors including bond angles.

XANES and pre-edge spectra. Interpretation of XANES or NEXAFS spectra is presently limited to a relatively qualitative level. However, very useful and unique structural and bonding information can be obtained from XANES (Bianconi, 1988). For example, the details of a XANES spectrum are quite sensitive to coordination geometry. Close similarity of XANES spectra for an unknown and a structurally well characterized compound indicates that the coordination environment of the absorber is similar in both compounds (see, e.g., Waychunas et al., 1983). The converse is also true. In addition, the energies of pre-edge and main-edge features can be correlated with the absorber's oxidation state when it is variable (Waychunas et al., 1983; Wong et al., 1984). For first-row transition elements, pre-edge spectral features caused by 1s to 3d electronic transitions contain useful information about the symmetry of the absorber's local environment. Specifically, the presence of a noticeable peak 4 to 5 eV below the main absorption edge usually indicates a regular site without a center of symmetry or a very irregular local coordination geometry. Similar though less intense features may occasionally arise from excitonic processes (Bianconi, 1988). The absence of such a feature in spectra from these elements provides important evidence that the absorber occupies a symmetrical site charac-terized by a center of symmetry. In addition to this qualitative information about site symmetry and oxidation state, XANES spectra can also yield information about interatomic distances between the absorber and neighboring atoms. Natoli (1983) has shown that delocalized, continuum features in a XANES spectrum should obey the relationship $(\Delta E)R^2$ = constant, where R is distance in Å and ΔE, for example, is the energy difference between the 3d band and the absorption edge crest for a 3d-transition element. This type of structural infor-mation is independent of and complementary to that provided by EXAFS analysis (see below). Continuum features in XANES spectra can show significant polarization dependence when data are taken on oriented single-crystals (Waychunas and Brown, 1990). This is also true for the EXAFS spectral region. Polarized XAS measurements are commonly made in SEXAFS and NEXAFS studies of molecular chemisorption on single-crystal surfaces (see Stöhr, 1988).

EXAFS spectra. In the single-scattering process producing EXAFS, the frequen-cies of backscattered photoelectron waves are approximately inversely proportional to the distances between the absorber and neighboring backscatterer atoms (i.e., a second-neighbor atom at twice the distance between the absorber and a first-neighbor atom will have a frequency approximately half that of the EXAFS contribution of the first neighbor). Frequency and distance are not truly proportional because the electron undergoes a phase shift due to interaction of the outgoing and backscattered waves with the potentials of the absorbing and backscattering atoms. Thus the raw distance must be corrected for the effects of phase shift, which is typically 0.2 to 0.5 Å and is characteristic of a particular absorber-backscatterer pair. The amplitude of the scattered wave is determined primarily by the number and backscattering cross-sections of the atoms in the vicinity of the absorber, modified by several effects discussed below.

When the background-subtracted EXAFS data [$\chi(E) = \{\mu(E) - \mu_0(E)\}/\mu_0(E)$] are converted to momentum space, $\chi(k)$, where $|k| = \{2m(E - E_0)h^2\}^{1/2}$, the extended fine structure is directly related to the sum of sinusoidal oscillations for each shell of neighboring atoms. In these equations $\mu(E)$ is the absorption due to the element in the sample, $\mu_0(E)$ is the normalized background absorption (physically corresponding to the absorption coefficient of the isolated atom), m is the reduced mass of the electron, h is Planck's constant divided by 2π, E is the energy of the incident x-ray beam, and E_0 is the threshold energy of the atom, which in theory is the energy above which the electron is no longer bound to its parent atom. The frequency of each oscillation produced by backscatterers in shell i is given by $\sin[2kR_{A-B} + f(k)_{A-B}]$, where R_{A-B} is the bond length between the absorber atom (A) and neighboring backscatterer atom (B) in shell i, and $f_i(k)_{A-B}$ is the phase-shift function specific to the atom pair A-B. The term $2kR_{A-B}$ accounts for the phase difference of a free electron making the return trip to the absorber from a neighboring atom; the additional phase shifts $f_i(k)_{A-B}$ are required to account for the potentials due to both the absorbing and backscattering atoms. The EXAFS function $\chi(k)$ is adequately modelled in the single-scattering, plane-wave approximation as the sum of scattering contributions from each shell i by Equation (5).

$$\chi(k) = \sum_i [S_i(k)N_i/kR_i^2] \, F_i(k) \, \exp[-2\sigma_i^2 k^2] \, \exp[-2R_i/\lambda(k)] \, \sin[2kR_i + f_i(k)] \quad . \tag{5}$$

Here $F_i(k)$ is the backscattering amplitude of the N neighboring atoms of type i at distance R_i. A Debye-Waller type factor, σ_i, accounts for thermal vibration (in the harmonic approximation) and static disorder (assuming Gaussian pair distribution). $S_i(k)$ is an amplitude reduction factor due to many-body relaxation effects of the absorber and multielectronic excitations (Stern and Heald, 1983). The term $\exp[-2R_i/\lambda(k)]$ accounts for inelastic losses in the scattering process with λ being the electron's mean-free path. In the formulation we use, $F_i(k)$ is parameterized as $c_1\exp(c_2k^2)k^{c_3}$, where the c_j's are adjustable. The additional phase shift experienced by the photoelectrons is parameterized as $f_i(k) = a_0 + a_1k + a_2k^2$, where the a_j's are adjustable. Thus each EXAFS wave scattered from the ith shell of N backscatterers is determined by the backscattering amplitude $N_iF_i(k)$, modified by four reduction factors ($S_i(k)$, the two exponential terms, the $1/kR_i^2$ distance dependence) and by the sinusoidal oscillation which includes a dependence on interatomic distance ($2kR_i$) and phase shift [$f_i(k)$].

<u>Analysis of EXAFS spectra.</u> Our purpose in analyzing EXAFS spectra is to extract quantitative estimates of the average distances between the absorber atom and its first-, second-, and possibly third-nearest neighbors, estimates of the average numbers of atoms in these shells, and estimates of the composition of the absorber's nearest neighbors. Quantitative estimates of interatomic distance between an absorber and its nearest neighbors require Fourier transformation of the $\chi(k)$ function, which produces a radial structure function (rsf). The lower k-space limit of the transform is generally taken above 3 Å^{-1}, because the single-scattering approximation used to describe EXAFS oscillations is not valid at small wave vector values, where multiple scattering dominates. The rsf is similar to a radial distribution function but contains only pair correlations involving the absorber and must be corrected for phase shift. The real part (modulus) of the Fourier transform consists of a sum of radial peaks located at R_i, which correspond to the distances between the absorber and successive shells of neighboring atoms, uncorrected for phase shift. The rsf also contains compositional information because the amplitudes of its peaks are dependent on the number and backscattering amplitudes of the nearest-neighbor atoms. In addition, compositional information comes from the phase shift which affects interatomic distance and is dependent on the types of absorber and backscatterer atoms. In practice for backscatterers of atomic number less than 36, the $\chi(k)$ function is multiplied by k^3 before Fourier transformation to weight the data approximately equally over the entire k range.

In order to demonstrate how structural information is obtained from an EXAFS spectrum, we will examine the sorption of Pb(II) on γ-Al$_2$O$_3$. We shall generate and discuss theoretical spectra for a hypothetical multinuclear sorption complex of two Pb^{2+} ions bound directly to oxygens on the γ-Al$_2$O$_3$ surface. This example will show that distances, number, and identities of second neighbors around the sorbed ion are essential parameters for defining the mechanism of chemisorption and identity of the complex.

The EXAFS spectra shown in Figure 11 represent simulated, background-subtracted Pb L$_{III}$ spectra for the hypothetical case of Pb(II) sorbed on γ-Al$_2$O$_3$. The L$_{III}$ edge of Pb occurs at approximately 13055 eV, which is used as the value for E_0. The

330

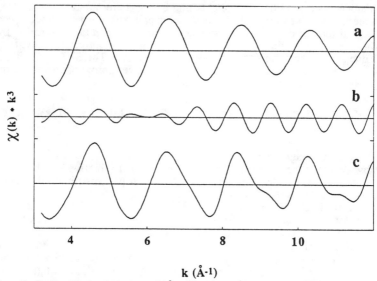

Figure 11. Simulated, background-subtracted, k^3-weighted frequency components of Pb L_{III}-EXAFS spectra for lead in a hypothetical dimeric complex such as may occur among adsorption complexes when lead sorbs on the γ-Al_2O_3 surface. The central Pb absorber is surrounded by four oxygens at a distance of 2.1 Å, a second Pb at 3.4 Å, and an Al in the γ-Al_2O_3 surface at 3.4 Å. Figure 11a is the $k^3\chi(k)_{Pb-O}$ function for Pb-O interactions, and Figure 11b is the $k^3\chi(k)_{Pb-Pb}$ function for the Pb-Pb interaction. Figure 11c is the simulated spectrum of the PbO_4Pb sorption complex produced by adding the $k^3\chi(k)$ functions for Pb-O and Pb-Pb interactions; it shows a "beat pattern" with shoulders at approximately 7, 9, and 11 Å^{-1}. From Brown and Parks (1989).

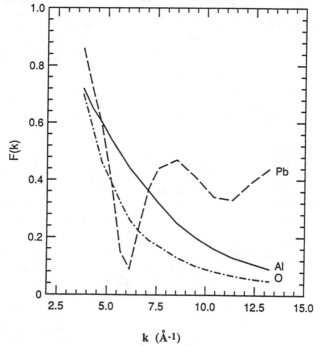

Figure 12. Calculated backscattering amplitudes, F(k), for Pb, Al, and O as a function of wave vector k using theoretical parameters from Teo and Lee (1979). From Brown and Parks (1989).

$k^3\chi(k)$ function, $k^3\chi(k)_{Pb-O}$, produced by a complex of one Pb atom surrounded by four oxygens at a distance of 2.1 Å, is shown in Figure 11a. It consists of a sinusoidal wave of a single frequency. The presence of one second-neighbor Pb atom at a distance of 3.4 Å from the central Pb atom results in a second $k^3\chi(k)$ function, $k^3\chi(k)_{Pb-Pb}$ (Fig. 11b), of higher frequency than $k^3\chi(k)_{Pb-O}$. Differences in the k-dependence of the amplitudes of these functions are due to differences in the k-dependence of the backscattering amplitudes of lead and oxygen (Fig. 12). When these functions are summed, a "beat pattern" results (Fig. 11c), which is seen as shoulders at k values of approximately 7, 9, and 11 Å^{-1}. The presence of such a beat pattern in a raw EXAFS spectrum is often indicative of second-neighbor metal atoms surrounding the central atom. Aluminum atoms would be expected among the second neighbors surrounding Pb, if Pb bonds directly to oxygens on the γ-Al$_2$O$_3$ surface. The $k^3\chi(k)_{Pb-Al}$ function for one second-neighbor Al atom at 3.4 Å from the central Pb is shown in Figure 13a, and a $k^3\chi(k)_{Pb-Pb}$ function for one second-neighbor Pb at 3.4 Å from the central Pb is shown in Figure 13b. The presence of a second-neighbor Al atom also would have contributed to the beat pattern of the summed spectrum (Fig. 11c) had it been included in the calculation. Its contribution would be weak at k greater than 8 Å because its backscattering amplitude is much less than that of Pb in this k-range (Fig. 12). Beat patterns in experimental EXAFS spectra can reveal the presence of second-neighbor metal atoms, but determination of exact numbers and identities requires further analysis of the EXAFS spectrum as described below.

Fourier transformation of the $k^3\chi(k)_{Pb-O}$, the $k^3\chi(k)_{Pb-Al}$, and the $k^3\chi(k)_{Pb-Pb}$ functions produces radial structure functions (Figs. 14a, 14b, and 14c, respectively) with peaks centered at approximately 1.6 Å, 2.9 Å, and 3.35 Å, corresponding to Pb-O, Pb-Al, and Pb-Pb distances, respectively. These distances are not corrected for phase shift. Thus they don't correspond to the distances of 2.1, 3.4 and 3.4 Å, respectively, used in the model calculations. These differences are caused by the effects of phase shifts for these atom pairs which emphasizes the need for accurate phase-shift information. The number of atoms in the first- and second-neighbor shells surrounding Pb can be extracted from the amplitudes of the peaks in the radial structure functions.

To derive accurate interatomic distances, coordination numbers, and atom identities from an EXAFS spectrum, it is necessary to calibrate the relationships between frequency and distance and between amplitude and the number and identity of coordinating atoms. This may be done using model compounds or theory. Model compounds must have known structures and contain the absorber and nearest-neighbor backscatterers of interest. For example, α-PbO, with Pb coordinated by four first-neighbor oxygen atoms, is a good model compound with which to estimate Pb-O phase-shift parameters and oxygen backscattering amplitude parameters for the Pb on γ-Al$_2$O$_3$ system. This compound is not suitable for estimating Pb-Pb phase shift parameters because of the high degree of disorder of second-neighbor lead relative to the central lead. Unfortunately there is no adequate model compound containing Pb with second-neighbor Pb or Al atoms, so theoretical phase-shift and amplitude parameters (e.g., Teo and Lee, 1979; McKale et al, 1988) must be used for Pb-Pb and Pb-Al pairs.

In practice, calibration of phaseshift and amplitude dependence on k using model compounds is accomplished by obtaining an EXAFS spectrum of the element of interest in a model compound. The peak in the radial structure function corresponding to the selected atom pair is isolated and back-Fourier transformed to produce the corresponding $\chi(k)$ contribution, $\chi(k)_{BT}$, for that atom pair alone. The parameterized form of Equation (5) is fit to $\chi(k)_{BT}$ after fixing interatomic distances, R_i, and coordination numbers, N_i, at values taken from the known crystal structure of that model compound. The parameters for phase shift and backscattering amplitude for the atom pair in question are adjusted along with σ_i as part of a least-squares fitting procedure which minimizes the variance S, expressed as S $= \Sigma\,[\chi(k)_{BT} - \chi(k)_p]^2$, where $\chi(k)_p$ is the parameterized form of Equation (5). Once these parameters are obtained, the radial structure function for the unknown compound is treated

332

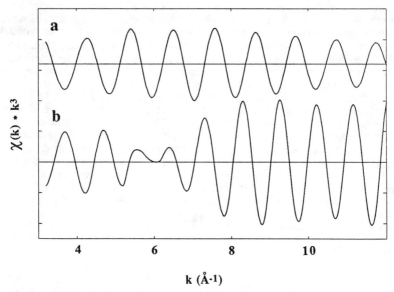

Figure 13. Simulated, background-subtracted, and k^3-weighted Pb L_{III} EXAFS spectra for (a) a lead atom with a single second-neighbor aluminum at 3.4 Å distance and (b) a lead atom with a single second-neighbor lead at 3.4 Å, illustrating the difference in amplitude versus k caused by Al and Pb second neighbors. From Brown and Parks (1989).

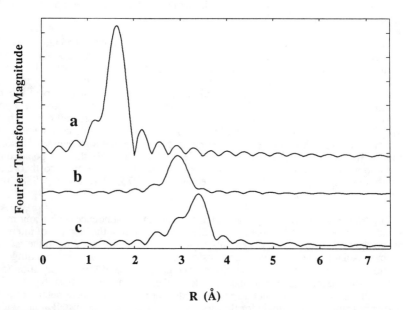

Figure 14. Fourier transformation of the $k^3\chi(k)_{Pb-O}$, the $k^3\chi(k)_{Pb-Al}$, and the $k^3\chi(k)_{Pb-Pb}$ functions in Figures 11 and 13, producing radial structure functions (rsf's) for (a) Pb-O, (b) Pb-Al, and (c) Pb-Pb atom pairs. The rsf's show peaks at approximately 1.6, 2.9, and 3.35 Å, respectively, before correction for phase shift, corresponding to distances of 2.1, 3.4, and 3.4 Å used in constructing the model. From Brown and Parks (1989).

in the same way, except that R_i, N_i, and σ_i are varied while phase-shift and amplitude parameters are fixed at the values refined from the model compound. This procedure will be demonstrated below in the discussion of EXAFS-derived results for sorption complexes. When differences in backscattering amplitudes and/or phase shifts of alternative atoms in the second-neighbor shell are sufficient, their identities can be determined by substituting various possible atoms during the fitting procedure. In the Pb on γ-Al_2O_3 case discussed above, the second-shell possibilities are oxygen, lead, and aluminum, which have backscattering amplitudes different enough that they can be distinguished (see Fig. 12).

Because reasonably accurate bond distances (at least ±0.03 Å) and coordination numbers (±15%) are needed to distinguish among possible second-neighhbor arrangements around a metal ion sorbed at a mineral/water interface, a few comments about the accuracy of EXAFS- or SEXAFS-derived bond distances and coordination numbers are required. A detailed study of the reliability of multi-shell EXAFS analysis by Holmes et al. (1988) on real and simulated data showed that careful analysis of high-quality data can result in accuracies of ±0.01 Å for first-shell distances and ±0.05 Å for fourth-shell distances. Intermediate shells should have accuracies between these two values. Coordination numbers are less accurate because of large correlations between Debye-Waller factors, coordination numbers, and inelastic terms in the EXAFS backscattered amplitude expression. Holmes et al. (1988) claim an error limit of ±0.2 atoms (~10%) for the nearest-neighbor shell around an absorber. The accuracy of coordination numbers decreases significantly with increasing shell number. In our analyses, we estimate accuracies of ±0.02 Å and ±0.03 Å, for first- and second-shell distances, respectively, and ±10% and ±20% for first- and second-shell coordination numbers, respectively. These estimates come from fits of the first and second shells around absorbers such as Co^{2+} in well-characterized model compounds using phaseshift and backscattered amplitude functions for both shells extracted from other similar model compounds.

Other in situ spectroscopic methods and selected applications

Magnetic resonance spectroscopies. There are several magnetic resonance methods (EPR, ENDOR, ESEEM, and NMR) that can provide valuable information on sorption complexes at solid/water interfaces. The most widely used is Electron Paramagnetic (or Spin) Resonance spectroscopy. EPR involves the absorption of radiation of microwave frequency by molecules possessing electrons with unpaired spins; placing the sample in a magnetic field results in resolution of degenerate spin states (see Howe, 1984; Calas, 1988, for reviews). EPR spectra are commonly plotted as the first derivative of the absorption curve versus the strength of the applied magnetic field. The EPR spectrum is sensitive to the local environment (approximately first neighbor) of the paramagnetic species and thus can be used to distinguish among different molecular arrangements of these species at surfaces (e.g., inner-sphere versus outer-sphere adsorption complexes). As an example, EPR spectra of Cu^{2+} (aq) and Cu^{2+} adsorbed on hydrous δ-Al_2O_3 and on hydrous TiO_2 are shown in Figure 15 (from Motschi, 1984). The signal from the aqueous Cu^{2+} (Fig. 15a), which would be similar to that from a fully-hydrated outer-sphere complex, is clearly distinguishable from that of the sorbed Cu^{2+} (Figs. 15b,c). The inferred inner-sphere molecular arrangements of the Cu^{2+} sorption complexes on the two surfaces are also shown (Figs. 15 d,e).

EPR is applicable only to paramagnetic ions and has been used primarily to study Cu^{2+} (McBride, 1976; 1978; 1982a; 1982b; 1985; 1986; McBride and Bouldin, 1984; McBride et al., 1984; Clark and McBride, 1984; Harsh et al., 1984; Bassetti et al., 1979; Zelewsky and Bemtgen, 1982; Motschi, 1983; Motschi, 1984; Motschi, 1987) and Mn^{2+} (e.g., Bleam and McBride, 1985), and V^{4+} (e.g., Martini et al., 1975; Motschi, 1987) at oxide/water interfaces, in addition, the substrate must be diamagnetic. The EPR studies of copper (II) on a series of aluminum oxides and clay minerals by McBride and coworkers have been used to infer oxidation state, coordination, hydrolysis, and orientation of the

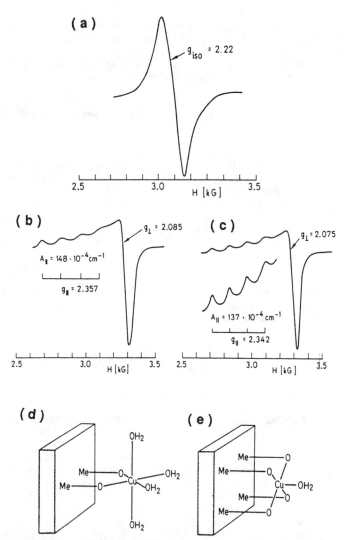

Figure 15. (a) EPR spectrum of Cu^{2+} (aq); (b) EPR spectrum of aqueous Cu^{2+} sorbed at the δ-Al_2O_3/water interface; (c) EPR spectrum of aqueous Cu^{2+} sorbed at the TiO_2/water interface; (d) schematic model of the inferred inner-sphere Cu^{2+} adsorption complex at the δ-Al_2O_3/water interface; (e) schematic model of the inferred inner-sphere Cu^{2+} adsorption complex at the TiO_2/water interface. After Motschi (1984).

adsorbate; to distinguish among different sorption mechanisms (mononuclear versus multi-nuclear complexes and three-dimensional precipitates); and to detect multiple adsorption sites.

ENDOR/ESEEM (Electron Nuclear Double Resonance/Electron Spin Echo Envelope Modulation) spectroscopies provide longer-range structural information than EPR, detecting weak interactions between the unpaired electron and neighboring magnetic nuclei up to separations of about 5 Å. ENDOR spectroscopy can be thought of as a combination of EPR and NMR because a nuclear transition is observed while the electron resonance is excited. ESEEM is also related to EPR and involves modulation of the electron spin echo by a series of pulses, which produces a decay curve whose shape depends on the arrangement and bonding of first- and second-neighbor atoms around the paramagnetic ion.

ENDOR measurements can yield interatomic distances if the interaction between the magnetic moments of the electron and the nucleus is dipolar in nature. For example, in situ EPR and ENDOR studies of Cu(II) sorbed on hydrous δ-Al_2O_3 (Motschi, 1983; Rudin and Motschi, 1984) showed that the Cu ion has a pseudo square planar coordination geometry and indicate the presence of protons of both axially and equatorially coordinated water molecules bonded to the Cu adion (Figs. 15 d,e). Estimates of the Cu-H and Cu-O(H_2O) distances (to ±0.1 Å) were made from the proton hyperfine couplings using a dipolar approximation. Similar indirect attempts to characterize the solvation geometry of Cu(II) sorbed on silica have been made using ESEEM spectroscopy (Ichikawa et al., 1981). Although capable of higher resolution than EPR, neither ENDOR nor ESEEM is as sensitive as EPR in terms of detection (see Mims, 1972; Motschi, 1987; Rudin and Motschi, 1984; Schweiger, 1982; and Mohl et al., 1988, for more details).

Nuclear magnetic resonance (NMR) spectroscopy appears to be less applicable in surface studies because the number of nuclei needed to observe a signal is large and the number of atoms on a surface is typically small. However, a few surface-NMR studies have been carried out, including [195]Pt on alumina (Slichter, 1985) and [13]CO on alumina-supported rhodium catalysts (Duncan et al., 1979, 1980).

FTIR and Raman spectroscopy. These methods can provide information on the vibrational states of molecules sorbed to surfaces which allows different molecules to be "fingerprinted". Infrared spectroscopy is one of the oldest and most sensitive methods used to study hydroxide groups and water molecules on oxide surfaces (e.g., Hair, 1967, and references therein). Fourier Transform Infrared (FTIR) spectroscopy (cf. Bell, 1984) provides improved signal-to-noise ratio relative to conventional dispersive IR methods as well as greatly reduced time for spectrum collection, thus allowing dynamic (time-resolved) studies. Diffuse-reflectance and photoacoustic spectra can also be collected using an FTIR spectrometer and appropriate sample cells. Although applications of FTIR spectroscopy to surfaces are mostly concerned with characterization of gas-phase molecules, including H_2O, chemisorbed at solid/gas interfaces (e.g., Hoffman and Knözinger, 1987; Scarano et al., 1988; Beebe et al., 1990; Echterhoff and Knözinger, 1990; Matsumoto and Kaneko, 1990), it is also used to characterize the solid/liquid interface, in situ (e.g., Tejedor - Tejedor and Anderson, 1986) as well as sorbed molecules at the solid/liquid interface (e.g., Chang and Weaver, 1990).

Normal Raman scattering applied to adsorbates on highly absorbing or highly reflective surfaces is a relatively insensitive method. However, the discovery of surface-enhanced Raman spectroscopy (SERS) by Fleischmann and co-workers (1974) has led to a highly surface-sensitive spectroscopic method that shows spectral enhancement of several orders of magnitude (Fleischmann et al., 1985; Campion, 1986; Weaver, et al., 1987; Cotton, 1988; Krasser, 1989). Although the origin of this enhancement is not yet thoroughly understood, it is thought that roughened substrates and/or tuning of the exciting radiation to a molecular electronic state, as in resonance Raman scattering, is necessary for the enhancement (Campion, 1986). Applications of SERS to adsorbed species at solid/water interfaces have not yet been made to mineral/water systems, but they do include studies of sorbates at electrode/aqueous interfaces (e.g., Corrigan et al., 1985; Weaver et al., 1987). Normal Raman scattering has been used to study the sorption of U(VI) on hydrous TiO_2, ZrO_2, and silica gel (Maya, 1982). This work indicates that the sorption process involves the displacement of the carbonate ligands from the soluble species, yielding a uranyl moiety, free of carbonate, that is strongly bound to the surfaces.

Mössbauer spectroscopy. Recoilless nuclear resonant γ-ray absorption in the transmission mode is most commonly used in mineralogical research to study the site geometry and oxidation state of iron in crystalline and amorphous solids. However, the emission of backscattered or "conversion" electrons or photons (fluorescent x-rays) following decay of an excited absorber nucleus can be used to obtain surface-sensitive

Mössbauer spectra (Tatarchuk and Dumesic, 1984). There are only eight practical "Mössbauer-active" isotopes, of which [57]Fe is the most important mineralogically and geochemically. Therefore, the technique is limited relative to the other in situ methods discussed above. An interesting application of conversion-electron Mössbauer spectroscopy (CEMS) to metal ion sorption at oxide/water interfaces is the study of aqueous [57]Co sorption on hematite (Ambe et al., 1986). They followed the sorption reaction as a function of pH and found evidence that Co forms strongly bonded, inner-sphere complexes under alkaline conditions.

XAS STUDIES OF CHEMISORPTION REACTION MECHANISMS
AT SOLID/LIQUID INTERFACES

Over the past three years, about a dozen synchrotron-based XAS studies have been made on cation or anion complexes sorbed at solid/water interface, most at less than monolayer coverages. All were done on high surface area powders (surface areas of 2 to 115 m[2]/g) that are base or acid titrated to achieve the desired uptake of ion from solution (typically ~95% uptake). The use of powders rather than single crystals, for minerals other than clays, makes detailed structural interpretation of sorbate-sorbent interactions difficult because the surface microtopography of the particles may vary and could differ significantly from that of single crystals prepared under more controlled conditions. However, it is necessary in order to achieve adequate concentrations of the sorbed species in the excitation volume illuminated by the synchrotron x-ray beam. The wet suspensions are centrifuged and about 90% of the supernatant removed to avoid possible interference of an EXAFS signal from any remaining probe ion in the supernatent with that from the sorbed species. The small amount of solution phase in the wet suspensions usually contains a small amount of the probe ion of interest; however, the amount is typically not sufficient to give a measurable EXAFS signal, as shown by EXAFS examination of the supernatant.

As mentioned earlier, there are certain sample requirements that must be met in order to obtain useful information on the structure and stoichiometry of adsorption complexes by XAS. These may be summarized as follows: (1) There must be sufficient sorbate concentration to produce x-ray fluorescence in the fluorescence-yield mode or a sufficiently high I_0/I_1 in the transmission mode of data collection to produce an EXAFS spectrum of adequate signal-to-noise ratio to observe frequencies due to second or more distant neighbors around an absorber. (2) The concentration of the element of interest must be very low in the bulk oxide or solution phase to maximize the contribution of the surface species to the total fluorescence or transmission. In practice, up to about 10% of the absorbing element may be present in the bulk solid or solution phase without causing significant intereference with the signal from the surface species. (3) The absorber must have a high enough K-edge energy to minimize the effects of matrix and water absorption of the fluorescence-yield or transmitted x-rays. This requirement makes the study of absorbers of atomic number less than ~22 (Ti: K-edge energy = 4966 eV) difficult or impossible with currently available XAS methods. Thus sorption of Na, Mg, or K, for example, at a solid/water interface cannot be studied by XAS. (4) There must be sufficient differences between the backscattering amplitudes and/or phases of the absorbing element (A) and elements in the substrate (B) to permit discrimination between A and B among second neighbors. This requirement is critical because of our desire to discriminate between mononuclear and multinuclear adsorption complexes, if present, to identify potential surface precipitates, and to detect absorption (i.e., the probe element of interest is incorporated in the bulk solid through a diffusion or dissolution-precipitation process) (see discussion below). Distinguishing between elements like Fe and Co in the second shell around a central Fe atom would be impossible because of the similarity in backscattering amplitudes and phases of Fe and Co. Thus, sorption of Co on Fe-oxide or Fe-hydroxide cannot be studied by XAS. The same is true of Fe sorbed on Mn-oxides. In practice, element A (in the sorbate) should differ from element B (in the sorbent) by at least five atomic numbers. (5) It is also essential that the systems chosen for study be as thoroughly characterized as possible using classical surface chemistry methods.

To illustrate how different sorption mechanisms are detected by EXAFS spectroscopy, consider the following sorption scenarios. (1) *Outer-Sphere Complex:* If the sorbing phase forms an outer-sphere complex, the EXAFS spectrum would not show a "beat" pattern due to interference of different frequency photoelectron waves (see Fig. 11) and the Fourier transform would show no evidence of second neighbors. (2) *Inner-Sphere Complex:* In this case the EXAFS spectrum would most likely show a "beat" pattern caused by the two shells of atoms, although in certain cases the phases and/or amplitudes associated with the second neighbors could cause partial or complete cancellation of the "beat". The Fourier transform should show a clear, second-shell feature, and fitting of the Fourier backtransform of this second-neighbor shell should reveal metal atoms characteristic of the substrate, indicating that the sorbing cation or anion is close enough to the metal surface to form chemisorption bonds to surface oxygens. This mechanism requires that one or more waters of hydration around the sorbing metal be lost from the first-coordination sphere of the sorbing metal during adsorption. Depending on the number and distances of second-neighbor metals obtained in the fitting process, the adsorbing metal could be present as a monodentate, bidentate, or multidentate complex. Geometric analysis of several possible complex-sorbent structural configurations is necessary to distinguish among these possibilities. If in the fitting process, metal atoms of the same type as in the adsorbing complex are detected in the second shell, this would be evidence for the presence of multinuclear adsorption complexes at the solid/water interface. (3) *Surface Precipitate:* If analysis of the second shell around the adsorbing atom shows a large number of second-neighbor atoms of the same type as the central absorber but no evidence for second-neighbor atoms characteristic of the substrate, this constitutes evidence for a three-dimensional surface precipitate. More distant metal neighbors of the same type may also be detected in the case of a surface precipitate if it is an ordered phase. In both cases, the EXAFS spectrum should show a "beat", and the Fourier transform should show a clear second-shell feature. (4) *Absorption:* In this case, a large number of second-neighbor metal atoms characteristic of the substrate would be detected, with no second neighbors of the same type as the sorbing ion. In favorable cases, EXAFS analysis is capable of distinguishing among these possibilities in addition to providing geometric details of the sorbing species.

Most of the XAS measurements for the sorption samples discussed below were made in the fluorescence-yield mode because of its higher sensitivity than the transmission mode for elements at low concentration. The presentation of results emphasizes those studies performed by the Surface and Aqueous Geochemistry Group at Stanford University. This summary follows and updates that of Brown et al. (1989), presenting new data for Co(II) sorption at kaolinite/water, silica/water, and calcite/water interfaces and for Pb(II) sorption at the goethite/water interface. In addition, new data from surface-sensitive, grazing-incidence EXAFS measurements of sorbed species at electrode/electrolyte interfaces are summarized.

<u>Co(II) on γ-Al$_2$O$_3$ and TiO$_2$ (rutile)</u>

The sorption of Co^{2+} at the γ-Al$_2$O$_3$/water and TiO$_2$(rutile)/water interfaces has been studied using conventional surface chemistry methods (James and Healy, 1972; Tewari et al., 1972), XPS (Tewari and McIntyre, 1974; Tewari and Lee, 1975), and most recently using XAS (Chisholm-Brause et al. 1989; Brown et al., 1989; Chisholm-Brause et al., 1990a,c). James and Healy (1972) successfully simulated extensive sorption density data with a model that assumes sorbed Co^{2+} and its mononuclear hydrolysis products retain their inner hydration shells. They observed an unusually rapid increase in sorption density with pH when the solution phase approached saturation with respect to Co(OH)$_2$ and suggested surface-induced precipitation as a possible explanation. Tewari and co-workers (Tewari and McIntyre, 1974; Tewari and Lee, 1975) interpreted similarity in the Co $2p_{3/2}$ binding energies of $Co(OH)_2$ and Co(II) sorbed at high surface coverage on γ-Al_2O_3 at 30°C as evidence for surface-induced precipitation of $Co(OH)_2$ in agreement with James and Healy. This interpretation of the XPS data was challenged (Briggs and Bosworth, 1977) based on the argument that the effect of Co spin states must be accounted for before the spectral similarity between a model compound and an unknown can be used to identify the unknown.

Comparison of Co K-edge XANES spectra for crystalline $Co(OH)_2$, aqueous Co^{2+} sorbed on γ-Al_2O_3, aqueous Co^{2+} sorbed on TiO_2 (rutile), and aqueous 12 mM $Co(NO_3)_2$ solution (Fig. 16) shows that the local environments of aqueous Co^{2+} sorbed on γ-Al_2O_3 and TiO_2 at similar surface coverages (~1.25 μmoles/m^2) are significantly different from each other. Furthermore, the spectra for these sorption samples are different from that of $Co(OH)_2$ and the aquo-complex of Co^{2+} in solution. These observations suggest that Co in the sorption samples has a local structural environment distinctly different from that in $Co(OH)_2$. If crystalline $Co(OH)_2$ is a reasonable structural model for a surface precipitate of Co hydroxide, then surface precipitation can be ruled out as the mode of accumulation for Co^{2+} sorbed on γ-Al_2O_3 or TiO_2 (rutile) under the conditions of these experiments. The lack of similarity of the XANES spectra for the sorption samples relative to that of Co^{2+}(aq) is strong evidence that Co does not form an outer-sphere complex at either oxide/water interface. Moreover, the broader main absorption edges for the two sorption samples could be due to a broader distribution of distances around Co relative to the Co^{2+}(aq) complex, which would be expected to have a shell of six water molecules with a single Co-O distance. Finally, none of these four spectra displays a significant 1s to 3d pre-edge feature, indicating that Co^{2+} in each of these samples is in a fairly regular, near-centrosymmetric local coordination environment, such as a slightly-distorted octahedron.

Background-subtracted EXAFS data for (a) solid $Co(OH)_2$, (b) aqueous Co^{2+} on γ-Al_2O_3, (c) aqueous Co^{2+} on kaolinite, (d) aqueous Co^{2+} on TiO_2, and (e) 12 mM $Co(NO_3)_2$ solution are shown in Figure 17. In order to extract quantitative estimates of the distance, number, and types of nearest neighbors surrounding sorbed Co ions, we must begin data analysis by Fourier transforming the EXAFS functions (Fig. 18). Least-squares fitting of the back-transforms of the first peak in the radial structure function (rsf) from the Co sorption data yielded six nearest-neighbor oxygens at 2.08±0.02 Å for each case. Fitting the second major peak in the rsf of (b) resulted in four Co atoms at 3.12± 0.03 Å and about one Al at 3.20±0.03 Å. A third major peak in the rsf at about 5.7 Å was fit to about one Co at 6.10±0.05 Å. These Co-Co distances are 0.05 to 0.25 Å shorter and the number of second-neighbor Co atoms are smaller than those in $Co(OH)_2$. Chisholm-Brause et al. (1990c) inferred that the one Co neighbor at 6.1 Å in (b) is co-linear with the Co(1)-Co(2) bond, by analogy with the strong Co feature at 6.35 Å in $Co(OH)_2$. In the latter compound, the 6 fourth-neighbor Co atoms are colinear with the Co(1)-Co(2) bond, whereas the 18 third-neighbor Co atoms are not and do not contribute significant amplitude to the $Co(OH)_2$ rsf in the 5.5 Å region. For Co^{2+} on TiO_2, fitting the second feature in the rsf centered at ≈ 3 Å resulted in 1.4 Co atoms at 3.36±0.03 Å and about one Ti atom at 3.31±0.03 Å (Figs. 19 a,b). Unlike samples (a), (b), or (c), no strong fourth-shell Co peak is present in (d) or (e).

Inclusion of metal ions of the oxide (Al or Ti) or clay mineral (Si or Al) in the second shell around the sorbed Co, together with second-neighbor Co, produced significantly better fits than only second-neighbor Co atoms (Fig. 19). This observation leads to the conclusion that Co is bonded directly to the oxide surface in cases (b) Co(II)/γ-Al_2O_3, (c) Co(II)/kaolinite, and (d) Co(II)/TiO_2. The presence of second-neighbor Co around the sorbed Co indicates the presence of multinuclear sorption complexes. To the best of our knowledge, this is the first direct structural evidence for multinuclear metal sorption

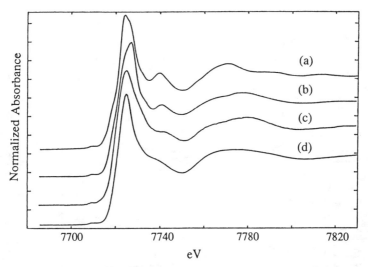

Figure 16. Comparison of Co K-XANES spectra for (a) Co(OH)$_2$, (b) Co sorbed on γ-Al$_2$O$_3$, (c) Co sorbed on TiO$_2$, and (d) 12 mM Co(NO$_3$)$_2$ aqueous solution. The lack of significant pre-edge features (4-5 eV below the main absorption edge) in any of the XANES spectra is indicative of a relatively undistorted, near-centrosymmetric site like an octahedron. From Brown et al. (1989).

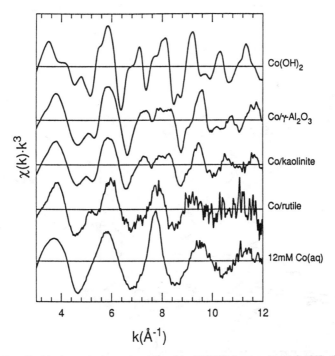

Figure 17. Normalized, background-subtracted, and k^3-weighted EXAFS functions for (a) Co(OH)$_2$ (i), (b) Co sorbed on γ-Al$_2$O$_3$, (c) Co sorbed on kaolinite, (d) Co sorbed on TiO$_2$ (rutile), and (e) 12 mM Co(NO$_3$)$_2$ aqueous solution. These data are Fourier transformed to yield radial structure functions (shown in Fig. 18), which are analyzed to determine bond distances and coordination numbers. From Chisholm-Brause et al. (1990c).

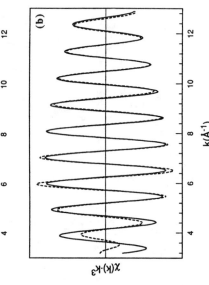

Figure 19. Comparisons of the EXAFS function generated by Fourier backtransforming the second radial structure function peak for Co sorbed on TiO2 (solid line) with a least-squares fit theoretical EXAFS function (dashed line) assuming (a) only Co atoms in the second shell and (b) Co and Ti atoms in the second shell around the central Co atom. For all sorption samples, significantly better fits were achieved by assuming that both Co and metal ions of the oxide (Al, Si, or Ti) were present in the second coordination shell. From Chisholm-Brause et al. (1990c).

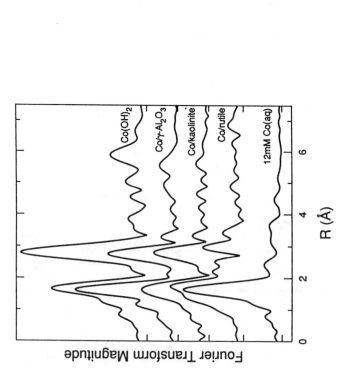

Figure 18. Radial structure functions (rsfs) for (a) Co(OH)2 (i), (b) Co sorbed on γ-Al2O3, (c) Co sorbed on kaolinite, (d) Co sorbed on TiO2 (rutile), and (e) 12 mM Co(NO3)2 aqueous solution. The first peak is attributed to first-neighbor oxygen atoms, and its amplitude reflects the number of first-neighbor oxygens surrounding Co. The second peak in (a), (b), (c), and (d) is due to second-neighbor atoms surrounding Co. Note the lower amplitude of this peak in (b), (c), and (d) relative to (a) and the absence of this peak in (e). A peak at about 6 Å due to fourth-shell Co atoms is present in (a) and, with lower amplitude, in (b) and (c). The fourth-shell peak is absent in (d) and (e). From Chisholm-Brause et al. (1990c).

complexes on oxide surfaces (Chisholm-Brause et al., 1990c), although clustering of Mn^{2+} on boehmite and goethite has been inferred from EPR measurements (Bleam and McBride, 1985). Comparison of the radial structure functions (rsf) and the fits from the three sorption samples (b,c, and d) with that of (a), the solid $Co(OH)_2$ (inactive), suggests that Co is present as a true adsorption complex and not as a precipitate. Furthermore, comparison of the rsf's and fits of (b), (c), and (d) with that of (e), the hexa-aquo Co(II) in the nitrate solution, indicates that the sorption complexes are the inner-sphere type because of the presence and absence, respectively, of second-neighbor Al or Ti atoms in the sorption samples and aqueous Co solution. The small number of Al or Ti nearest neighbors in the sorption complexes (b, c, or d) is strong evidence against significant diffusion (absorption) of Co into the oxides. More recent EXAFS data for aqueous Co(II) sorbed on γ-Al_2O_3 at lower surface coverages (0.25 and 0.61 µmoles/m^2) (Chisholm-Brause et al., 1990a) shows that as sorption density decreases, mononuclear, inner-sphere adsorption complexes form at the expense of multinuclear, inner-sphere complexes.

Several observations suggest that the properties of the solid/water interface influence the adsorption complex. For example, Co^{2+} adsorbs more strongly and in a different pH range on TiO_2 (rutile) than on α-SiO_2 (James and Healy, 1972). In addition, Ni^{2+} precipitates as an amorphous "active" hydroxide in the presence of α-SiO_2 surfaces, whereas in the presence of α-Al_2O_3, it nucleates as more ordered "inactive" $Ni(OH)_2$ (Ahmed, 1971). In the study of Chisholm-Brause et al. (1990c), the large number of second-neighbor Co atoms in the sorption complexes of Co^{2+} on γ-Al_2O_3 indicates the formation of multinuclear complexes containing an average of six cobalt atoms. Co multinuclear complexes of smaller size were found at the kaolinite/water and α-SiO_2/water interfaces (O'Day et al., 1990; see discussion below). In contrast, the small number of second-neighbor Co atoms in the sorption complexes of aqueous Co^{2+} on α-$SiO2$ and TiO_2 indicates smaller species, containing an average of 2-3 cobalt atoms. The surface coverages were roughly the same in all cases (1-3 µmoles/m^2). These results support the idea that the nature of the oxide surface influences the characteristics of the sorption complex formed.

The observations that TiO_2 surfaces favor formation of smaller adsorption complexes than γ-Al_2O_3 or kaolinite, with α-SiO_2 apparently favoring complexes closer in size to those on TiO_2, all at similar surface coverages, are related to differences in surface structure and bonding among these phases. The (100) or (111) (1x1) surfaces of γ-Al_2O_3 (a defect spinel structure like maghemite) should have both tetrahedrally- and octahedrally-coordinated Al atoms, with some oxygens (those typical of the bulk structure) four-coordinated by three octahedral Al and one tetrahedral Al and other oxygens (those terminating the Al octahedra or tetrahedra at the (100) surface) only one-coordinated to octahedral or tetrahedral Al. Any surface of α-SiO_2 should have only tetrahedrally-coordinated Si sites exposed with some of the oxygens two-coordinated by Si and others one-coordinated. The kaolinite (001) (1x1) surface could have either all octahedral Al sites or all tetrahedral Si sites exposed or more likely some combination of the two. In contrast, the (110) (1x1) surface of TiO_2 (Fig. 4) should have only octahedral Ti sites exposed, with some oxygens coordinated to three six-coordinated Ti and others one-coordinated to Ti. Although our EXAFS experiments were performed on high surface area powdered samples in aqueous suspensions, the surfaces listed are likely to be the most abundant ones. Because of differences in cation and anion coordination and differences in the covalency of metal-oxygens bonds, the residual charges on oxygen at these surfaces should be different, resulting in differences in surface reactivity of the different phases. Furthermore, the differences in atomic spacing on the different surfaces may exert an "epitaxial" effect on the geometry of the sorption complex formed. An analysis of these effects will be undertaken by Chisholm-Brause et al. (in preparation) and O'Day et al. (in preparation).

Co(II) on kaolinite and α-SiO_2 (quartz)

Recent XAS studies of the sorption mechanism of aqueous Co(II) on kaolinite

342

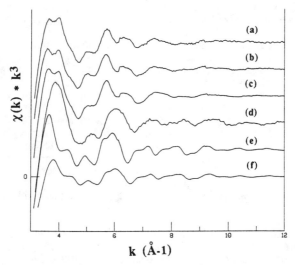

$\chi(k) * k^3$

k (Å⁻¹)

Figure 20. Comparison of normalized, background-subtracted, and k^3-weighted EXAFS spectra of (a) 480 ppm Co^{2+} sorbed at the calcite/water interface (initial Co concentration = 10^{-5}M); (b) 525 ppm Co^{2+} sorbed at the calcite/water interface (initial Co concentration = 10^{-5} M); (c) 2500 ppm Co in a natural calcite crystal; (d) 750 ppm Co^{2+} sorbed at the calcite/water interface (initial Co concentration = 10^{-4} M); (e) and (f) two different samples of $Co(OH)_2$ (inactive form). From Xu et al. (1990).

$(Al_2Si_2O_5(OH)_4)$ (O'Day et al., 1988, 1989; O'Day et al., 1990; Chisholm-Brause et al., 1990c) and on quartz (α-SiO_2) (O'Day et al., 1990) provide additional support for the conclusions reached above for aqueous Co(II) sorption on γ-Al_2O_3. This work, when compared with Co(II)/γ-Al_2O_3 results at different surface coverages, leads to the following conclusions (O'Day et al., 1990): (1) Below the solubilities of $Co(OH)_2$ (inactive) and $Co(OH)_2$ (active) and with surface coverages of 3 to 10 μmoles/m², aqueous Co(II) sorbs on kaolinite and quartz predominantly as two-dimensional, inner-sphere, multinuclear adsorption complexes. (2) Above the solubilies of the active and inactive forms of the solid hydroxide, Co (II) sorbs as a "large" (> 50 Co atoms), three-dimensional phase, but with local structure around Co different from that in crystalline $Co(OH)_2$ (i). This observation could be consistent with the precipitation of a disordered phase of $Co(OH)_2$ (i) or of small particles of $Co(OH)_2$ (i). (3) When compared with EXAFS results for Co(II) sorption on γ-Al_2O_3, these data suggest that surface hydrolysis and adsorption proceed from mononuclear adsorption complexes, through multinuclear adsorption complexes of increasing size as saturation is approached, to precipitation of a three-dimensional hydroxide phase as the solubility of the solid hydroxide is exceeded.

Co(II) on CaCO₃

An XAS study of the sorption mechanism of aqueous Co(II) on natural calcite cleavage surfaces (Xu et al., 1990) found that the Co(II) forms a solid solution with $CaCO_3$ under most of the experimental conditions examined. Figure 20 clearly shows the similarity of the EXAFS spectrum of the two lower concentration sorption samples (a) and (b) with that of 2500 ppm Co in a natural calcite. The higher concentration sorption sample (d) appears to have a local Co environment similar to that in $Co(OH)_2$ (i), suggesting that precipitation has occurred. Our observations for the two lower concentration sorption samples rule out the formation of a dominant precipitate phase with a local Co environment like that in $Co(OH)_2$ (i). Instead, they lead to the suggestion that <u>ab</u>sorption has occurred

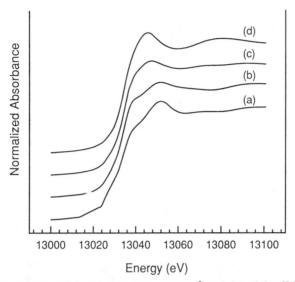

Figure 21. Pb L_{III}-XANES spectra for (a) PbO (orthorhombic), (b) β-Pb$_6$O(OH)$_6$(ClO$_4$)$_4$·H$_2$O, (c) 15 mM Pb sorbed at the γ-Al$_2$O$_3$/water interface, and (d) 15 mM Pb(NO$_3$)$_2$ aqueous solution. The presence of two shoulders on the absorption edge for sample (a) reflects the presence of second-shell metal atoms. The marked differences between the spectra for the sorption sample (c) and those of the model compounds (a, b and d) indicate that the local structural environments of Pb in these samples are different. The presence of two peaks in the region 13060-13100 eV of the spectra for samples (a), (b), and (c) indicates the presence of second-shell metal atom(s). From Brown et al. (1989) (cf. Chisholm-Brause et al., 1990b, Figs. 8a,b).

in these two samples. These spectra are currently being fit to structural models to provide more quantitative structural information and to determine if the Co is ordered or disordered in the solid solution or if Co second-neighbor clustering occurs.

<u>Pb(II) on γ-Al$_2$O$_3$ and α-FeOOH (goethite)</u>

Sorption of Pb^{2+} on γ-Al$_2$O$_3$ has been studied by Hohl and Stumm (1976) who found it necessary to assume two types of adsorption complexes, one involving a monodentate site and the other a bidentate site, in order to simulate experimental sorption data with a constant capacitance adsorption model. Davis and Leckie (1978), using an electrical triple-layer adsorption model to interpret the same experimental data, concluded that two monodentate-type complexes were needed and that the bidentate-type complex was unnecessary. The figures used by both groups to illustrate the sorption reactions imply inner-sphere complexes. Hayes and Leckie (1986a) later emphasized the necessity of distinguishing between inner- and outer-sphere complexes explicitly.

Recent XAS studies of Pb^{2+} sorption at the γ-Al$_2$O$_3$/water interface (Chisholm-Brause et al., 1989; Chisholm-Brause et al., 1990b) help clarify the sorption mechanism. XANES spectra are shown in Figure 21. The presence of a small shoulder at about 13050 eV in the derivative spectra of the sorption samples at the same energy as a major feature in the Pb model compounds (Fig. 22) suggests the presence of Pb among second-neighbor atoms in the 15 mM sorption sample. This feature is absent in the Pb^{2+}(aq) sample as it should be. Fourier transforms of the EXAFS (Fig. 23) clearly show the presence of a second shell around Pb^{2+} on γ-Al$_2$O$_3$. Fits of the back-transform of the first rsf peak yield one oxygen at 2.23±0.02 Å and two oxygens at 2.46±0.02 Å. Fits of back-transforms of the second rsf peak (Fig. 24) yield a second-neighbor Pb atom at 3.55± 0.03 Å and an average of about one Al atom at 3.77±0.03 Å, indicating inner-sphere bonding. The presence of only about one second-neighbor Pb atom indicates either dimeric complexes or

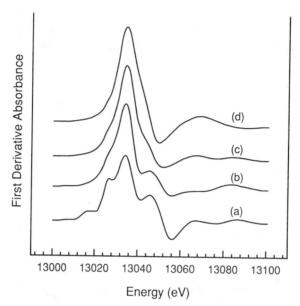

Figure 22. First derivatives of the Pb L_{III}-XANES spectra shown in Figure 21: Pb L_{III}-XANES spectra for (a) PbO (orthorhombic), (b) β-Pb$_6$O(OH)$_6$(ClO$_4$)$_4$·H$_2$O, (c) 15 mM Pb sorbed at the γ-Al$_2$O$_3$/water interface, and (d) 15 mM Pb(NO$_3$)$_2$ aqueous solution. The shoulders in the normalized edge spectra (Fig. 21) due to second-shell metal atoms are indicated by small peaks in the derivative spectra at ~13027 and 13045 eV. The two peaks in the 13060-13100 eV region of the normalized edge spectra for (a), (b), and (c) resulting from the first and second coordination shells around the central Pb absorber atom are clearly visible in the first derivative spectra of (a), (b), and (c). From Brown et al. (1989) (cf. Chisholm-Brause et al., 1990b, Figs. 8c,d).

a mixture of monomeric and a small fraction of multinuclear complexes. This and the presence of an Al in the second-neighbor shell indicate that the sorbed species is not a surface precipitate. The observed Pb-O and Pb-Al distances are consistent with a monodentate site in the sense that each Pb in the complex is bonded to a single surface oxygen site; bidentate and tridentate arrangements have much shorter Pb-Al distances than observed. In summary, these results provide direct, molecular-level evidence that Pb^{2+} forms inner-sphere complexes on γ-Al$_2$O$_3$, probably both mono- and multinuclear, involving an average of one surface site per Pb atom.

A similar XAS study of aqueous Pb(II) sorbed at the goethite/water interface (Roe et al., 1990) showed that at low coverages (2 mM Pb(II), corresponding to ~15% monolayer coverage), Pb forms monomeric, inner-sphere surface complexes. At higher coverages (up to 15 mM, corresponding to monolayer coverage), the EXAFS and XANES data are consistent with the formation of surface polymers of Pb, bonded to the surface by a few inner-sphere lead-oxygen bonds. No evidence was found for the formation of a surface precipitate or a solid solution at the highest surface coverages. The XAS results provide a molecular-level explanation for the pressure-jump relaxation results for lead ion adsorption at the goethite/water interface (Hayes and Leckie, 1986b).

Np(V) on α-FeOOH and U(VI) on ferric oxide-hydroxide gels

Isotopes such as ^{237}Np and ^{238}U are produced during plutonium processing and in nuclear power production; thus they are important radionuclides to be stored in high-level nuclear waste repositories. The transport of these isotopes and others in local ground-waters around a repository and their partitioning onto associated minerals must be under-

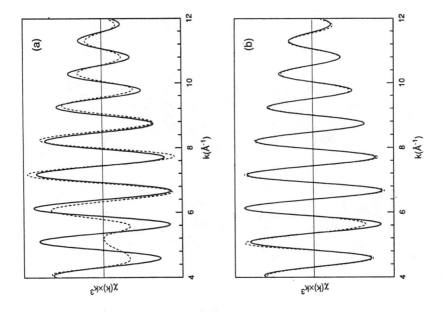

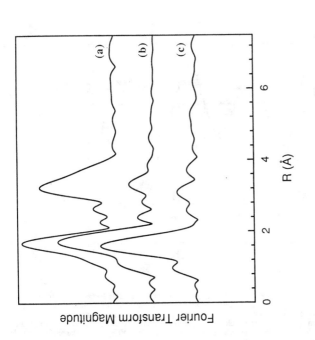

Figure 23 (left). Radial structure functions for Pb L$_{III}$-XANES spectra for (a) PbO (ortho-rhombic), (b) β-Pb$_6$O(OH)$_6$(ClO$_4$)$_4$·H$_2$O, and (c) 15 mM Pb sorbed at the γ-Al$_2$O$_3$/water interface. A small peak at about 3 Å in (b) indicates the presence of second-shell metal atoms. Fits of Fourier backtransform of this peak (Fig. 22) yielded a mixed shell of Pb and Al atoms. From Brown et al. (1989) (cf. Chisholm-Brause et al., 1990b, Fig. 7).

Figure 24 (right). Comparison of the EXAFS function generated by Fourier back-transforming the second radial structure function peak for 15 mM Pb sorbed at the γ-Al$_2$O$_3$/water interface (solid line) with a least-squares fit, theoretical EXAFS function (dashed line) assuming (a) only Pb atoms in the second shell and (b) Pb and Al atoms in the second shell around the central Pb atom. From Chisholm-Brause et al. (1990b).

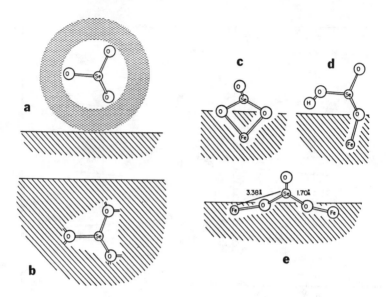

Figure 25. Schematic drawings of several of the possible geometric arrangements of selenite ions at the α-FeOOH (goethite)/water interface. The solid is indicated by the lined area, and the area above each surface is water. Model (e) is the one most consistent with the fit of the selenite on goethite EXAFS data, whereas model a is consistent with the fit of selenate on goethite. From Hayes et al. (1987).

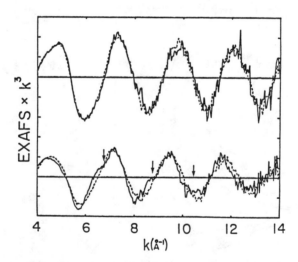

Figure 26. Background-subtracted, k^3-weighted Se K-EXAFS spectra for selenate sorbed on goethite (a) and selenite sorbed on goethite (b). Experimental data are shown as solid curves, and fits of these data assuming a single shell of four and three oxygen atoms, respectively, for selenate and selenite on goethite are shown as dashed curves. The arrows on spectrum (b) from selenite on goethite indicate features produced by a second shell of metal atoms surrounding Se. Spectrum (a) from selenate on goethite is fit well by a model assuming a single shell of four oxygens around Se. From Hayes et al. (1987).

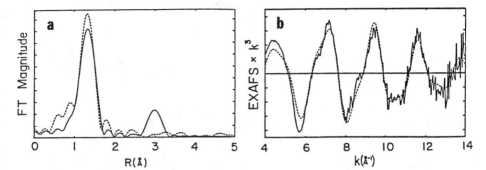

Figure 27. (a) Radial structure functions for selenite (solid curve) and selenate (dashed curve) sorbed on goethite. (b) Comparison of the EXAFS function generated by Fourier backtransforming the first and second radial structure function peaks for 5 mM selenite sorbed on goethite (solid line) with a least-squares fit theoretical EXAFS function (dashed line) assuming three oxygen atoms at 1.70 Å and two iron atoms at 3.38 Å around selenium. From Hayes et al. (1987).

stood at a fundamental level in order to reliably predict their far-field hydro-geochemical behavior. XAS studies of the sorption of Np(V) onto goethite (α-FeOOH) (Combes et al., 1990) and of U(VI) onto ferric oxide-hydroxide gels (Combes, 1988) indicate that one mode of adsorption under the solution conditions examined is the formation of inner-sphere complexes with neptunyl or uranyl moieties.

Selenium oxyanions on α-FeOOH

Elements other than radionuclides may also pose environmental or ecological hazards when they occur in sufficient concentrations. One such element is selenium. The sorption of selenium oxyanions on the common ferric iron oxy-hydroxide mineral, goethite (α-FeOOH) was studied by Hayes et al. (1987). Figure 25 illustrates several of the possible geometric arrangements of selenite ions at the goethite/water interface. Conventional surface chemistry studies have shown that selenite is sorbed more strongly than selenate on goethite, suggesting that the two ions are bonded differently (Hayes, 1987). In-situ Se K-edge fluorescence-yield EXAFS spectra of selenate and selenite sorption complexes are shown in Figure 26. The spectrum of selenate shows a single, damped sinusoidal wave typical of a single shell of four oxygen backscatterers, whereas the spectrum of selenite shows a distinctive "beat pattern" characteristic of more than one shell of backscatterers at different distances. Fourier transforms of these spectra (Fig. 27a) show a clear second-shell contribution in the case of selenite but none in the case of selenate. Fitting of the back-transformed spectra to one shell of four oxygens for selenate and two shells for selenite (one shell of three oxygens and one shell of two irons) gave good fits to the experimental EXAFS (Fig. 27b). The best-fit models are three oxygens at 1.70 Å and two irons at 3.38 Å around selenium in selenite ions at the goethite/water interface and four oxygens at 1.65 Å around selenium in selenate ions at this interface. These results indicate that selenate is sorbed as an outer-sphere complex (Fig. 25a) and that selenite is sorbed as an inner-sphere complex in a bidentate fashion (Fig. 25e). These structural results provide the first molecular-level explanation for the weak binding of selenate and the strong binding of selenite at the goethite/water interface.

Other XAS studies of sorption complexes at solid/water or solid/air interfaces

Other in situ EXAFS studies of sorbed species at solid/water interfaces include the following: (1) fluorescence-yield EXAFS study of $(AsO_4)^{3-}$ sorbed on powdered ferrihydrite (Waychunas et al., 1990); (2) a transmission EXAFS study of Ni^{2+} sorbed on silica gel (Bonneviot et al., 1988); and (3) a grazing-incidence EXAFS study of Pb sorbed on single-crystal Ag(111) (Samant et al., 1987, 1988). In addition to these in situ measurements, EXAFS studies have been performed on dried samples following sorption or electrochemical deposition experiments. These include (1) a grazing-incidence fluorescence-yield EXAFS study of copper on gold (Blum et al., 1986; Melroy et al., 1988); (2) transmission EXAFS studies of perrhenate, permanganate, molybdate, tungstate, and chromate on alumina (Mulcahy et al., 1990); (3) transmission EXAFS studies of phospho-molybdate on γ-Al$_2$O$_3$ and α-AlOOH (van Veen et al., 1990a, 1990b); and (4) a trans-mission EXAFS study of TiO$_2$ (rutile and anatase) on SiO$_2$ (Salama et al., 1990). Other EXAFS studies of interface regions include (1) a transmission EXAFS study of the changes in a Ni(OH)$_2$ electrode in contact with a concentrated alkali electrolyte solution as the electrode was being electrochemically oxidized (McBreen et al., 1987); (2) a trans-mission EXAFS study of a γ-alumina-supported CoMoS catalyst while hydro-desulfur-ization of benzothiophene was proceeding at high temperature (573 K) and pressure (7.3 MPa) (Boudart et al., 1985); (3) a glancing-angle EXAFS study of Cu, Ni, and Cr on Al (Heald et al., 1990); (4) a fluorescence-yield EXAFS study of the oxide film on iron and iron/chromium alloys (Kerker et al., 1990); (5) a time-resolved, energy-dispersive, in situ EXAFS study of the electrochemical inclusion of Cu and Fe in a conducting polymer (Guay et al, 1990); (6) an in situ transmission EXAFS study of Rh supported on γ-Al$_2$O$_3$, V$_2$O$_3$, Cr$_2$O$_3$, and MoO$_3$ (Johnston et al., 1990); (7) a fluorescence-yield EXAFS study of alumina-supported Mo$_2$(OAc)$_4$ and Li$_4$[Mo$_2$Me$_8$] (Evans et al., 1990); and (8) a trans-mission EXAFS study of a γ-Al$_2$O$_3$-supported Ir catalyst (Kampers and Koningsberger, 1990).

SEXAFS STUDIES OF MOLECULAR CHEMISORPTION
AT SOLID/VACUUM INTERFACES

As mentioned earlier, Surface Extended X-ray Absorption Fine Structure (SEXAFS) provides the same type of structural information as normal EXAFS; however, it requires an ultra-high vacuum (UHV) environment because electron detection methods are normally used instead of x-ray detection (see Stöhr, 1988, for discussion of various detection techniques used in SEXAFS measurements). This is both an advantage and a disadvantage. The UHV environment of the sample and detector currently precludes the possibility of studying sorption complexes at solid/water interfaces using SEXAFS. Therefore, SEXAFS is restricted to the study of solid/vacuum interfaces. However, the UHV environment makes it possible to study elements whose absorption edges occur in the soft x-ray - vacuum ultraviolet energy range (50 to ~2000 eV), which is not possible in an air or He environment. These include the K-edges of the second- and third-row elements (C, N, O, F, Na, Mg, Al, Si, etc.), many of which are of geochemical importance, as well as the L and M edges of heavier elements (e.g., Fe L$_{III}$). The enhanced detection sensitivity provided by the UHV environment also makes possible the study of sorption on clean, single-crystal surfaces. Studies of sorption complexes at solid/water interfaces using single-crystal samples are quite difficult at low coverages with conventional EXAFS methods, including fluorescence yield.

Although SEXAFS is not well suited for structural studies of clean surfaces, it can provide detailed information about molecules or ions chemisorbed at a solid/vacuum interface, and many studies of this type have been carried out over the past ten years (see Citrin, 1986, for a thorough review of SEXAFS applications through early 1986). Here we will highlight only a few of these applications.

One of the most thoroughly studied chemisorption systems by SEXAFS and other methods is oxygen on the (111) and (100) surfaces of Al metal (see Stöhr, 1988 for summary). Figure 28 shows three possible chemisorption sites for oxygen on the fcc (100) surface. The Al(111) surface would have three-fold "hollow" sites, rather than the four-fold "hollow" sites of the (100) surface, and would also have two-fold "bridge" sites as well as one-fold "atop" sites. Thus there are three possible sites on the Al(111) surface where oxygen can chemisorb. It would appear to be a relatively simple structural problem to sort out the site on which oxygen chemisorbs and at what Al-O distance using LEED; however, that was not the case. Three LEED studies of the (1x1) Al(111) surface exposed

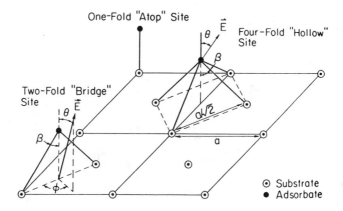

Figure 28. Chemisorption sites on an fcc (100) surface. From Stöhr (1988).

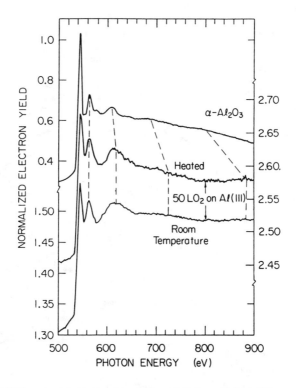

Figure 29. SEXAFS spectra at the K-edge of oxygen for α-Al$_2$O$_3$ (corundum) and 50-L oxygen on Al(111) at room temperature and after heating to 200°C for 10 min. From Norman et al. (1981).

350

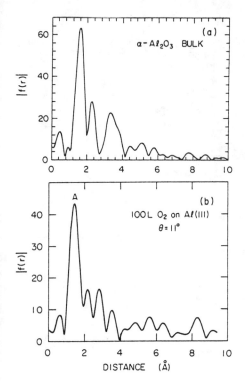

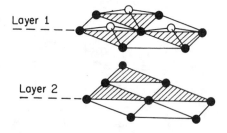

Figure 30. Fourier transforms of k^2-weighted oxygen K-SEXAFS spectra for (a) bulk α-Al_2O_3 (corundum) and (b) 100-L oxygen on Al(111) recorded at an 11° grazing-incidence angle. From Stöhr et al. (1980).

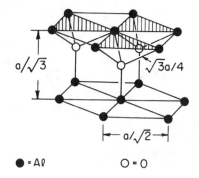

Figure 31. Structural models for (a) chemisorbed oxygen on Al(111) and (b) a subsurface, oxide-like phase that increases in volume with increasing oxygen surface coverage. From Norman et al. (1981).

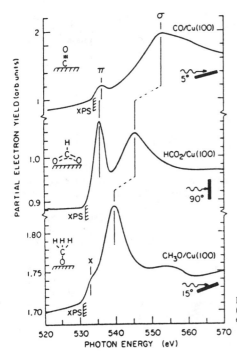

Figure 32. Oxygen K-NEXAFS spectra of CO, formate (HCO$_2$), and methoxy (CH$_3$O) on Cu(100). From Stöhr et al. (1983).

to 100-150 L of oxygen (Payling and Ramsey, 1980; Martinson et al., 1979; Yu et al., 1980) led to the conclusion that oxygen chemisorbs to the three-fold "hollow" site, resulting in an Al-O distance of about 2.20±0.04 Å. SEXAFS measurements made on similar samples (Johansson and Stöhr, 1979; Stöhr et al., 1980) (Figs. 29 and 30) gave an average Al-O distance of 1.79±0.05 Å, which was interpreted as being associated with Al chemisorbing in the three-fold "hollow" site but at a significantly different height above the surface than that indicated by LEED measurements. This rather significant discrepancy was finally cleared up by additional analyses of LEED intensities (Soria et al., 1981) which showed that the LEED intensity data were best accounted for by Al-O distances ranging from 1.80±0.02 Å to 1.83±0.02 Å, depending on oxygen exposure (90L to 150 L). Figure 31 contrasts the structural model derived for chemisorbed oxygen on Al(111) with a subsurface oxide-like phase that appears with increasing coverage. This test case established SEXAFS as a powerful new surface structural tool. Since this initial work, a number of SEXAFS studies of atomic sorption on single-crystal metal and semiconductor surfaces have been carried out (see Citrin, 1986, and Stöhr, 1988, for a summary).

More recently, SEXAFS and NEXAFS have been used to study the orientation of polyatomic molecules chemisorbed on clean metal surfaces. Some of this work is summarized by Stöhr (1984), Outka and Madix (1986), Stöhr and Outka (1987), and Outka and Stöhr (1988). Interest in these types of studies is driven by the desire to understand reaction intermediates in surface-catalyzed reactions. This work makes use of the strong polarization of the synchrotron x-ray beam in the plane of the storage ring by utilizing different incident glancing angles to determine molecular orientation. The near-edge structure contained in NEXAFS spectra is much more sensitive to different orientations of chemisorbed polyatomic molecules than the extended fine structure of SEXAFS. An example of this type of result (Stöhr et al., 1983) is shown in Figure 32, which illustrates variations in the position of the σ^* continuum resonance in NEXAFS spectra of CO, formate (HCO_2), and methoxy (CH_3O) chemisorbed on Cu(100). Distances can be obtained by an analysis of polarization-dependent NEXAFS spectra using the simple relationship discussed in the section on XANES and pre-edge spectra above. They also can be obtained by an analysis of the SEXAFS.

CONCLUSIONS AND OUTLOOK

The above examples show that useful, quantitative information on the structure and stoichiometry of sorption complexes at mineral/water interfaces can be obtained from x-ray absorption spectroscopy and several magnetic resonance methods (EPR, ENDOR, ESEEM). Careful analysis of XAS data provides unique information on the geometric details of sorption complexes that contain cations or anions of atomic number greater than ~22 (Ti). Such information makes it possible in favorable cases to distinguish among different molecular mechanisms of chemisorption. The current situation is certain to be improved by the completion of new synchrotron radiation sources (the Advanced Photon Source at Argonne National Laboratory and the Advanced Light Source at Lawrence Berkeley Laboratory) with several orders of magnitude more brightness than current sources. These new sources should permit EXAFS and x-ray scattering studies of chemisorbed species on surfaces at much lower surface coverages and on single-crystal surfaces as well as dynamical studies of sorption reactions. Less quantitative, although quite valuable information about sorption complexes at solid/water interfaces can also be obtained from vibrational spectroscopy methods (FTIR and SERS); however, FTIR and SERS have not been used extensively to characterize sorption complexes at mineral/water interfaces.

New methods such as scanning tunneling microscopy and atomic force microscopy are likely to provide us with a different perspective on the structure of surfaces, including defect sites, and the coverage and structure of sorption complexes on mineral surfaces (see, Hochella et al., 1989; Eggleston and Hochella, 1990; Hochella, this volume, for a discussion of these methods). At present, the insulating nature of many oxide-based minerals precludes atomic-resolution STM studies of their surfaces, but AFM holds the promise of providing such information for a wide range of mineral surfaces (e.g., Hochella et al, 1990). However, these methods are not likely to yield the geometric details provided by a carefully-performed surface EXAFS or scattering experiment. On the other hand AFM or STM can, in principle, look at individual sites or sorption complexes on a surface, rather than averaging over all sorption species of a selected cation or anion as EXAFS does. For example, recent atomic-resolution STM work on the structure of iodine on Pt(111) (Schardt et al., 1989) and of chlorine on Si(111)-(7x7) (Boland and Villarrubia, 1990) demonstrates the potential of these techniques. The most reasonable approach to the study of the structure and reactivity of mineral/water interfaces is to employ a variety of methods that provide complementary information, since no single characterization method gives a complete description of surface structure or the geometric details of sorption sites or sorption complexes. It is essential that the surface chemistry of sorption samples be well characterized by classical surface chemistry methods (e.g., adsorption isotherm measurements, BET surface area measurements, etc.) before they are studied by spectroscopic methods.

It would be very useful to have simple, atomic-level models for solid/water interfaces with which to predict local geometric arrangements, surface site charges, and bond strengths for clean and adsorbate-covered surfaces, much like we have for bulk crystalline materials. With such models the reactivity of a given surface for different sorbates could be predicted. However, we are not yet at the point in our study of mineral surfaces that we can draw generalizations of the type that led Pauling to develop his famous "rules" for complex ionic crystals, although such generalizations are now beginning to be made for metal surfaces (e.g., Citrin, 1986, 1987; Mitchell et al., 1986) and have been attempted for oxide surfaces (e.g., Ziolkowski, 1983).

ACKNOWLEDGMENTS

I thank Cathy Chisholm-Brause, Peggy O'Day, George Parks, and Ning Xu (all members of the Aqueous and Surface Geochemistry Group at Stanford University), Kim Hayes (now at the University of Michigan), and Larry Roe (now at Intel Corporation, Albuquerque, NM) for their contributions to this chapter through a number of collaborative EXAFS investigations of sorption processes at solid-water interfaces. All of us put in long hours collecting the data, and Cathy, Peggy, and Ning carried out all of the data reduction. Keith Hodgson (Department of Chemistry, Stanford University) is thanked for making his computer system available to us for some of the EXAFS data reduction. I also wish to thank my colleagues George Parks and Mike Hochella for numerous discussions concerning mineral surfaces and adsorption over the past few years and for critically reviewing this chapter. John Bargar (Stanford University) is thanked for his help with literature searches and for drafting Figure 2, which was modified after a similar figure prepared by Peggy O'Day. In addition, I am pleased to acknowledge the help of Brit Hedman (SSRL) with various beamline problems encountered during the many hours of data collection at SSRL and the advice and encouragement of Farrel Lytle (Boeing Corporation) during the past decade. Farrel's development of a simple and efficient x-ray fluorescence detector made feasible our EXAFS work on sorption complexes at low surface coverages. Much of the EXAFS work reported in this chapter has been supported by the National Science Foundation through grants EAR-8805440 and EAR-8513488. The EXAFS data reported herein were collected at the Stanford Synchrotron Radiation Laboratory and the National Synchrotron Light Source. SSRL is supported by DOE and the NIH biotechnology program. The staffs of both laboratories are thanked for their help over the past few years.

REFERENCES

Ahmed, A. (1971) The Sorption of Aqueous Ni(II) Species on Alumina and Quartz. Ph.D. Dissertation, Stanford University, Stanford, CA, 113 pp.

Ambe, F., Ambe, S., Okada, T., and Sekizawa, H. (1986) In situ Mössbauer studies of metal oxide-aqueous solution interfaces with adsorbed cobalt-57 and antimony-119 ions. In: Geochemical Processes at Mineral Surfaces, eds. J. A. Davis and K. F. Hayes, ACS Symposium Series 323, Washington, DC: Am. Chem. Soc., pp. 403-424.

Anderson, P.J., Horlock, R.F., and Oliver, J.F. (1965) Interaction of water with the magnesium oxide surface. Trans. Faraday Soc. 61, 2754-2762.

Ban, M.I., Van Hove, M.A., and Somorjai, G.A. (1987) The tilting of CO molecules chemisorbed on a Pt(110) surface: a molecular orbital study. Surface Sci. 185, 355-372.

Barrett, N.T., Greaves, G.N., Pizzini, S., and Roberts, K.J. (1990) The local atomic structure of the oxide coating on polished GaAs(100). Surface Sci. 227, 337-346.

Bassett, W.A. and Brown, G.E., Jr. (1990) Synchrotron radiation in the earth sciences. Ann. Rev. Earth Planet. Sci. 18, 387-447.

Bassetti, V., Burlamacchi, L. and Martini, G. (1979) Use of paramagnetic probes for the study of liquid adsorbed on porous supports. Copper (II) in water solution. J. Am. Chem. Soc. 101:19, 5471-5477.

Beebe, T.P., Jr., Crowell, J.E., and Yates, J.T., Jr. (1990) Infrared spectroscopic study of the rotation of chemisorbed methoxy species on an alumina surface. J. Chem. Phys. 92(8), 5119-5126.

Bell, A.T. (1984) Fourier-transform infrared spectroscopy in heterogeneous catalysis. In: Chemistry and Physics of Solid Surfaces V, Springer Series in Chemical Physics 35, eds. R. Vanselow and R. F. Howe, Berlin: Springer-Verlag, pp. 23-38.

354

Bianconi, A. (1988) XANES spectroscopy. In: X-ray Absorption: Principles, Applications, Techniques of EXAFS, SEXAFS, and XANES, eds. D.C. Koningsberger and R. Prins, New York: John Wiley & Sons, pp. 573-662.

Bleam, W.F. and McBride, M.B. (1985) Cluster formation versus isolated-site adsorption. a study of Mn(II) and Mg(II) adsorption on boehmite and goethite. J. Colloid Interface Sci. 103, 124-132.

Blum, L., Abruna, H.D., White, J., Gordon, J.G., II, Borges, G.L., Samant, M.G., and Melroy, O.R. (1986) Study of underpotentially deposited copper on gold by fluorescence detected surface EXAFS. J. Chem. Phys. 85, 6732-6738.

Boehm, H.P. (1971) Acidic and basic properties of hydroxylated metal oxide surfaces. Disc. Faraday Soc. 52 (Surface Chemistry of Oxides), 264-289.

Boland, J.J. and Villarrubia, J.S. (1990) Identification of the products from the reaction of chlorine with the silicon (111)-(7x7) surface. Science 248, 838-840.

Bonneviot, L., Clause, O., Che, M., Manceau, A., Decarreau, A., Villain, F., Bazin, D., and Dexpert, H. (1988) Investigation of the effect of pH on the structure of Ni^{2+} ions impregnated in a silica gel by EXAFS. Physica B 158, 43-44.

Bosio, L., Cortez, R., and Froment, M. (1984) ReflEXAFS studies of protective oxide formation on metal surfaces. In: EXAFS and Near-Edge Structure III, K.O. Hodgson, B. Hedman, and J.E. Penner-Hahn, eds., Springer Proc. Phys. 2, New York: Springer-Verlag, p. 484-486.

Boudart, M., Dalla Betta, R.A., Foger, K., Loffler, D.G., and Samant, M.G. (1985) Study by synchrotron radiation of the structure of a working catalyst at high temperatures and pressures. Science 228, 717-719.

Briggs, D. and Bosworth, Y.M. (1977) Comment on "Adsorption of Co(II) at the oxide-water interface. J. Colloid Interface Sci. 59, 194.

Brown, G.E., Jr. and Parks, G.A. (1989) Synchrotron-based x-ray absorption studies of cation environments in earth materials. Rev. Geophys. 27, 519-533.

Brown, G.E., Jr., Parks, G.A., and Chisholm-Brause, C.J. (1989) In situ x-ray absorption spectroscopic studies of ions at oxide-water interfaces. Chimia 43, 248-256.

Brown, G.E., Jr., Calas, G., Waychunas, G.A., and Petiau, J. (1988) X-ray absorption spectroscopy and its applications in mineralogy and geochemistry. In: Spectroscopic Methods in Mineralogy and Geology, ed. F. Hawthorne, Reviews in Mineralogy 18, 431-512.

Brown, I.D. (1977) Predicting bond lengths in inorganic crystals. Acta Crystallogr. B33, 1305-1310.

Brown, I.D. (1981) The bond-valence method: an empirical approach to chemical structure and bonding. In: Structure and Bonding in Crystals, Vol. II, eds. M. O'Keeffe and A. Navrotsky, New York: Academic Press, pp. 1-30.

Brown, I.D. (1987) Recent developments in the bond valence model of inorganic bonding. Phys. Chem. Minerals 15, 30-34.

Brown, I.D. and Altermatt, D. (1985) Bond-valence parameters obtained from a systematic analysis of the inorganic crystal structure database. Acta Crystallogr. B41, 244-247.

Brown, I.D. and Shannon, R.D. (1973) Empirical bond-strength-bond-length curves for oxides. Acta Crystallogr. A29, 266-282.

Calas, G. (1988) Electron paramagnetic resonance. In: Spectroscopic Methods in Mineralogy and Geology, ed. F. Hawthorne, Reviews in Mineralogy 18, 513-571.

Campion A. (1986) Raman spectroscopy of adsorbed molecules. In: Chemistry and Physics of Solid Surfaces VI, eds. R. Vanselow and R. Howe, Springer Series in Surface Sciences 5, Berlin: Springer-Verlag, pp. 262-283.

Chang, S.-C. and Weaver, M.J. (1990) In situ infrared spectroscopy of CO adsorbed at ordered Pt(110)-aqueous interfaces. Surface Sci. 230, 222-236.

Chisholm-Brause, C.J., Brown, G.E., Jr., and Parks, G.A. (1989) EXAFS investigation of aqueous Co(II) adsorbed on oxide surfaces in situ. Physica B 158, 646-648.

Chisholm-Brause, C.J., Brown, G.E., Jr., and Parks, G.A. (1990a) In situ EXAFS study of Co(II) sorption on γ-Al$_2$O$_3$ at different adsorption densities. In: XAFS VI, X-ray Absorption Fine Structure VI, ed. S.S. Hasnain, Chichester, U.K.: Ellis Horwood Ltd. Publishers (in press).

Chisholm-Brause, C.J., Hayes, K.F., Roe, A.L., Brown, G.E., Jr., Parks, G.A., and Leckie, J.O. (1990b) Spectroscopic investigation of Pb(II) complexes at the γ-Al$_2$O$_3$/water interface. Geochim. Cosmo-chim. Acta 54, 1897-1909.

Chisholm-Brause, C.J., O'Day, P.A., Brown, G.E., Jr., and Parks, G.A. (1990c) Synchrotron-based x-ray absorption evidence for cation polymerization at oxide-water interfaces. Nature. (submitted).

Chisholm-Brause, C.J., Roe, A.L., Hayes, K.F., Brown, G.E., Jr., Parks, G.A., and Leckie, J.O. (1989) XANES and EXAFS study of aqueous Pb(II) adsorbed on oxide surfaces. Physica B 158, 674-676.

Citrin, P.H. (1986) An overview of SEXAFS during the past decade. J. de Physique, Coll. C8, suppl. 12, vol. 47, 437-472.

Citrin, P.H. (1987) Correlation of SEXAFS-determined surface bond lengths and coordination numbers with simple bonding concepts. Surface Sci. 184, 109-120.

Clark, C.J. and McBride, M.J. (1984) Chemisorption of Cu(II) and Co(II) on allophane and imogolite. Clays and Clay Minerals 32, 300-310.

Combes, J.-M. (1988) Evolution de la Structure locales des Polymeres et Gels Ferriques lors de la Cristallisation de Oxydes de fer. Application au Piegeage de l'Uranium. Ph.D. Dissertation, University of Paris VI, 206 pp.

Combes, J.-M., Chisholm-Brause, C.J., Brown, G.E., Jr., Parks, G.A., Conradson, S.D., Eller, P.G., Triay, I., Hobart, D.E., and Meier, A. (1990) EXAFS spectroscopic study of neptunium (V) sorption at the α-FeOOH/water interface. Environ. Sci. Technol. (submitted).

Corrigan, D.S., Foley, J.K., Gao, P., Pons, S., and Weaver, M.J. (1985) Comparisons between surface-enhanced Raman and surface infrared spectroscopies for strongly perturbed adsorbates: thicyanate at gold electrodes. Langmuir 1, 616-620.

Cotton, T.M. (1988) The application of surface-enhanced Raman scattering to biochemical systems. In: Spectroscopy of Surfaces, eds., R.J.H. Clark and R.E. Hester, New York: John Wiley & Sons, pp. 91-153.

Davis, J.A. and Hayes, K.F. (1986) Geochemical processes at mineral surfaces: an overview. In: Geochemical Processes at Mineral Surfaces, eds. J. A. Davis and K. F. Hayes, ACS Symposium Series 323, Washington, DC: Am. Chem. Soc., pp. 2-18.

Davis, J.A. and Leckie, J.O. (1978) Surface ionization and complexation at the oxide/water interface: II. Surface properties of amorphous iron oxyhydroxides and adsorption of metal ions. J. Colloid Interface Sci. 67, 90-107.

Duncan, T.M., Yates, J.T., and Vaughan, R.W. (1979) ^{13}C NMR of CO chemisorbed on Rh dispersed on Al$_2$O$_3$. J. Chem. Phys. 71, 3129-3130.

Duncan, T.M., Yates, J.M., and Vaughan, R.W. (1980) A ^{13}C NMR study of the adsorbed states of CO on Rh dispersed on Al$_2$O$_3$. J. Chem. Phys. 73, 975-985.

Duriez, C., Chapon, C., Henry, C.R., and Rickard, J.M. (1990) Structural characterization of MgO(100) surfaces. Surface Sci. 230, 123-136.

Echterhoff, R. and Knözinger, E. (1990) FTIR spectroscopic characterization of the adsorption and desorption of ammonia on MgO surfaces. Surface Sci. 230, 237-244.

Eggleston, C.M. and Hochella, M.F., Jr. (1990) Scanning tunneling microscopy of sulfide surfaces. Geochim. Cosmochim. Acta 54, 1511-1517.

Evans, J., Gauntlett, J.T., and Mosselmans, J.F.W. (1990) Characterization of oxide-supported alkene conversion catalysts using x-ray absorption spectroscopy. Faraday Discuss. Chem. Soc. 89, (in press).

Fleischmann, M., Graves, P.R., and Robinson, J. (1985) Enhanced and normal Raman scattering from pyridine adsorbed on rough and smooth silver electrodes. J. Electroanal. Chem. 182, 73-85.

Fleischmann, M., Hendra, P.J., and McQuillan, A.J. (1974) Raman spectra of pyridine adsorbed at a silver electrode. Chem. Phys. Lett. 26, 163-166.

Gates, B.C. (1986) Molecular organometallic chemistry and catalysis on metal-oxide surfaces. In: Chemistry and Physics of Solid Surfaces VI, eds. R. Vanselow and R. Howe, Springer Series in Surface Sciences 5, Berlin: Springer-Verlag, pp. 49-71.

Gignac, W.J., Williams, R.S., and Kowalczyk, S.P. (1985) Valence and conduction band structure of the sapphire (102) surface. Phys. Rev. B32, 1237-1247.

Goodman, B.A. (1986) Adsorption of metal ions and complexes on aluminosilicate minerals. In: Geochemical Processes at Mineral Surfaces, eds. J. A. Davis and K. F. Hayes, ACS Symposium Series 323, Washington, DC: Am. Chem. Soc., pp. 342-361.

Gotoh, T., Murakami, S., Kinosita, K., and Murata, Y. (1981) Surface rumpling of $MgO(100)$ deduced from changes in RHEED Kikuchi patterns. I. Experimental. J. Phys. Soc. Japan 50, 2063-2068.

Guay, D., Tourillon, G., and Fontaine, A. (1990) Electrochemical inclusion of copper and iron species in a conducting polymer observed in situ using time-resolved x-ray absorption spectroscopy. Faraday Discuss. Chem. Soc. 89, (in press).

Gurman, S.J. (1989) Structural information in extended x-ray absorption fine structure (EXAFS). In: Synchrotron Radiation and Biophysics, ed. S.S. Hasnain, Chichester, U.K.: Ellis Horwood Ltd. Publishers, pp. 9-42.

Hair, M.L. (1967) Infrared Spectroscopy in Surface Chemistry. New York: Marcel Dekker, 315 pp.

Hall, W.K. (1986) Catalysis by molybdena-alumina and related oxide systems. In: Chemistry and Physics of Solid Surfaces VI, eds. R. Vanselow and R. Howe, Springer Series in Surface Science 5, Berlin: Springer-Verlag, pp. 73-105.

Harsh, J.B., Doner, H.E., and McBride, M.B. (1984) Chemisorption of copper on hydroxy-aluminum-hectorite: an electron spin resonance study. Clays and Clay Minerals 32, 407-413.

Haworth, A. (1990) A review of the modelling of sorption from aqueous solutions. Adv. Colloid Interface Sci. 32, 43-78.

Hawthorne, F.C. (ed.) (1988) Spectroscopic Methods in Mineralogy and Geology. Reviews in Mineralogy 18, 698 pp.

Hayes, K.F. (1987) Equilibrium, Spectroscopic, and Kinetic Studies of Ion Adsorption at the Oxide/Aqueous Interface. Ph.D. Dissertation, Stanford University, 260 pp.

Hayes, K.F. and Leckie, J.O. (1986a) Modeling ionic strength effects on cation adsorption at hydrous oxide/solution interfaces. J. Interface Colloid Sci. 115, 564-572.

Hayes, K.F. and Leckie, J.O. (1986b) Mechanism of lead ion adsorption at the goethite-water interface. In: Geochemical Processes at Mineral Surfaces, eds. J.A. Davis and K.F. Hayes, ACS Symposium Series 323, Washington, DC: Am. Chem. Soc., pp. 114-141.

Hayes, K.F., Roe, A.L., Brown, G.E., Jr., Hodgson, K.O., Leckie, J.O., and Parks, G.A. (1987) In situ x-ray absorption study of surface complexes: selenium oxyanions on α-FeOOH. Science 238, 783-786.

Heald, S.M., Barrera, E.V., and Chen, H. (1990) Glancing angle XAFS and x-ray reflectivity studies of transition-metal/aluminum interfaces. Faraday Discuss. Chem. Soc. 89, (in press).

Heiland, G. and Lüth, H. (1984) Adsorption on oxides. In: The Chemical Physics of Solid

Surfaces and Heterogeneous Catalysis, Vol. III, eds. D.A. King and D.D. Woodruff, Amsterdam: Elsevier, pp.137-219.

Heinrich, V.E. (1979) Ultraviolet photoemission studies of molecular adsorption on oxide surfaces. Prog. Surface Sci. 9, 143-164.

Heinrich, V.E. (1983) The nature of transition-metal-oxide surfaces. Prog. Surface Sci. 14, 175-200.

Heinrich, V.E. (1985) The surfaces of metal oxides. Rep. Prog. Phys. 48, 1481-1541.

Hochella, M.F., Jr. and Carim, A.F. (1988) A reassessment of electron escape depths in silicon and thermally grown silicon dioxide thin films. Surface Sci. 197, L260-L268.

Hochella, M.F., Jr., Eggleston, C.M., Elings, V.B., Thompson, M.S. (1990) Atomic structure and morphology of the albite {010} surface: an atomic-force microscope and electron diffraction study. Am. Mineral. 75, 723-730.

Hochella, M.F., Jr., Eggleston, C.M., Elings, V.B., Parks, G.A., Brown, G.E., Jr., Wu, C.M., and Kjoller, K. (1989) Mineralogy in two dimensions: scanning tunneling microscopy of semiconducting minerals with implications for geochemical reactivity. Am. Mineral. 74, 1233-1246.

Hoffman, R. (1988) Solids and Surfaces: A Chemist's View of Bonding in Extended Structures. New York: VCH Publishers, Inc., 142 pp.

Hoffman, P. and Knözinger, E. (1987) Novel aspects of mid and far IR Fourier transform spectroscopy applied to surface and adsorption studies on SiO_2. Surface Sci. 188, 181-198.

Hohl, H. and Stumm, W. (1976) Interaction of Pb^{2+} with hydrous γ-Al_2O_3. J. Interface Colloid Sci. 55, 281-288.

Holmes, D.J., Batchelor, D.R., and King, D.A. (1988) Surface structure determination by SEXAFS: The reliability of bond lengths and coordination numbers from multi-shell analyses. Surface Sci. 199, 476-492.

Hong, S.Y., Anderson, A.B., and Smialek, J.L. (1990) Sulfur at nickel-alumina interfaces. Surface Sci. 230, 175-183.

Howe, R.F. (1984) Magnetic resonance in surface science. In: Chemistry and Physics of Solid Surfaces V, Springer Series in Chemical Physics 35, eds. R. Vanselow and R. F. Howe, Berlin: Springer-Verlag, pp. 39-64.

Ichikawa, T., Yoshido, H., and Kevan, L. (1981) Electron spin echo studies of Cu^{2+} on silica surfaces: Interaction with water and ammonia adsorbates. J. Chem. Phys. 75, 2485-2485.

James, R.O. and Healy, T.W. (1972) Adsorption of hydrolyzable metal ions at the oxide/water interface. I. Co(II) adsorption on SiO_2 and TiO_2 model systems. J. Colloid Interface Sci. 40, 42-52.

Johansson, L.I. and Stöhr, J. (1979) Bonding of oxygen on Al(111): a surface extended x-ray absorption fine-structure study. Phys. Rev. Lett. 43, 1882-1885.

Johnston, P., Joyner, R.W., Pudney. P.D.A., Shpiro, E.S., and Williams, B.P. (1990) In situ studies of supported rhodium catalysts. Faraday Discuss. Chem. Soc. 89, (in press).

Kampers, F.W.H. and Koningsberger, D.C. (1990) EXAFS study of the influence of hydrogen desorption and oxygen adsorption on the structural properties of small irridium particles supported on Al_2O_3. Faraday Discuss. Chem. Soc. 89, (in press).

Kerker, M., Robinson, J., and Forty, A.J. (1990) In situ structural studies of the passive film on iron and iron/chromium alloys using x-ray absorption spectroscopy. Faraday Discuss. Chem. Soc. 89, (in press).

Kinniburgh, C.G. (1975) A LEED study of Mg(100): II. Theory at normal incidence. J. Phys. C8, 2382-2394.

Kinniburgh, C.G. (1976) A LEED study of MgO(100): III. Theory at off-normal

358

incidence. J. Phys. C9, 2695-2708.

Kiselev, V.F. and Krylov, O.V. (1985) Adsorption Processes on Semiconductor and Dielectric Surfaces I. Springer Series in Chemical Physics 32, Berlin: Springer-Verlag, 287 pp.

Kobayashi, H. and Yamaguchi, M. (1989) Ab initio MO study of adsorption of CO molecule on TiO_2 surfaces. Surface Sci. 214, 466-476.

Koningsberger, D.C. and Prins, R. (eds.) (1988) X-ray Absorption: Principles, Applications, Techniques of EXAFS, SEXAFS, and XANES. New York: John Wiley & Sons, 673 pp.

Krasser, W. (1989) Raman spectroscopy and surface chemistry. Chem. in Britain, 618-622.

Lad, R.J. and Heinrich, V.E. (1988) Structure of α-Fe_2O_3 single crystal surfaces following Ar^+ ion bombardment and annealing in O_2. Surface Sci. 193, 81-93.

Langmuir, I. (1916) The constitution and fundamental properties of solids and liquids. J. Am. Chem. Soc. 38, 2221-2295.

Langmuir, I. and J.B. Taylor (1932) The mobility of caesium atoms adsorbed on tungsten. Phys. Rev. 40, 463-464.

Lennard-Jones, J.E. (1932) Processes of adsorption and diffusion on solid surfaces. Trans. Farad. Soc. 28, 333-359.

Lundqvist, B.I. (1984) Chemisorption and reactivity of metals. In: Many-Body Phenomena at Surfaces, eds. D.C. Langreth and Suhl, Orlando, FL: Academic Press, pp. 93-144.

Lytle, F.W., Greegor, R.B., Sandstrom, D.R., Marques, E.C., Wong, J., Spiro, C.L., Huffman, G.P., and Huggins, F.E. (1984) Measurement of soft x-ray absorption spectra with a fluorescent ion chamber detector. Nucl. Instr. and Meth. 226, 542-548.

Mackrodt, W.C. (1988) Atomistic simulation of oxide surfaces. Phys. Chem. Minerals 15, 228-137.

Martini, G., Ottaviani, M.F., and Seravalli, G.L. (1975) Electron spin resonance study of vanadyl complexes adsorbed on synthetic zeolites. J. Phys. Chem. 79, 1716-1720.

Martinson, C.W.B., Flodström, S.A., Rundgren, J., and Westrin, P. (1979) Oxygen chemisorption on aluminum single crystals: site determination by LEED studies. Surface Sci. 89, 102-113.

Matsumoto, A. and Kaneko, K. (1990) Gradual change in the chemisorbed NO species on α-FeOOH. Langmuir 6, 1202-1204.

Maya, L. (1982) Sorbed uranium (VI) species on hydrous titania, zirconia, and silica gel. Radiochim. Acta 31, 147-151.

McBreen, J., O'Grady, W.E., Pandya, K.I., Hoffman, R.W., and Sayers, D.E. (1987) EXAFS study of the nickel oxide electrode. Langmuir 3, 428-433.

McBride, M.B. (1976) Origin and position of exchange sites in kaolinite: an ESR study. Clays and Clay Minerals 24, 88-92.

McBride, M.B. (1978) Copper(II) interactions with kaolinite: factors controlling adsorption. Clays and Clay Minerals 26, 101-106.

McBride, M.B. (1982a) Cu^{2+}-adsorption characteristics of aluminum hydroxide and oxyhydroxides. Clays and Clay Minerals 30, 21-28.

McBride, M.B. (1982b) Hydrolysis and dehydration reactions of exchangeable Cu^{2+} on hectorite. Clays and Clay Minerals 30, 200-206.

McBride, M.B. (1985) Influence of glycine on Cu^{2+} adsorption by microcrystalline gibbsite and boehmite. Clays and Clay Minerals 33, 397-402.

McBride, M.B. (1986) Paramagnetic probes of layer silicate surfaces. In: Geochemical Processes at Mineral Surfaces, eds. J.A. Davis and K.F. Hayes, ACS Symposium

Series 323, Washington, DC: Am. Chem. Soc., pp. 362-388.

McBride, M.B. and Bouldin, D.R. (1984) Long-term reactions of copper (II) in a contaminated calcareous soil. Soil Sci. Soc. Am. J. 48, 56-59.

McBride, M.B., Fraser, A.R., and McHardy, W.J. (1984) Cu^{2+} interaction with microcrystalline gibbsite. evidence for oriented chemisorbed copper ions. Clays and Clay Minerals 32, 12-18.

McKale, A.G., Veal, B.W., Paulikas, A.P., Chan, S.K., and Knapp, G.S. (1988) Improved ab-initio calculations for amplitude and phase functions for extended x-ray absorption fine structure spectros-copy. J. Am. Chem. Soc. 110, 3763-3768.

Melroy, O.R., Samant, M.G., Borges, G.L., Gordon, J.G. II, Blum, L., White, J.H., Albarelli, M.J., McMillan, M., and Abruna, H.D. (1988) In-plane structure of underpotentially deposited copper on gold(111) determined by surface EXAFS. Langmuir 4, 728-732.

Mims, W.B. (1972) Electron spin echos. In: Electron Paramagnetic Resonance, ed. S. Geschwind, New York: Plenum Press, Chapter 4.

Mitchell, K.A.R., Schlatter, S.A., and Sodhi, R.N.S. (1986) Further analysis of surface bond lengths measured for chemisorption on metal surfaces. Can. J. Chem. 64, 1435-1439.

Möhl, W., Schweiger, A., and Motschi, H. (1988) Modes of phosphate binding to Cu(II): investigations of the electron spin echo envelope modulation of complexes on surfaces and in solutions. (unpublished manuscript).

Motschi, H. (1983) Cu(II) bound to hydrous surfaces. EPR measurements to characterize surface coordination. Naturwissenschaften 70, 519-520.

Motschi, H. (1984) Correlation of EPR-parameters with thermodynamic stability constants for copper (II) complexes- Cu(II) as a probe for the surface complexation at the water/oxide interface. Colloids and Surfaces 9, 333-347.

Motschi, H. (1987) Aspects of the molecular structure in surface complexes: spectroscopic investigations. In: Aquatic Surface Chemistry, ed. W. Stumm, New York: John Wiley & Sons, pp. 111-125.

Mulcahy, F.M., Fay, M.J., Proctor, A., Houalla, M., and Hercules, D.M. (1990) The adsorption of metal oxyanions on alumina. J. Catalysis 124, 231-240.

Nakamatsu, H., Sudo, A., and Kawai, S. (1988) Relaxation of the MgO(100) surface studied by ICISS. Surface Sci. 194, 265-274.

Natoli, R. (1983) Near edge absorption structure in the framework of the multiple scattering model.. Potential resonance or barrier effects. In: EXAFS and Near Edge Structure, eds. A. Bianconi, L. Incoccia, and S. Stipcich, Springer Series in Chemical Physics, Vol. 27, Berlin: Springer-Verlag, pp. 43-47.

Norman, D., Brennan, S., Jaeger, R., and Stöhr, J. (1981) Structural models for the interaction of oxygen with Al(111) and Al implied by photoemission and surface EXAFS. Surface Sci. 105, L297-L306.

O'Day, P.A., Chisholm-Brause, C.J., Brown, G.E., Jr., and Parks, G.A. (1988) Characterization of Co(II) complexes at the kaolinite/water interface by x-ray absorption spectroscopy. (abstr.) EOS Trans. Am. Geophys. Union 69, p. 1482.

O'Day, P.A., Chisholm-Brause, C.J., Brown, G.E., Jr., and Parks, G.A. (1989) Synchrotron-based XAS study of Co(II) sorption complexes at mineral/water interfaces: effect of mineral surfaces. (abstr.) Symposium on Synchrotron Radiation in the Geological Sciences, Abstracts Vol., 28th Int'l Geol. Congress, Washington, DC, July 1989, p. 2-534.

O'Day, P.A., Brown, G.E., Jr., and Parks, G.A. (1990) EXAFS study of aqueous Co(II) sorption complexes on kaolinite and quartz surfaces. In: XAFS VI, X-ray Absorption Fine Structure VI, ed. S.S. Hasnain, Chichester, U.K.: Ellis Horwood Ltd. Publishers (in press).

360

Ohtani, H., Kao, C.-T., Van Hove, M.A., and Somorjai, G.A. (1986) A tabulation and clssification of the structures of clean solid surfaces and of adsorbed atomic and molecular monolayers as determined from low energy electron diffraction. Prog. Surface Sci. 23, 155-316.

Outka, D.A. and Madix, R.J. (1986) Structural characterization of molecules and reaction intermediates on surfaces using synchrotron radiation. In: Chemistry and Physics of Solid Surfaces VI, eds. R. Vanselow and R. Howe, Springer Series in Surface Sciences 5, Berlin: Springer-Verlag, pp. 133-167.

Outka, D.A., and Stöhr, J. (1988) Curve fitting analysis of near-edge core excitation spectra of free, adsorbed, and polymeric molecules. J. Chem. Phys. 88, 3539-32554.

Pauling, L. (1960) The Nature of the Chemical Bond. 3rd Ed., Ithaca, NY: Cornell University Press, pp. 505-562.

Payling, R. and Ramsey, J.A. (1980) Possible ordered structure in the adsorption of oxygen on Al(111). J. Phys. C Solid State Phys. 13, 505-515.

Petrovic, R., Berner, R.A., and Goldhaber, M.B. (1976) Rate control in dissolution of alkali feldspars-I. Study of residual feldspar grains by x-ray photoelectron spectroscopy. Geochim. Cosmochim. Acta 40, 537-548 .

Pisani, C., Dovesi, R., Nada, R., and Tamiro, S. (1989) Ab-initio Hartree-Fock perturbed cluster treatment of local chemisorption: isolated carbon monoxide on a periodic MgO(100) substrate. Surface Sci. 216, 489-504.

Razouk, R.I. and Mikhail, R.Sh. (1958) Hydration of magnesium oxide from the vapor phase. J. Phys. Chem. 62, 920-925.

Rhodin, T.N. and Ertl, G. (eds.) (1979) The Nature of the Surface Chemical Bond. Amsterdam: North-Holland Publishing Co., 405 pp.

Rhodin, T.N. and Gadzuk, J.W. (1979) Electron spectroscopy and surface chemical bonding. In: The Nature of the Surface Chemical Bond, eds. T.N. Rhodin and G. Ertl, Amsterdam: North-Holland Publishing Co., pp. 113-273.

Roe, A.L., Hayes, K.F., Chisholm-Brause, C.J., Brown, G.E., Jr., Parks, G.A., and Leckie, J.O. (1990) X-ray absorption study of lead complexes at α-FeOOH/water interfaces. Langmuir (accepted for publication).

Rudin, M. and Motschi, H. (1984) A molecular model for the structure of copper complexes on hydrous oxide surfaces: an ENDOR study of ternary Cu(II) complexes on δ-alumina. J. Colloid Interface Sci. 98, 385-393.

Salama, T.M., Tanaka, T., Ýamaguchi, T., and Tanabe, K. (1990) EXAFS/XANES study of titanium oxide supported on SiO_2: a structural consideration on the amorphous state. Surface Sci. 227, L100-L104.

Samant, M.G., Borges, G.L., Gordon, J.G., II, Melroy, O.R., and Blum, L. (1987) In situ extended x-ray absorption fine structure spectroscopy of a lead monolayer at a silver (111) electrode/electrolyte interface. J. Am. Chem. Soc. 109, 5970-5974.

Samant, M.G., Toney, M.F., Borges, G.L., Blum, L., and Melroy, O.R. (1988) In situ grazing incidence x-ray diffraction study of electrochemically deposited Pb monolayers on Ag(111). Surface Sci. 193, L29-L36.

Santoni, A., Tran-Thoai, D.B., and Urban, J. (1988) MgO(100) surface topology determined by surface extended electron energy loss fine structure. Solid State Comm. 68, 1039-1041.

Sayers, D.E., Lytle, F.W., and Stern, E.A. (1970) Point scattering theory of x-ray K absorption fine structure. Advan. X-ray Anal. 13, 248-271.

Sayers, D.E., Stern, E.A., and Lytle, F.W. (1971) New technique for investigating noncrystalline structures: Fourier analysis of the extended x-ray absorption fine structure. Phys. Rev. Lett. 27, 1204-1207.

Scarano, D., Zecchina, A., and Reller, A. (1988) IR study of CO adsorption on α-Cr_2O_3.

Surface Sci. 198, 11-25.

Schardt, B.C., Yau, S.-L. and Rinaldi, F. (1989) Atomic resolution imaging of adsorbates on metal surfaces in air: iodine adsorption on Pt(111). Science 243, 1050-1053.

Schindler, P.W. and Stumm, W. (1987) The surface chemistry of oxides, hydroxides, and oxide minerals. In: Aquatic Surface Chemistry, ed. W. Stumm, New York: John Wiley & Sons, pp. 83-110.

Schweiger, A. (1982) Electron nuclear double resonance of transition metal complexes with organic ligands. In: Structure and Bonding, Vol. 51, eds. M.J. Clark, J.B. Goodenough, P. Hemmerich, J.A. Ibers, C.K. Jorgensen, J.B. Neilands, D. Reinen, R. Weis, and R.J.P. Williams, Berlin: Springer-Verlag, pp. 1.

Slichter, C.P. (1985) NMR and surface structure. In: The Structure of Surfaces. Springer Series in Surface Sciences 2, Berlin: Springer-Verlag, pp. 84-89.

Somorjai, G.A. (1981) Chemistry in Two Dimensions: Surfaces. Ithaca, NY: Cornell University Press, pp. 126-175.

Somorjai, G.A. (1990) Modern concepts in surface science and heterogeneous catalysis. J. Phys. Chem. 94, 1013-1023.

Somorjai, G.A. and Van Hove, M.A. (1989) Adsorbate-induced restructuring of surfaces. Prog. Surface Sci. 30, 201-231.

Soria, F., Martinez, V., Munoz, M.C., and Sacedon, J.L. (1981) Structure of the initial stages of oxidation of Al(111) surfaces from low energy electron diffraction and Auger electron spectroscopy. Phys. Rev. B24, 6926-6935.

Sposito, G. (1986) Distinguishing adsorption from surface precipitation. In: Geochemical Processes at Mineral Surfaces, eds. J. A. Davis and K. F. Hayes, ACS Symposium Series 323, Washington, DC: Am. Chem. Soc., pp. 217-228.

Stern, E.A. and Heald, S.M. (1983) Basic principles and applications of EXAFS. In: Handbook on Synchrotron Radiation, Vol. 1b., ed. E.E. Koch, New York: North Holland, p. 955-1014.

Stöhr, J. (1984) Surface crystallography by means of SEXAFS and NEXAFS. In: Chemistry and Physics of Solid Surfaces V, eds. V. R. Vanselow and R. Howe, Springer Series in Chemical Physics, vol. 35, Berlin: Springer-Verlag, p. 231-275.

Stöhr, J. (1988) SEXAFS: everything you always wanted to know about SEXAFS but were afraid to ask. In: X-ray Absorption: Principles, Applications, Techniques of EXAFS, SEXAFS, and XANES, eds. D.C. Koningsberger and R. Prins, New York: John Wiley & Sons, pp. 443-571.

Stöhr, J. and Outka, D.A. (1987) Determination of molecular orientation on surfaces from the angular dependence of near-edge x-ray absorption fine-structure spectra. Phys. Rev. B36, 7891-7905.

Stöhr, J., Jaeger, R., and Brennan, S. (1982) Surface crystallography by means of electron and ion yield SEXAFS. Surface Sci. 117, 503-524.

Stöhr, J., Johansson, L.I., Brennan, S., Hecht, M., and Miller, J.N. (1980) Surface extended x-ray absorption fine structure study of oxygen interaction with Al(111) surfaces. Phys. Rev. B22, 4052-4065.

Stöhr, J. Gland, J.L., Eberhardt, W.E., Outka, D., Madix, R.J., Sette, F., Koestner, R.J., and Döbler, U. (1983) Bonding and bond lengths of chemisorbed molecules from near-edge x-ray absorption fine structure studies. Phys. Rev. Lett. 51, 2414-2417.

Suits, J. (ed.) (1962) The Collected Works of Irving Langmuir. Vols. 1-13, New York: Pergamon Press.

Tatarchuk, B.J. and Dumesic, J.A. (1984) Mössbauer spectroscopy: applications to surface and catalytic phenomena. In: Chemistry and Physics of Solid Surfaces V, Springer Series in Chemical Physics 35, eds. R. Vanselow and R. F. Howe, Berlin: Springer-Verlag, pp. 65-109.

362

Teo, B.K. (1986) EXAFS: Basic Principles and Data Analysis. Inorganic Chemistry Concepts 9, Berlin: Springer-Verlag, 349 pp.

Teo, B.K. and Lee, P.A. (1979) Ab initio calculation of amplitude and phase function for Extended X-ray Absorption Fine Structure (EXAFS) spectroscopy. J. Am. Chem. Soc. 101, 2815-2830.

Tewari, P.H. and Lee, W. (1975) Adsorption of Co(II) at the oxide-water interface. J. Colloid Interface Sci. 52, 77-88.

Tewari, P.H. and McIntyre, N.S. (1974) Characterization of adsorbed cobalt at the oxide-water interface. AIChE Symp. Series 71, 134-137.

Tewari, P.H., Campbell, A.B., Lee, W. (1972) Adsorption of Co^{2+} by oxides from aqueous solution. Can. J. Chem. 50, 1642-1648.

Tejedor-Tejedor, M.I. and Anderson, M.A. (1986) In situ attenuated total reflectance Fourier transform infrared studies of the goethite (α-FeOOH)-aqueous solution interface. Langmuir 2, 203-210.

Tsukada, M., Adachi, H., and Satoko, C. (1983) Theory of electronic structure of oxide surfaces. Prog. Surface Sci. 14, 113-174.

Urano, T., Kanaji, T., and Kaburagi, M. (1983) Surface structure of MgO(001) surface studied by LEED. Surface Sci. 134, 109-121.

Van Hove, M.A. (1979) Surface crystallography and bonding. In: The Nature of the Surface Chemical Bond, eds. T.N. Rhodin and G. Ertl, Amsterdam: North-Holland Publishing Co., pp. 275-311.

Van Hove, M.A. and Tong, S.Y. (eds.) (1985) The Structure of Surfaces. Springer Series in Surface Sciences 2, Berlin: Springer-Verlag, 435 pp.

Van Hove, M.A., Weinberg, W.H., and Chan., C.-M. (1986) Low Energy Electron Diffraction: Experiment, Theory, and Surface Structure Determination. Springer Series in Surface Science 6, Berlin: Springer-Verlag, 603 pp.

Van Hove, M.A., Wang, S.-W., Ogletree, D.F., and Somorjai, G.A. (1989) The state of surface structural chemistry theory, experiment, and results. In: Advances in Quantum Chemistry, vol. 20, eds. P.-O. Löwdin, J.R. Sabin, and M.C. Zerner, New York: Academic Press, Inc., pp. 1-184.

Vanselow, R. and Howe, R. (eds.) (1984) Chemistry and Physics of Solid Surfaces V. Springer Series in Chemical Physics 35, Berlin: Springer-Verlag, 554 pp.

van Veen, J.A.E., Hendriks, P.A.J.M., Romers, E.J.G.M., and Amdrea, R.R. (1990a) Chemistry of phosphomolybdate adsorption on alumina surfaces. 1. The molybdate/alumina system. J. Phys. Chem. 94, 5275-5282.

van Veen, J.A.E., Hendriks, P.A.J.M., Andrea, R.R., Romers, E.J.G.M., and Wilson, A.E. (1990b) Chemistry of phosphomolybdate adsorption on alumina surfaces. 2. The molybdate/phosphated alumina and phosphomolybdate/alumina systems. J. Phys. Chem. 94, 5282-5285.

Waychunas, G.A. and Brown, G.E., Jr. (1990) Polarized x-ray absorption spectroscopy of metal ions in minerals: applications to site geometry and electronic structure determination. Phys. Chem. Minerals (in press).

Waychunas, G.A., Apted, M.J., and Brown, G.E., Jr. (1983) X-ray K-edge absorption spectra of Fe minerals and model compounds: near edge structure. Phys. Chem. Minerals 10, 1-9.

Waychunas, G.A., Rea, B.B., Fuller, C.C., and Davis, J.A. (1990) Fe and As K-edge EXAFS study of arsenate $(AsO_4)^{3-}$ adsorption on "two-line" ferrihydrite. In: XAFS VI, X-ray Absorption Fine Structure VI, ed. S.S. Hasnain, Chichester, U.K.: Ellis Horwood Ltd. Publishers (in press).

Weaver, R.J., Corrigan, D.S., Gao, P., Gosztola, D., and Leung, L.-W.H. (1987) Some applications of surface Raman and infrared spectroscopies to mechanistic electrochemistry involving adsorbed species. J. Elec. Spec. Related Phenomena 45,

291-302.

Welton-Cook, M.R. and Berndt, W. (1982) A LEED study of the MgO(100) surface. J. Phys. C 15, 5691-5718.

Westall, J. and Hohl, H. (1980) A comparison of electrostatic models for the oxide/solution interface. Adv. Colloid Interface Sci. 12, 265-294.

Williams, R.H. and McGovern, I.T. (1984) Adsorption on semiconductors. In: The Chemical Physics of Solid Surfaces and Heterogeneous Catalysts, eds. D.A. King and D.P. Woodruff, Amsterdam: Elsevier, pp. 267-309.

Wong, J., Lytle, F.W., Messmer, R.P., and Maylotte, D.H. (1984) K-edge absorption spectra of selected vanadium compounds. Phys. Rev. B30, 5596-6510.

Wong, J. (1986) Extended x-ray absorption fine structure: a modern structural tool in materials science. Materials. Sci. Eng. 80, 107-128.

Woodruff, D.P. and Delchar, T.A. (1986) Modern Techniques of Surface Science. Cambridge: Cambridge Unviersity Press, 453 pp.

Xu, N., Brown, G.E., Jr., Parks, G.A., and Hochella, M.F., Jr. (1990) Sorption mechanism of Co^{2+} at the calcite-water interface. (abstr.) Abstr. Program Geol. Soc. Am. Ann. Mtg. 22 (in press).

Yu, H.L., Munoz, M.C., and Soria, F. (1980) On the initial stages of oxidation of Al(111) by LEED analysis. Surface Sci. 94, L184-L190.

Zangwill, A. (1988) Physics at Surfaces. Cambridge: Cambridge University Press, 454 pp.

Zelewsky, A. von and Bemtgen, J.M. (1982) Formation of ternary copper (II) complexes at the surface of silica gel as studied by EPR spectroscopy. Inorg. Chem. 21, 1771.

Ziolkowski, J. (1983) Advanced bond-strength model of active sites on oxide catalysts. J. Catalysis 84, 317-332.

MECHANISMS OF GROWTH AND DISSOLUTION
OF SPARINGLY SOLUBLE SALTS

INTRODUCTION

The precipitation and dissolution of sparingly soluble minerals play an important role in the establishment of geological equilibria (Davis and Hayes, 1986). There is therefore considerable interest in the elucidation of the mechanisms of these processes and this chapter introduces the theory and experimental kinetics methods including the most recent achievements in this area. Following a discussion of the calculation of the driving forces for crystallization and dissolution, several attractive kinetics models are discussed. The experimental methods are presented with particular reference to the constant composition (CC) technique (Tomson and Nancollas, 1978). Finally, some specific systems, such as the alkaline earth phosphates, fluorides, and carbonates, are presented in order to illustrate the methods used for deducing the mechanisms of a particular reaction.

THE DRIVING FORCES FOR GROWTH AND DISSOLUTION

Definition

The growth and dissolution of a crystalline phase can be regarded as two opposing processes. For an electrolyte crystal, $A_\alpha B_\beta$, in an aqueous solution, the reactions can be expressed by Equation (1).

$$\alpha A(aq.) + \beta B(aq.) \underset{\text{dissolution}}{\overset{\text{growth}}{=\!=\!=\!=\!=}} A_\alpha B_\beta(\text{lattice}) \quad , \tag{1}$$

where the ionic charges are omitted for clarity. For sparingly soluble salts, the equilibrium constant of reaction (1) is simply the reciprocal of the solubility product, K_s, which is defined by

$$K_s = a_{Ae}{}^\alpha a_{Be}{}^\beta \quad , \tag{2}$$

where a_{Ae} and a_{Be} are the equilibrium molar activities of ions A and B, respectively (see the list of symbols). When the crystal $A_\alpha B_\beta$ is brought into contact with a solution, its ability to grow or dissolve depends upon the value of the Gibbs free energy of the corresponding reaction. For the formation of one mole of crystalline $A_\alpha B_\beta$, the change in the Gibbs free energy, ΔG_m, is given by

$$\Delta G_m = -RT ln(I_p/K_s) \quad , \tag{3}$$

where R is the gas constant, T the absolute temperature and I_p the ionic activity product,

$$I_p = a_A{}^\alpha a_B{}^\beta \quad . \tag{4}$$

ΔG_m may be regarded as the thermodynamic driving force for crystal growth. However, the value of ΔG_m depends on the selection of the crystal formula unit. For example, calcium hydroxyapatite has been written either as $Ca_5(OH)(PO_4)_3$ or $Ca_{10}(OH)_2(PO_4)_6$, with ΔG_m values differing by a factor of two. In order to avoid such ambiguities in this chapter, ΔG_m

LIST OF SYMBOLS

a	Size of a growth unit
a_A	Molar activity of A ions
a_B	Molar activity of B ions
A	A cation
A_0	Initial surface area
A_D	Constant in Debye-Hückel equation
A_e	Pre-exponential term of homogeneous nucleation rate equation
A_t	Total surface area
$[A]_t$	Titrant concentration of A ions
b	Constant
B	An anion
$[B]_t$	Titrant concentration of B ions
C	Concentration in the bulk solution
C_{ad}	Concentration in the adsorption layer
C_e	Effective titrant concentration
C_i	Concentration adjacent to the adsorption layer
C_s	Solubility
C_t	Titrant concentration
C_w	Concentration of working solution
d	Diameter or edge length of a surface nucleus
dn	Increment of moles of precipitation
dv_t	Increment of volume of titrant addition
D	Diffusion coefficient
g	Number of growth units in a 3-dimensional nucleus
i	Number of growth units in a surface nucleus
$[I]$	Concentration of additive
$[I]_{ad}$	Additive concentration in the adsorption layer
I_p	Ionic activity product
j_a	Adsorption flux to a kink
j_b	Detachment flux from a kink
j_d	Desorption flux from a kink
j_f	Integration flux to a kink
j_k	Net integration flux to a kink
J	Crystal growth rate
J_0	Crystal growth rate in the absence of additives
J_n	Homogeneous nucleation rate
J_s	Surface reaction rate
J_v	Volume diffusion rate
k	Boltzmann constant
k_2	Rate constant of a parabolic rate law
k_{ad}	Adsorption controlled rate constant
k_{de}	Detachment controlled rate constant
k_{in}	Integration controlled rate constant
k_{sd}	Surface diffusion controlled rate constant
k_{vd}	Volume diffusion controlled rate constant
K_{ad}	Adsorption coefficient
K_s	Solubility product
K_t	Adsorption coefficient of additives on a terrace
M_0	Initial mass of crystals
M_t	Mass of crystals
n	Effective order of reaction
n_1	Density of single growth units on a surface
N_0	Avogadro constant

n_i Density of surface clusters containing i growth units.
r Radius of a 3-dimensional nucleus
r^* Radius of a critical 3-dimensional nucleus
$<r>$ Radius of a particle
R Gas constant
S Saturation ratio
S_a Specific surface area
S_m Maximum saturation ratio of a metastable solution
t Time
T Absolute temperature
v Linear growth rate
V_m Molar volume
V_t Volume of titrant addition
V_w Volume of working solution
x_0 Average distance between kinks
x_s Mean surface diffusion distance
X A cation
y_0 Step spacing in a spiral
y_j Ion activity coefficient
Y An anion
z Ionic charge

α Number of A ions in the formula unit $A_\alpha B_\beta$
β Number of B ions in the formula unit $A_\alpha B_\beta$
γ Edge free energy
δ Diffusion layer thickness
ΔG_c Thermodynamic driving force
ΔG_i Gibbs free energy for surface cluster formation
ΔG_i^* Gibbs formation energy of critical surface nuclei
ΔG_m Gibbs free energy change for growth
ΔG_n Gibbs free energy for formation of 3-dimensional nuclei
ΔG_n^* Gibbs formation energy of critical 3-dimensional nuclei
ΔG_s Surface free energy term in homogeneous nucleation
ΔG_v Gibbs free energy change in solution
ϵ Kink formation energy
Λ Adsorption layer thickness
μ Ionic strength
ν Number of ions in a formula unit
ν_a Adsorption frequency at a kink
ν_{ad} Adsorption frequency
ν_d Desorption frequency from a kink
ν_{de} Detachment frequency
ν_{in} Integration frequency
ρ Density
σ Relative supersaturation
σ_c Transition supersaturation from surface diffusion to adsorption
σ_i Relative supersaturation for surface processes
σ_L Value of σ_c at $x_s = a$
σ_γ Surface energy
θ Fraction of poisoned kinks
Ω Volume for adsorption at a kink

will be divided by the total number of ions in the formula unit, ν ($\nu = \alpha + \beta$), and the resultant value, ΔG_c, will be considered as the thermodynamic driving force.

$$\Delta G_c = \Delta G_m/\nu = -RT\ln S \quad , \tag{5}$$

where the saturation ratio,

$$S = (I_P/K_s)^{1/\nu} \quad . \tag{6}$$

When $S > 1$, ΔG_c is negative, the solution is supersaturated and growth may occur. In contrast, when $S < 1$, the reverse reaction (dissolution) may take place in the undersaturated solution. In kinetics studies, the state of the solution is often defined by the relative supersaturation, σ, given by Equation (7):

$$\sigma = S - 1 \quad . \tag{7}$$

When the solution composition has the same stoichiometry as the crystal, the saturation ratio may be approximated by Equation (8) in terms of the molarity of the solution, C, and the solubility value, C_s.

$$S = C/C_s \quad . \tag{8}$$

Equation (8) will be equivalent to Equation (6) if there is no ion association and complexation in the solution and the activity coefficients remain constant when the solute concentration varies from C to C_s.

Calculation

In solutions in which there are significant ion association and complexation, the free ion concentrations are considerably lower than the total stoichiometric values. In order to calculate S or σ using Equation (6) and (7), it is necessary to know concentrations of free lattice ions and their activity coefficients. The latter are functions of ionic strength, μ, given by Equation (9):

$$\mu = (1/2)\sum z_j^2 C_j \quad , \tag{9}$$

where C_j and z_j are the concentration and charge of ion species j, respectively.

For most cases involving mixed electrolytes, the empirical extension of the Debye-Hückel equation proposed by Davies (1962) may be used for the calculation of activity coefficients, y_j,

$$\lg y_j = -z_j^2 A_D[\mu^{1/2}/(1 + \mu^{1/2}) - 0.3\mu] \quad , \tag{10}$$

where A_D is a temperature dependent constant ($A_D = 0.5115$ at 25°C). Equation (10) generally offers satisfactory results at ionic strength less than about 0.1 M. At higher ionic strength, more complex equations such as those of Guggenheim (Guggenheim, 1935; Guggenheim and Turgeon, 1955) and Pitzer (Pitzer, 1973; Pitzer and Mayorga, 1973; Pitzer, 1986) may be used to estimate the activity corrections.

The free ion concentrations can be evaluated by using expressions for mass balance, electroneutrality, and equilibrium constants for the formation of any ion pairs or complexes in solution. The activity coefficients may be calculated using Equation (10) by iteration procedures involving the ionic strengths (I and Nancollas, 1972).

NUCLEATION

In a supersaturated solution, ions may form clusters or nuclei, which can be considered as crystal embryos. If the clusters have the same inner structure as large crystals, the capture of g growth units will reduce the Gibbs free energy of the system according to Equation (11):

$$\Delta G_v = -gkT \ln S, \tag{11}$$

where k is the Boltzmann constant. However, the formation of a cluster involves the creation of a solid-solution interface with finite surface energy σ_γ, resulting in an increase in the Gibbs free energy. If the cluster is spherical with a radius r, the surface energy term is given by Equation (12):

$$\Delta G_s = 4\pi r^2 \sigma_\gamma \ . \tag{12}$$

Assuming that each growth unit occupies a volume a^3, the nucleus will contain $(4/3)\pi r^3/a^3$ growth units. Thus the overall free energy of formation of a cluster will be (McDonald, 1962; 1963)

$$\Delta G_n = -(4/3)\pi r^3 kT \ln S/a^3 + 4\pi r^2 \sigma_\gamma \ . \tag{13}$$

A graphical representation of Equation (13) is given in Figure 1. The radius r^* given by Equation (14), corresponding to the maximum formation energy, ΔG_n^*, can be obtained by maximizing ΔG_n with respect to r.

$$r^* = \frac{2\gamma a}{kT \ln S} \ , \tag{14}$$

where γ is the edge free energy per growth unit composing the step which is related to the surface energy by Equation (15):

$$\gamma = \sigma_\gamma a^2 \ . \tag{15}$$

Substitution of r^* into Equation (13) yields the maximum Gibbs free energy for nucleus formation,

$$\Delta G_n^* = \frac{16\pi\gamma^3}{(kT \ln S)^3} \ . \tag{16}$$

ΔG_n^* may be considered as the activation energy for the nucleation process. Thus rate of nucleation, J_n, may be expressed in the form (Mullin, 1972):

$$J_n = A_e \exp(-\Delta G_n^*/kT) \ , \tag{17}$$

where A_e also depends on the supersaturation. A typical plot of J_n is given as a function of S in Figure 2. It can be seen that the nucleation rate is virtually zero below a certain supersaturation value, S_m, suggesting that the solution under these conditions may be stable for long periods without precipitation. The supersaturation range, $1 < S < S_m$, is the metastable zone, within which seeded crystal growth can be achieved without the complication of concomitant nucleation. Figure 3 illustrates the regions of under-, meta- and supersaturation of the solution. It should be noted, however, that foreign surfaces may induce nucleation in the metastable zone by a heterogeneous process.

370

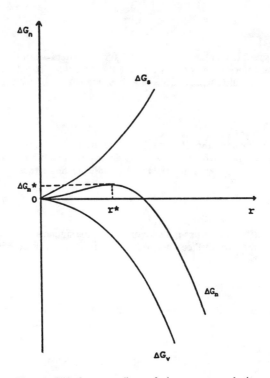

Figure 1. Gibbs free energy diagram for homogeneous nucleation.

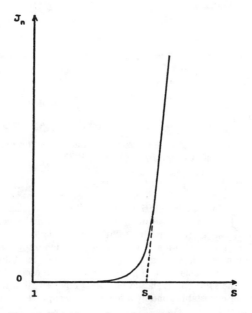

Figure 2. Nucleation rate J_n as a function of supersaturation ratio, S.

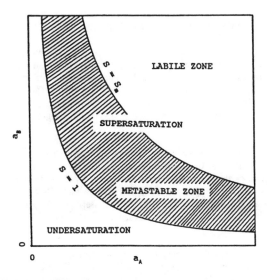

Figure 3. Solubility isotherm of symmetrical AB electrolyte. The dashed area represents the metastable zone.

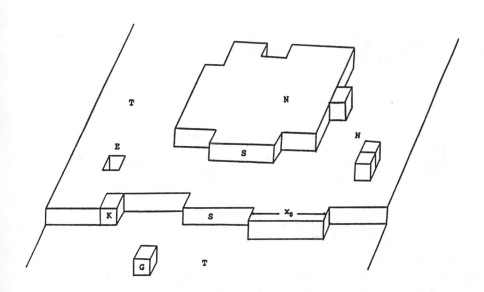

Figure 4. Schematic of the (001) surface of a simple cubic crystal. The steps (S) contain kinks (K) with an average distance X_0. On the terrace (T) there exist single growth units (G), clusters (N) and etch pits (E).

THE CRYSTAL-SOLUTION INTERFACE

Modern theories of crystal growth stem from considerations of crystal fine structure (Stranski, 1928). The most frequently used surface model is an unrelaxed (001) surface of a simple cubic crystal as depicted in Figure 4. The cubes are referred to as growth units which can be atoms, ions, or molecules of the crystal. On the surface, there are steps which contain a number of kinks (Frenkel, 1945; Burton and Cabrera, 1949). The area between two adjacent steps is referred to as a terrace on which some single adsorbed growth units, clusters, and etch pits may be found.

Ions adsorbed on the terraces are attached by only a single bond and would be expected to desorb readily into the solution phase. Kink sites on the surface offer the most stable configuration since the adsorbed ions are attached by three bonds. Therefore, crystals grow mainly by the addition of growth units at the kink sites (Nielsen and Christoffersen, 1982). In this section, the origins of these kinks and steps will be discussed.

Kink densities

To minimize the edge free energy, a step in equilibrium with solution will contain a certain number of kinks at T > 0 (Frenkel, 1945). The average spacing between kinks, x_0, can be calculated from Equation (18) based on statistical thermodynamics (Frenkel, 1945; Burton and Cabrera, 1949).

$$x_0 = (1/2)a[\exp(\epsilon/kT) + 2] \quad , \tag{18}$$

where a is the spacing between two adjacent growth units, and ϵ denotes the kink formation energy which can be approximated as the edge free energy, γ, at high ϵ values. For sparingly soluble salts with relatively high γ, x_0 may be written as

$$x_0 \approx (1/2)a \exp(\gamma/kT) \quad . \tag{19}$$

However, Equation (19), derived at equilibrium, may no longer be valid for a growing or dissolving step. Recently, the kink spacing x_0, derived kinetically for crystal steps with extremely high ϵ/kT values, has been shown to be dependent on S according to Equation (20) and (21) (Zhang and Nancollas, 1990):

$$x_0 = (1/2)aS^{-1/2}\exp(\gamma/kT) \qquad S > 1 \quad , \tag{20}$$

$$x_0 = (1/2)a(2 - S)^{1/2}\exp(\gamma/kT) \qquad S < 1 \quad . \tag{21}$$

As required, both Equation (20) and (21) reduce to Equation (19) when S is unity. Strictly speaking, Equation (20) and (21) may be significantly in error when $\epsilon/kT < 4$. In view of the high kink formation energy of sparingly soluble salts, and in order to account for their experimentally observed kinetics, Equation (20) and (21) will be used in this paper.

Steps from surface nucleation

The steps on a crystal surface can be created either by surface nucleation or the development of screw dislocation spirals. On a crystal surface in a supersaturated solution, the growth units have a tendency to form nuclei (Fig. 4), one growth unit thick, to lower the free energy of the system. Steps are thus created at the peripheries of these nuclei. The shape of the nucleus depends on the γ/kT value. At very low edge free energy, the nucleus is round while it becomes polygonized at very high γ/kT values (Burton et al., 1951). In the following discussion, the nuclei are assumed to be square shaped with the edge length, d. The Gibbs free energy for the formation of a surface nucleus is given by Equation (22):

$$\Delta G_i = 4\gamma i^{1/2} - ikT\ln S \quad , \tag{22}$$

where i $(= d^2/a^2)$ is the number of ions in the cluster. The first term on the right hand side of Equation (22) corresponds to the edge free energy of the cluster, and the second term to the decrease in chemical potential of the system. The density of the clusters on the surface (number per unit area) is given by Equation (23) (Hillig, 1966):

$$n_i = n_1 \exp(-\Delta G_i/kT) \quad , \tag{23}$$

where n_1 is the density of a single growth unit on the surface. The maximum Gibbs free energy for formation of a critical nucleus, $\Delta G_i{}^*$, is given by Equation (24):

$$\Delta G_i{}^* = \frac{4\gamma^2}{kT\ln S} \quad . \tag{24}$$

The edge length of the critical nucleus, d^*, is given by Equation (25):

$$d^* = \frac{2\gamma a}{kT\ln S} \quad . \tag{25}$$

It is interesting to note that the diameter of a circular critical nucleus is also given by Equation (25). Moreover, comparison of Equation (25) with Equation (14) indicates that the edge length of a critical two-dimensional nucleus is the same as the radius of a three-dimensional nucleus.

Clusters smaller than d^* are extremely unstable and are unlikely to form stable steps leading to crystal growth. Only those that overcome the energy barrier, $\Delta G_i{}^*$, can provide stable growth steps. According to Equation (23) and (24), the density of these clusters is very small at low driving forces, especially for sparingly soluble salts with very high edge free energies. Thus, it can be expected that the rate of crystal growth through two dimensional nucleation will be extremely small under these conditions.

Steps from screw dislocations

Most real crystal surfaces are intersected by screw dislocations as shown in Figure 5 (Frank, 1952), creating steps extending from the emergence point of the dislocation to the edge of the crystal surface. When such a surface is introduced into a supersaturated solution, the step can readily advance as growth units are continuously deposited on it. Since the angular velocity is greater near the dislocation, a screw axis is created and eventually the surface is covered by a series of steps in a spiral pattern as shown in Figure 6 (Frank, 1949; Burton et al., 1951). The dislocation therefore provides a perpetual step source for growth. The distance between two consecutive spiral steps, y_0, far from the dislocation center, is constant and is about 9.5 times of the d^* value (Cabrera and Levin, 1956; Budevski et al., 1975).

$$y_0 = \frac{19\gamma a}{kT\ln S} \quad . \tag{26}$$

In this derivation, some thermodynamic assumptions and mathematical simplifications have been made. Thus, Equation (26) is only applicable at very low supersaturation. However, due to the lack of suitable alternative models, this equation is often used well beyond its domain of theoretical validity.

374

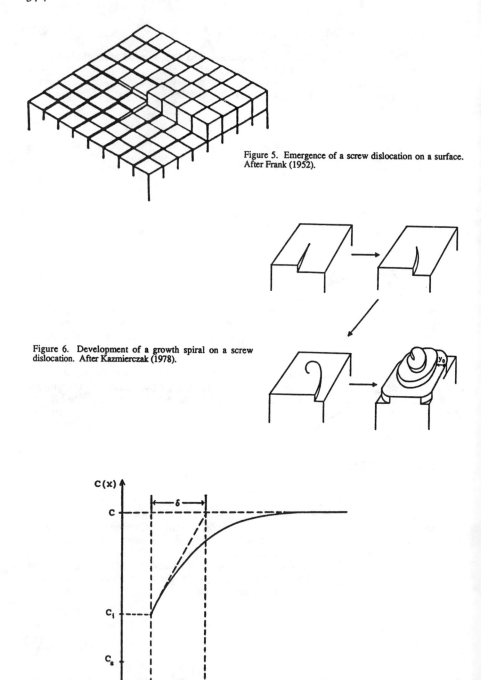

Figure 5. Emergence of a screw dislocation on a surface. After Frank (1952).

Figure 6. Development of a growth spiral on a screw dislocation. After Kazmierczak (1978).

Figure 7. Schematic representation of the concentration, C, as a function of distance, x, from a crystal surface. Λ is the thickness of the adsorption layer, δ is the diffusion layer thickness.

CRYSTAL GROWTH RATE LAWS

The linear growth rate of a crystal surface is defined as the displacement velocity in a direction normal to the surface (Nývlt et al., 1985). However, for the growth of polydispersed sparingly soluble electrolyte crystals suspended in a supersaturated solution, it is of greater interest for chemists and geochemists to express the rate in terms of the surface flux, which is defined as the number of moles of growth units deposited on unit area of the surface in unit time. The surface flux, J, is related to the linear growth rate, v, by Equation (27) (Nielsen, 1984):

$$v = V_m J \quad , \tag{27}$$

where V_m is the molar volume of the crystal.

The rate laws are equations which express the crystal growth rate as a function of the thermodynamic driving force, or supersaturation. In this section, the rate laws for the following rate determining processes will be discussed. In order for an ion in bulk solution to be incorporated at a kink site on a crystal surface step, it must (1) diffuse up to the surface, (2) enter the adsorption layer, (3) diffuse over the surface to a step and then to a kink site, and (4) integrate into the kink site. If steps are not readily available, surface nucleation has to occur and this becomes the most likely rate determining process.

Volume diffusion

A general concentration profile near a growing crystal surface is depicted in Figure 7, where C_i is the concentration adjacent to the adsorption layer whose thickness, Λ, is assumed to be of the dimension of the growth unit. When surface processes such as adsorption, surface diffusion, integration and surface nucleation, are extremely fast, the rate of crystal growth is governed by the rate at which growth units are transferred from bulk solution to the surface. The flux of these growth units, J, toward the surface is related to the concentration gradient, dC/dx, by Fick's first law, Equation (28) (Noyes and Whitney, 1897; Nernst, 1904):

$$J = -D\frac{dC}{dx} \quad , \tag{28}$$

where x is the distance from the surface, and D the diffusion coefficient. In the classic volume diffusion model, the surface reactions are assumed to be so fast that the equilibrium concentration can be established near the adsorption layer, with $C_i = C_s$. Thus the concentration gradient dC/dx can be replaced by $(C_s - C)/\delta$, where δ, shown in Figure 7, is defined as the diffusion layer thickness. Using Equation (7) and (8), the volume diffusion controlled rate becomes

$$J = k_{vd}\sigma \quad , \tag{29}$$

where

$$k_{vd} = DC_s/\delta \quad . \tag{30}$$

Adsorption and surface diffusion

If the transport of growth units in the solution to the surface is so rapid that no concentration gradients are established in the solution phase ($C_i = C$), the growth rate is controlled by processes such as adsorption, subsequent surface diffusion and integration. If a growth unit is accommodated by a kink immediately after diffusing to the step, the growth rate of a surface with a screw dislocation is controlled by adsorption and surface diffusion, and the corresponding rate law is given by Equation (31) (Burton et al., 1951; Ohara and Reid, 1973; Bennema and Gilmer, 1973; Nielsen, 1984):

$$J = k_{ad} \frac{\sigma lnS}{\sigma_c} tanh \frac{\sigma_c}{lnS} \quad . \tag{31}$$

where

$$k_{ad} = \nu_{ad} a C_s \quad , \tag{32}$$

and

$$\sigma_c = \frac{19\gamma a}{2kTx_s} \quad . \tag{33}$$

ν_{ad} represents the adsorption frequency and x_s is the mean displacement of a growth unit in the adsorption layer before desorption takes place. When $lnS > \sigma_c$, Equation (31) is transformed into a linear rate law given by Equation (34):

$$J = k_{ad}\sigma \quad . \tag{34}$$

In this case, upon entering the adsorption layer, a growth unit will readily find a kink site, and the reaction is controlled by adsorption. However, for $lnS << \sigma_c$, Equation (31) becomes

$$J = k_{sd}\sigma lnS \quad , \tag{35}$$

with

$$k_{sd} = \frac{2\nu_{ad}C_s x_s}{19(\gamma/kT)} \quad . \tag{36}$$

At very low supersaturation where $lnS \approx \sigma$, J has a parabolic dependence on the relative supersaturation. In this case, the surface diffusion of growth units to the steps is rate determining.

Integration

The growth rate is determined by integration of growth units into a kink site if the other kinetics steps involved in transferring these units to the vicinity of the kink site are much faster. The volume near the kink from which integration may occur is of the order $2a^3$ (Nielsen, 1984). In this volume, there are $2a^3 K_{ad}C$ moles of growth units, K_{ad} being the adsorption coefficient. If the integration frequency is ν_{in}, the forward flux toward a kink, j_f (moles per kink), will be given by Equation (37):

$$j_f = 2\nu_{in}a^3 K_{ad}C \quad . \tag{37}$$

However, the growth units may also detach from the kinks, resulting in a backward flux, j_b. Since at equilibrium, $j_b = j_f$, the backward flux can then be calculated from Equation (37) by substituting C_s for C. Therefore, with Equation (7) and (8), the net flux received by a kink, j_k ($= j_f - j_b$), is given by Equation (38):

$$j_k = 2\nu_{in}K_{ad}C_s a^3\sigma. \tag{38}$$

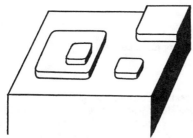

Figure 8. Schematic representation of the crystal surface growing by polynucleation.

For spiral growth, the kink density on the surface is $1/x_0 y_0$ and the growth rate in terms of surface flux is

$$J = j_k/x_0 y_0 \quad .\tag{39}$$

Using Equation (20) and (26) for x_0 and y_0, we obtain

$$J = k_{in}\sigma S^{\frac{1}{2}} ln S \quad ,\tag{40}$$

where

$$k_{in} = \frac{4\nu_{in}aK_{ad}C_s}{19(\gamma/kT)exp(\gamma/kT)} \quad .\tag{41}$$

It has been shown by Nielsen (1984) that Equation (40) can be approximated by a parabolic rate law, Equation (42), over a wide range of supersaturation.

$$J = k_{in}\sigma^2 \quad .\tag{42}$$

In this model, it is assumed that integration is only possible for growth units in contact with the crystal surface. This is reasonable because in the case of direct integration from solution, the activation energy will be greater than that from the adsorption layer where partial dehydration has already occurred (Bennema, 1967; Nielsen, 1984). Moreover, since the condition, $K_{ad} > 1$, is normally satisfied for electrolyte crystals in aqueous solution, the growth rate due to incorporation from the adsorption layer will be greater than that corresponding to capture from solution.

Surface nucleation

The growth of a defect-free crystal surface is realized by the spreading of two dimensional nuclei across the surface. When the crystal surface is very small, as soon as a critical nucleus forms, it may cover the entire surface before renucleation, as described by the so called mononucleation model (Ohara and Reid, 1973). However, on surfaces of a relatively large area, many critical nuclei may form simultaneously, even one upon another as shown in Figure 8. This has been referred as the polynuclear or birth and spread model (Hillig, 1966; Nielsen, 1984). If the surface nuclei adopt a square-like morphology and the rate of their expansion is controlled by integration of growth units into the kinks along the edges of the nuclei, the crystal growth rate through polynucleation is given by Equation (43) (Zhang, 1990):

$$J = k_e S^{7/6}\sigma^{2/3}(ln S)^{1/6}exp(-K_e/ln S) \quad ,\tag{43}$$

where

$$k_e \approx 7a\nu_{in}(K_{ad}C_s)^{4/3}(V_m)^{1/3}exp(-\gamma/kT) \quad ,\tag{44}$$

$$K_e = (4/3)(\gamma/kT)^2 \tag{45}$$

The numerical coefficient 7 in Equation (44) is an approximate value.

Combined mechanisms

As discussed above, crystal growth is realized through a series of consecutive processes with the slowest being the rate determine step. When the rates of these processes are comparable at equal driving forces, each may have a significant influence on the observed rate of crystal growth. When volume diffusion and surface reaction have comparable rates, the concentration adjacent to the adsorption layer, C_i, will be significantly different from the bulk concentration (see Fig.7). Thus the relative supersaturation for the surface process, σ_i (= $C_i/C_s - 1$), is only a portion of the overall driving force, σ. The remainder, $\sigma - \sigma_i$, is involved in volume diffusion. According to Equation (29), the volume diffusion flux, J_v, can then be written as

$$J_v = k_{vd}(\sigma - \sigma_i) \quad . \tag{46}$$

If the surface process is adsorption controlled, the surface reaction flux, J_s, will be given by Equation (47):

$$J_s = k_{ad}\sigma_i \quad . \tag{47}$$

Since $J_v = J_s$, the unknown quantity σ_i can be eliminated from Equation (46) and (47), yielding a rate law for combined volume diffusion and adsorption,

$$J = \frac{k_{vd}k_{ad}}{k_{vd} + k_{ad}}\sigma \quad . \tag{48}$$

When volume diffusion is much faster ($k_{vd} \gg k_{ad}$), Equation (48) reduces to Equation (34) for an adsorption controlled rate. On the other hand, when adsorption is much faster ($k_{ad} \gg k_{vd}$), the volume diffusion controlled rate law, Equation (29), is recovered.

If the surface process is controlled by integration or surface diffusion at very low supersaturation, the surface flux may be described by a parabolic rate law:

$$J_s = k_2\sigma_i^2 \quad , \tag{49}$$

where k_2 is the rate constant. Elimination σ_i from Equation (46) and (49) leads to Equation (50) for the combined volume diffusion and surface reaction (Nielsen, 1984),

$$J = \frac{2k_{vd}k_2\sigma^2}{2k_2\sigma + k_{vd}[1 + (1 + 4\sigma k_2/k_{vd})^{1/2}]} \quad . \tag{50}$$

At low supersaturation, $\sigma \ll k_{vd}/2k_2$, Equation (50) yields a parabolic dependence, while the linear volume diffusion rate law, Equation (29), is obtained at much higher supersaturation.

It is also possible that surface nucleation and spiral dislocation mechanisms may operate simultaneously since surface nuclei may form between the steps originating from a screw dislocation source. The rate is thus the sum of the two individual processes with the faster one dominating (Nielsen, 1984). At very low supersaturation, since surface nucleation is negligibly slow, the growth rate of a surface with a screw dislocation will be determined mainly by a spiral dislocation mechanism. However, at sufficiently high supersaturation, surface nucleation may dominate the overall growth rate.

CRYSTAL DISSOLUTION RATE LAWS

In contrast to growth, during dissolution, a growth unit will (1) detach from the kink, (2) diffuse away from the kink site, (3) desorb from the surface, and (4) diffuse into the bulk solution. The corresponding elementary mechanisms may thus be: (1) detachment; (2) surface diffusion; (3) desorption, and (4) volume diffusion. If there are no screw dislocations on the crystal surface, the dissolution rate may be determined by the formation of etch pits, one growth unit deep. For surface diffusion, desorption, and volume diffusion controlled dissolution, the rate equations are identical to the corresponding growth expressions but with negative σ values. However, for spiral dissolution controlled by detachment, since the kink density has a different dependence on S, the rate law is given by Equation (51) incorporating Equation (21).

$$J = k_{de}\frac{(S - 1)lnS}{(2 - S)^{1/2}} \quad , \tag{51}$$

where k_{de} is identical to k_{in} given by Equation (41).

Similarly, for polynuclear dissolution, with the rate of nucleus expansion controlled by detachment from kink sites,

$$J = k_e S^{2/3}\sigma^{2/3}(-lnS)^{1/6}(2 - S)^{-1/2}exp(K_e/lnS) \quad , \tag{52}$$

where k_e and K_e are given by Equation (44) and (45), respectively.

Traditionally, volume diffusion has been emphasized for dissolution (Noyes and Whitney, 1897). More recent experimental results, however, indicate that the dissolution rates of many sparingly soluble salts are determined by processes occurring at the crystal surfaces.

INFLUENCE OF ADDITIVES ON THE RATES

In nature, crystallization and dissolution almost always occur in the presence of ions and molecules other than those of the crystal lattice. It has long been known that very low concentrations of some foreign ions and molecules (hereafter, additives) can greatly affect the rates of crystal growth and dissolution (Buckley, 1951). In this section, the mechanisms by which the additives may influence these rates will be discussed.

Inhibitory effect of additives

Some additives form complexes with the lattice ions, thereby effectively removing them from solution and reducing the thermodynamic driving force for growth. In the following, the influence of additives on the rates of reaction is considered at constant thermodynamic driving forces. At very low concentrations, additive molecules are unlikely to significantly influence the volume transport process and their effect on the rate must be due to their participation in surface reactions when present in the adsorption layer. If the size of the additive molecule is similar to that of a growth unit, three types of adsorption sites, with different binding energies, can be distinguished (Davey, 1976; 1979). The binding will be strongest at kinks, moderate at steps, and weakest on crystal terraces. In the following, the inhibitory effects are discussed according to the position of the adsorbed additive molecules. Large inhibitor molecules, although their adsorption will not be so selective, may still be discussed in terms of the adsorption sites.

Inactivation of kink sites. When occupied by additive molecules, the kinks can be considered, at least temporarily, poisoned or inactivated (Ohara and Reid, 1973). Lattice ions can not be integrated into these kinks until the additive molecules are desorbed. In this case, the number of active kink sites is decreased, leading to a decrease in the integration controlled growth rate. Assuming that adsorption at kink sites occurs only for additive molecules already present in the adsorption layer, the flux, j_a, of inhibitors toward the kinks can be written as Equation (53):

$$j_a = (1 - \theta)\nu_a[I]_{ad}\Omega \quad , \tag{53}$$

where θ is the percentage of poisoned kinks, ν_a the adsorption frequency at a kink site, $[I]_{ad}$ the concentration of additives in the adsorption layer, and Ω the volume within which the additive molecule may jump directly to the kink site. The desorption flux of the additive molecules, j_d, from the kinks may be written as

$$j_d = \nu_d\theta/N_0 \quad . \tag{54}$$

where ν_d is the desorption frequency from a kink site, and N_0 is the Avogadro constant. At steady state, $j_a = j_d$, and Equation (55) is obtained:

$$\theta = \frac{\nu_a[I]_{ad}\Omega N_0}{\nu_a[I]_{ad}\Omega N_0 + \nu_d} \quad . \tag{55}$$

If the coverage of a terrace by additive molecules is very low, as anticipated in most cases, it follows that

$$[I]_{ad} = K_t[I] \quad , \tag{56}$$

where K_t is the adsorption coefficient of additives on the terrace, and $[I]$ is the inhibitor concentration in the bulk solution. Substitution of Equation (56) into Equation (55) will result in a Langmuir (1916) type adsorption at kink sites.

For an integration controlled process, the growth rate in the presence of additives, J, should be directly proportional to the fraction of active kinks. Thus

$$J = J_0(1 - \theta) \quad , \tag{57}$$

where J_0 is the rate in the absence of additives. From Equation (55), (56) and (57), one obtain

$$J = J_0/(b[I] + 1) \quad , \tag{58}$$

with the constant,

$$b = K_t\nu_a\Omega N_0/\nu_d \quad . \tag{59}$$

Thus a plot of J_0/J against $[I]$ will be linear according to Equation (60):

$$J_0/J = 1 + b[I] \quad . \tag{60}$$

Retardation of step movement. Even at an instant when all the kinks are covered by additive molecules, the steps may still advance as new kinks form on their uncovered parts. Thus it would appear that the stoppage of the step movement is possible only when it is completely covered by additives. Such intensive coverage may not be necessary since when the distance between two adsorbed molecules is less than the diameter of a critical nucleus, d*, step movement will be prevented (Sears, 1958; Albon and Dunning, 1962; Cabrera and Vermilyea, 1958). Even when the distance is greater than d*, the rate of step advancement will be reduced because of their high curvature. Since the size of a critical nucleus is larger at lower supersaturation, as indicated by Equation (25), it follows that inhibitors will be more effective at lower driving forces.

Reduction of the concentration of growth units on a terrace. Additives compete for adsorption sites with the lattice ions, thus reducing both the concentration of the latter in the adsorption layer and the subsequent growth rate. Considering multi-component adsorption, the concentration of growth units on the crystal surface, C_{ad}, is given by

$$C_{ad} = K_{ad}C/(1 + K_{ad}V_mC + K_tV_m[I]) \quad . \tag{61}$$

For sparingly soluble salts, the inequality, $K_{ad}CV_m << 1$, may be satisfied and Equation (61) becomes:

$$C_{ad} = K_{ad}C/(1 + K_tV_m[I]) \quad . \tag{62}$$

The rate of growth in the presence of inhibitor can therefore be written as Equation (63):

$$J = J_0/(1 + K_tV_m[I]) \quad . \tag{63}$$

Equation (61) is based on a Langmuir adsorption model although other types of adsorption may be possible (Giles et al., 1974). Unless the concentration of additives is very high, the reaction rate is not expected to be completely inhibited due to the decrease of C_{ad} by competitive adsorption.

In addition to the inhibition mechanisms discussed above, additive molecules on the crystal surface may serve as obstacles, preventing the growth units on the surface from diffusing to the steps. Such a mechanism will be more likely when the surface coverage is relatively high, as for the adsorption of macromolecules such as proteins and polyelectrolytes. Moreover, if growth is controlled by polynucleation, the rate can be markedly influenced if inhibitor molecules are distributed on the surface at a distance less than the diameter of a critical nucleus (Christoffersen et al., 1983).

In the above discussions, it has been emphasized that additives change the reaction rate by influencing the surface processes. However, it does not follow that additives would have no influence on an otherwise volume diffusion controlled reaction. Since the latter results from faster surface processes, when these are inhibited, the rate may become surface controlled. Further increase in additive concentration will therefore result in a decreased rate of growth.

Dual effects of additives

The influence of additives in inhibiting crystal growth has long been confirmed experimentally although it has been suspected that their adsorption may decrease the surface energy, leading to a rate increase (Mutaftsschiev, 1973). It is apparent from rate equations for surface controlled processes, Equation (33), (40), (41), (43), (44) and (45), that the rate is also indirectly proportional to the edge free energy. For crystals controlled by a polynuclear mechanism, an increased rate was indeed observed in the presence of some additives (Bliznakov and Kirkova, 1957; Troost, 1968). This was explained in terms of a lowering of the surface energy due to adsorption (Davey, 1979). However, since the surface underneath the additive molecules can no longer participate in the formation of surface nuclei, this explanation for rate promotion is only phenomenological. The promoting influence of additives on surface nucleation can be explained without invoking changes in the surface free energy. Frank pointed out that additive molecules on a perfect crystal surface may serve as sites for nucleation (Frank, 1952) and this idea has since been confirmed

382

by computer simulations (Weeks and Gilmer, 1979). It has been discovered that strongly adsorbed foreign ions of the same size as the lattice ions greatly enhance the surface nucleation. For the corresponding dissolution process, the adsorption of a foreign ion on the crystal surface may promote the detachment of neighboring lattice ions of the same charge due to coulombic repulsion, and the rate of etch pit formation will be increased (Cheng et al., 1984).

Although in some cases, experiments have shown that the crystal growth rate may be increased in the presence of additives, this situation is rare as compared with the long established inhibitory effect. This is due to the fact that, although the adsorption of some additive molecules may induce surface nucleation, their presence on the surface also retards the step advancement (Davey, 1979). However, it seems likely that promotion may occur at low surface coverage while inhibition will dominate at higher additive concentrations.

EXPERIMENTAL METHODS

Experimental approaches

In order to investigate the reaction mechanisms, it is important to study the growth or dissolution rates at different driving forces. For sparingly soluble salts growing in suspension, the rates can be determined directly by measuring the changes of the particle size as a function of time (Markovic and Komunjer, 1979) or indirectly from the measured changes in concentration of lattice ions in the solution. Titration methods may also be used to record the amount of lattice ions that must be added to maintain the concentrations at constant values during the reaction. These methods can be divided into three main categories.

Free drift method. One of the earliest experimental approaches is the "free drift" method (Davies and Jones, 1955; Nancollas, 1973), in which the rate of a reaction is obtained by following concentration changes as a function of time. With the development of rapidly responding sensors, the concentrations may be recorded continuously by means of potentiometric, conductometric, or photometric methods (Ebrahimpour, 1990). When such continuous monitoring methods are not available, the reaction can be followed by sampling the solution suspension, filtering rapidly and analyzing for lattice ions using accepted analytical techniques.

Potentiostatic method. For some systems, it is necessary and convenient to investigate the rate of reaction when the activity of one or more species is held at a constant value. For instance, for calcium phosphates and carbonates, the nature of the phases precipitating and rates of reaction are dependent upon solution pH at a given supersaturation. It is therefore an advantage to investigate these systems at a constant pH using traditional pH stat procedures (Whitnah, 1933; Jacobsen and Léonis, 1951). The pH is maintained constant by addition of acidic or basic titrant solutions controlled by a potentiostat incorporating glass and reference electrodes (Nancollas and Mohan, 1970). The reaction rate can be evaluated from the volume of titrant consumption with time, provided that the precipitate stoichiometry is known.

Constant composition method. In the conventional potentiostatic method, the activity of a single ion species is controlled by titrant addition while the concentration of other lattice ions are allowed to vary with time. This results in variations in the thermodynamic driving forces during the reaction. Although it would appear that the rate data for crystal growth as a function of supersaturation can be obtained from a single experiment, large errors are often involved in the determination of the rates, especially at low supersaturation. It is therefore desirable to use a method in which the activities of all the species in the solution can be maintained constant during the reaction. The Constant Composition (CC) method (Tomson and Nancollas, 1978) has several advantages as compared with the free drift and conventional potentiostatic techniques:

(1) The growth and dissolution rates can be determined more precisely, especially at very low driving forces.

(2) The experiments can be carried out at precisely known points on a phase diagram in order to avoid side reactions such as phase transformation (Nancollas, 1989).

(3) Relatively large amounts of material can be grown at a specific driving force for characterization of the solid phases (Tomson and Nancollas, 1978).

(4) The influence of additives on the reaction rate can be examined over extended time periods while maintaining constant concentrations of both additives and lattice ions (Budz et al., 1988).

(5) It enables changes in reactivity of the crystal surfaces to be investigated as a function of time at a given driving force (Zhang, 1990).

Titrant composition for CC experiments

In designing CC experiments, the most important problem is the calculation of the concentrations of the titrant solutions. During the reaction, the solution composition changes due to mass transfer between the solution and solid phases. The titrant must be able to compensate for these changes as well as concomitant dilution effects due to their addition. The diversity of systems that can be studied by the CC method makes it difficult to write generalized equations for the calculation of titrant concentrations, but in the following, the method will be illustrated for three typical systems.

Systems containing only lattice ions. In binary systems containing only lattice ions and solvent, the depletion of the solution concentration due to precipitation can be compensated by the addition of a titrant solution containing the same components but at a higher concentration. If a volume dV_t of the titrant is added to sustain solution composition when dn moles of solid precipitate, the titrant concentration, C_t, must satisfy the following mass balance relationship,

$$C_t dV_t = C_w dV_t + dn \quad , \tag{64}$$

where C_w is the concentration of the working solution in the reaction vessel and $C_w dV_t$ compensates for dilution by the titrant addition. Equation (64) simply states that the number of moles of lattice ions added to the solution, $C_t dV_t$, must be equal to the sum of the solution content, $C_w dV_t$, and the amount precipitated, dn. From Equation (64), the titrant concentration can be expressed as

$$C_t = C_w + C_e \quad , \tag{65}$$

where C_e $(= dn/dV_t)$ denotes the number of moles of precipitate formed per unit volume of added titrant. This will be referred to as the effective concentration of the titrant. The value of C_e is arbitrary and is optimally set so as to achieve the desired rate precision and reaction extent while maintaining the stability of the titrant solution.

During dissolution, any increase in lattice ion concentrations may be compensated by dilution, through the addition of a solution with $C_t < C_w$.

$$C_t = C_w - C_e \quad . \tag{66}$$

Pure solvent may also be used as the titrant. However, the use of a titrant solution with a finite C_t value is recommended at low dissolution rates since the increased volume of titrant consumption allows the rate to be determined with a greater precision.

The application of the CC method to crystal growth in these binary systems may be limited to substances with a significant temperature coefficient of solubility so that the metastable solution can be made at elevated or suppressed temperatures. However, this method may find a wide application in dissolution studies since the undersaturated solution can be prepared by dissolving known amounts of crystals in the solvent. The absence of supporting electrolytes also allows the use of more sensitive methods, such as conductometry, for controlling the reaction.

Systems involving supporting electrolytes. In this sightly more complex case, the growth of a binary electrolyte AB is studied in a supersaturated solution prepared by mixing of the solutions of soluble salts, AY and XB. The titrant may be prepared in the same manner but at higher concentrations and contained in separated burets to avoid precipitation. To conform with the mass balance requirements, the concentrations of A and B ions in the titrants are given by Equation (67) and (68):

$$[A]_t = 2[A] + C_e \quad , \tag{67}$$

$$[B]_t = 2[B] + C_e \quad , \tag{68}$$

where [A] and [B] denote the working solution concentrations and the factor 2 accounts for the dilution effect introduced by the use of two simultaneous titrants. The inert ions X and Y in the working solution will accumulate as the reaction proceeds, causing the ionic strength to increase. This difficulty can be overcome by the addition of supporting electrolyte XY to the working solution to maintain its concentration constant. Since C_e moles of XY will be released per liter of each titrant added, the concentration of XY contained in one of the two titrants is given by

$$[XY]_t = 2[XY] - C_e \quad . \tag{69}$$

Similarly, in the corresponding CC dissolution experiment, additional supporting electrolyte at a concentration $[XY]_t$ must be added to the titrant in order to avoid concentration decreases due to dilution by the titrants.

$$[XY]_t = 2[XY] + C_e \quad . \tag{70}$$

Although the addition of more supporting electrolyte may appear to be a disadvantage of the CC method, in many real situations, supporting electrolytes are present. Moreover, the ionic strength has been found to be an important factor influencing the reaction kinetics at a crystal-solution interface (Witkamp, et al., 1990).

Systems involving acid or base addition. An excellent example of such a system is found in the case of calcium phosphates. Suppose that the working solution is prepared by mixing KH_2PO_4, $CaCl_2$, KCl, and KOH and it is intended to study the growth rates of hydroxyapatite (HAP, $Ca_5(OH)(PO_4)_3$) at slightly acidic pH where $H_2PO_4^-$ is the dominating phosphate species. As HAP precipitates from the solution, H^+ ions will be released and KOH will be needed to maintain the pH constant. In order to avoid precipitation in the titrant solution, two titrant burets may be used, one containing $CaCl_2$ and KCl, and the other KH_2PO_4 and KOH. The titrant concentrations of $CaCl_2$ and KH_2PO_4 can be derived by using mass balance expressions for calcium and phosphate ions, respectively, according to the procedure used for Equation (65), (67) and (68),

$$[CaCl_2]_t = 2[CaCl_2] + 5C_e \quad , \tag{71}$$

$$[KH_2PO_4]_t = 2[KH_2PO_4] + 3C_e \quad . \tag{72}$$

In Equation (71) and (72), the numerical coefficients of C_e arise since precipitation of dn moles of HAP removes 5dn and 3dn moles of calcium and phosphate ions, respectively, from the solution. the KCl titrant concentration can be derived by balancing the number of moles of chloride ions before and after the addition of a volume dV_t of each titrant according to Equation (73):

$$(2[CaCl_2]_t + [KCl]_t)dV_t = (2[CaCl_2] + [KCl])(2dV_t) \quad . \tag{73}$$

Combination of Equation (71) and (73) gives the concentration of the supporting electrolyte concentration in the titrant,

$$[KCl]_t = 2[KCl] - 10C_e \quad . \tag{74}$$

Similarly, from the mass balance for potassium ion, we obtain

$$[KCl]_t + [KH_2PO_4]_t + [KOH]_t = 2([KCl] + [KH_2PO_4] + [KOH]) \quad . \tag{75}$$

Substitution of Equation (72) and (74) into Equation (75) yields the titrant concentration for KOH,

$$[\overset{\bullet}{K}OH]_t = 2[KOH] + 7C_e \quad . \tag{76}$$

For the corresponding dissolution reaction, KOH in the titrant is replaced by HCl. Since the solution is undersaturated, all the components can be contained in a single buret. The titrant concentrations can again be derived by solving simultaneous mass balance equations for total calcium, phosphate, chloride, and potassium ions, and the results are summarized in Equation (77)-(80):

$$[CaCl_2]_t = [CaCl_2] - 5C_e \quad ; \tag{77}$$

$$[KH_2PO_4]_t = [KH_2PO_4] - 3C_e \quad ; \tag{78}$$

$$[HCl]_t = - [KOH] + 7C_e \quad ; \tag{79}$$

$$[KCl]_t = [KCl] + [KOH] + 3C_e \quad . \tag{80}$$

If additives are introduced into the working solution, they should be included in the titrants at the same concentrations in order to avoid dilution effect.

Instrumentation for CC method

Constant Composition experiments involve the simultaneous addition of multiple titrants of precisely monitored volumes. Although this may be achieved with separately driven burets, it is usually preferable for them to be mechanically coupled. A thermostatted cell, a potentiometer incorporating ion selective and reference electrodes, a potentiostat containing an appropriate switching device (impulsomat), and a chart recorder (dosigraph) completes the instrumentation as shown in Figure 9 (Svehla, 1978). Before the reaction is initiated, the emf set point for the electrode system is selected on the impulsomat. During the reaction, the output of the potentiometer is constantly compared with the preset value. The difference, or the error signal, is amplified and relayed to the electronic switch. When the error signal exceeds a threshold value, the switch activates the motor-driven burets. In order to avoid overshooting the set point due to slow electrode response or inertia of the electronic switch, a pulse generator is usually built into the potentiostat (Svehla, 1978). This produces a square-wave signal which is then superimposed on the error signal, and the resultant is used to operate the electronic switch. If the frequency and amplitude of the superimposed square wave are properly chosen, the addition of titrant aliquot from the burets will be made intermittently when the signals are small and more continuously when the potential differs significantly from the set point value. Moreover, the effective titrant concentration and number of burets can also be changed in order to ensure optimal titrant addition rates.

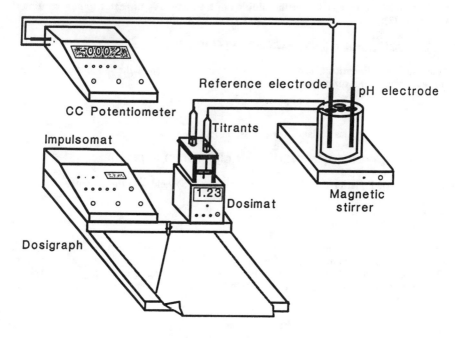

Figure 9. Apparatus for constant composition experiments.

Table 1. Effective orders, n, for different mechanisms

n	Probable mechanisms
1	Volume diffusion; Adsorption; Volume diffusion + Adsorption.
1-2	Combined mechanisms such as: Adsorption + Surface diffusion; Adsorption + Integration; Volume diffusion + Surface diffusion; Volume diffusion + Integration; Volume diffusion + Polynucleation.
2	Surface diffusion; Integration; Surface diffusion + Integration.
>2	Polynuclear growth; Polynuclear + Spiral growth. Polynuclear dissolution; Spiral dissolution controlled by surface diffusion and/or detachment; Polynuclear + Spiral dissolution.

Note: Except the case of n > 2, all the mechanisms are valid for both growth and dissolution.

Rate determination

In free drift experiments, the rate can be obtained from the measured concentration changes in solution as a function of time. The overall rate, J, can be calculated at any instant from the slope of the resultant C-t curve, dC/dt.

$$J = - \frac{V_W}{A_t} \frac{dC}{dt} ,$$
(81)

where V_w is the volume of the reaction solution and A_t is the instantaneous total surface area of the crystals. For potentiostatic and CC experiments, the rates of reaction can be calculated from the volume of titrant added, which can be converted to the amount of deposition or dissolution, as a function of time. For CC, the overall reaction rate is simply given by

$$J = \frac{C_e}{A_t} \frac{dV_t}{dt} ,$$
(82)

where (dV_t/dt) is the slope of the titrant volume-time curve recorded during the reaction.

During an experiment, the total surface area A_t changes with extent of growth or dissolution. If the crystals are monodisperse and retain their morphology during the reaction, A_t can be related to the initial value, A_0, by Equation (83) (Van Oosterhout and van Rosmalen, 1980; Barone et al., 1983):

$$A_t = A_0(M_t/M_0)^{2/3} ,$$
(83)

where M_0 and M_t are the mass of crystals present initially and at time t, respectively. A_0 can be calculated from the specific surface area, S_a, of the seed crystals by using Equation (84):

$$A_0 = S_a M_0 .$$
(84)

Although most seed crystals are not monodisperse, it has been shown that Equation (83) is a reasonable approximation when the particle size distribution is Gaussian (Barone et al., 1983). However, if the morphology changes during the reaction, Equation (83) may have significant error. In this case, certain corrections may be made using information obtained from scanning electron microscopy and surface area measurement at different extent of the reaction.

EXPERIMENTAL DETERMINATION OF REACTION MECHANISMS

The mechanisms of crystal growth and dissolution are usually interpreted from measured the reaction rates at different driving forces. These kinetics data can be confronted with theoretical models such as those discussed above to determine the most probable mechanism. It is common practice to fit the data to an empirical rate law such as Equation (85):

$$J = K\sigma^n .$$
(85)

The effective order, n, can be determined from a logarithmic plot of Equation (85). From the n value, the probable mechanism(s) can be deduced recognizing that one or more elementary processes may operate simultaneously. The probable mechanisms corresponding to experimental n values are summarized in Table 1.

For crystal growth, a value n > 2 indicates the participation of surface nucleation. On the other hand for dissolution, both polynucleation and spiral dissolution, with a rate controlled by detachment of growth units from a step, may yield an effective order higher than 2 (Zhang and Nancollas, 1990). However, it is still possible to distinguish between these two mechanisms using kinetics data at very low driving forces where the n value for spiral dissolution approaches 2 while polynucleation yields a much higher order. For both growth and dissolution, additional support for a polynucleation controlled mechanism may be gained by fitting the data to Equation (43) or (51) and obtaining the exponential constant K_e given in Equation (45). The estimated edge free energy γ/kT can then be compared with that obtained from homogeneous nucleation studies (Sangwal, 1989). A good agreement between these values can serve as evidence for polynucleation. If the two values differ significantly, the possible involvement of three dimensional, secondary, or heterogeneous nucleation on foreign surfaces may be considered.

For a second order reaction, it is difficult to distinguish between integration and surface diffusion mechanisms in the supersaturation range below the σ_c value given by Equation (33). The difficulty arises because independent x_s data are not presently available to allow estimation of the rate constants. However, advantage may be taken of the fact that the adsorption and surface diffusion controlled mechanism follows a parabolic rate law at $\sigma \ll \sigma_c$ and a linear rate law at $\sigma \gg \sigma_c$. Thus the experimental measurements may be extended to higher driving forces to determine whether the reaction changes to first order, reflecting an adsorption controlled process. Since x_s is not known, the change-over transition can not be predicted. However, if surface diffusion does not play any role ($x_s = a$), the change in effective order is unlikely to occur if lnS is less than σ_L given by Equation (86), obtained by substituting $x_s = a$ into Equation (33).

$$\sigma_L = \frac{19\gamma}{2kT} \ . \tag{86}$$

If a transition between second and first order occurs at a lower supersaturation value, it may indicate that $x_s > a$ and an important contribution due to surface diffusion. However, at $lnS > \sigma_L$, Equation (25) would indicate that the critical surface nucleus would be smaller than a single growth unit. In this situation, there is virtually no energy barrier for surface nucleation, and growth may no longer follow a spiral mechanism (Bennema, 1969). Furthermore, homogeneous nucleation may also occur at such high concentrations. Thus, in the supersaturation range where spiral growth predominates, a change in the effective reaction order from second to first (provided volume diffusion is not responsible for such a transition) may indicate that the rate is not entirely controlled by an integration process. On the other hand, even if there is no apparent change in the effective order under achievable experimental conditions, it may still be possible, when sufficient data are available, to distinguish between the surface diffusion and integration controlled mechanisms based on the different S functions given by Equation (35) and (40).

The involvement of volume diffusion can sometimes be verified by varying the stirring dynamics which influences the diffusion layer thickness δ (Nývlt et al., 1985). If the rate is sensitive to variations in stirring speeds, volume diffusion is likely to participate as a rate controlling process. However, an insensitivity to stirring may not rule out volume diffusion, since (1) at relatively high stirring speed, the diffusion layer thickness decreases only slightly with increasing shear at solution-crystal interface (Janssen-van Rosmalen, 1975); (2) for smaller crystals (less than $5\mu m$, depending on the density difference between crystal and solution), the suspended particles may be carried by the liquid flow so that any increase in the stirring speed will have little effect on the thickness of the diffusion layer (Nielsen, 1984).

In spite of these difficulties, the participation of volume diffusion as a rate determining step may still be investigated by comparing the experimental rate with that predicted using Equation (29) and (30). The diffusion layer thickness of small crystals suspended in an aqueous solution can be estimated by Equation (85) (Nielsen, 1980):

$$\delta = (5.74\mu m) <r>^{0.145} (\Delta\rho)^{-0.285} \quad , \tag{87}$$

where $<r>$ is the particle radius (μm) and $\Delta\rho$ (g/cm^3) the density difference between solid and solution. In deriving Equation (87), the ionic diffusion coefficient, D, is taken as 10^{-9} m^2/s, and this value may also be used for estimating the volume diffusion controlled rate constant in Equation (30). If the calculated rate is within one order of magnitude of the experimental value, the reaction will be strongly influenced by volume diffusion. However, if the former is significantly smaller, the rate is probably controlled by a surface process.

Activation energies, obtained from experiments at different temperatures, may also be used to differentiate between volume diffusion and surface controlled processes. The activation energy for volume diffusion, reflecting the temperature dependence of the diffusion coefficient, usually lies between 16 and 20 kJ/mol (Nielsen, 1984), while for a surface reaction the value may be in excess of 35 kJ/mol. For example, if a reaction, first order (n = 1) with rates sensitive to solution hydrodynamics, has an activation energy of less than 20 kJ/mol, it is safe to assume that it is overwhelmingly controlled by volume diffusion. However, if the reaction rate, insensitive to stirring speed changes, is much lower than the value estimated using Fick's law, and if the activation energy is higher than 35 kJ/mol, it is quite certain that an adsorption process predominates. In all other cases, both adsorption and volume diffusion mechanisms may participate for a first order reaction.

As shown in Table 1, various combined mechanisms can result in an effective order between 1 and 2. In this case, it is very important to fit the experimental data to the models to obtain important parameters such as the edge free energy and activation energy for comparison with values obtained using independent methods.

GROWTH AND DISSOLUTION OF SOME ALKALINE EARTH SALTS

Calcium phosphate

Increases in phosphate concentration in lakes and rivers near heavily populated areas have contributed to a resurgence of interest in chemical processes such as the precipitation and dissolution of phosphate salts. This provides a challenge to the chemist not only because of the presence of multiple components in the solution phase but also because of the numerous calcium phosphate phases that may be involved in the precipitation reactions. Typical solubility isotherms as a function of pH are given in Figure 10. Although the thermodynamically most stable phase is hydroxyapatite (HAP, $Ca_{10}(PO_4)_6(OH)_2$), it is now generally accepted that other phases such as amorphous calcium phosphate (ACP), dicalcium phosphate dihydrate (DCPD, $CaHPO_4 \cdot 2H_2O$), tricalcium phosphate (TCP, $Ca_3(PO_4)_2$), and octacalcium phosphate (OCP, $Ca_8H_2(PO_4)_3 \cdot 5H_2O$), as well as defect apatite may participate (Brown et al., 1962; Francis and Webb, 1971; Nancollas, 1982). In the spontaneous precipitation of calcium phosphate in highly supersaturated solutions, the formation of apatite is usually preceded by the precipitation of one or more precursor phases such as ACP in which no long range order can be detected by X-ray diffraction studies (Betts et al., 1981). The composition of this phase appears to depend upon the precipitation conditions and its formation may be followed by the nucleation of OCP which serves as a template for HAP (Brown et al., 1979; Nancollas, 1989).

As the acidity of the solution is increased, OCP or DCPD may be the appropriate precursor phases in accordance with Ostwald's rule of stages which states that the least stable salt with the highest solubility will always form first in a sequential precipitation reaction. This is illustrated by line AD in Figure 10. Starting at A, supersaturated with respect to all calcium phosphate phases, DCPD may form in the region AB, thereafter only TCP and HAP

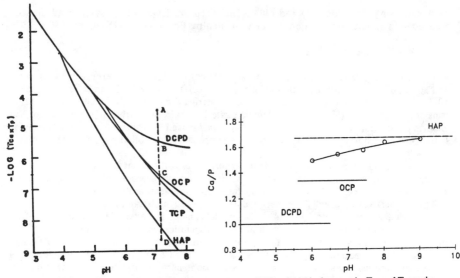

Figure 10 (left). Solubility isotherms for calcium phosphates at 37°C at 0.1 M ionic strength. T_{Ca} and T_P are the total calcium and phosphate concentrations. The isotherms are calculated at $(T_{Ca}/T_P) = 1, 1.33, 1.5$ and 1.67 for DCPD, OCP, TCP and HAP, respectively.

Figure 11 (right). Ranges of stability of calcium phosphate phases as determined by CC experiments showing the molar calcium/phosphate ratios as a function of pH. Modified from Nancollas and Zawacki (1989).

Table 2. Kinetics data for growth of alkaline earth fluorides.

Crystal	σ	n	μ(M)	Reference
MgF$_2$	0.4 - 0.8	1.8	0.06 - 0.18	a
	0.1 - 1.1	2.0	0.06 - 0.18	a
	1.0 - 1.6	5	0.15	b
CaF$_2$	0.37 - 1.94	1.8	0.025	c
	0.20 - 1.66	1.9	0.016 - 0.026	d
	1 - 3.4	3	0.012 - 0.015	e
SrF$_2$	0.09 - 1.03	2.0	0.06 - 0.15	f
	0.9 - 2.0	4.3	0.06 - 0.15	f
BaF$_2$	0.46 - 1.0	2.2	0.41	g

Note: All the studies were made at 25°C except reference d at 37°C.
References: a) Yoshikawa and Nancollas, 1983; b) Abdul-Rahman, 1988; c) Pérez, 1989;
d) Shyu and Nancollas, 1980; e) Christoffersen et al., 1988; f) Abdul-Rahman and Nancollas, 1984;
g) Barone et al., 1983.

Table 3. Kinetics data for the dissolution of alkaline earth fluorides at 25°C.

Crystal	$-\sigma$	n	E*	Reference.
MgF$_2$	0.1 - 0.8	2		a
	0.25 - 0.75	3.5	86	b
CaF$_2$		1	73	c
	0.25 - 0.75	0.8±0.2		d
SrF$_2$	0.1 - 0.3	1.7	72	a
	0.50 - 0.93	3.8		a

Note: E* = activation energy (kJ/mol)
References: a) Abdul-Rahman, 1988; b) Hamza and Nancollas, 1985;
c) Gardner and Nancollas, 1976; d) Christoffersen et al., 1988.

may form below C. Although the thermodynamic driving forces, represented in Figure 10, have to be favorable for a particular precipitation event, the appearance and stability of the phase actually formed will be much more dependent upon kinetic factors. Thus the less stable, more acidic calcium phosphate phases crystallize at rates considerably greater than that of HAP. The advantage of the CC method is that supersaturation can be maintained at any point on the solubility diagram (Figure 10) and the technique has demonstrated the formation of calcium deficient apatites having compositions dependent upon pH as shown in Figure 11 (Nancollas and Zawacki, 1989; Heughebaert et al., 1990; Zawacki et al., 1990). An important factor that must be taken into account in analyzing theses data is that the introduction of seed material into the supersaturated solutions may be accompanied by ion exchange and/or other surface modification processes, depending upon the surface area of the seed material. Thus the introduction of HAP crystallites of very high specific surface area into calcium phosphate solutions may be accompanied by a very rapid ion exchange reaction in which H^+ and Ca^{2+} ions are desorbed from the surface. If very high surface area solids are used, relatively large amounts of phosphate and calcium ions may be involved in these exchange processes and regulate the attainment of equilibrium.

Despite complications due to the formation of defect apatites, under well defined supersaturation conditions, it is possible to grow pure DCPD, or OCP on seed crystals of these phases. The growth rate of DCPD both at 25° and 37°C follows a parabolic rate law with an activation energy of 44 kJ/mol (Marshall and Nancollas, 1969; Salimi, 1985). However, at 37°C in the range of $0.45 < \sigma < 1.05$, growth of OCP crystals has an effective order of 4, suggesting a polynucleation mechanism (Salimi, 1985; Heughebaert and Nancollas, 1984). At higher driving forces $(1.11 < \sigma < 1.69)$, the rate of reaction is proportional to the supersaturation and is sensitive to stirring speed. Thus it appears that the reaction at higher driving forces may be diffusion controlled (Salimi, 1985). Moreover, it has also been found that at very low driving forces $(\sigma < 0.6)$, OCP growth kinetics can be interpreted in terms of an integration controlled spiral growth mechanism (Lundager Madsen, 1982; 1987). In the light of these results, three supersaturation regions may be identified for OCP growth (1) at very low supersaturation, the reaction follows a spiral dislocation mechanism, (2), at intermediate supersaturation, a polynuclear mechanism dominates the rate while (3), at even higher driving forces the surface reaction becomes so fast that the transport of ions from bulk solution to the surface may become rate determining. It is interesting that even simple ions such as magnesium significantly influence the growth rate of OCP, following a Langmuir type adsorption isotherm. However, this added ion has little influence on the growth of DCPD (Salimi et al., 1985).

While HAP dissolution follows the polynucleation mechanism (Christoffersen, 1980), dissolution of DCPD and OCP can be described by a combined volume diffusion and surface reaction mechanism, with the relative importance of these processes depending upon the pH and hydrodynamics (Nancollas and Marshall, 1971; Christoffersen and Christoffersen, 1988; Zhang, 1990). For both growth and dissolution of calcium phosphates, the rates show a marked dependence on pH at a constant driving force. Moreover, it is especially interesting that both the crystallization and dissolution rates of DCPD have minimum values at about pH = 5 (Salimi, 1985; Zhang, 1990).

Alkaline earth fluoride

The alkaline metal fluorides are becoming increasingly important since fluoride ions are introduced into the environment as industrial waste in the form of fluorite and fluorapatite. The possibility of alkaline earth fluoride precipitation, therefore, may be of considerable importance in many environmental systems. The rates of crystallization and dissolution have been subjected to extensive studies and some recent results are summarized in Tables 2 and 3.

It can be seen from Table 2 that the growth of Group II fluoride salts follows a parabolic rate law at low supersaturation consistent with a spiral mechanism (Yoshikawa et

392

al., 1984) while at higher driving forces, the effective orders are considerably greater than 2, indicating the participation of surface nucleation. All the growth rates are found to be insensitive to stirring speeds. The activation energy for CaF_2 growth is 61 kJ/mol (Gardner and Nancollas, 1976), also indicating a surface controlled process.

As shown in Table 3, the dissolution of MgF_2, CaF_2 and SrF_2 may also be surface controlled with activation energies of 86, 73, and 72 kJ/mol, respectively (Hamza and Nancollas, 1985; Gardner and Nancollas, 1982; Abdul-Rahman, 1988). For the dissolution of MgF_2, two independent studies have yielded n values of 2 and 3.5 in almost the same undersaturation ranges (Yoshikawa and Nancollas, 1983; Abdul-Rahman, 1988). The former indicates a detachment controlled mechanism while the latter suggests the involvement of polynucleation. This difference may be due to a lack of screw dislocation sources in the seed crystals used in the latter study and this suggestion is further supported by the significantly smaller experimental reaction rates.

Although an early free drift study of CaF_2 dissolution suggested a linear dependence on the relative undersaturation, the activation energy, 73 kJ/mol, appeared to be too high for a diffusion controlled reaction (Gardner and nancollas, 1976). More recently, Christoffersen et al (1988), using the CC method, found a dissolution order of about unity (0.8 ± 0.2) in the undersaturation range of $-0.8 < \sigma < -0.25$, but with a rate considerably less than that calculated for volume diffusion. It therefore follows that the dissolution of CaF_2 is probably surface controlled.

It is interesting to note that the dissolution of SrF_2 follows a parabolic rate law at lower undersaturation ($-0.3 < \sigma < -0.1$) while the reaction order changes to 3.8 at higher driving forces ($-0.93 < \sigma < -0.5$) (Abdul-Rahman, 1988). This suggests that the reaction is probably controlled by detachment of ions from kink sites. However, the possibility of involvement of surface nucleation at higher driving forces cannot be ruled out.

Calcium carbonate

Although numerous kinetics studies have been made of the dissolution of calcium carbonate using conventional kinetics methods, the section will only provide a brief description of the more recent CC investigations. Constant composition studies of the crystal growth of calcium carbonate (calcite) crystals have been made in the pH range 8.2-10 using a glass electrode to monitor the hydrogen ion activity (Kazmierczak et al., 1982). The rate of crystal growth follows a rate law parabolic in relative supersaturation with an activation energy of 39 kJ/mol and is insensitive to changes in stirring dynamics (Nancollas and Reddy, 1971). All these data point to a surface controlled mechanism and this conclusion has been confirmed by a subsequent CC investigation using both suspensions and rotating single crystal surfaces (Nancollas et al., 1981). More recently, Christoffersen and Christoffersen (1990) confirmed a spiral growth mechanism with cation integration as a possible rate controlling step.

A CC dissolution study of a natural calcite crystal mounted on a rotating disk indicates the involvement of both transport and surface control over a range of pH (Nancollas et al., 1983). At low pH, the dissolution process is controlled by the rate of hydration of carbon dioxide while at high pH, the rate is probably controlled by a surface process rather than by bulk diffusion.

ACKNOWLEDGMENT

We thank the National Institute of Health for grant DE03223 in support of the projects carried out in this laboratory.

REFERENCES

Abdul-Rahman, A. (1988) The Kinetics of Crystal Growth and Dissolution of Magnesium and Strontium Fluoride Salts. The Effect of Additives. Ph.D. Thesis, State University of New York at Buffalo.

Abdul-Rahman, A. and Nancollas, G. H. (1984) Crystal Growth of Strontium Fluoride from Aqueous Solution. J. Chem. Soc. Faraday Trans. 80, 217-224.

Albon, N. and Dunning, W. J. (1962) Growth of Sucrose Crystals: Determination of Edge Energy from the Effect of Added Impurity on Rate of Step Advance. Acta Crystallogr. 15, 474-476.

Barone, J. P., Nancollas, G. H. and Yoshikawa, Y. (1983) Crystal Growth as a Function of Surface Area. J. Crystal Growth 63, 91-96

Barone, J. P., Svrjeck, D. and Nancollas, G. H. (1983) The Crystal Growth of Barium Fluoride in Aqueous Solutions. J. Crystal Growth 62, 27-33.

Bennema, P. (1967) Analysis of Crystal growth Models for Slightly Supersaturated Solutions. J. Crystal Growth 1, 278-286.

Bennema, P. (1969) The Importance of Surface Diffusion For Crystal Growth from Solution. J. Crystal Growth 5, 29-43.

Bennema, P. and Gilmer, G. H. (1973) Kinetics of Crystal growth. In: Crystal Growth: An Introduction, ed., P. Hartman, North-Holland, Amsterdam, 263-327.

Betts, F., Blumenthal, N. C. and Posner, A. S. (1981) Bone Mineralization. J. Crystal Growth 53, 63-73.

Bliznakov. G. and Kirkova, E. (1957) Der Einfluß der Adsorption auf das Kristallwachstum. Zeits. Physik. Chem. 206, 271-280.

Brown, W. E., Smith, J. P., Lehr, J. R. and Frazier, A. W. (1962) Crystal Structure of Octacalcium Phosphate. Nature 196, 1048-1050.

Brown, W. E., Schroeder, L. W. and Ferris, J. S. (1979) Interlayering of Crystalline OCP and HAP. J. Phys. Chem. 83, 1385-1388.

Buckley, H. E. (1951) Crystal Growth, Wiley, New York.

Budevski, E., Staikov, G. and Bostanov, V. (1975) Form and Step Distance of Polygonized Growth Spiral. J. Crystal Growth 29, 316-320.

Budz, J. A., Lo Re, M., and Nancollas, G. H. (1988) The Influence of High- and Low-molecular-weight Inhibitors on Dissolution Kinetics of Hydroxyapatite and Human Enamel in Lactate Buffers: A Constant Composition Study. J. Dental Res. 67, 1493-1498.

Burton, J. J. (1977) Nucleation Theory. In: Statistical Mechanics, Part A: Equilibrium Techniques, ed., B. J. Berne, Plenum Press, New York, 195-234.

Burton, W. K. and Cabrera, N. (1949) Crystal Growth and Surface Structure. Faraday Disc. Chem. Soc. 5, 33-39.

Burton, W. K., Cabrera, N. and Frank, F. C. (1951) The Growth of Crystals and the Equilibrium Structure of Their Surfaces. Phil. Trans. Roy. Soc. London A243, 299-358.

Cabrera, N. and Levin, M. M. (1956) On the Dislocation Theory of Evaporation of Crystals. Phil. Mag. 1, 450-458.

Cabrera, N. and Vermilyea, D. A. (1958) The Growth of Crystal from Solution. In: Growth and Perfection of Crystals, eds., R. H. Doremus, B. W. Roberts and J. Turnbull, Wiley, New York, 393-408.

Cheng, V. K. W. and Coller, B. A. W. (1987) Monte Carlo Simulation Study on the Dissolution of a Train of Infinitely Straight Steps and of an Infinitely straight Crystal Edge. J. Crystal Growth 84, 436-454.

Cheng, V. K. W., Coller, B. A. W. and Powell, J. L. (1984) Kinetics and Simulation of Dissolution of Barium Sulphate. Faraday Disc. Chem. Soc., 77 (1984) 243-256.

Christoffersen, J. (1980) Kinetics of Dissolution of Calcium Hydroxyapatite. J. Crystal Growth 49, 29-44.

394

Christoffersen, J. and Christoffersen, M. R. (1988) The Kinetics of Crystal growth and Dissolution of Calcium Monohydrogen Phosphate Dihydrate. J. Crystal Growth 87, 51-61.

Christoffersen, J. and Christoffersen, M. R. (1990) Kinetics of Spiral Growth of Calcite Crystals. Determination of the Absolute Rate Constant. J. Crystal Growth 100, 203-211.

Christoffersen, J., Christoffersen, M. R., Christensen, S. B. and Nancollas, G. H. (1983) Kinetics of Dissolution of Calcium Hydroxyapatite. VI. The Effects of Adsorption of Methylene Diphosphate, Stannous Ions and Partly-Peptized Collegen. J. Crystal Growth 62, 254-264.

Christoffersen, J., Christoffersen, M. R., Kibalczyc, W. and Perdok, W. G. (1988) Kinetics of Dissolution and Growth of Calcium Fluoride and Effects of Phosphate. Acta Odontol Scand. 46, 325-336.

Davey, R. J. (1976) The Effect of Impurity Adsorption on the Kinetics of Crystal Growth from Solution. J. Crystal Growth 34, 109-119.

Davey, R. J. (1979) Control of Crystal Habit. In: Industrial Crystallization 78, eds., E. J. de Jong and S. J. Jancic, North-Holland, 169-183.

Davis, J. A. and Hayes, K. F. (1986) Geochemical Processes at Mineral Surface: An Overview. In: Geochemical Processes at Mineral Surfaces, eds., J. A. Davis and K. F. Hayes, American Chemical Society, 2-18.

Davies, C. W. (1962), Ion Association. Butterworths, London.

Davies, C. W. and Jones, A. L. (1955) The Precipitation of Silver Chloride from Aqueous Solutions. Part 2.—Kinetics of Growth of Seed Crystals. Trans. of Faraday Soc. 51, 812-817.

Ebrahimpour, A. (1990) Kinetics of Dissolution, Growth and Transformation of Calcium Salts and the Development of Spectroscopic and Dual Constant Composition Methods. Ph.D thesis, State University of New York at Buffalo, Ch.7.

Francis, M. D. and Webb, N. C. (1971) Hydroxyapatite Formation from Hydrated Calcium Monohydrogen Phosphate Precursor. Calcified Tissue. Res. 6, 335-342.

Frank, F. C. (1949) The Influence of Dislocations on Crystal Growth. Disc. Faraday Soc. 5, 48-54.

Frank, F. C. (1952) Crystal Growth and Dissolution. Advances in Physics 1, 91-109.

Frenkel, J. (1945) On the Surface Motion of Particles in Crystals and the Natural Roughness of Crystalline Faces. J. Phys. USSR. 9, 392-398.

Frenkel, J. (1946) Kinetic Theory of Liquids, Oxford Univ. Press, Oxford.

Gardner, G. L. and Nancollas, G. H. (1976) The Kinetics of Crystal growth and Dissolution of Calcium and Magnesium Fluoride. J. Dent. Res., 55, 342-352.

Giles, C. H., Smith, D. and Huitson, A. (1974) A General Treatment and Classification of the Solute Adsorption Isotherm. I. Theoretical. J. Colloid Interface Sci. 47, 755-765.

Guggenheim, E. A. (1935) The Specific Thermodynamic Properties of Aqueous Solutions of Strong Electrolytes. Phil. Mag. 19, 588-643.

Guggenheim, E. A. and Turgeon, J. C. (1955) Specific Interaction of Ions. Trans. Faraday Soc. 51, 747-761.

Hamza, S. M. and Nancollas, G. H. (1985) The Kinetics of Dissolution of Magnesium Fluoride in Aqueous Solution. Langmuir 1, 573-576.

Heughebaert, J. C. and Nancollas, G. H. (1984) The kinetics of Crystallization of Octacalcium Phosphate. J. Phys. Chem. 88, 2478-2481.

Heughebaert, J. C., Zawacki, S. J. and Nancollas, G. H. (1990) The Growth of Non-Stoichiometric Apatite from Aqueous Solution at 37°C. I. Methodology and Growth at pH = 7.4. J. Colloid Interface Sci. 135, 20-32.

Hillig, W. B. (1966) A Derivation of Classical Two-Dimensional Nucleation Kinetics and The Associated Crystal Growth Laws. Acta Metall. 14, 1868-1869.

I, Ting-po and Nancollas, G. H. (1972) EQUIL - A General Computational Method for the Calculation of Solution Equilibria. Anal. Chem. 44, 1940-1950.

Jacobsen, C. F. and Léonis, J. (1951) A Recording Auto-Titrator. Compt. rend trav. lab. Carlsberg Sér. chim. 27, 333-339.

Janssen-van Rosmalen, J., Bennema, P and Garside, J. (1975) The Influence of Volume Diffusion on Crystal Growth. J. Crystal Growth 26, 342-352.

Kazmierczak, T. K.(1978) Kinetics of Precipitation of Calcium Carbonate. Ph.D. Thesis, State University of New York at Buffalo.

Kazmierczak, T. K, Tomson, M. B. and Nancollas, G. H. (1982) The Crystal Growth of Calcium Carbonate. A controlled Composition Kinetics Study. J. Phys. Chem. 86, 103-107.

Langmuir, J. (1916) The Constitution and Fundamental Properties of Solids and Liquids. Part I. Solids. J. Am. Chem. Soc. 38, 2221-2295.

Lundager Madsen, H. E. (1982) Calcium Phosphate Crystallization. III. Overall Growth Kinetics of Tetracalcium Monohydrogen Phosphate. Acta Chem. Scand. A36, 239-249.

Lundager Madsen, H. E. (1987) Comments on "The Growth of Dicalcium Phosphate Dihydrate on Octacalcium Phosphate at 25°C by J. C. Heughebaert, J. F. De Rooij, and G. H. Nancollas". J. Crystal Growth 80, 450-452.

Markovic, M. and Komunjer, L. (1979) A New Method to Follow Crystal Growth by Coulter Counter. J. Crystal Growth 46, 701-705.

Marshall, R. W. and Nancollas, G. H. (1969) The Kinetics of Crystal Growth of Dicalcium Phosphate Dihydrate. J. Phys. Chem. 73, 3838-3944.

McDonald, J. E. (1962) Homogeneous Nucleation of Vapor Condensation. I. Thermodynamic Aspects. Am. J. Phys. 30, 870-877.

McDonald, J. E. (1963) Homogeneous Nucleation of Vapor Condensation. II. Kinetic Aspects. Am. J. Phys. 31, 31-41.

Mohan, M. S. and Nancollas, G. H. (1970) The Growth of Hydroxyapatite Crystals. Arch. Oral Biol. 15, 731-745.

Mullin, J. W. (1972) Crystallization. Butterworths, London.

Mutaftsschiev, B. (1977) Adsorption and Crystal Growth. In: Chemistry and Physics of Solid Surfaces, eds., R. Vanselow and S. Y. Tong, CRC Press, Cleveland, Ohio, 73-86.

Nancollas, G. H. (1973) The Crystal Growth of Sparingly Soluble Salts. Croat. Chem. Acta. 45 (1973) 225-231.

Nancollas, G. H. (1982) Phase Transformation During Precipitation of Calcium Salts. In: Biological Mineralization and Demineralization, ed., G. H. Nancollas, Springer-Verlag, New York, 79-99.

Nancollas, G. H. (1989) In Vitro Studies of Calcium Phosphate Crystallization. In: Biomineralization, eds., S. Mann and J. Webb and J. P. Williams, VCH Verlagsgesellschaft, Weinheim, 157-187.

Nancollas, G. H., Kazmierczak, T. F. and Schuttringer, E. (1981) A Controlled Composition Study of Calcium Carbonate Crystal Growth: The Influence of Scale Inhibitors. Corrosion 37, 76-81.

Nancollas, G. H. and Marshall, R. W. (1971) The Kinetics of Dissolution of Dicalcium Phosphate Dihydrate. J. Dent. Res. 50, 1268-1272.

Nancollas, G. H. and Reddy, M. M. (1971) Crystallization of Calcium Carbonate Part II. Calcite Growth Mechanisms. J. Colloid Interface Sci. 37, 824-830.

Nancollas, G. H., Sawada, K. and Schuttringer, E. (1983) Mineralization Reactions Involving Calcium Carbonates and Phosphates. In: Biomineralization and Biological Metal Accumulation, eds., P. Westbroek and E. W. de Jong, Reidel Publishing Co, Holland, 155-169.

Nancollas, G. H. and Zawacki, S. J. (1989) Calcium Phosphate Mineralization. Connective Tissue Research 21, 239-246.

Nernst, W. (1904), Theorie de Reaktionsgeschwindigkeit in Hetergenen Systemem. Z. Phys. Chem. 47, 52-55.

Nielsen, A. E. (1980) Transport Control in Crystal Growth from Solution. Croat. Chem. Acta 53, 255-279.

Nielsen, A. E. (1984) Electrolyte Crystal Growth Mechanisms, J. Crystal Growth 67, 289-310.

Nielsen, A. E. and Christoffersen, J. (1982), In: Biological Mineralization and Demineralization, ed., G. H. Nancollas, Springer-Verlag, New York, 37-77.

Noyes, A. A. and Whitney, W. R. (1897) Ueber die Auflösungsgeschwindigkeit von festen Stoffen in ihren eigenen Lösungen. Z. Phys. Chem. 23, 689-692.

Nývlt, J., Söhnel, O., Matuchová, M. and Broul, M. (1985) The Kinetics of Industrial Crystallization. Elsevier, Amsterdam.

Ohara, M. and Reid, R. C., Modeling Crystal Growth Rates from Solution, Prentice-Hall.

Pérez, L. (1989) The Kinetics of Crystallization, Flocculation and Phase Transformation of Some Alkaline-Earth Salts. Ph.D. Thesis, State University of New York at Buffalo.

Pitzer, K. S. (1973) Thermodynamics of Electrolytes. I. Theoretical Basis and General Equations. J. Phys. Chem. 77, 268-277.

Pitzer, K. S. (1986) Theoretical Considerations of Solubility with Emphasis on Mixed Aqueous Electrolytes. Pure Appl. Chem. 58, 1599-1610.

Pitzer, K. S. and Mayorga, G. (1973) Thermodynamics of Electrolytes. II. Activity and Osmatic Coefficients for Strong Electrolytes with One or Both Ions Univalent. J. Phys. Chem. 77, 2300-2308.

Salimi, M. H. (1985) The Kinetics of Growth of Calcium Phosphate. Ph.D thesis, State University of New York at Buffalo.

Salimi, M. H., Heughebaert, J. C. and Nancollas, G. H. (1985) The Crystal growth of Calcium Phosphates in the Presence of Magnesium Ions. Langmuir 1, 119-122.

Sangwal, K. (1989) On the Estimation of Surface Entropy Factor, Interfacial Tension, Dissolution Enthalpy and Metastable Zone-Width for Substances Crystallizing from Solution. J. Crystal Growth 97 (1989) 393-405.

Sears, G. W. (1958) Effect of Poisons on Crystal growth. J. Chem. Phys. 29, 1045-1048.

Shyu, L. J. and Nancollas, G. H. (1980) The Kinetics of Crystallization of Calcium Fluoride. A New Constant Composition Method. Croat. Chem. Acta 53, 281-289.

Stranski, I. N. (1928) Zur Theorie des Kristallwachstums. Z. Phys. Chem. 136, 259-278.

Svehla, G. (1978) Automatic Potentiometric Titration, Pergamon Press, New York. Ch.7

Tomson, M. B. and Nancollas, G. H. (1978) Mineralization Kinetics; A Constant Composition Approach. Science 200, 1059-1060.

Troost, S. (1968) Influence of Surface Active Agent on the Crystal Growth of Sodium Triphosphate Hexahydrate. J. Crystal Growth 3/4, 340-343.

van Oosterhout, G. W. and van Rosmalen, G. M. (1980) Analysis of Kinetic Experiments on Growth and Dissolution of Crystals in Suspension. J. Crystal Growth 48, 464-468.

Weeks, J. D. and Gilmer, G. H. (1979) Dynamics of Crystal Growth. Advan. Chem. Phys. 40, 157-228.

Whitnah, C. H. (1935) A Mechanically Operated Buret. Industrial and Engineering Chemistry (Analytical Edn.) 5, 352-354.

Yoshikawa, Y. and Nancollas, G. H. (1983) The Kinetics of Crystal Growth of Magnesium Fluoride. J. Crystal Growth 64, 222-228.

Yoshikawa, Y., Nancollas, G. H. and Barone, J. (1984) The Kinetics of Crystallization of Group II Fluoride Salts in Aqueous Solution. J. Crystal Growth 69, 357-361.

Zawacki, S. J., Heughebaert, J. C. and Nancollas, G. H. (1990) The Growth of Non-Stoichiometric Apatite from Aqueous Solution at 37°C. II: Effect of pH upon the Precipitated Phase. J. Colloid Interface. Sci. 135, 33-44.

Zhang, J. (1990) Theory of Crystal Growth. Kinetics of Dissolution and Transformation of Calcium Phosphates. Ph.D thesis, State University of New York at Buffalo.

Zhang, J. and Nancollas, G. H. (1990) Kink Density Along a Crystal Surface Step at Low Temperatures and under Non-Equilibrium Conditions. J. Crystal Growth, in press.

CHAPTER 10   W. H. CASEY & B. BUNKER

LEACHING OF MINERAL AND GLASS SURFACES DURING DISSOLUTION

INTRODUCTION

Dissolution is a complex process involving many distinct reaction steps and pathways. Different sites in a mixed-oxide mineral or glass structure separately react with the solution and simultaneously undergo hydration, hydrolysis, ion exchange, or condensation reactions. Each of these reactions affects, and is affected by, the chemistry and structure of the near-surface region of the solid. Furthermore, each of these parallel reactions implicitly involves the transport of reactive solutes to, and from, the reactive site.

Depending upon the relative rates of these reactions, a given material can dissolve uniformly, or can selectively react to form an extensive altered layer near the material-solution interface. As an example of an altered mineral surface, a transmission electron micrograph of labradorite feldspar after several hundred hours of reaction in 0.01 N HCl is shown in Figure 1. The mineral surface consists of a ~200 Ångstrom-thick layer which is dramatically depleted in aluminum, calcium, and sodium relative to the underlying feldspar. The leached layer is amorphous (no lattice fringes in electron diffraction), and is enriched in hydrogen and silicon relative to the unreacted labradorite.

In this chapter we provide a framework for understanding these leaching reactions in terms of the structure and chemistry of the solid and solution phases. A consistent theme in this chapter is that the corrosion of minerals and glasses can be understood at a simple level by treating the materials as inorganic polymers (e.g., Ray, 1978). Different reactive sites in the material are distinguishable from their function and bonding in the

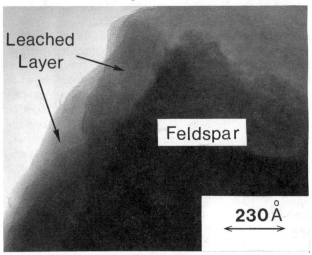

Leached Layer

Feldspar

230 Å

Figure 1. A high-resolution transmission electron photomicrograph of the surface of labradorite feldspar after leaching in a 0.01 N HCl solution for many hours. Photomicrograph by J. Banfield.

structure, and their reaction with the aqueous solution. Within this organizing theme, studies of glass leaching are very useful for unraveling mineral leaching reactions. Unlike minerals, small changes in glass composition cause continuous changes in the polymer structure which, if well characterized, can be directly related to the leaching behavior.

Few things are more fundamental to models of water-rock interaction than the reactive properties of mineral surfaces exposed to the aqueous solution. These models have taken on increased importance now that toxic materials are to be stored in geologic aquifers. The suitability of such storage is determined, in part, by predicting the chemical reactions which will take place between hypothetically released waste and rocks in the repository. The time scale for such predictions is suitably short (10-10,000 years) that disequilibrium pathways will dominate much of the important geochemistry.

If they are ever to become credible, these predictions must be based upon an understanding of the disequilibrium pathways by which toxic materials are immobilized. Leaching reactions are important to these pathways because they create a surface with chemical properties which are unique from the underlying mineral, and which can extend to depths of many unit cells. In many cases, these surfaces control the character of diagenesis.

THE STRUCTURE OF MIXED-OXIDE MINERALS AND GLASSES

The simplified structure of oxide minerals and glasses

There are some fundamental similarities between the structure of oxide glasses and minerals at a microscopic level. Both contain oxide groups(PO_4^{-3}, SiO_4^{-4}, AlO_4^{-5}, BO_4^{-5}) linked together into covalent polymers. The metal cations which form polymers are called network formers, and all tend to be found in tetrahedral or trigonal oxygen- coordination geometries (octahedral aluminum will be discussed later). Metal-oxygen bonds among oxide sites in these polymers have approximately a 50-60 percent covalent character.

These metal-oxide groups link to form a wide range of polymer structures. Forsterite, for example, contains only unconnected (Q^0) silicate tetrahedra with magnesium ions to balance the charge at nonbridging oxygens. The pyroxene chain contains only doubly connected (Q^2) tetrahedra (Fig. 2A); that is, there are no covalent crosslinks between the adjacent linear chains. Adjacent silicate chains in pyroxene are linked to one another via largely ionic bonds to alkali and alkaline-earth cations. Contrast this structure with amphibole (Fig. 2B), where every other tetrahedral unit has a bridging oxygen to an adjacent chain. Bridges between polymer chains are termed crosslinks, and are very important to the overall stability of an oxide material. The amphibole structure contains an equal mixture of Q^2 and Q^3 silicons, and an average crosslink density of 0.5.

The polymer notation becomes extremely useful when different oxides are compared. The structure of a hypothetical silicate glass, for example, is shown in Figure 3. The glass structure consists of linear silicate chains with a crosslink density of approximately 0.2; that is, the chains are bonded to each other at roughly one out of every five tetrahedra. Cesium and calcium ions are coordinated to nonbridging oxygens which have an unsatisfied, negative charge (this coordination may not be clear from the figure). Note that the same structural elements identified in minerals (non-bridging oxygens, bridging oxygens, crosslinks, etc.) are present in this oxide glass.

Metal cations which form ionic rather than covalent bonds to polymeric oxygens are called network modifiers. These network-modifying cations are commonly alkali (e.g., Na^+) or alkaline-earth (e.g., Ca^{+2}) cations. Bonds between these modifier cations and oxygen are ~80 percent ionic. Network-modifying cations perform two related functions in the structure of a glass or mineral: they coordinate to non-bridging oxygens branching from the polymer (as in Figs. 2 and 3) and they compensate the unbalanced charge stemming from isostructural substitutions in a polymer. In boro- and aluminosilicates, the unbalanced charge is caused by substitution of trivalent boron and aluminum for quatravalent silicon.

Network-modifying cations derive their name from their tendency to polymerize or depolymerize the structure of a glass. In other words, addition of modifier cations changes the Q-structure of the material. Sodium-borosilicate glass provides a useful example. The addition of Na_2O to borosilicate glass induces the following structural changes, depending upon the glass composition: (i) the concentration of non- bridging silicate sites increases if sodium enters the silicate network (Table 1); (ii) the concentration of trigonal boron sites decreases in favor of tetrahedral boron sites if sodium enters the borate network. The structural changes can be predicted from knowledge of the glass composition.

Reactive and unreactive sites in a structure

Mineral weathering rates vary dramatically with the mineral structure and composition. Those silicate materials which dissolve slowly tend to contain an extensively crosslinked (e.g., tectosilicate) structure (Goldich, 1938; Hurd, 1979). These materials also tend to be selectively leached by aqueous solutions, as there are sufficient unreactive bonds near the mineral surface to maintain integrity once the reactive constituents are removed (Murata, 1943; Petit et al., 1989). Conversely, silicates with a poorly connected fabric tend to dissolve rapidly and uniformly. These minerals contain few unreactive crosslinks between polymers.

Quantifying the relation between mineral structure and dissolution rate is difficult. Rates vary considerably with solution pH, as well as with mineral and solution compositions. Furthermore, most natural minerals are so impure that it is impossible to generate a truly homologous series of structures for study (e.g., Mg_2SiO_4, $MgSiO_3$, $Mg_3Si_2O_5(OH)_4$, etc). For these and other reasons, rates measured by different experimenters are commonly not comparable (see Fig. 7 in Bales and Morgan, 1985). The variation in rate with addition of silicate crosslinks to a mineral structure is poorly known.

The relation between bulk dissolution rates and structure is simple to illustrate with glasses. In borate glasses, the four-coordinated borons are much more resistant to hydrolysis than three-coordinated borons. The relationship of boron coordination to dissolution rates is illustrated for $Li_2 \cdot B_2O_3$ glass in Figure 4A. The rate initially decreases

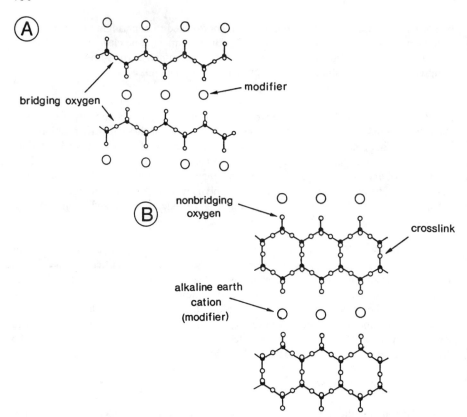

Figure 2. (A) Idealized structure of pyroxene. Filled circles correspond to silicon atoms and unfilled circles correspond to oxygens. This structure of linear tetrahedral chains is very common among nonsilicate oxide minerals and glasses. (B) Idealized structure of amphibole. In this case, the linear chains are bonded to one another via a covalent crosslink.

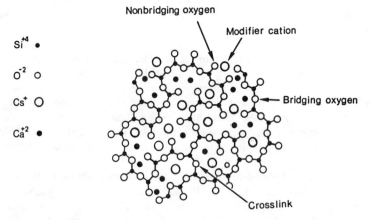

Figure 3. Idealized structure of a silicate glass. Note that the same structural elements in the silicate polymers shown in Figure 2 are also present in the glass. For our purposes, the main difference between crystalline phases and glass is that, for glasses, the repeated units in the polymer are arranged randomly.

Table 1: The mole-percentage of four- and three-coordinated silicon and boron in sodium borosilicate glass as a function of composition (Bunker et al., 1988). The site distributions are arranged in columns corresponding to the method of site determination (e.g., Raman spectroscopy).

[1]SILICON SITES

[2]Composition	Raman		^{29}Si-NMR		^{17}O-NMR	
	Q^3	Q^4	Q^3	Q^4	Q^3	Q^4
30·10·60	75	25	80	20	80	20
20·20·60	25	75	30	70	50	50
10·30·60	5	95	0	100	10	90

[3]BORON SITES

[2]Composition	Raman		^{11}B-NMR		^{17}O-NMR	
	N^3	N^4	N^3	N^4	N^3	N^4
30·10·60	25	75	20	80	0	100
20·20·60	35	65	25	75	50	50
10·30·60	70	30	65	35	75	25

[1]All silicon sites are coordinated to four oxygens. The Q-notation refers to the number of bridging oxygens attached to each tetrahedra.

[2]Glass compositions are reported as weight percent of the component oxides ($Na_2O \cdot B_2O_3 \cdot SiO_2$).

[3]There are either three (N^3) or four (N^4) oxygens coordinated to each boron.

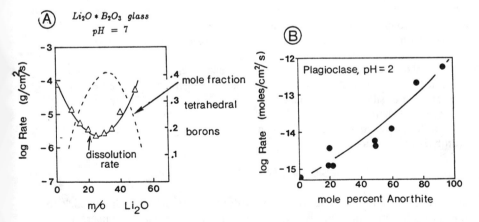

Figure 4. (A) Variation in the dissolution rate of B_2O_3-Li_2O glasses at 40°C and pH = 7.0. The minimum in dissolution rate at ~25 % Li_2O corresponds to the maximum fraction of tetrahedral boron sites (adapted from Velez et al., 1982 and Feller et al., 1982). The composition is given in mole percent of the oxides. (B) Variation in the dissolution rate of plagioclase at pH = 2 as a function of composition.

with increasing Li_2O content, reaches a minimum at approximately 25-30 mole percent Li_2O, and increases at higher Li_2O concentrations. This minimum rate at 25-30 mole percent Li_2O corresponds to the highest concentration of tetrahedrally coordinated boron groups. In this case, the most extensively crosslinked structure dissolves slowest.

A slightly different relation is observed for plagioclase (Fig. 4B). The dissolution rate in strong acid varies with the relative concentration of aluminate groups and silicate groups in the structure. In this case, the silicate groups are unreactive relative to aluminate groups. Thus the silicate-rich endmember (albite) dissolves most slowly. The relation between dissolution rate and composition is simpler than for $Li_2O \cdot B_2O_3$ glass because the plagioclase series exhibits only a binary solid solution. The variations in plagioclase composition are isostructural, unlike the borate glass where increases in Li_2O content changes the concentration of bridging bonds in the borate polymer.

The relation between material structure and leaching (as opposed to bulk dissolution rates) is not so straightforward. Several properties of a material structure play a role in determining the leaching characteristics: (i) the crosslink density of the structure, (ii) the transmissivity of that structure to solutes and water, and (iii) the reactivity of different structural sites near the mineral surface.

Consider the structures shown in Figure 5. The quartz structure (Fig. 5A) consists of completely linked silicate tetrahedra with very small structural pores. In the example shown in Figure 5A, we have arbitrarily added several aluminate groups to the quartz structure as tetrahedral substitutions. In a real mineral, these substitutions would be associated with monovalent interstitial ions to balance the charge.

The ratio of unreactive silicate groups to reactive aluminate groups is very large in this hypothetical mineral. One can imagine that the aluminate sites could be completely removed from the quartz without dramatically increasing the connected porosity of the mineral; that is, structural pores are small even after one or two of the tetrahedra are eliminated. Deep leaching of aluminum from quartz is impossible unless the crosslinked structure is made transmissive to solutes and water. Because the concentration of unreactive groups in quartz is so high, and pores are so small, leaching in quartz is limited to very near the mineral surface unless the surface is extensively fractured or very rough.

At the other structural extreme, deep leached layers are not observed on loosely connected molecular solids. As one can see in Figure 5B, forsterite has a high concentration of silicate groups, but a crosslink density of zero. The mineral consists of isolated silicate units separated by modifier cations (Mg^{+2} in this case). In an acid solution the silicate groups convert intact to silicic acid as the Mg-O bonds are protonated; no hydrolysis reaction is needed:

$$Mg_2SiO_4 + 4\,H^+ \rightarrow 2\,Mg^{+2} + H_4SiO_4^{\circ} \qquad (1)$$

Any leached layer on forsterite will be thin indeed as there are no bridging oxide bonds to maintain integrity once the alkaline-earth cations are removed. The mineral dissolves in a similar fashion as a molecular solid. It will not have an appreciable leached layer

403

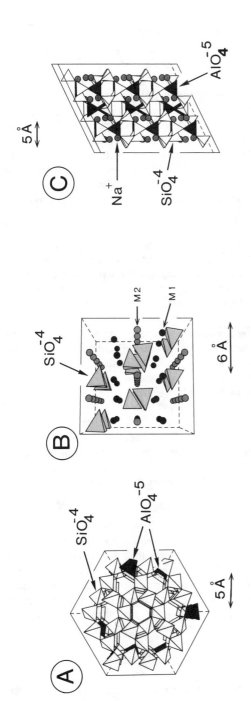

Figure 5. (A) The structure of quartz. Darkened polyhedral represent aluminate substitutions in the silicate framework. (B) The structure of forsterite. The polyhedra represent isolated (Q^0) silicate polyhedra. The circles represent magnesium ions in the M1 and M2 sites. (C) The structure of albite. Light and dark polyhedra represent silicate and aluminate groups respectively. The filled circles correspond to sodium ions.

unless nonbridging silanol ($\equiv$ Si-OH) groups condense upon dissolution into a framework of siloxane ($\equiv$ Si-O-Si$\equiv$) bridges.

Finally, examine the albite structure in Figure 5C. This structure contains exchangeable cations (sodium) along with two hydrolyzable sites in the oxide polymer (silicate and aluminate groups). The crosslink density of the material is as high as quartz. Unlike quartz, however, fully one third of the crosslinks in the albite structure are relatively reactive Al-O-Si bonds. Hydrolysis of these crosslinks significantly opens up the feldspar structure, allowing deep transport of solutes and water. (Imagine the albite structure in Figure 5C without the aluminate tetrahedra or the sodium atoms). The selective hydrolysis of aluminate groups proceeds to great depth in the mineral, causing a deep hydrogen- enriched zone (Fig. 1). Although the crosslink density is reduced by leaching, the residual silicate framework maintains structural integrity once other cations are removed.

Leaching reactions, particularly in acid solutions, have been well documented for phyllosilicate and tectosilicate minerals. Dealumination of zeolites by acid reaction, for example, is critical to production of hydrocarbon-cracking catalysts (Kerr, 1973), and has been studied for over a century. Selective leaching of phyllosilicate minerals has also been studied as a nonbauxitic source of aluminum (Aglietta et al., 1988).

Clay leaching is controlled by the mineral structure. As a class, clays consist of sheets of Q^3 silicon tetrahedra separated by layers of octahedral aluminum- or magnesium-hydroxide. Both the magnesium- hydroxide and aluminum-hydroxide sheets are unstable relative to the silicate structure in an acid solution. Thus clay leaching generally proceeds from the edges of the phyllosilicate inward (e.g., Gastuche, 1963; Newman and Brown, 1969), along the octahedral layer. Octahedrally coordinated aluminum ions are leached more rapidly in acids than aluminum in tetrahedral sites (Brindley and Youell, 1951; Man et al., 1990). Octahedral magnesium sheets can be preferentially dissolved from minerals in near-neutral and slightly basic pH solutions (e.g., Bales and Morgan, 1985; Lin and Clemency, 1981; Luce et al., 1972). In some cases this leaching produces a silica pseudomorph of the original grain (Pacco et al., 1976).

The effect of temperature

Rates of reactions which produce leached or hydrated layers on silicate minerals and glasses vary considerably with temperature. Extensive hydrated layers are observed on hydrothermally altered minerals which exhibit little or no alteration at lower temperatures (Petit et al., 1990). Because most natural minerals have experienced elevated temperatures during diagenesis, understanding reactions in leached layers is important to models of water-rock interaction, even at low temperature.

GENERAL REACTION MECHANISMS

The first step in understanding mineral or glass corrosion is to identify important sites in the structure, and to identify the reactivity of these sites with the aqueous solution.

The chemistry of the dissolved cations provides a useful guide to the reactivity of the different oxide sites near the leaching mineral surface. Those cations which exist as oxyacids in solution (e.g., $SiO_4^{-4} \rightarrow H_4SiO_4^{\circ}$) are removed from the mineral fabric by hydrolysis of the oxygen linkages between polyhedra. Cations which exist as solvated ions (e.g., Na^+) or simple ion-pairs in solution tend to be removed from the mineral by a coupled reaction involving ion-exchange for hydronium ions (to neutralize the charge) along with solvation (hydration).

It is useful to recognize three important classes of leaching reactions: hydration, ion-exchange, and hydrolysis. These classes are distinguished from one another based upon the extent to which water dissociates, the bonding character of the reactive site (ionic versus covalent), and whether or not mass is removed from the material.

Hydration

The effects of hydration range from simple penetration of water into a material to profound disruption of the structure. Factors which control the extent and rate of hydration include: (i) the size of structural pores in the material, (ii) the degree of covalent connectivity of the polymeric structure; and (iii) the hydration energies of cations ionically bonded to the structure. Most oxide glasses and minerals are not transmissive to water, therefore hydration does not extend to great depth without parallel reactions that relax the material structure.

Materials which are dramatically affected by hydration tend to be held together by ionic forces between polymers; that is, there are few covalent crosslinks. The ionic forces linking the structure together are diminished as ions between the polymers become solvated. In extreme cases, this hydration causes a material to dissolve into polymeric fragments, as is the case for hydration of metaphosphate glasses (Bunker et al., 1987). More commonly, hydration simply causes the polymeric structure to relax, such as in the case of swelling vermiculite.

In order to create a reacted layer by hydration, water must able to penetrate into the structure. To a first approximation, the size of structural pores in a solid controls the transmissivity of the material to molecular water. The relation between pore size and diffusivity can be illustrated by examining the transport of water in zeolites (Table 2). In zeolites with pores which are large (~7 Å) relative to a water molecule (2.8 Å), water diffusion is similar to the corresponding process in an aqueous solution;that is $D \sim 1 \times 10^{-5}$ cm^2/s and $E_a \sim 5$ kcal/mole where D is the diffusion coefficient and E_a is an activation energy. Continuous clusters of water exist in these large pores which have properties similar to bulk water (Rennie and Clifford, 1977).

Water diffusion is relatively slow ($D \sim 10^{-8}$ to 10^{-13} cm^2/s) for zeolites with pore spaces similar in size (2-6 Å) to the water molecule. Diffusion in these materials requires profound relaxation of the structure. Thus, activation energies for diffusion are intermediate to those observed for aqueous diffusion (~5-6 kcal/mole) and complete hydrolysis of the structure (8-20 kcal/mole).

406

Table 2: Tracer diffusion coefficients and activation energies for water in zeolites (Barrer, 1978) where: $D = D^{\circ}exp(-E_a/RT)$.

Zeolite	Pore Radius	D	D°	E_a	T C°
	Ångstroms	(cm²/s)	(cm²/s)	(kcal/mole)	
Analcime	2.2-2.4	$2.0x10^{-13}$	$1.5x10^{-1}$	17.0	46
Heulandite	2.4-7.8	$2.1x10^{-8}$	$7.6x10^{-1}$	11.0	45
Chabazite	3.7-4.2	$1.3x10^{-7}$	$1.2x10^{-1}$	8.7	45
Gmelinite	3.4-6.9	$5.8x10^{-8}$	$2.0x10^{-2}$	8.1	45
Na-X	~7.4	$2.1x10^{-5}$	-------	6.9	40
Ca-X	~7.4	$2.4x10^{-5}$	-------	6.8	40
Water	2.8	$3.9x10^{-5}$	$5.6x10^{-2}$	4.6	45

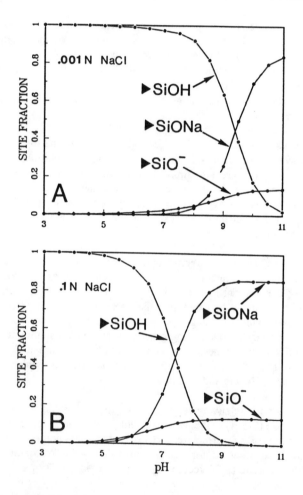

Figure 6. The calculated distribution of surface species on silica in 0.001 (A) and 0.1 (B) N NaCl solution. The distribution was calculated using the intrinsic equilibrium constants of Kent et al. (1986).

The material structure can be relaxed either by hydrolysis of the polymer network, or by the hydration and disruption of ionic bonds between polymers. Generally, penetration of water into tectosilicates requires complete disruption of at least some bridging bonds in the polymeric structure; that is, water diffusion is accompanied by hydrolysis. The diffusion coefficient for water in silica glass, for example, is less than 10^{-13} cm^2/s at 200°C, and diffusion has an activation energy of 10-20 kcal/mole (Wakabayashi and Tomozawa; 1989).

Ion-exchange reactions

The leached surface is commonly defined by replacement of modifier cations with hydrogen ions. Because the most-studied materials are simple glasses and minerals, the replacement generally involves alkali or alkaline-earth cations. Leaching of sodium silicate glass in acid, for example, involves exchange of hydronium at non-bridging silicate sites:

$$\geqslant Si-ONa + H_3O^+ = \geqslant Si-OH + Na^+ + H_2O \tag{2}$$

This exchange is rapid as long as solute transport to, and from, the site is unimpeded. Exchange reactions in this chapter are generally written to involve hydronium rather than hydrogen ion because the leaching reactions which concern us proceed after the mineral surface is extensively hydrated. There are cases, however, where ion exchange may involve hydrogen ions or the water molecule alone (e.g., Smets and Tholen, 1983).

Modeling these reactions is a nasty business. Most oxide materials contain several potential sites for ion exchange, and the relative abundance of these sites vary during dissolution as the structure of the mineral or glass is modified. Inferences about the exchange properties of structural groups must often be inferred from the properties of simple materials, or from reactions in solution.

A reaction similar to (2), for example, has been studied on silica (Kent et al., 1986). There is good reason to suspect that ion-exchange reactions actually affect the non-bridging sites, thus in this case, there is consistency between the modeling formalism and what actually proceeds on the surface. A calculated distribution of surface species is shown in Figure 6. Sodium (and other alkali ions, Dugger et al., 1964) completely desorbs from the nonbridging oxygens as the solution pH decreases from 11 to 7. This desorption virtually coincides with dissociation of the surface-hydroxyl group. The sodium complex ($\geqslant Si-ONa$) appears at pH-conditions where the neutral silanol ($\geqslant Si-OH$) group has dissociated to produce a negatively charged site: ($\geqslant Si-O^-$). Drawing an analogy between silica (Fig. 6) and the nonbridging sites in a leaching glass, one expects that alkali ions will desorb from these functional groups in dilute, near-neutral and acidic solutions.

The solution chemistry can also be used as a guide. For example, because bonding of alkali ions to nonbridging oxygen sites is largely electrostatic, ion exchange mimics the acid-base behavior of that nonbridging site. Silicic acid (H_4SiO_4), for example, has a dissociation constant of $\sim 10^{-10}$ and is fully protonated in near-neutral pH solutions. At pH > 9 silicic acid is partly dissociated. If the surface site behaves in a similar manner as the dissolved acid, one expects to find sodium-silicate surface complexes only at very basic conditions (e.g., Fig. 6). At these conditions, the sodium can electrostatically bond to a negatively charged, nonbridging site. Conversely, H_3PO_4 is a relatively strong acid, with a first dissociation constant of ~10^{-2}. One expects the nonbridging surface sites to be partly dissociated at relatively low pH. It is not surprising then that sodium is retained by nonbridging phosphate sites in a glass at much more acidic conditions than at the silicate sites (e.g., Bunker et al., 1983).

Some exchange reactions involve sites which have no dissolved analogue. Aluminosilicate and borosilicate glasses, for example, contain exchange sites at bridging and nonbridging oxygens. Reaction at the bridging sites commonly controls the overall exchange behavior of the solid (Isard, 1967), and yet cannot be studied in a simple material or by analogy with dissolved oxyacids. The leaching of sodium from an aluminosilicate glass, for example, proceeds via two potential pathways.

The first pathway is simple exchange:

$$\equiv \overset{\overset{\text{Na}}{\vdots}}{\text{Si–O–Al}} + H_3O^+ \ = \ \equiv \overset{\overset{\text{H}_3\text{O}}{\vdots}}{\text{Si–O–Al}} + Na^+ \qquad (3)$$

A second pathway requires that the bridging oxygen bond hydrolyzes simultaneous with exchange. The reaction is therefore irreversible, and the coordination of aluminum at the surface changes along with leaching of sodium (Barrer and Klinowski, 1975):

$$\equiv \overset{\overset{\text{Na}}{\vdots}}{\text{Si–O–Al}} + H_3O^+ = \equiv \text{Si–OH} + \overset{|}{\underset{\diagdown}{\text{Al}}} + Na^+ + H_2O \qquad (4)$$

The trigonal aluminum is unstable and rapidly hydrates to hexacoordination.

The relative importance of these two paths to alkali leaching in glasses is a subject of intense study (e.g., Smets and Lommen, 1982), but is unresolved. The two paths predict different pH-dependencies of alkali leaching. Ion-exchange coupled to hydrolysis of tetrahedral borate and aluminate groups proceeds only in acid solutions (pH < 3-5). Ion exchange without hydrolysis (Equation 3) proceeds at neutral and slightly basic pH conditions (Isard, 1967).

<u>Hydrolysis reactions</u>

The most important class of leaching reaction are those which create or destroy covalent metal-oxygen or metal-hydroxide bonds in the polymeric fabric. These are termed hydrolysis reactions since water typically dissociates to form hydroxyl groups:

$$\equiv\!M\!-\!O\!-\!M\!\!\leqq + H_2O = \equiv\!M\!-\!OH + HO\!-\!M\!\!\leqq \qquad (5)$$

In this reaction, M refers to a network-forming cation.

Rates of hydrolysis depend strongly on the character and structure of metal-oxygen bonds near the mineral surface. Consider the example of Group IIA metal oxides. In the bulk solid of BeO, oxygens are tetrahedrally coordinated to four metal atoms, and this mineral dissolves at a rate of $\sim\!10^{-15}$ moles/cm^2/s at pH = 3 (Furrer and Stumm, 1986). Magnesium and oxygen are octahedrally coordinated in MgO, which dissolves at a rate of $\sim\!10^{-8}$ moles/cm^2/s at pH = 3 (Vermilyea, 1969). The difference spans seven orders of magnitude!

In simple solutions, variations in hydrolysis rates with pH is controlled by the acid-base properties of bridging oxygens or hydroxyl groups at the mineral surface (see Wendt, 1973). These acid-base reactions are synonymous with the sorption of hydroxyl or hydrogen ions onto the oxide surface. These sorption reactions imbue the material with a net surface charge, which can be measured in a potentiometric titration or through studies of electrophoretic mobility (e.g., Furrer and Stumm, 1986). The variation in surface-charge concentration with solution pH is characteristic for a given oxide surface, just as the extent of protonation is characteristic of a dissolved oxyacid at a given pH. While the present discussion concerns acid-base chemistry, other ligands also profoundly change the rate and character of hydrolysis (Schindler and Stumm, 1987).

These observations can be related to hydrolysis kinetics by examining simple mechanisms. As a starting point we distinguish between S_N2 (Substitution-Nucleophilic bimolecular) and S_N1 (Substitution-Nucleophilic unimolecular) mechanisms of ligand exchange (see Basolo and Pearson, 1958). In ligand exchange, as in hydrolysis, the number of ligands coordinated to the metal site is typically similar before and after reaction.

A simple S_N1 hydrolysis mechanism is the acid/base-catalyzed depolymerization of M-O-M bonds (see Wendt, 1973). The mechanism proceeds by dissociation of the M-O bond followed by insertion of a water molecule. For linked, octahedrally coordinated metal sites:

$$\text{fast} \qquad (6)$$

$$\text{slow} \qquad (7)$$

$$\text{fast} \qquad (8)$$

In reactions (7) through (8), the $\oplus$ sign indicates a site which can accept a ligand; that is, the metal is coordinatively unsaturated.

The rate-controlling step, regardless of solution pH, is dissociation of the M-O bond to form a coordinatively undersaturated (five-coordinated in this case) metal site. The coordination sphere around the metal site becomes resaturated with ligands (increasing the coordination number from five to six in this case) by insertion of a water molecule after dissociation. Similar reactions could be written involving adsorbed hydroxyl rather than hydrogen ions.

The rapid acid-base reactions (Eqn. 6) control the rate of bond dissociation and account for the variation in hydrolysis rates with pH (Wendt, 1973). These reactions modify the distribution of electron charge in the M-O bond (Kawakami et al., 1984). Metal-oxygen bonds with considerable ionic character tend to dissociate more rapidly than strictly covalent bonds. Rates of exchange have been determined for some dissolved species (Eigen, 1962). For example, the rate of exchange of water between the solvent and the hydration sphere around an aquated Be^{+2} ion is three orders of magnitude slower than for the less covalent Mg^{+2} ion (Eigen, 1962). There is generally a profound consistency between the kinetics of ligand exchange of hydrolyzed solutes and the dissolution kinetics of oxide minerals.

In contrast, S_N2 mechanisms of hydrolysis proceed by association of ligands to metal sites, thereby increasing the coordination number of the site in an intermediate step. Consider the following depolymerization of a tetrahedrally coordinated metal site by water:

$$\equiv M-O-M\lessgtr + H_2O \rightarrow \equiv M-O-\overset{H_2O}{\underset{\diagdown}{\overset{|}{M}}}\lessgtr \rightarrow \equiv M-OH + HO-M\lessgtr \qquad (9)$$

The slow step in the reaction is formation of the five-coordinated intermediate from the usual, four-coordinated site.

These S_N2 reactions can involve any species which has sufficient electron density (i.e., is adequately nucleophilic) to intrude into the inner coordination sphere of the metal site and bond. Rates of these mechanisms vary with the strength of the nucleophile, and the hydroxyl ion is more nucleophilic than water. Therefore S_N2 reactions proceed rapidly in solutions where the hydroxide ion concentrations are appreciable. S_N2 mechanisms are also sensitive to the geometry of the association. Such nucleophilic hydrolysis is easier at cations which are undersaturated with ligands, such as trigonal boron. (Tetrahedral boron is coordinatively saturated.)

The primary tool for determining the acid-base properties of an oxide surface (titration for surface charge) is sensitive only to the net properties. One cannot, for example, unequivocally discriminate between hydrogen sorbed at the aluminum-, silicon-, and beryllium-oxygen sites in beryl ($Be_3Al_2Si_6O_{18}$). Furthermore, in most cases surface charge is attributed only to nonbridging hydroxyl sites, e.g., $\geq$ SiOH, $\geq$ SiO⁻, $\geq$ SiOH₂⁺,

while acid-base reactions also affect oxygens at highly coordinated sites. In fact, these reactions are most important, since bond cleavage between metal atoms is the very essence of dissolution.

It is worth reconsidering the simple oxides BeO and MgO to illustrate this point (see Boehm and Knozinger, 1983). BeO has a wurtzite structure with four beryllium atoms coordinated to each oxygen. The truncation of the tetrahedra results in acid-base sites coordinated to one, two and three beryllium atoms:

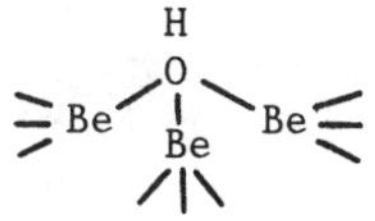

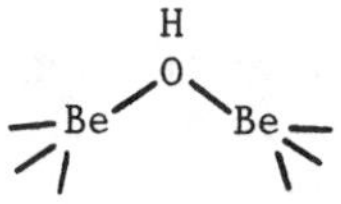

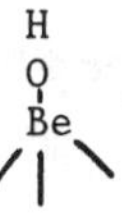

The distribution of these sites is not constant among the crystal faces of BeO. The sites also have different affinities for hydrogen ions, and can be resolved from one another spectroscopically (Tsyganenko and Filimonov, 1973).

Periclase (MgO) has the NaCl structure with oxygens octahedrally surrounded by six magnesium atoms. For this structure, the following three sets of hydroxyl groups are present, depending upon the number of Mg-O bonds (one, two or three) intersected by a cleavage plane:

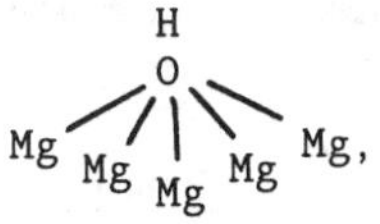

(Set #1)

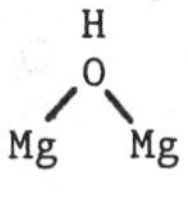

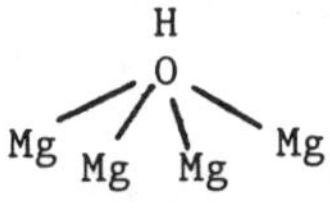

(Set #2)

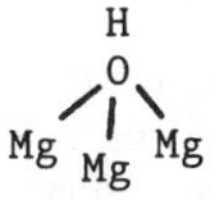

(Set #3).

Bonds from magnesium ions to oxygens within the mineral are not shown.

The acid-base chemistry becomes even more complicated if the solid contains a cation vacancy or a cation in more than one oxygen-coordination geometry. Consider the example of gamma-alumina, which has the following distinct hydroxyl groups on the (111) face (Knozinger and Ratnasamy, 1978):

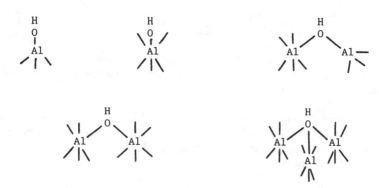

The acidity of a metal-hydroxide site varies considerably with the number of metal atoms coordinated to the oxygen. Hiemstra et al., (1989), for example, calculate that the deprotonation constant for an Al_2OH site is over eleven orders of magnitude larger than a simple AlOH site. Furthermore, the above acid-base sites on MgO, BeO and Al_2O_3 above were identified by considering only the crystallography. In aqueous solutions, the population of sites varies with the solution composition via rapid condensation reactions.

In general, hydrolysis rates vary with surface charge because the rate-enhancing acid-base reactions vary with pH in a similar fashion as those which account for surface charge. By analogy with the hydrolysis of dissolved cations, even simple oxide surfaces contain a rich acid-base chemistry. The pK's for protonation of oxygens on lead-hydroxide complexes, for example, span the entire measurable pH range from 0 to 14 (Baes and Mesmer, 1976; p. 364). The acid-base reactions on a mineral surface are no less complex.

The pH-dependence of leaching rates

As discussed above, the pH-dependence of hydrolysis is controlled by the acid-base properties and bonding in the M-O bond, and the mechanism of hydrolysis (see Wendt, 1973). Dissolution reactions proceed rapidly when the M-O bond is extensively coordinated to hydrogen or hydroxyl ions, and relatively slowly when the site is neutral. Thus leaching rates of network-forming cations tend to be rapid in strong acids and bases, and exhibit a characteristic minimum at more neutral pH-conditions.

To understand the pH-dependence of dissolution, one needs to: (i) identify the sites of acid-base reactions at a mineral surface, and (ii) estimate equilibrium constants for the reactions important to dissolution. For a complicated mineral, acid-base reactions at bridging oxygens control the pH-dependence of dissolution. As yet, however, there is no satisfactory method of predicting equilibrium constants for these reactions. Most workers use the relation between surface charge and pH on simple oxides (e.g., SiO_2, BeO, Al_2O_3) as a guide to acid-base reactions on a more complicated mineral surface. While this approach is somewhat successful (e.g., Bales and Morgan, 1985; Carroll-Webb and Walther, 1988; Guy and Schott, 1989), it is inadequate because not all acid-base sites on a complicated phase are represented in the simple mineral. Furthermore, the acid-base reactions which are important to dissolution do not necessarily correlate well with those which account for the surface charge.

Hiemstra et al., (1989) provide a method of predicting acid-base equilibrium constants at individual acid-base sites on a mineral surface. The method is based upon work by Kassiakoff and Harker (1938) and Ricci (1948). In this method, acid dissociation constants are separated into electrostatic and chemical terms. Electrostatic terms are calculated from the mineral crystallography, and by considering the formal charge on an exposed oxygen. The formal charge is defined as the valence of the oxygen (-2) plus the sum of the electrostatic bond strengths to that oxygen. Thus the formal charge on oxygens in the following sites:

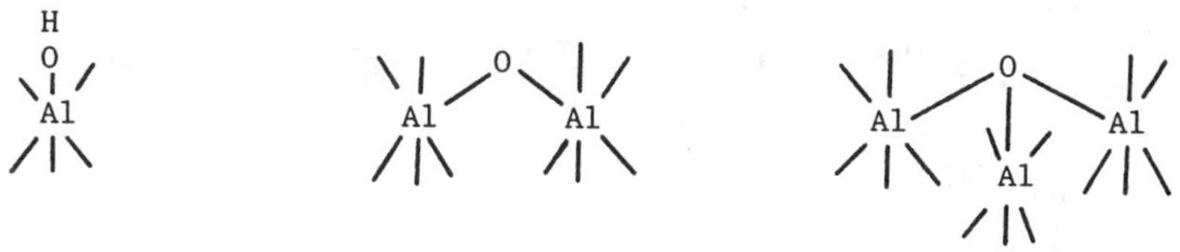

is -0.5, -1.0 and -0.5, respectively (Pauling, 1970). The chemical terms are estimated from the strengths of dissolved oxyacids and from knowledge of acid-base reactions on gibbsite, which has a particularly simple surface chemistry. With this empirical correlation, Hiemstra et al., (1989) reports pK's for protonation of the above sites of 10.0, 12.3, and -1.5, respectively.

The method, although promising, has several disadvantages. First, it can only be applied to simple oxide minerals (e.g., Hiemstra and van Riemsdijk, 1990), and not to the complicated linkages in minerals such as plagioclase, which include disparate cations

$$\text{e.g.,} \quad \geq\!\text{Al-O-Si}\!\leq$$

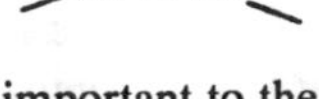

It is exactly these linkages which are important to the leaching behavior of a complicated mineral or glass. Secondly, the model relies upon crystallographic bond lengths while the surface of many glasses and minerals may be disordered. In this case, crystallography provides a poor guide to the atomic surface structure.

Insight into the pH-dependence of leaching can also be acquired from the chemistry of dissolved species. This approach works well with some conspicuous exceptions, such as silica. As discussed earlier, the minimum in dissolution rate commonly corresponds to that pH where there is little excess sorption of hydrogen or hydroxyls ions to the mineral surface. For many minerals, this condition is met at the isoelectric point.

Parks (1965) noted that isoelectric point also corresponds to pH where there are equal concentrations of positively and negatively charged metal complexes in solution. In other words, the isoelectric point of a oxide also commonly corresponds to the pH of minimum solubility. Quick examination of Baes and Mesmer (1976) shows that dissolved, polymeric complexes also become relatively abundant at this pH, at least in solutions with high total metal concentrations. Thus, the conditions which lead to polymerization of dissolved complexes also cause slow dissolution rates. (Polymerization can be the opposite of dissolution).

Rates of boron leaching from alkali-borate glass, for example, are slowest in the pH range 6-10, and increase dramatically at pH > 10 and at pH < 6. The dominant dissolved boron complex at pH < 9 is boric acid $(B(OH)_3^\circ)$ and at pH > 9, the dominant complex is the borate anion $(B(OH)_4^-)$. Dissolved polyborate species are not observed in either strong acids or at pH > 11.

Polyborate species are measurable, however, in the pH range 6 < pH < 11 and are relatively abundant at pH = 9. At these conditions there also exists a mixture of $B(OH)_3^\circ$ and $B(OH)_4^-$ complexes. The polyborate species form via an oxide bridge between the monomeric complexes:

$$B(OH)_3^\circ + B(OH)_4^- \rightarrow (HO)_2B-O-B(OH)_3^- + H_2O \qquad (10)$$

and ultimately forms dissolved ring structures.

The correlation between solution chemistry and leaching kinetics is good for borates because the dissolved borate species, including the polyborates, are similar to the structures in the glass. To ensure accuracy: (i) the acid-base chemistry of dissolved hydroxide complexes must resemble the hydrolyzable oxide bond in a solid structure; and (ii) other catalytic ligands are not present.

EXAMPLES

In previous sections we argued that modification of mineral and glass surfaces can be understood through stepwise analysis of the material structure and chemistry. In this section we apply this analysis to several simple examples. The first such example is chosen to be intentionally unfamiliar to geochemists.

Phosphate oxynitride glass

Phosphorus-oxynitride glass consists of linear phosphate chains linked to one another either by: (i) ionic bonds to modifier cations, or (ii) nitride anion linkages (Fig. 7). Thus there are three potential reaction sites: one ionic crosslink at nonbridging sites, and two covalent bridges

$$\geq P-O-P \leq \quad \text{and} \quad \geq P-N=P \leq$$

between phosphorus atoms (Fig. 7). The anhydrous material is impervious to water, but the transmissivity of the glass greatly increases with hydration.

Information about the relative rates of P-O-P hydrolysis versus hydration of the

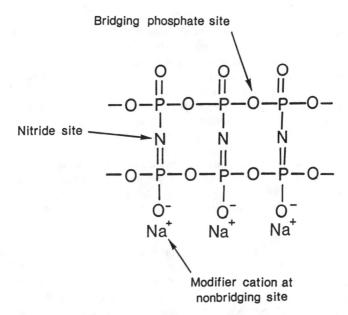

Bridging phosphate site

Nitride site

Modifier cation at
nonbridging site

Figure 7. The idealized structure of sodium-phosphate-oxynitride glass.

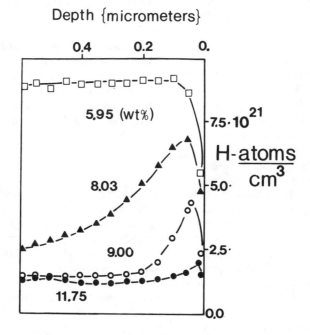

Depth {micrometers}

5.95 (wt%)

8.03

9.00

11.75

7.5·10²¹

H-atoms/cm³

5.0·

2.5·

0.0

Figure 8. Hydrogen depth profiles in sodium-phosphate-oxynitride glasses exposed to humid air (T = 30°C, 80% relative humidity, t = 1 hour) as a function of weight percent nitrogen content. Profiles show hydrogen concentration (atoms/cm³) as a function of depth from the glass interface. Background concentration of hydrogen in the glass is 1.5x10²⁰ H-atoms/cm³. The curves are labeled with respect to weight percent nitrogen.

modifier cation is derived from studies of nitrogen- free glass. In alkali-metaphosphate glass, which consist of linear PO_4^{-3} chains, the dominant dissolution mechanism is hydration and weakening of the ionic bond between chains. The hydration reaction is so rapid that the strength of metaphosphate glass virtually disappears upon exposure to humid air. Bridging-oxide sites within a phosphate polymer are nearly as inert to hydrolysis in near-neutral pH solutions as siloxane groups in silica (Crowther and Westman, 1956). Thus, phosphate chains tend to dissolve as intact polymeric fragments in an aqueous solution.

Oxynitride glass, however, contains a covalent nitride crosslink between polymers (Fig. 7). These nitride sites hydrolyze very slowly to ammonia, via a complicated reaction path (see Bunker et al., 1987). Therefore one expects that nitrided glasses would not be as susceptible to weakening by hydration as alkali phosphate glasses which contain no such covalent crosslink.

Hydrogen-depth profiles are shown in Figure 8 as a function of nitrogen concentration in nitrided sodium metaphosphate glass. These particular profiles were collecting using Elastic Recoil Detection Analysis (see Petit et al., 1990). All glasses were reacted for one hour under identical conditions, thus the amount of hydrogen in the glass (Fig. 8) reflects the rate of corrosion. As expected, rates of hydrogen penetration decrease with nitrogen content in the glass. Water has saturated the glass containing 5.95 percent nitrogen to a depth of greater than 0.6 micrometers. The glasses containing 8 and 9 percent nitrogen, however, show greatly reduced penetration by hydrogen. The only hydrogen in the glass containing 11.75 percent nitrogen corresponds to the bulk value in the unreacted glass ($\sim 1.5 \times 10^{21}$ H-atoms/cm^3).

In this example, alteration of the material surface was retarded by adding stable nitride crosslinks to the polymer structure. The addition of crosslinks to a polymer network can sometimes make it more, not less, reactive. Metaphosphate chains (each PO_4^{-3} unit bonded to two neighbors) are very stable relative to polymers where PO_4^{-3} units bond to three neighbors. The third phosphate cross-link is very reactive to water, as in the case of phosphorus-pentoxide (P_2O_5) which consists solely of triply-connected PO_4^{-3} tetrahedra and is a commercial desiccant.

Plagioclase

Although it is inherently unsatisfying to infer acid-base properties of specific sites on a complicated mineral surface from the simple oxides, sometimes this approach is unavoidable. There are no dissolved aluminosilicate complexes which can be used to unravel the acid-base chemistry of the Al-O-Si site in plagioclase. From studies of aluminosilicate minerals and gels, however, it is clear that the tetrahedral aluminum has less affinity for hydrogen ions than octahedral aluminum (Parks, 1967). The isoelectric point of tetrahedrally coordinated aluminum in a hypothetical Al_2O_3 is ~ 4.7, which is at much more acid conditions than octahedrally coordinated aluminum oxide (Parks, 1967).

With this reasoning, one suspects that tetrahedral aluminum-oxide sites on plagioclase may be protonated in moderately low pH-solutions. By analogy with SiO_2, the silicon-oxide sites in plagioclase will not be protonated until very acid conditions. The silica surface is not positively charged until the solution pH drops to much lower than 2.

This qualitative reasoning leads to predictions which are consistent with the observed leaching behavior of plagioclase. From consideration only of the isoelectric points of the component oxides, one expects Si-O-Si linkages in plagioclase to be very stable at 2 < pH < 7. The aluminate group, however, may be susceptible to acid-catalyzed hydrolysis at pH less than about 4. The hydrolysis of aluminate sites will open up the structure and release alkali- and alkaline-earth cations.

Depth profiles of hydrogen in the near-surface region of labradorite are shown in Figure 9 before and after reaction with solutions at varying pH. The depth profiles were determined via Elastic Recoil Detection Analysis (see Casey et al., 1988; 1989). The depth- distribution of hydrogen in the leached crystals at pH > 5 is somewhat similar to that in the unleached reference crystals. At these conditions, the mineral dissolves uniformly with no deep ion-exchange or selective hydrolysis reactions. At very high pH-conditions the inventory of hydrogen in the crystal is even slightly less than the unreacted reference crystal.

Conversely, at more acid conditions the surface is extensively leached of aluminum, calcium and sodium, and is enriched in silicon and hydrogen. This hydrogen enrichment corresponds to creation of the leached layer similar to that shown in Figure 1. One expects the structure of leached layers on plagioclase to vary dramatically with composition. In albite, for example, removal of aluminate groups from the structure still leaves an extensively connected structure of linked silicate tetrahedra. By contrast aluminum leaching from anorthite leaves only unconnected tetrahedra. Any detectable Al-depleted leached layer on anorthite requires condensation of silanol groups into a linked structure subsequent to leaching.

Beryl

Beryl provides an enlightening illustration of how stepwise analysis of a mineral structure can lead to clear hypotheses about the character of dissolution. The beryl structure (Fig. 10) consists of $Si_6O_{18}^{-8}$ rings separated by tetrahedral beryllate groups (BeO_4^{-6}) and octahedrally coordinated aluminum. The mineral contains large (~5 Å) structural pores parallel to the [001] crystallographic axis. From this information alone one expects that the leaching reactions may be strongly influenced by the mineral crystallography. Water and solutes will diffuse much more rapidly down these structural pores than in perpendicular directions.

Linkages among disparate cations in beryl provide potential sites for selective hydrolysis during dissolution,

e.g., $\geq$Be-O-Si$\leq$, $-$Al-O-Si$\leq$, $\geq$Be-O-Al$-$.

The rates of selective hydrolysis will be controlled by the character of bonding between cations and oxygen, and the acid-base properties of the oxygens. As is the case for

418

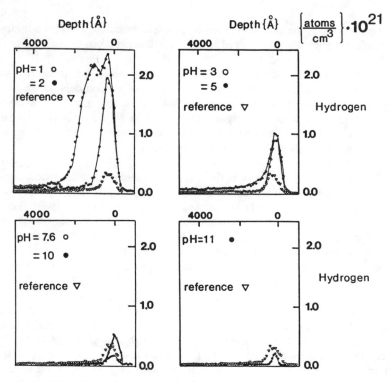

Figure 9. Depth profiles of hydrogen concentration in plagioclase feldspar as a function of solution pH (Casey et al., 1989).

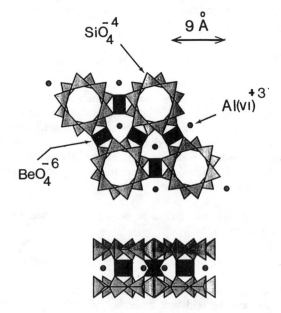

Figure 10. The structure of beryl. The [001] (top) and [010] (bottom) axes are perpendicular to the page.

plagioclase, however, the acid-base properties of the important linkages in beryl are unknown. We again must estimate the potential for selective leaching by comparing the isoelectric points of the component oxides.

By comparing isoelectric points of BeO, Al_2O_3 and SiO_2, one can surmise that beryllium-oxygen and aluminum-oxygen anions have a much higher affinity for hydrogen ions than the silicate groups. The mineral BeO, for example, has an isoelectric point in the range 9-10.5 (Parks, 1965), indicating that some oxygens are extensively coordinated to hydrogen ions (causing a net positive charge) in an acid solution (see also Baes and Mesmer, 1976; p. 96). $Al(VI)_2O_3$ has an isoelectric point at pH ~ 8.3 (Schindler and Stumm, 1987). The SiO_2 surface, however, is nearly neutral in the pH range 2-7 (Fig.6).

The discrepancy between isoelectric points of the component oxides suggests that beryllium- and aluminum-oxygen bonds in the mineral may be more susceptible to acid-catalyzed hydrolysis than silicon-oxygen bonds. Dissolution of beryl in acid should create a leached layer similar to that shown in Figure 1, but which is strongly controlled by the crystallographic orientation of the mineral. The silicate rings in beryl, however, are not linked to one another. Thus, no leached layer will form unless adjacent silicate rings link to one another after the beryllium and aluminum sites are leached away.

PROPERTIES OF THE LEACHED LAYER

At some early stage in the incongruent dissolution of minerals and glasses, the residual leached surface takes on a chemical life of its own; that is, the leached layer reacts with the solution somewhat independently of the bulk phase. Some interesting properties of the leached layers are introduced in this section.

Changes in cation coordination with leaching

By analogy with the dissolved oxyacid, one expects some network- forming cations to change oxygen coordination with solution pH. For example, boron exists in both trigonal and tetrahedral oxygen- coordination in unleached borosilicate glass. The relative concentration of borons in each of these sites can be measured spectroscopically, and results for two glass compositions is reported in Table 3. At pH < 9 boron is selectively leached from these glasses. For the 20•20•60 sodium borosilicate glass, over 95 percent of boron is removed during leaching at pH=1. The remaining boron sites in the leached glass, however, are overwhelmingly in trigonal coordination, as predicted from the solution chemistry. At pH of 9 and above, the distribution of boron species in the leached surface resembles that in the bulk glass (Table 3).

Similar changes in oxygen-coordination number with leaching are predicted for aluminum in some aluminosilicate minerals. Dissolved aluminum changes coordinated from tetrahedral $(Al(OH)_4^-)$ in basic solutions to the octahedrally coordinated aquo-complex $(Al•(H_2O)_6^{+3})$ in acid solutions. Thus any residual aluminum in the leached layer shown in Figure 1 is predicted to be octahedrally coordinated to oxygens.

420

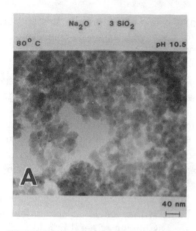

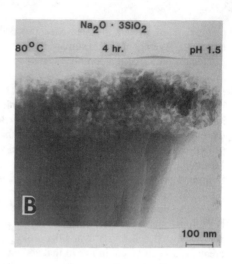

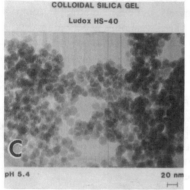

Figure 11. Transmission-electron photomicrographs of sodium-silicate glass after leaching in acidic (A) and basic (B) solutions. Repolymerization of silanol groups in the glasses has induced a porosity to the leached layer that is apparent in photomicrographs (A) and (B). These condensation reactions have proceeded so extensively that texture of the leached layers in (A) and (B) resembles colloidal silica (C).

Table 3: The speciation of boron in leached and unleached borosilicate glass. The trigonal sites are divided between axially symmetric (BO_3-S), axially asymmetric (BO_3-A) sites and boric acid (H_3BO_3). The tetrahedral sites are all reported as BO_4.

Conditions	BO_4	BO_3-S	BO_3-A	H_3BO_3
		[1]30•10•60		
unreacted	100	---	---	---
pH = 12	100	---	---	---
pH = 9	88.9	11.1	---	---
pH = 1	6	40	14	40
		[1]20•20•60		
unreacted	80.	11.4	8.6	---
pH = 12	80.	20.	---	---
pH = 9	81.8	18.9	---	---
pH = 1	8.7	17.4	26.1	47.8

[1]The glass compositions are reported as weight percent of the component oxides ($Na_2O•B_2O_3•SiO_2$).

Repolymerization of hydroxyl groups subsequent to leaching

Leached layers are commonly defined by high concentrations of total hydrogen (e.g., Bunker et al., 1988; Casey et al., 1988). Much of this hydrogen enters the structure to replace cations which have been removed. Thus an electrically neutral, leached feldspar must have enough hydrogen to balance the amount of removed charge, unless the structure completely reorganizes itself to eliminate nonbridging oxygens (see Reactions (9) and (10)). In some cases, repolymerization is so extensive that the leach layer separates into distinct water-rich and siloxane-rich phases (Tomazawa and Capella, 1983). Examples of such phase separation are shown in Figure 11. Condensation reactions have also been identified in leached tectosilicate minerals (Casey et al., 1988; 1989; Westrich et al., 1990; Hellman et al., 1990).

Repolymerization reactions profoundly affect the kinetics of leaching of alkali-silicate and alkali-borosilicate glasses. Alkali leaching of these glasses generally proceeds in two stages. In the first stage, reactive solutes (generally H_2O and H_3O^+) diffuse into the bulk glass, partly hydrolyze the structure and exchange with alkali ions. The hydrolysis reactions make the glass more transmissive to reactive solutes. During this stage of leaching, the concentration of a leached cation in solution, such as Na^+, increases nonlinearly with time in a batch experiment.

In the second stage of leaching, the silica-rich leached material repolymerizes into a porous silica network similar to that shown in Figure 11. Solute diffusion through this network is fast relative to hydrolysis of the unreacted glass. As long as the leached layer is not too thick, the overall rate of cation leaching is controlled by the rate of hydration of the bulk glass, which is nearly constant, and not the diffusion of solutes through the already modified glass. At this stage, the concentration of a leached cation in solution increases linearly with time in a bulk experiment. Ultimately, however, the leached layer becomes so thick that solute diffusion again controls the rate of the leaching reaction, or the entire particle is converted to silica gel.

Crazing and spallation of the leached layer

Leached layers exhibit textural and compositional differences depending upon the conditions of dissolution. In mildly corrosive solutions at low temperatures (20°C), for example, the leached layer on alkali-silicate glass is brittle and can develop an extensive network of intersecting cracks. Examples of such cracking (or crazing) are shown in Figure 12 for alkali-silicate glass and plagioclase.

Crazing is caused by the build up of stress within the leached layer. These stresses can build up while the material is immersed (Fig. 12-A) or can be caused by drying (Figs. 12-B, 12-C). The extent and character of crazing depends on the material composition, the solution composition, and the temperature. The crazing of an alkali- silicate glass, for example, correlates with the water content of the leached layer, which varies with temperature (Bunker et al., 1983).

In extreme cases, portions of the leached layer separate from the unreacted material.

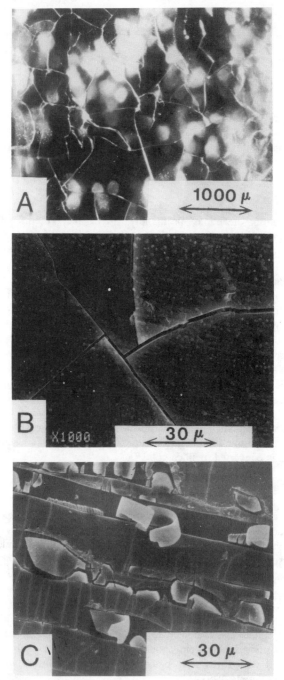

Figure 12. (A) The crazed surface of $Na_2O \cdot 3 \cdot SiO_2$ glass after reaction at pH = 7 for 1.3 h at 60°C. This photograph was taken through the leaching aqueous solution. Thus the crazing is not attributable to drying. Similar crazing has been reported on some leached mineral surfaces (Gastuche, 1963). (B) The surface of alkali silicate glass after drying. (C) The surface of plagioclase feldspar after reaction at pH=1 and drying at room temperature. Note that the leached layer is considerably cracked and portions have spalled away.

In Figure 12-C, we show the surface of leached feldspar after drying in air. As one can see, the leached surface is separating from the underlying mineral. Such spallation of leached surfaces may be important in weathering as leached minerals are subjected to wetting and drying cycles in an arid soil.

Chemisorption in the leached layer

The leached surfaces of minerals and glasses are notorious for concentrating solutes from the adjacent aqueous solution. There are three properties of a leached layer which account for this reactivity. First, leaching reactions consume hydronium ions. Therefore the pH of pore fluids in a very thick leached layer can be higher than in the bulk solution.

Secondly, solutes react with hydroxyl groups in the leached surface of a mineral or glass. Solutes which hydrolyze in the aqueous solution, for example, also condense hydroxyl groups in the leached layer into bridging bonds. The hydroxyl groups, however, must compete with other ligands in the solution for metal ions. Dissolved uranium, for example, interacts strongly with a leached glass surface only in a carbonate-free solution (Bunker, 1987). Thirdly, the leached layer consists largely of silica which has a negative surface charge in most natural waters. Positively charged colloids can enter and electrostatically bond to the leached layer.

These reactions can retard the rate leaching or dissolution by increasing the number of crosslinks, or by decreasing the transmissivity of the leached surface to reactive solutes. Some of the sorbing solutes which are important in natural waters include ferric iron and aluminum. Even trace impurities in an aqueous solution can dramatically affect the corrosion rates (Buckwalter and Pederson, 1982).

CONCLUSIONS

Most leachable minerals and glasses are best treated as polymeric oxides containing many structural elements. In order to understand the leaching or dissolution of these materials, one must first identify the mode of reaction of each structural element with the aqueous solution. It is particularly important to estimate the acid-base properties of reactive site. Traditional hydrolysis diagrams showing the predominance of dissolved hydroxide species can be useful in estimating the acid-base properties of a particular site. This approach is limited by the extent to which the important sites on a leachable mineral resemble dissolved hydroxide and oxide species.

Materials which selectively leach tend to contain a high concentration of covalent crosslinks in the structure. These crosslinks maintain structural integrity in the leached surface once the reactive constituents are removed. The rate and depth of leaching also depends upon the transmissivity of the material to undissociated water and solutes. Deep ion-exchange and hydration in most oxide materials requires partial relaxation of the structure by hydrolysis of reactive sites.

Once produced, the leached layer has unique reactive properties. Silanol groups in

424

the layer can repolymerize subsequent to the ion exchange. Trace solutes, organic acids and even colloids can react with the leached layer, and these reactions can dramatically affect the rate of dissolution and leaching. Finally, tensile stresses can build up in the leached layer which cause crazing and spallation of the leached material away from the unreacted surface.

ACKNOWLEDGMENTS

The authors would like to thank Drs. H.R. Westrich, C. Stockman, H. Stockman, and G.W. Arnold for valuable comments. This research was supported by the Department of Energy under contract DE-AC04-76DP00789.

REFERENCES

Aglietta, E.F., Porto-Lopez, J.M., and Pereira, E. (1988) Structural alterations in kaolinite by acid treatment. Appl. Clay Sci. 3, 155-163.
Baes, C.F. and Mesmer, R.E. (1976) The Hydrolysis of Cations. John Wiley, New York.
Bales, R.C. and Morgan, J.J. (1985) Dissolution kinetics of chrysotile at pH = 7 to 10. Geochim. Cosmochim. Acta 49, 2281-2228.
Barrer, R.M. (1978) Zeolites and Clay Minerals as Sorbents and Molecular Sieves. Academic Press, New York.
Barrer, R.M. and Klinowski, J. (1975) Hydrogen Mordenite and Hydronium Mordenite. J. Chem. Soc. Faraday Trans. 71, 690-698.
Basolo, F. and Pearson, R.G. (1958) Mechanisms of Inorganic Reactions-A Study of Metal-Complexes in Solution. John Wiley & Sons, New York.
Boehm, H.P. and Knozinger, H. (1983) Nature and estimation of functional groups on solids surfaces in: Catalysis Science and Technology, (J.R. Anderson and M. Boudart, eds.), Springer-Verlag, New York.
Brindley, G.W. and Youell, R.F. (1951) A chemical determination of tetrahedral and octahedral aluminum ions in a silicate. Acta Cryst. 4, 495-497.
Buckwalter, C.Q. and Pederson, L.R. (1982) Inhibition of nuclear waste glass leaching by chemisorption. J. Amer. Ceram. Soc. 65, 431-436.
Bunker, B.C. (1987) Waste glass leaching: chemistry and kinetics. Mat. Res. Soc. Symp. 84, 493-507.
Bunker, B.C., Arnold, G.W., Beauchamp, E.K. and Day, D.E. (1983) Mechanisms for alkali leaching in mixed-Na-K silicate glasses. J. Non-Cryst. Solids 58, 295-322.
Bunker, B.C., Arnold, G.W., Rajaram, M. and Day, D.E. (1987) Corrosion of phosphorus oxynitride glasses in water and humid air. J. Amer. Ceram. Soc. 70, 425-430.
Bunker, B.C., Tallant, D.R., Headley, T.J., Turner, G.L. and Kirkpatrick, R.J. (1988) The structure of leached sodium borosilicate glass. Phys. Chem. Glasses 29, 106-120.
Carroll-Webb, S. and Walther, J.A. (1988) A surface complex reaction model for the pH-dependence of corundum and kaolinite dissolution rates. Geochim. Cosmochim. Acta 52, 2609-2623.
Casey, W.H., Westrich, H.R. and Arnold, G.W. (1988) The surface of Labradorite feldspar reacted with aqueous solutions at pH = 2, 3, and 12. Geochim. Cosmochim. Acta 52, 2795-2807.
Casey, W.H., Westrich, H.R., Arnold, G.W. and Banfield, J.F. (1989) The surface chemistry of dissolving Labradorite feldspar. Geochim. Cosmochim. Acta 53, 821-832.
Crowther, J. and Westman, A.E.R. (1956) The hydrolysis of the condensed phosphates III: Sodium tetrametaphosphate and sodium tetraphosphate. Can. J. Chem. 34, 969-981.
Dugger, D.L., Stanton, J.H., Irby, B.N., McConnell, B.L., Cummings, W.W. and Maatman, R.W. (1964) The exchange of twenty metal ions with the weakly acidic silanol groups of silica gel. J. Phys. Chem. 68, 757-760.
Eigen, M. (1962) Fast elementary steps in chemical reaction mechanisms. Pure Appl. Chem. 6, 97-115.

425

Feller, S.A., Dell, W.J. and Bray, P.J. (1982) [10]B NMR studies of lithium borate glass. J. Non-Cryst. Solids 51, 21-30.

Furrer, G. and Stumm, W. (1986) The coordination chemistry of weathering I: Dissolution kinetics of δ-Al$_2$O$_3$ and BeO. Geochim. Cosmochim. Acta 50, 1847-1860.

Gastuche, M.C. (1963) Kinetics of acid dissolution of biotite: 1. Interfacial rate process followed by optical measurement of the white silica rim. Int'l. Clay. Conf., Stockholm 1, 67-83.

Goldich, S.S. (1938) A study of rock weathering. J. Geol. 46, 17-58.

Guy, C. and Schott, J. (1990) Multisite surface reaction versus transport control during the hydrolysis of a complex oxide. Chem. Geol. 78. 181-189.

Hellmann, R., Eggleston, C.M., Hochella, M.F. and Crerar, D.A. (1990) The formation of leached layers on albite surfaces during dissolution under hydrothermal conditions. Geochim. Cosmochim. Acta 54, 1267- 1281.

Hiemstra, T. and Van Riemdijk, W.H. (1990) Multiple activated complex dissolution of metal (Hydr)oxides: A thermodynamic approach applied to quartz. J. Coll. Interface Sci. 136, 132-149.

Hiemstra, T., Van Riemsdijk, W.H. and Bolt, G.H. (1989) Multisite proton adsorption modeling at the solid/solution interface of (hydr)oxides: A new approach. J. Coll. Interface Sci. 133, 91- 104.

Hurd, D.C., Fraley, C. and Fugate, J.K. (1979) Silica apparent solubilities and rates of dissolution and precipitation in Chemical Modeling in Aqueous Systems, E. A. Jenne, ed.; Amer. Chem. Soc. Symp. Ser. 93.

Isard, J.O. (1967) The dependence of glass-electrode properties on composition in: Glass electrodes for hydrogen and other cations, ed: G. Eisenman, Marcel Dekker.

Kawakami, H., Yoshida, S. and Yonezawa, T. (1984) A quantum-chemical approach to the generation of solid acidity in composite metal oxides. J. Chem. Soc. Faraday Trans. 2, 205-217.

Kent, D.B., Tripathi, V.S., Ball, N.B., and Leckie, J.O. (1986) Surface-complexation modeling of radionuclide adsorption in subsurface environments. Technical Report 294, Dep. Civil Engineering, Stanford University, Stanford, CA.

Kerr, G.T. (1973) Hydrogen zeolite Y, ultrastable zeolite Y, and aluminum-deficient zeolites. in: Molecular Sieves (eds. Meier and Uytterhoeven), Adv. Chem. Series 121, Amer. Chem. Soc., Washington, D.C.

Knozinger, H. and Ratnasamy, P. (1978) Catalytic aluminas: Surface models and characterization of surface sites. Catal. Rev.-Sci. Eng. 17, 31-70.

Kossiakoff, A. and Harker, D. (1938) The calculation of ionization constants of inorganic oxygen acids from their structures. J. Amer. Chem. Soc. 60, 2047-2055.

Lin, F-C. and Clemency, C.V. (1981) The dissolution kinetics of brucite, antigorite, talc and phlogopite at room temperatures and pressure. Amer. Mineral. 66, 801-806.

Luce, R.W., Bartlett, R.W. and Parks, G.A. (1972) Dissolution kinetics of magnesium silicates. Geochim. Cosmochim. Acta 36, 35-50.

Man, P.P., Peltre, M.J. and Barthomeuf, D. (1990) Nuclear magnetic resonance study of the de-alumination of an amorphous silica-alumina catalyst. J. Chem. Soc. Faraday Trans. 86, 1599-1602.

Murata, K.J. (1943) Internal structure of silicate minerals that gelatinize with acid. Amer. Mineral. 28, 545-562.

Newman, A.C.D. and Brown,G. (1969) Delayed exchange of potassium from some edges of mica flakes. Nature 223, 175-176.

Pacco, F., van Cangh, L., and Fripiat, J.J. (1976) Etude par spectroscopie infrarouge et resonance magnetique nucleaire de la distribution in homogene des groupes silinols d'un gel de silice fibreaux. Bull. Soc. Chim. France 7-8, 1021-1026.

Parks, G.A. (1965) The isoelectric points of solid oxides, solid hydroxides and aqueous hydroxo complex systems. Chem. Rev. 65, 177- 198.

Parks, G.A. (1967) Aqueous surface chemistry of oxides and complex oxide minerals; ch. 6 in: Equilibrium Concepts in Natural Water Systems, Adv. Chemistry Ser. 67, Amer. Chem. Soc.

Pauling, L. (1970) General Chemistry. Dover, New York.

Petit, J-C., Dran, J.-C. and Della Mea, G. (1990) Energetic ion beam analysis in the Earth Sciences. Nature 334, 621-626.

Petit, J-C., Dran, J-C., Paccagnella, A., and Della Mea, G.(1989) Structural dependence of crystalline silicate hydration during aqueous dissolution. Earth Planet. Sci. Lett. 93, 292-298.

Ray, N.H. (1978) Inorganic Polymers. Academic Press. 174 pp.

Rennie, G.K. and Clifford, J. (1977) Melting of ice in porous solids. J. Chem. Soc.-Faraday Trans. II, 73, 680-685.

Ricci, J.E. (1948) The aqueous ionization constants of inorganic oxygen acids. J. Amer. Chem. Soc. 70, 109-113.

Schindler, P.W. and Stumm, W. (1987) The surface chemistry of oxides, hydroxides and oxide minerals. Ch. 4: Aquatic Surface Chemistry, W. Stumm, ed., Wiley Interscience, New York

Smets, B.M.J. and Lommen, T.P.A. (1982) The leaching of sodium aluminosilicate glasses studied by secondary ion mass spectrometry. Phys. Chem. Glasses 23, 83-87.

Smets, B.M.J. and Tholen, M.G.W. (1983) The pH-dependence of the aqueous corrosion of glasses. J. Non-Cryst. Solids 49, 351-362.

Tomozawa, M. and Capella, S. (1983) Microstructure in hydrated silicate glass. J. Amer. Ceram. Soc. 66, C24-C25.

Tsyganenko, A.A. and Filimonov, V.N. (1973) Infrared spectra of surface hydroxyl groups and crystalline structure of oxides. J. Molec. Struct. 19, 579-589.

Velez, M.H., Tuller, H.L., and Uhlmann, D.R. (1982) Chemical durability of lithium borate glasses. J. Non-Cryst. Solids 49, 351-362.

Vermilyea, D.A. (1969) The dissolution of MgO and $Mg(OH)_2$ in aqueous solutions. J. Electrochem. Soc. 116, 1179-1183.

Wakabayashi, H. and Tomozawa, M. (1989) Diffusion of water into silica glass at low temperature. J. Amer. Ceram. Soc. 72, 1850-1855.

Wendt von, H. (1973) Die Kinetik typischer Hydrolysereaktionen von mehrwertigen Kationen. Chimia 27, 575-588.

Westrich, H.R., Casey, W.H. and Arnold, G.W. (1990) Oxygen isotope exchange in the leached layer of labradorite feldspar. Geochim. Cosmochim. Acta 53, 1681-1685.

CHAPTER 11

OXIDATIVE AND REDUCTIVE
DISSOLUTION OF MINERALS

INTRODUCTION

Over the history of the earth, organisms have initiated and maintained global chemical disequilibria. Products and by-products of biological processes are or have been thermodynamically unstable with respect to reaction with major constituents of the atmosphere and lithosphere. Reducible or oxidizable components of minerals have reacted with biological products and by-products under two fundamentally different regimes: the changing (i.e., non-steady-state) redox conditions of the evolving atmosphere and the modern redox condition, steady-state on a global average but subject to local (spatial or temporal) perturbations.

In the chemical evolution of the atmosphere and oceans from the reducing conditions of the primitive earth to the oxidizing conditions of the modern environment, reduced crustal minerals, e.g., those containing Fe(II), Mn(II), U(IV), S(-II) or S(-I), have reacted with free oxygen generated by photosynthesis. Garrels and Perry (1974) have speculated that the reaction of O_2 with Fe(II)carbonates, liberating CO_2 from the crustal matrix, was integrally involved in the establishment of modern atmospheric conditions. As discussed by Holland (1984), the occurrence of reduced minerals as detrital constituents of sediments or sedimentary rocks depends both on the level of atmospheric oxygen to which the minerals were exposed during weathering and the rates of mineral oxidation (and dissolution) reactions. Thus detrital uraninite, which is readily oxidized under modern atmospheric conditions, is rarely found in recent sediments but has been preserved in ancient sediments weathered under Pre-Cambrian levels of atmospheric oxygen. In contrast, detrital magnetite commonly survives exposure even to present levels of atmospheric oxygen during weathering, transport, and deposition due to its slow redox kinetics (Holland, 1984). Thus, over geologic time, oxidation of crustal minerals by free oxygen has profoundly influenced the composition of the atmosphere and of soils, waters, sediments, and sedimentary rocks.

On a global average, the modern environment may be considered to be nearly at steady-state with regard to proton and electron balances, as characterized by a present day atmosphere of approximately 20.9% O_2, 0.03% CO_2, 79.1% N_2 and a world ocean with pH ~ 8 and a redox potential of $E_H = 0.75$ V. Near steady-state redox conditions are maintained by coupled redox cycles. Coupling of the sulfur, sulfate S(VI)/sulfide S(-II), and carbon, carbonate C(IV)/organic carbon C(0), cycles influences the regulation of atmospheric oxygen (Garrels and Perry, 1974; Garrels and Lerman, 1984; Holland, 1984 and ref. cit.). In a heterogeneous environment, however, the balance of such redox cycles may be locally perturbed. Thus, both reduction and oxidation of minerals occur in modern environments and significantly affect the biogeochemical cycling of many elements. The effects of redox processes are due predominantly to the variation of metal solubility with oxidation state. Redox reactions of metals, such as Fe and Mn, in natural waters often result in the precipitation or dissolution of solid phases (Stumm and Morgan, 1981). The distribution of particle-reactive species (including other trace metals, phosphate, and organic micropollutants) can be markedly affected by the formation or elimination of surfaces that result from such redox reactions (Stauffer and Armstrong, 1986; Armstrong et al., 1987). In addition, some organic matter may be directly oxidized by reaction with reducible minerals such as Mn(III,IV) or Fe(III) oxides.

The effects of the redox reactions of minerals and the associated solid dissolution and precipitation reactions on the biogeochemical cycling of elements and the regulation of water composition depend on the rates of the various processes. Here, our goals are to

428

describe a general model for mineral dissolution that can be applied to redox-enhanced dissolution, to compare the predictions of this model with observations of dissolution kinetics in laboratory systems, and to evaluate the possibilities, difficulties, and implications of extrapolation of the model to the natural environment.

Objectives

The objectives of this chapter are:

(1) to briefly describe the redox conditions and important redox processes in natural systems;

(2) to review surface-controlled dissolution processes at the mineral-water interface and to emphasize that specific properties of the mineral surface, which depend on both surface chemistry and structure, must be considered;

(3) to demonstrate that a simple kinetic model for surface-controlled reductive and oxidative dissolution of minerals can be derived by considering surface coordination (surface complex formation with H^+, OH^-, metal ions, and ligands, including oxidants and reductants), lattice statistics, and activated complex theory;

(4) to illustrate that the surface complexation model may be generally applied to describe various dissolution pathways;

(5) to illustrate that, in oxygenation reactions of transition metal ions, the effect of complex formation with surface hydroxyl groups of minerals on reaction kinetics is comparable to the effect of complex formation with OH^- in solution;

(6) to examine a few case studies of the dissolution of Fe(III) (hydr)oxides, Mn(III,IV) oxides, UO_2, and FeS_2;

(7) to discuss the applicability of the surface-controlled dissolution model under field conditions and the extrapolation of laboratory dissolution rates to the field; and

(8) to consider the geochemical implications of redox-enhanced mineral dissolution processes.

The information provided should also form the basis for discussion of mineral dissolution (Casey and Bunker, Chapter 10, this volume), of surface-catalyzed redox processes (White, Chapter 12, this volume), and of heterogeneous photochemical processes at mineral surfaces (Waite, Chapter 14, this volume).

BACKGROUND

Redox processes in natural systems

Major redox couples and the distribution of redox-active species. For most of the biosphere, the controlling redox couple is the O_2/H_2O couple. The predominance of this redox couple arises from both the abundance and reactivity of O_2 (Stumm and Morgan, 1981). In systems with restricted exchange with the atmosphere or oxygenated waters (e.g., sediments, flooded soils, deep groundwaters, hypolimnetic waters), microbial oxidation of organic matter can result in depletion of O_2 and development of sub-oxic and anoxic conditions. Characteristic profiles of dissolved concentrations of redox-active species are observed at oxic-anoxic boundaries (Fig. 1). These profiles of dissolved inorganic species are paralleled by changes in the microbial communities (assemblages) with depth in the sediment and reflect the progression in terminal electron acceptors for the oxidation of organic carbon by microorganisms (i.e., O_2, NO_3^-, MnO_2, Fe(OH)$_3$, SO_4^{2-}, organic C in fermentation reactions). Figure 2 shows the important redox couples in natural systems (e.g., MnO_2/Mn^{2+} and Fe(OH)$_3$/Fe^{2+} under sub-oxic and SO_4^{2-}/H_2S or SO_4^{2-}/HS^- under anoxic conditions) as well as both the ranges of redox potential and pH that can support microbial activity and that have been observed in soils.

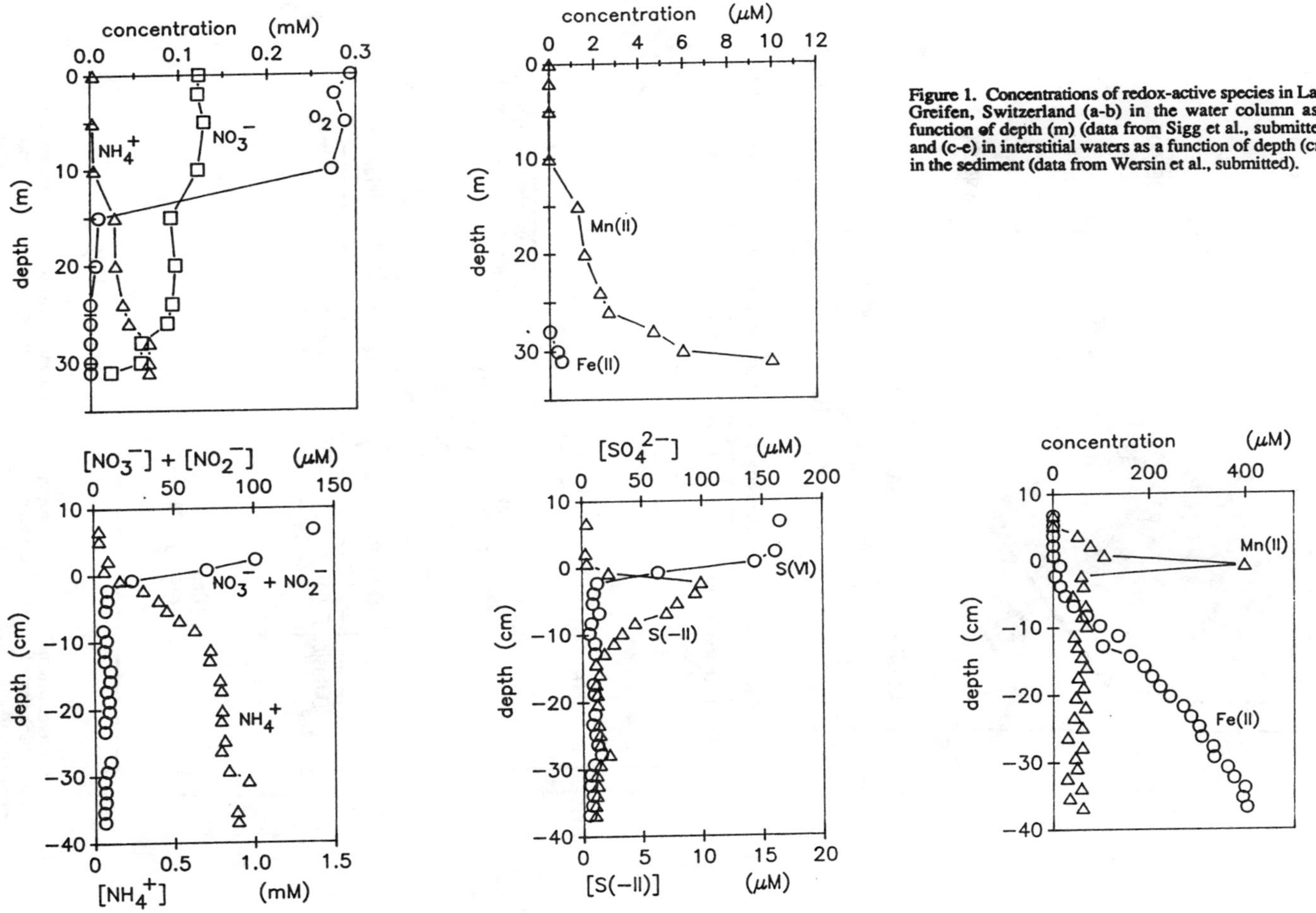

Figure 1. Concentrations of redox-active species in Lake Greifen, Switzerland (a-b) in the water column as a function of depth (m) (data from Sigg et al., submitted) and (c-e) in interstitial waters as a function of depth (cm) in the sediment (data from Wersin et al., submitted).

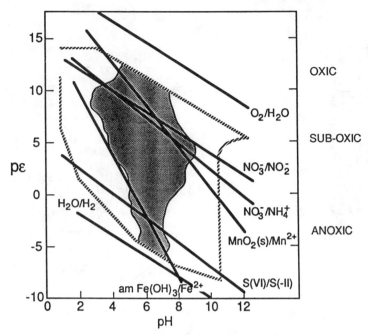

Figure 2. A pε–pH diagram showing the domain accessible to microorganisms (dashed perimeter) and that observed in soils, natural waters, and sediments (shaded area) with possible redox buffers (calculated for [reductant] = [oxidant]). [Adapted in part from Sposito (1989), based on data compiled by Baaș Becking et al. (1960)].

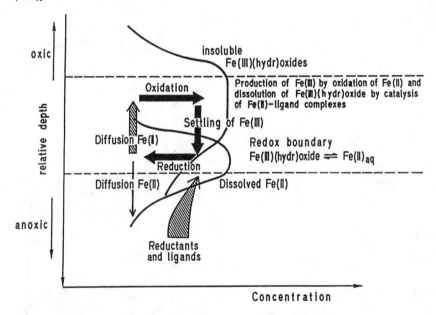

Figure 3. Transformations of Fe(II, III) at an oxic-anoxic boundary in a water column or sediment (modified from Davison, 1985). Peaks in the concentrations of solid Fe(III) (hydr)oxides and of dissolved Fe(II) coincide with the depth of maximum Fe(III) and Fe(II) production. Efficient dissolution of Fe(III) (hydr)oxides results from the combination of ligands and Fe(II) produced in the underlying anoxic zone (from Sulzberger et al., 1989).

Table 1. Examples of redox reactants in natural systems

REDUCTANT	ENVIRONMENT	REFERENCE
organic acids (e.g.- formate, acetate, oxalate, pyruvate, glycolate, lactate)	anoxic marine sediments	1-3
reduced S species		
H_2S, HS^-, polysulfides	anoxic waters	
pyrite, thiosulfate	and sediments	4, 5
SO_2, HSO_3^-	atmospheric waters	6, 7
humic substances/ dissolved organic matter (photochemical)	surface waters (photic zone)	8, 9
H_2O_2	surface waters (photic zone)	10-12
phenols	landfill leachates	13
	surface waters	14
	waste waters	15
Fe(II)	oxic-anoxic boundaries[a]	16
	surface waters (photic zone)[a]	17
	atmospheric waters[a]	6
As(III)	lake sediments	18, 19

OXIDANT	ENVIRONMENT	REFERENCE
O_2	oxygenated waters	20
H_2O_2	surface water (photic zone)	10-12
Fe(III) oxides	(see Table II)	
Mn(III,IV) oxides	(see Table III)	
free radicals (HO•, organic peroxy radicals)	surface water (photic zone)	21

REFERENCES: (1) Barcelona, 1980 (2) Sansone and Martens, 1982 (3) Sansone, 1986 (4) Jorgensen, 1983 and ref. cit. (5) Luther et al., 1986 (6) Behra and Sigg, 1990 (7) Hoffmann and Jacob, 1984 (8) Sunda et al., 1983 (9) Harvey and Boran, 1985 (10) Zika et al., 1985a (11) Zika et al., 1985b (12) Cooper and Zika, 1983 (13) Reinhard et al., 1984 (14) Wegman, 1979 (15) Leuenberger et al., 1983 (16) Sulzberger et al., 1989 (17) Sulzberger et al., 1990 (18) Huang et al., 1982 (19) Oscarson et al., 1981 (20) Stumm and Morgan, 1981 and ref. cit. (21) Hoigné, 1990 and ref. cit.

[a] plausibly important in this environment

Note: Some species can act as both reductants and oxidants in natural systems. For example, Fe(II) can be oxidized and Cu(II) reduced by H_2O_2 in seawater.

Rapid cycling of redox-active elements occurs immediately at <u>oxic-anoxic boundaries</u> in the water columnn as particulate oxides sink through the oxycline and are reduced and dissolved reduced species diffuse upwards and are oxidized (Fig. 3). Redox reactions involving solid phases may not go to completion in natural waters as indicated by the co-occurrence of Fe(II) and Fe(III) solid phases at the oxic-anoxic boundary in lakes (Buffle et al., 1989) and in anoxic freshwater sediments (Wersin et al., submitted).

<u>Specific reductants and oxidants</u>. Although the redox potential is controlled by a predominant redox couple, thermodynamically unstable species are often the proximate reactants in redox cycling in natural systems. In oxic waters, dissolved organic matter may be oxidized by reaction with free radicals such as HO• (Hoigné, 1990). In surface seawater,

Table 2. Examples of reductive dissolution of iron oxides in natural systems

ENVIRONMENT	REDUCTANT	IMPLICATIONS	REFERENCE
Oxic surface waters (photic zone)	H_2O/OH^-, humic substances/ dissolved organic matter- photochemical		1-5
		increased bioavailability of Fe	6-8
		of PO_4	9
	cell surface[a]	enhanced metal uptake[a]	10, 11
Atmospheric waters	[a]sulfite, Fe(II), organics-photochemical		12
Lake water column oxic-anoxic boundary	sulfide[a]		13
Lake sediment-water interface			14
		increased water column residence time for PO_4	15
		for PCB's	16
Anoxic marine sediments	sulfide[a]		17
Flooded soils			18
Anoxic freshwater sediments	direct microbial catalysis[a]		19

REFERENCES: (1) Collienne, 1983 (2) McKnight et al., 1988 (3) McKnight and Bencala, 1988 (4) Harvey and Boran, 1985 (5) Sulzberger et al., 1990 (6) Anderson and Morel, 1980 (6) Finden et al., 1984 (8) Rich and Morel, 1990 (9) Francko and Heath, 1982 (10) Price and Morel, 1990 and ref. cit.(11) Jones et al., 1987 (12) Behra and Sigg, 1990 (13) de Vitre et al., 1988 (14) Davison, 1985 and ref. cit. (15) Stauffer and Armstrong, 1986 (16) Armstrong et al., 1987 (17) Canfield and Berner, 1987 (18) Sposito, 1989 (19) Lovley and Phillips, 1988

[a]hypothesized

Table 3. Examples of reductive dissolution of manganese oxides in natural systems

ENVIRONMENT	REDUCTANT	IMPLICATIONS	REFERENCE
surface seawater	humic substances/ matter- photochemical	increased Mn bioavailability	1, 2
sub-oxic waters (Black Sea)			3
anoxic hypolimnetic waters			4
sediment-water interface			5
anoxic freshwater sediments	sulfide[a], thiols[a], direct microbial catalysis[a]		6-8
lake sediments	As(III)		9, 10
polluted soils or sediments	phenols[a]	degradation of organic pollutants[a]	(see Table I)

REFERENCES: (1) Sunda et al. (1983) (2) Sunda and Huntsman, 1988 (3) Tebo et al., 1990. (4) Sigg et al., in prep. (5) Davison, 1985 (6) Lovely and Phillips, 1988 (7) Burdige and Nealson, 1986 (8) Mopper and Taylor, 1986.(9) Huang et al., 1982 (10) Oscarson et al., 1981

[a]hypothesized

Table 4. Examples of oxidative dissolution of minerals in natural systems

MINERAL	ENVIRONMENT	OXIDANT	IMPLICATIONS	REFERENCES
Pyrite	salt marsh sediments	biologically-mediated[a]	rapid S, Fe cycling, thiol production	1, 2
	acid-mine drainage	Fe(III)		3
	leaching of mine tailings	chemoautotrophic bacteria[a]	solubilization of Fe(II) and other trace metals	4
Fe(II) silicates	ocean hydrothermal systems	O_2		5
	exposed bedrock-atmospheric weathering	O_2		6
U(IV) oxides	exposed detrital minerals-atmospheric weathering	O_2		7
	subsurface- ore deposits	O_2		8
	subsurface- buried nuclear waste	O_2		9, 10

REFERENCES: (1) Luther et al., 1986 (2) Luther and Church, 1988 (3) Singer and Stumm, 1970 (4) Francis et al., 1989 and ref. cit. (5) Seyfried and Mottl, 1982 (6) Giovanoli et al., 1988 (7) Holland, 1984 (8) Drever, 1988 (9) Garisto and Garisto, 1986 (10) Ollilla, 1989

[a]hypothesized

the dominant oxidation pathway for Fe(II) is predicted to be reaction with H_2O_2 rather than with O_2 (Moffet and Zika, 1987). Table 1 lists some of the important redox reactants in natural systems.

Role of the biota. It may be noted that many of the reductants and oxidants listed in Table 1 are metabolic products or undergo extensive biological cycling. The biota, from microbes to vascular plants, participate both indirectly and directly in redox cycling. Biological processes that indirectly affect redox cycling and redox-enhanced dissolution include: (1) release of redox-active substances, either exudates or decomposition intermediates, especially those that can specifically adsorb to mineral surfaces, (2) maintenance of redox microenvironments at cell surfaces (Richardson et al., 1988) or within microbial colonies (Carpenter and Price, 1976; Paerl and Prufert, 1987), (3) stabilization of soils (Schwarzman and Volk, 1989), and (4) microfracturing of mineral grains. Microorganisms, both bacteria and phytoplankton, can also directly catalyze reduction of metal oxides (Lovley and Phillips, 1988; Arnold et al., 1988) or metal complexes (Jones et al., 1987) and oxidation of Mn^{2+} (Sunda and Huntsman, 1987) or metal sulfides (Silver and Torma, 1974; Francis et al., 1989). Under natural conditions, however, the exact nature of microbial mediation is not always clear.

Reductive and oxidative dissolution of minerals. As is clear from the above discussion, many elements undergo extensive redox cycling in natural systems. For some elements, particularly Fe, Mn, U, and S, precipitation and dissolution of solid phases commonly accompany changes in oxidation state. In natural systems, Fe(III) and Mn(III,IV) oxides are subject to reductive dissolution and Fe(II) silicates and sulfides and U(IV) oxides to oxidative dissolution. Some examples of reductive and oxidative dissolution of minerals in natural systems, the environmental regimes under which such processes occur, important oxidants or reductants, and some of the biogeochemical implications of redox-enhanced dissolution are given in Tables 2-4.

Some theoretical background

Dependence of the rate of surface-controlled mineral dissolution on surface structure. We would first like to provide the reader with a qualitative understanding of the subject of dissolution kinetics. The most important reactants in mineral dissolution are H_2O, H^+, OH^- ligands and, in case of reducible or oxidizable minerals, reductants and oxidants.

The dissolution of a mineral is a sum of chemical and physical reaction steps. If the chemical reactions at the surface are slow in comparison with transport (diffusion) processes, dissolution kinetics will be controlled by one step in the overall chemical process occurring at the mineral surface. Then, rates of transport of reactants or products between the bulk solution and the surface can be neglected in describing the overall dissolution rate. The rates of dissolution of many minerals, especially under conditions encountered in nature, have been shown to be surface-controlled (Petrovic et al., 1976; Berner and Holdren, 1979). Thus the dissolution reaction may be represented schematically by two steps:

surface sites + reactants (H^+, OH^-, ligands, reductants, or oxidants)

$$\xrightarrow{\text{fast}} \text{surface species} \qquad (1)$$

$$\text{surface species} \xrightarrow[\text{metal detachment}]{\text{slow}} \text{metal}_{(aq)} \qquad (2)$$

Although both of these steps certainly consist of a series of elementary reactions, the rate law for surface-controlled dissolution is based on the assumptions that the sorption of reactants to surface sites, Step (1), is fast and that the subsequent detachment of the metal species from the surface of the crystalline lattice into the solution, Step (2), is slow and thus rate-limiting. The assumption of rapid formation of surface complexes is supported by experimental evidence (Hachiya et al., 1984; Hayes and Leckie, 1986; Wehrli et al., in press). Coordination of reactants (H^+, OH^-, ligands, reductants or oxidants) at the surface may allow polarization, weakening and breaking of metal-oxygen bonds or reduction or oxidization of one of the constituents of the surface of the lattice. Figure 4 shows a few examples of surface configurations that enhance or inhibit dissolution. For slow metal detachment (i.e., when Step (2) is rate-limiting), the rate law for dissolution will show a dependence on the concentration (activity) of the appropriate surface species.

$$\text{Dissolution rate} \propto \{\text{surface species}\} \tag{3a}$$

As described by Wieland et al. (1988), this conclusion (Eqn. 3a) is also consistent with transition state theory. The surface species formed by the interaction of H^+, OH^-, or ligands with surface sites can be considered as the precursor of the activated complex.

$$\text{Dissolution rate} \propto \{\text{precursor of the activated complex}\} \tag{3b}$$

The concentration of the surface species (or of the activated complex precursor) (Eqn. 3a or 3b) can usually be estimated from the number of surface sites and the extent of surface protonation or deprotonation or the surface concentration of ligands, reductants or oxidants (Furrer and Stumm, 1986).

Surface-controlled dissolution kinetics and the surface complexation model. The mechanistic formulation of rate laws requires knowledge of the structure and concentration of the reactive surface species (precursor of the activated complex). Surface complexation models, based on the studies of oxides in aqueous solution, incorporate both the principles of coordination chemistry in homogeneous solution and electric double layer theory (Schindler and Kamber, 1968; Stumm et al., 1970; Yates et al., 1974; Davis et al., 1978; Kummert and Stumm, 1980; Sigg and Stumm, 1980). Observed dissolution kinetics have been successfully explained by the surface complexation model for simple oxides and some silicates (Furrer and Stumm, 1983, 1986; Pulver et al., 1984; Stumm and Furrer, 1987; Sulzberger et al., 1989; Stumm and Wieland, 1990; Stumm et al., in press). These authors formulated mechanistic rate laws, i.e., rate as a function of concentrations of surface species, to replace more empirical rate laws, i.e., rate as a function of dissolved concentrations.

Central to the surface complexation model is the role of surface functional groups, present on all hydrous inorganic solids. The functional groups on the surface depend on the mineral constituents; thus, they may be, in addition to -OH-groups (on oxides and hydroxides, -SH (sulfides), -S-S (disulfides like pyrite), or $-CO_2OH$ (carbonates). Deprotonated surface groups behave as Lewis bases and their interaction with H^+ and metal ions (including metallic oxidants and reductants) can be understood as competitive complex formation. Similarly, the central ion of the mineral surface can exchange its structural functional groups (-OH ions in case of hydrous oxides) for other ligands (anions or the conjugate bases of weak acids). Thus, the surface of a mineral can be treated, in a statistical approach, as a polymeric acid (or base) which undergoes acid-base and coordination reactions; these reactions may include redox reactions with (surface) coordinated reductants or oxidants.

436

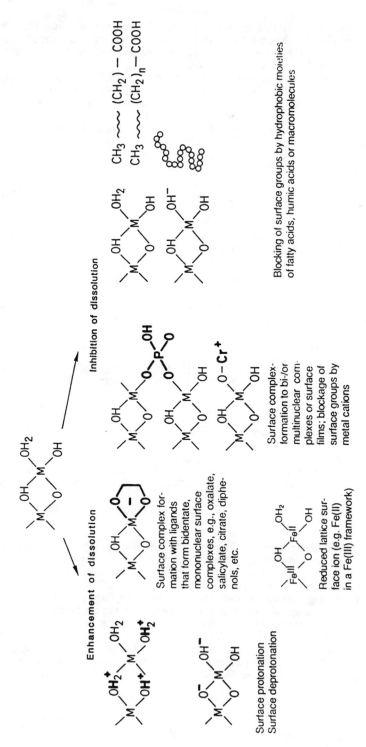

Enhancement of dissolution

Inhibition of dissolution

Surface complex formation with ligands that form bidentate, mononuclear surface complexes, e.g., oxalate, salicylate, citrate, diphenols, etc.

Surface complexation to bi-/or multinuclear complexes or surface films; blockage of surface groups by metal cations

Blocking of surface groups by hydrophobic moieties of fatty acids, humic acids or macromolecules

Surface protonation
Surface deprotonation

Reduced lattice surface ion (e.g. Fe(II) in a Fe(III) framework)

Figure 4. Effect of protonation, complex formation with ligands and metal ions, and reduction on dissolution rate. These structures are schematic short hand notations given here to illustrate the principal features of the surface complexes. They do not reveal, however, the structural properties or the coordination numbers of the oxides under consideration; charges given are relative.

The following concepts are characteristic of all surface complexation models (Dzombak and Morel, 1990):

(i)   sorption takes place at specific coordination sites;
(ii) sorption reactions can be described by mass law equations;
(iii) surface charge results from the sorption reaction itself; and
(iv) the effect of surface charge on sorption can be taken into account by
      applying a correction factor derived from the electric double layer theory
      to the mass law constants for surface reactions.

Dzombak and Morel have lucidly illustrated that inorganic sorption data for hydrous oxides can be interpreted thermodynamically in a consistent way. Data are readily amenable to computer calculations.

The nature (i.e., inner-sphere or outer-sphere complexes) and structure (i.e., mono-dentate or bidentate, mononuclear or binuclear complexes – cf. Fig. 4) of surface species cannot be predicted unequivocally from thermodynamic data alone. These characteristics must be inferred from combinations of thermodynamic data with linear-free energy comparisons between solute and surface complex stability constants (Schindler and Stumm, 1987), ionic strength dependence of these constants (Hayes and Leckie, 1987), and observations of reactivity (for example, in dissolution reactions). Although it is often difficult to use spectroscopic techniques because of the presence of water, significant progress has been made recently by using electron spin resonance, electron nuclear double resonance spectroscopy, electron spin echo envelope modulation, Fourier transform in-frared spectroscopy, and EXAFS (extended X-ray absorption fine structure) (Motschi, 1987; Tejedor-Tejedor and Anderson, 1986; Brown et al., 1989).

Reactivity of surface species and redox reactions at mineral surfaces. The specific characteristics of surface species determine the rates of reactions at the mineral-water inter-face. For example, ligands that can form bidentate complexes on the surface are more effi-cient than monodentate ligands in promoting oxide dissolution (Furrer and Stumm, 1986). As discussed by Wehrli et al. (1989), oxidants or reductants can form both inner-sphere and outer-sphere surface complexes and the nature of the surface complex affects the reac-tivity of the surface-bound reactant. Because the surface OH-groups (or deprotonated O-groups) act as σ-donor ligands, a surface-bound Cu(II) ion experiences a higher electron density at the metal center and its redox potential (with regard to $Cu(I)_{aq}$) is lower than that of Cu^{2+} in solution. Fe(II) bound inner-spherically to an oxide mineral is a much stronger reductant than Fe^{2+}_{aq}. It is also likely that the ligand exchange kinetics (e.g., exchange of coordinated water) is facilitated for surface-bound metals as compared to the aqueous ionic species.

Table 5 compares the catalysis of redox reactions by ligands in solution with catalysis by surface functional groups (>M-OH). As shown, electron transfer may occur by either outer-sphere or inner-sphere pathways. In the outer-sphere reaction (Table 5, Reaction (a)), electrostatic models can be used to calculate the stability of the ion pair and the rate con-stant for the electron transfer can often be estimated from calculations of the change of bond distances in the coordination sphere upon electron transfer (Marcus theory, see Cannon, 1980). Inner-sphere redox processes (Reaction (b)) involve specific chemical interactions. Chemical models must replace electrostatic models.

Electron transfer reactions between dissolved metal ions typically occur by such reac-tion sequences. As discussed by Wehrli (1990), electron transfer from Fe^{2+}_{aq} to Fe^{3+}_{aq} is facilitated by hydrolysis of Fe^{3+}; the OH^- -bridge permits the formation of a precursor complex,

$$FeOH^{2+}_{aq} + Fe^{2+} \longrightarrow [Fe^{III} - OH - Fe^{II}]^{4+} \quad ,$$

Table 5. Mechanisms of redox reactions: comparison of the effect of ligands and of surface functional groups (modified from Wehrli et al., 1989)

---

1) HOMOGENEOUS LIGANDS

Outer-sphere. Initial formation of ion pair followed by rate-determining electron transfer, ET (electrostatic models)

$$A^+ + B \rightleftarrows A^+...B \rightleftarrows A...B^+ \rightleftarrows A + B^+$$
$$\text{diffusion} \qquad \text{ET} \qquad \text{diffusion}$$

(a)

Inner-sphere. Specific chemical interaction (inner-sphere complex formation involving a bridging ligand X) is followed by ET. (Chemical models)

$$AX^+ + B \rightleftarrows AX^+B \rightleftarrows AXB^+ \rightleftarrows A + XB^+$$
$$\text{complex formation} \qquad \text{ET} \qquad \text{dissociation}$$

(b)

Any of the three steps may determine the reaction rate. The electronic structure of the bridging ligand is critical in the electron transfer step.

---

2) HETEROGENEOUS LIGANDS

$$\text{ET}$$
$$>M\text{-}OH + HRed_{aq} \rightleftarrows >MRed \rightleftarrows >M'Ox \rightleftarrows M'_{aq} + Ox_{aq}$$
$$\text{where Red = reductant, Ox = oxidant (Ox + e}^- \rightleftarrows \text{Red)}$$
$$\text{and M' = reduced metal}$$

(c)

For reaction at the hydrous iron(III) oxide surface:

$$>Fe^{III}OH + Fe^{2+} \rightleftarrows (>Fe^{III}OFe^{II})^+ + H_3O^+ \rightleftarrows (>Fe^{II}OFe^{III})^+$$
$$\text{adsorption} \qquad \text{ET}$$
$$(>Fe^{II}OFe^{III})^+ + H_3O^+ \rightleftarrows (>Fe^{II}OH_2) + Fe^{III}(OH)^{2+}$$
$$\text{desorption}$$

(d)

For reactions at the hydrous iron(III) oxide surface with a bridging organic ligand L (e.g. oxalate):

$$>FeOH + HL^- \rightleftarrows >FeLH + OH^-$$
$$>FeLH + Fe^{2+} \rightleftarrows >FeLFe^+ + H^+$$
$$>Fe^{III}LFe^{II+} \rightleftarrows >Fe^{II}LFe^{III+}$$
$$\text{ET}$$
$$>Fe^{II}LFe^{III+} + H_2O \rightleftarrows >Fe^{II}OH_2 + Fe^{III}L^+$$

(e)

---

which undergoes electron transfer to form the successor complex $[Fe^{II} - OH - Fe^{III}]^{4+}$. This species subsequently dissociates into $Fe^{2+}_{aq} + FeOH^{2+}_{aq}$ (Haim, 1983). The accelerating effect of bridging ligands on electron transfer in solution has been demonstrated by Taube (1979) for complexes such as

$$(NH_3)_5 \, Co^{III}\text{-}X\text{-}Cr^{II} \quad ,$$

where X is a bridging ligand, e.g., Cl^- or oxalate, that permits the transfer of electrons and thus facilitates the reduction of Co(III) by Cr(II).

Similar reaction sequences may be postulated for heterogeneous electron transfer reactions. If inner-sphere complexes are formed (Table 5, Reaction (c)), any of the three steps may determine the reaction rate, as in homogeneous solution. Stone and Morgan (1987) have analyzed such three-step processes. Adsorption equilibria and the structure of the precursor complexes strongly affect heterogeneous redox kinetics.

Electron transfer at the hydrous iron(III) oxide surface can also be facilitated by bridging ligands. This is illustrated, for the general case, in Table 5, Reaction (e); hydroxo bridges (Reaction (d)) are a special case. The role of surface hydroxo bridges in the oxidation of U^{IV} by solid PbO_2 has been elucidated in ^{18}O tracer studies (Gordon and Taube, 1962). Both oxygen atoms in the product UO_2^{2+} originate from the reactant PbO_2. Thus, electron transfer must occur through the surface hydroxo bridges, $(Pb^{IV}\text{-}O)_2U^{IV}_{aq}$.

Application of surface complexation model to surface-controlled dissolution: model assumptions. Thus far we have described the basic concepts of surface-controlled dissolution and the surface complexation model, focusing particularly on the central importance of the concentration and reactivity of surface species. Explicit rate laws for surface-controlled dissolution can be formulated on the basis of the surface-complexation model. This formulation involves the following assumptions:

(i) Surface processes are rate-controlling. Diffusion of dissolved species to or from the mineral surface or within pores is assumed to be rapid relative to dissolution.

(ii) Back reactions, such as adsorption and precipitation, are neglected.

(iii) Surface sites with different activation energy (and thus different rates of reaction) exist due to the presence of steps, kinks, pits, etc. and surface defects.

(iv) Dissolution reactions proceed in parallel at surface sites with varying reactivity, but the overall dissolution is dominated by reaction at the most abundant of the most active sites (i.e., the sites of fastest dissolution).

(v) The active sites for dissolution are continuously regenerated. Thus, the mole fraction of active sites is constant during steady-state dissolution.

(vi) Detachment, taken to be the rate-limiting step in dissolution, may be considered as the decomposition of an activated surface complex. The precursor of the activated complex is a surface metal center which has been destabilized (relative to others on the surface of the crystal lattice) by formation of a surface complex, protonation of neighboring hydroxyl groups, reduction or oxidation to a more labile oxidation state, or a combination of these processes.

(vii) Formation of the precursor complex is rapid; the precursor complex is regenerated and its concentration is constant during dissolution.

(viii) The concentration of the precursor complex is small relative to the total concentration of surface sites.

(ix) The dissolution rate is proportional to the concentration (activity) of the precursor complex.

Although we postulate that the mole fraction of active sites remains constant during dissolution, it should be noted that some uncertainty remains concerning the role of the active sites in the dissolution processes. Crystal defects, which have little effect on the thermodynamic properties of the solid, could markedly affect the dissolution process. In recent experimental studies, however, dissolution rates changed only by a factor of two despite large variations in point defect densities (Casey et al., 1988; Schott et al., 1989).

The distribution of sites with different activation energies (and thus different reaction rates) on the surface can be represented by the conventional kink and step model of active sites. The overall rate of surface reactions (e.g., dissolution) reflects the contributions of parallel reactions occurring at different active sites.

Dissolution may be inhibited by substances that block the active sites by adsorption either of cations, such as VO^{2+}, Al^{3+}, Cr(III), or anions, such as phosphate, silicate or certain polymeric ligands (cf. Fig. 4). Not all surface sites have to be blocked, but only those active sites (e.g., dislocations and/or kinks) that determine the kinetics of surface reactions; very small concentrations of inhibitors may have a pronounced effect. (The same idea applies to certain features of corrosion control, especially the passivity of oxide films.)

The assumption of a constant distribution of active sites during dissolution is supported by numerical simulations of dissolution processes (Wehrli, 1989). The simulations indicate that the formation of steady-state morphologies might "override" the kinetic influence of point defects. The effect of reaction at distinct crystallographic planes on mineral dissolution has been discussed by Wehrli et al. (1990). Spectroscopic measurements of surface complexes during ligand-promoted dissolution also support the assumption that the concentration of the precursor complex is constant (Hering, unpub.).

Application of the surface complexation model: a generalized rate law for dissolution. For dissolution reactions under conditions far from equilibrium and for which the concentrations of the reactants (H^+, OH^-, ligands, reductants, or oxidants) in solution are not depleted during the course of the reaction, a steady-state dissolution rate is predicted by the surface complexation model. The (apparent) zero-order dissolution rate is proportional to the concentration of the precursor complex, that is,

$$R = k \, C_p^s = k \, x_a \, P_j S \quad , \tag{4}$$

where R = dissolution rate (mol m^{-2}s^{-1}), k = rate constant (s^{-1}), C_p^s = precursor concentration (mol m^{-2}), x_a = mole fraction of the active sites, P_j = probability that a specific site has the coordinative arrangement of the precursor complex, and S = total concentration of surface sites (mol m^{-2}). For a given oxide, the mole fraction of active sites (x_a) may be considered constant and this term included in the rate constant. Then, the reaction rates for different surface-controlled dissolution rates of (hydr)oxides may be expressed as shown in Table 6.

Application of the surface complexation model: a descriptive example. Here we will use the example of Fe(III) (hydr)oxide dissolution to illustrate the general applicability of the surface complexation model. Several possible pathways for Fe(III)(hydr)oxide dissolution are outlined in Figure 5. The structures in this figure are highly schematic. They are not intended to give the details of structural or coordinative arrangements but simply to illustrate the presence of surface hydroxyl groups and of oxo- and hydroxo-bridges and to indicate correct relative charges in a convenient way. This schematic presentation illustrates the diversity of reaction pathways. Steady-state dissolution kinetics are expected if the original surface structure is restored after detachment of the surface iron atom. In this case, the dissolution rate will be a function of the concentration of the appropriate surface species (cf. Table 6).

The first two pathways (a) and (b) show, respectively, the influence of H^+ and of surface complex-forming ligands. The fast binding of either or both of these reactants to the surface, which results in the weakening of the bonds in the proximity of a surface Fe(III) center, is followed by a slow detachment of the surface Fe(III) species into solution. The corresponding rate laws are given in Table 6 for R_H and R_L. These two pathways are discussed in detail (Furrer and Stumm, 1986; Zinder et al., 1986) and will not be further discussed here except to recall, as a background for the reductive dissolution, that bidentate surface complexes (surface chelates) are particularly efficient in enhancing dissolution.

Reductive dissolution mechanisms are illustrated in pathways (c)-(e). Reductants adsorbed to the hydrous oxide surface can readily exchange electrons with an Fe(III) surface

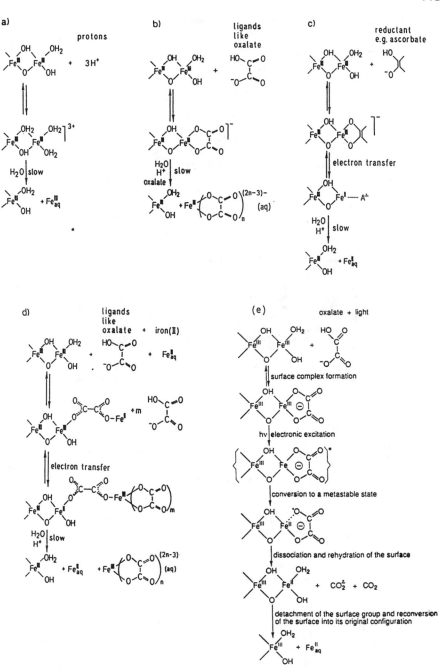

Figure 5. Schematic representation of the various reaction modes for the dissolution of Fe(III)(hydr)oxides: (a) by protons; (b) by bidentate complex formers that form surface chelates; (c) by reductants such as ascorbate that can form surface complexes and transfer electrons inner-spherically; (d) catalytic dissolution of Fe(III)(hydr)oxides by Fe(II) in the presence of a complex former and; (e) light-induced dissolution of Fe(III)(hydr)oxides in the presence of an electron donor such as oxalate. In all of the above examples, surface coordination controls the dissolution process (adapted from Sulzberger et al., 1989).

Table 6. Rate laws for the dissolution of oxide minerals [1,2]

REACTANT	EXPERIMENTAL RATE LAW
1) H^+ (acids)	$R_H = k_H^{'}(C_H^s)^j$
2) OH^- (bases)	$R_{OH} = k_{OH}^{'}(C_{OH}^s)^j$
3) L (ligands)	$R_L = k_L^{'}(C_L^s)$
4) Re (reductant)	$R_{Re} = k_{Re}^{'}(C_{Re}^s)$
5) Ox (oxidant)	$R_{ox} = k_{ox}^{'}(C_{Ox}^s)$
6) H_2O	$R_{H_2O} = k_{H_2O}^{'}$

1) The total rate can often (in absence of complications) be expressed as the sum of the individual rate laws, e.g., the pH-dependent rate of dissolution may be expressed as $R_T = R_H + R_{OH} + R_{H_2O}$ where C_H^s, C_{OH}^s, C_L^s, C_{Re}^s, C_{Ox}^s, are, respectively, the surface concentrations (mol m^{-2}) of H^+, OH^-, L, reductant and oxidant. The exponent j corresponds to the valence of the central ion (e.g., j = 2, 3 for BeO, Al_2O_3, respectively). The surface concentration of a species is related by an adsorption equation (surface complex formation mass law) to its solute concentration, e.g., $c_x^s \approx kc_x^a$ (where a<1). Thus if the rate law is written in terms of solute concentrations (activities) $R=kc_x^a$, the rate shows a dependence on the fractional order of the reactant in solution, c_x.

2) The rate constants k' are experimentally determined rate constants (with units h^{-1} $mol^{-(j-1)}$ $m^{2(j-1)}$); they include factors that cannot be determined individually.

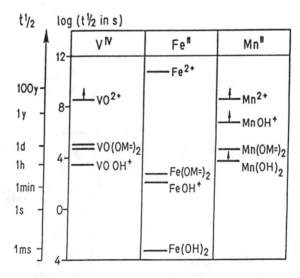

Figure 6. Effects of hydrolysis and adsorption on the oxygenation half-lives of transition metal ions. Arrows indicate lower limit. Data represent order of magnitude estimates because surface area concentrations were not determined (from Wehrli and Stumm, 1989)

center. Those reductants, such as ascorbate, that form inner-sphere surface complexes are especially efficient. The electron transfer leads to an oxidized reactant (often a radical) and a surface Fe(II) atom. The Fe(II)-O bond in the surface of the crystalline lattice is more labile than the Fe(III)-O bond and thus the reduced metal center is more easily detached from the surface than the original oxidized metal center. The corresponding rate law for pathway (c) is given by R_{Re} in Table 6. (The details of this and other reductive pathways for Fe(III) oxide dissolution will be discussed below under CASE STUDIES.)

Pathway (d) in Figure 5 provides a possible explanation for the efficiency of a combination of a reductant and a complex former in promoting fast dissolution of Fe(III) (hydr)oxides. In this pathway, Fe(II) is the reductant. In the absence of a complex former, however, Fe^{2+} does not transfer electrons to the surface Fe(III) of an Fe(III) (hydr)oxide to any measurable extent . The electron transfer occurs only in the presence of a suitable bridging ligand (e.g., oxalate). As illustrated in Figure 5 (d), a ternary surface complex is formed and an electron transfer, presumably inner-sphere, occurs between the adsorbed Fe(II) and the surface Fe(III). This is followed by the rate-limiting detachment of the reduced surface iron. In this pathway, the concentration of $Fe(II)_{aq}$ remains constant while the concentration of disolved Fe(III) increases; thus $Fe(II)_{aq}$ acts as a catalyst for the dissolution of Fe(III) (hydr)oxides.

Although thermodynamically favorable, reductive dissolution of Fe(III) (hydr)oxides by some metastable ligands (even those, such as oxalate, that can form surface complexes) does not occur in the absence of light. The photochemical pathway is depicted in Fig. 5(e). In the presence of light, surface complex formation is followed by electron transfer via an excited state (indicated by *) either of the iron oxide bulk phase or of the surface complex. (Light-induced reactions are discussed in Chapter 14.)

Several pathways may contribute to the overall dissolution reaction. Over the course of the dissolution, the relative importance of the contributing pathways may change, possibly due to accumulation of some reactant. Thus the overall rate law may not be simply the sum of the individual ones. This effect is apparent in the case of autocatalysis of photochemical reductive dissolution of iron(III)oxides by Fe(II) (cf. Fig. 5(e)) (Sulzberger et al., 1989; Sulzberger, 1990; Siffert and Sulzberger, in prep.)

Catalysis of redox reactions by mineral surfaces. Although this topic is discussed in more detail in Chapter 12, we present here one important example of redox reactions catalyzed by oxide surfaces, the oxidation of surface-bound Fe(II) by O_2. As mentioned previously, the redox characteristics (redox potential and redox kinetics) of transition metals (Fe^{2+}, Mn^{2+}, and VO^{2+}) are affected by complexation either by ligands in solution or by surface functional groups.

For the oxygenation of Fe(II) in solution (at pH > 5), the empirical rate law shows a second-order dependence on OH⁻ concentration (Stumm and Lee, 1961), that is,

$$- \frac{d\,[Fe(II)]}{dt} = k\,[Fe(II)]\,[OH^-]^2\,[O_2]\ . \tag{5}$$

This dependence is consistent with predominant reaction of the hydrolyzed species, $Fe(OH)_2$, within the pH range of interest, or

$$- \frac{d[Fe(II)]}{dt} = k'\,[Fe(OH)_2]\,[O_2]\ . \tag{6}$$

Similarly, the rate of oxygenation of Fe(II) bound to the surface hydroxyl groups of a hydrous oxide can be expressed in terms of the surface species (Tamura et al., 1980; Wehrli and Stumm, 1989; Wehrli, 1990). Thus,

$$- \frac{d[Fe(II)]}{dt} = k''\{Fe(OM<)_2\}[O_2] \; . \tag{7}$$

Coordination of Fe(II) either by OH⁻ in solution or by surface hydroxyl groups facilitates electron transfer to O_2. The rate of oxygenation depends strongly on the speciation of Fe(II); the overall rate law reflects the contributions of the Fe(II) species. Then,

$$- \frac{d[Fe(II)]}{dt} = (k_0[Fe^{2+}]+k_1[Fe(OH)^+]+k_2[Fe(OH)_2]+k''\{Fe(OM<)_2\}) \, [O_2] \; . \tag{8}$$

Figure 6 compares the half lives of V(IV), Fe(II) and Mn(II) species, including surface-bound metals, with respect to oxidation by oxygen (Po_2 = 0.2 atm). In each case, the relative effects of hydrolysis and adsorption are significant and similar.

Table 7. Examples of reductive dissolution – laboratory studies

MINERAL	REDUCTANT	ADDITIONAL LIGANDS	REFERENCES
[(a) iron oxides – reductant alone]			
hematite	ascorbate	–	1-3
goethite	ascorbate	–	1
goethite	phenols	–	4
goethite	S(-II)	–	5
goethite	dithionite	–	6
hematite	dithionite	–	6
magnetite	ascorbate	–	7
[(b) iron oxides – reduced metal and a ligand]			
hematite	Fe(II)	oxalate	8, 9
goethite	Fe(II)	oxalate	3, 9
magnetite	Fe(II)	oxalate	10
magnetite	Fe(II)	NTA	11
magnetite	Fe(II)	EDTA	12, 13
magnetite	V(II)	picolinate	14, 15
hematite	V(II)	picolinate	14, 15
(probably am-FeOOH)	Ti(III)	citrate/EDTA	16
[(c) iron oxides – (non-metal) reductant and a ligand]			
hematite	ascorbate	oxalate	2, 3
iron oxide/clays, soils	dithionite	citrate	17
[(d) iron oxides – photochemical]			
hematite	hv/ oxalate		18
hematite	hv/ bisulfite		19
[(e) manganese oxides]			
Mn(III,IV) oxides	phenols	–	20-24
	(non-aromatic) organic acids		21-25
	U(IV)$_{aq}$	–	26
birnessite	As(III)		27-29
birnessite	Se(IV)		29
[(f) other]			
Co(III) oxide	hydroquinone	–	24
NiFe$_2$O$_4$	V(II)	picolinate	30

REFERENCES: (1) Zinder et al., 1986 (2) Banwart, 1989 (3) Suter et al., submitted (4) LaKind and Stone, 1989 (5) Pyzik and Sommer, 1981 (6) Torrent et al., 1987 (7) dos Santos et al., in press (8) Siffert, 1989 (9) Suter et al., 1988 (10) Blesa et al., 1987 (11) del Valle Hidalgo et al., 1988 (12) Blesa et al., 1984 (13) Borghi et al., 1988 (14) Segal and Sellers, 1980 (15) Allen et al., 1988 (16) Hudson and Morel, 1989 (17) Mehra and Jackson, 1960 (18) Siffert and Sulzberger, in prep.(19) Faust and Hoffmann, 1986 (20) Stone and Morgan, 1984a (21) Stone and Morgan, 1984b (22) Stone, 1987a (23) Ulrich and Stone, 1989 (24) Stone and Ulrich, 1989 (25) Stone, 1987b (26) Gordon and Taube, 1962 (27) Oscarson et al., 1983a (28) Oscarson et al., 1983b (29) Scott, in prep. (30) Segal and Sellers, 1982

Table 8. Examples of oxidative dissolution – laboratory studies

MINERAL	OXIDANT	REFERENCES
[(a) sulfides]		
pyrite		(review) 1
	O_2	2-5
	Fe(III)	2,3
	H_2O_2	2
PbS	O_2	6
[(b) Fe(II) silicates]		
bronzite	O_2	7
fayalite	O_2	7
basalt	O_2	8
augite	Fe(III), O_2	9
biotite	Fe(III), O_2	9
hornblende	Fe(III), O_2	9
[(c) U(IV) oxides]		
uraninite	O_2	10
UO_2 (nuclear fuel)	O_2	11
	O_2/ electrochemical	12, 13
[(d) other]		
Cr_2O_3	MnO_4^-	14
NiFeCr-oxides	MnO_4^-	15

REFERENCES: (1) Lowson, 1982 (2) McKibben and Barnes, 1986 (3) Moses et al., 1987 (4) Nicholson et al., 1988 (5) Nicholson et al., 1990 (6) Hsieh and Huang, 1989 (7) Schott and Berner, 1983 (8) White et al., 1985 (9) White and Yee, 1985 (10) Grandstaff, 1976 (11) Ollilla, 1989 (12) Shoesmith et al., 1988 (13) Shoesmith et al., 1989 (14) Segal and Williams, 1986 (15) O'Brian et al., 1987.

CASE STUDIES

Many general reviews of reductive and oxidative dissolution are available (e.g., Valverde and Wagner, 1976; Segal and Sellers, 1984; Stone and Morgan, 1987). Tables 7 and 8 list some examples of the extensive laboratory studies of redox-enhanced dissolution. Here, we will discuss both reductive dissolution (for iron and manganese oxides) and oxidative dissolution (for pyrite, iron silicates, and uranium oxides) in terms of the surface complexation model.

Reductive dissolution of iron oxides

Dissolution of iron oxides is markedly enhanced by reducing agents such as organic reductants, inorganic reductants, reduced metals in combination with organic ligands, or combinations of organic reductants and ligands. Although redox reactions at mineral surfaces clearly influence dissolution rates, it becomes increasingly difficult to define the speciation of all redox-reactive species (particularly surface species) in complicated systems.

Reaction with a reductant. In the simpler cases of dissolution by an organic or inorganic reductant, surface-controlled dissolution may be unambiguously demonstrated. In the reduction of hematite by ascorbate (at pH 3), linear (steady-state) dissolution rates were observed (Fig. 7a). The dissolution rate showed saturation at high total ascorbate concentrations, i.e., a fractional-order dependence on total ascorbate, consistent with a first-order dependence of the rate on the surface concentration of ascorbate as shown in Figure 7b-d (Banwart, 1989; Suter et al., submitted). The dissolution rate, R, is a Langmuir-type function of the concentration of ascorbate in solution:

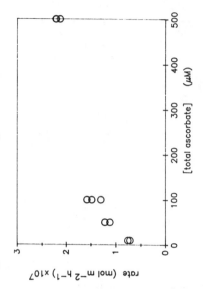

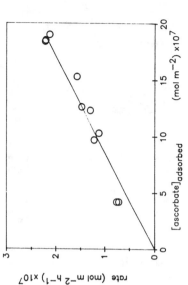

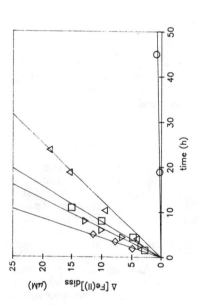

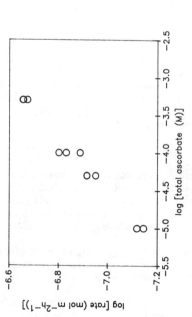

Figure 7. Reductive dissolution of hematite (0.61 g/L) at pH 3 in the presence of ascorbate. (a) Dissolved concentrations of Fe(II) relative to the value at $t = 0$ (Δ[Fe(II)]$_{\text{dissolved}}$) as a function of time for initial ascorbate concentrations: (o) 0, (Δ) 10, (II) 50, (∇) 100, (◊) 500 μM (adapted from Sulzberger et al., 1989), (b) dissolution rate vs. total ascorbate concentrations showing saturation at high total ascorbate, (c) logarithmic transform of dissociation rate vs. total ascorbate showing fractional-order dependence on total ascorbate, and (d) dissolution rate vs. adsorbed ascorbate showing first-order dependence on surface ascorbate (data from Banwart, 1989).

$$R = k \frac{K^s S [HA^-]}{1 + K^s [HA^-]} \quad , \tag{9}$$

where K^s is the equilibrium constant of the adsorption equilibrium, $[HA^-]$ is the dissolved concentration of ascorbic acid, k is the rate constant (h^{-1}) and S (mol m^{-2}) is the maximum capacity of the α-Fe_2O_3 surface for the adsorption of ascorbate, S={>FeA^-} + {>$FeOH$}. The Langmuir quotient in Equation (10) gives the concentration (mol m^{-2}) of the surface bound ascorbate

$$\{>FeA^-\} = \frac{K^s S [HA^-]}{1 + K^s [HA^-]} \tag{10}$$

for the following equilibrium:

$$>FeOH + HA^- \rightleftarrows > FeA^- + H_2O \quad . \tag{11}$$

Thus, as shown in Figure 7d, the rate is directly proportional to the concentration of surface-bound ascorbate,

$$R = k \{>FeA^-\} \quad . \tag{12}$$

Dissolution of goethite by S(-II) has also been interpreted as showing fractional-order dependence on total dissolved sulfide concentration (Pyzik and Sommer, 1981). In both these cases, a fractional-order dependence on proton concentration was also observed indicating a dependence of the rate on surface protonation.

Linear dissolution rates have also been observed for reaction of goethite with phenolic reductants. Phenolic substitution patterns significantly affected dissolution rates; dissolution was enhanced by chelating substituents (such as ortho-carboxylate or hydroxy groups) and electron-donating substituents and retarded by electron-withdrawing substituents. Substituent effects were ascribed to a rate-limiting electron transfer step. The agreement between the measured ratios of the oxidized phenolic product (p-benzoquinone) to Fe^{2+} and predicted stoichiometry for the reaction of goethite with hydroquinone was also taken to indicate that release of ferrous iron from the oxide surface (in the pH range from 2.0 to 4.6) was not rate-limiting (LaKind and Stone, 1989).

Inhibition of reductive dissolution by aluminum (either sorbed on or incorporated into iron oxides) has been observed both for reaction with ascorbate (Banwart, 1989; Suter et al., submitted) and dithionite (Torrent et al., 1987). This inhibition is consistent with a decrease in the concentration of surface complexes between Fe(III) and the reductant due to blocking of the surface by aluminum.

<u>Reaction with a reduced metal and a ligand.</u> Iron oxide dissolution is accelerated by combinations of reduced metals with (non-reducing) organic ligands. Some examples are given in Table 7, part (b). Because of rapid dissolution kinetics and low concentrations of (intermediate) surface complexes, concentrations of reactive surface species cannot be determined directly. Several mechanisms for such reactions have been suggested based on the observed dependence of the dissolution rate on total concentrations of reactive species; these observations for reactions with Fe(II) and an organic ligand Y (Y = oxalate, NTA, or EDTA) are depicted schematically in Figure 8. Obviously, the rate is strongly influenced by both Fe(II) and ligand concentrations. Reactions of hematite and goethite with Fe(II)/oxalate in our laboratory showed saturation of the dissolution rate with total Fe(II) but a direct dependence of the rate on the total oxalate concentration. This behavior has been attributed to an electron transfer step involving the surface species,

448

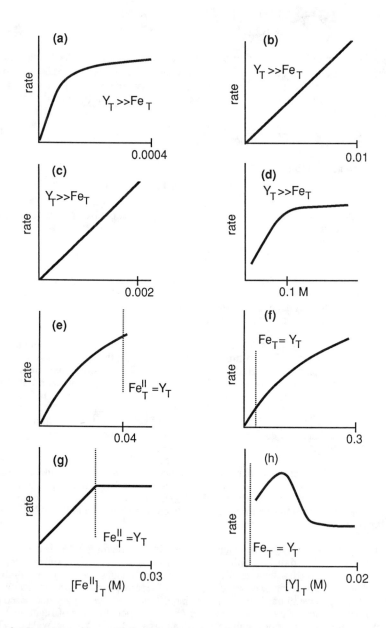

Figure 8. Dependence of iron oxide dissolution rates on total Fe(II) and ligand, Y, concentrations (a,b) for hematite, Y = oxalate (Siffert, 1989), and for magnetite (c,d) Y = oxalate (Blesa et al., 1987b), (e,f) Y = NTA (del Valle Hidalgo et al., 1988), (g,h) Y = EDTA (Borghi et al., 1988) (vertical dotted lines indicate concentrations at which $Fe^{II}_T = Y_T$.).

$$>Fe^{III}-Ox-Fe^{II}-Ox_n^{(2n-1)-} \quad .$$

The direct dependence of the rate on total oxalate concentration can be rationalized if the surface concentration of this species does not reach a saturating value within the experimental range of oxalate concentrations (Siffert, 1989; Suter et al., 1988, submitted). In contrast, the reaction of magnetite with Fe(II)/oxalate showed the opposite pattern – saturation of the dissolution rate with total oxalate and a linear dependence on total Fe(II). In a seeming paradox, both of these patterns of reactivity are consistent with a mechanism involving the proposed surface species. This is because of the different concentration ranges of total oxalate in these experiments. Consider the following surface species and equilibria between them:

$$>Fe^{III}Y + Fe^{II} \quad \rightleftharpoons \quad >Fe^{III}YFe^{II} \ ;$$
$$(I)$$

$$>Fe^{III}YFe^{II} + Y \quad \rightleftharpoons \quad >Fe^{III}YFe^{II}Y \ .$$
$$(I) \qquad\qquad\qquad (II)$$

Both species are favored at high total Fe^{II}, but species I is more favored at relatively lower Y concentrations as shown in Figure 9. [Calculations of species concentrations were made with the (arbitrarily chosen) stability constants for species I and II given in Table 9.] Under these conditions, species II may reach a maximum concentration with increasing total Fe(II) (note that this concentration is much less than the total surface site concentration). At high total Y, the concentration of species II continues to increase with increasing total Fe(II). Figure 10a shows the concentration of species II as a function of total Fe(II) at two total oxalate concentrations; a saturating concentration of species II is predicted at low total oxalate (consistent with our observations) but not at high total oxalate (consistent with observations of Blesa and co-workers). The concentration of species II is also predicted to reach a maximum value and in fact decrease as a function of total oxalate (due to the competition between solution oxalate and surface oxalate for Fe(II)) as shown in Figure 10b. Such an effect was indeed observed by Blesa and co-workers in the reaction of Fe(II)/EDTA with magnetite at high EDTA concentrations. Although such speculative arguments cannot prove the proposed reaction pathway, the observed dependence of dissolution rates on bulk solution parameters is at least consistent with a surface-controlled dissolution mechanism.

In contrast, the extremely rapid reaction of hematite with V(II)picolinate$_3$ most probably proceeds by an outer-sphere mechanism. The rate of the reaction approaches diffusion-limitation despite the relative substitution inertness of V^{2+} (Segal and Sellers, 1980). In the reaction of V(II)picolinate$_3$ with $NiFe_2O_4(s)$, a linear dependence of the rate on the V(II) concentration has been observed (Segal and Sellers, 1982).

Reaction with a (non-metal) reductant and a ligand. Rates of reductive dissolution can be significantly enhanced by organic ligands as has been shown for iron oxides with the reagent combinations dithionite/citrate (Mehra and Jackson, 1960) and ascorbate/oxalate (Zinder et al., 1986; Banwart et al., 1989). As discussed above (for the Fe(II)/oxalate reaction), accurate assessment of surface concentrations of reactive species is difficult when dissolution is rapid. Figure 11 shows the dissolution rates of hematite for the ascorbate/oxalate reaction as a function of total concentrations of oxalate and ascorbate and also the sorption isotherms for the individual reactants. Dissolution rates were linearly related to total oxalate concentrations over the range in concentrations examined. The experimental range in concentration, however, barely reached that required to saturate the surface (even in the absence of a competing sorbing ligand). Thus, the observed dependence is consistent with reaction of a surface oxalate species although this cannot be proven based on the

450

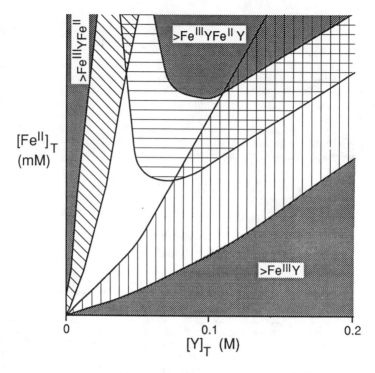

Figure 9. Calculated distribution of (hypothetical) surface species as a function of total Fe(II) and total ligand (Y = oxalate) concentrations (stability constants given in Table 9).

Table 9. Stability constants used to calculate
concentrations of surface species [1,2]

SPECIES	LOG β
[solution species]	
HY	3.97
H_2Y	5.01
Fe(II)Y	3.52
Fe(II)Y_2	5.85
[surface species]	
>Fe(III)Y	8.3
>Fe(III)YFe(II)	12.8
>Fe(III)YFe(II)Y	14.4

[1] Calculated for total concentration of surface
sites = 1.8 µM; pH = 3; ionic strength 0.005M
(no electrostatic effects considered).

[2] Stability constants given as $\beta = [H_2Y] / [H]^2 [Y]$.
Stability constants for solution species (Y = oxalate) from
Martell and Smith (1977) corrected for ionic strength.
Stability constants for surface species chosen arbitrarily.

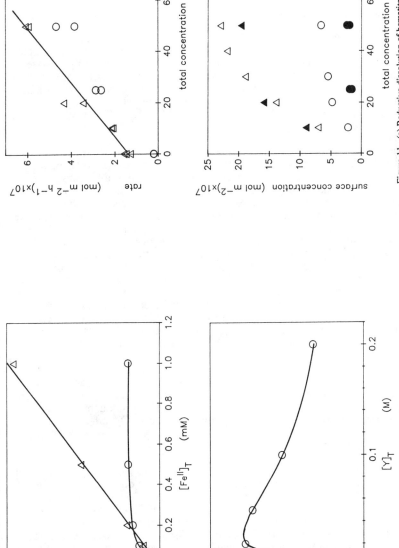

Figure 11. (a) Reductive dissolution of hematite by ascorbate at pH 3 in the presence of oxalate as a function of total concentrations of (o) ascorbate or (Δ) oxalate. (b) Surface concentrations of oxalate (Δ,▲) and ascorbate (o,●) as a function of total concentration. Open symbols: only one ligand present. Closed symbols: both ligands present (▲) total ascorbate = 100 μM, (●) total oxalate = 50 μM (data from Banwart, 1989).

Figure 10. Calculated concentrations of (hypothetical) ternary surface species, >Fe^{III}YFe^{II}Y, (a) as a function of total Fe(II) for 0.005 M (o) or 0.2 M (Δ) total oxalate, and (b) as a function of total ligand where Y = oxalate (stability constants given in Table 9).

available data. In contrast, the dependence of the dissolution rate on the total ascorbate concentration was clearly non-linear indicating that reductive dissolution by ascorbate in the presence of oxalate (as well as without oxalate – cf. Fig. 7) involves reaction of a surface ascorbate species (Banwart, 1989; Banwart et al., 1989). The accelerating effect of oxalate suggests that the rate-limiting step in the reductive dissolution by ascorbate is detachment rather than electron transfer since oxalate is more likely to facilitate detachment of a reduced iron center than to participate directly in the electron transfer from ascorbate to a surface iron atom.

Reductive dissolution of manganese oxides

The dissolution of Mn oxides in the presence of organic reductants (particularly phenols) has been studied extensively by Stone and co-workers (for references, see Table 7). These experiments demonstrate that the dissolution rate is controlled by the rate of chemical reactions on the mineral surface rather than diffusive processes. Increased stirring rates had no effect on the dissolution rate, activation energies were too high to be attributed to transport processes, and, in some cases, dissolution was inhibited in the presence of inorganic species (Ca^{2+} and HPO_4^{2-}) that sorb to mineral surfaces (Stone and Morgan, 1984a).

A significant effect of the structure of the organic reductant on the dissolution rate was observed. For chloro-substituted phenols, the dissolution rates (normalized for the extent of phenol adsorption) followed the pattern para-substituted > ortho-substituted >> meta-substituted for monochlorophenols and para, ortho-disubstituted >> para, meta-disubstituted >> meta, meta-disubstituted for dichlorophenols, consistent with substituent effects on the stability of the phenol oxidation intermediate, i.e., the phenoxy radical. On this basis, the authors suggest that the rate-limiting step involves an electron transfer reaction at the mineral surface rather than a detachment step (Ulrich and Stone, 1989). They support their conclusion by observations of the relative concentrations of the products released into solution in the reaction of MnO_2 with hydroquinone. The ratio of products (p-benzoquinone : Mn^{2+}) initially exceeded the 1:1 ratio predicted from stoichiometry, gradually decreasing to approach the stoichiometric value. This discrepancy could be explained by sorption of Mn^{2+} on the oxide surface. Quantitative agreement between values for total Mn (i.e., sorbed + dissolved) calculated from reaction stoichiometry and from measured Mn^{2+} and independent sorption isotherms suggest that Mn(II) sorption/desorption is fast compared to other surface chemical reactions such as electron transfer (Stone and Ulrich, 1989).

Studies of isotopic exchange in the reaction of MnO_2 with the inorganic reductant U^{4+} have shown that both of the oxygen atoms in the product uranyl ion (UO_2^{2-}) are derived from the solid oxidant (Gordon and Taube, 1962). This result demonstrates that this reaction proceeds by an inner-sphere electron transfer at the MnO_2 surface.

Oxidative dissolution of pyrite

Pyrite dissolution is enhanced in the presence of oxidants such as O_2, Fe(III), and H_2O_2. With properly cleaned pyrite, linear dissolution rates are observed in oxygenated suspensions (McKibben and Barnes, 1986; Moses et al., 1987). Surface-controlled dissolution is evidenced by an activation energy in the range of 60 to 90 kJ/mole and non-linear dependence of the dissolution rate on O_2 concentration (McKibben and Barnes, 1986; Moses et al., 1987; Nicholson et al., 1988). Non-linear dissolution kinetics were observed for reaction with Fe(III) and H_2O_2 because of depletion of the oxidant during the course of the dissolution. Oxidative dissolution by Fe(III), in contrast to O_2, showed a marked pH dependence; fractional-order dependence on Fe(III) concentration was observed (McKibben and Barnes, 1986; Moses et al., 1988).

The observed kinetics of pyrite oxidation have been interpreted by Luther (1987) according to molecular orbital theory. Facile pyrite oxidation by Fe^{3+} may be explained by formation of an Fe(III) surface complex with the persulfido ligands of the pyrite:

$$>Fe^{II}\text{-}S\text{-}S\text{-}Fe^{III}(H_2O)_5]\ ^{3+}$$

followed by inner-sphere electron transfer. In contrast, O_2 cannot form a bridged complex with the pyrite persulfido groups and thus electron transfer must proceed through an outer-sphere mechanism (Luther, 1987).

Oxidative dissolution of uranium(IV) oxides

The dissolution of UO_2 is also thermodynamically favored under oxidizing conditions as demonstrated both by equilibrium calculations and solubility measurements (Garisto and Garisto, 1986; Ollilla, 1989). Grandstaff (1976) has shown that the oxidative dissolution of uraninite in aqueous solution is surface-controlled. The observed linear dissolution rates were not affected by stirring rates and were proportional to mineral surface area. First-order, rather than fractional-order, dependencies on oxygen partial pressure and pH were reported; dissolution was not studied, however, for pH less than 4 or O_2 partial pressures greater than 0.2 atm. Langmuir-type dependence of the rate on dissolved bicarbonate (at pH 8.3) was observed. Dissolution was inhibited by the presence of non-uranium cations in the uraninite samples and by dissolved organic matter. Studies of the anodic oxidation of UO_2 (nuclear fuel) in aqueous solution have shown formation of successive surface films. The observed solid formation sequence was

$$UO_2 \rightarrow U_4O_9 \rightarrow U_3O_7 \rightarrow U_2O_5 \rightarrow U_3O_8 \rightarrow UO_3\cdot 2H_2O\ \ .$$

In solutions containing carbonate or phosphate, formation of uranyl carbonate or phosphate surface layers was observed (Garisto and Garisto, 1986 and ref. cit.; Shoesmith et al., 1988; 1989).

Oxidative dissolution of iron(II) silicates

Dissolution of Fe(II) silicates under oxic conditions clearly demonstrate the importance and complexity of chemical reactions at mineral surfaces. Initial incongruent dissolution of bronzite was attributed to formation of a thin (<10 Å) surface leached layer, depleted in Fe relative to Si (Schott and Berner, 1983). Initial incongruent dissolution of basalt was ascribed to reactions of a mixture of mineral phases (White et al., 1985). Long-term linear dissolution kinetics were observed for bronzite under anoxic conditions and low-pH, oxic conditions (Schott and Berner, 1983). Comparison of the final concentrations of dissolved iron in (initially) low-pH hornblende dissolution experiments reported by White and Yee (1985) suggest enhancement of dissolution under oxic conditions. At pH 6, the formation of a surface layer consisting of Fe(III) in a silicate matrix indicates the oxidation of surface Fe(II). Silicate mineral dissolution under oxic conditions does not, however, result in increased dissolved concentrations of Fe(III) but rather of Fe(II). White and Yee (1985) have demonstrated that under these conditions aqueous Fe(III) is reduced by the Fe(II) silicate to give Fe(II) in solution and have suggested that Fe^{2+} is adsorbed onto silicate mineral surfaces especially at higher pH. In the dissolution of bronzite under oxic conditions at pH 6, the observation of parabolic rather than linear dissolution kinetics was attributed to inhibition of dissolution by the surface Fe(III)/silicate layer. Thus while oxidation of silicate-bound Fe(II) with concomitant reduction of O_2 or Fe(III) occurs at the silicate mineral surface, these redox reactions do not result in enhancement of dissolution at neutral or slightly acidic pH (though there may be some enhancement at pH 1). Oxidative dissolution of Fe(II) silicates by NO_3^- occurs at low rates at pH 2-7. The mechanism is complex; the reaction appears to be catalyzed by FeOOH(s) (Postma, 1990).

DISCUSSION

Applicability and limitations of the surface complexation model

Laboratory studies. Surface control of reductive and oxidative dissolution has been clearly demonstrated in many instances, as discussed under CASE STUDIES above. Generally, the predictions of the surface complexation model (linear dissolution kinetics and fractional-order dependence on the dissolved concentrations of reactants) have been confirmed.

Thus, with the surface complexation model as a conceptual base, we may examine the discrepancies, real or apparent, between model predictions and laboratory observations. Some of these discrepancies arise from experimental artifacts. Initial rapid dissolution kinetics (followed by steady-state dissolution) generally occurs as a result of the presence of fine particles produced by mineral grinding and can be eliminated by appropriate sample preparation (Berner and Holdren, 1977, 1979; Berner et al., 1980; Holdren and Berner, 1979; Schott et al., 1981; Eggleston et al., 1989). Incongruent dissolution is likely to be observed in heterogeneous samples, such as soils, due to rapid dissolution of amorphous phases (Schnoor, 1990). An apparent decrease in dissolution rates at later reaction times may occur as an artifact if some cation released in the dissolution is re-precipitated as a secondary mineral phase (Schott and Petit, 1987 and ref. cit.).

In some cases, however, there are real limitations on the applicability of the surface complexation model. This is most obvious if the dissolution rate approaches the rate of solute diffusion as has been observed for some extremely soluble minerals, e.g., calcite dissolution at low pH (Berner and Morse, 1974). Initial incongruent dissolution may be observed during the development of a surface leached layer. As discussed by Schnoor (1990), incongruent dissolution of silicates is observed as base cations (Ca^{2+}, Mg^{2+}, Na^+, and K^+) are depleted preferentially to Si and Al. Growth of such leached layers, however, does not appear to continue during further dissolution for which steady-state, congruent dissolution is observed (Berner et al., 1980; Berner and Schott, 1982; Schott and Berner, 1983; Hellmann et al., 1990). Formation of extensive surface coatings or stable leached layers on dissolving minerals could result in inhibition of long-term dissolution. In this case, slow diffusion through the precipitate coating would control the overall dissolution rate. This mechanism has been invoked to explain the significant decrease in pyrite oxidation rates over approximately 100 days; accumulation of iron oxides on the pyrite surface was observed (Nicholson et al., 1990). Dissolution of iron silicates under oxic conditions was inhibited by formation of a surface layer consisting of Fe^{3+} in a silicate matrix (Schott and Berner, 1983). Thus, in some instances, physical processes (particularly diffusion through surface layers on solids) must be considered.

Applicability to field observations. The utility of the surface complexation model in interpreting field weathering rates is supported by laboratory weathering experiments with natural soil samples. Linear, congruent dissolution kinetics have been observed in soil column experiments; initial, incongruent dissolution could be explained by rapid dissolution of amorphous phases and short-term processes such as ion-exchange (Schnoor, 1990). The development of surface leached layers is also important for field samples, but comparison of soil grains with minerals subjected to laboratory dissolution experiments indicates that continued growth of surface leached layers on the time scale of environmental weathering can be discounted (Schott and Berner, 1983).

A major complication in the application of the surface complexation model to describe and interpret weathering in the field is the difficulty in evaluating the wetted surface area of reacting minerals and the mass of wetted soil under field conditions, crucial parameters in the estimation of field weathering rates. Generally, estimated field weathering rates have been reported to be between 10 and 100 times slower than mineral dissolution rates determined in the laboratory (Velbel, 1986; Schnoor, 1990) but this may be due to inaccurate estimation of the contact area.

As discussed by Schnoor (1990), the dependence of field weathering rates on mass-normalized flow rate indicates a shift from hydrologic control (at low flow) to surface-controlled dissolution (at high flow); weathering rates at high flow are comparable to laboratory experiments. Hydrologic control of dissolution might arise from unsaturated flow through soil macropores or inhibition of aluminosilicate dissolution due to high Al concentrations in soil macropores.

The importance of microbial processes in the natural environment also complicates the extrapolation of laboratory weathering rates to the field. For example, the rate-determining step in the oxidative dissolution of pyrite in acidic mine drainage waters is not surface-controlled oxidation of S(-II) to SO_4^{2-} and release of Fe(II) but the subsequent oxygenation of Fe(II) in solution. Thus, the overall rate of pyrite dissolution is insensitive to the mineral surface area but is influenced by the microbial catalysis of Fe(II) oxygenation. At the relatively low pH values typically encountered under conditions of pyrite dissolution, the oxidation of Fe(II) is slow unless catalyzed by autotrophic bacteria (Singer and Stumm, 1970; Stumm-Zollinger, 1972). In this case, although the surface complexation model may be used to predict the rate of specific reactions at the mineral surface, these are not the reactions that determine the <u>overall</u> rate of the reaction in the natural environment.

If microbial catalysis of redox reactions is important, extrapolation of rates from abiotic laboratory systems may significantly underestimate actual environmental rates. For example, environmental rates of manganese oxidation exceed the abiotic rates for this reaction in homogeneous solution (Diem and Stumm, 1984) by many orders of magnitude and even exceed the abiotic rates for the oxidation catalyzed by mineral surfaces (Davies and Morgan, 1989). This effect has been ascribed both to indirect microbial catalysis, by maintenance of high oxygen partial pressures and high pH in microenvironments near cell surfaces (Richardson et al., 1988), and to direct microbial mediation (Sunda and Huntsman, 1987).

Dissolution and its reverse: precipitation

Dissolution and precipitation are interrelated in two different ways. First, incongruent dissolution is often accompanied by the formation of a new solid phase. For example, weathering of feldspar is often coupled to the formation of kaolinite; oxidative dissolution of an Fe(II) bearing silicate may lead to the formation of a surface layer of $Fe(OH)_3$ (s); and reductive dissolution of MnO_2(s) by Fe(II) may cause the precipitation of $Fe(OH)_3$ (s). Thus, dissolution and precipitation may be kinetically coupled. Second, dissolution and precipitation of a particular solid phase are mechanistically interdependent. According to the principle of detailed balancing, the ratio of the forward and reverse rate constants is equal to the equilibrium constant. As discussed by Lasaga (1981), this principle of microscopic reversibility (i.e., detailed balancing) applies only to simple reaction mechanisms and must be applied with caution in the case of more complex processes. If we consider a dissolution reaction where the rate is controlled by desorption (detachment) of a central ion, then the rate of the corresponding precipitation pathway would be controlled (in the sense of heterogeneous nucleation) by an adsorption (incorporation) of the corresponding central ion. The validity of the principle of microscopic reversibility in the case of reductive or oxidative dissolution has not been ascertained; we would suggest this as an interesting question for future investigation. The study of heterogeneous nucleation processses should be a valuable adjunct to the study of dissolution processes. Schneider and Schwyn (1987) have illustrated that the morphological properties of Fe(III) (hydr)oxides reflect the hydrolytic pathway of their formation in synthetic, biological and aqueous media. They have also shown that the mechanistic pathways of formation (e.g., by hydrolysis of Fe(III) or by hydrolysis induced by oxidation of Fe(II)) can be understood, at least in part, by studying the depolymerization of polynuclear iron species by H+, ligands and reductants.

Remarkable progress has been made recently in the study of biomineralization. In this process, inorganic elements are extracted from the environment and selectively precipitated by organisms. Many oxidizable and reducible minerals such as magnetite, goethite and pyrite are formed at intracellular sites. Usually, templates consisting of suitable biological macromolecules serve as a matrix for the heterogeneous nucleation of bulk mineralized structures such as bone, teeth and shells. Biological control mechanisms are reflected not only in the type of the mineral phase formed but also in its morphology and crystallographic orientation (Mann, 1988). The understanding of the mechanisms of biomineralization processes may also be of great value in assessing dissolution pathways for biologically-precipitated minerals.

Some geochemical implications

Obviously the cycling of an element through the various global reservoirs (atmospheres, oceans, soils, sediments) depends on many physical, chemical, and biological processes. In all these reservoirs, solids are present with high surface area to volume ratios. Surface-controlled reactions are of great significance in regulating the composition of the reservoirs and influencing the mass fluxes of elements from one reservoir to another. What are the roles of thermodynamics, of surface kinetics, and of the biota in controlling the geochemical cycles of elements? With some exceptions, there is not yet sufficient information available to give a satisfactory answer to this question. Above all, we must remember – and we have already given a few examples – that the cycles are interdependent and that they are often kinetically coupled (Lasaga, 1981). We may illustrate this in a simplified way by elaborating on certain aspects of the iron(II,III) cycle, its interdependence with the cycles of P, S, heavy metals, O_2, and C, and the role of the biota and light.

The iron cycle shown in Figure 12 illustrates some redox processes typically observed in soils, sediments and waters, especially at oxic-anoxic boundaries. The cycle includes the reductive dissolution of iron(III) oxides by organic ligands (which, in surface waters, may also be photocatalyzed) and the oxygenation or surface catalyzed oxidation of Fe(II) by oxygen. As has already been mentioned, the oxidation of Fe(II) to Fe(III) (hydr)oxides is accompanied by the binding of reactive compounds (heavy metals, phosphate, or organic compounds). The reduction of the ferric (hydr)oxides is accompanied by the release of these substances. Phosphate sorption on surfaces of iron(III) oxides in surficial sediments can influence the concentration of phosphate in interstitial waters, which in turn regulates the flux of phosphate to the overlying water. Van Cappellen and Berner (1988) show in their mathematical models for the diagenesis of P in marine sediments that the pore water and solid sediment profiles of P depend on (i) the amount of reactive organic matter, biogenic mineral debris and hydrous ferric oxides and their characteristic depth scales for diagenetic remobilization and (ii) the mode and rate of fluorapatite precipitation. Krom and Berner (1981) suggest that the release of reactive P associated with hydrous ferric oxides closely tracks changes in redox conditions.

As is shown schematically in Figure 3, a rapid turn-over of iron takes place at the oxic-anoxic boundary in the water column. The principal reductant is the biodegradable biogenic material that settles down into the underlying waters. Within the depth-dependent redox gradient, concentration peaks of solid Fe(III) and dissolved Fe(II) develop (Davison, 1985), the peak of Fe(III) overlapping the peak of Fe(II). Organic acids or phenolic compounds can be generated in the water column by microbial degradation of biopolymers or, in the sediments, through fermentation processes (which typically occur at redox potentials around -0.2 V at pH ~7). These compounds may then participate in the reductive dissolution of Fe(III) (hydr)oxides, possibly by the Fe(II) catalyzed pathway discussed earlier (cf. Fig. 5(d) and CASE STUDIES). The sequence of diffusional transport of Fe(II), oxidation to

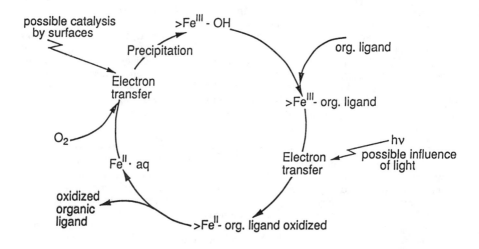

Figure 12. The dissolution of Fe(III)(hydr)oxides promoted by oxidizable organic ligands and the subsequent reoxidation of Fe(II) by O_2 plays an important role in soils, sediments, and waters and causes relatively rapid cycling of electrons and of reactive elements (e.g., organic carbon, oxygen, trace metals, and phosphate) at the oxic-anoxic boundary. This cycle can also occur, photochemically induced, in oxic surface waters. On overall balance, the iron cycle mediates the progressive oxidation of organic matter by oxygen.

insoluble $Fe(OH)_3$, and subsequent settling and reduction to dissolved Fe(II) typically occurs within a relatively narrow redox-cline.

On overall balance the cycle shown in Figure 12 represents a mediation (by iron) of the oxidation of organic matter by oxygen. This oxidation may be important both in the degradation and polymerization of organic matter in soils and waters. The interdependence of the iron cycle with that of other redox cycles is obvious; e.g., Fe(II) can reduce Mn(III,IV) oxides and HS⁻ is an efficient reductant for hydrous Fe(III) oxides.

Many of the processes mentioned above occur also in soils. Microorganisms and plants produce a larger number of biogenic acids. Oxalic, maleic, acetic, succinic, tartaric, ketogluconic and p-hydroxybenzonic acids have been found in top soils with oxalic acid, the most abundant, in concentrations as high as 10^{-5}–10^{-4} M (in soil water). The downward vertical displacement of Al and Fe observed during podzolisation of soils can be explained by considering the effect of pH and of complex formers on the solubility of iron and aluminum oxides and on their dissolution rates. The reductive dissolution of Fe(III) (hydr)oxides is also of importance in iron uptake by higher plants.

Somewhat surprisingly, the iron cycle is also important in atmospheric waters. Redox reactions, particularly the oxidation of SO_2, NO, and reduced C-compounds, occur in the atmosphere on an immense scale. Although liquid water constitutes less than 10^{-6} % of the troposphere, chemical processes in the aqueous phase (clouds, fog, rain) are of great relevance. Transition metals such as Fe, Mn and Cu and their redox pairs Fe(III)/Fe(II), Mn(IV,III)/Mn(II) and Cu(II)/Cu(I) are important as catalysts and reductants. Jacob et al. (1989) and Jacob and Hoffmann (1983) have suggested that Fe(III) acts either as a catalyst or as a direct oxidant for S(IV). Behra and Sigg (1990) have reported that a large fraction of the total Fe in fog water is present as dissolved Fe(II). The concentration of Fe(II) increased both with decreasing pH and exposure to light; maximum concentrations of 0.2 mM were reported. These authors assume that Fe(III) (hydr)oxide is reduced by sulfite, organic compounds, and free radicals formed photochemically during daytime. These pH- and light-dependent reduction reactions can be fast enough to outcompete reoxidation of Fe(II) by O_2.

Thus, atmospheric depositions may be a source of dissolved Fe(II) as well as particulate Fe(III) (present in wind-borne dust as coatings on minerals or colloidal oxide particles) to surface waters. Photo-reduction and dissolution of Fe(III) (hydr)oxides, either in atmospheric waters prior to deposition or in surface waters after deposition, increases the bioavailability of iron to phytoplankton. The speciation and redox cycling of Fe is of particular significance in areas where primary productivity is iron-limited. Martin et al. (1990) have recently suggested that phytoplankton productivity is limited by iron availability in off-shore Antarctic waters which, although rich in the major nutrients (N, P, and Si), are iron-poor. These authors have also suggested that variations, between glacial and inter-glacial times, in atmospheric iron inputs to these regions and the consequent changes in primary productivity may have been of sufficient magnitude to influence atmospheric CO_2 concentrations, yet another aspect of the interdependence of global Fe and C cycles.

CONCLUDING REMARKS

Redox reactions often enhance the dissolution rate of reducible and oxidizable minerals and thus significantly influence chemical weathering processes and the global hydrogeochemical cycles of many elements. Changes in oxidation state affect mineral solubility; some oxides (such as Cr_2O_3, V_2O_3) become more soluble upon oxidation. For others, particularly Fe(III) and Mn(III,IV) oxides, the reduction of surface metal centers results in easier detachment of the reduced metal ions from the lattice surface because the reduced metal-oxygen bond is more labile than the non-reduced metal-oxygen bond. Many of these dissolution processes are controlled by a reaction step at the surface (and not by a transport step). The biota assist the weathering processes by providing organic ligands (which are often potential reductants) and acids; the geochemical cycling of electrons is enhanced by redox gradients established by photosynthesis.

The surface reactivity depends on the surface species and their structural identity. A general rate law for reductive or oxidative surface-controlled dissolution of minerals can be derived by considering surface coordination. Functional groups present at the surface of minerals (-OH on oxides and silicates, -SH and -S-S on sulfides and polysulfides, and -CO_2OH on carbonates) can interact (chemically) with H^+, metal ions and ligands; oxidants and reductants can form surface complexes and exchange electrons with the central ions of the surface lattice. Electron transfer reactions can occur more readily if the redox couple is linked by a bridging ligand. Usually the dissolution rate is found to be proportional to the concentration of the surface bound (specifically adsorbed) reductant or oxidant.

The cycling of iron as it occurs in natural systems (waters, sediments, soils and atmosphere) is used to illustrate the various processes, including photocatalyzed reactions, that involve iron minerals and that are mediated by surfaces. In this cycle, reductive dissolution is accompanied by the release of reactive elements (which were adsorbed to the iron(III) oxide surfaces) and subsequent oxidation and precipitation are accompanied by sorption of reactive compounds. These processes affect metabolic transformations in soils and marine and lacustrine water-sediment systems, the cycling of elements (oxygen, trace metals, phosphate, sulfur, organic carbon), and biological productivity.

ACKNOWLEDGMENTS

We thank L. Sigg and P. Wersin for providing unpublished data and P. Brady, J. Morgan, and B. Wehrli for their helpful comments on the manuscript. Our research on weathering kinetics was supported by the Swiss National Science Foundation.

REFERENCES

Allen, G.C., Kirby, C., and Sellers, R.M. (1988) The effect of the low-oxidation-state metal ion reagent trispicolinatovanadium(II) formate on the surface morphology and composition of crystalline iron oxides. J. Chem. Soc. Faraday Trans. 1, 84, 355-364.

Anderson, M.A. and Morel, F.M.M. (1980) Uptake of Fe(II) by a diatom in oxic culture medium. Mar. Biol. Lett. 1, 263-268.

Armstrong, D.E., Hurley, J.P., Swackhammer, D.L., and Shafer, M.M. (1987) Cycles of nutrient elements, hydrophobic organic compounds, and metals in Crystal Lake: Role of particle-mediated processes. In: Sources and Fates of Aquatic Pollutants, R.A. Hites and S.J. Eisenreich, eds., Amer. Chem. Soc. Adv. Chem. Ser. 216, pp. 491-518.

Arnold, R.G., DiChristina, T.J., and Hoffmann, M.R. (1988) Reductive dissolution of Fe(III) oxides by Pseudomonas sp. 200. Biotechnol. Bioeng. 32, 1081-1096.

Baas Becking, L.G.M., Kaplan, J.R., and Moore, D. (1960) Limits of the natural environment in terms of pH and oxidation - reduction potential. J. Geol. 68, 243-284.

Banwart, S. (1989) The reductive dissolution of hematite (α-Fe_2O_3) by ascorbate. Ph.D. dissertation. Swiss Federal Institute of Technology, Zürich, Switzerland.

Banwart, S., Davies, S., and Stumm, W. (1989) The role of oxalate in accelerating the reductive dissolution of hematite (α-Fe_2O_3) by ascorbate. Coll. Surf. 39, 303-309.

Barcelona, M.J. (1980) Dissolved organic carbon and volatile fatty acids in marine sediment pore waters. Geochim. Cosmochim Acta 46, 575-89.

Behra, P. and Sigg, L. (1990) Evidence for redox cycling of iron in atmospheric water. Nature 344, 419-421.

Berner R.A. and Holdren, G.R. Jr. (1977) Mechanism of feldspar weathering: Some observational evidence. Geology 5, 369-372.

Berner, R.A, and Holdren, G.R. Jr. (1979) Mechanism of feldspar weathering – II. Observations of feldspars from soils. Geochim. Cosmochim. Acta 43, 1173-1186.

Berner, R.A. and Morse, J.W. (1974) Dissolution kinetics of calcium carbonate in seawater: IV. Theory of calcite dissolution. Amer. J. Sci. 274, 108-134.

Berner, R.A. and Schott, J. (1982) Mechanisms of pyroxene and amphibole weathering. II. Observations of soil grains. Amer. J. Sci. 282, 1214-1231.

Berner, R.A., Sjöberg, E.L., Velbel, M.A., and Krom, M.D. (1980) Dissolution of pyroxenes and amphiboles during weathering. Science 207, 1205-1206.

Blesa, M.A., Borghi, E.B., Maroto, A.J.G., and Regazzoni, A.E. (1984) Adsorption of EDTA and iron-EDTA complexes on magnetite and the mechanism of dissolution of magnetite by EDTA. J. Coll. Int. Sci. 98, 295-305.

Blesa, M.A., Marinovich, H.A., Baumgartner, E.C., and Maroto, A.J.G. (1987) Mechanism of dissolution of magnetite by oxalic acid-ferrous ion solutions. Inorg. Chem. 26, 3713-3717.

Borghi, E.B., Regazzoni, A.E., Maroto, A.J.G., and Blesa, M. (1988) Reductive dissolution of magnetite by solutions containing EDTA and Fe[II]. J. Coll. Int. Sci. 130, 299-310.

Brown, G.E.Jr., Parks, G.A., and Chisholm-Brause, C.J. (1989) In-situ X-ray absorption spectroscopic studies of ions at oxide-water interfaces. Chimia 43, 248-256.

Buffle, J., de Vitre, R.R., Perret, D. and Leppard, G.G. (1989) Physico-chemical characteristics of a colloidal iron phosphate species formed at the oxic-anoxic interface of a eutrophic lake. Geochim. Cosmochim. Acta 53, 399-408.

Burdige, D.J. and Nealson, K.H. (1986) Chemical and microbiological studies of sulfide-mediated manganese reduction. Geomicrobiology J. 4, 361-87.

460

Canfield, D.E. and Berner, R.A. (1987) Dissolution and pyritization of magnetite in anoxic marine sediments. Geochim. Cosmochim. Acta 51, 645-659.

Cannon, R.D. (1980) Electron Transfer Reactions. Butterworth, London.

Carpenter, E.J. and Price, IV, C.C. (1976) Marine Oscillatoria (Trichodesmium): Explanation for aerobic nitrogen fixation without heterocysts. Science 191, 1278-1280.

Casey,W.H., Westrich, H.R. and Arnold G.W. (1988) Surface chemistry of labradorite feldspar reacted with aqueous solutions at pH = 2, 3 and 12. Geochim. Cosmochim. Acta 52, 2795-2807.

Collienne, R.H. (1983) Photoreduction of iron in the epilimnion of acidic lakes. Limnol. Oceanogr. 28, 83-100.

Cooper, W.J.and Zika R.G. (1983) Photochemical formation of hydrogen peroxide in surface and groundwaters exposed to sunlight. Science 220, 711-2.

Davies, S.H.R. and Morgan, J.J. (1989) Manganese(II) oxidation kinetics on metal oxide surfaces. J. Coll. Int. Sci. 129, 63-77.

Davis, J.A., James, R.O., and Leckie, J.O. (1978) Surface ionization and complexation at the oxide/water interface. I. Computation of electrical double layer properties in simple electrolytes. J. Coll. Int. Sci. 63, 480-499.

Davison, W. (1985) Conceptual models for transport at a redox boundary. In: Chemical Processes in Lakes, W. Stumm, ed., Wiley-Interscience, New York, pp. 31-53.

de Vitre, R.R., Buffle, J., Perret, D. and Baudet, R. (1988) A study of iron and manganese transformations at the $O_2/S(-II)$ transition layer in a eutrophic lake (Lake Bret, Switzerland): A multimethod approach. Geochim. Cosmochim. Acta 52, 1601-1613.

del Valle Hidalgo, M., Katz, N.E., Maroto, A.J.G., and Blesa, M.A. (1988) Dissolution of magnetite by nitrilotriacetatoferrate(II). J. Chem. Soc., Faraday Trans. 1, 84, 9-18.

Diem, D. and Stumm, W. (1984) Is dissolved Mn^{2+} being oxidized by O_2 in the absence of Mn-bacteria or surface catalysts? Geochim. Cosmochim. Acta 48, 1571-1573.

dos Santos, M., Morando, P.J., Blesa M. A., Banwart, S. and Stumm, W. The reductive dissolution of iron oxides by ascorbate. J. Coll. Int. Sci., in press.

Drever, J.I. (1988) The Geochemistry of Natural Waters, 2nd ed., Prentice Hall, New York, 437 pp.

Dzombak, D. and Morel, F. (1990) Surface Complexation Modeling: Hydrous Ferric Oxide, Wiley-Interscience, New York.

Eggleston, C.M., Hochella, Jr., M.F. and Parks, G.A. (1989) Sample preparation and aging effects on the dissolution rate and surface composition of diopside. Geochim. Cosmochim. Acta 53, 797-804.

Faust, B.C. and Hoffmann, M.R. (1986) Photo-induced reductive dissolution of hematite (α–Fe_2O_3) by bisulfite. Env. Sci. Technol. 20, 943-948.

Finden, D.A.S., Tipping, E., Jaworski, G.H.M. and Reynolds, C.S. (1984) Light-induced reduction of natural iron(III) oxide and its relevance to phytoplankton. Nature 304,783-784.

Francis, A.J., Dodge, C.J., Rose, A.W., and Ramirez, A.J. (1989) Aerobic and anaerobic microbial dissolution of toxic metals from coal wastes: mechanisms of action. Environ. Sci. Technol. 23, 435-441.

Francko, D.A. and Heath, R.T. (1982) UV-sensitive complex phosphorus: Association with dissolved humic material and iron in a bog lake. Limnol. Oceanogr. 27, 564-569.

Furrer, G. and Stumm W. (1983) The role of surface coordination in the dissolution of δ-Al_2O_3 in dilute acids. Chimia 37, 338-341.

Furrer, G. and Stumm W. (1986) The coordination chemistry of weathering: I. Dissolution kinetics of δ-Al_2O_3 and BeO. Geochim. Cosmochim. Acta 50, 1847-1860.

Garisto, N.C. and Garisto, F. (1986) The dissolution of UO_2: A thermodynamic approach. Nucl. Chem. Waste Manag. 6, 203-211.

Garrels, R.M. and Lerman, A. (1984) Coupling of the sedimentary sulfur and carbon cycles – An improved model. Amer. J. Sci. 284, 989-1007.

Garrels, R.M. and Perry, E.A. (1974) Cycling of carbon, sulfur, and oxygen through geologic time. In: The Sea, vol 5, ed. E.D. Goldberg, Wiley-Interscience, New York, pp. 303-336.

Giovanoli, R., Schnoor, J.L., Sigg, L., Stumm, W., and Zobrist, J. (1988) Chemical weathering of crystalline rocks in the catchment area of acidic Ticino lakes, Switzerland. Clays Clay Minerals 36, 521-529.

Gordon, G. and Taube, H. (1962) Oxygen tracer experiments on the oxidation of aqueous uranium(IV) with oxygen-containing oxidizing agents. Inorg. Chem. 1, 69-75.

Grandstaff, D. E. (1976) A kinetic study of the dissolution of uraninite. Econ. Geol. 71, 1493-1506.

Hachiya, K., Sasaki, M., Ikeda, T., Mikami, N. and Yasunga, T. (1984) Static and kinetic studies of adsorption-desorption of metal ions on a γ-Al_2O_3 surface. II. Kinetic study by means of pressure jump technique. J. Phys. Chem. 88, 27-31.

461

Haim, A. (1983) Mechanisms of electron transfer reactions: The bridged activated complex. Progr. Inorg, Chem. 30, 273-357,

Harvey, G.R, and Boran, D.A. (1985) Geochemistry of humic substances in seawater. In: Humic Substances in Soil, Sediment and Water, G.R.Aicken, D.M. McKnight, R.L Wershaw, and P. MacCarthy, eds., Wiley-Interscience, New York, pp. 233-47.

Hayes, K.F. and Leckie J.O. (1986) Mechanism of lead ion adsorption at the goethite-water interface. J. Amer. Chem. Soc. 323, 114-141

Hayes, K.F. and Leckie, J.O. (1987) Modelling of ionic strength effects on cation adsorption at hydrous oxide/solution interfaces. J. Coll. Int. Sci. 115, 564-572.

Hellmann, R., Eggleston, C.M., Hochella, Jr., M.F. and Crerar, D.A. (1990) The formation of leached layers on albite surfaces during dissolution under hydrothermal conditions. Geochim. Cosmochim. Acta 54, 1267-1281.

Hoffmann, M.R. and Jacob, D.J. (1984) Kinetics and mechanisms of the catalytic oxidation of dissolved sulfur dioxide in aqueous solution: An application to nighttime fog water chemistry. In: SO_2, NO, and NO_2 Oxidation Mechanisms: Atmospheric Considerations, J.G. Calvert, ed., Acid Precipitation Ser., vol. 3, pp. 101-172, Butterworth Pub., Boston, Massachusetts.

Hoigné, J. (1990) In: Aquatic Surface Chemistry, W. Stumm, ed., Wiley-Interscience, New York.

Holdren, G.R. and Berner, R.A. (1979) Mechanism of feldspar weathering – I. Experimental studies. Geochim. Cosmochim. Acta 43, 1161-1171.

Holland, H.D. (1984) The Chemical Evolution of the Atmosphere and Oceans, Princeton University Press, Princeton, N.J. 582 pp.

Hsieh, Y.H. and Huang, C.P. (1989) The dissolution of PbS(s) in dilute aqueous solutions. J. Coll. Int. Sci. 131, 537-549.

Huang, P.M., Oscarson, D.W., Liaw, W.K., and Hammer, U.T. (1982) Dynamics and mechanisms of arsenite oxidation by freshwater lake sediments. Hydrobiologia 91-92, 315-322.

Hudson, R.J.M. and Morel, F.M.M.(1989) Distinguishing between extra- and intracellular iron in marine phytoplankton. Limnol. Oceanogr. 34, 1113-1120.

Jacob, D., Gottlieb, E.W., and Prather, M.J. (1989) Chemistry of a polluted cloudy boundary layer. J. Geophys. Research 94, 12975.

Jacob, D. and Hoffmann M.R. (1983) A dynamical model for the production of H^+, NO_3^-, and SO_4^{2-} in urban fog. J. Geophys. Res. 88, 6611-6621.

Jones, G.J., Palenik, B.P., and Morel, F.M.M. (1987) Trace metal reduction by phytoplankton: the role of plasmalemma redox enzymes. J. Phycol. 23, 237-244.

Jørgensen, J.J. (1983) Processes at the sediment water interface. In: The Major Biochemical Cycles and Their Interactions, B.Bolin and R.B. Cook, eds., SCOPE 21, pp. 477-509.

Krom, M.D. and Berner, R.A.(1981) The diagenesis of P in nearshone marine sediments. Geochim. Cosmochim. Acta 45, 207-216.

Kummert, R. and Stumm, W. (1980) The surface complexation of organic acids on hydrous $\gamma-Al_2O_3$. J. Coll. Int. Sci. 75, 373-385.

LaKind, J.S. and Stone, A.T. (1989) Reductive dissolution of goethite by phenolic reductants. Geochim. Cosmochim. Acta 53, 961-971.

Lasaga, A.C. (1981) Rate laws of chemical reactions. In: A.C. Lasaga and R.J. Kirkpatrick, eds., Kinetics of Geochemical Processes. Rev. in Mineral. 8, 1-68.

Leuenberger, C., Caney, R., Graydon, J.W., Molnar-Kubica,E., Giger, W. (1983) Persistent organic chemicals in pulp-mill affluents: occurrence and behavior in a biological treatment plant. Chimia 37, 345-354.

Lovley, D.R. and Phillips, E.J.P. (1988) Novel mode of microbial energy metabolism: organic carbon oxidation coupled to dissimilatory reduction of iron or manganese. Appl. Environ. Microbiol. 54, 1472-1480.

Lowson, R.T. (1982) Aqueous oxidation of pyrite by molecular oxygen. Chem. Rev. 82, 461-497.

Luther, III, G.W. (1987) Pyrite oxidation and reduction: Molecular orbital theory considerations. Geochim. Cosmochim. Acta 51, 3193-3199.

Luther, III, G.W. and Church, T.M. (1988) Seasonal cycling of sulfur and iron in porewaters of a Delaware salt marsh. Mar. Chem. 23, 295-309.

Luther, III, G.W., Church, T.M., Scudlark, J.R., and Cosman, M. (1986) Inorganic and organic sulfur cycling in salt-marsh pore waters. Science 232, 746-749.

Mann, S. (1988) Molecular recognition in biomineralization. Nature 332, 119-124.

Martell, A.E. and Smith, R. (1977) Critical Stability Constants, vol. 3, Plenum Press, New York.

Martin, J.H., Gordon, R.M., and Fitzwater, S.E. (1990) Iron in antarctic waters. Nature 345, 156-158.

McKibben, M.A. and Barnes, H.L. (1986) Oxidation of pyrite in low temperature acidic solutions: Rate laws and surface textures. Geochim. Cosmochim. Acta 50, 1509-1520.

462

McKnight , D.M. and Bencala, K.E. (1988) Diel variations in iron chemistry in an acidic stream in the Colorado Rocky Mountains, U.S.A. Arctic Alpine Res. 20, 492-500.

McKnight, D.M., Kimball, B.A., and Bencala, K.E. (1988) Iron photoreduction and oxidation in an acidic mountain stream. Science 240, 637-640.

Mehra, O.P. and Jackson, M.L. (1960) Iron oxide removal from soils and clays by dithionite-citrate systems buffered with sodium bicarbonate. Clays Clay Minerals 7, 317-327.

Moffett , J.W. and Zika, R.G. (1987) Reaction kinetics of hydrogen peroxide with copper and iron in seawater. Environ. Sci. Technol. 21, 804-810.

Mopper, K. and Taylor, B.F. (1986) Biochemical cycling of sulfur: Thiols in coastal marine sediments. In: Organic Marine Chemistry, M.L. Sohn, ed., Amer. Chem Soc. Symp. Ser. 305, pp. 324-39.

Moses, C., Nordstrom, D.K., Herman, J., and Mills, A.L. (1987) Aqueous pyrite oxidation by dissolved oxygen and by ferric iron. Geochim. Cosmochim. Acta 51, 1561-1571.

Motschi, H. (1987) Aspects of the molecular structure in surface complexes; spectroscopic investigations. In: Aquatic Surface Chemistry, W. Stumm, ed., Wiley-Interscience, New York, pp. 111-125.

Nicholson, R.V., Gillham, R.W., and Reardon, E.J. (1990) Pyrite oxidation in carbonate-buffered solution: 2. Rate control by oxide coatings. Geochim. Cosmochim. Acta 54, 395-402.

Nicholson, R.V., Gillham, R.W., and Reardon, E.J. (1988) Pyrite oxidation in carbonate-buffered solution: 1. Experimental kinetics. Geochim. Cosmochim. Acta 52, 1077-1085.

O'Brian, A.B., Segal, M.G., and Williams, W. J. (1987) Kinetics of metal oxide dissolution: Oxidative dissolution of chromium from mixed nickel-iron-chromium oxides by permanganate. J. Chem. Faraday Trans. 1, 83, 371-382.

Ollilla, K. (1989) Dissolution of UO_2 at various parametric conditions: a comparison between calculated and experimental results. Mat. Res. Symp. Proc. 127, 337-342.

Oscarson, D.W., Huang, P.M., Liaw, W.K., and Hammer, U.T. (1983a) Kinetics of oxidation of arsenite by various manganese dioxides. Soil Sci. Soc. Amer. J. 47, 644-648.

Oscarson, D.W., Huang, P.M., Hammer, U.T., and Liaw, W.K. (1983b) Oxidation and sorption of arsenite by manganese dioxide as influenced by surface coatings of iron and aluminum oxides and calcium carbonate. Water, Air, Soil Poll. 20, 233-244.

Oscarson, D.W., Liaw, W.K., and Huang, P.M. (1981) The kinetics and components involved in the oxidation of arsenite by freshwater lake sediments. Verh. – Int. Ver. Theor. Angew. Limnol. 21, 181-186.

Paerl, H.W.and Prufert, L.E. (1987) Oxygen-poor microzones as potential sites of microbial N_2 fixation in nitrogen-depleted aerobic marine waters. Appl. Environ. Microbiol. 53, 1078-1087.

Petrovic, R., Berner, R.A. and Goldhaber, M.B. (1976). Rate control in dissolution of alkali feldspars – I. Study of residual feldspar grains by X-ray photoelectron spectroscopy. Geochim. Cosmochim. Acta 40, 537-548

Postma, D. (1990) Kinetics of NO_3^- reduction by detrital Fe(II) silicates. Geochim. Cosmochim. Acta 54, 903-908.

Price, N.M. and Morel, F.M.M. (1990) Role of extracellular enzymatic reactions in natural waters. In: Aquatic Chemical Kinetics, W. Stumm, ed., Wiley, New York, pp. 235-257.

Pulver, K., Schindler, P.W., Westall, J.C. and Grauer, R. (1984) Kinetics and mechanism of dissolution of bayarite (γ-Al(OH)$_3$) in HNO$_3$-HF solutions at 298.2 K. J. Coll. Int. Sci. 101, 554-564.

Pyzik, A.J. and Sommer, S.E. (1981) Sedimentary iron monosulfides: kinetics and mechanism of formation. Geochim. Cosmochim. Acta 45, 687-698.

Reinhard, M., Goodman, N.L., and Barker, J.F. (1984) Occurrence and distribution of organic chemicals in two landfill leachate plumes. Environ. Sci. Technol. 18, 953-961.

Rich, H.W. and Morel, F.M.M. (1990) Availability of well-defined colloids to the marine diatom Thalassiosira weissflogii. Limnol. Oceanogr. 35, 000.

Richardson, L.L., Aguilar, C., and Nealson, K.H. (1988) Manganese oxidation in pH and O_2 microenvironments produced by phytoplankton. Limnol. Oceanogr. 33, 352-363.

Sansone, F.J. and Martens, C.S. (1982) Volatile fatty acids cycling in organic-rich marine sediments. Geochim. Cosmochim. Acta 46, 1575-89.

Sansone, F.J. (1986) Depth distribution of short chain organic acid turnover in Cape Lookout Bight sediments. Geochim. Cosmochim. Acta 50, 99-105.

Schindler, P.W. and Kamber, H.R. (1968) Die Acidität von Silanolgruppen. Helv. Chim. Acta 51, 1781-1786.

Schindler, P.W. and Stumm, W. (1987) The surface chemistry of oxides, hydroxides and oxide minerals. In: Aquatic Surface Chemistry, W. Stumm, ed., Wiley-Interscience, New York, 83-110.

Schneider, W. and Schwyn, B.(1987) In: Aquatic Surface Chemistry, W. Stumm ed., Wiley-Interscience, New York, pp. 167-194.

Schnoor, J.L. (1990) Kinetics of chemical weathering: A comparison of laboratory and field weathering rates. In: Aquatic Chemical Kinetics, W. Stumm, ed., Wiley-Interscience, New York, pp. 475-504.

Schott, J. and Berner, R.A. (1983) X-ray photoelectron studies of the mechanism of iron silicate dissolution during weathering. Geochim. Cosmochim. Acta 47, 2233-2240.

Schott, J. and Petit, J.C. (1987) New evidence for the mechanisms of dissolution of silicate minerals. In: Aquatic Surface Chemistry, W. Stumm, ed., Wiley-Interscience, New York, pp. 293-315.

Schott, J., Berner, R.A., and Sjöberg, E.L. (1981) Mechanism of pyroxene and amphibole weathering - I. Experimental studies of iron-free minerals. Geochim. Cosmochim. Acta 51, 2123-2135.

Schott, J., Brantley, S., Crerar, D., Guy, C., Borcsik, M., and Willaime, C. (1989) Dissolution kinetics of strained calcite. Geochim. Cosmochim. Acta 53, 373-382.

Schwarzmann, D.W. and Volk, T. (1989) Biotic enhancement of weathering. Nature 340, 457.

Scott, M.J., Reactions of As(III) and Se(IV) at goethite and birnessite surfaces. Ph.D. dissertation. California Institute of Technology, Pasadena, CA, in prep.

Segal, M.G. and Sellers, R.M. (1980) Reactions of solid iron(III) oxides with aqueous reducing agents. J. Chem. Soc. Comm., 991-993.

Segal, M.G. and Sellers, R.M. (1982) Kinetics of metal oxide dissolution. J. Chem. Soc. Faraday Trans. 1, 78, 1149-1164.

Segal, M.G. and Sellers, R.M. (1984) Redox reactions at solid-liquid interfaces. Adv. Inorgan. Bioinorgan. Mechanisms 3, 97-129.

Segal, M.G. and Williams, W.J. (1986) Kinetics of metal oxide dissolution: oxidative dissolution of chromium(III) oxide by potassium permanganate. J. Chem. Faraday Trans. 1, 82, 3245-3254.

Seyfriend, W.E.J. and Mottl, M.J. (1982) Hydrothermal alteration of basalt by seawater under seawater dominated conditions. Geochim. Cosmochim. Acta 46, 985-1002.

Shoesmith, D.W., Sunder, S., Bailey, M.G., and Wallace, G.J. (1988) Anodic oxidation of UO_2. V. Electrochemical and X-ray photoelectron spectroscopic studies of the film-growth and dissolution in phosphate-containing solutions. Can. J. Chem. 66, 259-265.

Shoesmith, D.W., Sunder, S., Bailey, M.G., and Wallace, G.J. (1989) The corrosion of nuclear fuel (UO_2) in oxygenated solutions. Corrosion Sci. 29, 1115-1128.

Siffert, C. (1989) L'Effet de la Lumière sur la Dissolution des Oxydes de Fer(III) dans les Mileux Aqueux. Ph.D. dissertation, Swiss Federal Institute of Technology, Zürich, Switzerland.

Siffert, C. and Sulzberger, B., in preparation.

Sigg, L., Johnson, C.A., and Kuhn A., Redox conditions and alkalinity generation in a seasonally anoxic lake (Lake Greifen). Mar. Chem., submitted.

Sigg, L. and Stumm, W. (1980) Chimie des surfaces d'oxydes en milieu aqueux. In: Géochimie des Interactions, Y. Tardi /S.A.R. Elément ed., Tarbes, France, 27-48.

Silver, M. and Torma, A.E. (1974) Oxidation of metal sulfides by Thiobacillus ferrooxidans grown on different substrates. Can. J. Microbiol. 20, 141-147.

Singer, P.C. and Stumm, W. (1970) Acid mine drainage – the rate limiting step. Science 167, 1121-1123.

Sposito, G. (1989) The Chemistry of Soils , Oxford Univ. Press, Oxford, England, 277 pp.

Stauffer, R.E. and Armstrong, D.E. (1986) Cycling of iron, manganese, silica, phosphorous, calcium, and potassium in two stratified basins of Shagwa Lake, Minn. Geochim. Cosmochim. Acta 50, 215-229.

Stone, A.T. (1987a) Reductive dissolution of manganese(III/IV) oxides by substituted phenols. Environ. Sci. Technol. 21, 979-988.

Stone, A.T. (1987b) Microbial metabolites and the reductive dissolution of manganese oxides: Oxalate and pyruvate. Geochim. Cosmochim. Acta 51, 919-925.

Stone, A.T. and Morgan, J.J. (1984a) Reduction and dissolution of manganese(III) and manganese(IV) oxides by organics. 1. Reaction with Hydroquinone. Environ. Sci. Technol. 18, 450-456.

Stone, A.T. and Morgan, J.J. (1984b) Reduction and dissolution of manganese(III) and manganese(IV) oxides by organics: 2. Survey of the Reactivity of Organics. Environ. Sci. Technol. 18, 617-624.

Stone, A.T. and Morgan J.J. (1987) Reductive dissolution of metal oxides. In: Aquatic Surface Chemistry, W. Stumm, ed., Wiley-Interscience, New York.

Stone, A.T. and Ulrich, H.-J, (1989) Kinetics and reaction stoichiometry in the reductive dissolution of manganese(IV) dioxide and Co(III) oxide by hydroquinone. J. Coll. Int. Sci. 132, 509-522.

Stumm, W. and Lee, G.F. (1961) Oxygenation of ferrous iron. Industrial and Eng. Chem. Acta 53, 143-146.

Stumm, W. and Furrer, G. (1987) The dissolution of oxides and aluminum silicates: Examples of surface-coordination-controlled kinetics. In: Aquatic Surface Chemistry, W. Stumm, ed., Wiley-Interscience, New York, pp. 197-219.

Stumm, W. and Morgan, J.J. (1981) Aquatic Chemistry, 2nd ed., Wiley Interscience, New York.

464

Stumm, W. and Wieland, E. (1990) Dissolution of oxide and silicate minerals: Rates depend on surface speciation. In: Aquatic Surface Chemistry, W. Stumm ed., Wiley-Interscience, New York.

Stumm, W., Huang, C.P. and Jenkins, S.R. (1970) Specific chemical interaction affecting the stability of dispersed systems. Croat. Chem. Acta 53, 291-312.

Stumm, W., Sulzberger, B., and Sinniger, J. The coordination chemistry of the oxide-electrolyte interface: the dependence of surface reactivity (dissolution, redox reactions) on surface structure. Croatica Chim. Acta, in press

Stumm-Zollinger, E. (1972) Die bakterielle Oxidation von Pyrit. Arch. Mikrob. 83, 110-119.

Sulzberger, B. (1990) Photoredox reactions at hydrous metal oxide surfaces; a surface coordination chemistry approach. In: Aquatic Chemical Kinetics, W. Stumm, ed., Wiley-Interscience, New York.

Sulzberger, B. and Siffert, C. (in preparation).

Sulzberger, B., Suter, D., Siffert, C. and Banwart, S. (1989) Dissolution of Fe(III) (hydr)oxides in natural waters; laboratory assessment on the kinetics controlled by surface coordination. Mar. Chem. 28, 127-144.

Sulzberger, B., Schnoor, J.L., Giovanoli, R., Hering, J.G., and Zobrist, J. (1990) Biogeochemistry of iron in an acidic lake. Aquatic Sciences 52, 56-74.

Sunda, W.G. and Huntsman, S.A. (1987) Microbial oxidation of manganese in a North Carolina estuary. Limnol. Oceanogr. 32, 552-564.

Sunda, W.G. and Huntsman, S.A. (1988) Effect of sunlight on redox cycles of manganese in the southwestern Sargasso Sea. Deep-Sea Res. 35, 1297-1317.

Sunda, W.G., Huntsman. S.A., and Harvey, G.R. (1983) Photoreduction of manganese oxides in seawater and its geochemical and biological implications. Nature 301, 234-236.

Suter, D., Siffert, C., Sulzberger, B. and Stumm, W. (1988) Catalytic dissolution of iron(III) (hydr)oxides by oxalic acid in the presence of Fe(II). Naturwiss. 75, 571-573.

Suter, D., Banwart, S., and Stumm, W. (submitted) The reductive dissolution of hydrous iron(III) oxides by ligands. Langmuir. Tamura, H., Kawamura, S., and Nagayama, M. (1980) Acceleration of the oxidation of Fe^{2+} ions by Fe(III)-oxyhydroxides. Corrosion Sci. 20, 963-971.

Taube, H. (1970) Electron transfer reactions of complex ions in solution. Academic Press, New York.

Tebo, B.M., Lewis, B.L., and Landing, W.M. (1990) The kinetics of manganese(II) oxidation in the Black Sea. EOS Trans. Amer. Geophys. Union 71, 173.

Tejedor-Tejedor, M.I. and Anderson, M.A. (1986) "In situ" attenuated total reflection Fourier transform infrared studies of the goethite ($FeOOH$)-aqueous solution interface. Langmuir 2, 203-210.

Torrent, J., Schwertmann, U., and Barron, V. (1987) The reductive dissolution of synthetic goethite and hematite in dithionite. Clay Minerals 22, 329-337.

Ulrich, H.-J. and Stone, A.T. (1989) Oxidation of chlorophenols adsorbed to manganese oxide surfaces. Environ. Sci. Technol. 23, 421-428.

Valverde, N. and Wagner, C. (1976) Considerations on the kinetics and the mechanism of the dissolution of metal oxides in acidic solutions. Ber. Bunsenges. Physik. Chem. 80, 330-333.

van Cappellen, P. and Berner, R.A. (1988) A mathematical model for the early diagenesis of phosphorus and fluorine in marine sediments: apatite precipitation. Amer. J. Sci. 288, 289-333.

Velbel, M.A. (1986) The mathematical basis for determining rates of geochemical and geomorphic processes in small forested watersheds by mass balance: Examples and implications. In: Rates of Chemical Weathering of Rocks and Minerals, S. Colman and D.P. Dethier, eds., pp. 439-451.

Wegman, R.C.C. and Hofstee, A.W.M. (1979) Chlorophenols in surface waters of the Netherlands (1976-1977). Water Res. 13, 651-657.

Wehrli, B. (1989) Monte Carlo simulations of surface morphologies during mineral dissolution. J. Coll. Int. Sci. 132, 230-242.

Wehrli, B. (1990) Redox reactions of metal ions at mineral surfaces. In: Aquatic Chemical Kinetics, W. Stumm, ed., Wiley-Interscience, New York.

Wehrli, B. and Stumm, W. (1989) Vanadyl in natural waters: Adsorption and hydrolysis promote oxygenation. Geochim. Cosmochim. Acta 53, 69-77.

Wehrli, B., Sulzberger, B., and Stumm, W. (1989) Redox processes catalyzed by hydrous oxide surfaces. Chem. Geol. 78, 167-179.

Wehrli, B., Ibric, S. and Stumm, W., Adsorption kinetics of vanadyl(IV) and chromium(III) to aluminum oxide: Evidence for a two-step mechanism. Coll. and Surf., in press

Wehrli, B., Wieland, E., and Furrer, G. (1990) Chemical mechanisms in the dissolution kinetics of minerals; the aspect of active sites. Aquatic Sci. 52, 3-31.

Wersin, P., Höhener, P., Giovanoli, R., and Stumm, W. Early diagenetic influences on iron transformations in a freshwater lake sediment. Chem. Geol., submitted.

White, A.F. and Yee, A. (1985) Aqueous oxidation-reduction kinetics associated with coupled electron-cation transfer from iron-containing silicates at 25°C. Geochim. Cosmochim. Acta 49, 1263-1275.

465

White, A.F., Yee, A., and Flexser, S. (1985) Surface oxidation-reduction kinetics associated with experimental basalt-water reaction at 25°C. Chem. Geol. 49, 73-86.

Wieland, E, Wehrli, B., and Stumm, W. (1988). The coordination chemistry of weathering: III. A generalization on the dissolution rates of minerals. Geochim. Cosmochim. Acta 52, 1969-1981.

Yates, D.E., Levine, S., and Healy, T.(1974) Site-binding model of the electrical double layer at the oxide/water interface. J. Chem. Soc., Faraday Trans. 1, 70, pp. 1807-1818.

Zika, R.G., Moffett, J.W., Cooper, W.J., Petasne, R.G., and Saltzman, E.S. (1985a) Spatial and temporal variations of hydrogen peroxide in gulf of Mexico waters. Geochim. Cosmochim. Acta 49, 1173-84.

Zika, R.G., Saltzman, E.S., Cooper, W.J. (1985b) Hydrogen peroxide concentrations in the Peru upwelling area. Mar. Chem. 17, 265-75.

Zinder, B., Furrer, G. and Stumm, W. (1986) The coordination chemistry of weathering: II. Dissolution of Fe(III) oxides. Geochim. Cosmochim. Acta 50, 1861-1870.

HETEROGENEOUS ELECTROCHEMICAL REACTIONS ASSOCIATED WITH OXIDATION OF FERROUS OXIDE AND SILICATE SURFACES

INTRODUCTION

Iron is the most abundant element in the bulk earth and ranks behind only O, Si, and Al as the fourth most abundant element in the earth's crust. Within the crust, Fe is dominated by the Fe^{3+} and Fe^{2+} oxidation states. The average Fe^{3+}/Fe^{2+} ratio for primary igneous rocks is reported to be 0.53 compared to 1.35 in their weathered counterparts represented by sedimentary rocks (Ronov and Yaroshevsky, 1971). Therefore oxidation of ferrous iron, and consequent electron transfer to the aquatic, atmospheric, and biologic environments, is a significant geochemical process. The present chapter reviews the electrochemistry involved in the oxidation of structurally bound Fe^{2+} in oxides and silicates and in the resulting electron transfer to solution. A general review of this topic has not been available and the purpose of the paper is bring together the results of recent studies addressing these heterogeneous electron transfer processes under ambient environmental conditions. Related redox processes, including oxidative and reductive dissolution of minerals (Hering and Stumm, 1990), adsorption involving redox reactions (Bancroft and Hyland, 1990), and photo-redox processes (Waite, 1990), are reviewed elsewhere in this volume.

Structural oxidation of Fe^{2+} and related electron transfer can be represented by the simple expression:

$$[Fe^{2+}] \rightarrow [Fe^{3+}] + e^- \quad , \tag{1}$$

using the convention in equations that solid state species are bracketed and aqueous and sorbed species are unbracketed. This half cell reaction is termed heterogeneous because it involves transfer of the electron from one phase, a solid, to another phase, a solution. The reaction represents a free energy minimization in response to the presence of an electron acceptor, for example O_2, at the solution-mineral interface and/or in the aqueous media. This review will be confined to redox processes in which the Fe species in Equation 1 remains generally immobile within the parent mineral structure.

Minerals containing oxidizable Fe can be broadly classified as oxides, ortho and chain silicates, micas, and clays. The most common minerals within these groups, which in total contain the bulk of Fe^{2+} exposed to oxidation at the earth's surface, are listed in Table 1. Under each mineral group, discussions will include (1) structural distributions of Fe^{2+} and Fe^{3+} in the lattice, (2) solid state oxidation mechanisms, (3) applicable dissolution reactions, (4) heterogeneous electron transfer, and finally, (5) reactions involving aqueous electron acceptors.

The oxidation reaction, represented by Equation 1, can occur as a precursor to dissolution when solid state electron transfer is kinetically more rapid than structural disintegration. Overall dissolution processes are extensively reviewed elsewhere in this volume (Casey and Bunker, 1990; Hering and Stumm, 1990), and will be addressed only to the extent to which they are either coupled to or compete with the solid state electron transfer reactions. However, because reaction kinetics of silicate and oxide minerals are usually relatively slow in comparison to aqueous redox reactions, oxidative electron transfer mechanisms described by Equation 1 may be extremely important,

TABLE 1. Average crustal abundances of mineral phases considered in this paper. Also listed are mean average Fe compositions and ranges in Fe compositions for each mineral (Ronov and Yaroshevsky, 1971).

Mineral	Volume %	Mean Wt. %Fe	Range Wt. %Fe
Olivine	2	10-30	0-55
Pyroxenes	8	12	0-42
Amphiboles	5	15	0-39
Biotite	4	15	12-33
Magnetite + ilmentite	2	72/37	-

TABLE 2. Band gap energies of some binary semiconductor compounds (eV). Adapted from Shuey (1975).

SiO_2 (quartz)	11.0
TiO_2 (rutile)	3.0
$FeTiO_3$ (ilmenite)	2.8
Fe_2O_3 (hematite)	2.34
FeS_2 (pyrite)	1.2
MnO_2 (pyrolusite)	2.6
Fe_3O_4 (magnetite)	0.10

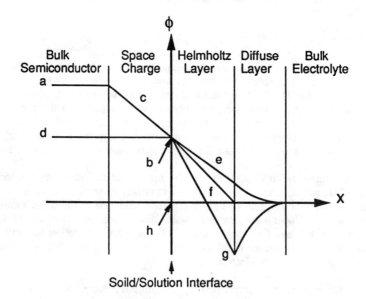

Figure 1. Electrochemical potential (E) as a function of distance (x) in an oxide semiconductor-electrolyte system: (a) bulk semiconductor potential, (b) solid/solution interface potential, (c) space charge potential, (d) flat band potential, (e) double layer potential in a dilute electrolyte, (f) double layer potential in a concentrated electrolyte, (g) reverse polarity resulting from specific cation adsorption, and (h) solid-solution potential at the ZPC (adapted from Diggle, 1973).

dominating the surface chemistry of Fe-containing minerals, and acting both as electron donors and electrochemical buffers in natural and perturbed surface and groundwater systems.

SOLID STATE ELECTROCHEMISTRY

Processes describing coupled structural oxidation and heterogeneous electron transfer are subsets of the more general topic of solid state electrochemistry. Although a review of this topic is clearly outside the scope of this paper, a brief summary will serve to integrate a number of specific observations regarding heterogeneous redox processes associated with ferrous oxides and silicates.

Solids can be classified as metals, insulators, or semiconductors based on their electrochemical characteristics, specifically the electron energy gap between the valance and conduction bands (Diggle, 1973). If the valance and conduction bands overlap, the solid is considered to be a metal. Under such conditions, the Fermi level, below which most or all electron energy levels are filled and above which they remain unoccupied, is well within the valence band. If a positive energy gap exists between the valence and conduction bands, the solid is a non-metal. A solid can be considered an insulator if the band gap is positive and > 5 eV, and a semiconductor if the energy gap is < 5 eV. In a semiconductor, the Fermi level lies somewhere in the band gap between the valence and conduction bands. Several oxide minerals considered in the present review have band gaps representative of semiconductors (Table 2). As suggested by the large band gap of quartz, non-Fe containing silicates are generally considered insulators. However, as will be documented in the following sections, electron transfer does occur in the conduction bands of ferrous silicates and, therefore, no distinction will be made between the oxides and Fe silicates as to semiconducting or insulating properties.

The net electrochemical potential (E) of the Fe^{2+} oxide or silicate is typically negative or cathodic relative to the aqueous state since the net electron transfer must be from the mineral to the solution (Eqn. 1) The variation of E as a function of distance x in an oxide electrode in contact with an aqueous electrolyte can be represented by the schematic diagram in Figure 1 (Diggle, 1973). In a semiconductor, the bulk solid state potential (Fig. 1, line a) is separated from solid/solution interface potential (point b) by the potentials within the space charge region (line c). The solid/solution interface is a plane created by the termination of the semiconductor lattice at the solution boundary. The potential at this surface is established both by the surface structure of the solid phase and adsorbates from solution. The space charge region retains the original structural framework as the bulk semiconductor but the energy band gaps in this region are influenced or "bent" by the potentials residing in the solid-solution interface and therefore indirectly by the electrolyte chemistry. The space charge region is analogous to the double layers in solution except the charge is electronic rather than ionic in nature. The thickness of the space charge region, termed the Debye depth, increases with decreasing electron charge concentration and mobility ($\rightarrow$ insulator) and decreases with increasing charge concentration and mobility ($\rightarrow$ metal). The specific charge distribution, n, within the space charge region can be defined by the Boltzmann relationship (Diggle, 1973),

$$n = n_b \exp[(\phi_s - \phi_b)e/kT] \quad , \qquad (2)$$

where n_b is the electron charge distribution in the bulk phase, ϕ_s is the electrostatic potential at the solid-solution/interface where the electron concentration is n, ϕ_b is the electrostatic potential in the bulk semiconductor where the space charge is zero, e is the electronic charge, and kT is the kinetic energy due to thermal activation at temperature T where k is the Boltzmann constant. The electrostatic potential ϕ at a distance x into a semiconductor is defined by the Poisson equation,

$$d^2\phi/dx^2 = q/(\varepsilon \cdot \varepsilon_o) \quad , \tag{3}$$

where q is the charge density, ε is the dielectric constant of the solid and ε_o is the dielectric constant of a vacuum. Electrical balance requires that the charge on the solid/solution interface be balanced by the charge in the space charge region. If no excess charge is trapped in the surface state, the space charge region becomes constant and equal to the bulk semiconductor electrode potential at the flat-band potential (line d), a situation analogous to the zero point of charge in the electrolyte double layer.

Solid state redox reactions are strongly influenced by the nature of Helmholtz and diffuse layer interactions. The Fermi levels in the solid, represented by the electrochemical potentials in the space charge region (Eqn. 3) are fixed in part by the electrochemical potential of the solution at the solid/aqueous interface. This induced potential in the space charge region is therefore the driving force that permits structural oxidation reactions of the form of Equation 1 to occur. The electrical potential in the Helmholtz and diffuse layers can vary significantly depending on the charge potential at the solid/solution interface in addition to the electrolyte composition of the bulk solution. For a dilute aqueous solution, the potential generally decreases across the Helmholtz and diffuse layers to zero relative potential in the bulk solution (line e). At high electrolyte concentrations, the charge differential may be compressed to within the Helmholtz region (line f). In some cases, specific sorption of cations can create inverted potential distributions in the double layer (Fig. 1, g). Finally, at the ZPC, the potential at the solid/solution interface will equal that of the bulk electrolyte (point h).

The potential distributions shown in Figure 1 can produce charge transfer between the solid and the solution. Electrical conduction in the solid can occur either as electrons in normally empty conduction bands or as holes in normally full valency bands (Morrison, 1980). Most oxides are extrinsic semiconductors in which charge carriers are produced from defect or vacancy structures in the solid or from trace impurities. If these sites act as donors, the conduction is mainly by electrons and the material is an n-type semiconductor. If the sites are electron acceptors, charge transfer occurs principally by holes or structural vacancies and the oxide is a p-type semiconductor. Oxidation reactions of the type described by Equation 1 can also result from intrinsic semiconduction in which the number of electrons and holes produced are equal as a result of electron removal from the valance band. As will be documented in considerable detail, the electron transfer processes in ferrous oxides and silicates is relatively simple and rapid, involving electron hopping between adjacent Fe^{3+} and Fe^{2+} lattice atoms. The conduction of holes is considerably more complex and often rate limiting, involving processes including proton transfer between oxygen and hydroxyl species and interlayer and structural cation diffusion.

The potential distribution in the aqueous double layers controls the transport of electrons between the solid and solution phases. Free electrons do not exist to any extent in solution and therefore either inner-sphere or outer-sphere electron transfer must occur between the solid interface and adsorbed ligands as described by the heterogeneous electron transfer theory of Marcus (1965). Diffusion of electrochemically active reactant or product species through the Helmholtz and diffuse layers can also be rate-limiting under certain circumstances. Finally, aqueous species in the bulk solution

TABLE 3. Common Fe^{2+}-containing oxide minerals and their structural symmetries.

Magnetite	$Fe^{3+}[Fe^{2+}Fe^{3+}]O_4$	Cubic
Ilmenite	$Fe^{2+}Ti^{4+}O_3$	Rhombohedral
Chromite	$Fe^{2+}[Cr^{3+}_2]O_4$	Cubic
Ulvöspinel	$Fe^{2+}[Fe^{2+}Ti^{4+}]O_4$	Cubic

must serve as the ultimate electron acceptors for solid state oxidation. The availability of such acceptors depends on the speciation and complexation of the aqueous solution. The concentration of non-electrochemically active species can affect the electrolyte concentration and the thickness of the diffuse layer, in addition to supplying exchangeable ligands to the solid/solution interface.

OXIDATION OF FERROUS-CONTAINING OXIDES

Ferrous oxide minerals can be structurally classified as possessing either a cubic spinel structure or rhombohedral symmetry (Table 3). In the "normal" spinel structure, trivalent ions occupy the majority of the octahedral sites. However in inverse spinels such as magnetite, the most common ferrous oxide mineral, sites are randomly occupied by equal numbers of Fe^{2+} and Fe^{3+} atoms, while the tetrahedral sites are occupied only by Fe^{3+}. The preference of Fe^{3+} for the tetrahedral sites may be a result of its smaller ionic radius, 0.064 nm, relative to 0.074 nm for Fe^{2+}. Mössbauer spectroscopy indicates that there is a very rapid exchange of electrons between Fe^{2+} and Fe^{3+} atoms in the octahedral sites so that the net charge on each Fe atom can be considered +2.5 (Gedikogla, 1983).

Continuous solid solutions exist between magnetite and ulvöspinel (Table 3) with most solid solution models assuming a coupled substitution in which Ti^{4+} replaces Fe^{3+} in the octahedral site and Fe^{2+} replaces Fe^{3+} in the tetrahedral site. The magnetite end member is normally slightly metal deficient with vacancies, $\varnothing$, occurring in the octahedral sites. The simplest ionic model for iron-deficient magnetite is $Fe^{3+}[Fe^{3+}_{1+2/3x}Fe^{2+}_{1-x} \varnothing_x]O_4$ where the distribution of the vacancy sites are localized in the octahedral positions (Jolivet and Tronc, 1988). Oxidized magnetites range from 0 < x < 1, which corresponds to a continuous solid solution with maghemite (γFe_2O_3), an Fe-deficient spinel polymorph of hematite. The maghemite structure can be partly stabilized by occupation of the vacant octahedral cation sites with H^+ up to a limiting composition of γ-$Fe_2O_3 \cdot 1/5H_2O$ (Swaddle and Oltmann, 1980). These sites may also be partially exchangeable with K^+ and Na^+. Maghemite is thermodynamically metastable with respect to hematite (α-Fe_2O_3) and its formation is dependent on particle size, temperature, and solution chemistry.

Ilmenite, the second most common Fe^{2+} oxide, possesses a rhombohedral structure similar to hematite with some distortion of the oxygen layers (Table 3). The continuous solid solution can be expressed as $Fe^{3+}_{2-2x}Fe^{2+}_xTi_xO_3$ where x is the mole fraction of ilmenite. Mössbauer studies (Shirane et al., 1962) and electrical conductivity measurements (Ishikawa, 1958) indicate that the intermediate solid solutions contain Fe^{2+} and Fe^{3+} in amounts approximately proportional to x and (2-2x) respectively, indicating that Ti is present only as Ti^{4+}. Along the triad axis in ilmenite, pairs of Ti atoms alternate with pairs of Fe^{2+} atoms, which necessitates an uneven distribution of Fe^{2+} and Ti in successive cation layers. In Ti-rich hematite, the Fe^{2+} impurity is randomly distributed.

Solid state electron transfer

Magnetite can exhibit both n-type and p-type semiconductor characteristics with a Fermi level in a low-mobility spin-polarized 3d band. Above the Verwey transition, magnetite contains separate Fe 3d sub-bands for the octahedral and tetrahedral iron atoms (Shuey, 1975). Two possible spin directions and two types of orbital orientation (t_{2g} and e_g) make for eight sub-bands even without considering lattice distortions or cation substitution. Half the bands are fully occupied but in only one spin direction. These bands are all rather narrow, so the electrons must be considered highly localized. In addition, there is one electron per Fe_3O_4, partly filling the opposite-spin band for octahedral site and t_{2g} symmetry. The low band gap energy (0.10 eV, Table 2) is the reason that magnetite exhibits the lowest resistivity of any oxide mineral ($\sim 5 \cdot 10^{-5}$ ohm-m). If the concentration of conduction band electrons is one per unit cell of magnetite, such a resistivity corresponds to an electron mobility of 0.1 $cm^2 \cdot V^{-1}$, a magnitude expected for electron hopping at a frequency equal to the atomic vibration frequency (Shuey, 1975).

In contrast to magnetite, ilmenite ($FeTiO_3$) has a much higher resistivity with a corresponding band gap energy of 2.8 eV (Table 2). The formation of solid solutions with hematite produces p-type semiconduction (Ishikawa, 1958) with the resistivity decreasing from 1 to 0.1 ohm-m as the hematite component increases from 5 to 50%. The electrical conductivity in the basal plane of ilmenite can be viewed as electron transfer within the Fe layers which Mössbauer spectroscopy has shown to be very fast. A simple ionic model does not predict conductivity along the c-axis, which agrees qualitatively with the large electrical anisotropy of ilmenite.

Solid state oxidation

Electrochemical processes involving oxidation of Fe^{2+} in oxide minerals are generally much better understood than oxidation in the silicates due to several factors: simpler structure, greater electrical conductivity, and corresponding synthetic analogues which are important corrosion products of steel and other alloys. In summarizing earlier experimental work on hydrothermal oxidation of magnetite to maghemite by atmospheric O_2, Lindsley (1976) pointed out that oxygen can not be added to the existing structure which consists of a nearly cubic-close-packed oxygen framework. However, free oxygen at the surface can be reduced by electrons derived from within the magnetite by the reaction $[Fe^{2+}] \rightarrow Fe^{3+} + e^-$ (Eqn. 1). The charge within the original crystal will no longer be balanced, necessitating the removal of 2 Fe^{3+} for every 3 oxygens reduced at the surface. The removal is accomplished by diffusion of the Fe^{3+} through the oxygen framework to the surface of the crystal. Once at the surface, experimental work has shown that the expelled Fe^{3+} and surface O^{2-} can either be removed to solution, be deposited epitaxially as hematite on the surface of the crystal, or be deposited as an extension of the original spinel structure, thus forming new maghemite by the idealized reaction:

$$2[Fe_1^{2+} Fe_2^{3+}]O_4 + 1/2O_2 \rightarrow 3\gamma[Fe_2^{3+}]O_3 \quad . \tag{4}$$

Experimental studies by Gallagher et al. (1968) at ambient room temperatures indicate that the solid state magnetite-maghemite transformation was topotatic with minimal changes in external crystal dimensions, surface area, and X-ray diffraction lines for magnetite-maghemite compositions down to 25% Fe^{2+}. Jolivet and Tronc (1988) achieved the complete oxidation of colloidal magnetite to maghemite at ambient temperature in anoxic perchloric acid solutions (pH 2). Such a reaction, in the absence of O_2 but in the presence of H^+, can be written as:

$$[Fe_1^{2+} Fe_2^{3+}]O_4 + 2H^+ \rightarrow [Fe_2^{3+}]O_3 + Fe^{2+} + H_2O \quad , \tag{5}$$

where one quarter of the structural O in magnetite is combined to form water with the concurrent release of Fe^{2+} to solution.

Studies on the oxidation of titanomagnetite, representative of the magnetite-ulvöspinel solid solution $Fe^{2+}_{1+x} Fe^{3+}_{2-2x}Ti^{4+}_xO_4$, at ambient temperature indicate decreases in structural Fe^{2+} and increases in structural Ti and O. (Akimoto et al., 1984). With an increasing degree of oxidation, electron mobility shifts from the interaction between $Fe^{2+} \leftrightarrow Fe^{3+}$ atoms to reaction between to $Ti^{3+} \leftrightarrow Ti^{4+}$ atoms which decreases the electrical conductivity (Lastovickova and Kropacek, 1983). Two solid state oxidation mechanisms have been proposed for titanomagnetite; either the substitution of O into the lattice, forming a cation deficient product (Lastovickova and Kropacek, 1983),

$$[Fe^{2+}]TiO_4 + (z/2)O \rightarrow [(1-z)Fe_1^{2+} zFe_1^{3+} (3z/8)\varnothing]TiO_{3+z/2} \quad , \tag{6}$$

or oxidation by the migration of Fe and other cations out of the lattice (Akimoto et al., 1984; Furuta et al., 1985):

$$[Fe^{2+}]TiO_4 \rightarrow [(1-z)Fe_1^{2+} 2z/3Fe_1^{3+} z/3\varnothing]TiO_4 + 2z/3e^- + z/3Fe^{2+} \quad . \tag{7}$$

Detailed microprobe analysis of naturally weathered marine basalts demonstrated that the Ti/O ratios were essentially constant, while with increasing oxidation, the $(Fe^{2+}+Fe^{3+})$/Ti ratios decreased indicating that cation migration is the dominant process (Furuta et al., 1985).

A number of studies have characterized weathering of natural ilmenite. Although ilmenite exists as a solid solution with hematite, the immobility of Ti precludes conversion of ilmenite to hematite. Initial stages of ilmenite oxidation involve the conversion of Fe^{2+} to Fe^{3+} without substantial loss of total Fe from the structure (Pakharykov et al., 1976). The Mössbauer spectra of some naturally weathered ilmenites indicate a range in Fe^{2+}/(total Fe) ratios of 0.90 to 0.15. Other studies of more extensively oxidized ilmenites show a progressive decrease in the total Fe/Ti ratio and the conversion of ilmenite first to hexagonal pseudorutile and, with further loss of Fe, to tetragonal rutile (TiO_2). Temple (1966) first recognized that Fe must be mobilized from the structure during this process and invoked Fe^{3+} migration through the ilmenite lattice.

Oxide electrode processes

As discussed in the previous sections, the electrochemical response of an oxide surface to an electrolyte represents the sum of three principal factors: (1) solid state electrochemical reactions and electron transfer, (2) interfacial electron transfer mechanisms dependent on the nature and coordination of surface ligands, and (3) diffusional transport through the double layer to the bulk solution (Diggle, 1973; Bard and Faulkner, 1980). The oxidation and reduction reactions in solution can be studied by construction of an electrode with an oxide/solution interface. Most electrode studies on ferrous oxides have either characterized the self-induced potential (SIP) of the electrode relative to the standard hydrogen electrode or have measured DC or AC currents generated by the oxide electrode from application of an external potential.

The SIP is measured under open current conditions at which the anodic current i_a and cathodic current i_c are equal to the exchange current i_o and the net current i is zero (i.e., $i = i_a + i_c = 0$). The condition such that $i_o = i_a = i_c = 0$ is met only for a chemical system in thermodynamic equilibrium. A number of workers (Engell, 1956; Allen et al., 1979a; White and Hochella, 1989) have measured the SIP for magnetite and ilmenite and found them to vary between -50 and +300 mV relative to the standard hydrogen electrode depending on the solution composition. As indicated by the examples of freshly polished magnetite electrodes (Fig. 2a), the SIP of ferrous oxides initially increase rapidly with time when placed in either oxic or anoxic solutions and ultimately approach only sightly higher potentials in the oxic case. Much larger positive potentials occur as a function of decreasing pH (Fig. 2b). Such potential variations may be produced in part by electrochemical changes within the Helmholtz layer, which, as in the ideal case of a platinum electrode, are reversible. However, as demonstrated by the hysteresis of the forward and backward pH titration, the oxidation reactions are clearly irreversible with the higher residual potentials signifying irreversible oxidation of the magnetite and ilmenite surfaces.

A simple yet powerful technique commonly used in electrochemistry is the application of an external potential which polarizes the oxide electrode, thereby generating a net current between the solid and solution ($i_a + i_c \neq 0$). Figure 3 shows the dynamic polarization of a magnetite electrode in 0.1 N NaCl solutions at pH 3 and 7. Significant features include strong cathodic and anodic currents generated from the respective decomposition of water to H_2 and O_2 (potentials shown by vertical dashed lines). The corrosion current for magnetite crosses zero at approximately +400 mV at pH 3 (Fig. 3a) and +100 mV at pH 7 (Fig. 3b); voltages which correspond to the SIP at which cathodic and anodic reactions for the magnetite electrode must be equal. A strong corrosion current is generated at approximately +200mV when the magnetite electrode is polarized slightly cathodic relative to the SIP in pH 3 solution (Fig. 3a). This potential corresponds to the maximum dissolution rate of magnetite in acidic solutions and has been investigated in considerable detail by a number of workers in corrosion chemistry including Engell (1956), Sukhotin et al. (1976), and Haruyama and Masamura (1978).

Elegant electrochemical models have been developed over the last fifty years to quantify the relationship between potential and current in oxide electrodes. In a simplistic situation, the reaction at the electrode surface can be written as,

$$R \leftrightarrow ne^- + Ox \quad , \tag{8}$$

where R is the reduced species (i.e., Fe^{2+}), Ox is the oxidized species (i.e., Fe^{3+}), and n is the number of electrons involved in the change of oxidation state. The relationship between the net current i resulting from an applied potential, and the exchange current i_o at the SIP, can be described by the current overpotential equation. The general form of this equation is (Bard and Faulker, 1980)

$$i = i_o \left[\frac{C_o(0,t)}{C_o^*} e^{-\alpha n f \eta} - \frac{C_r(0,t)}{C_r^*} e^{(1-\alpha)nf\eta} \right] \quad , \tag{9}$$

where C_o and C_r are the respective oxidant and reductant concentrations at time t and distance zero from the surface; C_o^* and C_r^* are the respective concentrations in the bulk electrolyte; α is the electron transfer coefficient; $f = F/RT$ where F is the Faraday constant; and η is the overpotential defined as the difference between the applied potential, E, and SIP. Obviously when $\eta = 0$, $i = 0$.

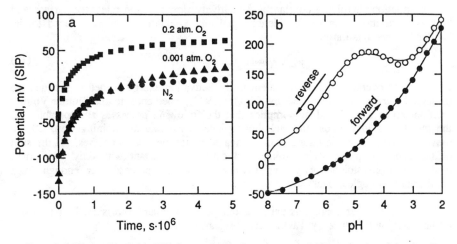

Figure 2. Self-induced potentials (SIP) for a magnetite electrode measured (a) as a function of time in oxic and anoxic 0.1N NaCl solutions at pH 7 and (b) as a function of pH in anoxic 0.1N NaCl solutions. The arrows show the direction of irreversible pH titration using HCl and NaOH solutions (White, unpublished data).

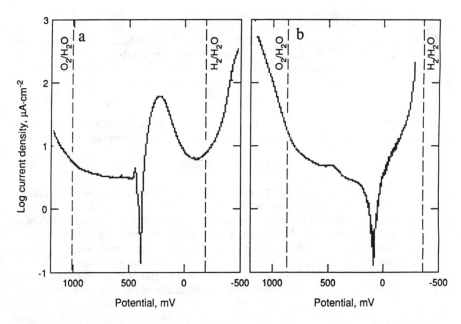

Figure 3. Dynamic polarization of magnetite indicating current response to applied potential in 0.1N NaCl solutions at (a) pH 3.0 and (b) pH 7.0. Vertical dashed lines correspond to stability limits of water (White, unpublished data).

For situations involving constant and relatively slow reaction rates, the surface and bulk concentrations can be considered equal and Equation 9 can be reduced to the Bultler-Volmer relationship:

$$i = i_0[e^{-\alpha n f \eta} - e^{(1-\alpha)n f \eta}] \quad , \tag{10}$$

in which the overpotential is equivalent to the activation energy needed to produce the measured current. As discussed by Diggle (1973), obedience to the Bultler-Volmer relationship in a semiconductor implies that the following processes are operable: (1) the transfer of ions across the Helmholtz plane is the rate determining step, (2) the movement of anions and cations can be treated as independent processes, (3) the sum of the chemical potentials of anions and cations is constant with only very small deviations from stoichiometry, and (4) the dissolution proceeds far enough from equilibrium to neglect backward reactions.

Two end member conditions exist for the Bultler-Volmer expression. If η is small, Equation 10 will reduce by Taylor series expansion to

$$i = i_0(-n f \eta) \quad , \tag{11}$$

in which the net current is linearly related to the overpotential in a narrow range near the SIP. The ratio $-\eta/i$ has the dimensions of resistance and is often called the charge transfer resistance (Mansfield, 1976). Applying Faraday's law, $N = Q/nF$, permits the calculation of the rate of electron transfer in moles$\cdot$s^{-1} at the SIP:

$$dN/dt = i_0/nF \quad . \tag{12}$$

An example of the application of the preceding derivations is shown in Figure 4 for a plot of overpotential versus measured current for magnetite electrodes in a 0.1 N NaCl solution at pH 7. As indicated in Figure 4, the overpotential relationship is linear only very near the SIP which is 100 mV and at positive overpotentials representing the anodic limb of the dynamic polarization curve. The values for the charge transfer resistance and dN/dt (Eqn. 12) are calculated to be 5.3 k$\cdot\Omega\cdot$cm^{-2} and 1.2×10^{-13} moles$\cdot$cm$^{-2}\cdot$s^{-1} respectively.

If the applied overpotentials are large, one of the bracketed terms in Equation 10 becomes negligible and the Butler-Volmer equation can be reduced to:

$$\eta = (RT/\alpha nF)\ln i_0 - (RT/\alpha nF)\ln i \quad . \tag{13}$$

This expression is a form of the Tafel relationship in which the measured current increases exponentially with the applied overpotential. The Tafel relationship has been observed for magnetite in the presence of acidic solutions (Engell, 1956; Sukhotin et al., 1976; and Haruyama and Masamura, 1978). As demonstrated in Figure 5 (from Haruyama and Masamura, 1978), the positive branch of the linear plot is strongly affected by pH while the negative branch is not. This relationship can be explained by the reactions $[O^{2-}] + 2H^+ \rightarrow H_2O$ and $[Fe^{2+}] \rightarrow Fe^{2+}$. The composition of the magnetite surface layer would then be anion-deficient due to cation removal during the anodic reaction, at stoichiometric equilibrium at the SIP, and cation deficient during the cathodic reaction as a function of anion removal into solution. Diggle (1973) pointed out that because the application of anodic polarization increased the magnetite reaction rate, the limiting reaction must be the removal of the anion, i.e., O^{2-}, as a hydroxide

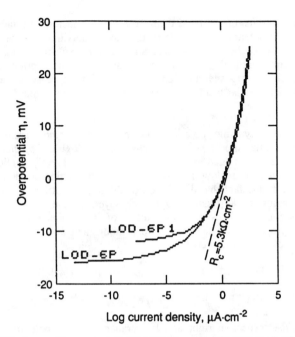

Figure 4. Applied overpotential versus measured current for two magnetite electrodes in a 0.1N NaCl solution at pH 7.0. Slope of the dashed line correspondes to the calculated polarization resistance (White, unpublished data).

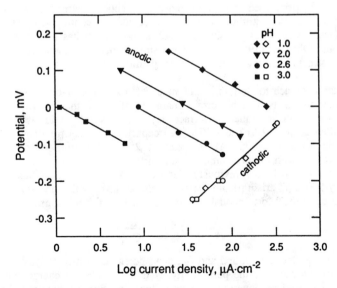

Figure 5. Polarization curves for magnetite in perchlorate solutions of various pHs (Haruyama and Masamura, 1978).

from the surface. In terms of the band theory for semiconductors, the fact that the rest potential or flat band potential lies anodic relative to the maximum dissolution potential indicates that increasing cation removal occurs as the semiconductor bands bend upward (anodic polarization) until degeneracy occurs at some point prior to the freely dissolving potential. An opposite situation is observed for n-type semiconductors, such as hematite, in which the flat band potential is cathodic relative to the freely dissolving potential.

The fact that the dynamic polarization scan of the same magnetite electrode produced only a minimal anodic peak in solutions of pH 7 (Fig. 3b) indicates that the reaction mechanisms investigated in acidic solutions are not directly applicable to more neutral pH conditions. The slow rate of magnetite dissolution at neutral pH most probably results from the fact that dissolution of the oxygen lattice is rate determining causing the formation of a passive maghemite layer. In the case of ilmenite, the reaction is further inhibited by the insolubility of Ti in the structure. White and Hochella (1989) proposed that the respective cathodic and anodic half cell reactions controlling redox of magnetite were

and

$$[Fe_1^{2+} Fe_2^{3+}]O_4 + 8H^+ + 2e^- \rightarrow 3Fe^{2+} + 4H_2O \ , \tag{14}$$

$$3[Fe_1^{2+} Fe_2^{3+}]O_4 \rightarrow 4\gamma[Fe_2^{3+}]O_3 + Fe^{2+} + 2e^- \ , \tag{15}$$

which represent the respective right and left branches of the potentiodynamic limbs surrounding the SIP of magnetite at pH 7 (Fig. 3b). The net reaction at the SIP, in the absence of O_2, is Equation 5. The experimental work of Jolivet and Tronc (1988) confirmed that the reaction stoichiometry for Fe^{2+} and H^+ was exactly 1 to 2 during the conversion of colloidal magnetite to maghemite. The observed decrease in the magnetite reaction rate with pH can also be attributed to rate control by Equation 5. Anodic polarization of magnetite relative to the SIP was investigated by Allen et al. (1979a, 1980) who performed short term transient potential experiments which showed that the charge generation exhibited a parabolic relationship controlled by H^+ diffusion into the surface. As previously discussed, such a mechanism has also been proposed by Swaddle and Oltmann (1980) to adjust the stoichiometry of magnetite due to Fe^{2+} loss from the structure during oxidation to maghemite.

An alternate approach to studying solid state Fe^{2+} oxidation involves the use of AC impedance measurements which permit characterization of the capacitance of the electrode related both to aqueous surface layer effects and oxide semiconductor properties. Allen et al. (1979b) identified two capacitance peaks for magnetite in 1M perchloric acid solutions (Fig. 6a). A peak at 250 mV corresponded to faradaic reactions associated with the dissolution of the surface which releases Fe and electrons to solution at low pH as previously described. A second capacitance peak at approximately 900 mV, arising from the space charge layer within the semiconductor, was found to obey the Mott-Schottky theory defined by the relationship,

$$\phi - \phi_o = \frac{k}{e} \cdot T + \frac{e\varepsilon N}{\delta\pi} \cdot \frac{1}{C^2} \ , \tag{16}$$

where the ϕ and ϕ_o are the applied and flat band potentials respectively, δ is distance, e is the electron charge, ε is the permittivity, N is the number of charge carriers, and C is the capacitance. This equation predicts a linear relationship between the applied potential and one over the square of the capacitance (Fig. 6b). The two linear segments, with differences in slope, were interpreted by Allen et al. (1979b) to correspond to a

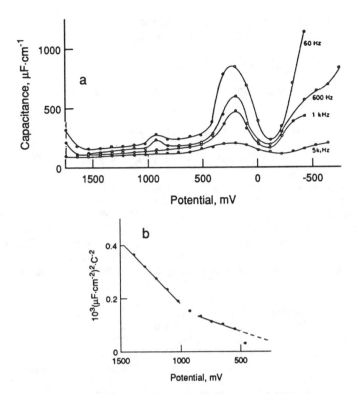

Figure 6. AC impedance measurements on a magnetite electrode in a 1 M NaClO$_4$ solution at pH 1. (a) Capacitance-potential curve, (b) Mott-Schottky plot of potential versus the inverse square of the capacitance at a frequency of 1kHz (Haruyama and Masamura, 1978).

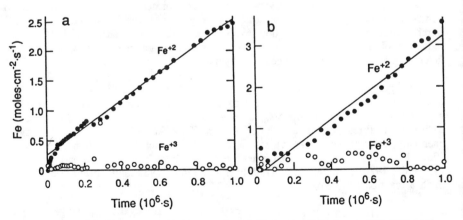

Figure 7. Fe speciation associated with (a) magnetite and (b) ilmenite dissolution at pH 3 (White and Hochella, 1989; and unpublished data).

change in conduction mechanism from the intrinsic p-semiconduction of Fe_3O_4 at or beneath the surface to the n-type semiconduction of γ-Fe_2O_3 at the surface.

Heterogeneous redox reactions

Congruent chemical dissolution of magnetite and ilmenite can be described by the reactions:

$$[Fe_i^{+2} Fe_2^{3+}]O_4 + 8H^+ \rightarrow 2Fe^{3+} + Fe^{2+} + 4H_2O \quad , \quad (17)$$

and
$$[Fe^{2+}]TiO_3 + 2H^+ \rightarrow Fe^{2+} + TiO_{2_{(aq)}} + H_2O \quad . \quad (18)$$

However, congruent dissolution has been reported only at high H^+ activities including 0.5 N HCl and 0.5 to 2.0 N H_2SO_4 in the case of magnetite (Sidhu et al., 1981; Bruyere and Blesa, 1985) and 9 M H_2SO_4 in the case of ilmenite (Barton and McConnel, 1978). At lower H^+ activities, the reaction of both ilmenite and magnetite produce only linear increases in aqueous Fe^{2+} with respect to time (Fig. 7) along with a concurrent uptake of H^+. The lack of aqueous Fe^{3+} and Ti release indicates that magnetite and ilmenite react both incongruently and heterogeneously with the aqueous solution.

Based on the polarization data, the release of Fe^{2+} and consumption of H^+ in solution (Eqn. 5) can described by two coupled redox reactions which involve the concurrent reductive dissolution of the magnetite structure (Eqn. 14) and the structural conversion of magnetite to maghemite (Eqn. 15). Both reactions involve heterogeneous electron transfer; in the anodic case by electron transfer from the solution and in the cathodic case by electron transfer to the solution. In the presence of aqueous species which act as electron donors or acceptors, Equations 14 and 15 can become decoupled from each other and recoupled with a corresponding aqueous half-cell reaction. An example involves the concurrent reduction of dissolved O_2 to form water:

$$3[Fe^{2+}Fe_2^{3+}]O_4 + 1/2O_2 + 2H^+ \rightarrow 4\gamma[Fe_2^{3+}]O_3 + Fe^{2+} + H_2O \quad , \quad (19)$$

which represents the solid state topotatic transformation of magnetite to maghemite (Lindsley, 1976). Although reactions described by Equations. 5 and 19 are superficially similar, they are mechanistically very different. The former involves partial dissolution of magnetite including the loss of structural oxygen, whereas the second reaction is a purely heterogeneous redox reaction involving the oxidation of structural Fe^{2+} and the reduction of dissolved O_2 to form water. A similar reaction can be written for the heterogeneous reduction of oxygen by ilmenite.

Reductive dissolution involving transition metals.   Based on the preceding discussion, the presence of multivalent aqueous transition metals, which can function as electron acceptors or donors, is expected to have a profound influence on electrochemical reactions involving Fe^{2+} oxides. The effects of aqueous transition metal chemistry, in particular aqueous Fe^{2+}, on decoupled reductive dissolution of magnetite (Eqn. 14) at low pH have been extensively documented in the corrosion literature (Valverde, 1976; Sidhu et al., 1981; Jolivet and Tronc, 1988; and Borghi et al., 1989). Such a reaction can be written as

$$[Fe_i^{2+} Fe_2^{3+}]O_4 + 8H^+ + 2^*Fe^{+2} \rightarrow 3Fe^{2+} + 2^*Fe^{+3} + 4H_2O \quad , \quad (20)$$

where the asterisks signify the aqueous oxidation of Fe^{2+} to Fe^{3+}. As discussed by

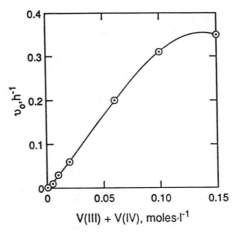

Figure 8. Dissolution rate υ_o of magnetite as a function of equal ratio concentrations of V^{3+} and VO^{2+} at pH 1 and 50°C (Valverde, 1976).

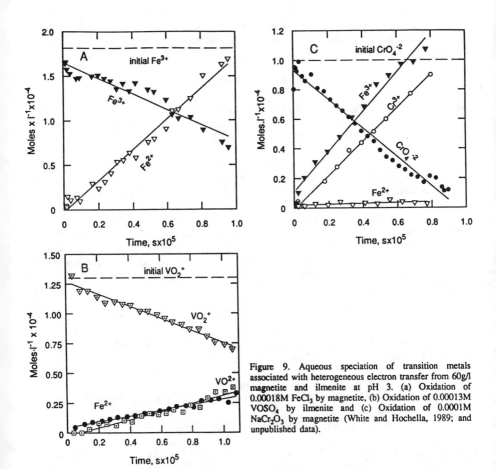

Figure 9. Aqueous speciation of transition metals associated with heterogeneous electron transfer from 60g/l magnetite and ilmenite at pH 3. (a) Oxidation of 0.00018M $FeCl_3$ by magnetite, (b) Oxidation of 0.00013M $VOSO_4$ by ilmenite and (c) Oxidation of 0.0001M $NaCr_2O_3$ by magnetite (White and Hochella, 1989; and unpublished data).

Blesa et al. (1984), an increase of aqueous and surface Fe^{2+}, acting as electron donors, shifts the structural Fe species on the magnetite surface in the direction:

$$[Fe^{3+}] + Fe^{2+}_{adsorb} \rightarrow [Fe^{2+}] + Fe^{+3}_{adsorb} \quad , \tag{21}$$

which results in an increase in the release rate of the less strongly bound Fe^{2+} ion from the magnetite structure. This reaction becomes autocatalytic; as the amount of Fe^{2+} released to solution increases, the reductive dissolution increases.

Compounds such as EDTA (Valverde, 1976; Blesa et al., 1984; Borghi et al., 1989) or oxalate (Baumgartner et al., 1983) complex with Fe^{2+} and accelerate the magnetite dissolution rate under experimental conditions. This is illustrated by the reaction:

$$[Fe^{3+}](C_2O_4) + Fe^{2+}(C_2O_4)_{adsorb} \rightarrow [Fe^{2+}](C_2O_4) + Fe^{3+}(C_2O_4)_{adsorb} \quad , \tag{22}$$

with the subsequent release of Fe^{2+} to solution. The complexing carboxylic ligand increases the sorptive potential of the ferrous species, thus facilitating the inner-sphere electron transfer between the adsorbed and structural Fe states.

Valverde (1976) demonstrated that a linear relationship existed between the rate of magnetite dissolution and aqueous concentrations of equal molar V(IV)/V(III) at pH 1 and 50°C as shown in Figure 8. The corresponding reductive dissolution of magnetite can then be written:

$$[Fe^{2+}_3 Fe^{3+}_2]O_4 + 4H^+ + 2V^{3+} \rightarrow 3Fe^{2+} + 2VO^{2+} + 2H_2O \quad , \tag{23}$$

where the dependence of the dissolution rate on the sum of the V(IV) and V(III) valence species results from the incomplete establishment of equilibrium between Fe ions in different valence states on the Fe_3O_4 surface and the V couple in solution.

Oxidative electron transfer. Heterogeneous electron transfer associated with the oxidation of magnetite to maghemite (Eqn. 15) will proceed in the presence of transition metals that act as aqueous electron acceptors as opposed to aqueous electron donors as in the case of reductive dissolution just discussed. The extent of this effect, however, is less well documented. Engell (1956) was the first to report a decrease of Fe^{3+} in solution in the presence of magnetite. An example of the linear rate of Fe^{3+} reduction on magnetite surfaces at pH 3 is shown in Figure 9a from the work of White and Hochella (1989). Such reduction can be described by a heterogeneous reaction which couples the solid state oxidation of magnetite to maghemite to the aqueous reduction of Fe^{3+} to Fe^{2+}:

$$3[Fe^{2+}_; Fe^{3+}_2]O_4 + 2^*Fe^{3+} \rightarrow 4\gamma[Fe^{3+}_2]O_3 + 2^*Fe^{2+} + Fe^{2+} \quad . \tag{24}$$

In this reaction, two Fe^{3+} atoms, indicated by the asterisks, are reduced in aqueous solution for every Fe^{2+} atom released from the solid.

Recent unpublished data (Fig. 9b) indicate that magnetite is capable of reducing VO_2^+ (V(V)) to VO^{2+} (V(IV)) by oxidative electron transfer according to the reaction:

$$3[Fe^{2+}_; Fe^{3+}_2] + 2VO_2^+ + 4H^+ \rightarrow 4\gamma[Fe^{3+}_2] + Fe^{2+} + 2VO^{2+} + 2H_2O \quad . \tag{25}$$

As demonstrated in Figure 9b, VO_2^+ decreases linearly with time in the presence of magnetite while VO^{2+} exhibits a linear increase. In addition, a 1:2 stoichiometry exists between Fe^{2+} produced from the oxidation of magnetite to maghemite and VO_2^+ consumed by electron transfer as predicted by Equation 25. The fact that a comparable 1:2 stoichiometry does not exist between the Fe^{2+} and VO^{2+} indicates that the bivalent vanadate is more strongly adsorbed. A comparison of Equations 23 and 25 underscores the importance of heterogeneous reactions in buffering aqueous redox speciation. In the case of vanadium, only the VO^{2+} species is stable in solution in the presence of magnetite. The more reduced species, V^{3+} is oxidized to VO^{2+} by reductive dissolution of magnetite while the more oxidized species, VO_2^+ is reduced to VO^{2+} by the solid state oxidation of magnetite to maghemite.

The heterogeneous reduction of chromate (Cr(VI)) to Cr(III) in the presence of magnetite and ilmenite was investigated by White and Hochella (1989). In the case of ilmenite, the reaction can be represented as:

$$9[Fe^{2+}]TiO_3 + 16H^+ + 2CrO_4^{2-} \rightarrow 9[2/3Fe^{3+}]TiO_3 + 3Fe^{2+} + 2Cr^{3+} + 8H_2O \quad . \quad (26)$$

The linear rates of reduction of CrO_4^{2-} to Cr^{3+} in the presence of ilmenite is shown in Figure 9c as a function of time. Unlike the case for V reduction, the Fe^{2+} released during structural oxidation is not stable in the presence of the residual chromate, and is subsequently oxidized via the aqueous reaction:

$$3Fe^{2+} + 8H^+ + CrO_4^{2-} \rightarrow 3Fe^{3+} + Cr^{3+} + 4H_2O \quad . \quad (27)$$

This reaction accounts for the increase in Fe^{3+} in the pH 3 experiment shown in Figure 9c. The increasing Cr^{3+} results from both heterogenous reduction (Eqn. 26) and homogeneous reduction (Eqn. 27). Based on the required 3 to 1 stoichiometry of Equation 27, aqueous oxidation of Fe^{2+} can not explain the amount of CrO_4^{2-} reduced in solution.

OXIDATION OF FERROUS ORTHO AND CHAIN SILICATES

Ortho and chain silicates are common rock-forming minerals, of which olivines, pyroxenes, and amphiboles contain the bulk of the Fe^{2+} in the earth's crust (Table 1). Surprisingly, little information is available on the electrochemical mechanisms associated with oxidation of Fe^{2+} in these minerals under ambient temperature and pressure conditions.

Orthorhombic olivine minerals are composed of independent SiO_4 tetrahedra linked by divalent atoms in six-fold coordination. In the continuous solid solution between fayalite (Fe_2SiO_4) (Table 4) and forsterite (Mg_2SiO_4), some degree of ordering occurs in the octahedral sites with the larger Fe^{2+} atoms displaying a sight preference for the M(1) sites. Most primary Fe-rich olivines contain less than 2% Fe^{3+} although relatively rare ferrifayalite ($Fe^{2+}Fe^{3+}_2Si_2O_8$) does occur. In such olivines, Fe^{3+} is ordered into the M(1) position and Fe^{2+} into the M(2) position requiring the formation of structural vacancies in the octahedrons such as $[3Fe^{2+}] \Leftrightarrow [2Fe^{3+}, \varnothing]$ (Deer et al., 1982).

The pyroxene structure consists of the linkage of SiO_4 tetrahedra at two of the four corners to form continuous chains of the composition $(SiO_3)_n$. The orthorhombic subgroup consists essentially of the compositional series $MgSiO_3$-$FeSiO_3$, representing enstatite-orthoferrosilite (Table 4). Mössbauer and optical spectral studies indicate that Fe^{2+} prefers the M(2) octahedral sites in orthopyroxenes while Mg preferentially occupies the M(1) sites (Rossman, 1980). The lower site energies of the M(2) sites result in greater cation mobility and are responsible for the preferential loss of Fe^{2+}

TABLE 4. Common Fe^{2+}-containing ortho and chain silicates

Olivine	
Fayalite-Forsterite	Fe_2SiO_4 - Mg_2SiO_4
Orthorhombic pyroxenes	
Enstatite-Orthoferrosilite	$MgSiO_3$-$FeSiO_3$
Monoclinic pyroxenes	
Diopside-Hedenbergite	$(Ca,Mg)Si_2O_6$-$(Ca,Fe)Si_2O_6$
Augite-Ferroaugite	$(Ca,Na,Mg,Fe^{2+},Mn,Fe^{3+},Al,Ti)_2(Si,Al)_2O_6$
Pigeonite	$(Mg,Fe^{2+},Ca)(Mg,Fe^{2+})Si_2O_6$
Aegirine-Aegirine-augite	$(Na,Fe^{3+})Si_2O_6$-$(Na,Ca)(Fe^{3+},Fe^{2+},Mg)Si_2O_6$
Amphiboles	
Hornblende	$(Na,K)_{0.1}Ca_2(Mg,Fe^{2+},Fe^{3+},Al)_5(Si_{6.7}Al_{2.1})O_{22}(OH,F)_2$
Ferroactinolite	$Ca_2(Mg,Fe^{2+})_5 (Si_8O_{22})(OH,F)_2$

relative to Mg during aqueous dissolution and weathering (Schott and Berner, 1983). Fe^{2+} has been observed in both the M(1) and M(2) octahedral sites for clinopyroxenes. Fe^{3+} is a common component in clinopyroxenes in basaltic flows and can be situated in both the M(1) sites in the octahedral position and in the tetrahedral sites. Relatively rare ferric pyroxenes, such as monoclinic acmite ($NaFe^{3+}Si_2O_6$), also can be produced from crystallization of peralkaline syenites or under higher temperature metamorphic conditions.

Amphiboles are composed of $(Si,Al)O_4$ tetrahedra linked to form chains which have double the width of those in pyroxenes and have the composition $(Si_4O_{11})_n$. The amphibole structure allows great flexibility for substitution in the octahedral sites, and minerals of the group exhibit an extremely wide range of chemical composition (Table 4). Fe^{2+} in amphiboles can occupy the M(1), M(2), M(3), and/or 6-coordinated M(4) octahedral sites. The ordering pattern between sites is complex and varies as a function of specific mineral composition and structure. The behavior of Fe^{2+} in amphiboles is generally antipathetic to that of Mg. Unlike for pyroxenes, there is no convincing evidence for tetrahedrally-coordinated Fe^{3+} in amphiboles. Detailed analyses have indicated strong ordering of Fe^{3+} into the M(2) site with lesser amounts of Fe^{3+} in M(1) and/or M(3) sites (Hawthorne, 1981a).

Solid state electron transfer

Of the chain and orthosilicates, only the olivines have been investigated extensively as to bulk electrical conductivity, which was found to increase with increasing temperature, pressure and Fe composition. The range of conductivities for Fa_{42} to Fa_{100} varied between 10^{-10} and 10^{-7} $ohm^{-1} \cdot cm^{-1}$ at 1 bar pressure and 25°C (Mao, 1973). The anisotropy relative to crystallographic direction is low. Brandley et al. (1964) initially proposed that electrical conductivity was created by electron transfer or hopping according the solid state reaction as expressed in Equation 1. This mechanism is supported by the observation that the conductivity of olivine decreases with the reduction of Fe^{3+} to Fe^{2+} under low O_2 fugacity (Duba and Nicholls, 1973). Littler and Williams (1969) have also attributed electrical conductivity at ambient temperature in the amphibole mineral crocidolite to the above electron hopping mechanism.

Several electronic absorption spectroscopy studies have documented the presence of an intervalence electron charge interaction (IVCT) in the visible spectrum between Fe^{2+} and Fe^{3+} in ortho and chain silicates. In pyroxenes, the IVCT effect is strongly a function of Fe^{3+} substitution. For example, in lunar samples formed under extremely low O_2 fugacities and no Fe^{3+}, IVCT effects are observed only between Fe^{2+} and Ti^{4+}

(Burns et al., 1976). In calcic and alkali amphiboles, the intervalence charge-transfer bands for the $Fe^{2+}_{M(1)} \rightarrow Fe^{3+}_{M(2)}$ interaction are shown to be greater than the $Fe^{2+}_{M(3)} \rightarrow Fe^{3+}_{M(2)}$ due to closer cation distances in the structure (Hawthorne, 1981b).

Solid state oxidation

In contrast to olivines and pyroxenes, for which few data are available, oxidation of amphiboles has been intensively studied, principally in relation to concurrent dehydroxylation at high temperature. For an idealized amphibole, such a reaction is:

$$[Fe^{2+}_7]Si_8O_{22}[OH]_2 + 1/2O_2 \rightarrow [Fe^{3+}_2, Fe^{2+}_5]Si_8O_{22}[O]_2 + H_2O \quad . \qquad (28)$$

This reaction is similar to that proposed for low temperature clay oxidation as will be discussed latter. Solid state diffusion of O_2 into the amphibole structure is not considered to be a viable mechanism. Instead, as initially proposed by Addison et al. (1962), structural oxidation of Fe occurs by the same electron hopping mechanism as discussed in the preceding section. This process can be written as the reaction sequence:

$$[Fe^{2+}] \rightarrow e^- + [Fe^{3+}] \quad ,$$

$$[Fe^{3+}]_{surface} + e^- \rightarrow [Fe^{2+}]_{surface} \quad ,$$

and
$$[Fe^{2+}]_{surface} + H^+ + 1/4O_2 \rightarrow [Fe^{3+}]_{surface} + 1/2H_2O \quad , \qquad (29)$$

in which the surface Fe acts as a catalyst undergoing successive reduction by electrons from the underlying lattice, and subsequently undergoing reoxidation at the surface from atmospheric oxygen. This process is the same as positive hole conduction, involving transfer of electrons from Fe^{2+} atoms to neighboring Fe^{3+} atoms. The corresponding proton transfer associated with Equation 29 can be described by the concurrent reactions:

$$[OH^-] \rightarrow [O^{2-}] + H^+ \quad ,$$

$$[O^{2-}]_{surface} + H^+ \rightarrow [OH^-]_{surface} \quad ,$$

and
$$[OH^-]_{surface} + OH^- \rightarrow [O^{2-}] + H_2O \quad , \qquad (30)$$

which is a surface catalytic reaction involving the protonation and deprotonation of the amphibole surface. In amphiboles, the cations are arranged in ribbons, and electron transfer may take place along this structure (Addison et al., 1962). The presence of other cations such as Mg in the ribbon blocks the transfer process and, accordingly, Mg-rich amphiboles tend to be more resistant to oxidation.

Mössbauer spectra, which supports the scenario that dehydroxylation is accomplished by oxidation, show extra Fe^{3+} doublets in the spectra which have been assigned to Fe^{3+} at the M(1) and M(3) sites. Comparisons of the Mössbauer spectra of primary and oxidized amphiboles also suggest that considerable cation disorder results from oxidation, including the loss of Fe^{2+} from the M(3) sites and Fe^{3+} from the M(2) sites (Phillips et al., 1986). Oxyhornblende, the end member resulting from the above hydrothermal process, often has Fe^{3+}/(total Fe) ratios approaching unity.

486

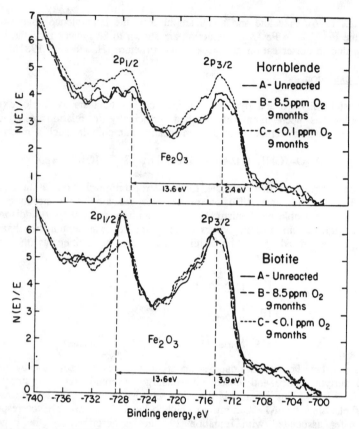

Figure 10. XPS spectra for $2P_{1/2}$ and $2P_{3/2}$ Fe bonding on the surfaces of unreacted and reacted hornblende and biotite. The respective 2.4 and 3.9 eV shifts are the result of surface charging. Vertical dashed lines are the peak positions for αFe_2O_3 with an energy separation of 13.6 eV (White and Yee, 1985).

Similar oxidation processes are expected to occur at ambient temperatures, but at much slower rates. Comparisons of naturally weathered and unweathered hornblende unit cell parameters showed no significant differences, indicating that the removal of Fe^{2+} occurs principally from the M(1) and M(3) sites which results in only minor structural perturbations (Goodman and Wilson, 1976). The Mössbauer spectra did show that the extent and widths of the Fe^{3+} peaks increased and that oxidation affected the electric field gradient at the Fe^{3+} sites. Both the net effect of Fe^{2+} loss and oxidation in the bulk hornblende must be relatively small since the overall ratio of Fe^{3+}/(total Fe) does not change appreciably during the initial stages of weathering.

In contrast to the bulk Fe compositions just discussed, X-ray photoelectron spectroscopy (XPS) indicates that surficial Fe in both experimentally and naturally weathered Fe silicates is extensively oxidized. For example, Figure 10a shows the binding energies for the $Fe_{2p1/2}$ and $Fe_{2p3/2}$ photoelectron peaks of hornblende surfaces reacted for 9 months under both oxic and anoxic solutions at pH 5. The peak positions compare closely to those of hematite, indicating the complete oxidation of the surface Fe (White and Yee, 1985). Comparable XPS binding energy shifts of between 1.0 and 1.2 eV have been observed for surface Fe relative to energy shifts for Fe in naturally weathered grain interiors of olivine, pyroxene, and amphiboles (Table 5).

Dissolution processes

A significant number of researchers have investigated the behavior of Fe during the dissolution of olivine at ambient temperatures (Sanemasa et al., 1972; Schott and Berner, 1983; Siegel and Pfannkuch, 1984; and Grandstaff, 1986). Except for reactions under oxic conditions leading to the precipitation of ferric oxyhydroxides, the release of Fe^{+2} proceeded congruently with respect to Mg and Si and the stoichiometry of the olivine phase (Fig. 11a). This condition is ascribed by Murphy and Helgeson (1987) to be both a consequence of the isolation of the silica tetrahedra from one another in the olivine orthosilicate structure and to the only slight energy differentials controlling Fe^{2+} and Mg ordering between the M(1) and M(2) sites.

In contrast, Siever and Woodford (1979) documented incongruent release of Fe^{2+} from pyroxene (bronzite) surfaces which later oxidized either in solution or on the silicate surface and subsequently precipitated as $Fe(OH)_3$. Schott and Berner (1983) also found that while Mg release from bronzite was generally congruent relative to Si, the initial release rates of Fe^{2+} were much more rapid (Fig. 11b). Murphy and Helgeson (1987) proposed that dissolution of orthopyroxenes proceeded as

$$[Fe^{2+}_{(2)} \, M_{(1)}Si_2O_6] + H^+ \rightarrow [H^+_{(2)} \, M_{(1)}Si_2O_6]^- + Fe^{2+} \quad , \qquad (31)$$

with preferential release of Fe^{2+} from the less energetically favorable M(2) site caused by an exchange reaction with absorbed H^+ and the subsequent release of Fe^{2+} to solution. The breakdown of the silicate structure would then occur by the entry of a second H^+ into the vicinity of the exchanged site or near an M(2) site bearing another Fe atom, thus further destablizing the structure. Murphy and Helgeson (1987) speculated that slower rates of hydrolysis accompanying surface oxidation of Fe^{2+} may be the result of the formation of a more stable Fe^{3+}-bearing structure which inhibits H^+ penetration and retards the rate of silicate detachment. White and Yee (1985) demonstrated that Fe^{2+} is the dominant aqueous Fe species released during dissolution of hornblende under both oxic and anoxic conditions over a pH range of 1 to 7. As would be expected, Fe^{2+} concentrations decrease in solution with time under oxic neutral pH conditions due to the precipitation of ferric oxyhyroxide.

Coupled electron-cation transport

The preceding data indicate that although the surfaces of both experimentally and naturally weathered Fe-containing olivines, pyroxenes, and amphiboles are oxidized to Fe^{3+}, the principal species released by dissolution is Fe^{2+}. This apparent discrepancy, as originally pointed out by White and Yee (1985), must be reconciled by a heterogeneous redox process occurring between the silicate surface and the solution:

$$[Fe^{2+}, 1/zM^{z+}] \rightarrow [Fe^{3+}] + 1/zM^{z+} + e^- \quad , \qquad (32)$$

where M is a cation of charge z. For an idealized pyroxene of the composition $[Fe,Ca]Si_2O_6$, for example, the coupled electron-cation reaction can be written as:

$$[Fe^{2+}, 1/2Ca] \rightarrow [Fe^{3+}] + 1/2Ca^{2+} + e^- \quad , \qquad (33)$$

where charge balance in the oxidized surface phase is maintained by the formation of a vacancy site. The resulting oxidized Fe^{3+} on the surface can also undergo subsequent hydrolysis and dissolution by the reaction:

488

TABLE 5. XPS data for the $Fe_{2p3/2}$ photoelectron showing
elemental ratios and binding energies for the surfaces
and interiors of pyroxenes and amphiboles from soils.
The weathered surfaces were ultrasonically cleaned prior
to analysis (from Berner and Schott, 1982).

	Fe/Si		Fe (eV)	
	Surface	Interior	Surface	Interior
Pyroxene				
Bronzite	1.61	1.65	609.7	608.3
Diopside	0.95	1.85	609.9	608.5
Hypersthene	0.36	0.46	609.6	608.5
Augite	2.44	2.55	609.7	608.7
Amphibole				
Hornblende	3.10	2.80	609.8	608.7

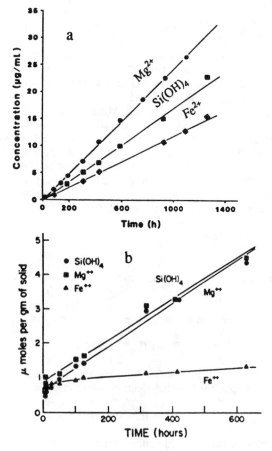

Figure 11. Aqueous speciation associated with (a) olivine dissolution in oxic 0.1M KCl at pH 3.2 (Grandstaff, 1986), and (b) bronzite dissolution in oxic pH 1 solution (Schott and Berner, 1983).

$$[Fe^{3+}] + 3H^+ \rightarrow [3H^+] + Fe^{3+} \quad . \tag{34}$$

Coupling this reaction with the half cell reaction describing the reduction of aqueous Fe at the mineral surface results in the corresponding hydrolysis half cell reaction:

$$[Fe^{3+}] + 3H^+ + e^- \rightarrow [3H^+] + Fe^{2+} \quad . \tag{35}$$

If no reducible species exist in aqueous solution, the two half cell reactions (Eqns. 33 and 35) are directly coupled and the sum of the reactions will produce congruent cation dissolution:

$$[Fe^{2+}_? \ 1/2Ca] + 3H^+ \rightarrow [3H^+] + Fe^{2+} + 1/2Ca^{2+} \quad . \tag{36}$$

If the hydrolysis reaction (Eqn. 35) is slow and rate controlling relative to the solid state oxidation reaction (Eqn. 33), residual Fe^{3+} will be present on the surface at the same time as Fe^{2+} is released to solution. As previously discussed, these respective conditions are confirmed by aqueous and XPS analyses of Fe.

The above sequence of reactions can be been studied in greater detail by using an extraneous electron acceptor in aqueous solution which decouples the half cell reaction for the structural oxidation of Fe^{2+} from the dissolution of Fe. As discussed by White and Yee (1985), this effect can be demonstrated most readily by the addition of Fe^{3+} to solution. In such a case, the electron transfer associated with the structural oxidation of Fe is no longer directly coupled to the release of Fe^{3+} into solution but can be described for the idealized pyroxene by the reaction:

$$[Fe^{2+} \ 1/2Ca] + {}^*Fe^{3+} \rightarrow [Fe^{3+}] + 1/2Ca^{2+} + {}^*Fe^{2+} \quad . \tag{37}$$

The Fe species designated by the asterisks represents the extraneous Fe component added to solution. As predicted by Equation 37, aqueous Fe^{3+} is reduced to Fe^{2+} in the presence of augite and hornblende (Fig. 12). In the case of the augite experiment, the initial $FeCl_3$ was tagged with ${}^{59}Fe$. Beta counting indicated a constant count rate in solution confirming the aqueous reduction of Fe^{3+} to Fe^{2+}. This result precludes the possibility of simple exchange of Fe^{3+} and Fe^{2+} on the mineral surface.

Addition of Fe^{3+} to the aqueous solutions significantly increased the rate of Ca release from augite (Fig. 13a) relative to rates in solutions containing no Fe^{3+} (Fig. 13b). Rates of Mg and Na release also increased although to a lesser extent reflecting more closely the actual mineral stoichiometry and suggesting that other cations besides Ca^{2+} may act as charge compensators in Equation 37. As predicted, however, the concentrations of Ca released and Fe^{3+} consumed were approximately equivalent. The addition of aqueous Fe^{3+} had the effect of decoupling the redox reaction rate from the hydrolysis reaction rate. In fact, as indicated by the delay in the rate of H^+ uptake, the hydrolysis reaction (Eqn. 35) was strongly suppressed by the redox reaction (Eqn. 33) and commenced only after Fe^{3+} was completely consumed in solution.

Dissolved oxygen reduction. The reduction of dissolved O_2 (DO) by ferrous silicates has been documented for granite (Torstenfelt et al., 1983), basalt (White et al., 1985) and for individual ferrous silicate phases (White and Yee, 1985). An example of O_2 uptake by augite and hornblende as a function of pH and time, is shown in Figure 14. A coupled heterogeneous redox reaction involving the reduction of O_2 to H_2O can be

490

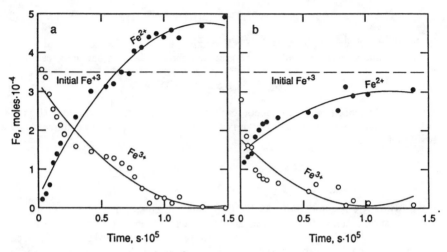

Figure 12. Heterogeneous reduction of Fe^{+3} in the presence of 30 g/l (a) hornblende and (b) augite at pH 3.0 (White, unpublished data).

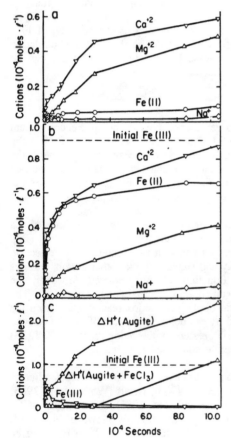

Figure 13. Changes in aqueous chemistry at pH 3.0 during heterogeneous oxidation of augite. (a) Major cation release from a starting solution of deionized water. (b) Major cations release from a starting solution of 0.0001 molar $FeCl_3$. (c) Comparisons of rates of H^+ decrease (ΔH^+) in deionized water and $FeCl_3$ solutions. Also shown is the corresponding rate of Fe^{3+} loss from the $FeCl_3$ solution (White and Yee, 1985).

written as:

$$[Fe^{2+}_i \; 1/zM^{z+}] + 1/4O_2 + H^+ \rightarrow [Fe^{+3}] + 1/zM^{z+} + 1/2H_2O \quad . \tag{38}$$

Within the stability field of ferric oxyhydroxides, the uptake of O_2 shown in Figure 14 can also occur by the aqueous redox reaction:

$$Fe^{2+} + 1/4O_2 + 3/2H_2O \rightarrow [FeOOH] + 2H^+ \quad , \tag{39}$$

in which ferrous ion produced by ferrous silicate dissolution (Eqn. 36) is first oxidized in solution and then precipitated. White and Yee (1985) presented evidence indicating that reduction rates for DO based on structural oxidation are more rapid than those based on aqueous Fe^{2+} oxidation and subsequent oxyhydroxide precipitation. The rates of O_2 uptake at neutral pH are much faster than Fe^{2+} release rates and increased dramatically with increasing H^+ concentrations beyond the point at which aqueous Fe^{3+} became undersaturated with respect to the oxyhydroxide phase (Fig. 14). In addition, secondary ion mass spectroscopy (SIMS) failed to detect ^{18}O enrichment on the mineral surfaces containing dissolved $^{18}O_2$, implying that the tagged O_2 did not precipitate on the mineral surfaces as an oxyhydroxide phase.

Actinide reduction on olivine and basalt.   The effects of heterogeneous redox processes on neptunium retardation and sorption on Fe^{2+}-silicate rocks and minerals have been addressed by Susak et al. (1983) and Meyer et al. (1984) in relation to predicting transport characteristics of nuclear waste in geologic repositories. The redox chemistry of Np is characterized by three stable valence states, Np^{4+}, NpO_2^+, and NpO_2^{2+}. The (V) and (VI) ligands have minimal tendencies to sorb or react with mineral surfaces while Np^{4+} is strongly sorbed and immobilized at near-neutral pH. Addition of basalt and olivine to such solutions resulted in the significant loss of NpO_2^{2+} and increases of NpO_2^+ in solution (Fig. 15) and Np^{+4} on the mineral surfaces.

In the case of NpO_2^{2+} reduction in the presence of olivine, Susak et al. (1983) proposed the following sequence of reactions:

$$NpO_2^{2+} + 4H^+ \rightarrow Np^{6+}_{adsorbed} + 2H_2O \tag{40}$$

$$Np^{6+}_{adsorbed} + [Fe^{2+}SiO_3] \rightarrow Np^{5+}_{adsorbed} + [Fe^{3+}SiO_3] \tag{41}$$

$$Np^{5+}_{adsorbed} + 2H_2O \rightarrow NpO_2^+ + 4H^+ , \tag{42}$$

where Equation 41 involves the heterogeneous electron exchange between Fe^{2+} in the olivine structure with Np^{6+} sorbed on the mineral surface. Meyer et al. (1984) ran experiments using initial NpO_2^+ solutions in which the aqueous electrochemical potential was stabilized using a porous platinum electrode. The results indicated that NpO_2^+ was being reduced to Np^{4+} on the mineral surfaces of crushed basalt. The reduction rate increased with increasing pH indicating that NpO_2^+ must first be sorbed on the substrate before reduction can proceed.

Nitrate reduction.   Postma (1990) recently investigated the potential of nitrate reduction by Fe^{2+} on the surfaces of pyroxenes (augite) and amphiboles (arvedsonite). The experimental results demonstrated that the behavior of Fe and the rate of nitrate reduction were both strongly pH dependent (Fig. 16). At pH 1, both Fe^{2+} and Fe^{3+} were released into solution, suggesting congruent dissolution without corresponding nitrate reduction. Nitrate loss and subsequent nitrite gain were observed in higher pH

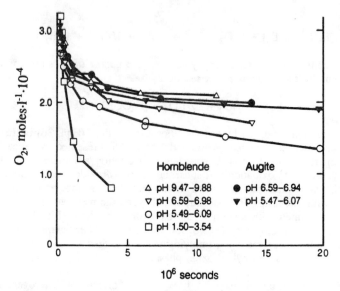

Figure 14. Dissolved O_2 decreases with time from solutions of differing pH containing 80 g/l augite and hornblende (White and Yee, 1985).

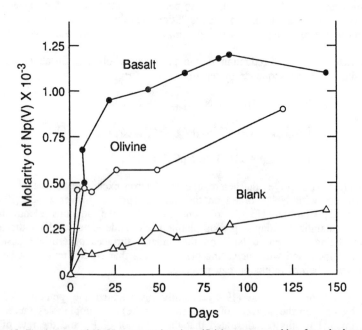

Figure 15. The increase of Np(V) concentrations in artificial seawater resulting from the heterogeneous reduction of Np(VI) as a function time at an initial pH of 4. Long term increase in the blank solution results from alpha decay of ^{237}Np (Susak et al., 1983).

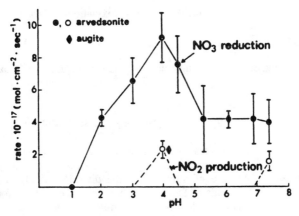

Figure 16. Rates of nitrate reduction and nitrite production in the presence of pyroxene and amphibole (arvedsonite) at differing pH (Postma, 1990).

solutions. Nitrate reduction rates were comparable for both the pyroxene and amphibole and reached maximum rates at pH 4. Quantification of the nitrite production rates was not possible due to decomposition under the experimental conditions. In terms of oxidative electron transfer, the half cell describing nitrate reduction to nitrite,

$$NO_3^- + 2H^+ + 2e^- \rightarrow NO_2^- + H_2O \quad , \tag{43}$$

can be combined with the solid state oxidation half-cell (Eqn. 32) to produce the reaction:

$$2[Fe^{2+}, 1/zM^{z+}] + NO_3^- + 2H^+ \rightarrow 2[Fe^{3+}] + NO_2^- + 2/zM^{z+} + H_2O \quad . \tag{44}$$

The onset of nitrate reduction at pH 2 also coincides with a significant decrease in both Fe^{2+} and total Fe aqueous concentrations. This was interpreted by Postma (1990) as signifying the catalytic effect of ferric oxyhyroxide precipitation on nitrate reduction although an exact mechanism was not proposed. As indicated by Postma, heterogeneous nitrate reduction may have important applications to problems such as agricultural contamination of aquifers, but additional work is needed to clarify the reactions involved.

OXIDATION OF MICAS

The crystal structure of sheet silicates, including micas, consists of two basic units: the Si tetrahedron formed by a Si ion surrounded by four O^{2-} ions in tetrahedral coordination, and the Al octahedron formed by an Al ion surrounded by four O^{2-} and two OH^- ions in octahedral coordination. In micas, one tetrahedral sheet separates two octahedral sheets. Trioctrhedral biotite and annite and dioctahedral glauconite are common micas containing significant amounts of Fe. Generally Fe^{2+} and Fe^{3+} are randomly distributed among the octahedral sites although short-range bonding preferences of Mg for F and Fe^{2+} for OH may result in small scale domains in the octahedral sites as demonstrated by magnetic resonance studies of Sanz and Stone (1983). Fe^{3+} occasionally occurs in the tetrahedral sites but both Fe^{2+} and Fe^{3+} only rarely occur in interlayer positions. The layer charge in a mica arises by some combination of four primary mechanisms: (1) substitution of Al^{3+} or Fe^{3+} for Si^{4+} in the tetrahedral sites, (2) substitution of Fe^{2+} for Fe^{3+} in the octahedral positions, (3) vacancies in the octahedral positions, or (4) dehydroxylation of OH⁻ to O²⁻.

<u>Oxidation of structural Fe</u>

A number of studies have documented coupled structural Fe^{2+} oxidation and dehydroxylation of biotite at temperatures >200°C. For an ideal annite, the net reaction can be written as:

$$K_2[Fe^{+2}_3](AlSi_3)O_{10}[OH]_2 + 1/2O_2 \rightarrow K_2[Fe^{2+}Fe^{3+}_2](AlSi_3)O_{10}[O]_2 + H_2O \quad . \quad (45)$$

where Fe occupies octahedral sites and K interlayer sites. Sanz and Stone (1983) showed that the rate of Fe oxidation was dependent on the O_2 partial pressure. As previously discussed in the case of amphiboles, the diffusion of O_2 into the lattice structure is not a viable mechanism. Sanz and Stone proposed that O_2 forms charged oxygen species on the surface which promote the migration of electrons and protons from the underlying lattice structure. Based on Mössbauer data and an activated complex model, Ferrow (1987) demonstrated that at ambient temperature, Fe^{2+} in the M(2) site in biotite was oxidized preferentially to the M(1) site due to the shorter distance between the M(2)-M(2) shared edge than the M(2)-M(1) shared edge. This shorter distance, in addition to shared symmetry of the M(2) sites, favors electron charge transfer across the M(2)-M(2) edge, and consequently, preferred oxidation.

Scott and Amonette (1988) proposed that the formation of oxybiotite phases at ambient temperature involves four processes: (1) the oxidation of Fe^{2+} to Fe^{3+} in the structure, (2) a decrease in K content, (3) an increase in the structural H_2O content, and (4) the increase in the ion exchange capacity. The link between structural Fe^{2+} oxidized in the octahedral site and K released from the interlayer position to aqueous solution can be ascribed to the reaction:

$$K[Fe^{2+}] \rightarrow [Fe^{3+}] + K^+ + e^- \quad . \quad (46)$$

A direct 1:1 stoichiometric link between these processes would ultimately result in a oxybiotite with no layer charge. Observations by Coleman et al. (1963) and others however indicate an increase in cation exchange capacity of biotite during weathering. This apparent contradiction led Scott and Amonette (1988) to propose three additional oxidation reactions which occurs simultaneously to maintain much of the layer charge despite the occurrence of oxidation. These reactions are:

$$[Fe^{2+}(OH)_{2 \, octa}] \rightarrow [Fe^{2+}O(OH)_{octa}] + H^+ + e^- \quad (47)$$

$$3[Fe^{2+}_{octa}] \rightarrow 2[Fe^{3+}_{octa}] + Fe^{3+} + 3e^- \quad (48)$$

$$[Fe^{2+}_{octa}] + H_2O \rightarrow [Fe^{3+}_{octa}(OH^-_{inter})] + H^+ + e^- \quad , \quad (49)$$

which respectively reflect: (1) a dehydroxylation process similar to high temperature conditions (Eqn. 45), (2) the ejection of additional octahedral cations from the mica structure, and (3) a gain of hydroxyls in vacant interlayer sites. In theory, both deprotonation and octahedral cation ejection should increase the selectivity of mica for K by eliminating or changing the structural orientation of the structural hydroxyls, and thus lessen the amounts of K released during weathering. The addition of structural Fe^{3+} should also enhance the resistance of the mica to dissolution by increasing the Fe bonding strength. Calculations indicate that the oxide bonding energy of octahedrally coordinated Al (and presumably Fe^{3+}) is roughly twice that of the divalent cation in

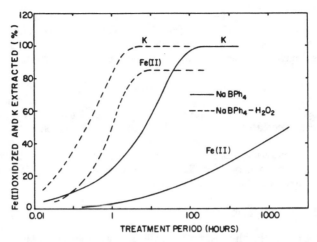

Figure 17. K extracted from and structural Fe^{2+} oxidized in lepidiomelane mica at 25°C with oxic NaCl and sodium tetraphenyl boron (NaBPh$_4$) solutions that did or did not contain H_2O_2 and were exposed to air (Scott and Amonette, 1988).

silicates such as micas (Voskresenskaya and Coliazo, 1983). This effect is supported by the observation that oxybiotites, in which structural Fe^{2+} has been oxidized and hydroxyls deprotonated, are very resistent to additional weathering.

A number of studies have investigated the effects of oxidizers on the valence state of structural Fe and the artificial weathering of Fe micas (Gilkes et al., 1973). The most commonly employed agents include O_2, OCl^-, H_2O_2, and Br_2. The rate of oxidation of Fe-micas by these agents is strongly related to the concurrent expansion of the mica structure. Aqueous solutions such as NaCl and NaBPh$_4$, which can exchange interlayer K and therefore expand the mica, result in relatively rapid oxidization of structural Fe^{2+} by dissolved oxygen at ambient temperatures (Scott and Amonette, 1988) Figure 17, adapted from their work, indicates that the oxidation of Fe^{2+}, and the concurrent release of K, occurs almost as rapidly in the presence of O_2 as H_2O_2. Using the degree of K extraction by sodium tetraphenyl boron (NaBPh$_4$) as a measure of the mica expansion, the data also indicate that structural Fe^{2+} is not oxidized as fast as it is exposed.

In contrast, the inhibition of mica expansion by the presence of aqueous K has also been shown to greatly decrease or prevent the oxidation of Fe^{2+} (Gilkes and Young, 1982). Thus, the penetration of O_2 and other oxidants are strongly dependent on diffusion in or along the interlayer region of the mica rather than dependent on diffusion through the tetrahedral or octahedral structure. Unlike expandable smectites and vermiculites, in which structural Fe^{3+} can be reduced by dithionite, hydrazine, and sulfide, there is no experimental evidence that Fe^{3+} can be reduced in micas.

Dissolution

Most studies of the experimental dissolution of Fe micas have been performed under strongly acidic conditions, or else have used chelating agents or strong oxidizers. Biotite reacted with strong acids exhibits congruent release of Fe, Al, and K and retention of Si, resulting in the formation of visible leached layers (Gastuche, 1963). Other studies have related the effects of natural organic chelators, such as fulvic acids, to accelerated dissolution of Fe from biotite (Schnitzer and Kodama, 1976). The organics may have the dual effects of complexing with surface Fe, thus reducing the structural bonding energy, and preventing the reprecipitation of secondary surface Fe oxyhydroxides which could act as diffusion barriers to chemical transport.

In experimental weathering at neutral pH, White and Yee (1985) documented, using XPS, that the Fe $2p_{1/2}$ and Fe $2p_{3/2}$ photoelectron peak intensities remained relatively constant on the basal plane of biotite and were dominated by Fe^{3+} (Fig. 10b), while the K_{1s} XPS peak decreased in intensity at the onset of dissolution. Thus, the surface dissolution of the biotite reflects the processes documented for the bulk material discussed in the previous section, i.e., the oxidation of structural Fe^{2+} to Fe^{3+} and the release of K to solution (Eqn. 46). The data also indicated that Fe^{2+} is released to solution under both oxic and anoxic solutions.

Heterogeneous redox reactions

As in the case of chain silicates, Fe^{2+} release from an oxidized surface of mica must involve a heterogeneous redox reaction. Such a process is also required based on the reaction as proposed by Scott and Amonette (1988) for biotite oxidation (Eqn. 46) in which both K and an electron are transferred from the solid to solution, maintaining electrical balance in both phases (Eqn. 46 is a specific form of Eqn. 32). A concurrent hydrolysis reaction, which removes the oxidized structural Fe and reduces it at the solution interface, can be written as

$$[Fe^{3+}] + 3H^+ + e^- \rightarrow [3H^+] + Fe^{2+} \quad . \tag{50}$$

The occurrence of this reaction is supported both by the decrease in aqueous hydrogen activity during dissolution, and the observed protonation of micas during weathering processes (Mackintosh et al., 1972). Under conditions in which no external redox couple exists, electrochemical balance requires that the oxidation and hydrolysis reaction be coupled such that

$$K[Fe^{2+}] + 3H^+ \rightarrow [3H^+] + K^+ + Fe^{2+} \quad . \tag{51}$$

In the presence of an external redox couple, half cell reactions described by Equations 46 and 50 become decoupled in a manner analogous to that previously described for amphiboles and pyroxenes. White and Yee (1985) initially tested this hypotheses by addition of aqueous Fe^{3+} to differing solution/mineral ratios of muscovite and biotite (Fig. 18). Slight increases and decreases occurred respectively for aqueous Fe^{2+} and Fe^{3+} in the presence of muscovite. In contrast, addition of Fe^{3+} to comparable masses of biotite resulted in dramatic decreases in Fe^{3+} and increases in Fe^{2+}, demonstrating the importance of solid state Fe speciation in effecting aqueous redox speciation. Eary and Rai (1989) also demonstrated the effective reduction of Fe^{3+} in the presence of biotite. The net redox processes in the case of an extraneous source of Fe^{3+} can be defined by combining the aqueous half cell reaction describing aqueous Fe^{3+} reduction with Equation 50 to produce:

$$K[Fe^{2+}] + {}^*Fe^{3+} \rightarrow [Fe^{3+}] + K^+ + {}^*Fe^{2+} \quad , \tag{52}$$

where the asterisks are used to denote Fe initially added to solution. The release of K is significantly increased by the addition of aqueous Fe^{3+} (Figs. 19a and 19b) and exhibits a 1:1 stoichiometry with the Fe^{2+} increase as predicted by Equation 51. In the absence of a reducible aqueous species, the rate of H^+ uptake was balanced by the sum

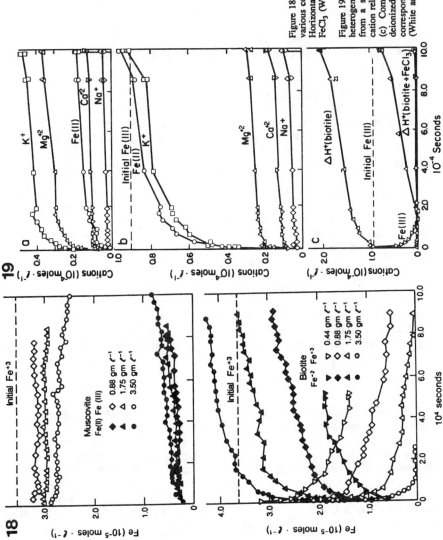

Figure 18. Reduction of Fe³⁺ with time in the presence of various concentrations of muscovite and biotite at pH 3.0. Horizontal dashed lines are the initial concentrations of FeCl₃ (White and Yee, 1985).

Figure 19. Changes in aqueous chemistry at pH 3.0 during heterogeneous oxidation of biotite. (a) Major cation release from a starting solution of deionized water. (b) Major cation release from a starting solution of 0.0001 M FeCl₃. (c) Comparisons of rates of H⁺ decrease (ΔH⁺) in deionized water and FeCl₃ solutions. Also shown is the corresponding rate of Fe³⁺ loss from the FeCl₃ solution (White and Yee, 1985).

of the release rates of K^+, Fe^{2+}, Mg^{2+}, and Ca^{2+} during the hydrolysis reaction (Fig. 19c). In contrast, no H^+ uptake occurs in the presence of Fe^{3+} until Fe^{3+} has been consumed. This indicates, as in the case for augite reaction (Fig. 13c), that the heterogeneous redox reaction effectively suppresses silicate hydrolysis.

Reduction of transition metals. Eary and Rai (1989) investigated chromate (CrO_4^{2-}) reduction by biotite as functions of pH and time in the presence of atmospheric oxygen (Fig. 20). The effects of stirring rates were found to be minimal and reaction rates were proportional to the biotite surface area. Rates of chromate reduction were found to be strongly dependent on both pH and major ion concentrations in solution. Eary and Rai (1989) proposed that the low chromate reduction rates at pH > 6.0 may be due to low aqueous Fe^{2+} concentrations caused by slower dissolution rates and accelerated Fe^{2+} oxidation by dissolved oxygen. At lower pH, the reduction rate was found to be proportional to $(H^+)^n$ were n = 0.96. Changes in anion type, in order of $SO_4^{2-} > Cl^- > ClO_4^- > PO_4^{2-}$, were shown to have considerable effect on increasing both the rates of chromate reduction and the rates of Fe^{+2} release from the biotite. Eary and Rai (1989) attributed this effect to an increasing affinity for complexation of Fe^{2+} on the biotite surface. The reduction of chromate in the presence of biotite occurred by the reaction

$$3[Fe_2^{2+} K^+] + HCrO_4^- + 6H^+ \rightarrow 3[Fe^{3+}] + CrOH^{2+} + 3H_2O + 3K^+ \quad , \quad (53)$$

where the reaction is written with Cr species that are relevant to acidic solutions. As pointed out by these workers, the above reaction is complicated by the fact that hexavalent Cr can also be reduced by aqueous Fe^{2+} which is produced by the coupled hydrolysis reaction (Eqn. 27).

Reduction of organics. Halogenated aliphatic compounds have been shown to be chemically transformed by reductive dehalogenation and elimination. Such degradation involves electron transfer reactions that remove one or more halogens, forming halides, radical intermediates, and organic products which are less oxidized. An example of reductive elimination investigated by Kriegman and Reinhard (1989) is the transformation of hexachloroethane (HCA) to tetrachloroethylene (PCE):

$$Cl_3C\text{-}CCl_3 + 2e^- \rightarrow Cl_2C=CCl_2 + 2Cl^- \quad . \quad (54)$$

These workers demonstrated that both HCA and carbon tetrachloride (CTET) could be chemically transformed in the presence of biotite and vermiculite at pH 7-8 and 50°C (Fig. 21a). They proposed the following reaction for describing the heterogeneous electron transfer:

$$2[Fe_2^{2+} 1/zM^{z+}] + Cl_3C\text{-}CCl_3 \rightarrow 2[Fe^{3+}] + Cl_2C=CCl_2 + 2Cl^- + 2/zM^{z+} \quad . \quad (55)$$

The corresponding rate of increase in the dehalogenated product, PCE, from HCA is shown in Figure 21b. For CTET, the expected product is chloroform (CF) although less than 10-15% of the CTET reacted to form CF, indicating other secondary byproducts. Although homogeneous reduction with aqueous HS^- was very slow, HS^- in the presence of biotite and vermiculite greatly accelerated the heterogeneous reduction of halogenated compounds. Although the exact mechanism is still under investigation, Kriegman and Reinhard (1989) have proposed that structural oxidation may be accompanied by a catalytic reaction in which HS^- regenerates the Fe^{2+} sites on the biotite and vermiculite surfaces by the reaction:

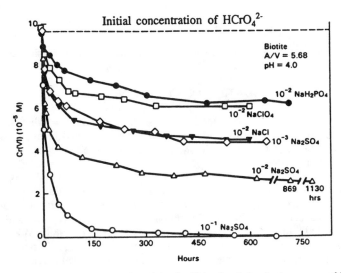

Figure 20. Chromate reduction as a function of time in pH 4 oxic solutions in the presence of biotite (Eary and Rai, 1989).

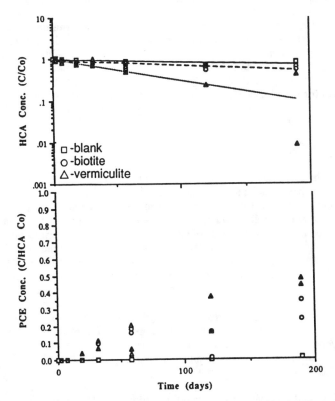

Figure 21. Heterogeneous reduction of haloaliphatic organics in the presence of biotite and vermiculite at pH 8 and 50°C. (a) Decreases in hexachloroethane with time. (b) Increases in tetrachloroethylene (Kriegman and Reinhard, 1990).

$$2[Fe^{3+}] + 2H_2S \rightarrow HSSH + 2[Fe_2^{2+} H^+] \quad , \tag{56}$$

where HSSH is a polysulfide which can become further oxidized to sulfate.

OXIDATION OF CLAY MINERALS

Clays are the only class of minerals considered in this review which commonly approach thermodynamic stability under ambient temperature and pressure conditions. The distribution of Fe in their structure should therefore more closely reflect the redox conditions of the surficial environment in which the clays formed. Clays have a phyllosilicate structure similar to that of mica consisting of Si tetrahedrons and Al octahedrons. Clays organized with one tetrahedral and one octahedral sheet are termed 1:1 clays whereas clays comprised of one tetrahedral sheet separating two octahedral sheets are termed 2:1 clays. The most common 1:1 clays, including kaolinite, dickite, and halloysite, have total Fe concentrations below 1% (Jepson, 1988). Mössbauer and electron spin resonance (ESR) techniques indicate only limited substitution of Fe^{3+} for Al in the octahedral sites which is probably associated with crystal defects (Fysh et al., 1983).

Smectites and chlorites are the most common Fe-containing 2:1 clays. They exhibit a wide range of structural Fe concentrations resulting from the substitution of Fe^{2+} for Mg and Al in the octahedral positions and Fe^{2+} and Fe^{3+} for Al in the tetrahedral sites. Dioctahedral smectites usually contain very little Fe^{2+} whereas the trioctahedral clays can be very rich in both Fe^{2+} and Fe^{3+} (Stucki, 1988). The theoretical compositions for the end member tetrahedral and octahedral layers for dioctahedral clays are, respectively, ferripyrophyllite ($Fe^{3+}Si_3$) and nontronite ($Fe^{3+}_4(Al_2Si_6)$). Corresponding end members for the trioctahedral clays are Fe^{2+}-talc and $Fe^{2+}_6Si_8$ respectively. Of these clays, nontronite is the most common, occurring as a weathering product of basaltic glass, paragonite, and augite.

Structural oxidation

Based on the preceding description of the clay structure, Fe^{2+} in both the tetrahedral and octahedral sites could undergo oxidation. The calculated redox potential required to oxidize Fe^{2+} to Fe^{3+}, when adjacent to Al in the octahedral sheet, is 4.4 eV based on the quantum mechanical model of Aronowitz et al. (1982). This result suggests that Fe is easily reduced but difficult to oxidize. Most studies of experimental and natural clay systems suggest, however, that the potential is much lower then 4.4 eV and that oxidation and reduction of structural Fe in smectites is largely reversible (Lear and Stucki, 1985). Experimental studies have concentrated on methods of reducing structural Fe^{3+} in smectite using strong aqueous reducing agents such as Na-dithionite, hydrazine, sodium hydrosufide, and benzidine which generally have limited relevance to natural redox conditions. Based on studies using $Na_2S_2O_4$ on nontronite, Roth and Tullock (1973) proposed that reduction first involved the transfer of the electron from the reducing agent to the basal surface of the clay layer. This step was then followed by structural oxidation of Fe^{3+}, accompanied or immediately followed by the loss of a structural H_2O and reprotonation by H^+ from solution.

Qualitative field data demonstrate rapid oxidation of Fe^{2+} in fresh smectites exposed to air (Kohyama et al., 1973). Structural oxidation of Fe has been shown to strongly affect layer charge and cation exchange capacities of clays. Smectite layers carry a negative charge of 0.4 to 1.2 per unit cell. Experimental studies by Stucki et al. (1984)

found that both the layer charge and cation exchange capacity decreased as octahedral Fe in dioctahedral smectite was oxidized from Fe^{2+} to Fe^{3+}. However, the relationship between charge and Fe^{3+} content did not obey a 1:1 linear relationship as expected.

Dissolution

As discussed by Stucki (1988), the dissolution rate of smectite increases with the substitution of Fe^{2+} for Al in the octahedral positions. Novák and Cícel (1978) found that $t_{1/2}$, the time to dissolve half the smectite structure in strong HCl, was related to the Fe^{2+} and Mg content by the relationship:

$$t_{1/2} = 3.95 - 1.96 \, Fe^{2+} - 2.30 \, Mg \; . \tag{57}$$

Chemical reduction of Fe^{3+} has also been shown to increase the dissolution of total Fe from the smectite structure (Rozenson and Heller-Kallai, 1976; Lear and Stucki, 1985). These results imply that the oxidization of structural Fe^{2+} generally increases the thermodynamic stability of clays in the presence of aqueous solution. This is confirmed by estimates of higher free energies and lower solubilities of Fe^{3+} phyllosilicates relative to Fe^{2+} phyllosilicates as calculated by Tardy (1990) from the thermodynamic properties of constituent elements and oxides (Table 6).

Heterogeneous reduction reactions

The potential for heterogeneous electron transfer during the oxidation of structural Fe^{2+} would appear significant for Fe-rich clays based on their reactivity, large surface areas and high exchange capacities. Surprisingly little research has been conducted on redox processes involving oxidation of Fe^{2+} in clay minerals. The reversible redox reaction scheme proposed by Stucki (1988) can be rewritten to describe electron transfer to solution during the oxidation of a Fe^{2+} smectite such that

$$[Fe^{2+}] \rightarrow [Fe^{3+}] + e^- \tag{58}$$

$$H_2O + [O^{2-}] \rightarrow [2(OH)^-] \tag{59}$$

and

$$[OH^-] \rightarrow [O^{2-}] + H^+ \; . \tag{60}$$

Thus, the oxidation of Fe in dioctahedral smectites involves at least two processes. The first is the oxidation of the octahedral Fe site (Eqns. 1 and 58) and the second involves sequential hydroxylation and deprotonation reactions. The overall electron transfer reaction can be written as the half-cell:

$$[Fe_2^{2+} OH^-] + H_2O \rightarrow [Fe_2^{3+} (OH^-)_2] + H^+ + e^- \; . \tag{61}$$

The preceding reactions assume three conditions: (1) Fe is not leached from the clay structure, (2) no change occurs in the structural charge or cation exchange capacity, and (3) the clay is hydroxylated from solution. However as previously indicated, experimental data indicate that surface charge does in fact change as a function of oxidation and reduction. Therefore, the stoichiometry between Fe oxidation and reduction and hydroxylation or dehydroxylation must be different than the simple 1:1 relationship suggested in Equation 61.

Reduction of transition metals. Several workers have investigated the effects of transition metal sorption on the oxidation state of structural Fe in smectites and

502

TABLE 6. Comparison of free energies calculated from
elements (kJ/mol) and constituent oxides (cJ/mol) and
solubility products of Fe^{2+} and Fe^{3+} phyllosilicates
(after Tardy, 1990).

	ΔG_f°	ΔG_{ox}°	$\log K_{sp}$
Talc			
$[Fe^{2+}]_3Si_4O_{10}(OH)_2$	-4470.0	-76.2	7.24
$[Fe^{3+}]_2Si_4O_{10}(OH)_2$	-4371.2	+20.0	-13.70
Celadonite			
$K[Fe^{2+}]_{2.50}Si_4O_{10}(OH)_2$	-4696.3	-258.3	11.75
$K[Fe^{3+}]_{1.67}Si_4O_{10}(OH)_2$	-4613.7	-185.4	-7.00
Biotite			
$K[Fe^{2+}]_3AlSi_3O_{10}(OH)_2$	-4799.8	-301.4	23.67
$K[Fe^{3+}]_2AlSi_3O_{10}(OH)_2$	-4699.1	-212.1	.09

TABLE 7. Fe^{2+} contents of vermiculitized biotites exposed
to various 0.1N ionic solutions (After Sayin, 1982).

Exchange ion	salt used	Fe^{2+} (wt%)
Mg^{2+}	$MgCl_2$	9.05
Ni^{2+}	$NiCl_2$	9.05
Zn^{2+}	$ZnCl_2$	9.32
Cu^{2+}	$CuCl_2$	1.76
Cu^{2+}	$CuSO_4$	2.59

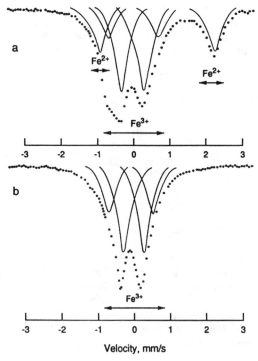

Figure 22. Computer fitted Mössbauer spectra of (a) untreated montmorillonite, and (b) Cu^{2+} exchanged
montmorillonite (Rozenson and Heller-Kallai, 1976).

vermiculites. Using Mössbauer spectroscopy, Rozenson and Heller-Kallai (1976) demonstrated that octahedrally coordinated Fe^{2+}, situated in the M(1) and M(2) sites of montmorillonite, could be oxidized by aqueous Cu^{2+}. Figure 22 compares the Mössbauer spectrum for untreated Wyoming montmorillonite and montmorillonite treated with 0.1 N $CuCl_2$ for 24 hours. The untreated clay exhibits two doublets corresponding to octahedrally coordinated Fe^{3+} in the M(1) and M(2) sites and one doublet corresponding to octahedrally coordinated Fe^{2+} in M(1) or M(2) sites. After treatment, the Fe^{2+} doublet is clearly missing indicating quantitative conversion to Fe^{3+}. Treatment with comparable concentrations of other cations including Li, Na, K, Mg, and Ca in oxic solutions did not produce structural Fe reduction leading the workers to propose that Cu is acting as a catalyst in the oxidation of structural Fe.

The electrochemical interaction between sorbed Cu^{2+} and structural Fe was also investigated in vermiculite by Sayin (1982). After exchanging interlayer K by $BaCl_2$ and $MgCl_2$, the mineral vermiculite was exposed to several divalent cation solutions listed in Table 7 for 8 hours at 70°C. As indicated in Table 7, both cupric chloride and cupric sulfate solutions significantly decreased the Fe^{2+} component. The authors demonstrated that the oxidation of structural Fe^{2+} by Cu^{2+} was electrochemically possible based on the standard aqueous potentials for Cu and Fe but concluded that Cu^+ is not stable as an interlayer species and would be reoxidized by O_2. Hence Cu^+ ions acted as transitory electron carriers, the ultimate electron acceptors being dissolved O_2 in solution. As Fe^{2+} is oxidized, electroneutrality is maintained either by ejection of interlayer cations, deprotonation of the octahedral OH groups or by loss of octahedral Fe.

The reaction scheme described by Sayin (1982) was based on the sequential reactions:

$$[Fe_s^{2+} OH] + Cu^{2+} \rightarrow [Fe_s^{3+}\ O^{2-}] + Cu^+ + H^+ , \qquad (62)$$

$$Cu^+ + H^+ + 1/4O_2 \rightarrow Cu^{2+} + 1/2H_2O \quad . \qquad (63)$$

The overall reaction for vermiculite oxidation would therefore be

$$[Fe_s^{2+} OH^-] + 1/4O_2 \rightarrow [Fe_s^{3+}\ O^{2-}] + 1/2H_2O , \qquad (64)$$

which is essentially identical to the reaction proposed for high temperature dehydroxylation of amphiboles (Eqn. 28).

CONCLUSIONS

The preceding review presents an attempt at synthesizing available information concerning heterogeneous electron transfer processes controlled by the in situ oxidation of structural Fe^{2+} to Fe^{3+} in oxide and silicate minerals. While such mechanisms may potentially be very important in controlling and buffering redox conditions in water-rock systems at ambient temperatures and pressures, this subject has only recently been investigated by the geochemical community.

The presence of Fe^{3+} in the near surface and bulk phases of both naturally and experimentally weathered oxides and silicates has been demonstrated by analytical methods including XPS and Mössbauer spectroscopy. Direct structural and thermodynamic information concerning how the higher valence charge and smaller diameter of the ferric ion is accommodated in the mineral is minimal. However for nearly all the ferrous oxides and silicates considered, ferric polymorphs do exist which represent reasonable analogues to structural oxidation. Examples include magnetite-maghemite, fayalite-ferrifayalite, hornblende-oxyhornblende, and biotite-oxybiotite. These oxidized phases most often accommodate Fe^{3+} by creating structural vacancies in the octahedral positions. In the case of amphiboles and phyllosilicates, charge balances can also be partly compensated for either by dehydroxylation of the OH sites or by addition of hydroxyl ions to the structure. Except for clays, the formation of these oxidation products is not reversible, reflecting the fundamental thermodynamic instability of most primary Fe^{2+} oxide and silicate minerals at ambient conditions.

Electron transport from oxidation sites to the solid/solution interface is dependent on the band gap energies and the concentration of conduction band electrons per unit cell of an oxide or silicate. The potential in the space charge region in the solid is induced by the electrolyte potential which is the driving force causing oxidation. For Fe^{3+} and Fe^{2+} atoms occupying adjacent sites in oxides and silicates, Mössbauer spectroscopy and other techniques indicate that the electron transfer is very rapid. However, as the distance between electrochemically active species increases, such as by separation in aluminosilicate structures, the electrical conductivity dramatically decreases which explains large variations in conductivity between minerals. The rates of structural oxidation exhibit smaller variations than does conductivity because the electron transfer must be coupled with the migration of positive charge to preserve electrical neutrality in the mineral structure.

The mechanism of the co-migrating positive charge is specific to the mineral phase. In the case of ferrous oxides, Fe is the only mobile cation. For biotite, Fe oxidation and electron transfer are coupled to K depletion from the interlayer site. For augite, electron transfer is coupled principally with Ca mobility from octahedral sites. An alternate mechanism of positive charge transport for clays and amphiboles is the deprotonation of the interstitial hydroxyl sites. Such a process would lead to coupled transport of an electron or hydrogen ion from the structure. Transport of cations through the crystal structure must be viewed as the probable rate-limiting step in solid state oxidation. In fact, coupled cation-electron transfer is difficult to justify at all based on extremely low solid state diffusion rates for cations at ambient temperatures. However, the formation of ferric oxide and silicate phases, containing abundant cation vacancy sites, could substantially alter pathways by which positive charge diffuses to the surface. Also, in the case of phyllosilicates, interlayer cation diffusion would be substantially faster than solid state diffusion.

Structural oxidation and surface dissolution are parallel reaction pathways which minimize ferrous oxide and silicate surface free energies. In the case of ferrous oxides, structural oxidation and reductive dissolution both release Fe^{2+} to solution and are distinguished only by the fact that dissolution reactions involve the structural loss of oxygen whereas structural oxidation does not. However, the response of ferrous oxides to electrochemically active aqueous species, which decouple the two reactions, is significantly different. The rate of reductive dissolution is accelerated by the presence of electron donors capable of reducing Fe^{3+} to more mobile Fe^{2+} on the surface. Structural oxidation is accelerated by the presence of aqueous electron acceptors which permit in situ oxidation of Fe^{2+} to Fe^{3+}. For ferrous silicates, structural oxidation and hydrolysis reactions compete for the removal of cations from the structure. In the case of oxidation, cation release is coupled to electron transport whereas in dissolution, cations are exchanged for H^+ at the silicate surface. As documented for biotite and augite, the redox process dominates over the hydrolysis process until electron acceptors, such as Fe^{3+}, are depleted in solution.

Structural oxidation and coupled heterogeneous electron transfer can control redox-sensitive speciation of transition metals such as Fe, Cr, V, and Cu, actinides such as Np, in addition to promoting the reduction of nitrate and halogenated aliphatic compounds including hexachloroethane and carbon tetrachloride. An important distinction is whether or not reduction of the electron acceptor is promoted by aqueous Fe^{2+}. If such a situation exists, the relative importance of the heterogeneous and homogeneous reduction is simply directly proportional to the respective reaction rates. For the examples of dissolved oxygen and chromate reduction, electron transfer from oxidation of magnetite and ilmenite is faster than electron transfer resulting from reductive dissolution. If the electrochemical potential of the electron acceptor is larger than for the Fe^{2+}/Fe^{3+} couple, as in the case of vanadate, the reaction can only proceed as a result of solid state electron transfer.

Heterogeneous electron transfer from the in situ oxidation of ferrous oxides and silicates has important ramifications for the electrochemistry of geologic and hydrologic systems. In terms of availability of electron donors, the concentrations of electrochemically active Fe in the solid state are generally much greater than in the coexisting aqueous phase allowing the creation of dynamic redox buffers in poorly poised aqueous systems. In addition, the electron transfer is driven by essentially irreversible oxidation kinetics, contributing to commonly observed redox disequilibrium in the aqueous phase. Finally, the oxidation potential in the solid state is greater than for the corresponding aqueous Fe couple and is therefore capable of reducing a wider range of redox sensitive species. While the preceding review documents the importance of these reactions in controlling the speciation of a number of environmentally sensitive compounds under experimental conditions, coupled structural oxidation and electron transfer processes in natural and perturbed geochemical systems remains to be to be to be characterized.

ACKNOWLEDGEMENTS

The author expresses thanks to Maria Peterson, Alex Blum, and Kirk Nordstrom of the U.S. Geological Survey and to Michael Hochella of Stanford University for providing suggestions and comments which significantly enhanced the paper. Appreciation is also expressed to Andy Yee of Lawrence Berkeley Laboratory who performed portions of unpublished work presented in the paper.

REFERENCES

Addison, C. C. , W. W. Addison, G. H. Neal, and J. H. Sharp (1962) Amphiboles. Part 1, The oxidation of crocidolite. J. Chem Soc. 1468-1471.

Akimoto, T., H. Kinoshita, and T. Furuta (1984) Electron probe microanalysis study on processes of low-temperature oxidation of titanomagnetite. Earth Plant. Sci. Letters 71, 263-278.

Allen, P. D., N. A. Hampton, and G. J. Bignold (1979a) The electrochemistry of magnetite. Part 1. The electrochemistry of Fe_3O_4/C discs-potentiodynamic experiments. J. Electroanal. Chem. 99, 299-309.

506

Allen, P. D., N. A. Hampton, and J. F. Tyson (1979b) The differential capacitance of magnetite. Surface Tech. 9, 395-400.

Allen, P. D., N. A. Hampson, and G. Bignold (1980) The electrochemical dissolution of magnetite Part II, The oxidation of bulk magnetite. J. Electroanal. Chem. 223-233.

Aronowitz, S., L. Coyne, J. Lawless, and J. Rishpon (1982) Quantum-chemical modeling of smectite clays. Inorg. Chem. 21, 2589-2593.

Bancroft, G. M. and M. M. Hyland (1990) Spectroscopic methods for the study of adsorption/reduction reaction of metal complexes on sulphide minerals. In: Mineral-Water Interface Geochemistry, M. F. Hochella Jr. and A. F. White eds., Reviews in Mineral. 23 (this volume).

Bard, A. J. and L. R. Faulkner (1980) Electrochemical Methods, Fundamentals and Applications. John Wiley and Sons, New York, 718p.

Barton, A. F. M. and S. R. McConnel (1978) Rotating disc dissolution rates of ionic solids, Part 3-natural and synthetic ilmenite. J. Chem. Soc. Faraday Trans. 89, 234-245.

Baumgartner, E., M. A. Bleasa, H. A. Marinovich, and A. J. G. Marto, (1983) Heterogeneous electron transfer as a pathway in the dissolution of magnetite in oxalic acid solutions. Inorg. Chem. 22, 2224-2226.

Berner, B. and J. Schott (1982) Mechanism of pyroxene and amphibole weathering II: Observations of soil grains. Amer. J. Sci. 282, 1214-1225.

Blesa, M. A., E. B. Borghi, A. J. Marto, and A. E. Regazzoni (1984) Adsorption of EDTA and iron-EDTA complexes on magnetite and the mechanism of dissolution of magnetite by EDTA. J. Colloid Interface Sci. 98, 295-300.

Borghi, E. B., A. E. Regazzoni, A. J. G. Maroto, and M. A. Blesa (1989) Reductive dissolution of magnetite by solutions containing EDTA and Fe^{II}. J. Colloid Interface Sci. 130, 301-310.

Brandley, R. S., A. K. Jamil, and D. C. Munro (1964) The electrical conductivity of olivine at high temperatures and pressures. Geochim. Cosmochim. Acta 28, 1669-1678.

Bruyere, V. I. and M. A. Blesa (1985) Acidic and reductive dissolution of magnetite in aqueous sulfuric acid. J. Electroanal. Chem. 182, 141-156.

Burns, R.G., K. M. Parkin, B. M. Loeffler, I. S. Leung, and R. M. Abu-Eid (1976) Further characterization of spectral features attributable to titanium on the moon. Proc. 7th Lunar Conf., 2571-2578.

Casey, W. H., and B. Bunker (1990) The leaching of mineral and glass surfaces during dissolution. In: Mineral-Water Interface Geochemistry, M. F. Hochella Jr. and A. F. White ed., Reviews in Mineral. 23 (this volume).

Coleman, N. T., F. H. le Roux, and J. G. Cady (1963) Biotite-hydrobiotite-vermiculite in soils. Nature 198, 409-410.

Deer, W. A., R. A. Howie, and J. Zussman, (1982) Rock-Forming Minerals, Vol. 1A Orthosilicates. Longman Press, London, 919 p.

Diggle, J. W. (1973) Oxide and Oxide Surfaces Vol. 2 The Anodic Behavior of Metals and Semiconductors. J. W. Diggle, ed., N. Marcel Dekker, New York 385p.

Duba, A. and I. A. Nicholls (1973) The influence of oxidation state on the electrical conductivity of olivine. Earth Planet. Sci. Letters 18, 59-64.

Eary, L. E. and D. Rai (1989) Kinetics of chromate reduction by ferrous ions derived from hematite and biotite at 25°. Amer. J. Sci. 289, 180-213.

Engell, H. J. (1956) Über die Auflösung von Oxyden in verdünnten Säauren. Zeit. physik. Chemie 7, 158-181.

Ferrow, E. (1987) Mössbauer and X-ray studies on the oxidation of annite and ferriannite. Phys. Chem. Minerals 14, 270-275.

Furuta, T. , M. Otsuki, and T. Akimoto (1985) Quantitative electron probe microanalysis of oxygen in titanomagnetites with implications for oxidation processes. J. Geophys. Res. 90, 3145-3150.

Fysh, S. A., J. D. Cashion, and P. E. Clark (1983) Mössbauer effect studies of Fe in kaolinite. Clays Clay Minerals 31, 285-292.

Gallagher, K. J., W. Feitnechtm and U. Mannweiler (1968) Mechanism of oxidation of magnetite to γ-Fe_2O_3. Nature 217, 1118-1121.

507

Gastuche, M. C. (1963) Kinetics of acid dissolution of biotite I. Interfacial rate process followed by optical measurement of the white silica rim. Proc. Int'l. Clay Conf., Stockholm, 67-76.

Gedikogla, A. (1983) Mössbauer study of low temperature oxidation of natural magnetite. Script. Metal. 17, 45-48.

Gilkes, R. J., R. C. Young, J. P. Quirk (1973) The oxidation of octahedral iron in biotite. Clays Clay Minerals, 20, 303-315.

Gilkes, R. J. and R. C. Young (1982) Artificial weathering of oxidized biotite; IV, The inhibitory effect of potassium on dissolution rates. Proc. Soil Science Soc. Amer. 38, 529-532.

Goodman, B. A. and Wilson, M. J. (1976) A Mössbauer study of the weathering of hornblende. Clay Minerals 11, 153-163.

Grandstaff, D. E. (1986) The dissolution rate of forsteritic olivine from Hawaiian beach sand. In: Rates of Chemical Weathering of Rocks and Minerals, S. M. Colman and D. P. Dethier, eds. Academic Press, Orlando Florida 41-59.

Haruyama, S. and K. Masamura (1978) The dissolution of magnetite in acidic perchlorate solutions. Corrosion Sci. 18, 263-274.

Hawthorne, F. C. (1981a) Amphibole spectroscopy. In: Amphiboles and Other Hydrous Pyriboles - Mineralogy. D.R. Veblen, ed., Reviews in Mineral. 9a, 103-139.

Hawthorne, F. C. (1981b) Crystal chemistry of the amphiboles In: Amphiboles and Other Hydrous Pyriboles- Mineralogy. D.R. Veblen ed. Reviews in Mineral. vol. 9a, 1-95.

Hering, J. G., and W. Stumm (1990) Oxidative and reductive dissolution of minerals. In: Mineral Water Interface Geochemistry, M. F. Hochella Jr. and A. F. White eds., Reviews in Mineral. 23 (this volume).

Ishikawa, Y. (1958) Electrical properties of $FeTiO_3$-Fe_2O_3. J. Phys. Soc. Japan 13, 37-42.

Jepson, W. B. (1988) Structural iron in kaolinites and in associated ancillary minerals. In: Iron in Soils and Clay Minerals. J. W. Stucki, B.A. Goodman, and U. Schertmann, eds., D. Reidel Publishing Co., Dordrecht, Netherlands, 467-536.

Jolivet, J. and E. Tronc (1988) Interface electron transfer in colloidal spinel iron oxide. Conversion of Fe_3O_4 to Fe_2O_3 in aqueous medium. J. Coll. Interface. Sci. 125, 688-701.

Kohyama, N., S. Shimoda, and T. Sudo (1973) Iron-rich saponite (ferrous and ferric forms). Clays Clay Minerals 21, 229-237.

Kriegman, M. R., and Reinhard, M. (1989) Electron transfer reactions of haloaliphatics and ferrous iron bearing minerals. Abst. Amer. Chem. Soc. Annual Mtg. Miami, Florida.

Lastovickova, M and V. Kropacek (1983) Electrical conductivity of Fe-Ti-O minerals in connection with oxidation processes. J. Geomag. Geoelectr. 35, 777-786.

Lear, P. R. and J. W. Stucki (1985) The role of structural hydrogen in the reduction and reoxidation of iron in nontronite. Clays Clay Minerals. 33, 539-545.

Lindsley, D. H. (1976) The crystal chemistry and structure of oxide minerals as exemplified by Fe-Ti oxides. In: Oxide Minerals D. Rumble, ed., Reviews in Mineral. 3 1-L52.

Littler, J. G. F. and J. M. Williams (1969) Electrical conductivity measurements in amphibole minerals. J. Chem. Soc. 6368.

Mackintosh, E. E., D. G. Lewis, and D. J. Greenland (1972) Dodecylammonium-mica complexes-- II. Characterization of the reaction products. Clays Clay Minerals 20, 125-134.

Mansfield, F. (1976) The polarization resistance technique for measurement of corrosion currents. In: Advances in Corrosion Science. M. G. Fontana and R. W. Staehle, eds., Plenum Press, New York.

Marcus, R. A. (1965) The theory of electron-transfer reactions. VI. Unified treatment for homogeneous and electrode reactions. J. Chem. Phys. 43, 679-691.

Mao, H. K. (1973) Electrical and optical properties of the olivine series at high pressure. Carnegie Inst. Washington Ann. Rept. Dir. Geophys. Lab. 1972-1973, 552-554.

Meyer, R. E., W. D. Arnold, and F. I. Case (1984) Valence effects on adsorption: Laboratory controls of valence state. NRC Nuclear Waste Geochemistry '83. Proc. of the Nuclear Regulatory Commission NUREG/CP-0052, 535.

Morrison, S. R. (1980) Electrochemistry at Semiconductor and Oxidized Metal Electrodes, Plenum Press, New York, 526 p.

508

Murphy, W. M., and H. C. Helgeson (1987) Thermodynamic and kinetic constraints on reaction rates among minerals and aqueous solutions. III. Activated complexes and the pH-dependence of the rates of feldspar, pyroxene, wollastonite, and olivine hydrolysis. Geochim. Cosmochim. Acta 51, 3137-3153.

Novák, I. and B. Cícel (1978) Dissolution of smectites in hydrochloric acid. II. Dissolution rate as a function of crystallochemical composition. Clay Minerals 26, 341-344.

Pakharykov, N. M., M.M. Kazvchenko, V. V. Solov'ev, V.K. Shakhanov, and V. D. Tyan (1976) Supergene alteration of ilmenite in the Karaotekel deposits, Kazakhstan. Lithol. Mineral. Res. 11, 95-98.

Phillips, M. W., R. K. Popp, and A.A. Pinkerton (1986) Structural investigation of oxidation -dehydroxylation in hornblende. Trans. Amer. Geophys. Union EOS 67, 1270.

Postma, D. (1990) Kinetics of nitrate reduction by detrital Fe(II)-silicates. Geochim. Cosmochim. Acta 54, 903-908.

Ronov, A.B., and A.A. Yaroshevsky (1971) Chemical composition of the earth's crust In: The Earth's Crust and Upper Mantle. P.J. Hart ed. Amer. Geophys. Union, Washington, D.C.37-57.

Rossman, G. R. (1980) Pyroxene spectroscopy. In: Pyroxenes. C.T. Prewitt, ed., Reviews in Mineral. 7, 93-116.

Rozenson, I. and L. Heller-Kallai (1976) Reduction and oxidation of Fe^{+3} in dioctahedral smectites- III. Oxidation of octahedral iron in montmorillionite. Clay Minerals 26, 88-92.

Roth, C. B., and R. J. Tullock (1973) Deprotonation of nontronite resulting from chemical reduction of structural ferric iron. Proc. Int'l. Clay Conf., Madrid. J. M. Serratosa ed. Div. Ciencias C. S. I. C., Madrid.

Sanemasa, I., M. Yoshida, and T. Ozawa (1972) The dissolution of olivine in aqueous solutions of inorganic acids. Bull. Chem. Soc. Japan 45, 1741-1746.

Sanz, J. J and W. E. Stone (1983) NMR applied to minerals -- IV. Local order in the octahedral sheet of micas: Fe-F avoidance. Clay Minerals 18, 187-192.

Sayin, M. (1982) Catalytic action of copper on the oxidation of structural iron in vermiculated biotite. Clays Clay Minerals 30, 287-290.

Schnitzer, M. and H. Kodama (1976) The dissolution of micas by fulvic acid. Geoderma 15, 381-391.

Schott, J. and R. A. Berner (1983) X-ray photoelectron studies of the mechanism of iron silicate dissolution during weathering. Geochim. Cosmochim. Acta 47, 2233-2240.

Scott, A. D. and J. Amonette (1988) Role of iron in mica weathering. In:Iron in Soils and Clay Minerals, J. W. Stucki, B. A. Goodman, and U. Schwertmann eds. D. Reidel Publishing Co., Dordrecht, The Netherlands.

Shirane, G, D. E. Cox, and S. L. Ruby (1962) Mössbauer study of isomer shift, quadrapole interaction, and hyperfine structure in several oxides containing ^{57}Fe. Phys. Rev. 125, 1158-1165.

Shuey, R. T. (1975) Semiconducting ore minerals. Developments in Economic Geology 4, Elsevier, New York. 415 p.

Sidhu, P. S., R. J. Gilkes, R. M. Cornell, A.M. Posner, and J. P. Quirk (1981) Dissolution of iron oxides and oxyhydroxides in hydrochloric and perchloric acids. Clays Clay Minerals 29, 269-276.

Siegel, D. I. and H. O. Pfannkuch (1984) Silicate mineral dissolution at pH 4 and near standard temperature and pressure. Geochim. Cosmochim. Acta 48, 197-201.

Siever, R. and N. Woodford (1979) Dissolution kinetics and weathering of mafic minerals. Geochim. Cosmochim. Acta 43, 717-724.

Stucki, J.W., D. C. Golden, and C. B. Roth (1984) Effects of reduction and reoxidation of structural iron on the surface charge and dissolution of dioctahedral smectites. Clay Minerals 32, 350-356.

Stucki, J. W. (1988) Structural iron in smectites. In:Iron in Soils and Clay Minerals, J. W. Sucki, B. A. Goodman, and U. Schwertmann eds., D. Reidel Publishing Company, Dordrecht, The Netherlands.

Sukhotin, A. M., E. A. Gankin, and A. I. Khentov (1976) Electrochemical properties of ferrous oxides. Protect. Metals 12, 38.

Susak, N. J., A. Friedman, S. Fried, and J. C. Sullivan (1983) The reduction of neptunium (VI) by basalt and olivine. Nuclear Technology 63, 266-270.

Swaddle, T. W., and P. Oltmann (1980) Kinetics of magnetite-maghemite-hematite transformation, with special reference to hydrothermal systems. Canadian. J. Chem. 58, 1763-1772.

Tardy, Y. (1990) Gibbs free energy of formation of hydrated and dehydrated clay minerals. Chem. Geology 84, 255-258.

Temple, A. K. (1966) Alteration of ilmenite. Econ. Geol. 81, 695-714. Torstenfelt, B., B. Allard, B. Johnson, and T. Iltner (1983) Iron content and reducing capacity of granites and bentonites. Sversk Kamslanlefersoyning AB Avdelning KBS Tech Rept. 83-36, 9 p.

Valverde, N. (1976) Investigations on the rate of dissolution of metal oxides in acidic solutions with additions of redox couples and complexing agents. Ber. Bunsenges. Phys. Chem. 80, 333-340.

Voskresenskaya, N. T. and H. Coliazo (1983) Kinetics and mechanisms of dissolution of some layer silicates in a weathering crust. Geochem. Int'l. 20, 40-50.

Waite, T. D. (1990) Photo-redox processes at the mineral-water interface. In: Mineral-Water Interface Geochemistry, M. F. Hochella Jr. and A. F. White eds., Reviews in Mineral. 23 (this volume).

White, A. F. and A. Yee (1985) Aqueous oxidation-reduction kinetics associated with coupled electron-cation transfer from iron-containing silicates. Geochim. Cosmochim. Acta 49, 1263-1275.

White, A.F., A. Yee, and S. Flexser (1985) Surface oxidation-reduction kinetics associated with experimental basalt/water reaction at 25°C. Chem. Geol. 49, 73-86.

White, A. F., and M. F. Hochella, Jr. (1989) Electron transfer mechanisms associated with the surface dissolution and oxidation of magnetite and ilmenite, Proc. 6th Int'l. Symp. Water-Rock Interaction 765-768.

CHAPTER 13   G. M. BANCROFT & M. M. HYLAND

SPECTROSCOPIC STUDIES OF ADSORPTION/REDUCTION REACTIONS OF AQUEOUS METAL COMPLEXES ON SULPHIDE SURFACES

The class of minerals known as the sulphides are found in virtually all rock types, of all ages throughout the world. This is a result of their formation from a wide range of temperature and pressure regimes, in magmatic, metamorphic and sedimentary processes (Park and McDiarmid, 1975). They are the primary ore source of most of the base metals. In view of their ubiquity, it is not surprising that reactions occurring between sulphide minerals and aqueous solutions play an important role in many technological and natural geochemical processes.

Three broad classes of sulphide/aqueous interfacial reactions of geochemical interest are oxidation, adsorption and reduction. Sulphide oxidation by aqueous agents has received considerable attention from researchers interested in the problem of acid mine drainage, the processing of sulphide ores, and the biogeochemical cycling of sulphur (McKibben and Barnes, 1986).

Adsorption reactions at sulphide surfaces have been recognized as a mechanism for the control of metal concentrations in reducing aqueous environments (James and Parks, 1975; Morse et al., 1987). The low levels of heavy metals, such as mercury and lead, in anoxic fluids are believed to be a result of their scavenging by sulphide particulates (Dyrssen et al. 1984; Bacon et al., 1980). On the technological side, the adsorption of mercury on sulphides has been studied as a method of reducing levels of that element in industrial wastewater (Jean and Bancroft, 1986; Brown et al., 1979).
Adsorption reactions also play an extremely important role in the mineral processing of sulphides. The reaction of sulphides with organic collectors, such as xanthate, and activators, such as copper ions, is used to control the hydrophobicity/hydrophilicity of the surface and thus effect their separation from the other constituents of the ore (Fuerstenau et al., 1985).

It has recently been emphasized, (initially by Sakharova and coworkers, 1967, 1976, 1978 and Bancroft and Jean, 1982) that adsorption/reduction reactions at sulphide surfaces may be very important for the concentration of noble metals such as Au and Ag in hydrothermal regimes (see Brief Review below). Traditional wisdom has always emphasized that ore deposition results from a large reduction in solubility of aqueous gold or noble metal complexes due to a change in temperature, pressure or solution composition (Seward, 1973, 1984; Romberger, 1986, 1988) and until recently there has been little mention of the importance of existing or growing sulphide surfaces for concentrating metals. Seward (1973, 1984) for example, suggested two possibilities (which could be generalized) for the deposition of gold transported as bisulphide complexes by alkaline, sulphur-rich reducing fluids. First, gold precipitation can be induced by a drop in pH or a decrease in sulphur fugacity. Secondly, in solutions undersaturated in $Au(HS)_2^-$, gold may coprecipitate with other metal sulphides (Seward, 1984). Where chloride complexes are responsible for gold transport, a change in temperature, pH or reduction may be responsible for gold deposition (Henley, 1973; Romberger, 1986, 1988). Studies of hydrothermal metal deposition require knowledge of all the complex processes involved; how the fluids are formed, what gold complexes are responsible for mobilization, and how the solubility may change. Recent thermodynamic modelling studies (Romberger, 1986, 1988) are very useful in providing a basis for discussion, but there still are far too few laboratory studies (which are exceedingly difficult to perform at high temperature and pressure) to test the thermodynamic models.

Experimentally, it has become far easier in the past decade to examine solution/surface reactions to obtain information on metal deposition. In addition, there is now evidence from natural sulphides that such surface reactions could be generally important for noble metal ore formations (Bakken et al., 1989; Starling et al., 1989). To understand the solution/surface reaction, it is crucial to be able to probe specifically the outermost atomic layers of the solid. Surface sensitive techniques such as X-ray photoelectron spectroscopy (XPS) [initially called ESCA (Electron Spectroscopy for Chemical Analysis) by Siegbahn et al., 1967] and Auger electron spectroscopy (AES) have been extremely useful in many scientific areas, including geochemistry. For an excellent review, see Hochella (1988).

When studying the reaction of one metal complex with one sulphide mineral, the researcher must try to answer some of the following fundamental questions:

1) What is the oxidation state and chemical environment of the metal on the sulphide surface during the reaction? What metal species is adsorbed and what is the reducing agent? What happens to the ligands bound to the aqueous metal complex during adsorption/reduction?

2) How much of the metal species is adsorbed/reduced? What is the rate of adsorption/reduction? If reduction of the metal does not occur, can this be explained?

3) If the metal species is reduced on the surface, what is the reducing agent, and what is it oxidized to? What is the rate of oxidation?

4) Where does nucleation and growth of the metal species begin? Where is the metal distributed on the surface, and what is the nature (e.g. amorphous or crystalline) of the metal species?

Of course, it is also important to characterize the solution, i.e., to determine the metal complexes present before and after reaction, in order to decide which complexes are adsorbed and/or reduced. It is equally important to compare the reactions of a variety of analogous aqueous metal complexes with a variety of chemically, and even physically different sulphide minerals.

Modern surface analytical techniques are capable of providing answers to many of the above questions. In this chapter we examine in great detail the reactions of Au, Pd, Ag and Hg complexes (in particular $AuCl_4^-$ and $PdCl_4^=$) with sulphides such as pyrite, galena and sphalerite, to demonstrate how XPS and other surface probes can be used to characterize adsorption/reduction reactions at mineral surfaces. One cannot expect a single technique to provide all the answers, or to be applicable to all systems, and therefore a multi-disciplinary approach is absolutely necessary. This chapter will also deal with AES, Raman Spectroscopy, Scanning Tunneling Microscopy (STM), Secondary Ion Mass Spectroscopy (SIMS) and Mössbauer Spectroscopy, which have been used to provide complementary information and to confirm observations from XPS.

We shall begin with a review of laboratory studies of reactions of metal complexes with sulphide minerals, and then move on to a discussion of XPS, and briefly, AES, as a background for the discussion of the application of these techniques to selected metal complex/sulphide systems.

BRIEF REVIEW OF LABORATORY STUDIES

It has been known for more than a century that metallic sulphides are capable of precipitating precious metals from solution. Skey (1871) and Palmer and Bastin (1913) qualitatively assessed the efficiency of a variety of sulphides in removing gold and silver from solutions of their chloride complexes. Palmer and Bastin (1913) noted that the nature of the metallic deposit varied depending on the precipitating mineral. Both groups recognized the application of their findings to the formation of ore deposits.

Despite this early start, there have been relatively few studies in the intervening years. In the late 1960s Sakharova and co-workers initiated a series of electrochemical experiments on gold, silver and gold/silver codeposition on pyrite, pyrrhotite and galena (Sakharova and Lobacheva, 1967, 1978; Sakharova et al., 1975, 1976, 1980, 1981). The reduction process was studied using the sulphide mineral as the working electrode and measuring the current as a function of voltage when the electrodes were immersed in solutions of $AgNO_3$ or $HAuCl_4$ or a mixture of the two. Voltage vs. log current plots indicate at what potentials reduction occurs. The influence of pH, anion composition and the efficiency of the sulphide were studied. Unfortunately, the systems are very complex, with more than one reaction contributing to the measured current, and therefore, these studies were more descriptive than interpretive. Some of their results were quite interesting. They found, for example, that pyrite has a greater capacity to reduce and precipitate gold than either galena or pyrrhotite. Also, when pyrite and galena or pyrite and pyrrhotite were coupled to form a microgalvanic system, at low concentrations of $HAuCl_4$, metallic gold was deposited on the less readily oxidized mineral (pyrite, in both cases), and visible oxidation of the other electrode was noted.

In the 1980s, research from several groups showed that adsorption mechanisms can be an effective means of removing gold from solution, but no surface work was performed in most of these studies. Gold adsorption onto various oxide minerals has been studied by Nechayev (1984) and Nechayev and Nikolenko (1986a,b) in order to understand the immobilization of secondary gold, solubilized from primary deposits under oxidizing conditions. Seward (1984) has proposed that sulphides will play an important role in the adsorption of gold under conditions encountered in hydrothermal systems. Renders and Seward (1989) have studied the transport of gold as Au(I) thio complexes (probably the most important Au transporting complex) and the adsorption of these complexes on As and Sb sulphides. This system closely mimics those found in active geothermal regions. They determined that Au(HS) was the adsorbing species and [197]Au

Mössbauer indicated that gold was present on the sulphide in the Au(I) oxidation state. In their estimation, As_2S_3 and Sb_2S_3 were capable of concentrating gold from solution by four or five orders of magnitude, sufficient to account for the levels of gold found in active geothermal systems.

The first detailed surface studies of metal reduction by sulphides were those of Jean and Bancroft (Bancroft and Jean, 1982; Jean and Bancroft, 1985) who studied the reaction of Au complexes with a number of sulphides. Hiskey and co-workers and Buckley et al. have conducted detailed electrochemical studies of the reduction of Ag^+ on pyrite (Hiskey et al., 1987; Hiskey and Pritzker, 1988; Buckley et al., 1989). Their interest in this system arises from the need to find an agent to enhance oxidation of pyrite for mineral processing applications. There is disagreement in these studies about the nature of Ag (either Ag, Ag_2S or Ag_2O) deposited on the pyrite. Reduction is inhibited by the presence of a passivating layer if the FeS_2 surface is allowed to oxidize before contacting the Ag^+ solution. There was also an indication that the reduction was kinetically controlled by an electrochemical surface reaction. Hyland then studied the reaction of Au complexes in more detail (Hyland, 1989; Hyland and Bancroft, 1989) and studied the reaction of Pd and Hg complexes with sulphides (Hyland and Bancroft, 1990a,b). The Pd studies were undertaken because geochemists have recognized recently that palladium is not as inert as assumed, and that aqueous processes will contribute to the geochemical cycling of palladium (Cousins, 1973; Fuchs and Rose, 1974; McCallum, 1976; Mountain and Wood, 1988). In the following discussion, we will concentrate on describing the recent surface results from our laboratory on the reaction of Au and Pd complexes with sulphides.

SURFACE STUDIES OF METAL COMPLEX/SULPHIDE SYSTEMS

The aqueous metal complex/sulphide systems discussed below have been studied using XPS as the primary technique. In a pedogogical chapter such as this, it seems worthwhile to spend some time introducing XPS, and in less detail AES, as a background to the following discussion. We will then show how XPS, with support from Raman Spectroscopy can be used to provide chemical information about the reacted sulphide surface, and answer many aspects of the first three questions in the Introduction. Then, in the fourth section, we will turn to SEM and AES results to obtain spatial information on the adsorbed/reduced metal species. Then, in the fifth section, we will briefly comment on desorption and electrochemistry experiments to delineate the nature of surface species. We also comment on two very recent techniques, EXAFS and STM, which should be extremely useful for obtaining chemical and spatial information on sulphide surfaces.

An introduction to X-ray photoelectron spectroscopy (XPS)

The most useful general surface technique for delineating the chemistry of the surface is XPS or ESCA (Siegbahn et al., 1967; Carlson, 1975). In the XPS technique, a source of X-rays (usually Al Kα at 1486 eV, but now often monochromatized synchrotron radiation) is directed at a sample, and the X-rays ionize all the atoms on the surface:

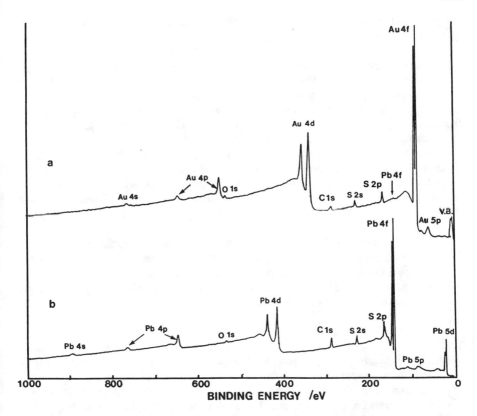

Figure 1. Broadscan spectra of cleaved PbS plates (a) after a 10 h reaction with a 20 ppm Au solution of KAuCl$_4$, and (b) an unreacted clean PbS surface (Hyland and Bancroft, 1989).

$$A_{surface} + h\upsilon \longrightarrow A^+_{surface} + e \ , \tag{1}$$

where e is the ejected photoelectron.

With the intensity of the normal sources, multiply charged atoms are not formed (i.e., $A^+_{surface}$ relaxes to $A_{surface}$ before the next photon hits A). The kinetic energy of the ejected electron(s) in Equation 1 is given approximately by the Einstein photoelectric equation

$$h\upsilon = E_B + E_k \ , \tag{2}$$

where $h\upsilon$ is the known X-ray energy, E_B is the ionization or binding energy of the electron(s), and E_k is the kinetic energy of the photoelectron(s). Using an electrostatic or magnetic analyzer to measure E_k, we can, in principle, obtain the binding energy for all electrons in all elements, except H, if $h\upsilon$ is large enough. In practice, the electron kinetic energies are scanned, for example, by varying the voltage on the plates of a hemispherical electrostatic analyzer, and a plot is obtained of the number of photoelectrons versus kinetic energy (or binding energy using Equation 2 to convert kinetic energy to binding energy, Fig. 1). High vacuum conditions are necessary: both to minimize surface contamination, and to minimize the collisions of photoelectrons with gas molecules which would reduce signal intensity.

Broadscan spectra (1000 to 0 eV binding energy, ~486 to ~1486 eV kinetic energy) taken with Al Kα radiation (Fig. 1) illustrate a number of important properties and uses of this spectroscopy. First, at least one peak or doublet appears for every element except H. Considering the spectrum of cleaved PbS in Figure 1b, it is immediately apparent that a number of peaks appear for the Pb and S, and these are listed in Table 1. The binding energies for these peaks are characteristic of the element, although they can vary by up to about 10 eV depending on the chemical state of the element (ie. the chemical shift, see below). Other core level electrons for Pb and S at larger binding energies (such as the S 1s and Pb 1s electrons at binding energies of ~2470 eV and 88,000 eV respectively) are of course, not excited by Al Kα radiation. In addition to Pb and S peaks, C 1s and O 1s peaks appear from submonolayer hydrocarbon and oxygen vacuum contaminants.

The presence of these C and O 1s peaks immediately emphasizes the extreme surface sensitivity of the technique. Because photoelectrons will be scattered by other nuclei and electrons (and Equation 2 will no longer hold), a primary photoelectron has to be ejected near the surface to be detected. This surface sensitivity is perhaps seen more clearly using the following equation (Bancroft et al., 1979),

$$I = I_o[1 - \exp(-d/\wedge)] \quad , \tag{3}$$

where I = intensity (area) of a peak for thickness d; I_o = intensity of the peak for an infinitely thick sample; and $\wedge$ is the mean free-path, or escape depth for primary photoelectrons in Å. The mean free path is normally in the 10-20 Å region (Hochella, 1988). For example, if $d = 2\wedge$, then $I = 86\% \ I_o$, and if $d = 3\wedge$, $I = 95\% \ I_o$. Thus XPS analyzes just the first ~50 Å of the surface, and XPS is especially useful for looking at reactions on surfaces.

The intensity of one or more elemental peaks is normally used for semiquantitative or quantitative analyses of the surface. For example, in Figure 1b, the Pb 4f and S 2p peaks are normally used for Pb and S analyses on surfaces because these peaks appear to be the most intense and narrow elemental peaks. Before looking at the use of these peak areas in more detail, a number of questions might be asked about Figure 1b. For example, first, why do many electronic levels (e.g. Pb 5d, Pb 5p, Pb 4f) give rise to doublets; second, why are some peaks much more intense than others (e.g., the Pb 4f peaks are much more intense than the other Pb peaks); and third why are some peaks much narrower than others (e.g., the Pb 4f peaks are obviously much narrower than the Pb 4d peaks)? First, removal of a p, d or f electron always gives a doublet in the spectrum, and the separation of the two peaks is the so-called spin-orbit splitting. This spin orbit splitting results from a coupling of the orbital (L) and spin (S) angular moments in the ion states. Thus, removal of a core s, p, d or f electron results in ion states with respectively $s^1(L = 0, S = \frac{1}{2})$, $p^5(L = 1, S = \frac{1}{2})$, $d^5(L = 2, S = \frac{1}{2})$ and $f^{13}(L = 3, S = \frac{1}{2})$ electron states.

In the j - j coupling scheme, the total electron angular momentum is the sum or difference of the orbital and spin angular moments ($J = L + S$, and $L - S$). This gives rise to the spin orbit split, $P_{3/2}$ and $P_{1/2}$, $D_{5/2}$ and $D_{3/2}$, and $F_{7/2}$ and $F_{5/2}$ states. The lower case letters are normally used as in Table 1, following one electron terminology. Note that the spin-orbit splitting increases for heavy elements relative to light elements (e.g., the Pb 4f splitting of ~5 eV is resolved, while the S 2p splitting of ~1 eV is not readily resolved). Also, the spin orbit splitting generally increases as the binding energy increases (e.g. the Pb 5d, 5p, 4f, 4d and 4p splittings are ~3 eV, 24 eV, 5 eV, 22 eV and 119 eV, respectively, Table 1).

The heights of the elemental peaks are very different for two reasons. First, as seen in Table 1, the cross section for ionization (the peak area is proportional to the cross section) are very different for different energy levels (e.g., the Pb $4f_{7/2}$ and $4d_{5/2}$ cross sections of 12.73 and 13.02 respectively are much larger than the $5p_{3/2}$ cross section of 1.33). Second, the peak widths obviously vary considerably. For example, the Pb 4f peaks are much narrower than the Pb 4d or 4p peaks. Thus the Pb 4d peaks appear much smaller than the Pb 4f peaks, even though the Pb $4d_{5/2}$ cross section (and area) is larger than the Pb $4f_{7/2}$ cross section (Table 1). Generally, core levels of an element with relatively small binding energy will be the most intense and also the narrowest (usually either the 1s, 2p, 3d or 4f level with the smallest binding energy). A few brief comments may illuminate why this is true. The intensity, I, depends on the magnitude of the dipole matrix element for excitation of an electron (Berkowitz, 1979),

$$I \alpha \int \phi_0(r_1) \underline{r}_1 \phi_f^*(r_1)d\tau_1 \quad , \tag{4}$$

where ϕ_0 and ϕ_f are the one electron wave functions for the core electron and the outgoing photoelectron respectively and $\underline{r}_1$ is the dipole operator. By the nature of the dipole matrix element, the allowed transitions are $\Delta\ell = \pm 1$, ie., d electrons can be excited to p or f waves with the $\ell + 1$ transition (d → f) being generally the most intense. The above dipole integral can be regarded as an overlap between the initial d function and the outgoing f wave, and there will generally be greater overlap between the continuous outgoing wave and outer core orbitals (e.g. S 2p and Pb 4f). However, cross sections vary greatly with the photon energy (Berkowitz, 1979; Yeh and Lindau, 1985); but at >1000 eV photon energies, the outer 1s, 2p, 3d or 4f core levels are generally, but by no means always, the most intense.

Before turning to linewidths, it is interesting to note that the intensity ratios of spin orbit doublet peaks are very close to the m_j multiplicity (m_j = J, J-1...-J) of the levels. Thus, $P_{3/2}$ has four m_j levels (+3/2, +1/2, -1/2, -3/2) while $P_{1/2}$ has two such levels (+1/2, -1/2) leading to a peak intensity ratio of close to 2:1 for the sulphur $2p_{3/2}:2p_{1/2}$ ratio (Table 1). Note that this leads to a $F_{7/2} : F_{5/2}$ intensity ratio of ~8/6 (1.33), and a $D_{5/2} : D_{3/2}$ ratio of ~6/4 (1.50). The Pb 4f, 4d and 4p spin orbit intensities reflect the above ratios rather well in Figure 1b and Table 1 and p, d, and f atomic levels are often readily confirmed in an XPS spectrum by this ratio of spin-orbit intensities.

Turning now to the linewidths, the overall observed linewidth (Γ observed) can be written as a first approximation:

TABLE 1. Binding Energies (eV), Cross Sections, and Line Widths (eV) for Core Levels of a Number of Elements Which are Important for Sulphide Surface Studies[a]

Atomic Number/ Element	Energy Level	Approximate[b] Binding Energy	Cross[c] Section	Inherent[d] Line Width
6 C	C 1s*	285	1.0	< 0.1
8 O	O 1s*	540	3.1	< 0.1
16 S	S $2p_{3/2}$*	160	1.11	< 0.1
	S $2p_{1/2}$	161	0.57	< 0.1
	S 2s	230	1.43	broad
17 Cl	Cl $2p_{3/2}$*	200	1.51	narrow
	Cl $2p_{1/2}$	202	0.78	narrow
	Cl 2s	270	1.69	broad
26 Fe	Fe $3p_{3/2,1/2}$	53	1.10,0.57	broad
	Fe 3s	91	0.75	V. broad
	Fe $2p_{3/2}$*	707	10.82	narrow
	Fe $2p_{1/2}$	720	5.60	narrow
	Fe 2s	845	4.57	V. broad
46 Pd	Pd $4p_{3/2}$	58	1.24	V. broad
	Pd $4p_{1/2}$	64	0.64	V. broad
	Pd 4s	97	0.60	V. broad
	Pd $3d_{5/2}$*	368	9.48	narrow
	Pd $3d_{3/2}$	374	6.56	narrow
	Pd $3p_{3/2}$	532	7.63	V. broad
	Pd $3p_{1/2}$	560	3.83	V. broad
	Pd 3s	672	2.81	V. broad
82 Pb	Pb $5d_{5/2}$	18	1.50	< 0.3
	Pb $5d_{3/2}$	21	1.11	< 0.3
	Pb $5p_{3/2}$	83	1.33	V. broad
	Pb $5p_{1/2}$	106	0.53	V. broad
	Pb 5s	147	0.54	V. broad
	Pb $4f_{7/2}$	137	12.73	< 0.1
	Pb $4f_{5/2}$	142	10.01	< 0.1
	Pb $4d_{5/2}$	412	13.02	V. broad
	Pb $4d_{3/2}$	434	8.87	V. broad
	Pb $4p_{3/2}$	643	6.33	V. broad
	Pb $4p_{1/2}$	762	2.12	V. broad
	Pb 4s	890	1.96	V. broad

(a) The Hg and Au cross sections and linewidths are very similar to the Pb values.

(b) Wagner et al. (1979); Vaughan (1985).

(c) Cross sections in units of 13,600 barns ie. the C 1s cross section is 13,600 barns. Scofield, 1976.

(d) Narrow, < 0.4 eV; broad, > 1.0 eV, V. broad, > 5 eV. Gupta et al. (1977), McGuire (1972, 1974).

$$\Gamma^2_{observed} = \Gamma^2_{instrumental} + \Gamma^2_{solid} \qquad (5)$$

$$\text{where: } \Gamma^2_{instrumental} = \Gamma^2_{source} + \Gamma^2_{analyzer}$$

$$\text{and } \Gamma^2_{solid} = \Gamma^2_{inherent} + \Gamma^2_{extra} .$$

For most modern electron analyzers, the analyzer contribution to the instrument width is small and $\Gamma_{instrumental} \simeq \Gamma_{source}$: where Γ_{source} is ~ 1.0 eV for a non-monochromatized Al Kα source, ~ 0.5 eV for a monochromatized Al Kα source, but can be as small as 0.1 eV for new monochromatized synchrotron radiation sources. Γ_{extra} arises from many effects [differential broadening, surface chemical shifts, crystal field effects, multiplet splitting, vibrational splittings (Carlson et al., 1975)], but often this term is smaller than $\Gamma_{inherent}$, and $\Gamma_{solid} \approx \Gamma_{inherent}$. The inherent lifetime width (Γ_i) of the hole state is given by the Heisenberg uncertainty principle $\Gamma_i \tau = h/2\pi$ which in units of eV is given by

$$\Gamma_i = \frac{4.56 \times 10^{-16}}{t_{1/2}} \quad (\tau = \frac{t_{1/2}}{0.693}). \qquad (6)$$

Thus, if $t_{1/2}$ (the half-life of the hole state) is long, Γ_i is small, but if $t_{1/2} \sim 10^{-16}$ sec, $\Gamma_i > 1$ eV.

What controls the lifetime of the hole? The lifetimes are controlled by the Fluorescence and Auger rates which determine how quickly the electrons fill the hole created by photoejection (Carlson, 1975; McGuire, 1972, 1974). For core electrons, the lifetimes and thus the linewidths are controlled mainly by Coster-Kronig transitions, an Auger process (see next section) in which the hole and the electron filling the hole are from the same principle quantum number i.e., a 3d electron fills a 3p hole, or a 2p electron fills a 2s hole. Because of the greater overlap between subshells of the same quantum number, Coster-Kronig rates are fast and natural widths are wide. Note that it is not possible to have a Coster-Kronig transition to 1s, 2p, 3d or 4f hole states, and these peaks are almost always the narrowest (Table 1) and are often < 0.3 eV wide (Table 1). Thus, the source width Γ_{source} often controls $\Gamma_{observed}$, and it is extremely important to decrease Γ_{source} to obtain the best resolution for the C 1s, O 1s, S $2p_{3/2}$, Cl $2p_{3/2}$, Pd $3d_{5/2}$, Pb $5d_{5/2}$ and Pb $4f_{7/2}$ levels (Table 1). Both for intensity and linewidth, then, the above levels are used for surface analyses and they are asterisked in Table 1.

Using monochromatized synchrotron radiation, it is now possible to obtain total core level linewidths on metals and semiconductors of ~ 0.2 eV, and recent gas phase Si 2p spectra show that total core level linewidths of ~ 0.1 eV can be readily obtained (Bozek et al., 1990, to be published). These narrow sources have not yet been used for examining sulphide surfaces, but undoubtedly such studies will be performed soon and greatly increased chemical sensitivity will be obtained.

Semiquantitative, and even quantitative, analysis is possible using the measured peak areas, A, the theoretical cross sections (σ) in Table 1, and the kinetic energy E_k. The peak area can be written as:

$$A_x \propto n_x \, \sigma_x \, E_k^y, \tag{7}$$

where: n_x is the number of x atoms on the surface and E_k^y reflects the sensitivity of the lens/electron analyzer used. For most of the spectra taken in our laboratory, $y = 0.7$. To eliminate the proportionality constant, usually ratios (Eqn. 8) are used:

$$\frac{n_x}{n_y} = \frac{\sigma_y}{\sigma_x} \frac{(K.E.)_y^{0.7}}{(K.E.)_x^{0.7}} \frac{A_x}{A_y}. \tag{8}$$

Looking at the PbS spectrum in Figure 1b, it is apparent that the Pb $4f_{7/2}$ peak is about an order of magnitude more intense than the S 2p peaks. Because the E_k of both lines are very similar, and $\sigma_{Pb\ 4f_{7/2}} : \sigma_{S\ 2p_{3/2}} \approx 10{:}1$ (Table 1), it is easily seen that $n_{Pb}{:}n_S \approx 1$ as expected from the stoichiometry of the compound. For a variety of reasons, the atomic ratios can vary by up to 50% from the stoichiometric ratios, but the variation of atomic ratios such as: $n_{Pb}{:}n_S$ when the PbS surface is reacted with aqueous metal complexes is extremely useful for following surface reactions.

An introduction to Auger electron spectroscopy (AES), and a comparison of techniques

The core hole produced when an atom is bombarded with an energetic source, either X-rays or electrons, may be filled by two processes: X-ray flourescence or Auger decay. In fluorescence, the core hole is filled by an outer electron with emission of an X-ray. Competing with X-ray fluorescence is a radiationless Auger transition in which an outer electron is ejected (Auger electron) when the core hole is filled by another electron. For example, if a 1s (or K) electron is ejected, a 2p (or $L_{2,3}$) electron can fill the hole and the energy given off in this transition can cause another 2p ($L_{2,3}$) electron (the Auger electron) to be ejected. The Auger electron is labelled with reference to the three electrons involved in its production; in the above example it is a KLL Auger electron.

In Auger electron spectroscopy (also called Scanning Auger Microscopy, SAM) an electron beam is used as the primary source, and the spectrum is a plot of number of electrons produced as a function of kinetic energy. Each element will have a unique set of Auger peaks at characteristic kinetic energies, which may be used to identify, quantify and provide chemical state information about the elements present on the surface. As with XPS, the surface sensitivity of AES is due to the small escape depth of Auger electrons. Because of the large increasing background in a Auger spectrum, they are generally presented in the derivative mode, which makes the Auger peaks more apparent.

Although it is possible to derive chemical state information from Auger spectra, AES is not as powerful as XPS in this regard. A change in chemical environment may cause peak shapes and positions to change, but these are less readily predictable than in XPS, and a 'fingerprint' approach to chemical state identification is usually taken. Also, AES linewidths are normally broader than XPS widths.

The major advantage of AES comes from the ability to finely focus the primary electron source, and thus to obtain excellent spatial resolution. It is possible, for example to analyze < 1 μm features on a surface, or to map adsorbate distributions across an area

TABLE 2.   A Comparison of the Major Spectroscopic Techniques for Surface Analysis of Sulfides

Technique	Advantages	Disadvantages
XPS or ESCA	1. Very surface sensitive (< 50 Å). 2. Reasonably good chemical sensitivity (will improve). 3. Semiquantitative analysis - both elemental and different chemical species of same element. 4. Relatively easy to obtain spectra of all solids. 5. X-ray beam relatively non-destructive.	1. High vacuum technique. 2. Are we looking at a changed species due to vacuum and X-rays? 3. Poor spatial resolution (0.1×0.1 mm) 4. Vacuum contaminates.
Auger (AES)	1. Very surface sensitive (< 50 Å) 2. Good spatial resolution (~ 1000 Å or better) and good elemental images. 3. Semiquantitative elemental analyses. 4. Relatively easy to get spectra and images of semiconductor sulfides.	1. High vacuum technique. 2. Electron bombardment is more destructive than X-rays. 3. Chemical sensitivity is generally not good.
Raman	1. High resolution spectra can be obtain in solution. 2. Very good chemical sensitivity for sulfur species.	1. Not surface sensitive. 2. Laser beam destructive. 3. Often not simple to interpret.
SEM	1. Very high spatial resolution (< 100 Å). 2. Easy to use and obtain images and EDX analysis for most elements.	1. Poor surface sensitivity for elemental analyses 2. Electron beam relatively destructive.

Note: EXAFS, STM, AFM, and Photoelectron Microscopy are new powerful techniques. For a discussion see G.E. Brown and M.G. Hochella in this volume.

as small as a few microns square. Of course it is not possible with AES to determine distributions on an atomic level. For this, techniques such as Scanning Tunneling Microscopy, with very high spatial resolution are needed. The combination of surface sensitivity and spatial resolution gives AES a clear advantage over other mapping techniques, such as EDX, where the signal from a thin surface layer will be lost in the strong signal from the substrate. Hochella et al. (1986) gave an excellent review of some geological applications of AES.

The major advantages and disadvantages of XPS and AES for sulphide surface studies (and indeed most surface studies) are briefly compared with the advantages and disadvantages of Raman spectroscopy and SEM in Table 2. Both XPS and AES are extremely surface sensitive with analyses depths of ~ 50 Å, whereas Raman spectroscopy and EDX analysis on the SEM are much less surface sensitive. XPS gives probably the most general surface chemical information of any of these techniques, but Raman and infrared spectroscopies can be used to confirm results in many cases. Raman spectroscopy has the great advantage that analyses of the surface can be done in situ. The other techniques all require high or ultrahigh vacuum conditions for analyses and it is sometimes possible that the chemistry of the surface will change when the sample is removed from solution and washed, during the pumpdown to high vacuum, or due to the impinging X-ray or electron beam. XPS and Raman, although chemically very useful, give minimal spatial information, although photoelectron microscopy is becoming useful now. High spatial resolution is obtained by AES and SEM, while other techniques such as SIMS can also be used to obtain spatial information at poorer resolution than the AES or SEM techniques.

The Chemistry of the surface from XPS and Raman spectroscopies

When sulphide surfaces are reacted with dilute aqueous solutions of Au, Hg and Pd complexes, the broadscan spectra (Figs. 1-3) are very useful in characterizing the reaction qualitatively. In Figure 1a, a 20 ppm Au solution of $KAuCl_4$ has been reacted with a cleaved PbS surface. It is immediately obvious that the Au peaks (note how similar they are to the corresponding Pb peaks but at lower binding energy as expected for the lower atomic number Au) dominate the spectrum, and no Cl or K peaks are apparent. The S 2p and S 2s peaks are still relatively intense, while the Pb peaks can barely be detected. The usual small C 1s and very small O 1s peaks from adsorbed contaminants are also present. A layer of Au metal must have formed on the PbS, because no Cl 2p and O1s peaks from Cl or $H_2O(OH^-)$ ligands are present. Because S is still showing, the Au metal layer is either: uniform and thin (several monolayers), or non-uniform and much thicker. The absence of Pb peaks shows that a sulphur-rich, Pb-depleted layer forms at the surface. The SEM results in the next section immediately answer the question of uniformity of the Au layer.

Figure 2 shows the spectra of sulphide plates reacted with $HgCl_2$ solutions. Although, the Hg 4f peaks are the most intense, there are two qualitative differences between these spectra and the Au spectra shown in Figure 1. First, the metal peaks from the sulphide are relatively intense, especially for PbS (Fig. 2b) and FeS_2 (Fig. 2c). This shows that only a very thin layer of mercury species (adsorbed?) is on the sulphide surface. Second, in all three cases, a substantial amount of Cl is on the surface indicating that Cl has been adsorbed along with the Hg (probably as an $HgCl_x$ species).

In Figure 3, pyrite (FeS_2) and hematite (Fe_2O_3) plates have been reacted with Na_2PdCl_4 solutions. Obviously, much more Pd is on the pyrite than hematite surface, but the Pd uptake is much more similar to Hg uptake on sulphides (Fig. 2c) than the Au uptake on sulphides (Fig. 1a). Thus, a relatively thin layer of Pd is present on the surface (note the intense Fe 2p and S 2p peaks), and a small Cl 2p peak is present showing the presence of chlorine. Again, no cations are adsorbed.

To obtain much more chemical information from the XPS peaks, it is important to obtain narrow scan spectra of both the metal and sulphur levels. Initial Au spectra for solid $KAuCl_4$, AuCN, and Au metal standards using a nonmonochromatized source are shown in Figure 4. This figure (and the numbers in Table 3) shows that the Au $4f_{7/2}$ binding energy increases by ~ 1 eV for each unit increase in Au oxidation state. Thus

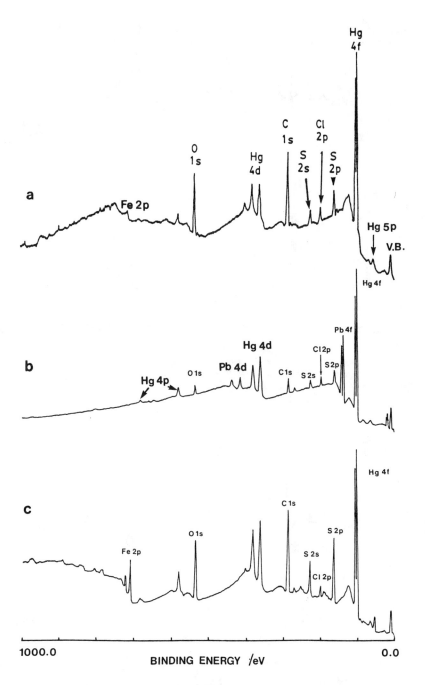

Figure 2. Broadscans of polished: (a) pyrrhotite; (b) galena; and (c) pyrite plates after reaction with 100 ppm mercury (as HgCl₂) solution at pH 3.0 for 2 hours (Hyland and Bancroft, 1990b).

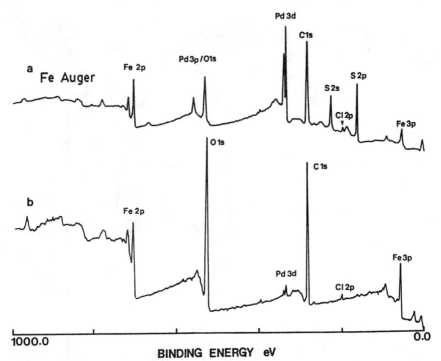

Figure 3. Broadscans of (a) pyrite; and (b) hematite plates after 16 h in solutions of Na_2PdCl_4 (40 ppm Pd), pH 4.0, 1000 ppm Cl⁻. Much more palladium is deposited on pyrite plates relative to hematite (Hyland and Bancroft, 1990a).

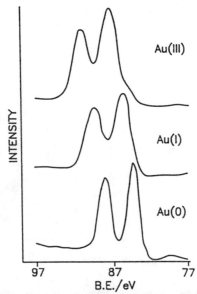

Figure 4. Au 4f narrow scan spectra using a non-monochromatized source in the McPherson ESCA-36 spectrometer for $KAuCl_4$ (upper trace), AuCN (middle trace) and elemental gold (lower trace) (Jean and Bancroft, 1985).

TABLE 3. Au $4f_{7/2}$ Binding Energies (eV) Referenced to the Metal Au $4f_{7/2}$ line at 84.0 eV[a]

Species	Reacted With	Ox State	B.E./eV Au $4f_{7/2}$	ΔB.E.
$KAu(Cl)_4^-$	Standard	Au(III)	87.2	3.2
AuCN	Standard	Au(I)	85.2	1.2
$Na_3Au(S_2O_3)_2$	Standard	Au(I)	85.1	1.1
$Au_{11}L_7X_3$	Standard	$Au^{\delta+}$[b]	84.7	0.7
Au metal	Standard	Au(0)	84.0	0
FeS	$KAu(Cl)_4$	Au(0)	84.0	+ 0.0
FeS_2	$KAu(Cl)_4$	Au(0)	84.1	+ 0.1
ZnS	$KAu(Cl)_4$	Au(0)	83.9	- 0.1
PbS	$KAu(Cl)_4$	Au(0)	84.0	0.0
$FeCuS_2$	$KAu(Cl)_4$	Au(0)	84.3	+ 0.3
SiO_2	$KAu(Cl)_4$	Au(III)	87.0	+ 3.0
Fe_2O_3	$KAu(Cl)_4$	Au(III)	86.9	+ 2.9
Fe_3O_4	$KAu(Cl)_4$	Au(III)	86.7	+ 2.7
FeS	$KAu(CN)_2Br_2$	Au(I)	85.1	+ 1.1
PbS	$KAu(CN)_2Br_2$	Au(I)	84.7	+ 0.7
FeS_2	$Na_3Au(S_2O_3)_2$	$Au^{\delta+}$[b]	84.4	+ 0.4
PbS	$Na_3Au(S_2O_3)_2$	Au(0)	84.0	0.0

[a] From Jean and Bancroft (1985) and Hyland (1989).

[b] δ^+ designates a fractional oxidation state.

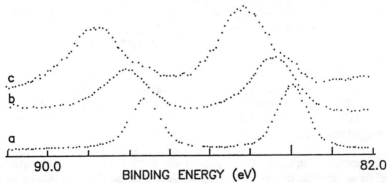

Figure 5. Au 4f narrow scan spectra using a monochromatized source in the Surface Science Laboratories SSX-100 spectrometer: a) Au foil; b) FeS_2 plate reacted with a $Na_3Au(S_2O_3)_2$ standard (c) solid $Na_3Au(S_2O_3)_2$ (Hyland, 1989).

the binding energies for Au(0), Au(I) and Au(III) are 84.0 eV, ~85.0 eV and ~87.0 eV respectively. Higher resolution Au 4f spectra can be obtained with a monochromatized source as shown in Figure 5. Once again, the Au(I) standard, solid $Na_3 Au(S_2O_3)_2$, gives a Au $4f_{7/2}$ binding energy of ~85.0 eV. The chemical shift arises because a rearrangement of valence electrons changes the potential felt by the core electrons. Specifically, loss of valence electrons on going from Au(0) to Au(III) deshields the Au 4f electrons from the nuclear positive charge, and the 4f electrons are "pulled in" more towards the nucleus thus increasing the 4f binding energy.

When Au complexes are reacted with oxides or sulphides, the Au $4f_{7/2}$ chemical shift is extremely useful for characterizing the Au species on the surface. In all cases in Table 3, the sulphides reduce KAuCl$_4$ to Au metal, whereas gold remains as Au(III) on the oxides. Thus, the Au 4f binding energy for KAuCl$_4$ solutions reacted with sulphides is 84.1±0.2 eV; whereas the Au 4f binding energy for KAuCl$_4$ solutions reacted with oxides is 86.8±0.2 eV. The less reactive Au(III) complex $Au(CN)_2Br_2^-$ is only partially reduced to Au(I) (the Au $4f_{7/2}$ binding energy is 84.9±0.2 eV). The Au(I) species $Au(S_2O_3)_2^{3-}$ is either: partially reduced on FeS$_2$ forming a gold cluster in which the formal oxidation state of the Au is between 0 and 1 (labelled Au$^{\delta+}$ in Table 3), or completely reduced on PbS to Au metal. Thus, in Figure 5, the FeS$_2$ surface reacted with $Au(S_2O_3)_2^{3-}$ yields a Au 4f spectrum with Au $4f_{7/2}$ binding energies intermediate between Au(0) and $Au(S_2O_3)_2^{3-}$, whereas the Au $4f_{7/2}$ binding energy on PbS of 84.0 eV shows that Au metal is present. Note that the Au 4f linewidths in Figure 5 (~1 eV) are much narrower in this spectrum than in Figure 4 (> 1.5 eV) because the spectra in Figure 5 were taken with the narrower monochromatized source. Somewhat surprisingly the lack of any Au 4f peak showed that the very inert Au(CN)$_2^-$ species was not even adsorbed (Jean and Bancroft, 1985).

The peak areas provide additional information on the reactions (Tables 4 and 5). When AuCl$_4^-$ is reacted with FeS$_2$ (Table 4), PbS (Table 5 and Fig. 1) and (Zn,Fe)S (Hyland and Bancroft, 1989), the Au 4f intensity increases rapidly (Au is rapidly adsorbed and reduced), while a metal-deficient sulphide surface is formed. This is especially apparent for PbS in Table 5, where after 10 hours reaction of a 20 ppm Au solution of AuCl$_4^-$, the atomic ratios of Au:Pb:S are ≈2.5:0.02:1. The results from both S 2p and S 2s peaks in Table 4 and 5, show that the reliability is normally about ± 25%. However, for $Au(S_2O_3)_3^{3-}$, the two very different types of reactions on FeS$_2$ and PbS are confirmed by the atomic ratios in Tables 4 and 5. Thus, on FeS$_2$, very little Au goes down on pyrite, and the Fe:S ratios remain quite constant. In contrast, on PbS, the behavior of $Au(S_2O_3)_2^{3-}$ is very similar to that of AuCl$_4^-$: Au:S ratios are large, and a Pb depleted sulphide surface is formed. These results, and others, show that simple adsorption of the Au complexes results in low Au coverage (probably monolayer), with little surface reaction at the sulphide surface; while adsorption/reduction leads to large Au coverages (many monolayers), with a radically altered sulphide surface. In general, then, these surface reactions are highly dependent on both the natures of the surface and the aqueous complex; and XPS results pin-point some important features quickly.

The reaction of aqueous PdCl$_4^{2-}$ solutions with sulphides (Hyland and Bancroft, 1990a) highlight some of the important features of this type of reaction, and emphasize some of the complexities. In this study, the reactions of 40 ppm Pd solutions of Na$_2$PdCl$_4$ on FeS$_2$, PbS and (Zn,Fe)S were studied as a function of pH (from 1.5 to 5.0) and using

TABLE 4. A Comparison of Au:Fe:S Surface Atomic Ratios for FeS$_2$ Plates Reacted for Increasing Lengths of Time with 20 ppm Au Solutions as AuCl$_4^-$ and Au(S$_2$O$_3$)$_2^{3-}$ based on peak areas (a)

		Au 4f	Fe 2p$_{3/2}$	S 2p
AuCl$_4^-$	0.0	--	0.17	1.0
	1 min	0.08	0.23	1.0
	30 min	0.97	0.17	1.0
	1 hr	1.10	0.23	1.0
	3 hr	3.04	0.21	1.0
	24 hr	5.10	0.06	1.0

		Au 4f	Fe 2p$_{3/2}$	S 2s
AuCl$_4^{-(b)}$	0.0	--	0.20	1.0
	1 min	0.09	0.24	1.0
	30 min	1.08	0.19	1.0
	1 hr	1.15	0.24	1.0
	3 hr	3.44	0.24	1.0
	24 hr	7.06	0.09	1.0

		Au 4f	Fe 2p$_{3/2}$	S 2s
Au(S$_2$O$_3$)$_3^{3-}$	0.0	0.01	0.20	1.0
	30 min	0.01	0.21	1.0
	1 hr	0.01	0.28	1.0
	5 hr	0.04	0.32	1.0
	10 hr	0.04	0.25	1.0
	24 hr	0.08	0.34	1.0
	48 hr	0.08	0.34	1.0
	1 week	0.05	0.25	1.0

(a) Data taken from: Hyland and Bancroft (1989) and Hyland (1989).

(b) Recalculated using the S 2s peak rather than the S 2p$_{3/2}$ peaks.

TABLE 5. A Comparison of Au:Pb:S Surface Atomic Ratios for PbS Plates Reacted for Increasing Lengths of Time with 20 ppm Au Solutions as AuCl$_4^-$ and Au(S$_2$O$_3$)$_2^{3-}$ based on peak areas$^{(a)}$

		Au 4f	Pb 4f	S 2p
AuCl$_4^-$	0.0	--	1.09	1.0
	1 min	0.56	0.50	1.0
	30 min	2.92	0.10	1.0
	1 hr	3.40	0.13	1.0
	10 hr	2.26	0.01	1.0

		Au 4f	Pb 4f	S 2s
AuCl$_4^{-(b)}$	0.0	--	1.36	1.0
	1 min	0.67	0.60	1.0
	30 min	3.00	0.10	1.0
	1 hr	3.74	0.14	1.0
	10 hr	2.82	0.03	1.0

		Au 4f	Pb 4f	S 2p
Au(S$_2$O$_3$)$_3^{3-}$	0.0	--	1.36	1.0
	30 min	1.39	1.10	1.0
	1 hr	1.98	0.99	1.0
	5 hr	1.79	0.42	1.0
	10 hr	2.43	0.47	1.0
	24 hr	3.28	0.32	1.0
	48 hr	2.25	0.48	1.0
	72 hr	8.70	0.27	1.0

(a) Data taken from: Hyland and Bancroft (1989); and Hyland (1989).

(b) Recalculated using the S 2s peak rather than the S 2p$_{3/2}$ peaks.

TABLE 6. Pd 3d$_{5/2}$ Binding Energies (eV)^a of Reacted Minerals and Selected Standards^b

Reacted Minerals	Pd(0)	Pd A	Pd B	Pd C	Pd D
FeS$_2$	335.5	336.7	337.7	339.2	337.3
PbS	335.5	336.5	337.8	--	--
(Zn,Fe)S	335.5	336.7	337.9	--	--
Fe$_2$O$_3$					337.3
Standards					
Pd metal	335.5				
PdS	336.6				
PdS$_2$	336.9				
Na$_2$PdCl$_4$	337.9				
K$_2$PdCl$_4$	338.1				
PdCl$_2$	338.4				
PdCl$_2$	338.2				
PdO.nH$_2$O	337.1				
PdO.nH$_2$O	337.0				
PdCl$_x$ on oxides	338.2				

^{a} ± 0.1 eV, referenced to Pd 3d$_{5/2}$ of Pd metal at 335.5 eV.

^{b} From Hyland and Bancroft (1990a).

[Cl⁻]/[Na₂PdCl₄] ratios of ≈ 15,000, 160, and 11. By UV spectrophotometry, these Cl⁻ concentrations yielded mainly the solution species $PdCl_4^{2-}$, $PdCl_3(H_2O)^-$ and $PdCl_2(H_2O)_2$ respectively. Figure 3 shows that palladium does not deposit on pyrite plates as fast as gold does; and that little palladium is taken up by hematite. The Pd 3d XPS spectra (Fig. 6, Table 6) clearly indicate the variety of Pd species deposited on all surfaces. On hematite, the Pd $3d_{5/2}$ binding energy was measured consistently at 337.3 eV, as expected. This is a shift of 2 eV above Pd metal (Pd $3d_{5/2}$ = 335.5 eV) and corresponds to PdO.nH₂O (Table 6). In contrast, a number of Pd species are distinguished on pyrite, (Fig. 6), and the relative amount of each was strongly dependent on the reaction conditions; pH, [Cl⁻] and time. In the initial stages of the reaction for 40 ppm Pd, (with 1000 ppm Cl⁻, pH 4.2), the Pd $3d_{5/2}$ spectra are dominated by a peak at 336.7 eV, labelled Pd A in the diagram. The pronounced asymmetry on the high binding energy side of the peak implies the presence of other palladium species, and two other doublets (Pd B and Pd C) at 337.9 eV and 339.2 eV, respectively have been fit to the spectrum. The Pd $3d_{5/2}$ binding energies of Pd A, Pd B, and Pd C are all within the range found for Pd(II) compounds, (Table 6) and thus these species have not undergone reduction. Comparison of the Pd A and Pd B with standard Pd compounds shows that Pd A corresponds to a Pd sulphide, either PdS or PdS_2; and Pd B corresponds to a $PdCl_x$ species adsorbed on the sulphide. XPS intensities once again show that these Pd(II) species only build up to less than a monolayer.

In addition to Pd A, Pd B and Pd C, there is a small shoulder at lower energy in Figure 6a which grows with time. After 18 hours of reaction, (Fig. 6c), this peak which corresponds to the Pd metal standard (Fig. 6d, Table 6), predominates. After monolayer coverage of mostly PdS or PdS_2, palladium starts to be reduced to Pd metal.

The proportion of these species varies greatly depending on both [Cl⁻] (Table 7, Fig. 7) and pH. Thus, if the [Cl⁻] concentration is very high as in Table 7a, little Pd is taken up by the FeS_2, and no Pd is reduced to Pd(0). As the [Cl⁻] concentration is increased, much more Pd is taken up, and the majority of the Pd becomes reduced. It is apparent that the $PdCl_4^{2-}$ species does not readily adsorb or react whereas $PdCl_2(H_2O)_2$ reacts readily with the sulphide, and leads to reduction. The reduction can thus be controlled by the concentration of chloride. Unlike the $AuCl_4^-$ reactions, substantial Cl⁻ goes down on the surface - as $PdCl_x$ species (doublet B).

Similarly, the percent Pd(0) detected at pH 1.5, 3.0, 4.0, and 5.0 for the low [Cl⁻] solutions varies markedly (Hyland and Bancroft, 1990a). At the high and low pH values, little or no Pd(0) is detected at all reaction times. At pH 3 and 4, the reduction is obviously favored. At the high pH, the formation of PdO.H₂O [or Pd(OH)₂] reduces the concentration of the "active" species, $PdCl_2(H_2O)_2$, thus inhibiting reduction to Pd(0). The lack of reduction at low pH suggests that the electroactive species is $PdCl_2(H_2O)_2(OH)^-$ which is converted to $PdCl_2(H_2O)_2$ at low pH.

$$PdCl_2(H_2O)(OH)^- + H^+ \longrightarrow PdCl_2(H_2O)_2 \ . \tag{9}$$

The rate of reduction would then be proportional to the concentration of Pd $Cl_2(H_2O)(OH)^-$.

The above gold and palladium results have given specific answers to most of the first

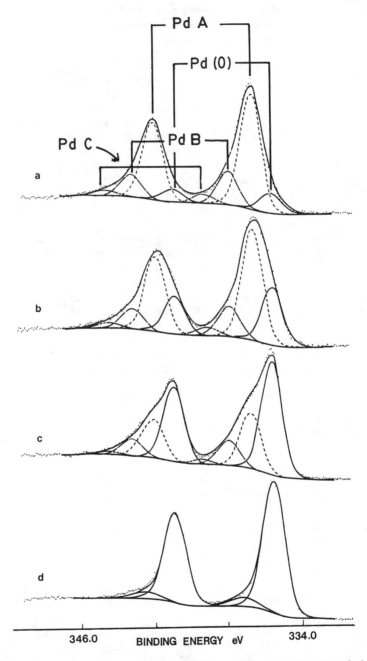

Figure 6. Curve-resolved Pd 3d spectra of polished pyrite plates using the SSX-100 monochromatized source after: (a) 1 hour; (b) 4 hours; and (c) 18 hours in solutions of Na_2PdCl_4 (40 ppm Pd), pH 4.0, 1000 ppm Cl^-. Palladium foil standard is shown in (d). The spectra show the four palladium species detected on reacted pyrite plates. The dots are the original data, the dashed curves are the individual peaks and the solid lines represent the composite fits (Hyland and Bancroft, 1990a).

530

TABLE 7. Surface Atomic Compositions[a,b] for FeS$_2$ Plates in Na$_2$PdCl$_4$ Solutions, pH 3.0 and [Cl$^-$]/[Na$_2$PdCl$_4$] of: (a) 15,000; (b) 11

(a) [Cl$^-$]/[Na$_2$PdCl$_4$] = 15,000

	% Pd(0)[c]	Pd$_t$	:	Pd A-C	:	Cl	:	S
0.5 hr	0	0.13	:	0.13	:	0.02	:	1.0
1 hr	0	0.04	:	0.04	:	0.01	:	1.0
3 hr	0	0.05	:	0.05	:	0.02	:	1.0
6 hr	0	0.06	:	0.06	:	0.07	:	1.0
12 hr	0	0.08	:	0.08	:	0.02	:	1.0
18 hr	0	0.06	:	0.06	:	0.05	:	1.0
24 hr	0	0.08	:	0.08	:	0.04	:	1.0
2 days	0	0.10	:	0.10	:	0.05	:	1.0
14 days	0	0.11	:	0.11	:	0.06	:	1.0

(b) [Cl$^-$]/[Na$_2$PdCl$_4$] = 11

	% Pd(0)[c]	Pd$_t$	:	Pd A-C	:	Cl	:	S
0.5 hr	17	0.16	:	0.13	:	0.05	:	1.0
1 hr	27	0.19	:	0.14	:	0.07	:	1.0
3 hr	42	0.30	:	0.17	:	0.08	:	1.0
6 hr	44	0.41	:	0.21	:	0.15	:	1.0
12 hr	62	0.94	:	0.30	:	0.29	:	1.0
18 hr	72	1.71	:	0.36	:	0.42	:	1.0
24 hr	69	1.68	:	0.40	:	0.40	:	1.0
2 days	70	3.07	:	0.71	:	0.61	:	1.0
14 days	61	1.21	:	0.41	:	0.20	:	1.0

[a] Calculated using Pd 3d, Cl 2p and S 2p broadscan peak areas.
[b] From Hyland et al. (1990a).
[c] From curve-fit Pd 3d spectra.

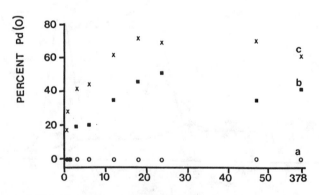

Figure 7. A plot of the percent Pd(0) detected on reacted polished pyrite plates as a function of time for different [Cl$^-$]/[Na$_2$PdCl$_4$] ratios: (a) 15,000 (o); (b) 160 (■); (c) 11 (x). The percent Pd(0) was determined from the curve-resolved Pd 3d peaks (see Fig. 6) (Hyland and Bancroft, 1990a).

two sets of questions in the INTRODUCTION involving the metal species. Turning now to the third questions, can the reducing agent be identified and the oxidation reaction be followed? Initially (Jean and Bancroft, 1985), the reducing agent was not identified with XPS when $AuCl_4^-$ was reacted with sulphides. The expected reaction of S^{2-} going to SO_4^{2-},

$$S^{2-} + 4H_2O \longrightarrow SO_4^{2-} + 8H^+ + 8e \ , \tag{10}$$

was not evident as no SO_4^{2-} was detected on the sulphide surfaces. With higher resolution spectra, and high sensitivity solution analyses for SO_4^{2-}, we have confirmed that S^{2-} is indeed the reducing agent, polysulphides are formed on most sulphide surfaces and SO_4^{2-} is present in solution. Sulphur 2p results are summarized in Table 8, and representative S 2p spectra are shown in Figures 8-10. Qualitatively, the oxidation of sulphide is seen in Figure 8 for (Zn,Fe)S reacted with a 20 ppm Au solution of $KAuCl_4$. The cleaved surface initially gives a clean well-resolved S 2p doublet (A) with a S $2p_{3/2}$ binding energy of 161.4 eV corresponding to S^{2-} (Table 8). The sulphide in PbS and FeS_2 give S $2p_{3/2}$ binding energies of 160.7 and 162.4 eV (Fig. 9, Table 8). The linewidths in these spectra are all less than 1 eV, and this resolution from a monochromatized instrument is essential for following the change in S chemistry on the surface. After reacting (Zn,Fe)S in a control solution with no $AuCl_4^-$ for one hour (Fig. 8b), a hint of a peak at ~ 164 eV is seen, and another doublet has been fit (C). However, the sulphide surface is quickly modified in the $AuCl_4^-$ solutions. After 10 minutes reaction, peaks C are very noticeable; while after one hour reaction, peaks C are the dominant peaks. Peaks B have to be fit in this spectrum (and other spectra) to obtain satisfactory fits to the data. Similar spectra are obtained for PbS and (Zn,Fe)S reacted with $PdCl_4^{2-}$ at intermediate [Cl⁻] where reduction of Pd(II) to the metal occurs. In the PbS spectrum (Fig. 9b), peaks B are the most intense. Pyrite is much less reactive than the other two sulphides. The reaction of $PdCl_4^{2-}$ with FeS_2 gives no obvious change in the S 2p spectrum; but reaction of FeS_2 with $KAuCl_4$ does lead to a similar three doublet spectrum (Hyland and Bancroft, 1989). Electrochemical oxidation of FeS_2 by applying a potential of + 0.7 volts relative to the standard calomel electrode (SCE) (Mycroft et al., 1990) also gives a very similar spectrum (Fig. 10a) again with three doublets; and oxidation of S^{2-} on a gold electrode (Buckley et al., 1987) also gives a three doublet spectrum. Table 8 shows that the S $2p_{3/2}$ binding energies for species A, B, and C are rather similar: 161.4 (±0.1) eV for A (except for $PdCl_4^{2-}$ on PbS), 162.2 (±0.2) eV for B, and 163.2 (±0.1) eV for C.

As discussed previously for the Au 4f peaks, an increase in binding energy corresponds to an increase in oxidation state. Table 8 shows that sulphide, with a formal oxidation state of -2, gives a S $2p_{3/2}$ binding energy of ~ 161 eV; while S(0) gives a S 2p binding energy of 163.7 eV, an increase again of ~ 1 eV per unit increase in binding energy. Obviously doublets B and C correspond to S in an oxidation state of between -2 and 0. The only possible species are polysulphides such as S_x^{2-} (x = 3,4,5,6). There is no evidence at all for the presence of S-O surface species such as $S_2O_3^{2-}$, SO_3^{2-} or SO_4^{2-} with sulphur oxidation states of +2, +4 and +6 respectively, which would give much higher binding energy peaks. Although SO_4^{2-} is not on the surface, a considerable amount of SO_4^{2-} normally appears in solution (especially for pyrite) (Hyland and Bancroft, 1989, 1990a).

TABLE 8. S $2p_{3/2}$ Binding Energies (eV) of Unreacted and Reacted Minerals and Related Compounds Referenced to the Metal Au $4f_{7/2}$ Line at 84.0 eV[*]

	Unreacted	Reacted with	Species after Reaction		
			A[†] M-$\underline{S}$-S	B[†] M-S-$\underline{S}$	C S-S
FeS$_2$	162.4	AuCl$_4^-$	161.5	162.4	163.2
		Au(S$_2$O$_3$)$_2^{3-}$		162.4	
		PdCl$_4^{2-}$		162.5	
		electrochemical oxidation (Fig.12)	161.4	162.5	163.3
PbS	160.7	AuCl$_4^-$	161.3	162.2	163.1
		Au(S$_2$O$_3$)$_2^{3-}$	161.3	162.4	163.1
		PdCl$_4^{2-}$	160.7	162.0	163.2
(Zn,Fe)S	161.4	AuCl$_4^-$	161.5	162.2	163.1
		PdCl$_4^{2-}$	161.5	162.2	163.2
oxidation of S^{2-} on Au			161.4	162.5	163.1
Cu$_3$(S$_4$)$_3^{3-}$			162.1		163.1
Pt(S$_5$)$_3^{2-}$			162.0	162.9	163.3
Rh(S$_5$)$_3^{3-}$			161.7	162.8	163.4
S(0)					163.7

[*] Data taken from Hyland and Bancroft (1989); Hyland and Bancroft (1990a); Hyland (1989); and Mycroft et al. (1990).

[†] Note that the original S 2p peak of unreacted PbS and (Zn,Fe)S overlaps strongly with peak A of the polysulfide. Similarly the S 2p peak of unreacted FeS$_2$ is at the same position as peak B of the polysulfide.

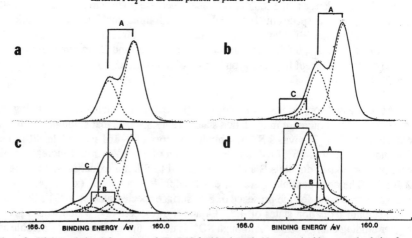

Figure 8. High resolution S 2p spectra of (Zn,Fe)S freshly cleaved, (a); reacted with a control solution for one hour, (b); reacted with a 20 ppm Au solution (KAuCl$_4$) for ten minutes, (c); and, one hour, (d). The dots represent the original data, the dashed curves are the individual peaks from the curve-fitting procedure and the solid lines are the composite fit (Hyland and Bancroft, 1989).

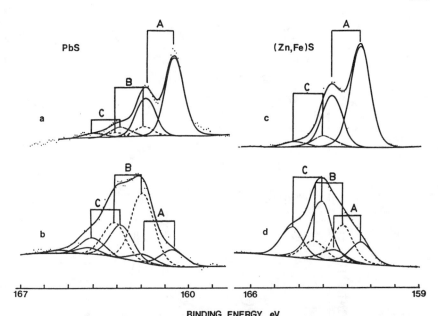

Figure 9. Curve-resolved high resolution S 2p spectra of cleaved galena plates after one hour on: (a) in a control solution; and (b) in a solution of Na_2PdCl_4 (40 ppm Pd), pH 4.0, 1000 ppm Cl^-. Shown in (c) and (d) are S 2p of (Zn,Fe)S plates reacted under the same conditions as (a) and (b), respectively (Hyland and Bancroft, 1990a).

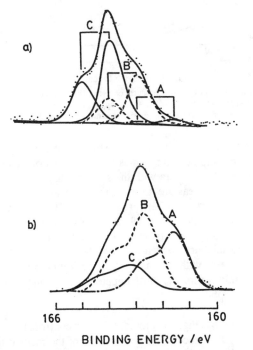

Figure 10. High resolution S 2p spectra of polysulphides from electrochemical oxidation of FeS_2. See binding energies in Table 4 (Mycroft et al., 1990).

534

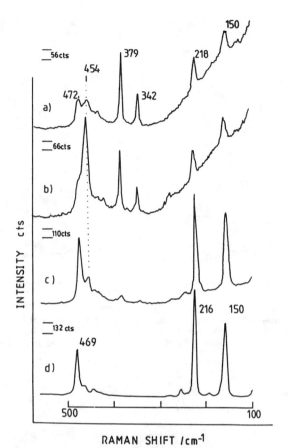

Figure 11. In situ Raman spectra of oxidation products on pyrite electrode surfaces. Laser excitation line 514.5 nm, laser power at sample 45 mW, slit width 90 μm, applied potential 700 mV(SCE). (a)-(c), surface products after 1, 2.2, 23 hours, respectively; (d), ex situ spectrum of sublimed elemental sulphur (Mycroft et al., 1990).

The polysulphide surface species have been confirmed in two ways. First, the S 2p XPS spectra of known polysulphide compounds (at the bottom of Table 8 and Fig. 10b) show very similar three doublet spectra with the same binding energies (Buckley et al., 1988). However, the relative areas are not the same, and we cannot characterize what polysulphide we have on the surface. There are at least two structurally distinct sulphides in each polysulphide. For example in S_5^{2-}, which forms a ring with two S bonded to the metal atom, three distinct sulphurs are present, with the lower binding energy (A) corresponding to the two sulphurs bonded to the metal (Buckley et al., 1988). The S $2p_{3/2}$ binding energy increases with increasing distance from the metal, and approaches the value for elemental sulphur for the third sulphur from the metal (C). This sulphur only has bonds to other sulphurs.

In situ Raman spectroscopy (Fig. 11) provides the best confirmation of polysulphides, and also confirms that the XPS spectra are not artifacts of the method ie. caused by the vacuum and/or X-ray damage. After one hour of anodic oxidation (Mycroft et al., 1990), new features in the spectrum can be seen in addition to the initial pyrite peaks at 379, and

342 cm^{-1}. The new peaks at 450-460 cm^{-1} in Figure 11a increase sharply in intensity after 2.2 hours reaction (Fig. 11b), and can readily be attributed to polysulphides based on previous Raman studies on inorganic polysulphides (Mycroft et al., 1990 and references therein). Since no iron polysulphides were known which could be used as representative standards, it was not possible to positively identify the specific polysulphide formed. However, the sharpness of the polysulphide peak at 452-460 cm^{-1} in Figure 11b suggests that a single polysulphide species was present, rather than a mixture. After 23 hours (Fig. 11c), elemental sulphur (peaks at 150, 216 and 470 cm^{-1}) becomes the predominant surface component, as shown by the similarity of this spectrum with that of elemental sulphur (Fig. 11d). The kinetics of the sulphide oxidation can be followed beautifully with this technique. The Raman technique analyzes the first ~ μm of the surface, and thus the rapid disappearance of pyrite peaks indicates that rather thick layers are grown quickly. Somewhat surprisingly, the metastable polysulphides are rather stable both in solution and in air. Solution analyses show that indeed SO_4^{2-} is also formed, but is found only in solution. The SO_4^{2-} is formed as a result of further oxidation of the sulphur, or by a separate mechanism via $S_2O_3^{2-}$ which we also do not see on the surface (Mycroft et al., 1990).

These surface studies have given considerable insight into the reduction process and answered much of question 3 in the Introduction. We know now that, as initially expected, S^{2-} is the major reducing agent. We also know the fate of the sulphide, and can follow the rate of sulphide oxidation directly both in solution with Raman spectroscopy, and ex-situ, with XPS.

Spatial distribution of metal species from SEM and Auger studies

In the previous section of this chapter, we have answered many important questions about the surface chemistry of these reactions. However, XPS and Raman spectroscopies have given no spatial information, and no answers to question 4 in the INTRODUCTION which centre on where and how the metal species are distributed on the surface. To obtain spatial information, we now turn to SEM and Auger studies.

In our original study (Jean and Bancroft, 1985), several polished sulphide plates were examined by SEM before and after reaction with 100 ppm KAuCl$_4$ solutions, and chemically analyzed with the energy dispersive X-ray (EDX) system. Figure 12 shows the secondary electron images of a blank polished sphalerite plate (A) and a plate reacted with the gold solution (B). The micrograph (Fig. 12b) shows the presence of agglomerates of gold balls distributed unevenly on the surface of the reacted plate. The balls have diameters of several thousand angstroms after only 1 minute of reaction. These gold balls grow in size with reaction time until the gold reaches several layers think and almost covers the surface after a few days. Using a small beam size (400 Å), we analyzed the Au balls and the "clean" regions of the reacted plates. The X-ray spectrum of the agglomerates shows the characteristic X-ray lines of Au with only traces of Zn from the background. [The EDX analysis depth is of the order of 10,000 Å (1 μm).] A close examination of Figure 12b suggests that agglomerates grew on every specific sites, preferentially along edges of crystals. Often pockets of abnormally high concentration of gold were found on the surface (Jean and Bancroft, 1985). In such cases, gold seems to have deposited abundantly at a few sites. Even after gold almost covers the surface,

536

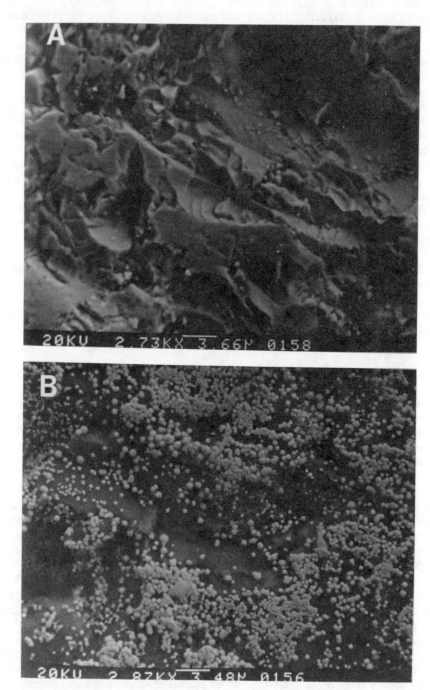

Figure 12. SEM micrographs of of an unreacted polished sphalerite (A) and sphalerite reacted with a 100 ppm gold solution (KAuCl₄) (B) for one minute at pH 5.6 (Jean and Bancroft, 1985).

the layer is not uniform, but rather disordered with agglomerates of various sizes. This leaves a variety of holes that would still allow the solution to come in contact with the sulphide surface.

On freshly fractured surfaces of (Zn,Fe)S (Fig. 13, see also Fig. 8) (Hyland and Bancroft, 1989) the Au metal is again very unevenly distributed. There is not as much gold present on the fractured (Zn,Fe)S surface after few minutes of reaction as in the polished surface after one minute of reaction (Fig. 13b). The surface reacted for two days is severely corroded (Fig. 13c), and the Au balls are more difficult to distinguish. X-ray powder diffraction of the surface material shows the expected peaks for metallic gold plus a high background resulting from the amorphous polysulphide surface.

We showed previously (Tables 3 and 5) that the reaction of $Au(S_2O_3)_3^{3-}$ on PbS produced Au metal; whereas the Au(I) was not completely reduced on FeS_2. In Figure 14, the secondary electron (a,c) and Au Auger images (b,d respectively) are shown for PbS plates reacted for 10 hours in 20 ppm Au solutions of $Na_3Au(S_2O_3)_2$. The Au metal is more uniform in Figure 14b,d than in Figures 12 and 13, but gold is still deposited in a non-uniform manner as can be seen best by the changes in signal brightness in the Au Auger maps (Fig. 14b,d). In some region, gold could be detected as a film on the surface (Fig. 14a). In general, high concentrations of gold on the PbS surfaces could not be correlated with any visible features such as cracks, pits or fracture lines - mainly because the PbS surface reacts quickly to produce a polysulphide layer depleted in Pb.

The SEM and Auger studies of the sulphide plates reacted with $PdCl_4^{2-}$ show important spatial features of these reactions. In Figure 15, a polished pyrite plate has been reacted with $PdCl_4^{2-}$ at the high chloride concentrations where the predominant solution species is $PdCl_4^{2-}$. Under these conditions, only Pd(II) species are present (Table 7a). The secondary electron image (Fig. 15a) does not detect any appreciable palladium (although the surface roughness create contrast); but the surface sensitive Auger technique shows that the Pd(II) species are adsorbed relatively uniformly. In oxidized regions, or regions where aluminosilicate inclusions are present (Fig. 16), the Pd Auger intensity (Fig. 16b) is very low. The O and S Auger maps (Fig. 16c and 16d respectively) confirm the presence of an oxide inclusion. The Pd Auger intensity here is not as uniform as in Figure 15b.

Similarly, when aqueous $HgCl_2$ is reacted with sulphides, the Hg is not reduced to the metal, but forms Hg(II) adsorbed species on the surfaces (Hyland and Bancroft, 1990b). The distribution of adsorbed species as shown by the Auger maps for cleaved PbS surfaces is somewhat similar to that shown in Figures 15 and 16. After one hour of reaction with a 100 ppm Hg solution of $HgCl_2$, sub-monolayer coverage of Hg and Cl is evident. Already, a Pb depleted sulphide surface is evident by the much lower intensity of the Pb image compared to the S image. Some areas of the surface do have higher levels of mercury than others, indicating that sorption occurs to a greater extent in some regions.

538

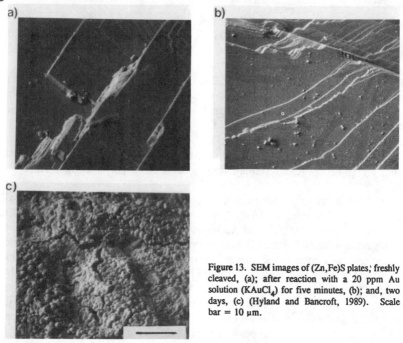

Figure 13. SEM images of (Zn,Fe)S plates, freshly cleaved, (a); after reaction with a 20 ppm Au solution (KAuCl₄) for five minutes, (b); and, two days, (c) (Hyland and Bancroft, 1989). Scale bar = 10 µm.

Figure 14 (below). Auger maps of cleaved PbS plates after 10 hours in Na₃Au(S₂O₃)₂ solutions (20 ppm Au). Secondary electron images of two plates are shown in (a) and (c), while the Au Auger map of (a) and (c) are shown in (b) and (d) respectively. In some regions, (a), gold was visible as a film on the surface. (Hyland, 1989)

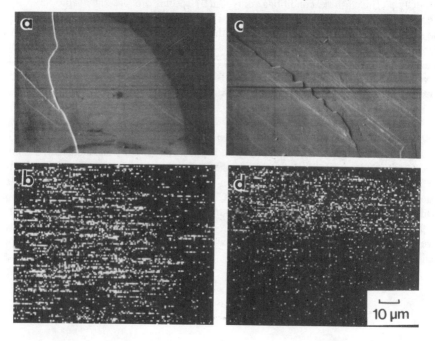

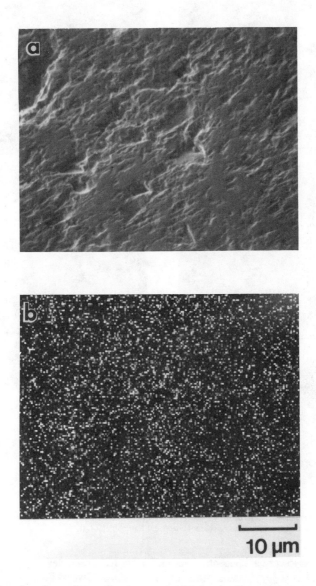

Figure 15. Secondary electron image, (a), of a reacted polished pyrite plate with only adsorbed Pd(II) (Pd A-C), and the corresponding Pd Auger map, (b), showing the uniformity of palladium distribution (Hyland and Bancroft, 1990a).

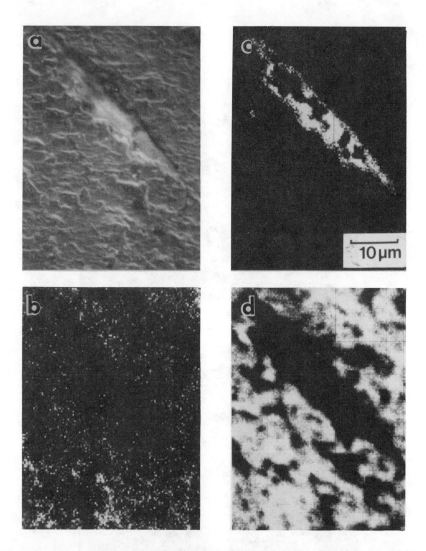

Figure 16. Secondary electron image (a) of a reacted polished pyrite with a silicate inclusion and the corresponding Auger maps: (b) Pd; (c) O; and (d) S. Pd is not detected within the inclusion, suggesting palladium adsorption is specific to the sulphide (Hyland and Bancroft, 1990a).

In contrast, at lower Cl⁻ concentrations and longer reaction times, Pd metal is formed and its distribution is <u>very</u> nonuniform. Figure 17 shows several interesting features at varying SEM resolutions. The Pd reduction is much slower than the Au reduction, and this leads to beautiful well-formed 1-10 μm Pd crystals (Fig. 17b) which cluster as shown in the poorer resolution photograph Figure 17a. Even smaller crystals on other parts of the surface are found growing preferentially along edges (Fig. 17c), pits (Fig. 17d and e), and fracture sites (Fig. 17f). Clearly, such sites on the surface are preferred sites for Pd(0) reduction.

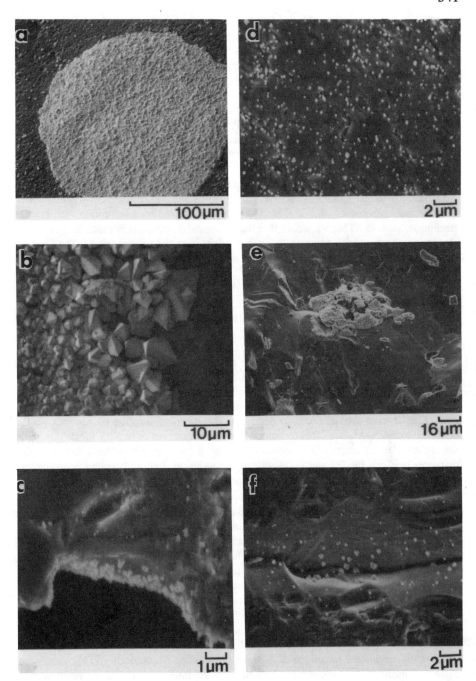

Figure 17. Secondary electron images of polished pyrite plates with Pd(0) crystals. The well formed crystals shown in (b), tend to cluster as in (a); and form preferentially at edges (c), pits (d), (e); and fracture sites (f) (Hyland et al., 1990b).

The morphology of the Pd metal varies with the sulphide. For example, the SEM images (Fig. 18a-d) and Pd Auger map (Fig. 18e) of freshly fractured and reacted galena plates show the expected corrosion of the surface as a result of Pd leaching. The Pd metal on galena takes the form of small spherical modules, usually less than 1 μm in diameter in contrast to larger pyramidal crystals on pyrite.

Finally, we have recently observed very different crystallization behavior for pyrite plates reacted with AgNO₃ solutions (Fig. 19) (Scaini and Bancroft, unpublished). The Ag metal crystallizes very differently than seen for the Au and Pd, and layered, filigree formation is apparent.

The SEM and Auger results do answer part of question 4 in the INTRODUCTION. Adsorption of metal species is relatively uniform on the Auger scale of ~ 1 μm. It is quite probable, however, that if these systems were studied by a technique with atomic resolution, such as STM, we would find that the metal distribution is much less uniform. Reduction occurs at high energy sites such as edges, pits and fracture sites. The form of the metal growth varies greatly depending on both the metal reduced and the sulphide substrate.

Other techniques for obtaining surface information

A number of other techniques have been useful, and should be useful in the future, for obtaining surface information on these reactions, and we will comment briefly on these techniques. Extended X-ray Absorption Fine Structure (EXAFS) has become an invaluable technique for determining bond distances in crystalline and amorphous materials, and bond distances for adsorbates on surfaces. Recently, Chisholm-Brause et al. (1990) have used EXAFS to determine the mode of sorption of aqueous Pb(II) complexes onto γ-Al₂O₃. This technique has two advantages over XPS for determining the chemical environment of the sorbed Pb. First, it provides direct measurement of the distances between the Pb and neighbouring atoms in the first and second coordination shells, and numbers and identities of the atoms in these shells. Second, it can often be applied in-situ, while the aqueous γ-Al₂O₃ is still in contact with the aqueous Pb solution (Chisholm-Brause et al., 1990). For example, the EXAFS spectra of sorbed lead on γ-Al₂O₃ yielded one oxygen atom at 2.23 Å and two oxygens at 2.46 Å. Also, there was one second-neighbour lead atom at 3.45 Å and one second-neighbour aluminum atom at 3.72 Å. This technique could be very useful for looking both at the heavy metal and sulphur EXAFS. However, if more than one metal species is present (as for $PdCl_4^{2-}$ adsorption onto sulphides), the EXAFS spectra will not be resolvable for the different species. XPS would have to be used in conjunction with EXAFS to insure that only one metal species is present on the surface.

For determining the site(s) where the initial metal atoms are absorbed and/or reduced, the recently developed technique of Scanning Tunneling Microscopy (STM) should be extremely useful (see the chapter by Hochella, this volume). This technique yields atomic resolution, and it should be possible to image sorbed/reduced heavy metal atoms on sulphides. Already Hochella et al. (1989) and Eggleston and Hochella (1990) have obtained atomic images of PbS and FeS₂, although they are not easy to interpret. In the STM images of galena {100} surfaces, atomic vacancies have been observed, and these may be the sites where initial metal nucleation takes place in our aqueous reaction.

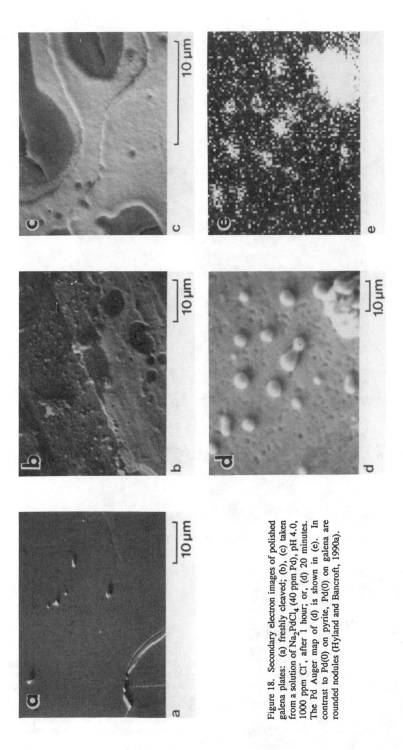

Figure 18. Secondary electron images of polished galena plates: (a) freshly cleaved; (b), (c) taken from a solution of Na$_2$PdCl$_4$ (40 ppm Pd), pH 4.0, 1000 ppm Cl$^-$, after 1 hour; or, (d) 20 minutes. The Pd Auger map of (d) is shown in (e). In contrast to Pd(0) on pyrite, Pd(0) on galena are rounded nodules (Hyland and Bancroft, 1990a).

544

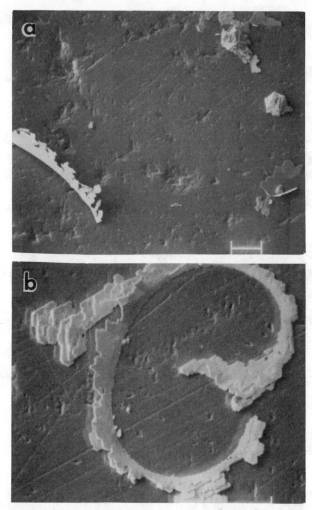

Figure 19. Secondary electron images of polished pyrite plates reacted with a 20 ppm Ag solution of AgNO$_3$ for 48 hours. Scale bars = 10 μm (Scaini and Bancroft, unpublished).

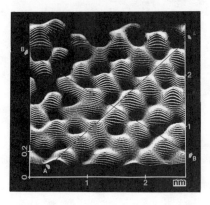

Figure 20. STM image of a fresh pyrite fracture surface taken at ~ 19.5 mV sample bias under oil. The frequency of peaks along the line labeled "A" is about 3.9 Å, and the frequency of peaks along the line labeled "B" is about 5.5 Å; within drift and calibration error this arrangement corresponds to that expected for a {101} surface of pyrite if the peaks correspond to Fe sites (Eggleston and Hochella, 1990).

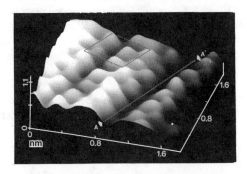

Figure 21. STM image of pyrite {100} growth surface taken at -42.4 mV sample bias under oil. This image was taken in constant current mode; tip height is displayed. The surface structure is arranged in "ranges" and "valleys" parallel to the line marked A-A´, which is parallel to a principal cystallographic axis. A 5.4 × 5.4 Å surface cell, as expected for termination of the bulk pyrite structure, is marked. However, the atomic arrangement does not correspond to a simple termination of the pyrite structure (Eggleston and Hochella, 1990).

Figures 20 and 21 shows STM images of pyrite fracture surfaces (Eggleston and Hochella, 1990). Because pyrite fractures conchoidally, there is no a priori knowledge of the crystallographic orientation of the surface. However, the frequency of peaks along the line labelled A in Figure 20 is 3.9 Å, and along B is 5.5 Å. Within error, these distances correspond to those expected for Fe atoms lying on a {101} surface of pyrite.

Figure 21 is an image obtained from a growth surface of pyrite which has been exposed to air for several years (Eggleston and Hochella, 1990). A square 5.4 × 5.4 Å (the size expected for unreconstructed {100} pyrite) is marked. The positions of peaks in the image do not correspond to the expected atomic position for a termination of the pyrite structure on {100}. They may well be imaging both S and Fe atoms, or imaging Fe atoms in an Fe_2O_3-like surface alteration product. Surface techniques are required to monitor the surface composition, and more development of STM is required to interpret images unambiguously. However, there can be little doubt that STM will be the method to look at the initial stages of sorption/reduction.

Other more classical methods combined with surface science techniques should be useful for determining the chemistry of these reactions. Desorption studies can be very helpful for determining the chemical nature of the adsorbed species - especially when a number of different metal species are on the surface eg Pd A, Pd B and Pd C in the reaction of $PdCl_4^{2-}$ with sulphides. An example of the qualitative use of desorption studies is provided by the desorption of adsorbed Hg on sulphides (Hyland, 1989; Hyland et al., 1990b). In these studies, sulphide plates were immersed in 100 ppm Hg solutions of $HgCl_2$ at ~pH 2 for several hours. This resulted in relatively uniform surface coverage of Hg with atomic ratios on PbS, for example, of Cl:Hg:Pd:S of 0.7:1.4:0.6:1. A detailed XPS analysis showed that there were two or three different Hg species on the surface. The plates were them immersed in 0.1 M or 1 M solutions of NaCl, Na_2SO_3, $Na_2S_2O_3$ and NaCN. The Hg:S atomic ratios were obtained by XPS using the Hg 4f and S 2p lines before and after reaction with the above solutions. The preliminary results

(Fig. 22) on pyrrhotite showed that the different ligands removed varying amounts of Hg in the order $Cl^- < SO_3^{2-} < S_2O_3^{2-} < CN^-$. Further more detailed results with just Cl^-, $S_2O_3^{2-}$ and CN^- ligands with Hg on FeS_2 and PbS plates showed the same trend. In this case, Cl^- removed little or no Hg; $S_2O_3^{2-}$ removed about one-half of the Hg, and all the Cl^-; and CN^- removed all Hg from PbS and >90% of the Hg from FeS_2 in less than 10 minutes.

The desorbing strengths of CN^-, $S_2O_3^{2-}$ and Cl^- will be proportional to their Hg^{2+} complexing ability. This can be estimated by calculating the standard free energies ΔG° of the complexation reactions,

$$Hg^{2+} + aL^{b-} \longrightarrow HgL_a^{2-ba} , \qquad (11)$$

where a represents the number of ligands involved, and b the charge. ΔG° for the formation of $HgCl_4^{2-}$, $Hg(S_2O_3)_3^{4-}$, and $Hg(CN)_4^{2-}$ are -86, -175 and -236 kJ/mol, respectively (Hyland, 1989). Cyanide then is expected to be the strongest desorbing agent, chloride the weakest, and thiosulphate intermediate - as seen experimentally. The desorption studies show immediately that there were at least two distinct Hg species on the surface using the XPS results and atomic ratios. Hyland et al. (1990b) showed that $S_2O_3^{2-}$ removes two Hg species in which Hg is bound to chloride, (one is $Hg_3S_2Cl_2$), while CN^- removes the sorbed HgS species as well. By using ligands, whose complexing strength varies over a smaller range, we feel that the desorption studies should be useful for determining the upper and lower limits of the free energy of formation of the surface species ($\Delta G^\circ_{f,surface}$), and thus provide important information regarding the nature of the surface species. In this method, $\Delta G^\circ_{f,surface}$ is bracketed by assuming equilibrium conditions for the reactions of surface species with ligands which do, and do not, desorb them. The results are not given here because lack of sufficient data did not allow us to narrowly delineate the upper and lower limits. The method however, is described in Hyland (1989). In addition, desorption studies should be generally useful for confirming that only one species is adsorbed on a surface; so that spectroscopic studies, such as EXAFS, are not unduly complicated by spectral overlap.

Finally as outlined in the REVIEW section of this chapter, many groups have used electrochemical methods to study the reaction of heavy metals on sulphides, but only the very recent studies of Hiskey et al. (1987), Hiskey and Pritzker (1988), and Buckley et al. (1989) have combined electrochemical and surface science measurements to look at the reactions of heavy metal ions with pyrite. In these studies, Ag_2O, Ag_2S and Ag metal were detected under different conditions. It is our hope that the $AuCl_4^-$ and $PdCl_4^{2-}$ reactions can be studied in greater detail by combining electrochemistry with surface techniques. By varying the sample potentials, we hope to stabilize different Au or Pd intermediates or stable species on the surface and study these using XPS and Raman spectroscopies.

Summary of mechanisms

A considerable body of chemical and physical information has been obtained for the sulphide reactions, and a number of workers have noted (e.g., Jean and Bancroft, 1985) the similarity of these reactions with what is often termed electrocrystallization. The steps

547

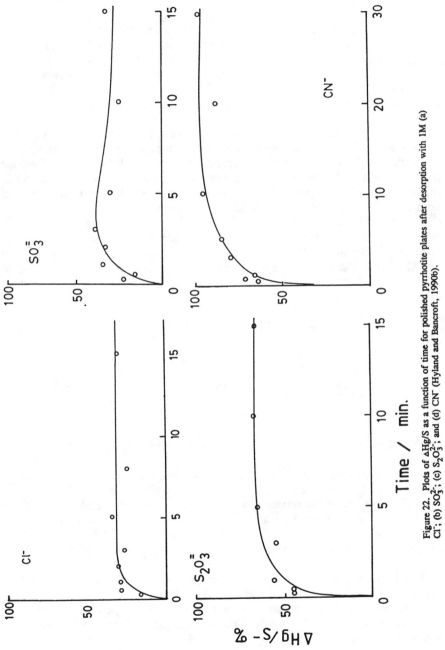

Figure 22. Plots of ΔHg/S as a function of time for polished pyrrhotite plates after desorption with 1M (a) Cl⁻; (b) SO₃²⁻; (c) S₂O₃²⁻; and (d) CN⁻ (Hyland and Bancroft, 1990b).

548

involved in the reactions on semiconductor sulphides are very similar to
those involved in electrocrystallization (Fig. 23), and involve a number of steps
(Greef et al., 1985).

1) diffusion of ions in solution to the sulphide surface with sorption;
2) electron transfer accompanied by partial or complete loss of ligands, resulting in so-
   called ad-atoms;
3) surface diffusion of ad-atoms, with subsequent clustering of ad-atoms to form critical
   nuclei;
4) development of crystallographic and morphological characteristics of the metal deposit

With regard to the first diffusion and sorption step, the mechanism can be at least
partially deduced. Our results have shown that the surface sulphides are necessary to
initiate the adsorption/reduction of metal species. For example, the amount of adsorption
is always much smaller on oxides or silicates than sulphides, and we have proven that
sulphide is the reducing agent. Therefore, we have proposed a two step mechanism (like
steps 1 and 2 above). The first step is the adsorption of metal complex onto the sulphide
(Fig. 24a). For the reactions of $AuCl_4^-$ on sulphides, reduction is so fast that no
intermediates in the electron transfer stage (Fig. 24b) are observed. However, for $PdCl_4^{2-}$
, $Au(S_2O_3)_2^{3-}$ and $Au(CN)_2Br_2^-$ non-reduced or partially reduced intermediates were seen
by XPS. The fact that $Au(CN)_2^-$ is not adsorbed indicates that at least one unstable ligand
must be present in order to vacate a site for the surface bond. The reaction can be written
as

$$ML_4 + Sf \longrightarrow L_3MSf + L \quad \text{(Fig. 24A)} \qquad (12)$$

or

$$ML_4 + 2Sf \longrightarrow \overset{\displaystyle Sf}{\underset{\displaystyle Sf}{\overset{\displaystyle /}{\underset{\displaystyle \backslash}{L_2M}}}} + 2L \;, \qquad (13)$$

where: M = Au, Pd, etc., Sf = surface site, L = ligand.

Ligand substitution on Au(III) or Pd(II) almost always takes place by an associative
mechanism (SN_2) (Hall et al., 1976). The same type of mechanism could occur with
the sulphide groups of the surface acting as a ligand. For species like $Au(CN)_2^-$ the
AuCN bond is too stable and cannot be displaced by the surface, which explains why
no adsorption occurred for $Au(CN)_2^-$. The postulate of Renders and Seward (1989) that
Au(HS) is the adsorbed species rather than $Au(HS)_2^-$ is also consistent with this argument.

Metal sulphides are often observed on the surface, (e.g., PdS or PdS_2); but they
appear to be present as a sorbed surface layer rather than a precipitate despite the
predicted thermodynamic favorability. Equations 14 and 15 describe the reactions
(Hyland and Bancroft, 1990a) between FeS_2 and aqueous Pd(II) species resulting in the
formation of PdS_2 and PdS:

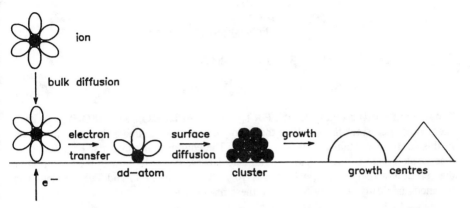

Figure 23. Some of the steps involved in the electrocrystallization of a metal on a substrate of a different material such as carbon (Greef et al., 1985).

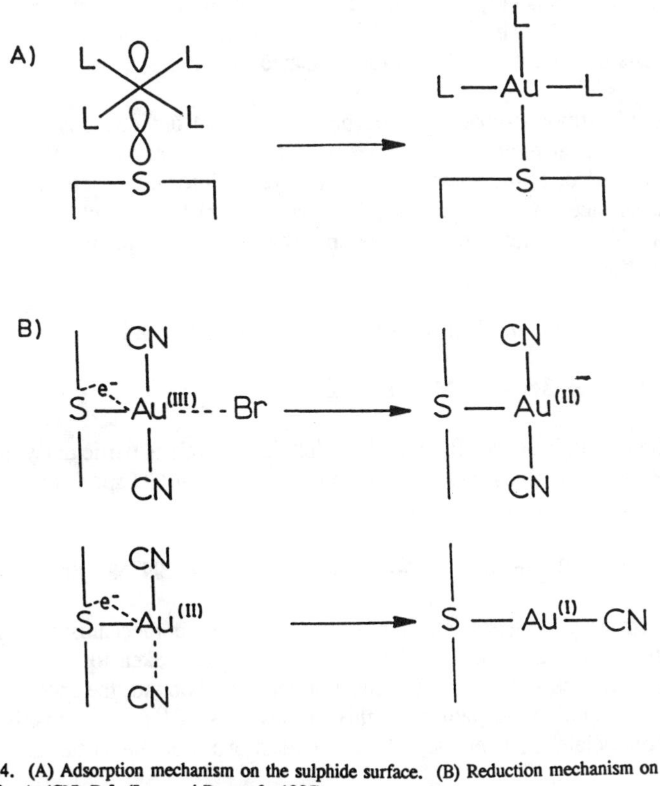

A)

B)

Figure 24. (A) Adsorption mechanism on the sulphide surface. (B) Reduction mechanism on the sulphide surface for $Au(CN)_2Br_2^-$ (Jean and Bancroft, 1985).

$$PdX_m^{2-m} + FeS_2 \longrightarrow PdS_2(s) + Fe^{2+} + mX^- \quad ; \tag{14}$$

$$PdX_m^{2-m} + FeS_2 + 3/2\ O_2(g) + H_2O \longrightarrow$$
$$PdS(s) + Fe^{2+} + SO_4^{2-} + mX^- + 2H^+ \tag{15}$$

The aqueous Pd(II) species, $PdCl_4^{2-}$, $PdCl_3(H_2O)^-$, $PdCl_2(H_2O)_2$ and $Pd(OH)_2(aq)$, have been considered since these will be the major solution species under the conditions used in this study (Mountain and Wood, 1988). The equilibrium constants, K_{eq}, for these reactions (Hyland et al., 1990a) were all calculated to be $> 10^{16}$, with the exception of the reaction with $Pd(OH)_2$, which was 10^2. Therefore, PdS_2 and PdS formation is thermodynamically strongly favored in these reactions.

The formation of palladium sulphide is limited to no more than a few monolayers on the surface, since the signal from the underlying pyrite is still very strong. As well, we would expect to detect a decrease in the pyritic iron:sulphur surface ratios if, as predicted from Equations 14 and 15, iron is leached from the surface. However the ratios for reacted and unreacted plates are the same. Further, Pd atomic absorption analysis (AA) of $PdCl_4^{2-}$ solutions reacted for two weeks with $FeS_2^=$ powders (160 ppm Pd, 4000 ppm Cl$^-$, pH 4.0) showed that at equilibrium $>90\%$ of the Pd remained in solution. Since Equations (14) and (15) are not limited by thermodynamics, they must be under kinetic control. The fact that palladium sulphides do form on other sulphide minerals suggests that the kinetic limitation is related to pyrite.

The initial adsorption probably occurs preferentially at defects, pits or edges, but STM studies and much higher resolution Auger studies are required to pin-point the adsorption sites. Also, at the present time, we have no information on the role of SH groups on the surface. Certainly, it appears that a metal-OH complex adsorbs more readily than a M-X complex, and it is possible that S-H-M interactions are important before the S-M bond is formed.

The second step in the mechanism is the reduction of the metal,

$$L_3M - Sf + 3e \longrightarrow M - Sf + 3L^- \quad . \tag{16}$$

The reaction between $KAu(CN)_2Br_2$ and the sulphide minerals is particularly interesting and is depicted in Figure 24b. We saw that in that case, a Au-CN species was found on the surface. The reaction can be written as

$$Au^{III}(CN)_2Br_2^- + Sf + 2e \longrightarrow (CN)Au^I - Sf + CN^- + 2Br^- \quad . \tag{17}$$

This reaction strongly suggests that a reductive elimination mechanism is in effect. Indeed the reaction indicates that one Au-CN bond must be broken for the reduction to occur. If reduction occurred without breakage of one Au-CN bond, the species $Au(CN)_2^-$ would be formed. Due to the stability of this species ($K \simeq 10^{39}$), it is unlikely that this species would dissociate at a later stage. Thus the reactions could be written as (Fig. 24b)

$$(CN)_2BrAu^{III} - Sf + 1e \longrightarrow (CN)_2Au^{II} - Sf + Br^- \quad ; \tag{18}$$

$$(CN)_2Au^{II} - Sf + 1e \longrightarrow (CN)Au^{I} - Sf + CN^- \quad . \qquad (19)$$

The reduction of Au(III) and Pd(II) is known to proceed by one- or two-electron steps (Hall et al., 1976).

Why are some of the complexes reduced and not others? From a thermodynamic perspective, this is readily rationalized by the E^o values for the reduction of the metal complex to the metal. For four of the reactions considered here, the E^o values are (Skibsted and Bjerrum, 1977):

$$AuCl_4^- + 3e \longrightarrow Au(0) + 4Cl^- \qquad E^o = + 1.00 \text{ V}$$

$$PdCl_4^{2-} + 2e \longrightarrow Pd(0) + 4Cl^- \qquad E^o = + 1.5 \text{ V}$$

$$Au(S_2O_3)_2^{3-} + e \longrightarrow Au(0) + 2 S_2O_3^{2-} \quad E^o = + 0.15 \text{ V}$$

$$Au(CN)_2^- + e \longrightarrow Au(0) + 2 CN^- \qquad E^o = - \; 0.48 \text{ V}$$

We have shown that the reducing agent is sulphide, and the overall oxidation reactions are:

$$S^{2-} + 4 H_2O \longrightarrow SO_4^{2-} + 8H^+ + 8e \quad E^o = \; -0.150 \text{ V},$$

$$\text{or} \quad S^{2-} \longrightarrow S^o + 2e \qquad\qquad E^o = \; +0.48 \text{ V}.$$

Obviously, $AuCl_4^-$ and $PdCl_4^{2-}$ should be very easily reduced, while $Au(CN)_2^-$ will be reduced with difficulty. The very different E^o values, for the metal complexes, are related to the metal-ligand bond strength which increases from Cl^- to CN^-. Similarly, Watt and Cunningham (1963) found from electrochemical measurements that the tendency for electrodeposition for aqueous square planar complexes of Pt(II) and Pd(II) and Au(III) decreases as the metal-ligand bond strengths increases.

However, although there is qualitative agreement between the ease of reduction, and free energies of reaction, there are a number of experimental results that show that kinetic factors are just as important. For example, at high Cl^- concentration, $PdCl_4^{2-}$ is stabilized rather than other $PdCl_x(H_2O)_{4-x}^{2-x}$ species. We showed earlier that $PdCl_4^{2-}$ is not reduced on sulphides yet the ΔE^o values for reduction of $PdCl_4^{2-}$ with S^{2-} are very favorable thermodynamically. Or if we consider that the reduction of $PdCl_x(H_2O)_{4-x}^{2-x}$ occurs at the expense of pyrite oxidation to give sulphate,

$$7.5 \; PdCl_x (H_2O)_{4-x}^{2-x} + FeS_2 + (4+x)H_2O \longrightarrow$$

$$7.5 \; Pd(0) + Fe^{3+} + 2SO_4^{2-} + 7.5 \; x \; Cl^- + 16 \; H^+,$$

then calculations (Hyland and Bancroft, 1990a) show that the reduction is favored for all $PdCl_x(H_2O)_{4-x}^{2-x}$. For example, at $[Cl^-] = 1.48$ M and pH = 3.0, the resulting $[Fe^{3+}]/[PdCl_x(H_2O)_{4-x}^{2-x}]$ will be $> 10^7$. Thus, there should be almost no $PdCl_x(H_2O)_{4-x}^{2-x}$

552

unreduced. Since experimental evidence shows that reduction is suppressed at much lower chloride concentrations than thermodynamics would predict, the effect of changing chloride must be related to kinetic parameters. These kinetic parameters certainly are not clear at the present time. Nor is it clear why palladium sulfides have to form on the surface before reduction to Pd metal occurs to an appreciable extent. Both PdS and PdS_2 are thermodynamically stable under our laboratory conditions.

Turning to the third and fourth stages in the mechanism in Figure 23 - surface diffusion and growth - the SEM and Auger results in the previous section of this paper certainly show that these processes are important in forming aggregates and crystals. Both processes have been modelled extensively for electrocrystallization (Greef et al., 1985). These results show that clustering should occur along defects, edges, and kinks - and this is what is found in many of our studies and in natural specimens (next section).

This discussion shows that we understand the adsorption/reduction reactions on sulphides in a qualitative manner. However, there is a great deal more to do: to identify the initial adsorption intermediate(s) and the S^{2-} sites on which adsorption takes place; and to understand the kinetic parameters which control the reduction, surface diffusion, and crystal growth. Some of the techniques mentioned in the previous section should be useful, and many more metal complexes need to be studied after reaction with a number of sulphides using existing and novel techniques. This is an exciting new area of study, and we have barely scratched the surface!

GEOCHEMICAL IMPLICATIONS OF THE LABORATORY STUDIES

As outlined in the INTRODUCTION, there has been little mention until recently of the importance of adsorption/reduction reactions for concentrating heavy metals. However, in the last few years there has been substantial evidence that these surface reactions are indeed important geochemically, and we discuss some of this evidence for natural gold deposits below.

The nature of Au in natural pyrites from SEM, SIMS and Mössbauer studies

Excellent reviews of the nature of gold in natural sulphides have been very recently provided by Cabri et al. (1989) and Cook and Chryssoulis (1990). The techniques normally used for detecting Au are bulk in nature and fail to determine the size of the Au metal particles, or whether the Au substitutes in the sulphide lattice, or is adsorbed. In order to understand the nature of Au deposits, and to process them successfully, we must obtain a better understanding of the nature of the Au in natural sulphides.

Recent spectroscopic studies show that Au can be present as very small discrete mineral inclusions or chemically combined with the sulphide. [197]Au Mössbauer spectroscopy appears to be the most promising, bulk technique for distinguishing chemically bound Au from metallic Au (Marion et al., 1986; Wagner et al., 1986; Cathelineau et al., 1988). Fig. 25 shows the Mossbauer spectrum of Au in arsenopyrites. Mössbauer spectroscopy is a very useful method for distinguishing oxidation states and chemical environments for many elements such as Fe, Sn and Au (Bancroft, 1973) using

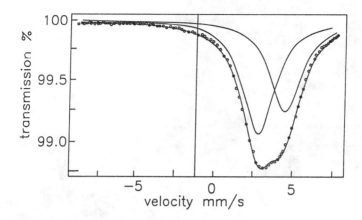

Figure 25. Mössbauer spectrum of Au-bearing arsenopyrite. The vertical line indicates the position of pure metallic gold absorption peak (Cathelineau et al., 1988).

the isomer shift (the shift of the peaks on the x axis) and the quadrupole splitting (the splitting of the two peaks). In the case of [197]Au Mössbauer, metallic Au with an isomer shift of -1.23 mm/sec; (vertical line in Figure 25) and no quadrupole splitting (Wagner et al., 1986; Cathelineau et al., 1988) can be readily distinguished from the doublet with an isomer shift (the mean of the two peaks) of close to +4 mm/sec. Most of the arsenopyrites gave more of the Au doublet at +4 mm/sec than the Au metal singlet. The doublet is due to Au which is chemically bonded, and suggested (Cathelineau et al., 1988) the presence of Au-X (X = As) bonds due to either solid solution within the arsenopyrite lattice or to inclusions having too small sizes to be detected by conventional methods. Both electron probe measurements (detection limit 200 ppm) and Secondary Ion Mass Spectrometry (SIMS) results (detection limit < 0.5 ppm) show a general correlation of Au with As, suggesting a Au-As complex or that the substitution of gold into that structure is facilitated by As (Cook and Chryssoulis, 1990).

It is generally assumed that the chemically bonded Au coprecipitated with the sulphide. However, Renders and Seward's recent work on the adsorption of Au(HS) complexes on As_2S_3 and Sb_2S_3 (Seward, 1984; Renders and Seward, 1989) strongly suggests that adsorption of the Au complex onto growing sulphides could account for the enhanced concentrations of Au in many hydrothermal deposits. Renders and Seward (1989) mention that a [197]Au Mössbauer spectrum of As_2S_3 saturated with AuHS detected only Au(I) - S bonds. Their result is entirely consistent with the adsorption mechanism given in the results in this chapter and in Figure 23: most of the adsorbed metal species that we have observed in the laboratory are bonded directly to the surface sulphide groups. Obviously, it is now critical to compare the Mössbauer spectra of adsorbed laboratory samples with those of natural samples. Mössbauer spectroscopy should be able to detect the difference between Au-As and Au-S bonds.

Very recently, several studies have reported Au metal on sulphides which could be observed with transmission electron microscopy (TEM) (Bakken et al., 1989) or SEM

554

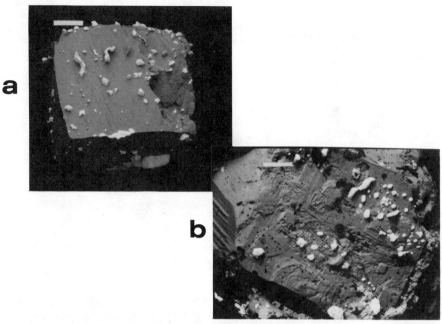

Figure 26. Back-scattered electron (BSE) images of gold and telluride on pyrite. (a), Gold and altaite (PbTe) on pyrite crystal faces and edges, Lennox Mine. Scale bar = 100 μm. (b), Gold on hydrothermally etched pyrite, Golden Valley Mine. Scale bar = 100 μm (Starling et al., 1989).

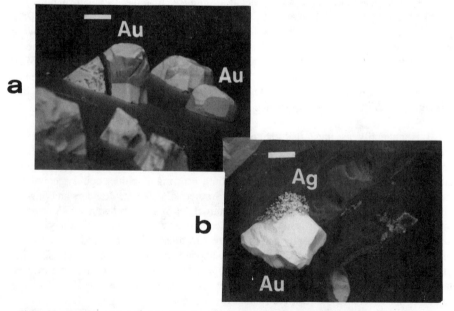

Figure 27. Precious metals and altaite (PbTe) on pyrite, Lennox Mine. (a), Higher magnification image of the right-central portion of Figure 1a; gold and altaite grains displaying planar basal surfaces adjacent to, the pyrite; BSE image, scale bar 10 μm. (b), Gold and filigree silver on pyrite; BSE image, scale bar = 20 μm (Starling et al., 1989).

(Clough and Craw, 1989; Starling et al., 1989). The form and distribution of the metal on natural sulphides is not unlike what we have seen in the laboratory studies, and suggests that the metal was deposited by adsorption/reduction rather than precipitation or coprecipitation. For example, Bakken et al. (1989) characterized colloid gold particles of <200 Å in diameter on pyrite in Carlin-type ores. Perhaps the best evidence to date of adsorption/reduction reactions on sulphides is given recently by Starling et al. (1989). They studied individual pyrite grains separated from four different African load gold deposits using optical and scanning electron microscopy, and looked at the textural relationships of late gold on the surface of pyrite. In the backscattered electron images (Figs. 26-27), gold metal occurs as native grains adhering to the pyrite grains (Fig. 26a,b), and the Au is often concentrated along crystal edges (Fig. 26a) or pits. The habit of native metal varies from good crystalline forms (Fig. 27a) of diameter ~10 μm to subhedral grains (Fig. 27b) and to subspherical and domal grains (Fig. 26). Other pyrite crystals exhibit a preferential concentration of gold on fracture surfaces. Notice also in Figure 27b, that the Ag occurs as spongeform aggregates and delicate filigree growths in contact with the pyrite. Their observation indicate that a "significant proportion" of the gold in all four African deposits was formed after crystallization of the pyrite and that adsorption/reduction was the important mechanism for Au and Ag formation.

ACKNOWLEDGEMENTS

We are very grateful to the N.S. McIntyre and the staff at Surface Science Western, University of Western Ontario, London, Canada, for their help and cooperation, and to J.R. Mycroft for a critical reading of this manuscript.

REFERENCES

Bacon, M.P., Brewer, P.G., Spencer, D.W., Murray, J.W. and Goodland, J. (1980) Lead 210, polonium-210, manganese and iron in the Cariaco Trench, Deep Sea Res. 27A, 119-135.

Bakken, B.M., Hochella, M.F. Jr., Marshall, A.F. and Turner, A.M. (1989) High resolution microscopy of gold in unoxidized ore from the Carlin mine, Nevada. Econ. Geol. 84, 171-179.

Bancroft, G.M. (1973) In Introduction to Mossbauer Spectroscopy for Inorganic Chemists and Geochemists. Halstead Press, London.

Bancroft, G.M., Brown, J.R. and Fyfe, W.S. (1979) Advances in, and applications of, X-ray photoelectron spectroscopy (ESCA) in Mineralogy and Geochemistry. Chem. Geol. 17, 227-243.

Bancroft, G.M. and Jean, G.E. (1982) Gold deposition at low temperatures on sulphide minerals. Nature 298, 730-731.

Berkowitz, J. (1979) Photoabsorption, photoionization and photoelectron spectroscopy. Academic Press, New York.

Bozek, J.D., Cutler, J.N., Bancroft, G.M., Coatsworth, L.L., Tan, K.H., Yang, D.S. and Cavell, R.G. (1990) High resolution molecular gas phase photoelectron spectra of core levels using synchrotron radiation. Chem. Phys. Lett. 165, 1-5.

Brown, J.R., Bancroft, G.M., Fyfe, W.S. and McLean, R.A.N. (1979) Mercury removal from water by iron sulphide minerals. An Electron Spectroscopy for Chemical Analysis (ESCA) Study. Environ. Sci. Technol. 13, 1142-1144.

Buckley, A.N., Hamilton, I.C. and Woods, R. (1987) An investigation of the sulphur(-II)/sulphur(0) system on gold electrodes. J. Electroanal. Chem. 216, 213-227.

556

Buckley, A.N., Wouterlood, H.J., Cartwright, P.S. and Gillard, R.D. (1988) Core electron binding energies of platinum and rhodium polysulphides. Inorg. Chim. Acta 143, 77-80.

Buckley, A.N., Wouterlood, H.J. and Woods, R. (1989) The interaction of pyrite with solutions containing silver ions. J. Applied Electrochem. 19, 744-757.

Cabri, L.J., Chryssoulis, S.L., DeVilhers, J.P.R., Laflamme, J.H.G., and Buseck, P.R. (1989) The nature of "invisible" gold in arsenopyrite. Canadian Mineral. 27, 353-362.

Carlson, T.A. (1975) Photoelectron and Auger electron spectroscopy. Plenum Press, New York.

Cathelineau, M., Boiron, M.C., Holliger, P., Marion, P. (1988) Gold-rich arsenopyrites, crystal chemistry, gold location and state, physical and chemical conditions of crystallization. Bicentennial Gold 88, 235-240, Extended Abstracts, Morphet Press, Melbourne.

Chisholm-Brause, C.J., Hayes, K.F., Roe, A.L., Brown, G.E. Jr., Parks, G.A. and Leckie, J.O. (1990) Spectroscopic investigation of Pb(II) complexes at the γ-Al$_2$O$_3$/water interface. Geochim. Cosmochim. Acta, in press.

Clough, D.M., Craw, D (1989) Authigenic Gold-Marcasite Association: Evidence for nugget growth by chemical accretion in fluvial gravels, Southland, New Zealand. Econ. Geo. 84, 953-958.

Cook, N.J., and Chryssoulis, S.L. (1990) Concentration's of "invisible gold" in the common sulfides. Canadian Mineral. 28, 1-16.

Cousins, C.A. (1973) Notes on the geochemistry of the platimun group elements. Geol. Soc. Amer. Trans. 76, 77-81.

Dyrssen, D., Hall, P., Haraldsson, C., Iverfeldt, A., and Westerlund, S. (1984) In Complexation of Trace Metals in Natural Waters. (eds. C.J.M. Kramer and J.C. Duinker), Martinius Nijhoff/Dr. W. Junk Publishers, The Hague, 239-245.

Eggleston, C.M. and Hochella, M.F. Jr. (1990) Scanning tunneling microscopy of sulphide surfaces. Geochim. Cosmochim. Acta 54, in press.

Fuchs, W.A. and Rose, A.W. (1974) The geochemical behavior of platinum and palladium in the weathering cycle in the Stillwater complex, Montana. Econ. Geol. 69, 332-346.

Fuerstenau, M.C., Miller, J.D. and Kuhn, M.C. (1985) Chemistry of flotation. Society of Mining Engineers, New York.

Greef, R., Peat, R., Peter, L.M., Pletcher, D. and Robinson, J. (1985) Instrumental methods in electrochemistry. Ellis Norwood Ltd., Chichester, U.K. and Halstead Press, New York.

Gupta, R.P., Tse, J.S. and Bancroft, G.M. (1980) Core level ligand field splitting in photoelectron spectra. Philos. Trans. Royal Soc. 293, 535-569.

Hall, A.J. and Satchel, D.P.N. (1976) The observation of an intermediate in the reaction between Tetrachloro-gold(III) and thiocyanate ions. J. Chem. Soc. Chem. Commun. 163-164.

Henley, R.W. (1973) Solubility of gold in hydrothermal chloride solutions. Chem. Geol. 11, 73-87.

Hiskey, J.B., Phule, P.P. and Pritzker, M.D. (1987) Studies on the effect of addition of silver on the direct oxidation of pyrite. Metallurgical Trans. 18B, 641-647.

Hiskey, J.B. and Pritzker, M.D. (1988) Electrochemical behavior of pyrite in sulfuric acid solutions containing silver ions. J. Appl. Electrochem. 18, 484-490.

Hochella, M.R. Jr., Harris, D.W. and Turner, A.M. (1986) Scanning Auger microscopy as a high resolution microprobe for geologic materials. Amer. Mineral. 71, 1247-1257.

Hochella, M.F. Jr., Eggleston, C.M., Elings, V.B., Parks, G.A., Brown, G.E. Jr, Wu, C.M. and Kjoller, K. (1989) Mineralogy in two dimensions: Scanning tunneling microscopy of semiconducting minerals with implications for geochemical reactivity. Amer. Mineral. 74, 1235-1248.

Hochella, M.F. Jr. (1988) Auger electron and X-ray photoelectron spectroscopies. Ch. 13 in Spectroscopic Methods in Mineralogy and Geology, Reviews in Mineralogy, 18, 573-637. Mineralogical Society of America

Hyland, M.M. (1989) X-ray photoelectron spectroscopic studies of the interaction of aqueous metal complexes with sulphide minerals, Ph.D. Thesis, University of Western Ontario, London, Ontario.

Hyland, M.M. and Bancroft, G.M. (1989) An XPS study of gold deposition at temperatures on sulphide minerals: Reducing agents. Geochim. Cosmochim. Acta 53, 367-372.

Hyland, M.M. and Bancroft, G.M. (1990a) Palladium sorption and reduction on sulphide mineral surfaces: An XPS and AES study. Geochim. Cosmochim. Acta 54, 117-130.

Hyland, M.M., Jean, G.E. and Bancroft, G.M. (1990b) XPS and AES studies of Hg(II) sorption and desorption reactions on sulphide minerals. Geochim. Cosmochim. Acta 54, in press.

James, R.O. and Parks, G.A. (1975) Adsorption of zinc(II) at the cinnabar (HgS)/H$_2$O interface. Amer. Inst. Chem. Eng. Symp. Series 150, 71, 157-164.

557

Jean, G.E. and Bancroft, G.M. (1985) An XPS and SEM study of gold deposition at low temperatures on sulphide mineral surfaces: Concentration by adsorption/reduction. Geochim. Cosmochim. Acta 49, 979-987.

Jean, G.E. and Bancroft, G.M. (1986) Heavy metal adsorption by sulphide mineral surfaces. Geochim. Cosmochim. Acta 50, 1455-1463.

Marion, P., Regnard, J.R. and Wagner, F.E. (1986) Study of the chemical state of gold in auriferous sulfides by [199]Au Mössbauer spectroscopy: first results. C. R. Acad. Sci. II 302, 571-574.

Marion, P., Boiron, M.C., Halliger, Ph, and Marion, Ph. (1988) Gold-rich arsenopyrites: crystal chemistry, gold location and state, physical and chemical condition of crystallization. Proceed. Bicentennial Gold 1988, p. 235-240.

McCallum, M.E., Louks, R.R., Carlson, R.R. Cooley, E.F. and Dourge, T.A. (1976) Platinum metals associated with hydrothermal copper ores of the New Rambler mine, Medicine Bow Mountains, Wyoming. Econ. Geol. 71, 1429-1450.

McGuire, E.J. (1972) Atomic M-shell Coster-Kronig, Auger and radiative rates and fluorescence yields for Ca-Th. Phys. Rev. A5, 1043-1047.

McGuire, E.J. (1974) Atomic N-shell Coster-Kronig, Auger and radiative rates and fluorescence yields for $38 \leq Z \leq 103$. Phys. Rev. A9, 1840-1851.

McKibben, M.A. and Barnes, H.L. (1986) Oxidation of pyrite in low temperature acidic solutions: Rate law and surface textures. Geochim. Cosmochim. Acta 50, 1509-1526.

Morse, J.W., Millero, F.J., Cornwell, J.C. and Rickard, D. (1987) The chemistry of the hydrogen sulfide and iron sulphide systems in natural waters. Earth-Science Rev. 24, 1-42.

Mountain, B.W. and Wood, S.A. (1988) Chemical controls on the solubility, transport and deposition of platinum and palladium in hydrothermal solutions: A thermodynamic approach. Econ. Geol. 83, 492-510.

Mycroft, J.R., Bancroft, G.M., McIntyre, N.S., Lorimer, J.W. and Hill, I.R. (1990) Detection of sulphur and polysulphides on electrochemically oxidized surfaces by X-ray photoelectron spectroscopy and Raman spectroscopy. J. Electroanalyt. Chem. and Interfacial. Electrochem., in press.

Nechayev, Ye.A. (1984) The effects of solution composition on the adsorption of gold(III) chloride complexes on haematite. Geochem. Int. 21, 87-93.

Nechayev, Ye.A. and Nikolenko, N.V. (1986a) Adsorption of gold(III) chloride complexes on alumina silica and Kaolin. Geochem. Int. 23, 32-37.

Nechayev, Ye.A. and Nikolenko, N.V. (1986b) Effects of surface charge on the adsorption of gold(III) chloride complexes on oxides. Geochem. Int. 23, 142-146.

Palmer, C. and Bastin, E.S. (1913) Metallic minerals as precipitants of silver and gold. Econ. Geology 8, 140-170.

Park, C.F., and McDiarmid, R.A. (1975) Ore Deposits. 3rd Edition, W.H. Freeman and Co., San Francisco 522 pages.

Renders, P.J. and Seward, T.M. (1989) The adsorption of thio gold(I) complexes by amorphous As_2S_3 and Sb_2S_3 at 25 and 90°C. Geochim. Cosmochim. Acta 53, 255-267.

Romberger, S.B. (1986) The solution chemistry of gold applied to the origin of hydrothermal deposits in Gold in the Western Shield: Montreal, Canadian Institute of Mining and Metallurgy Special Vol 38, (ed. L.A. Clark) 168-186.

Romberger, S.B. (1988) Geochemistry of gold in hydrothermal deposits: U.S. Geol. Surv. Bull. 1857-A, A9-A25.

Sakharova, M.S. Lobachera, I.K. (1967) Electrochemical study of the process of deposition of gold on sulfides. Geol. Rudn. Mest. 9, 45-55. Abstract translated - Econ. Geol. 64, 591-592.

Sakharova, M.S., Batrakova Yu, A., Ryakhovskaya, S.K. (1975) Investigation of electrochemical interactions between sulphides and (gold)-bearing solutions. Geochem. Int. 12, 84-89.

Sakharova, M.S., Batrakova Yu, A., Ryakhovskaya, S.K. (1976) The effect of anion composition of solution on coprecipitation of gold and silver on sulfides. Geochem. Int. 13, 160-166.

Sakharova, M.S. and Lobacheva, I.K. (1978) Microgalvanic systems involving sulfides and gold-bearing solutions, and characteristics of gold deposition. Geochem. Int. 15, 152-157.

Sakharova, M.S., Batrakova, Yu, A. and Pascikhova, T.V. (1980) Laboratory data on the effects of solution composition on the deposition of native silver. Geochem. Int. 30-36.

Sakharova, M.S., Batrakova, Yu, A., Ryakhovskaya, S.K. (1981) The effects of pH on the deposition of gold and silver from aqueous solution. Geochem. Int. 18, 28-34.

558

Scofield, J.H. (1976) Hartree-Slater subshell photoionization cross sections at 1254 and 1486 eV. J. Electron Spectrosc. Related Phenom. 8, 129-137.

Siegbahn, K., Nordling, C.N., Fahlman, A., Nordberg, R., Hamrin, K., Hedman, J., Johnsson, G., Bermark, T., Karlsson, S.E., Lindgren, I., Lindberg, B. (1967). ESCA: Atomic, molecular and solid state structure studied by means of electron spectroscopy. Almqvist and Wiksells: Uppsala, Sweden.

Seward, T.M. (1973) Thio complexes and the transport of gold in hydrothermal ore solutions. Geochim. Cosmochim. Acta 37, 379-399.

Seward, T.M. (1984) The transport and deposition of gold in hydrothermal systems. In Gold 82: The Geology Geochemistry and Genesis of Gold Deposits (ed. R.P. Foster) 165-181. A.A. Balkema Press.

Skey, W. (1871) On the capability of certain sulphides to form the negative pole of a galvanic circuit or battery; also on the electromotive power of metallic sulphides, Trans and Proc. New Zealand Institute Vol. III.

Skibsted, L.H. and Bjerrum, J. (1977) Studies on gold complexes III. The standard electrode potentials of Aqua gold ions. Acta Chem. Scandinavia A31, 155-156.

Starling, A., Gilligan, J.M., Carter, A.H.C., Foster, R.P., and Sanders, R.A. (1989) High temperature hydrothermal precipitation of precious metals on the surface of pyrite. Nature 340, 298-300.

Vaughan, D. ed. (1985) X-ray Data Booklet, Center for X-ray Optics, Lawrence Berkeley Laboratory, Livermore, Calif.

Wagner, F.E., Marion, Ph. and Regnard, J.R. (1986) Mössbauer study of the chemical state of gold in gold ores. GOLD 100. Preceedings of the International Conference on Gold. Vol. 2: Extractive Metallurgy of gold. Johannesburg, SAMM, 435-443.

Wagner, C.D., Riggs, W.M., Davis, L.E., Moulder, J.F. and (ed.) Muilenberg, G.E. (1979) Handbook of X-ray Photoelectron Spectroscopy. Perkin-Elmer Corp., Eden Prairie,

Watt, G.W. and Cunningham, J.A. (1963) Mechanism of electrodeposition from aqueous solutions of square planar complexes. J. Electrochem. Soc. 110, 716-723.

Yeh, J.J. and Lindau, I (1985) Subshell photoionization cross sections. Atomic Nuclea Data Tables 32, 1-155.

PHOTO-REDOX PROCESSES

AT THE MINERAL-WATER INTERFACE

INTRODUCTION

In this chapter, electron transfer reactions (redox processes) occurring at mineral - water interfaces that are either induced or assisted by light are considered. The possibility of using semiconducting minerals such as hematite or rutile in the conversion of solar energy into fuels (for example, by the splitting of water into H_2 and O_2) has resulted in intense research activity in this area over the last twenty years (Kalyanasundaram et al., 1986; Gratzel, 1988). More diverse applications of photochemical, heterogeneous redox processes are now apparent and include the selective synthesis of organic compounds at semiconductor surfaces (Fox, 1988; Pichat and Fox, 1988), the catalytic degradation of toxic organic and inorganic species in waters and wastes (Pelizzetti et al., 1988; Serpone et al., 1988a) and the catalysed precipitation of precious metals (Serpone et al., 1988b). It is also becoming increasingly clear that, in addition to the critical role of light in photosynthesis, light may induce transformations in natural systems that are important in the geochemical cycling of the elements. Many such processes are heterogeneous and involve the catalysis by or transformation of minerals (Waite and Morel, 1984a; McKnight et al., 1988; Sunda and Huntsman, 1988). For example, it now appears that in some marine waters, the solubility of iron, a vital nutrient to the primary producers, may be mediated by the light-assisted dissolution of iron oxides (Martin and Gordon, 1988; Martin and Fitzwater, 1988; Martin et al., 1990).

Every photochemical reaction is initiated by the absorption of radiation (the First Law of Photochemistry). Indeed, the Stark-Einstein law states that if a species (the chromophore) absorbs radiation, then one particle is excited for each quantum of radiation absorbed (Wayne, 1970). Photons of radiation of specific frequency v are associated with a fixed energy ε given by the relation:

$$\varepsilon = hv \quad , \tag{1}$$

where h is called Planck's constant. The excitation energy per mole of chromophore, E, is easily obtained by multiplying the molecular excitation energy, ε, by Avogadro's number, N; i.e.

$$E = Nhv = \frac{Nhc}{\lambda} \quad , \tag{2}$$

where c is the velocity of light and λ the wavelength of the exciting radiation. Thus, 300 nm radiation corresponds to approximately 400 kJ mol^{-1} or 4 eV while 700 nm radiation corresponds to approximately 180 kJ mol^{-1} or 1.8 eV.

A molecule which has absorbed a quantum of radiation becomes "energy-rich" or "excited" in the absorption process. Typically, absorption in the wavelength region of photochemical interest (the ultraviolet-visible region covering the wavelength range of approximately 250 - 700 nm) leads to electronic excitation of the absorber. Absorption at longer wavelengths usually leads to the excitation of vibrations or rotations of a molecule in its ground electronic state (Wayne, 1970).

The wavelength dependency of absorption, the nature of the excited species and, to some extent, the subsequent reactions of these excited species are dependent on the nature of the chromophore. Three major chromophore types involving minerals are considered to be particularly important: (i) absorption of light by the mineral bulk resulting in production of positive and negative charge carriers (holes and electrons) which subsequently induce electron transfer reactions at the mineral-water interface, (ii) light absorbing species located at mineral surfaces and for which the mineral is intimately involved in photo-induced (or enhanced) electron transfer, and (iii) light absorbing species adsorbed to mineral surfaces for which the underlying mineral modifies the photochemistry of the chromophore but for which the mineral does not participate in the electron transfer process. Mechanistic aspects of photoprocesses associated with each of these chromophore types are considered below and examples of photoprocesses involving these chromophores are presented.

PHOTO-REDOX PROCESSES INVOLVING ABSORPTION BY MINERALS

Electronic structure and optical properties of minerals

Obtaining an understanding of the electronic structure of solids is fundamental to understanding the phenomenon of light absorption by minerals. However, the electronic structure of solids is complex and cannot readily be accounted for by any one theory. Indeed, a number of theories (or models) which supplement each other have evolved over the last sixty years including crystal field theory, molecular orbital theory, band theory and the bond orbital model. Briefly, in the crystal field theory it is assumed that the interactions between neighbouring atoms are sufficiently weak that each electron remains localised at a discrete atomic position, while in molecular orbital theory, it is assumed that the solid possesses significant covalence, necessitating the sharing of electrons over more than one atom. Atomic nuclei are assumed to interact strongly in the band theory resulting in almost complete electron delocalisation over the solid. In this theory, the crystal structure rather than the individual atomic groupings is assumed to have a dominant effect on electron energy. In the bond orbital model, features of molecular orbital theory and band theory are combined (Marfunin, 1979).

A complete description of the electronic structure of a mineral can only be achieved using band theory since this theory enables analysis of the effect of the lattice (and the position of the electron in this lattice) on total electron energy (Marfunin, 1979). However, the accurate calculation of electron energy bands in solids with more than a few atoms per unit cell is often infeasible. In addition, in many minerals (particularly the transition metal oxides and silicates) the electrons are relatively localised and are not easily described by band theory (since the high degree of localisation removes the translational symmetry of the lattice) (Sherman, 1985). In such cases, a "cluster" molecular orbital approach in which the electronic structure of the solid is approximated by that of a finite atomic cluster which represents a basic structural entity of the mineral has been found to be more appropriate (Tossell and Gibbs, 1977). Molecular orbital approaches for minerals and band theory are discussed in a little more detail below.

Molecular orbital theory. Molecular orbitals (MO) are one electron wave functions describing the electron behaviour in the field of two or more nuclei and are obtained typically as the linear combination of atomic orbitals (LCAO):

$$\psi_{MO} = \psi_A \psi_B \quad . \tag{3}$$

Near the atom A the electron behaviour is approximated by the atomic orbitals ψ_A, while near the atom B, by ψ_B. The wave functions ψ_A and ψ_B are the s, p, d and f atomic orbitals. The square ψ_{MO}^2 of the wave function gives the probability of finding an electron in element volume dv. In order to normalise to unity the probability of finding an electron in all space of a given wave function, a normalising constant N is introduced such that:

$$N^2 \int \psi_{MO}^2 dv = 1 \quad .$$ (4)

To take into account the contribution of every AO (ψ_A and ψ_B) in MO, coefficients c_i are introduced:

$$\psi_{MO} = N(c_1\psi_A c_2\psi_B) \quad .$$ (5)

It can be shown that pair combinations of s, p and d atomic orbitals are restricted to two types: σ and π molecular orbitals, each of which can be bonding (σ^b, π^b) and antibonding (σ^*, π^*). The overlapping of atomic orbitals by lobes gives the σ molecular orbitals which are symmetric to rotation about the bond direction (z-axis). The σ orbitals are made of overlapping of $s-s, s-p_z, s-d_{z^2-y^2}, s-d_{z^2}$ atomic orbitals and also of overlapping $p_z - p_z, p_z - d_{z^2}, p_z - d_{x^2-y^2}$. The π orbitals arise from overlapping of $p_x - p_x, p_y - p_y, p_x - d_{xz}$ (and similar) atomic orbitals and are not symmetric with respect to rotation about the bond direction. The π bonds are always far less stable than σ bonds.

Antibonding σ^* and π^* molecular orbitals are formed from the same pairs of atomic orbitals as a result of their subtraction rather than their addition. Atomic orbital overlapping by lobes of the same sign leads to a bonding molecular orbital while that of opposite sign to an antibonding orbital. In bonding orbitals, the electron density increases in the region between the two nuclei (Fig. 1), the molecular orbital energy is less than either of the atomic orbitals and the electron spins are antiparallel (attractive). In antibonding orbitals the electron density between two nuclei diminishes to zero, the MO energy is higher than either of the AO energies and the electron spins are parallel (repulsive).

Ligand atomic orbitals, not having a type of orbital among the metal atomic orbitals with which they could overlap (as stipulated by symmetry conditions), are called nonbonding orbitals. They do not form molecular orbitals.

Molecular energy level diagrams may be constructed from the atomic orbitals under the following conditions: (1) the atomic orbitals should have comparable energy, (2) they must overlap appreciably, (3) they must have the same symmetry type, (4) each pair of atomic orbitals forming a bonding molecular orbital must also generate an antibonding orbital, (5) each of the atomic orbitals contributes in greater or lesser degree to all the molecular orbitals of the same symmetry type.

The symmetry types of the one electron atomic and molecular orbitals are dependent on their positions with respect to the ligands. Thus, for an octahedral complex (O$_h$ point group), s, p and d atomic orbitals will adopt the following symmetry types:

$$s \rightarrow a_{1g};\ p \rightarrow t_{1u};\ d_{xy}, d_{yz}, d_{xz} \rightarrow t_{2g};\ d_{x^2-y^2}, d_{z^2} \rightarrow e_g \quad ,$$ (6)

where g (gerade or even) and u (ungerade or uneven) indicate orbitals with and without a center of symmetry respectively. The molecular orbitals are obtained by combining the central ion atomic orbitals with the ligand group orbitals having the same symmetry. For example, for an octahedral complex, there will be three σ bonding molecular orbitals a_{1g}

562

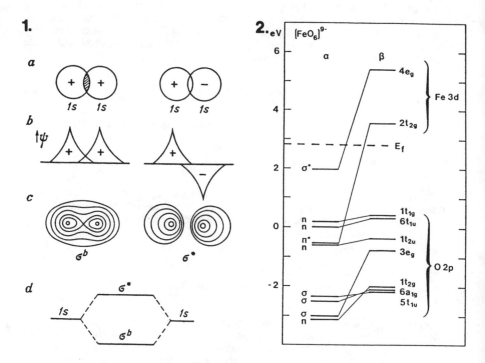

Figure 1. Bonding (σ^b) and antibonding (σ^*) σ molecular orbitals. (a) Overlapping of s orbital boundary surfaces of two atoms; (b) overlapping of the wave functions; (c) contours of constant electron density of σ^b and σ^* molecular orbitals; and (d) corresponding molecular orbital scheme (Marfunin, 1979).

Figure 2. Molecular orbital diagram of $(FeO_6)^{9-}$ obtained from SCF-Xα-SW calculations. The symbols α and β refer to spin-up and spin-down electron orbitals respectively. E_f refers to the fermi level, below which all orbitals are occupied (Sherman, 1985).

Table 1. Calculated optical spectra for iron oxides containing the octahedral $(FeO_6)^{9-}$ cluster (Sherman, 1985).

One electron transition	Energy (kK)[*]	Comments
$6t_{1u}^\beta \rightarrow 4e_g^\beta$	51.9	LMCT
$1t_{2u}^\beta \rightarrow 2t_{2g}^\beta$	43.6	LMCT
$6t_{1u}^\beta \rightarrow 2t_{2g}^\beta$	38.1	LMCT
$2t_{2g}^\alpha \rightarrow 2t_{2g}^\beta$	29.3	
$4e_g^\alpha \rightarrow 4e_g^\beta$	25.4	Ligand field
$4e_g^\alpha \rightarrow 2t_{2g}^\beta$	11.1	Ligand field

[*] $1kK = 1000\ cm^{-1} = 0.124\ eV$

(made from s + six p_x), t_{1u} (p + two p_z), $e_g(d_{x^2-y^2}, d_{z^2}$ + four p_z), two π bonding molecular orbitals t_{1u} (p + two p_x and two p_y), t_{2g} (d_{xy}, d_{xz}, d_{yz} + two p_x and two p_y), corresponding antibonding molecular orbitals and also two nonbonding ligand orbitals (t_{2u} and t_{1g}).

Calculation of molecular orbital energies are based on solution of the one electron Schrödinger equation for each orbital ψ:

$$\nabla^2\psi + \frac{8\pi^2 m}{h^2}(E-V)\psi = 0 \quad , \tag{7}$$

where E is the total orbital energy; V is the potential energy; m is the electron mass; and h is Planck's constant. For complex inorganic systems such as minerals it is essential that computational simplifications be introduced. Schemes such as the Extended Huckel Molecular Orbital (EHMO) method and the Self Consistent Field-Xα-Scattered Wave (SCF-Xα-SW) method have been used to this end (Tossell and Gibbs, 1977).

The SCF-Xα-SW method has been used particularly successfully in modelling the electronic structure of iron oxides and hydroxides (Tossell et al., 1974; Tossell, 1978; Sherman, 1985; Sherman and Waite, 1985), manganese oxides (Sherman, 1984), nontronites and iron-bearing smectites (Sherman and Vergo, 1988). In this method, solutions of the Schrödinger equation are obtained for volume-averaged forms of the potential within three separate regions: (1) spheres centered on the atomic nuclei, (2) an intersphere region, and (3) the region outside an outer sphere enclosing all the atomic spheres.

The molecular orbital diagram obtained by Sherman (1985) for the octahedral $(FeO_6)^{9-}$ cluster using the SCF-Xα-SW method is shown in Figure 2 for both spin-up (α) and spin-down (β) forms. Calculated optical transitions for this system are given in Table 1 and have been shown to correspond closely to the observed transitions in octahedral iron oxides (Sherman and Waite, 1985). Of particular note from these computations is the recognition that O → Fe charge transfer transitions in Fe^{3+} oxides (and silicates) occur at much higher energies than previously thought. The visible region absorption edge, which gives the iron oxides their red to yellow colours, does not result from ligand to metal charge transfer transitions but is a consequence of very intense Fe^{3+} ligand field and Fe^{3+}-Fe^{3+} pair transitions. The ligand field transitions in hematite are particularly intense compared to the other iron oxides because of increased magnetic coupling resulting from the unique face-sharing arrangements between FeO_6 polyhedra in this oxide (Sherman, 1985).

Band theory. Transition metal oxides, such as those of iron and manganese which exhibit significant ionic character, are preferentially modelled using a cluster molecular orbital approach. Other minerals such as periclase (MgO) and the chalcogenides (which includes sulfides, selenides, tellurides and arsenides) are considerably more covalent and their electronic structure is more conveniently represented via band theory.

Electron behaviour in crystals (as in atoms and molecules) can be described by the Schrödinger equation. Consider first the motion of the free electron, i.e. an electron which does not interact with the nucleus or other electrons. In this case the potential energy $V = 0$ and the Schrödinger equation can be written:

$$\nabla^2\psi + \frac{8\pi^2 m}{h^2}E\psi = 0 \quad , \tag{8}$$

where E now represents the kinetic energy of the electron ($mv^2/2$ for electron of velocity v). Using de Broglies expression for the wavelike behaviour of electrons ($mv=h/\lambda$), the electron energy E becomes $E = h^2/2m\lambda^2$ yielding:

$$\nabla^2\psi + k^2\psi = 0 \quad , \tag{9}$$

where $k = 2\pi/\lambda$ is the wave vector. Solution of the Schrödinger equation for the free electron is given by the function $\psi_k(r)=\exp(ikr)$ (i.e. by a sinusoidal wave of wavelength λ and wave vector k). For electrons occurring in the periodic field of lattice atoms, this expression is modified to:

$$\psi_k(r) = \exp(ikr)U_k(r) \quad , \tag{10}$$

where $U_k(r)$ is the periodic function (with the periodicity of the crystal lattice) modulating the electron motion and reflecting the perturbing effect of the periodic field of the lattice atoms. Equations of this form representing solutions to Schrödinger's equation for an electron in a crystal lattice are known as Bloch functions.

The electron motion in a crystal can be considered as diffraction by the crystal lattice with the diffraction phenomenon described by the Bragg equation:

$$n\lambda = 2d \sin\phi \quad , \tag{11}$$

where λ is the wavelength of the electrons, n is an integer, d is the interplanar distance of the lattice and ϕ is the angle between the lattice plane and the beam direction. At normal incidence of the beam to the lattice plane, $n\lambda = 2d$ or $n\pi/d = 2\pi/\lambda = k$. This condition means that the wave vector k is determined not only as the inverse of the wavelength values $(2\pi/\lambda)$ but also as the inverse of the interplanar distances in the lattice (π/d), i.e. the k-space is the space of the reciprocal lattice. The Bragg equation also indicates that when $n\lambda = 2d$ (or $k = n\pi/d$), electrons are reflected by the lattice planes. This means that, within the mineral bulk, electrons cannot exist as continuous moving waves, i.e. there must be forbidden values of kinetic energy corresponding to the condition $k = n\pi/d$. Thus, the k-space (i.e. a diagram with k_x, k_y and k_z axes) is divided into concentric regions formed by polyhedra with the electron energy within them changing continuously but exhibiting discontinuities (energy jumps associated with potential energy changes) at the faces. The regions within these polyhedra are called Brillouin zones with the shape of these polyhedra determined by the lattice symmetry (Marfunin, 1979). The term "Brillouin zone" is typically restricted to the first or "reduced" zone.

The electron wave functions in a crystal change from point to point within the Brillouin zone according to the different symmetry types at different points of the zone. Consider for example the band structure for periclase, MgO. This mineral possesses a face-centered cubic lattice for which the first Brillouin zone is a truncated octahedron. At characteristic points Λ, E, Δ, K, Ξ etc., the states formed from $2p$ orbitals of oxygen with an admixture of $3s$ orbitals of magnesium are described not as $2p$ but as Λ_{15} at the point Λ, as $E_1E_2E_3$ at the point E etc. The corresponding energies at these points are different and change from point to point quasicontinuously (quasi because at each point between Λ and E and between E and K etc. there is a discrete set of levels) (Fig. 3).

The whole set of states (at all points of the Brillouin zone) arising, for instance, from the $2p$ orbital, forms a band which may be referred to as a $2p$-like band; other bands form from $3s$, $3p$ orbitals etc. ($3s$-, $3p$-like bands). The $2s$- and $2p$-like bands are occupied and are called valence bands; the $3s$- and $3d$-like bands are unoccupied and are called conduction bands. These bands are separated by an energy gap E_g. The states composing these bands can be compared with molecular orbitals: valence bands are analogous to bonding molecular orbitals, conduction bands to antibonding ones (Marfunin, 1979). The energy band scheme for periclase and the optical transitions that may occur between the top levels of the valence band and the bottom levels of the conduction band are shown in Figure 3.

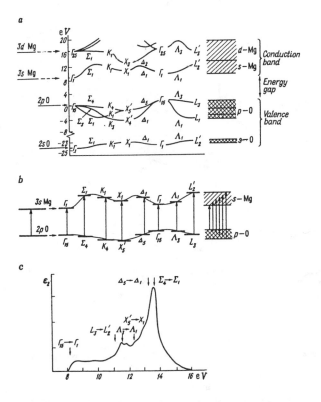

Figure 3. Energy band scheme and reflectance spectrum of periclase, MgO. (a) Energy band structure of MgO; (b) Optical transitions between top levels of valence band and bottom levels of conduction band in characteristic points of Brillouin zone; and (c) Assignment of reflectance spectrum (Marfunin, 1979).

The conduction band of solids is normally much broader than the valence band because of the greater overlap of the orbital wave functions. Bands associated with higher excited states normally overlap the conduction band so there are allowed electronic energy levels from the bottom of the conduction band to infinite energy (Morrison, 1980). In a perfect crystal, there are no allowed energy levels for electrons in the band gap.

Three types of solids are distinguished according to the size of the band gap (Marfunin, 1979): (1) the solid is considered to be a nonmetal if the band gap is more than approximately 5 eV; they are transparent in the near UV-visible region and are dielectrics (insulators) with mostly ionic bond type; (2) the solid is considered to be a semiconductor if the band gap is less than approximately 5 eV; for these materials, the absorption edge typically lies between the IR and near UV regions; (3) the solid is considered to be a metal if the valence and conduction bands overlap significantly (with the term semimetal sometimes used to describe solids in which the valence and conduction bands only overlap slightly). The nature of solids may also be defined in terms of the Fermi energy level - the energy of the highest occupied molecular orbitals of the solid. To a first aproximation, all energy levels below the Fermi level are occupied by electrons while all levels above are unoccupied. Thus, in a metal the Fermi level is well within the valence band while for a semiconductor, the Fermi level lies in the band gap between valence and conduction bands.

566

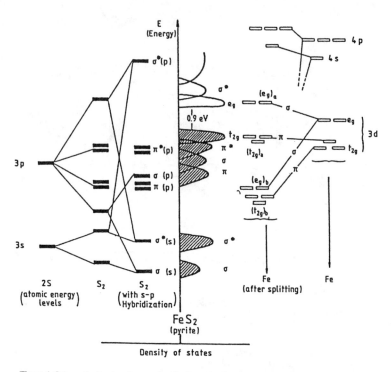

Figure 4. Schematic density of states distribution of pyrite type compounds (Ennaoui et al., 1986).

As mentioned earlier, chalcogenides exhibit more covalency than oxides (particularly in the case of sulfides because the ionisation potential of the S $3p$ electrons is significantly lower than that for the O $2p$ electrons; Jaegermann and Tributsch, 1988) and as such are better described by band theory. As an example, consider the dichalcogenide, pyrite FeS_2. The electronic structure can be understood in terms of an interaction between S_2 molecules (with s-p hybridization) and iron atoms with $3d$-states split in the octahedral environment into t_{2g} and e_g levels. Owing to this interaction, there is a further splitting of energy levels into bonding and antibonding states that accounts for the observed semiconducting properties of pyrite (Fig. 4). In accord with X-ray Photoelectron Spectroscopic (XPS) data (Ennaoui et al., 1986), the upper region of the valence band of pyrite consists mainly of iron t_{2g} states on top of sulfur states towards lower energies. A similar but inverted situation is found in the conduction band with iron e_g levels mainly forming the edge with increasing density of sulfur states at higher energy.

Semiconducting minerals

Of major interest in this chapter are minerals for which the absorption of light occurs in the near UV/visible spectral region and as a result of which, electron transfer processes at the mineral-water interface are induced or enhanced. According to the band gap criterion given above, such minerals are considered to be semiconductors. Typically, a semiconductor conducts neither as well as a metal such as copper, nor as poorly as an insulator like halite. The electrical conduction in semiconducting minerals is due to the motion of free charge carriers which may be of two types: electrons in the normally empty conduction band or holes in the normally full valence band. Three sources of free carriers may be distinguished (Shuey, 1975): deviation from stoichiometry, trace elements in solid solution, and thermal excitation across the band gap. When the last source of free carriers dominates, the number of electrons and holes is equal since holes are produced as a result of removal (excitation) of electrons from the valence band. In such cases, the material is considered to be an intrinsic semiconductor and includes minerals such as arsenopyrite (FeAsS) and pyrolusite (β-MnO_2).

Most semiconducting minerals are extrinsic, that is, the carriers are due to donor or acceptor defects which represent nonstoichiometry and/or the presence of point defects. Donor impurities donate electrons to the conduction band while acceptor impurities accept electrons from the valence band (which is equivalent to donating holes to the valence band). When donors are the main impurities present in minerals, the conduction is mainly by way of electrons and the material is called an n-type semiconductor. Similarly, if acceptors are the major impurities present conduction is mainly by way of holes and the material is called a p-type semiconductor. Some solids including ZnO, TiO_2, V_2O_5, $CuFeS_2$ and MoO_3 always occur as n-type while others such as Cu_2O, NiO and Cr_2O_3 always occur as p-type. A large group of minerals including α-Fe_2O_3, GaP, GaAs and FeS_2 can be either n or p-type.

In the dark, under thermal equilibrium, the chemical potential of the electron is equal to that of the hole and corresponds to the Fermi energy level of the solid. The concentration of electrons in the conduction band, n, is related to the energy of the bottom of this band, E_c, to the effective density of states, N_c, in this band and the Fermi energy E_F by

$$\frac{n}{N_C - n} = \exp\left(-\frac{E_F - E_C}{kT}\right) \quad . \tag{12}$$

For $n \ll N_c$,

$$E_F = E_C + kT \ln\left(\frac{n}{N_C}\right) \quad . \tag{13}$$

Analogous equations describe the concentration of holes in the valence band, p:

$$\frac{p}{N_V - p} = \exp\left(\frac{E_V - E_F}{kT}\right) \quad . \tag{14}$$

For $p \ll N_v$,

$$E_F = E_V - kT \ln\left(\frac{p}{N_V}\right) \quad . \tag{15}$$

In an intrinsic semiconductor, the Fermi level will lie at a potential about halfway between the valence and conduction bands. In n-type material, the Fermi level will lie just below the conduction band edge while in p-type material, it will be poised just above the valence band edge. While still the subject of some debate (Khan and Bockris, 1985; Scherson et al., 1985), the Fermi level of the semiconductor (under dark, equilibrium conditions) is established by the redox potential of the electrolyte with which the semiconductor is in contact (Morrison, 1980).

Under equilibrium conditions, the concentrations of electrons and holes are constrained by an expression derived from the product of the above conditions; i.e.

$$n \times p = (N_c - n)(N_V - p)\exp\left(\frac{-E_C - E_V}{kT}\right) \quad , \tag{16}$$

where E_c-$E_v = \Delta E_g$, the width of the band gap. For $n \ll N_c$ and $p \ll N_v$, this expression simplifies to:

$$n \times p = N_C N_V \exp\left(-\frac{E_C - E_V}{kT}\right) \quad , \tag{17}$$

and indicates that if E_g is large, the product np is small, indicating why at equilibrium both n and p cannot be large. Thus if n is high, perhaps because there is a high density of donors, and if E_g is large, then p must be small.

Effects of illumination. Upon absorption of light of energy equal to or higher than the band gap of the semiconductor, electrons are excited from the valence band to the conduction band and equilibrium between electron and holes is no longer preserved. Under such conditions, the chemical potential of the electron will be different from that of the hole (Gerischer, 1980). As a result, the Fermi level splits into two "quasi-Fermi" levels, $_nE_F$ for the electron and $_pE_F$ for the hole, where

$$_nE_F = E_C + kT \ln\left(\frac{n^*}{N_C - n^*}\right) \quad , \tag{18}$$

$$_pE_F = E_V - kT \ln\left(\frac{p^*}{N_V - p^*}\right) \quad . \tag{19}$$

Denoting the excess concentrations by Δn^* and Δp^*, we can write $n^* = n + \Delta n^*$ and $p^* = p + \Delta p^*$ where n and p are the equilibrium concentrations of electrons and holes respectively. Clearly, drastic changes in free energy under illumination can only be expected for carriers (c) with a low equilibrium concentration such that $\Delta c^* \gg c$. These are the minority carriers of a doped (extrinsic) semiconductor. In an intrinsic or very low doped material, the deviation in free energy can be large for both types of electronic carriers.

Charge carrier mobility. While light may significantly modify the number of charge carriers, the effect of these charge carriers on redox reactions at the mineral-water interface will obviously be dependent on the efficiency of migration of electrons or holes to (or away from) the solid surface. The theory of the mobility of delocalised charge carriers within semiconducting materials is discussed at length elsewhere (see for example Nag, 1972) thus only certain aspects are reviewed here.

If an energy band is either completely empty or completely filled with electrons, that band will not contribute to the electronic conductance of the solid since none of the electrons can gain energy in response to an applied field. However, conductance can occcur in any partially filled band. If the bands are narrow, arising from orbitals that do not appreciably overlap, then the electrons or holes do not move as freely; i.e., they have lower mobility.

Of major interest is the tendency for recombination of electron-hole pairs to occur within the semiconductor bulk at point defects and dislocations. Such "recombination centers" will normally be fully occupied by majority carriers thus the rate-limiting step will be that of minority carrier capture.

When minority carriers are photoproduced, the average distance the minority carrier diffuses before deexcitation (recombination) is termed the diffusion length: L_p if the minority carriers are holes; L_n if electrons. The diffusion length is given by the expression (Morrison, 1980):

$$L_{p,n} = (D_{p,n}\tau)^{1/2} \quad , \tag{20}$$

where D is the diffusion coefficient for the minority carrier and τ is its bulk lifetime. Taking advantage of the optical absorption of conduction band electrons in colloidal TiO_2 particles and using picosecond laser techniques, Rothenberger et al. (1985) showed that electron-hole pairs had a lifetime of 30 ns corresponding, in these particles, to a diffusion length of 2.2 x 10^{-5} cm.

It is now well recognised that specific defects in semiconducting materials create energy levels within the band gap which facilitate recombination of the electron-hole pair. Efforts to determine the distribution of these levels, let alone to identify them with particular structural defects, have been fraught with difficulty and a large number of models have been developed to explain a bewildering array of experimental results (Davis, 1984). In addition to specific defects, amorphous semiconductors possess intrinsic disorder which lead to the occurrence of localised states in tails at the band edges (Davis, 1984). Such states may again be expected to aid recombination and thus lower the efficiency of transport of charge carriers to the semiconductor surface.

Interfacial electron transfer

The efficiency and rate of electron transfer at the semiconductor-liquid interface is dominated by the charge characteristics at the interface. Charge at the interface will be mediated by the electronic energy levels at the surface of the solid ("surface states"). Both nature and properties of surface states and their impact on surface charge are discussed below as is the influence of the surface charge characteristics on the kinetics of interfacial electron transfer.

Surface states. Surface states on "clean" surfaces are classified as intrinsic and arise because of the disturbance to lattice periodicity by the surface. Extrinsic surface states are formed by adsorbates on the surface and are related to the interaction of the semiconductor with another phase.

Intrinsic surface states of two different types have been identified: covalent surface states (Shockley surface states) and ionic surface states (Tamm states) (Morrison, 1977; Many et al., 1965). The surface of relatively ionic semiconductors (for example, the iron oxides) will have "ionic surface states" or "Tamm states" on its clean surface. Consider, for example, a lattice oxygen ion, O_2^-. At the surface, this ion will be surrounded by less than its normal complement of cations. Thus the energy of the electrons on this surface O_2^- will be higher than that of the bulk (valence band) electrons simply because of the lower electrostatic attraction from the neighbouring cations (Fig. 5). Because they are occupied when the surface is neutral, these are donor states (Morrison, 1980). Analogously, the unoccupied energy levels associated with the surface cation can be acceptor states of lower energy than the conduction band simply because the surface cation has fewer neighbouring anions than the bulk cation.

Shockley states occur in more covalent solids and arise because of the lack of neighbouring atoms at the surface with which to share valence electrons resulting in the formation of "dangling bonds". These bonds provide surface states that can act either as donor surface states (the dangling electron being excited to the conduction band, leaving a positively charged surface) or as acceptor surface states (electrons being captured to pair with the electrons in the dangling bonds, resulting in a negatively charged surface). A significant overlap occurs between neighbouring dangling orbitals causing splitting of the resulting "molecular orbitals" into bonding and antibonding states (Fig. 5). Shockley states will be broadened either because the orbitals overlap with each other or overlap with the bands of the solid. Surface heterogeneity can also give rise to a spectrum of energy levels appearing as a band (Morrison, 1980).

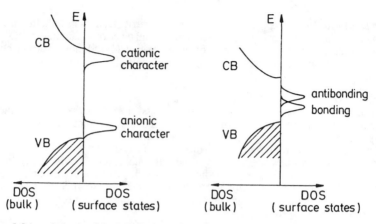

Figure 5. Schematic density of states distribution of Tamm (left) and Shockley (right) surface states for ionic and covalent solids respectively where the Y-axis represents potential (E) and the X-axis represents distance from the solid-liquid interface (Jaegermann and Tributsch, 1988).

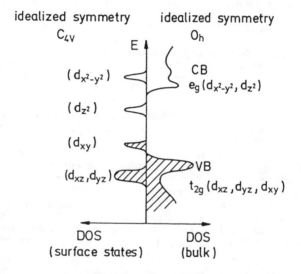

Figure 6. Schematic density of states distribution of surface states on (100) faces of pyrite type compounds showing dangling bonds of metal d-character (Jaegermann and Tributsch, 1988).

In transition metal semiconductors, coordination bonding via directed metal d-states within the solid may also lead to dangling bonds of metal d-character at the surface for certain crystal plane terminations. As an example, the possible electronic surface states formed on the (100) face of FeS_2-type compounds are schematically indicated in Figure 6. The surface states are derived from a reduction of the idealized octahedral coordination of the metal in the bulk to a square pyramid (idealized C_{4v} symmetry) at the surface. As a consequence, the degeneracy of the t_{2g} and e_g states is broken. The occupancy of the different

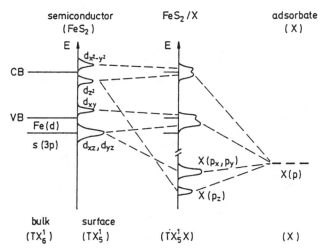

Figure 7. Hypothetical changes of energetic positions of surface states of FeS₂ as a result of coupling with the electronic states of an adsorbate (Jaegermann and Tributsch, 1988).

levels obtained may deviate from the bulk. In extreme cases, a change from a low spin configuration to a high spin configuration might be possible (a distinct possibility for FeS_2; Jaegermann and Tributsch, 1988).

The electronic surface states of real (clean) semiconductor surfaces deviate to some extent from the idealized schemes outlined above due to possible relaxation and reconstruction of the surface. The surface atoms rearrange to find a relative minimum of the surface free energy. The process most likely to occur involves a change in bonding distance and/or bonding angles without major migration of surface atoms. This process may, in some instances, occur at room temperature but may be kinetically hindered and require heating to proceed to completion (Jaegermann and Tributsch, 1988).

Extrinsic surface states arise as a result of extrinsic sources such as defects or adatoms of the semiconductor constituents at steps or dislocations on the surface or strongly interacting adsorbates which lead to new electronic states at the interface (Nakabayashi and Kawai, 1988). It should be noted that intrinsic and extrinsic states exclude each other. Intrinsic surface states are related to surface sites of increased reactivity and it is with these states that adsorbates will preferentially react with the resultant formation of new electronic surface states. The energy of these new states is given by a bonding-antibonding coupling of intrinsic surface states and adsorbate electron states as schematically shown in Figure 7 for FeS_2 surfaces. For surface metal sites in transition metal compounds, the adsorbates can be considered as extra ligands of different ligand field parameters to the lattice anions.

Space charge layer, band bending and electron transfer. With surface states accepting or donating electrons in response to the presence of an electroactive species in solution or application of an external bias voltage (by mounting an ohmic contact at the back of the electrode and inserting a second electrode in solution), electronic equilibrium between the surface and the bulk of the semiconductor demands that the trapped surface charge and the space charge within the so-called semiconductor space charge layer balance each other. Due to the relatively low concentration of mobile charge carriers in the semiconductor, the space charge layer spreads into the bulk of the crystal. As a consequence, the potential within the semiconductor bulk varies continuously leading to a bending of energy bands (Fig. 8). Of course, if excess charge carriers are not trapped in surface states, the interior of the semiconductor is at a constant potential. Under these "flat band" conditions, there is no space charge layer and the electronic levels are at constant energy up to the surface.

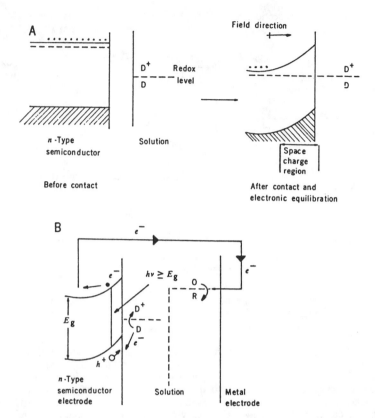

Figure 8. (A) Formation of a space charge region (a depletion layer) on equilibration of an n-type semiconductor with a solution containing a redox couple D/D⁺ The direction of the local electric field is also shown. In (B), the direction of electron flow occurring on photolysis of an n-type semiconductor is shown (Bard, 1980).

Figure 8 refers to n-type semiconductors. For p-type materials, analogous considerations apply with holes being the mobile charge carriers and immobile, negatively charged acceptor states forming the excess space charge within the depletion layer. The bands bend downwards in this case (Fig. 9).

The depletion layer at the semiconductor-solution interface, sometimes referred to as a Schottky barrier, plays an important role in light induced charge separation on semiconducting electrodes. The local electrostatic field present in the space charge layer serves to separate the electron pairs generated by illumination of the semiconductor. For n-type materials, the direction of the field is such that holes that survive recombination migrate to the surface where they may participate in chemical reactions while the electrons, in the case of a two electrode system, drift through the bulk to the back contact of the semiconductor and subsequently through the external circuit to the counter electrode (Fig. 8B). The opposite holds for p-type semiconductors (Fig. 9B).

It is of particular interest to consider the space charge layer in particulate semiconductors. In spherical particles the potential drop across the space charge layer can be obtained from the solution of the linearized Poisson-Boltzman equation (Albery and Bartlett, 1984); i.e.,

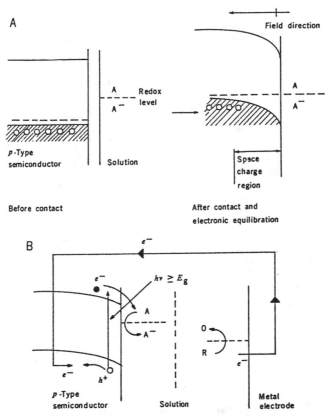

Figure 9. (A) Formation of a space charge region on equilibration of a p-type semiconductor with a solution containing a redox couple A/A⁻. In (B), the direction of electron flow occurring on photolysis of a p-type semi-conductor is shown (Bard, 1980).

$$\Delta\phi = \left(\frac{kT}{6e}\right)\left(\frac{r-(r_o-W)}{L_D}\right)^2\left(1+\frac{2(r_o-W)}{r}\right) \quad , \tag{21}$$

where r is the radius at which the potential drop $\Delta\phi$ is being determined, r_o is the particle radius, W is the depletion layer thickness and L_D is the Debye length:

$$L_D = (\varepsilon_o\varepsilon T/2e^2N)^{1/2} \quad , \tag{22}$$

where N is the concentration of ionised dopants, expressed in number of ions per cm³.

Following Gratzel (1988), a graphical illustration of the potential distribution for an n-type semiconducting particle which is in equilibrium with a solution containing a redox system for which the Fermi level is E_F is shown in Figure 10. Two limiting cases are considered particularly important for light induced electron transfer for semiconducting dispersions. If the size of the particles is much larger than the depletion layer width ($r_o \gg W$), the condition $r_o \approx r$ holds within the depletion layer and the preceding equation simplifies to:

$$\Delta\phi = \frac{kT}{2e}\left(\frac{r-(r_o-W)}{L_D}\right)^2 \quad . \tag{23}$$

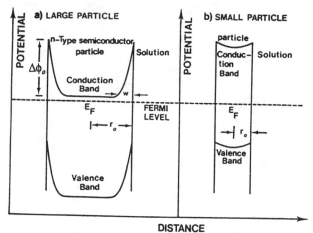

Figure 10. Space charge layer formation in a large (a) and small (b) semiconductor particle in equilibrium with a solution redox system for which the Fermi level is E_F. The small particle depletes almost completely of charge carriers. Hence, its fermi potential is located approximately in the middle of the band gap and the band bending is negligibly small (Gratzel, 1988).

For $r = r_0$, one obtains:

$$\Delta\phi_o = \frac{kT}{2e}\left(\frac{W}{L_D}\right)^2 \quad , \tag{24}$$

where $\Delta\phi_o$ corresponds to the total potential drop within the semiconductor particle. This expression is identical to that applicable for planar electrodes (Gratzel, 1988).

For very small (colloidal) semiconductor particles, the condition $W = r_o$ holds and the preceding general expression for $\Delta\phi$ reduces to :

$$\Delta\phi = \frac{kT}{6e}\left(\frac{W}{L_D}\right)^2 \quad . \tag{25}$$

From this equation and Figure 10 it is apparent that the electrical field in colloidal semiconductors is usually small and that high dopant levels are required to produce a significant potential difference between the surface and the center of the particle. The charge carrier concentration in undoped semiconducting colloids will be small and the band bending in such particles would be expected to be negligible (Gratzel, 1988). Clearly, if such a particle depletes almost entirely of charge to solution and the particles are too small to develop a space charge layer, the potential difference resulting from the transfer of charge from the semiconductor to the solution must drop in the Helmholtz layer. As a consequence, the position of the band edges of the semiconductor particle will shift. As shown in Figure 10, the Fermi level in such a case will be located approximately in the middle of the band gap.

Since the band bending is small in colloidal semiconductors, charge separation occurs via diffusion. The absorption of light leads to the generation of electron-hole pairs in the particle which are oriented in a spatially random fashion along the optical path (Gratzel, 1988). These charge carriers subsequently recombine or diffuse to the surface where they may undergo reactions with catalysts deposited onto the particle surface or suitable solutes. When

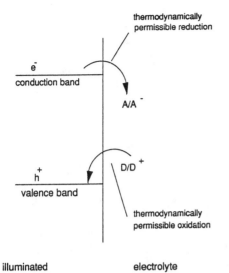

Figure 11. Thermodynamic requirements for electron transfer deduced from a comparison of the oxidation potentials of electron donors and acceptors in solution with the flat band potentials of the semiconductor. The vertical axis in this figure represents potential (Pichat and Fox, 1988).

Table 2. Band gap energies and wavelength of light equivalent to band gap for a variety of semiconductors.

Semiconductor	Band Gap (eV)	Equivalent Wavelength (nm)	Semiconductor	Band Gap (eV)	Equivalent Wavelength (nm)
ZrO_2	· 5.0	248	CdS	2.4	516
Ta_2O_5	4.0	310	α-Fe_2O_3	2.34	530
SnO_2	3.5	354	ZnTe	2.3	539
$KTaO_3$	3.5	354	$PbFe_{12}O_{19}$	2.3	539
$SrTiO_3$	3.4	365	GaP	2.3	539
Nb_2O_5	3.4	365	$CdFe_2O_4$	2.3	539
ZnO	3.35	370	CdO	2.2	563
$BaTiO_3$	3.3	376	$Hg_2Nb_2O_7$	1.8	689
TiO_2	3.0-3.3	376-413	$Hg_2Ta_2O_7$	1.8	689
SiC	3.0	376	CuO	1.7	729
V_2O_5	2.8	443	PbO_2	1.7	729
Bi_2O_3	2.8	443	CdTe	1.4	885
$FeTiO_3$	2.8	443	GaAs	1.4	885
PbO	2.76	449	InP	1.3	954
WO_3	2.7	459	Si	1.1	1127
$WO_{3-x}F_x$	2.7	459	β-HgS	0.54	2296
$YFeO_3$	2.6	476	β-MnO_2	0.26	4768
$Pb_2Ti_{1.5}W_{0.5}O_{6.5}$	2.4	516			

*From Nozik (1978) and Strehlow and Cook (1973)

electron donors and acceptors are both adsorbed on a photocatalyst, oxidative and reductive exchanges may occur on the same surface. Each particle can be pictured as a "short-circuited" photoelectrochemical cell, where the semiconductor electrode and counter electrode have been brought into contact (Bard, 1980).

The likelihood of reaction with electron donors and acceptors can be assessed from thermodynamic considerations. As shown in Figure 11, as long as a donor's oxidation potential is less positive than the valence band edge and the acceptor's reduction potential is less negative than the conduction band edge, electron transfer at the illuminated interface will be thermodynamically permissible. Since typical redox potentials for common organic and inorganic redox couples often lie between the band edges of easily accessible semiconductors, many electron transfer reactions will be feasible, at least based on thermodynamic criteria (Fox, 1983; Pichat and Fox, 1988).

576

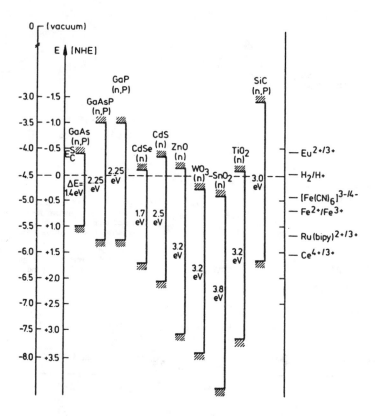

Figure 12. Band edge position of several semiconductors in contact with an aqueous electrolyte of pH 1 (Gratzel, 1989).

Factors influencing semiconductor reactivity

The electrochemical potentials of the valence and conduction bands of a semiconductor determine the amount of energy required to initially produce separated charge carriers and the redox potentials of adsorbed couples in relation to these band edges are critical determinants of the thermodynamic feasibility of a photochemical reaction proceeding at a semiconductor surface. The band edge positions of a semiconductor can be established by a variety of physical and electrochemical methods (Nozik, 1978). The optical and electrochemical properties of a wide range of semiconductors are reviewed by Harris and Wilson (1978) and the band gaps and band gap positions of selected semiconductors shown in Figure 12 and Table 2.

A wide range of factors may mediate the efficiency of generation of separated charge at the surface of semiconducting electrodes or powders on illumination with light and the yield of products generated in subsequent redox reactions at the solid-liquid interface. Some of these factors are discussed briefly below.

Morphology. Analogous photoelectrochemical events occur whether the semiconductor is formulated as a single crystal, a polycrystalline electrode, as a powder or as a finely divided colloid (Pichat and Fox, 1988). However, a variety of factors including crystallinity and surface area of the semiconductor may significantly alter the lifetime of generated charge carriers and the extent and rate of reaction at the solid-liquid interface.

For example, Sclafani et al. (1990) have shown that TiO_2 particles prepared by a variety of methods differed widely in their ability to photocatalytically degrade phenol. An increasing degree of crystallinity was shown to increase the degree of photodegradation. The decrease in hydroxylation of surface groups (particularly as influenced by high temperature treatment) was considered the most likely reason for the lowered activity of rutile phases prepared at high temperature (OH^- groups effectively trap holes).

Iron oxides exist in many crystal structures and stoichiometries and provide good examples for the assessment of effects of structure and associated physicochemical properties on semiconductivity. The semiconducting properties of different forms of these oxides have been investigated by a number of workers including Yeh and Hackerman (1977), Hardee and Bard (1978), Stramel and Thomas (1986), Somorjai and Salmeron (1986), and Leland and Bard (1987). The studies of Leland and Bard (1987) are of particular interest and involved (i) the estimation of the quasi-Fermi level for electrons and the standard heterogeneous rate constants for electron transfer in "mediated" charge collection experiments in which sodium tartrate served as the irreversible hole scavenger and $Fe(CN)_6^{3-}$ or $Ru(NH_3)_6^{3+}$ as electron acceptors; and (ii) estimation of the relative rate of photogenerated hole reactions by measurement of the rate of photooxidation of oxalate and sulfite.

Leland and Bard (1987) obtained large differences in $_nE_F$ for the various iron oxide polymorphs and attributed the differences to the mediation of electron transfer by intra-band gap states (or surface states) rather than by the conduction band level. Studies of the effect of potential of the charge collecting electrode on the photocurrent indicated that large overpotentials (i.e. potentials significantly more positive than $_nE_F$) were required to generate the small currents observed. Such an observation is indicative of slow heterogeneous electron transfer most likely as a result of extensive electron-hole pair recombination. Observed wide variation in heterogeneous rate constants for the different oxides demonstrates that crystal structure plays an important role in determining electron transfer kinetics. An observed square root dependency of steady-state photocurrent on light intensity is also indicative of extensive recombination of charge carriers (Leland and Bard, 1987). Large variations observed in photooxidation rates of oxalate and sulfide for the different forms of iron oxide has been attributed to differences in electronic and structural properties because there is no correlation between these rates and physicochemical parameters such as hydrodynamic diameter, band gap or BET surface areas.

Size. Size effects in photoreactions on dispersed catalysts are well-documented (e.g. Reidel, 1984; Lenoir et al., 1988), although their interpretation is far from complete. Farin and Avnir (1988) report that many heterogeneous reactions obey the scaling relation:

$$v = kR^{D_R - 3} \quad , \tag{26}$$

where v (mol/time.mL) is the initial reaction rate with the surface, $2R$ is the particle size, k is a constant and D_R, the reaction dimension, is a characteristic parameter of the heterogeneous reaction. D_R carries information on the sensitivity of the process to changes in the geometric parameters (particle size in this case) and on the distribution, location and nature of the subset of the reactive surface sites. A possible interpretation of D_R is that it reflects a fractal pattern of distribution of active sites (Farin et al., 1989). A number of examples of the applicability of the above relationship are presented and discussed by Farin et al. (1989). Two of these examples are given here.

Consider firstly the photocatalytic hydrogenation of propene by dispersions of TiO_2 (anatase) containing small amounts of Al_2O_3 (Anpo et al., 1988) (pure anatase is not an active photocatalyst for this reaction). The TiO_2 particle size decreases with increasing percentage of Al_2O_3 with a corresponding increase in specific photocatalytic activity (yield/TiO_2 content) (Table 3). This data closely follows the relationship in Equation 26 with $D_R = 0.26 \pm 0.12$. This very low value of D_R indicates a weak dependence of the

Table 3. Specific activity of TiO_2 for propene photohy-
drogenation on Ti-Al binary oxide catalysts (Farin et al.,
1989).

Particle size (Å)	Photocatalytic Activity (mmol/(h.g of TiO_2))
160	0.7
145	0.9
113	1.7
100	2.5
90	3.0
80	5.0

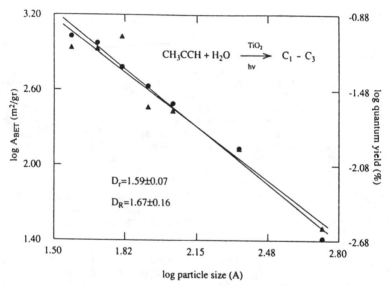

Figure 13. The dependence on TiO_2 particle size of TiO_2 surface area (Δ, left axis) and quantum yield (⊕, right axis) of the photocatalytic reaction of CH_3CCH with water (Farin et al., 1989).

photocatalytic ability on particle size. Farin et al. (1989) note that the corners and edges of a crystallite grow very slowly with increase in crystallite size and suggest that the low D_R indicates that only parts of the crystal edges are active in this case. Farin et al. (1989) suggest that the main role of Al_2O_3 as a cocatalyst is to introduce a small percentage of TiO_2 crystal imperfections, in which the exposed active Ti ion is not part of the ordered anatase crystalline structure.

A second example concerns the size effect found by Anpo et al. (1987) in the photocatalytic hydrogenation of methyl acetylene with water on dispersed titania (anatase and rutile). In this case, information is available not only on size effects on activity but also on the Ar BET surface area, A. This enables one to compare the fractal dimension, D_r, of the surface available for physisorption to the subset of reactive surface sites, as characterized by D_R. The former is given by (Avnir et al., 1985; Van Damme et al., 1988):

$$A \propto R^{D_r - 3} \quad , \tag{27}$$

while the relation between D_R and D_r is given by (Avnir, 1989; Farin and Avnir, 1987):

$$v \propto A^{(D_R-3)(D_r-3)} \quad , \tag{28}$$

which is a fractal interpretation of the power law:

$$v \propto A^m \quad , \tag{29}$$

in which m is the reaction order in surface area. As can be seen from the transformed data in Figure 13, $D_r \sim D_R$ ($m \sim 1$) indicating that the areas exposed for physisorption and reaction are the same.

Substitutional doping. Substitutional doping has been widely used, especially in attempts at photodissociating water, to extend the photosensitivity of large band gap semiconductors to the visible range and to control the position of the Fermi level. In some cases, the voluntary introduction of impurities stabilizes the semiconductor against corrosion (Pichat and Fox, 1988).

Substitution of Ti^{4+} in TiO_2 by cations of higher valency such as Nb^{5+} ions introduces extra valence electrons into the lattice (one per substituted cation). In this case, Nb^{5+} ions behave as complexation sites for electron donors. The opposite occurs when substituting Ti^{4+} by cations of lower valency such as Cr^{3+}. In this case, Cr^{3+} ions act as complexation sites for electron acceptors. Isoelectronic substitution, such as Zr^{4+} for Ti^{4+} changes the energetic position of the conduction band edge (Pichat and Fox, 1988; Bin-Daar et al., 1983).

Experimentally, semiconductors can be doped by sintering of the required salts at high temperature. In this way, materials of low surface area are obtained. Homogeneously doped powders of high surface area can be prepared in a flame reactor (Pichat and Fox, 1988).

Surface modification. Metals like Pt, Rh, Ru and Pd may be incorporated in the surface of a semiconductor crystal, or in that of a dispersed semiconductor crystallite, for two major reasons: (i) to introduce catalytic centers which accelerate a critical step of a reaction of interest, and (ii) to increase the separation of photoproduced electron-hole pairs (Heller, 1986). Platinum deposition is particularly influential in increasing the evolution of H_2 because of the lowered overpotential for this reaction in the presence of this catalyst (Kiwi et al., 1982). The issue of increasing charge separation, at least on an n-type semiconductor, relates to the ability of the metal particles to effectively "drain" electrons from the semiconductor (i.e., to function as a cathode).

Table 4. Photocatalytic decomposition of 2-propanol on illumination with a 500-W high pressure mercury arc lamp for 24 h of 3 mg of TiO_2 suspended in 8 mL of 2-propanol containing 0.2 mL of water. The data represent time averaged values for 24 h (Yoneyama et al., 1989).

Catalyst	Production rate (μmol h^{-1})	
	H_2	$(CH_3)_2CO$
TiO_2	0.26	0.28
TiO_2/clay	1.34	1.40
Pt/TiO_2	183	172
Pt/TiO_2/clay	697	661

In addition to the deposition of metals, semiconductors may be optimised for specific tasks by the deposition of a variety of substances on the semiconductor surface including organic dyes (Ryan and Spitler, 1988) and polymeric stabilizing agents (Nosaka and Fox, 1987). Thus, polyvinyl alcohol stabilizes colloidal TiO_2 suspensions from precipitation without quenching photoactivity (Nishimoto et al., 1985). Gratzel et al. (1989) report that tungstosilicate $(SiW_{12}O_{40})^{4-}$ deposited on the surface of TiO_2 markedly alters the catalytic properties of the titania semiconductor. In this study, the oxidation of CH_4 was found to procede to CO (a useful synthesis gas in the petrochemical industry) rather than to CO_2 (the major product in the absence of the tungstosilicate).

Attachment to supports. Attachment of particulate semiconductors to supports is an effective means of dispersal and may enable modification of the photochemical properties of the semiconductor. For example, dispersal of semiconductor particles into macroscopic gel arrays produces a porous material with high surface area. With multiple layers formed by this sol-gel method, sensitizers can be strongly adsorbed for extended wavelength response (Kalyanasundaram et al., 1987). Semiconductors can also be dispersed into cationic or anionic vesicles, either on the inside or the outside of the heterogeneous aggregate (Rafaeloff et al., 1985; Degani and Willner, 1986). Polymeric films can similarly be employed as supports, particularly if the semiconductor can be introduced by cation exchange (Meissner et al., 1983; Mau et al., 1984). Cellophane has been used as a support for CdS (Milosavljevic and Thomas, 1986).

Semiconducting colloids such as TiO_2, RuO_2, RuS_2 and CdS have also been stabilized against aggregation and precipitation by intercalation in clays and used in photocatalysis (Van Damme et al., 1986; Yoneyama et al., 1989). The results of studies by Yoneyama et al. (1989), in which the photocatalytic properties of microcrystalline TiO_2 incorporated in the interlayer space of montmorillonite, are particularly noteworthy. The pillared TiO_2 microcrystallites showed an approximately 0.58 eV blue shift in absorption compared with TiO_2 particles in accord with the quantum size effect (Anpo et al., 1987). The excited electronic states of the pillared TiO_2 were determined to be 0.36 V more negative than that of TiO_2 powder particles. Photocatalytic activities of the pillared TiO_2 were significantly greater than those of the TiO_2 powder particles for decomposition of 2-propanol to give acetone and hydrogen (Table 4) and of n-carboxylic acids with up to eight carbons to give the corresponding alkanes and carbon dioxide. The increased photocatalytic activity of the pillared semiconductor has been accredited in part to its higher surface area and particularly to the higher excited electronic states (Yoneyama et al., 1989).

Zeolites are also attractive candidates for supports for photoactive semiconducting particles. Their aluminum silicate frameworks render them highly stable under ambient conditions, and their varied structural porosity allows for controlled variance in the introduction or growth of the semiconductor particle (Liu and Thomas, 1989; Fox and Pettit, 1989). Fox and Pettit (1989) found that encapsulated CdS powders modified with an appropriate hydrogen evolution catalyst (Pt or ZnS) evolved hydrogen at nearly the same rate as large particles adsorbed onto an inert silica support but were considerably more stable to mechanical disruption than the externally supported semiconductor.

Intercalation of foreign species into semiconductors. Intercalation of semiconducting layer type transition metal dichalcogenides by a variety of organic molecules, alkali metals or "3d" transition metals provides a powerful way to finely tune the electron occupation of the relatively narrow "d" bands in these solids (Friend and Yoffe, 1987). These transition metal dichalcogenides are highly anisotropic solids, sometimes referred to as "two-dimensional" solids, and the intercalant molecules which are electron donors enter between the layers. This can result in profound changes in the electronic properties of the host lattice.

Applications of geochemical significance

The major motivation for investigation of semiconducting electrodes and powders has been the potential for developing an efficient water-splitting process capable of generating oxygen and hydrogen. Most of this work has been done on wide band gap semiconductors such as TiO_2 and $SrTiO_3$ loaded with noble metal catalysts. Excitation of the semiconductor by ultraviolet light generates electron-hole pairs; i.e.,

$$TiO_2 \xrightarrow{h\nu} TiO_2 \, (e_{cb}^- + h^+) \ .$$

The electrons are trapped by the noble metal deposit where hydrogen is subsequently evolved:

$$2e_{cb}^- + 2H^+ \rightarrow H_2 \ .$$

The concomitant hole reaction involves water oxidation to oxygen:

$$4h^+ + 2H_2O \rightarrow O_2 + 4H^+ \ .$$

Aspects of these processes have been presented in many articles and reviews (e.g. Nozick, 1978; Harris and Wilson, 1978; Gratzel, 1986; Gratzel, 1988) and are not considered further here. Rather, attention is focussed in this article either on reactions involving semiconducting minerals that are of some geochemical interest or on reactions providing added insight into the nature of naturally occurring semiconducting minerals.

Photodissolution of semiconducting minerals. The photo-generated holes and electrons in semiconductors are generally characterized by strong oxidizing and reducing potentials, respectively. Instead of being injected into the electrolyte to drive redox reactions, these holes and electrons may oxidize or reduce the semiconducor itself and cause decomposition (or phase transformation). Thermodynamic and kinetic aspects of the photodissolution of semiconductors have been reviewed by Gerischer (1980). Essential aspects of the thermal (i.e., non-photochemical) dissolution of minerals are reviewed elsewhere in this volume (Hering and Stumm, 1990).

Following Gerischer (1980), and specifying a semiconductor as binary compound MA, the reaction,

$$MA + ze^- + solv \rightarrow M + X_{solv}^{z-} \ ,$$

represents corrosion of the host crystal by its conduction band electrons, resulting in reduction of the cation and dissolution of the anion. Similarly the reaction,

$$MA + zh^+ + solv \rightarrow M_{solv}^{z+} + X \ ,$$

represents corrosion of the host crystal by its valence band holes, resulting in oxidation of the anion and dissolution of the cation. If the redox potentials associated with these reactions are $_nE_{decomp}$ and $_pE_{decomp}$ respectively, then (somewhat simplistically) electrons would be expected to be corrosive if $_nE_{decomp}$ lies below the conduction band edge (i.e., $_nE_{decomp} < E_c$) and holes would be expected to be corrosive $_pE_{decomp}$ lies above the valence band edge (i.e., $_pE_{decomp} > E_v$). These conditions are summarised in Figure 14.

It should be recognised that the above conditions are thermodynamically approximate and only give an indication of the likelihood of decomposition occurring. In addition, kinetic factors may modify the thermodynamic perception markedly. For example, even when decomposition of the semiconductor is thermodynamically feasible, it may be negligible if a competing thermodynamically allowed redox reaction in solution is rapid (Nozik, 1978;

582

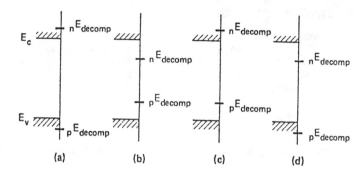

Figure 14. Possible positions for the thermodynamic energy levels for electron-induced or hole-induced corrosion relative to the band edges. If condition (a) exists, the semiconductor is relatively stable; condition (b) implies instability if either electrons or holes reach the surface; conditions (c) and (d) imply instability to holes and to electrons, respectively (Morrison, 1980).

Gerischer, 1980). Also, structural influences such as surface orientation, structural defects, etc., must be taken into account (Morrison, 1980).

Considerable interest has been shown in the light-induced decomposition of semi-conducting iron oxides both because of their small band gap (and thus their potential for excitation by solar radiation) and because of their importance in nature both as source of iron for plant nutrition and as adsorbers for other species of geochemical interest. The role of organic ligands including hydroxycarboxylates (Waite and Morel, 1984b; Waite, 1986; Cornell and Schindler, 1987; Cunningham et al., 1988; Litter and Blesa, 1988; Siffert, 1989; Sulzberger, 1990), a glycollate (Cunningham et al., 1985) and thiocarboxylates (Waite and Torikov, 1987) in inducing or enhancing the dissolution of a variety of iron oxides (hematite, maghemite, magnetite, lepidocrocite, goethite, ferrihydrite) have been reported as has the ability of bisulfite to induce the photodissolution of hematite (Faust and Hoffmann, 1986; Faust et al., 1989). For example, illumination of α-Fe_2O_3 with visible light ($\lambda = 504$ nm) in the presence of S(IV) results in an approximately 40-fold increase in the rate of dissolution of iron oxide over that observed in the dark (Fig. 15). Similarly, photolysis of suspensions of γ-FeOOH in the presence of relatively low concentrations of citric acid results in a significant enhancement in rate of dissolution of this oxyhydroxide at pH $\leq$ ca.5 (Fig. 16).

The primary photoprocess leading to dissolution of the oxide in these cases may well be the light induced generation of electron-hole pairs with the hole oxidizing the ligand at the oxide surface. The conduction band electrons may induce reduction of lattice Fe(III) to Fe(II) with subsequent bond breakage and, depending on the affinity of Fe(II) for the oxide surface, release to solution (Fig. 16). Alternatively (or, in addition), the conduction band electrons may induce phase transformation within the oxide lattice. Thus, in studies of the dynamics of charge transfer from reduced methylviologen radicals to colloidal Fe(III) oxides and oxyhydroxides, Mulvaney et al. (1988a,b) reported that a fraction of the electrons transferred to α-Fe_2O_3 colloids were able to migrate into the particle interior and form stable Fe_3O_4. Other workers have previously indicated that redox transformations at iron oxide surfaces may result in phase transformations (Tamaura et al., 1983; Tronc et al., 1984a,b). Further details of structural changes associated with (non-photochemical) electron transfer within minerals may be found elsewhere in this volume (White, 1990).

Rather than generation of electron-hole pairs as a result of light absorption by the bulk semiconducting oxide, a primary photoprocess whereby light induces transfer of electrons from the adsorbed ligand to the metal centers of the underlying substrate may be envisaged. Such ligand to metal charge transfer (LMCT) transitions are well documented for a wide range of ferric carboxylate complexes in homogeneous solution (Balzani and Carassiti,

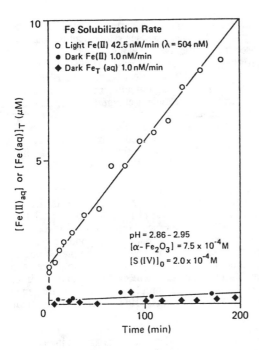

Figure 15. Photochemical and thermal dissolution of hematite in anoxic suspensions containing S(IV) (Faust and Hoffmann, 1986).

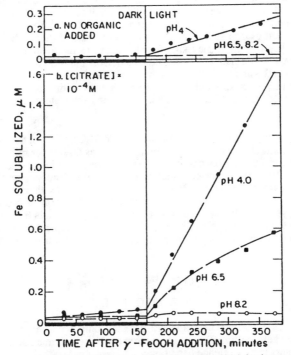

Figure 16. Dissolution of 5μM γ-FeOOH under dark and light conditions in a) the absence, and b) the presence of 10^{-4}M citrate. Light source: simulated solar spectrum of total intensity 300 μEinsteins cm^{-2} min^{-1} (Waite and Morel, 1984).

584

1970) and might be expected to occur at similar energies for "surface located" complexes. Processes involving light absorption by surface located species rather than the bulk solid are discussed in more detail in a later section. While the wavelength dependency of these primary processes might be expected to differ somewhat, the form of the kinetic expressions for dissolution of the solid substrates are essentially identical (Litter and Blesa, 1988; Sulzberger, 1990). For example, both processes are dependent upon intimate contact of the ligand with the underlying solid substrate to ensure efficient charge transfer. Indeed, the rate of dissolution of iron oxides exhibits a rectangular hyperbolic dependency on ligand concentration expected for a process dependent on a rapid preliminary adsorption step (Fig. 17). Not surprisingly, the rate determining step in the photoinduced or photoassisted solubilization of iron oxides appears to be that of ferrous iron detachment rather than ligand adsorption or interfacial electron transfer (Waite and Morel, 1984b; Waite, 1986).

In many systems, photochemical processes may simply complement existing thermal processes and may exhibit both homogeneous and heterogeneous components. Thus, the thermal dissolution of iron oxyhydroxides in the presence of ligands such as oxalate and EDTA involves a preliminary, rapid formation of surface complexes $>Fe^{III}$-L, a slow decomposition of these surface complexes through parallel paths corresponding to acid and reductive dissolution, and a fast reductive reaction between Fe^{II}-L complexes and Fe^{III} surface centers, conducive to a strong acceleration when Fe_{aq}^{2+} reaches appropriate values. In the presence of light, both the slow and fast processes are accelerated due to the additional formation of Fe^{2+} by photolysis of the oxide surface complexes, and by homogeneous photolysis of aqueous Fe^{III}-L (Litter et al., 1990; Sulzberger, 1990; Cornell and Schindler, 1987).

Diurnal variation in the concentrations of dissolved iron and manganese in marine and fresh waters have been attributed to the absorption of light by iron and manganese oxides or by complexes located at their surfaces (Sunda et al., 1983; Waite and Morel, 1984; Waite et al., 1988; Hong and Kester, 1986; McKnight et al., 1988). In addition, light is certain to be a major determinant of iron speciation in atmospheric water droplets (Behra and Sigg, 1990). The possibility that light may significantly increase the availability of iron and manganese to organisms has also been suggested (Anderson and Morel, 1983; Finden et al., 1984; Sunda, 1988) and is an effect that may have particular significance in aquatic systems where iron availability is considered to be the key factor limiting primary productivity (Entsch et al., 1983; Martin and Gordon, 1988; Martin and Fitzwater, 1988; Martin et al., 1990).

Hydrogen peroxide production. Hydrogen peroxide, H_2O_2, is one of the most powerful oxidants present in natural waters. Concentrations as high as 60 μM for rainwater and 250 μM for cloudwater samples have been reported. Peroxide concentrations in the range 10-300 nM are more typical for surface marine waters and low micromolar concentrations of H_2O_2 have been reported for freshwaters (Waite, 1990 and references therein). A potential mechanism of production in natural waters (and particularly in low organic content aerosols) involves the photolysis of naturally occurring semiconducting minerals (such as the oxides of iron, titanium and zinc) with light of energy above the band gap energy (Bahnemann et al., 1987). H_2O_2 can then be generated via two different pathways once electron-hole pairs are produced; i.e.,

$$O_2 + 2 e_{CB}^- + 2 H_{aq}^+ \rightarrow H_2O_2 ,$$

$$2H_2O + 2 h_{VB}^+ \rightarrow H_2O_2 + 2 H_{aq}^+ .$$

Appreciable yields of hydrogen peroxide are detected only when appropriate electron donors, D, are added before illumination. The elctron donor, which must be adsorbed to the particle surface, reacts with a valence band hole:

$$D + h^+ \rightarrow D^+ .$$

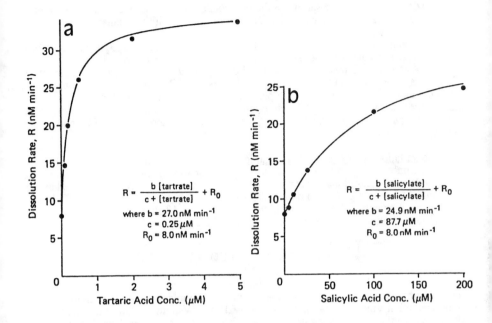

Figure 17. Dependence of rate of dissolution of 5μM γ-FeOOH in pH 4.0, 0.01M NaCl on concentration of a) tartaric acid, and b) salicylic acid. Fitted parameters obtained for rectangular hyperbolic model are given. Light source: Mercury arc lamp with 365 nm bandpass filtering (Waite 1986).

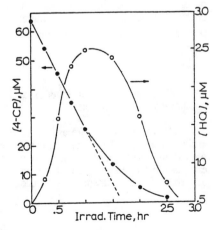

Figure 18 (right). Heterogeneous (TiO₂) photocatalyzed degradation of 4-chlorophenol (4-CP) (●) and the formation of hydroquinone (HQ) (O) as a function of irradiation time. Initial [4-CP] = 63 μM, initial pH = 5.85 (El-Akabi et al., 1989).

Electron donors bound to the surface of the semiconductor particles interfere with e^-/h^+ pair recombination, allowing conduction band electrons to react with molecular oxygen.

Kormann et al. (1988) obtained steady-state concentrations of H_2O_2 in excess of 100 μM on illumination of suspensions of ZnO and report quantum yields for peroxide formation of ca. 15% at 330 nm. However, only low hydrogen peroxide concentrations were found on TiO_2 suspensions, presumably because of the high stability of Ti-peroxo complexes. Illumination of desert sand resulted in peroxide formation with the attainment of a steady state concentration of ca. $2 \times 10^{-7}M$ although thermal degradation of H_2O_2 appeared to be relatively rapid, possibly as a result of a "Fenton-type" reaction between H_2O_2 and Fe(II) present in the desert sand.

   Degradation of organic and inorganic pollutants. Semiconducting minerals may induce the photocatalysed transformation of organic and inorganic species both at natural and pollutant concentrations. Essential aspects of this process are aptly demonstrated by the results of Al-Ekabi et al. (1989) on the TiO_2-mediated degradation of a variety of chlorophenols used as herbicides and fungicides. Thus, Figure 18 illustrates the course of the degradation reaction of 4-chlorophenol (4-CP) (63 μM),

$$ClC_6H_4OH + (13/2)O_2 \xrightarrow[TiO_2]{h\nu} 6CO_2 + 2H_2O + HCl \ ,$$

and the formation and subsequent degradation of its major intermediate, hydroquinone (HQ) with irradiation time. Direct photolysis of 4-CP by UV-light (≥ 250 nm) yields eleven detected organic intermediates, some of which undergo no further degradation (Pichat, 1988). By contrast, UV illumination of an aqueous slurry of 4-CP and TiO_2 produces six detected intermediates, all of which further degrade to CO_2 and HCl.

The rate of degradation of 4-CP follows a Langmuir-Hinshelwood expression:

$$r_o = \frac{k_r K C_o}{1 + K C_o + \sum_{i=1}^{n} K_i C_i} \ , \tag{30}$$

where r_o is the reaction rate, k_r is the reaction rate constant, K is the adsorption coefficient for the primary reactant, C_o is the initial concentration of the primary reactant, i is the number of intermediates (including the solvent) and C_i and K_i are the concentrations and adsorption coefficients for intermediates of interest. For high concentrations of 4-CP (in this case ≥ 13 μM), $KC_o \gg 1 + \sum_{i=1}^{n} K_i C_i$ and Equation 30 reduces to the simple zero order rate expression,

$$C_o - C = k_r t \ , \tag{31}$$

and follows partial order kinetics at lower concentrations of 4-CP (indicating some competition for surface sites).

The formation of HQ as the major intermediate suggests the involvement of •OH radicals (Fig. 19). Hydroxyl radicals have been implicated as the reactive species in the photocatalysed degradation of many organic compounds (Turchi and Ollis, 1990 and references therein) and may be formed either (i) via reaction of valence band holes with either adsorbed H_2O or with surface OH$^-$ groups; i.e.,

$$h_{VB}^+ + H_2O(ads) \rightarrow •OH + H^+ \ ,$$

$$h_{VB}^+ + OH^-(surf) \rightarrow •OH \ ,$$

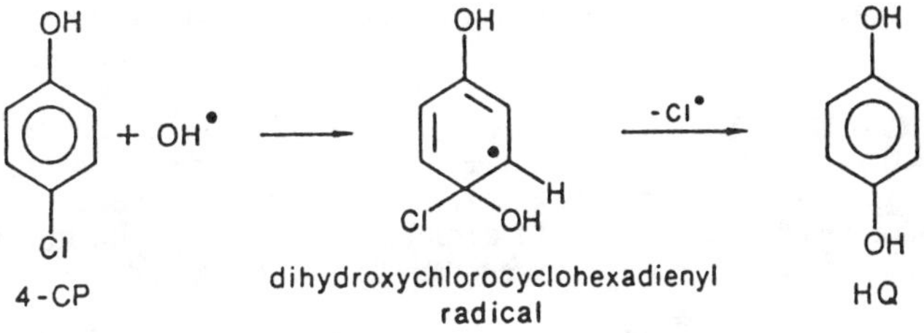

Figure 19. Generation of hydroquinone via interaction of •OH with 4-CP (El-Akabi et al., 1989).

or (ii) via H_2O_2 from $O_2^{\bullet-}$. It is generally accepted that surface-adsorbed oxygen delays the electron-hole pair recombination process by trapping the conduction band electron as superoxide ion which subsequently disproportionates to H_2O_2; i.e.,

$$O_2 + e_{CB}^- \rightarrow O_2^{\bullet-} \rightarrow H_2O_2 \ .$$

Cleavage of H_2O_2 by a variety of pathways may then yield •OH; i.e.,

$$H_2O_2 + e_{CB}^- \rightarrow \bullet OH + OH^- \ ,$$

$$H_2O_2 + O_2^{\bullet-} \rightarrow \bullet OH + OH^- + O_2 \ ,$$

$$H_2O_2 \xrightarrow{h\nu} 2\bullet OH \ .$$

In addition to the possibilities for degradation of organic pollutants in waters and wastes, some consideration has been given to the use of photolyzed semiconductor suspensions in the oxidation of inorganic contaminants. Particular attention has been given to the oxidation of cyanide, CN^-, to the less toxic forms of cyanate, OCN^-, and thiocyanate, SCN^- (Serpone et al., 1988a).

PHOTO-REDOX PROCESSES INVOLVING INTERFACIAL CHARGE TRANSFER TO THE MINERAL SUBSTRATE

Charge injection into semiconductors

The generation of separated electron-hole pairs at the semiconductor-liquid interface using band-gap excitation has been discussed above. However, most of the effective, stable (i.e. non-corrosive) semiconductors exhibit large band gaps requiring energy in the UV - near UV spectral region. In terms of both (i) utilization of solar energy to productive ends and (ii) the effect of light in influencing the geochemical cycling of the elements, heterogeneous systems which absorb solar radiation in the near UV to visible region (i.e., 300 - 700 nm) are of major interest.

An alternative process by which charge transfer processes may be photo-initiated at the solid-liquid interface involves the excitation of a surface-located species with subsequent transfer of either energy or electrons to the semiconductor (Fox, 1986; Nakabayashi and Kawai, 1988). Charge (or energy) transfer by this route is particularly effective if the sensitizing chromophore at the semiconductor surface is generated as a result of complexation of surface cations. However, electron and energy transfer have also been reported in cases where the adsorbed species is a chromophore in its own right. Both cases are considered in more detail below. It should be noted that the distinction between these two

588

cases is relatively fine since electron transfer will occur via surface states in both cases, the energy of which will be influenced by the adsorption and/or complexation reactions at the surface.

<u>Charge transfer via adsorbed chromophores.</u> A semiconductor may be activated by electron or energy transfer from a photoexcited chromophore if the chromophore's ground state reduction potential (and hence the energy of its lowest unoccupied molecular orbital) lies above that of the conduction band of the solid (Fox, 1986). This situation is effectively equivalent to that of oxidative hole capture by an adsorbed electron donor.

While examples of the sensitization of wide band gap semiconducting solids by visible light induced excitation of adsorbed chromophores are not as numerous as that of direct excitation of the semiconductor, a number of cases have been reported (see for example Watanabe et al., 1977; Hawn and Armstrong, 1978; Matsumura et al., 1981; Kamat and Fox, 1983). The early study by Watanabe et al. (1977) in which highly efficient N-deethylation of rhodamine B occurs on impact of visible light when adsorbed to CdS (but not in homogeneous solution) clearly demonstrates the essential elements of this process. As shown in Figure 20, absorption of visible light by the adsorbed dye molecule creates an excited (singlet) state at an energy 0.1-0.2 eV above the conduction band edge of CdS. Electron transfer from this excited state to the conduction band of CdS initiates the photochemical N-deethylation of adsorbed rhodamine-B (represented below as $[>N-C_2H_5]_{ads}$) via formation of a dye radical cation; i.e.,

$$[>N-C_2H_5]_{ads} \xrightarrow{h\nu} [>N-C_2H_5]_{ads}^{*} ,$$

$$[>N-C_2H_5]_{ads}^{*} \rightarrow [>N^{+}-C_2H_5]_{ads} + e_{cond}^{-} .$$

The conduction band electron, e_{cond}^{-}, is most likely trapped by surface bound oxygen to give the superoxide radical anion, according to:

$$e_{cond}^{-} + O_{2\,ads} \rightarrow O_2 \bullet^{-} .$$

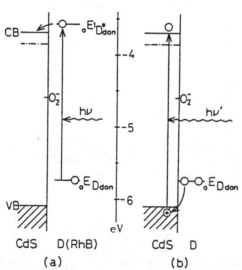

Figure 20. Schematic diagram of the initiation steps for the photochemical N-deethylation of rhodamine B on CdS by (a) excitation of adsorbed dye, and (b) excitation of CdS. $_{o}E1_{Ddon}^{*}$ and $_{o}E_{Ddon}$ denote the electron transfer energy terms of the excited singlet state and of the ground state respectively of rhodamine B in polar solvents. The perpendicular scale is the electronic energy vs. vacuum (Watanabe et al., 1977).

Watanabe et al. (1977) suggest that the N-deethylation most likely occurs via reaction between the dye radical cation and the superoxide radical anion.

For comparitive purposes, it is instructive to also present the preliminary steps that would occur on direct excitation of the semiconductor by band gap radiation (see Fig. 20):

$$CdS \xrightarrow{h\nu} h_{val}^+ + e_{cond}^- \, ,$$

$$h_{val}^+ + [>N-C_2H_5]_{ads} \rightarrow [>N^{+\cdot}-C_2H_5]_{ads} \, .$$

Gratzel (1988) points out that surface attachment of the sensitizer is usually essential. If the dye is merely dissolved in the electrolyte, the excited state will be so rapidly quenched (typically within 10^{-8} seconds) that charge transfer following diffusion to the semiconductor surface will be very inefficient. Gratzel (1988) also notes that it is typically only the first monolayer of adsorbed molecules that can transfer charge with thicker layers tending to be insulating.

Charge transfer via photoactive surface complexes. As indicated above, the complexation of surface cations may give rise to the formation of chromophores acting as sensitizers for the semiconductor. An effect of this type has been reported by Houlding and Gratzel (1983) in which 8-hydroxyquinoline was shown to undergo a chelation reaction with colloidal TiO$_2$ resulting in the formation of a yellow complex that was effective in sensitizing the visible light induced generation of hydrogen from water. Similarly, Brandli et al. (1970) established that o,o'-dihydroxy azo dyes, such as 1-[(2-hydroxy phenyl) azo]-2-naphthol, form a 1:1 complex at the surface of ZnO powders and, in so doing, sensitize the semiconductor to visible light.

More recently, Gratzel and coworkers (Vrachnou et al., 1987; Desilvestro et al., 1988) have investigated the visible light induced sensitization of TiO$_2$ by surface complexation with Fe(CN)$_6^-$. While the ferrocyanide anion shows little optical absorption above 360 nm, it forms a charge transfer complex with Ti^{IV} at the TiO$_2$ surface which shifts the onset of the photo-response of this semiconductor from 400 to 700 nm. Similarly, Frei et al. (1990) report that complexation of Ti^{4+} at the TiO$_2$ surface by phenylfluorone (PF) results in sensitization of the semiconductor to visible light. Pulsed laser studies indicate that charge injection from the electronically excited surface chelate into the conduction band of the colloid:

$$PF-TiO_2 \xrightarrow{h\nu} PF^+ + TiO_2(e_{cb}^-)$$

is a very rapid reaction occurring in less than 10^{-8} secs. By contrast, the recombination of the conduction band electron with the oxidized chelate is a much slower process. Interestingly, Frei et al. (1990) show that the presence of adsorbed PF significantly increases (by about 100 times) the rate of transfer of conduction band electrons to other electron acceptors present in solution (methylviologen in their case) over that observed in the absence of chelator (Gratzel, 1989) (Fig. 21). Indeed, the acceleration of the redox reaction by the chelator is so significant that the interfacial redox reaction can successfully compete with the recapture of the conduction band electron by the parent cation radical PF$^+$. Frei et al. (1990) rationalise this dramatic enhancement of interfacial electron transfer in terms of removal by chelation of intra band gap surface states present at the TiO$_2$ surface as a result of the partial coordination of lattice Ti^{4+} ions to water molecules (Howe and Gratzel, 1985).

Gratzel (1988) notes that although the sensitizer-initiated excitation of a semiconductor such as TiO$_2$ is functionally similar to the operation of a conventional n-type photoanode, the surface complex (and adsorbed chromophore) sensitized semiconductor operates by electron injection and is therefore a majority carrier device. Thus, the high recombination

590

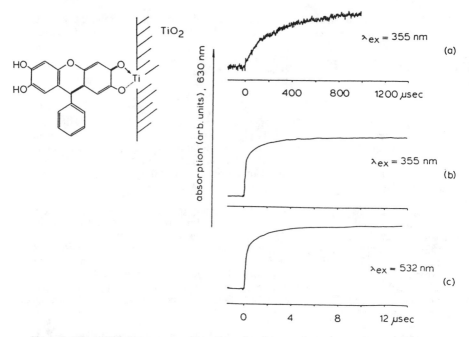

Figure 21. Effect of surface chelation on the kinetics of electron transfer from the conduction band of TiO$_2$ to methylviologen. Oscillograms showing the temporal behaviour of the 630 nm absorbance after laser excitation of aqueous solutions (pH 4.85) containing colloidal TiO$_2$ (1 g/L) and 10^{-3}M MV^{2+}: (a) bare TiO$_2$ particles, excitation wavelength 355 nm (i.e. band gap excitation); (b) PF-chelated TiO$_2$ colloid (PF concentration 3 x 10^{-5}M), excitation wavelength 355 nm (i.e. band gap excitation); (c) PF-chelated TiO$_2$ colloid (PF concentration 3 x 10^{-5}M), excitation wavelength 532 nm (i.e. surface chelate excitation) (Frei et al., 1990).

losses due to disorder in a semiconductor structure where the photoexcited electroactive charge carriers are holes are not encountered in this case.

A process involving excitation of a surface complex followed by charge transfer to the solid semiconductor may also be involved in the photodissolution of iron and manganese oxides in the presence of adsorbed carboxylic acids though, as discussed earlier, the observed phenomena may equally be accounted for by electron-hole pair generation within the bulk solid.

PHOTO-REDOX PROCESSES INVOLVING CHROMOPHORES ADSORBED ON OR INCORPORATED IN MINERAL SUBSTRATES

In this section, consideration is given to photo-redox processes involving chromophores adsorbed on or incorporated in mineral substrates. In these cases, the underlying (or encompassing) substrate may modify or mediate the photo-process but does not accept or donate charge itself. Of particular interest are the effects of mineral surfaces including silica and aluminosilicates (clays) on the photochemistry of adsorbed chromophores and this issue is discussed in some detail below. Of less geochemical relevance (but considerable industrial interest) are the photochemical and photophysical properties of chromophores incorporated within zeolite cavities. A little more detail on this topic is presented at the end of this section.

It should be noted that two complementary goals have motivated studies of photo-processes of molecules in the adsorbed state; (i) to observe and understand how surface interactions modify the excited state and subsequent reactions, and (ii) to use photochemistry to probe the nature of the underlying solid surface. Both goals are relevant to varying degrees in the material presented here.

Photoprocesses of chromophores adsorbed to silica and clay surfaces

There is now considerable interest in the influence of inorganic supports (in the form of colloidal dispersions, or as granules, powders or gels) on the photochemistry of adsorbed molecules. Indeed, there is growing evidence that specific features of the surface geometry and chemistry can dramatically affect both properties of the excited state and complete reaction pathways (Oelkrug et al., 1986; Thomas, 1987).

Photoprocesses on particulate silica and silica gel. Particles of silica represent an interesting substrate for photochemical investigation because the surface silanol groups are typically deprotonated over a considerable portion of the pH range and the resulting surface potentials are high ($\approx$ -170 mV) (Iler, 1979). The negative surface charge can aid quite significantly in separation of photoredox products. Such a phenomenon has been clearly demonstrated by Calvin, Willner and coworkers using silica-bound sensitizers such as $Ru(bpy)_3^{2+}$ and viologen salts (e.g. propylviologen sulfonate, PVS) as acceptor relays (Willner et al., 1981a; Willner et al, 1981b; Degani and Willner, 1983). Thus, steady-state excitation of $Ru(bpy)_3^{2+}$ in the presence of colloidal SiO_2 results in an emission spectrum similar to that observed in homogeneous aqueous phase. Addition of PVS° quenches the luminescence of the excited $Ru(bpy)_3^{2+}$:

$$^*Ru(bpy)_3^{2+} + PVS^\circ \xrightarrow{k_q} Ru(bpy)_3^{3+} + PVS\bullet^-$$

at a rate similar to that in homogeneous solution ($k_q = 1.5 \times 10^9$ $M^{-1}s^{-1}$). However, the back-electron-transfer reaction:

$$Ru(bpy)_3^{3+} + PVS\bullet^- \xrightarrow{k_b} Ru(bpy)_3^{2+} + PVS^\circ$$

occurs at a 100-fold slower rate in the presence of SiO_2 than in the homogeneous phase. In the presence of silica, the negatively charged reduced acceptor, $PVS\bullet^-$, is repelled by the negatively charged interface while the oxidized sensitizer, $Ru(bpy)_3^{3+}$, is still associated with the SiO_2 particle (Fig. 22).

Another example of the effect of negatively charged colloidal silica is given by the photo-physical behaviour of adsorbed uranyl ions (Wheeler and Thomas, 1984). The emission is much longer lived ($t_{1/2} \approx 440$ μsec) and more intense in the presence of silica than in water ($t_{1/2} \approx 11$ μsec). In addition, quenching by anionic species such as I^-, Br^- and SCN^- is inhibited by a factor of 10^3-10^4 due to adsorption of the uranyl ion on the negatively charged silica particle.

Studies have also been conducted using larger silica particles possessing a porous structure capable of incorporating guest chromophores (Wheeler and Thomas, 1983; Birenbaum et al., 1989). For example, Birenbaum et al. (1989) have investigated the photoinduced charge transfer interactions between pyrene (Py) and diethylaniline (DEA) on porous silica possessing varying average pore size distributions. These workers found that fluorescence quenching of $^1Py^*$ by DEA was completely inhibited for monolayer coverage of the silica surface by DEA, presumably as a result of the strong adsorption interactions of the n-electrons of DEA with the acidic protons of the surface silanol groups.

592

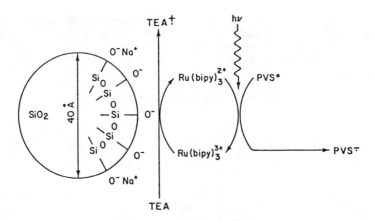

Figure 22. Schematic depicting function of negatively charged SiO₂ particles in assisting in separation of products formed in photosensitized reactions (Willner et al., 1981b).

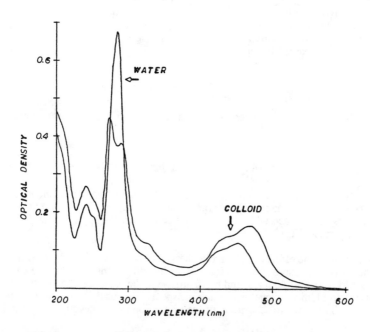

Figure 23. Absorption spectrum of 8.5 x 10⁻⁶M Ru(bpy)₃²⁺ in water and adsorbed on 1g of montmorillonite/L with a montmorillonite colloid (1 g/L) in the reference cell (Della Guardia and Thomas, 1983).

An effective charge transfer process was only observed when an excess of free (nonadsorbed) DEA molecules became available with the process then being markedly affected by the average pore size and the surface fractal dimension of the silica.

Extensive studies of photoprocesses have also been reported for chromophores associated with silica gel. Much of this work has been reviewed by Kalyanasundaram (1987) and will not be covered here.

Photoprocesses on clay minerals. Clays (in their various forms) may be considered a higher form of silica with unique structural features. For example, the swelling clays, known as smectites, possess layer lattice structures containing Si and Al in which two-dimensional oxyanions are separated by layers of (exchangeable) hydrated cations. In addition to the exchangeable cations, which may be almost any kind of organic, inorganic or organometallic cation, the interlayer space is able to accommodate water and/or a wide variety of organic species. This "intercalation" process does not occur in nonswelling clays such as kaolinite and any adsorbed molecules in this type of clay are confined to the external surface.

Transition metal complexes such as $Ru(bpy)_3^{2+}$ are readily introduced onto smectite clays such as hectorite or montmorillonite and the photochemistry of this and other incorporated complexes have been the subject of several investigations (Krenske et al., 1983; Della Guardia and Thomas, 1983; Ghosh and Bard, 1984; Schoonheydt et al., 1984; Nakamura and Thomas, 1985; Nakamura and Thomas, 1986). Some of the implications of intercalation of chromophores within smectites are exemplified by the effects on $Ru(bpy)_3^{2+}$. Thus, the absorption and luminescence of $Ru(bpy)_3^{2+}$ undergo significant perturbations in clays (Fig. 23). The charge transfer absorption band shifts toward the red (maxima at 472, 462 and 465 nm in montmorillonite, hectorite and kaolinite respectively) with about 50% increase in the absorption cross-section. The intensity of the π-π^* band near 285 nm is significantly reduced in intensity (and splits into two bands). The differences in the absorption spectra have been attributed to interaction between the $Ru(bpy)_3^{2+}$ complex and the clay surface. Possible explanations include distortion of the bipyridine ligands due to steric constraints and an enhancement in the metal-to-ligand charge transfer in the ground state (Kalyanasundaram, 1987). Double exponentials have been fitted to the emission decay profiles for this complex in the presence of clays and rationalized in terms of two types of adsorption sites on the clay particle: on the outer surface and intercalated between the layers.

As discussed above for silica substrates, other molecules may be added to clay suspensions which quench the fluorescence of excited $Ru(bpy)_3^{2+}$ via electron transfer reactions. The quenching kinetics are dependent, among other factors, on the extent of interaction of the quenching agent with the surface. Thus, Cu^{2+} is strongly adsorbed and quenches at a lower rate than in aqueous solution thus providing an estimate of the degree of movement of Cu^{2+} ions on the clay surface. Other quenching agents such as dimethylanaline and nitrobenzene are minimally adsorbed to the clay surface and exhibit enhanced quenching kinetics over that observed in aqueous solution. The kinetics of these latter reactions indicate that the reactive quencher molecules are adsorbed around the $Ru(bpy)_3^{2+}$ in a zone-like manner, rather than being adsorbed randomly throughout the system. This tends to indicate that the sites of adsorption are not uniform on the clay surface but occur in regions.

Photoprocesses in zeolites

General features of zeolites. Zeolites (from the Greek words zeo, "to boil", and lithos, "stone") are synthetic or natural minerals that often expel water so violently when heated that they appear to boil (Turro, 1986). Zeolites are a special class of crystalline aluminosilicates with a well-defined framework structure within which there are cavities, pores or channels of varying dimensions. The fundamental building blocks of the framework are SiO_4 and AlO_4 tetrahedra that are arranged in such a way that each of the four oxygen atoms is shared with another silica or alumina tetrahedron. Each AlO_4 unit bears a net negative charge (Al is normally trivalent, so the fourth bond to oxygen results in a negative charge on the Al atom) which must be compensated by the presence of cations such as H^+, Na^+, Ca^{2+}, etc. These cations are generally mobile, because they are ionically and not covalently bound to the framework structure. The ratio of Si to Al atoms in zeolites may be varied from about 1 to essentially infinity. The greater the number of Al atoms, the greater the number of cations and sites which may adsorb water (i.e. Si/Al ratio small implies strongly hydrophilic and vice versa). Obviously, the Si/Al ratio will also be an indicator of size, shape and topological characteristics of the internal void space.

In X, Y and A zeolites, the Si and Al tetrahedra link together to form cubooctahedron, the so-called sodalite unit, which may stack in various ways and contain various amounts of Si and Al. X, Y and A zeolites have Si/Al ratios of 0.7-1.2, 1.0-1.5 and 1.5-3.0 respectively. For example, the sodium zeolite NaX (faujasite type) has a Si/Al ratio of about 1.5 (and is thus strongly hydrophilic) and contains relatively large cavities (roughly spherical "super-cages" of approx. 13 Å diameter) which are connected by pores (roughly circular "windows" of approx. 8 Å diameter). At the other extreme, the framework of the ZSM class involves a pentasil rather than a sodalite building block and has a Si/Al ratio of 20 or greater (and is thus very hydrophobic). The pentasil building blocks link together to form chains or sheets which generate a void space topology which consists of long tubular channels of diameter approx. 6 Å and lengths of approx. 50 Å. In both these systems, the ratio of internal to external surface area is very large (>1000) providing a strong driving force for occupancy of the internal surface for any molecules that have the size/shape characteristics which allow entry to the interior (Turro, 1986).

Photochemistry of inorganic ions exchanged into zeolites. This topic has been reviewed by Kalyanasundaram (1987) and only selected examples of zeolite-mediated photoprocesses involving exchanged ions are given here. For example, Ag^+- and Ti^{3+}-exchanged zeolites have been found to be sensitive to visible light and have been adapted to photochemical and thermal water-splitting cycles (Jacobs et al., 1977; Kuznicki and Eyring, 1978). Ag^+-zeolites release molecular oxygen (O_2) from adsorbed water on visible light photolysis with the formation of molecularly dispersed Ag°:

$$Ag^+\text{-}Y \underset{+H_2O}{\overset{h\nu}{\rightarrow}} Ag^\circ\text{-}Y + H^+ + O_2 \ .$$

The reduced zeolites are capable of thermal reduction of water to H_2 and oxidation of Ag° at temperatures $\geq 600°C$:

$$Ag^\circ\text{-}Y \underset{+H_2O}{\overset{\Delta}{\rightarrow}} Ag^+\text{-}Y + H_2 + OH^- \ .$$

The zeolite system loses the reversibility after several cycles, presumably due to the sintering of Ag or due to loss of OH groups during the thermal treatment. Ti^{3+}-exchanged zeolites evolve H_2 upon photolysis:

$$Ti^{3+}\text{-}Y \underset{+H_2O}{\overset{h\nu}{\rightarrow}} Ti^{4+}\text{-}Y + H_2 + OH^- \ .$$

However, unlike the Ag-exchanged zeolites, thermolysis of oxygenated zeolite does not evolve O_2 in any significant amounts.

Potential applications of uranyl ion (UO_2^{2+})-exchanged zeolites in solar energy conversion have also received some scrutiny (Suib et al., 1981; Suib et al., 1984). Isopropanol has been found to quench the luminescence of incorporated UO_2^{2+} ions with a dual quenching pattern corresponding to quenching at different sites in the zeolite (internal and external regions). Bulk photolysis in isopropanol lead to the formation of molecular H_2 and acetone which was sustained for over 300 hours:

$$(CH_3)_2CHOH \underset{UO_2^+ - zeolite}{\overset{h\nu}{\rightarrow}} (CH_3)_2CO + H_2 \ ;$$

<u>Photochemistry of organic molecules in zeolites.</u> Particular attention in recent years has focussed on the ability of zeolites to modify and to control the reaction channels available to organic molecules adsorbed on internal and external zeolite surfaces. The shape/size selectivity of zeolites is clearly demonstrated by the ability of pentasil zeolites to separate o- and p-xylene. The approx. 6 Å diameter of ZSM channels allows relatively free diffusion of benzene (kinetic diameter approx. 5 Å) throughout the internal surface. However, the shape of o-xylene is such that there is no orientation which the molecule can assume and achieve a kinetic diameter ≤ 6 Å. As a result, o-xylene is kinetically excluded from diffusion into the channels of ZSM zeolites. On the other hand, p-xylene may achieve an orientation in which its long axis lies along the axis of the ZSM channel, and thereby achieve an effective kinetic diameter (in this orientation) that is equal to benzene (Turro, 1986).

The ramifications of the size/shape selectivity of zeolites is demonstrated by the ability of ZSM zeolites to dramatically alter the product distribution on photolysis of o- and p-dibenzylketones (denoted 0-ACOB and p-ACOB respectively; see Fig. 24) from that obtained in homogeneous solutions. Thus, in homogeneous solution, the photolysis of o-ACOB and p-ACOB leads to similar results; i.e.,

$$\text{o-ACOB} \xrightarrow[-CO]{hv} pA\bullet + B\bullet \rightarrow \underset{25\%}{oAoA} + \underset{50\%}{oAB} + \underset{25\%}{BB} \ ,$$

$$\text{p-ACOB} \xrightarrow[-CO]{hv} pA\bullet + B\bullet \rightarrow \underset{25\%}{pApA} + \underset{50\%}{pAB} + \underset{25\%}{BB} \ .$$

The ratio of the symmetrical (AA + BB) to asymmetrical (AB) coupling products is 1:1, a result consistent with complete randomization of the initial geminate radical pairs into free radicals and consistent with 0% cage effect (Fig. 24). Photolysis of o-ACOB and p-ACOB adsorbed on ZSM zeolits results in strikingly different product distributions:

$$\text{p-ACOB} \xrightarrow[ZSM]{hv} pAB \ ,$$

$$\text{o-ACOB} \xrightarrow[ZSM]{hv} oAoA + BB \ .$$

In the case of the p-ACOB, the cage effects (see Fig. 24) approach 100%, whereas for o-ACOB the cage effects approach -100%; i.e., p-ACOB leads to pAB as the major product, whereas o-ACOB leads to oAoA and to BB as the major products.

CONCLUSIONS

In this chapter, electron transfer reactions occurring at mineral-water interfaces that are either induced or assisted by light have been considered. Such reactions have been divided into three types depending on the nature of the chromophore: (i) those involving absorption of light by the mineral bulk resulting in production of positive and negative charge carriers (holes and electrons) which subsequently induce electron transfer reactions at the mineral-water interface; (ii) those involving light absorbing species located at mineral surfaces and for which the mineral is intimately involved in photo-induced (or enhanced) electron transfer, and iii) those involving light absorbing species adsorbed to mineral surfaces for which the underlying mineral modifies the photochemistry of the chromophore but for which the mineral does not participate in the electron transfer process.

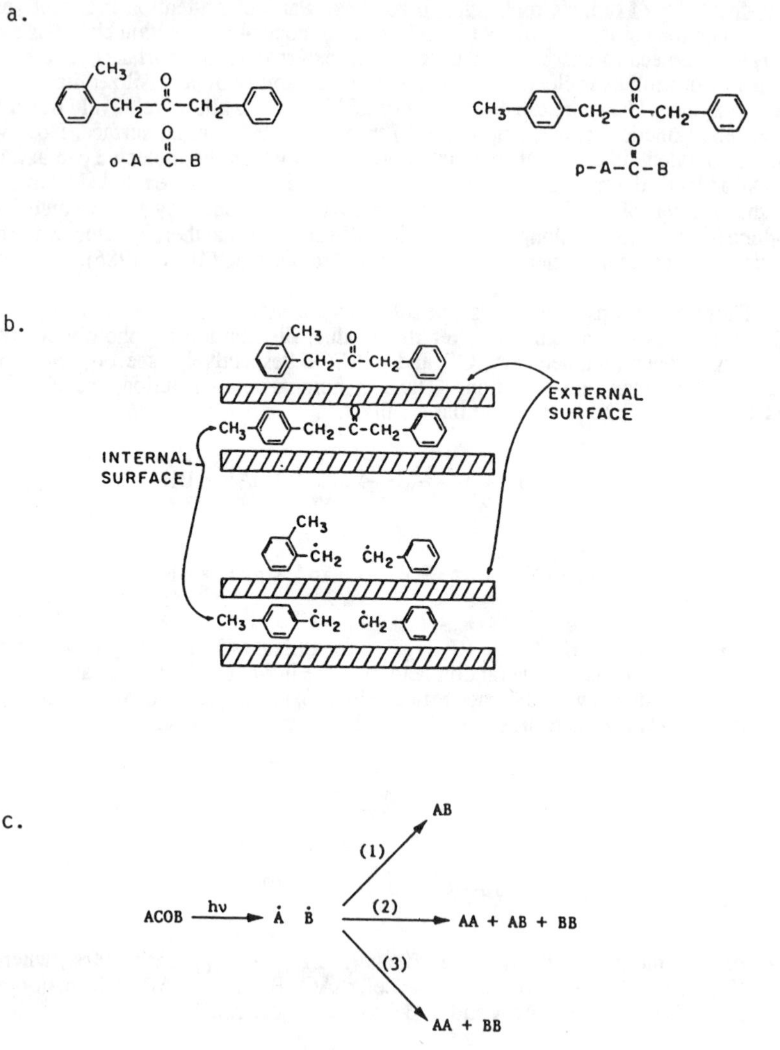

(1) Geminate Combination;

Limit = 100% Cage = 100% AB

(2) Random Combination;

Limit = 0% Cage = AA (25%) + AB (50%) + BB (25%)

(3) Sorting Combination;

Limit = -100% Cage = AA (50%) + BB (50%)

$$\%\ \text{CAGE} = \frac{AB - [AA + BB]}{AA + AB + BB}$$

Figure 24. (a) Representation of o-methyl dibenzylketone as o-ACOB and p-methyl dibenzylketone as p-ACOB; (b) Schematic representation of the adsorption of o-ACOB (on the external surface) and of p-ACOB (on the internal surface) of ZSM zeolites. Photolysis produces oA and B radicals on the external surface, and produces pA and B radicals on the internal surface; (c) General scheme for possible combinations of radical pairs (Turro, 1986).

The semiconducting minerals for which electron-hole pairs are generated on absorption of light of sufficient energy (i.e. greater than the band-gap) have been (and continue to be) of major interest because of their potential for capture and utilisation of solar energy and their ability to catalyse the degradation of adsorbed toxicants. The electronic properties of such minerals are most completely described by band theory since this theory enables analysis of the effect of the lattice on total electron energy. However, in many minerals (particularly the transition metal oxides and silicates) the electrons are relatively localised and are not easily described by band theory. In such cases, a "cluster" molecular orbital approach such as the Self Consistent Field-Xα-Scattered Wave method is more appropriate.

While light significantly modifies the number of charge carriers in semiconducting minerals, a number of additional factors influence the efficiency of migration of charge carriers to the mineral surface and subsequent electron transfer at the semiconductor-liquid interface. Of particular importance is the influence of mineral form, substitutional doping and surface modification on surface electronic states. The feasibility of any proposed electron transfer process at the solid-liquid interface will be dependent on the energy of these surface states. A variety of processes involving semiconducting minerals including the transformation of adsorbed species, the production of reduced oxygen species at the mineral surface and the photo-dissolution of the mineral are all of geochemical significance.

While most attention in the literature to date has been focussed on photo-processes involving direct absorption of light by the mineral, increasing attention is now being given to charge transfer processes at the solid-liquid interface involving excitation of surface-located species which subsequently transfer energy or electrons to the underlying (semiconducting) mineral. Charge (or energy) transfer by this route is particularly effective if the sensitizing chromophore at the mineral surface is generated as a result of complexation of surface cations. However, electron and energy transfer have also been reported in cases where the adsorbed species is a chromophore in its own right. The natural organic acid mediated photo-dissolution of minerals such as iron and manganese oxides may best be described by such processes.

Minerals such as silica and certain aluminosilicate clay minerals, while not accepting or donating charge themselves, may influence the efficiency of charge transfer within chromophores adsorbed to their surfaces through charge effects or through steric effects. In addition, selective adsorption of compounds capable of quenching the excited state of adsorbed molecules may also influence the efficiency and extent of photoprocesses at mineral surfaces. While such phenomenon may well occur in nature, investigation of such processes under more idealised conditions has also been of considerable use in probing the nature of the underlying mineral surface.

REFERENCES

Al-Ekabi, H., Serpone, N., Pelizzetti, E., Minero, C., Fox, M.A. and Draper, R.B. (1989) Kinetic studies in heterogeneous photocatalysis. 2. TiO_2-mediated degradation of 4-chlorophenol alone and in a three-component mixture of 4-chlorophenol, 2,4-dichlorophenol, and 2,4,5-trichlorophenol in air-equilibrated aqueous media. Langmuir 5, 250-255.

Albery, W.J. and Bartlett, P.N. (1984) The transport and kinetics of photogenerated carriers in colloidal semiconductor electrode particles. J. Electrochem. Soc. 131, 315-325.

Anderson, M.A. and Morel, F.M.M. (1983) The influence of aqueous iron chemistry on the uptake of iron by the coastal diatom Thalassiosira weissflogii. Limnol. Oceanogr. 27, 789-813

Anpo, M., Kawamura, T., Kodama, S., Maruya, K. and Onishi, T. (1988) Photocatalysis on Ti-Al binary metal oxides: Enhancement of the photocatalytic activity of TiO_2 species. J. Phys. Chem. 92, 438-440.

Anpo, M., Shima, T., Kodama, S. and Kubokawa, Y. (1987) Photocatalytic hydrogenation of CH_3CCH with H_2O on small-particle TiO_2: Size quantization effects and reaction intermediates. J. Phys. Chem. 91, 4305-4310.

598

Avnir, D. (1989) The Fractal Approach to Heterogeneous Chemistry: Surfaces, Colloids, Polymers. Wiley, Chichester, England.

Avnir, D., Farin, D. and Pfeifer, P. (1985) Surface geometric irregularity of particulate materials: The fractal approach. J. Colloid Interface Sci. 103, 112-123.

Bahnemann, D.W., Hoffmann, M.R., Hong, A.P. and Kormann, C. (1987) Photocatalytic formation of hydrogen peroxide. In "The Chemistry of Acid Rain: Sources and Atmospheric Processes", ACS Symposium Series 349, American Chemical Society, Washington, DC, pp. 120-132.

Balzani, V. and Carassiti, V. (1970) Photochemistry of Coordination Compounds. Academic Press, New York.

Bard, A.J. (1980) Photoelectrochemistry. Science 207, 139-144.

Behra, P. and Sigg, L. (1990). Evidence for redox cycling of iron in atmospheric water droplets. Nature 344, 419-421.

Bin-Daar, G., Dare-Edwards, M.P., Goodenough, J.B. and Hamnett, A. (1983) New anode materials for photoelectrolysis. J. Chem. Soc., Faraday Trans. 1, 79, 1199-1213.

Birenbaum, H., Avnir, D. and Ottolenghi, M. (1989) Surface geometry and pore size effects on photoinduced charge-transfer interactions between pyrene and diethylaniline on silica surfaces. Langmuir 5, 48-54.

Brandli, R., Rys, P., Zollinger, H., Oswald, H.R. and Schweitzer, F. (1970) Über die spektrale sensibilisierung von zinkoxid: I. Sorptions-verhalten von o,o'-dihydroxyazofarbstoffen. Helv. Chim. Acta 53, 1133-1145.

Cornell, R.M. and Schindler, P.W. (1987) Photochemical dissolution of goethite in acid/oxalate solution. Clays Clay Min. 35, 347-352.

Cunningham, K.M., Goldberg, M.C. and Weiner, E.R. (1985) The aqueous photolysis of ethylene glycol adsorbed on goethite. Photochem. Photobiol. 41, 409-416.

Cunningham, K.M., Goldberg, M.C. and Weiner, E.R. (1988) Mechanisms for aqueous photolysis of adsorbed benzoate, oxalate, and succinate on iron oxyhydroxide (goethite) surfaces. Environ. Sci. Technol. 22, 1090-1097.

Davis, E.A. (1984) Electronic properties of non-crystalline semiconductors. In "Physics and Chemistry of Electrons and Ions in Condensed Matter", J.V. Acrivos, N.F. Mott and A.D. Yoffe (Eds), NATO ASI Series C, Vol. 130, D. Reidel Publ. Co., Dordrecht, The Netherlands, pp. 145-164.

Degani, Y. and Willner, I. (1983) Photoinduced hydrogen evolution by a zwitterionic diquat electron acceptor. The functions of SiO_2 colloid in controlling the electron transfer process. J. Am. Chem. Soc. 105, 6228-6233.

Degani, Y. and Willner, I. (1986) Photoinduced hydrogenation of ethylene and acetylene in aqueous media: the functions of palladium and platinum colloids as catalytic charge relays. J. Chem. Soc., Perkin Trans. II, 37-42.

Della Guardia, R.A. and Thomas, J.K. (1983) Photoprocesses on colloidal clay systems. Tris(2,2'-bipyridine)ruthenium(II) bound to colloidal kaolin and montmorillonite. J. Phys. Chem. 87, 990-998.

Desilvestro, J., Pons, S., Vrachnou, E. and Gratzel, M. (1988) Electrochemical and FTIR spectroscopic characterization of ferrocyanide-modified TiO_2 electrodes designed for efficient photosensitization. J. Electroanal. Chem. 246, 411-422.

Ennaoui, A., Fiechter, S., Jaegermann, W. and Tributsch, H. (1986) Photoelectrochemistry of highly quantum efficient single-crystalline n-FeS_2 (pyrite). J. Electrochem. Soc. 133, 97-106.

Entsch, G., Sim, R.G. and Hatcher, B.C. (1983) Indications from photosynthetic components that iron is a limiting nutrient in primary producers on coral reefs. Mar. Biol. 73, 17-30.

Farin, D. and Avnir, D. (1987) Reactive fractal surfaces. J. Phys. Chem. 91, 5517-5521.

Farin, D. and Avnir, D. (1988) The reaction dimension in catalysis on dispersed metals. J. Am. Chem. Soc. 110, 2039-2045.

Farin, D., Kiwi, J. and Avnir, D. (1989) Size effects in photoprocesses on dispersed catalysts. J. Phys. Chem. 93, 5851-5854.

Faust, B.C. and Hoffmann, M.R. (1986) Photo-induced dissolution of α-Fe_2O_3 by bisulfite. Environ. Sci. Technol. 20, 943-948.

Faust, B.C., Hoffmann, M.R. and Bahnemann, D.W. (1989) Photocatalytic oxidation of sulfur dioxide in aqueous suspensions of α-Fe_2O_3. J. Phys. Chem. 93, 6371-6381.

Finden, D.A.S., Tipping, E., Jaworski, G.H.M. and Reynolds, C.S. (1984) Light-induced reduction of natural iron (III) oxide and its relevance to phytoplankton. Nature 309, 783-.

599

Fox, M.A. (1983) Organic heterogeneous photocatalysis: chemical conversions sensitized by irradiated semiconductors. Accounts Chem. Res. 16, 314-321.

Fox, M.A. (1986) Charge injection into semiconductor particles - importance in photocatalysis. In "Homogeneous and Heterogeneous Photocatalysis", E. Pelizzetti and N. Serpone (Eds), NATO ASI Series C, Vol. 174, D. Reidel Publ. Co., Dordrecht, The Netherlands, pp. 363-383.

Fox, M.A. (1988) Photocatalytic oxidation of organic substrates. In "Photocatalysis and Environment: Trends and Applications", M. Schiavello (Ed.), NATO ASI Series C, Vol. 237, Kluwer Academic Publ., Dordrecht, The Netherlands, pp. 445-467.

Fox, M.A. and Petit, T.L. (1989) Photoactivity of zeolite-supported cadmium sulfide: Hydrogen evolution in the presence of sacrificial donors. Langmuir 5, 1056-1061.

Frei, H., Fitzmaurice, D.J. and Gratzel, M. (1990) Surface chelation of semiconductors and interfacial electron transfer. Langmuir 6, 198-206.

Friend, R.H. and Yoffe, A.D. (1987) Electronic properties of intercalation complexes of the transition metal dichalcogenides. Advances in Physics 36, 1-94.

Gerischer, H. (1980) Photodecomposition of semiconductors: thermodynamics, kinetics and application to solar cells. Faraday Disc. Chem. Soc. 70, 137-151.

Ghosh, P.K. and Bard, A.J. (1984) Photochemistry of tris(2,2'-bipyridyl)ruthenium(II) in colloidal clay suspensions. J. Phys. Chem. 88, 5519-5526.

Gratzel, M. (1986). Dynamics of interfacial electron transfer reactions in colloidal semiconductor systems and water cleavage by visible light. In "Homogeneous and Heterogeneous Photocatalysis", E. Pelizzetti and N. Serpone (Eds), NATO ASI Series C, Vol. 174, D. Reidel Publ. Co., Dordrecht, The Netherlands, pp. 91-110.

Gratzel, M. (1988) Solar energy harvesting. In "Photoinduced Electron Transfer", M.A. Fox and M. Channon (Eds), Part D, Elsevier, Amsterdam, Chapter 6.3, pp. 394-440.

Gratzel, M. (1989) Heterogeneous Photochemical Electron Transfer. CRC Press, Boca Raton, Florida.

Gratzel, M., Thampi, K.R. and Kiwi, J. (1989) Methane oxidation at room temperature and atmospheric pressure activated by light via polytungstate dispersed on titania. J. Phys. Chem. 93, 4128-4132.

Hardee, K.L. and Bard, A.J. (1978) Semiconductor electrodes. V. The application of chemically vapour deposited iron oxide films to photosensitized electrolysis. J. Electrochem. Soc. 123, 1024-1026.

Harris, L.A. and Wilson, R.H. (1978) Semiconductors for photoelectrolysis. Ann. Rev. Mater. Sci. 8, 99-134.

Hawn, D.D. and Armstrong, N.R. (1978) Electrochemical adsorption and covalent attachment of eryhrosin to modified tin dioxide electrodes and measurement of the photocurrent sensitization to visible wavelength light. J. Phys. Chem. 82, 1288.

Heller, A. (1986) Metallic catalysts on semiconductors: transparency and electrical contact properties. In "Homogeneous and Heterogeneous Photocatalysis", E. Pelizzetti and N. Serpone (Eds), NATO ASI Series C, Vol. 174, D. Reidel Publ. Co., Dordrecht, The Netherlands, pp. 303-315.

Hering, J.G. and Stumm, W. (1990) Oxidative and reductive dissolution of minerals. In "Mineral-Water Interface Geochemistry", M.F. Hochella and A.F. White (Eds), Reviews in Mineralogy 23 (this volume).

Hong, R. and Kester, D.R. (1986) Redox states of iron in the offshore waters of Peru. Limnol. Oceanog. 31, 512-524.

Houlding, V.H. and Gratzel, M. (1983) Photochemical H_2 generation by visible light. Sensitization of TiO_2 particles by surface complexation with 8-hydroxyquinoline. J. Am. Chem. Soc. 105, 5695-5696.

Howe, R.F. and Gratzel, M. (1985) EPR observation of trapped electrons in colloidal TiO_2. J. Phys. Chem. 89, 4495-4499.

Iler, R.K. (1979) The Chemistry of Silica, 2nd Ed. Wiley, New York.

Jacobs, P.A. and Uytterhoeven, J.B. (1977) Cleavage of water over zeolites. J. Chem. Soc. Chem. Comm. 128-129.

Jaegermann, W. and Tributsch, H. (1988) Interfacial properties of semiconducting transition metal chalcogenides. Prog. Surf. Sci. 29, 1-167.

Kalyanasundaram, K. (1987) Photochemistry in Microheterogeneous Systems. Academic Press, Orlando, Florida.

Kalyanasundaram, K., Gratzel, M. and Pelizzetti, E. (1986) Interfacial electron transfer in colloidal metal and semiconductor dispersions and photodecomposition of water. Coord. Chem. Rev. 69, 57-125.

Kalyanasundaram, K., Vlachopoulous, N., Krishnan, V., Monnier, A. and Gratzel, M. (1987) Sensitization of TiO_2 in the visible light region using zinc porphyrins. J. Phys. Chem. 91, 2342-2347.

600

Kamat, P.V. and Fox, M.A. (1983) Photosensitization of TiO_2 colloids by erythrosin B in acetonitrile. Chem. Phys. Lett. 102, 379-384.

Khan, S.U.M. and Bockris, J.O'M. (1985) Reply to comment on "Electronic states in solution and charge transfer". J. Phys. Chem. 89, 555-556.

Kiwi, J., Kalyanasundaram, K. and Gratzel, M. (1982) Visible light induced cleavage of water into hydrogen and oxygen in colloidal and microheterogeneous systems. Structure and Bonding 49, 37-125.

Kormann, C., Bahnemann, D.W. and Hoffmann, M.R. (1988) The photocatalytic production of H_2O_2 and organic peroxides in aqueous suspensions of TiO_2, ZnO, and desert sand. Environ. Sci. Technol. 22, 798-806.

Krenske, D., Abdo, S., Van Damme, H., Cruz, M. and Fripiat, J.J. (1983) Photochemical and photocatalytic properties of adsorbed organometallic compounds. 1. Luminescence quenching of tris (2,2'-bipyridine) ruthenium(II) and -chromium(III) in clay membranes. J. Phys. Chem. 84, 2447-2457.

Kuznicki, S.M. and Eyring, E.M. (1978) "Water splitting" by titanium exchanged zeolite A. J. Am. Chem. Soc. 100, 6790-6791.

Leland, J.K. and Bard, A.J. (1987) Photochemistry of colloidal semiconducting iron oxide polymorphs. J. Phys. Chem. 91, 5076-5083.

Lenoir, P.M., Sassoon, E.R. and Kozak, J.J. (1988) Computer simulation of photochemical water cleavage systems. 2. Effects of particle size and ion concentration on the production of hydrogen from water mediated with a colloidal ctalyst. J. Phys. Chem. 92, 2526-2536.

Litter, M.I., Baumgartner, E.C., Urruturia, G.A. and Blesa, M.A. (1990) Photodissolution of iron oxides III. The interplay of photochemical and thermal processes in maghemite/carboxylic acid systems. Environ. Sci. Technol. (in press).

Litter, M.I. and Blesa, M.A. (1988) Photodissolution of iron oxides I. Maghemite in EDTA solutions. J. Colloid Interface Sci. 125, 679-687.

Liu, X. and Thomas, J.K. (1989) Formation and photophysical properties of CdS in zeolites with cages and channels. Langmuir 5, 58-66.

Many, A., Goldstein, Y. and Grover, N.B. (1965) Semiconductor Surfaces, North Holland, Amsterdam.

Marfunin, A.S. (1979) Physics of Minerals and Inorganic Materials. Springer-Verlag, Berlin.

Martin, J.M. and Fitzwater, S.E. (1988) Iron deficiency limits phytoplankton growth in the northeast Pacific subarctic. Nature 331, 341-343.

Martin, J.M. and Gordon, R.M. (1988) Northeast Pacific iron distributions in relation to phytoplankton productivity. Deep-Sea Res. 355, 177-196.

Martin, J.M., Gordon, R.M. and Fitzwater, S.E. (1990) Iron in Antarctic waters. Nature 345, 156-158.

Matsumura, M., Mitsuda, N., Yoshizawa, N. and Tsubomura, H. (1981) Photocurrents in the ZnO and TiO_2 photoelectrochemical cells sensitized by xanthene dyes and tetraphenylporphines. Effect of substitution on the electron injection processes. Bull. Chem. Soc. Japan 54, 692-695.

Mau, A.W.H., Huang, C.B., Kakuta, N., Bard, A.J., Campion, A., Fox, M.A., White, J.M. and Webber, S.E. (1984) H_2 photoproduction by nafion/CdS/Pt films in H_2O/S^{2-} solutions. J. Am. Chem. Soc. 106, 6537-6542.

McKnight, D.M., Kimball, B.A. and Bencala, K.E. (1988) Iron photoreduction and oxidation in an acidic mountain stream. Science 240, 637-640.

Meissner, D., Memming, R. and Kastening, B. (1983) Light-induced generation of hydrogen at CdS-monograin membranes. Chem. Phys. Lett. 96, 34-37.

Milosavljevic, B.H. and Thomas, J.K. (1986) Photochemistry of compounds in the constrained medium cellulose. 7. The effect of temperature on photoinduced electron transfer from tris (2,2'-bipyridine) ruthenium(II) to methylviologen solubilized in cellophane. J. Am. Chem. Soc. 108, 2513-2517.

Morrison, S.R. (1977) The Chemical Physics of Surfaces. Plenum Press, New York.

Morrison, S.R. (1980) Electrochemistry at Semiconductor and Oxidized Metal Electrodes. Plenum Press, New York.

Mulvaney, P., Cooper, R., Grieser, F. and Meisel, D. (1988) Charge trapping in the reductive dissolution of colloidal suspensions of iron(III) oxides. Langmuir 4, 1206-1211.

Mulvaney, P., Swayambunathan, V., Grieser, F. and Meisel, D. (1988) Dynamics of interfacial charge transfer in iron(III) oxide colloids. J. Phys. Chem. 92, 6732-6740.

Nag, B.R. (1972) Theory of electrical transport in semiconductors. Sci. Solid State 3, 1-227.

Nakabayashi, S. and Kawai, T. (1988) Electron transfer at interfaces. In "Photoinduced Electron Transfer", M.Á. Fox and M. Channon (Eds), Part B, Elsevier, Amsterdam, Chapter 3.3, pp. 599-641.

Nakamura, T. and Thomas, J.K. (1985) Photochemistry of materials adsorbed on clay systems. Effect of the nature of the adsorption on the kinetic description of the reactions. Langmuir 1, 568-573.

Nakamura, T. and Thomas, J.K. (1986) The interaction of alkylammonium salts with synthetic clays. A fluorescence and laser excitation study. J. Phys. Chem. 90, 641-644.

Nishimoto, S., Ohtani, B., Shirai, H. and Kagiya, T. (1985) Photolysis of aqueous poly(vinyl alcohol) solution by heterogeneous TiO_2/Pt catalyst. J. Polymer Sci., Polymer Lett. Ed. 23, 141-145.

Nosaka, Y. and Fox, M.A. (1987) Effect of charge of polymeric stabilizing agents on the quantum yields of photoinduced electron transfer from photoexcited colloidal semiconductors to adsorbed viologens. Langmuir 3, 1147-1150.

Nozik, A.J. (1978) Photoelectrochemistry: Applications to solar energy conversion. Ann. Rev. Phys. Chem. 29, 189-222.

Oelkrug, D., Flemming, W., Fullemann, R., Gunther, R., Honnen, W., Krabichler, G., Schafer, M. and Uhl, S. (1986) Photochemistry on surfaces. Pure Appl. Chem. 58, 1207-1218.

Pelizzetti, E., Minero, C., Pramauro, E., Serpone, N. and Borgarello, E. (1988). Photodegradation of organic pollutants in aquatic systems catalyzed by semiconductors. In "Photocatalysis and the Environment: Trends and Applications", M. Schiavello (Ed.), NATO ASI Series C, Vol. 237, Kluwer Academic Publ., Dordrecht, The Netherlands, pp. 469-498.

Pichat, P. (1988) Powder photocatalysts: characterization by isotopic exchanges and photoconductivity; potentials for metal recovery, catalyst preparation and water pollutant removal. In "Photocatalysis and the Environment: Trends and Applications", M. Schiavello (Ed.), NATO ASI Series C, Vol. 237, Kluwer Academic Publ., Dordrecht, The Netherlands, pp. 399-424.

Pichat, P. and Fox, M.A. (1988) Photocatalysis on semiconductors. In "Photoinduced Electron Transfer", M.A. Fox and M. Channon (Eds), Part D, Elsevier, Amsterdam, Chapter 6.1, pp. 242-302.

Rafaeloff, R., Tricot, Y.-M., Nome, F. and Fendler, J.H. (1985) Colloidal catalyst coated semiconductors in surfactant vesicles. In situ generation of rhodium-coated cadmium sulfide particles in dioctadecyldimethylammonium halide surfactant vesicles and their utilization in photosensitized charge separation and hydrogen generation. J. Phys. Chem. 89, 533-537.

Reidel, M. (1984) Photoelectrochemistry, Photocatalysis and Photoreactions. D. Reidel Publ. Co., Dordrecht, The Netherlands.

Rothenberger, G., Moser, J., Gratzel, M., Serpone, N. and Sharma, D.K. (1985) Charge carrier trapping and recombination dynamics in small semiconductor particles. J. Am. Chem. Soc. 107, 8054-8059.

Ryan, M.A. and Spitler, M.T. (1988) Light-initiated surface modification of oxide semiconductors with organic dyes. Langmuir 4, 861-867.

Scherson, D., Ekardt, W. and Gerischer, H. (1985) Comments on "Electronic states in solution and charge transfer". J. Phys. Chem. 89, 554-555.

Schoonheydt, R.A., DePauw, P., Vliers, D. and DeSchryver, F.C. (1984) Luminescence of tris(2,2'-bipyridine)ruthenium(II) in aqueous clay mineral suspensions. J. Phys. Chem. 88, 5113-5118.

Sclafani, A., Palmisano, L. and Schiavello, M. (1990) Influence of the preparation methods of TiO_2 on the photocatalytic degradation of phenol in aqueous dispersion. J. Phys. Chem. 94, 829-832.

Serpone, N., Borgarello, E. and Pelizzetti, E. (1988a) Photoreduction and photodegradation of inorganic pollutants I. Cyanides. In "Photocatalysis and the Environment: Trends and Applications", M. Schiavello (Ed.), NATO ASI Series C, Vol. 237, Kluwer Academic Publ., Dordrecht, The Netherlands, pp. 499-526.

Serpone, N., Borgarello, E. and Pelizzetti, E. (1988b) Photoreduction and photodegradation of inorganic pollutants II. Selective reduction and recovery of Au, Pt, Pd, Rh, Hg and Pb. In "Photocatalysis and the Environment: Trends and Applications", M. Schiavello (Ed.), NATO ASI Series C, Vol. 237, Kluwer Academic Publ., Dordrecht, The Netherlands, pp. 527-566.

Sherman, D.M. (1984) The electronic structures of manganese oxide minerals. Am. Mineral. 69, 788-799.

Sherman, D.M. (1985) The electronic structures of Fe^{3+} coordination sites in iron oxides; applications to spectra, bonding and magnetism. Phys. Chem. Minerals 12, 161-175.

Sherman, D. and Waite, T.D. (1985) Electronic spectra of Fe^{3+} oxides and oxide hydroxides in the near IR to near UV. Am. Mineral. 70, 1262-1269.

Sherman, D.M. and Vergo, N. (1988) Optical (diffuse reflectance) and Mossbauer spectroscopic study of nontronite and related Fe-bearing smectites. Am. Mineral. 73, 1346-1354.

Shuey, R.T. (1975) Semiconducting Ore Minerals. Elsevier, Amsterdam.

602

Siffert, C. (1989) L'effet de la lumiere sur la dissolution des oxydes de fer(III) dans les milieux aqueux, PhD Thesis, ETH, Zürich.

Somorjai, G.A. and Salmeron, M. (1986) Surface properties of catalysts. Iron and its oxides: surface chemistry, photochemistry and catalysis. In "Homogeneous and Heterogeneous Photocatalysis", E. Pelizzetti and N. Serpone (Eds), NATO ASI Series C, Vol. 174, D. Reidel Publ. Co., Dordrecht, The Netherlands, pp. 445-478.

Stramel, R.D. and Thomas, J.K. (1986) Photochemistry of iron oxide colloids. J. Colloid Interface Sci. 110, 121-129.

Strehlow, W.H. and Cook, E.L. (1973) Compilation of energy band gaps in elemental and binary compound semiconductors and insulators. J. Phys. Chem. Ref. Data 2, 163-199.

Suib, S.L., Bordeianu, O.G., McMohan, K.C. and Psaras, D. (1981) In "Inorganic Reactions in Organised Media", S.L. Holt (Ed.), ACS Symposium Series No. 117, American Chemical Society, Washington, D.C.

Suib, S.L., Kostapapas, A. and Psaras, D. (1984) Photoassisted catalytic oxidation of isopropyl alcohol by uranyl-exchanged zeolites. J. Am. Chem. Soc. 105, 1614-1620.

Sulzberger, B. (1990) Photoredox reactions at hydrous metal oxide surfaces: a surface coordination chemistry approach. In "Aquatic Chemical Kinetics - Reaction Rates of Processes in Natural Waters", W. Stumm (Ed.), Wiley Interscience, New York.

Sunda, W.G. (1989) Trace metal interactions with marine algae. In "Marine Photosynthesis", R. Alberte and R.T. Barber (Eds), Oxford University Press, New York.

Sunda, W.G. and Huntsman, S.A. (1988) Effect of sunlight on redox cycles of manganese in the southwestern Sargasso Sea. Deep-Sea Res. 35, 1297-1317.

Sunda, W.G., Huntsman, S.A. and Harvey, G.R. (1983) Photoreduction of manganese oxides in seawater and its geochemical and biological implications. Nature 301, 234-236.

Tamaura, Y., Ito, K. and Katsura, T. (1983) Transformation of γ-FeO(OH) to Fe_3O_4 by adsorption of iron(II) ion on γ-FeO(OH). J. Chem. Soc. Dalton Trans. 189-194.

Thomas, J.K. (1987) Characterization of surfaces by excited states. J. Phys. Chem. 91, 267-276.

Tossell, J.A. (1978) Self-consistent-field-X α study to one-electron energy levels in Fe_3O_4. Phys. Rev. B17, 484-487.

Tossell, J.A. and Gibbs, G.V. (1977) Molecular orbital studies of geometries and spectra of minerals and inorganic compounds. Phys. Chem. Minerals 2, 21-57.

Tossell, J.A., Vaughan, D.J. and Johnson, K.H. (1974) The electronic structure of rutile, wustite and hematite from molecular orbital calculations. Am. Mineral. 59, 319-334.

Tronc, E. and Jolivet, J.-P. (1984) Exchange and redox reactions at the interface of spinel-like iron oxide colloids in solution: Fe(II) adsorption. Adsorption Sci. Technol. 1, 247-251.

Tronc, E., Jolivet, J.-P., Lefebvre, J. and Massart, R. (1984) Ion adsorption and electron transfer in spinel-like iron oxide colloids. J. Chem. Soc., Faraday Trans. 1, 80, 2619-2629.

Turchi, C.S. and Ollis, D.F. (1990) Photocatalytic degradation of organic water contaminants: mechanisms involving hydroxyl radical attack. J. Catalysis 122, 178-192.

Turro, N.J. (1986) Photochemistry of organic molecules in microscopic reactors. Pure Appl. Chem. 58, 1219-1228.

Van Damme, H., Bergaya, F. and Challal, D. (1986) Photocatalysis over clay supports. In "Homogeneous and Heterogeneous Photocatalysis", E. Pelizzetti and N. Serpone (Eds), D. Reidel Publ. Co., Dordrecht, The Netherlands, pp. 479-508.

Van Damme, H., Levitz, P., Gatineau, L., Alcover, J.F. and Fripiat, J.J. (1988) On the determination of the surface fractal dimension of powders by granulometric analysis. J. Colloid Interface Sci. 122, 1-8.

Vrachnou, E., Vlachopoulos, N. and Gratzel, M. (1987) Efficient visible light sensitization of TiO_2 by surface complexation with $Fe(CN)_6^{4-}$. J. Chem. Soc., Chem. Comm. 868-870.

Waite, T.D. (1986) Photoredox chemistry of colloidal metal oxides. In "Geochemical Processes at Mineral Surfaces", J.A. Davis and K.F. Hayes (Eds), ACS Symposium Series No.323, American Chemical Society, Washington, D.C., pp. 426-445.

Waite, T.D. (1990) Photochemistry of colloids and surfaces in natural waters and water treatment. In "Chemistry of Colloids and Surfaces in Natural Waters and Water Treatment", R. Beckett (Ed.), Plenum Press, New York.

Waite, T.D. and Morel, F.M.M. (1984a) Photoreductive dissolution of colloidal iron oxides in natural waters. Environ. Sci. Technol. 18, 860-868.

Waite, T.D. and Morel, F.M.M. (1984b) Photoreductive dissolution of colloidal iron oxide: effect of citrate. J. Colloid Interface Sci. 102, 121-137.

Waite, T.D. and Torikov, A. (1987). Photo-assisted dissolution of colloidal iron oxides by thiol-containing compounds. J. Colloid Interface Sci. 119, 228-235.

Waite, T.D., Wrigley, I.C. and Szymczak, R. (1988) Photo-assisted dissolution of a colloidal manganese oxide in the presence of fulvic acid. Environ. Sci. Technol. 22, 778-785.

Watanabe, T., Takizawa, T. and Honda, K. (1977) Photocatalysis through excitation of adsorbates. 1. Highly efficient N-deethylation of rhodamine B adsorbed to CdS. J. Phys. Chem. 81, 1845-1851.

Wayne, R.P. (1970) Photochemistry. Butterworths, London.

Wheeler, J. and Thomas, J.K. (1983) Photochemistry in porous colloidal silica particles. J. Phys. Chem. 86, 4540-4544.

Wheeler, J. and Thomas, J.K. (1984) Photochemistry of the uranyl ion in colloidal silica solution. J. Phys. Chem. 88, 750-754.

White, A.F (1990) Heterogeneous electron transfer mechanisms associated with structural oxidation in ferrous oxides and silicates. In "Mineral-Water Interface Geochemistry", M.F. Hochella and A.F. White (Eds), Reviews in Mineralogy 23 (this volume).

Willner, I., Otvos, J.W. and Calvin, M. (1981a) Photosensitized electron-transfer reactions in colloidal SiO_2 systems: charge separation at a solid-aqueous interface. J. Am. Chem. Soc. 103, 3203-3205.

Willner, I., Yang, J-M., Laane, C., Otvos, J.W. and Calvin, M. (1981b) The function of SiO_2 colloids in photoinduced redox reactions. Interfacial effects on the quenching, charge separation and quantum yields. J. Phys. Chem. 85, 3277-3282.

Yeh, L-S.R. and Hackerman, N. (1977) Iron oxide semiconductor electrodes in photoassisted electrolysis of water. J. Electrochem. Soc. 124, 833-836.

Yoneyama, H., Haga, S. and Yamanaka, S. (1989) Photocatalytic activities of microcrystalline TiO_2 incorporated in sheet silicates of clay. J. Phys. Chem. 93, 4833-4837.